PLANT BIOACTIVES AND DRUG DISCOVERY

Wiley Series in Drug Discovery and Development

Binghe Wang, Series Editor

A complete list of the titles in this series appears at the end of this volume.

PLANT BIOACTIVES AND DRUG DISCOVERY

Principles, Practice, and Perspectives

Edited by

VALDIR CECHINEL-FILHO
Universidade do Vale do Itajaí (UNIVALI)
Itajaí-SC, Brazil

WILEY
A JOHN WILEY & SONS, INC., PUBLICATION

Published by John Wiley & Sons, Inc., Hoboken, New Jersey
Published simultaneously in Canada

Library of Congress Cataloging-in-Publication Data:

Plant bioactives and drug discovery : principles, practice, and perspectives / edited by Valdir Cechinel-Filho.
p. ; cm. – (Wiley series in drug discovery and development)
Includes bibliographical references and index.
ISBN 978-0-470-58226-8 (cloth)
I. Cechinel-Filho, Valdir. II. Series: Wiley series in drug discovery and development.
[DNLM: 1. Plants, Medicinal–metabolism. 2. Drug Discovery. 3. Plant Extracts–pharmacology. QV 766]
615.3'21–dc23 2011041434

Printed in the United States of America

10 9 8 7 6 5 4 3 2 1

CONTENTS

CONTRIBUTORS

A. Douglas Kinghorn, Division of Medicinal Chemistry and Pharmacognosy, The Ohio State University, Columbus, OH, USA

Alan L. Harvey, University of Strathclyde, Glasgow, UK

Alberto Boffi, Sapienza Università di Roma, Rome, Italy

Alberto Macone, Sapienza Università di Roma, Rome, Italy

Alessandra Bonamore, Sapienza Università di Roma, Rome, Italy

Andrea Calcaterra, Sapienza Università di Roma, Rome, Italy

Angélica G. Couto, Universidade do Vale do Itajaí (UNIVALI), Itajaí-SC-Brazil

Arturo San Feliciano, Department of Pharmaceutical Chemistry, University of Salamanca, Salamanca, Spain

Bruno Botta, Sapienza Università di Roma, Rome, Italy

Carlos A. M. Fraga, Faculty of Pharmacy, Universidade Federal do Rio de Janeiro Rio de Janeiro, RJ, Brazil

Chung Ki Sung, Division of Medicinal Chemistry and Pharmacognosy, The Ohio State University, Columbus, OH, USA

David G. I. Kingston, Department of Chemistry, Virginia Polytechnic Institute and State University, Blacksburg, VA, USA

David J. Newman, National Cancer Institute-Frederick, Frederick, MD, USA

Dayana R. Gouveia, Departamento de Física e Química, Universidade de São Paulo, (USP), Ribeirão Preto – SP, Brazil

Eliezer J. Barreiro, Faculty of Pharmacy, Universidade Federal do Rio de Janeiro, Rio de Janeiro, RJ, Brazil

Elisaldo A. Carlini, Universidade Federal de São Paulo (UNIFESP), São Paulo, Brazil

Elisa Petenatti, National University of San Luis, Argentina

Emerson F. Queiroz, University of Geneva and University of Lausanne, Switzerland

Esther del Olmo, Department of Pharmaceutical Chemistry, University of Salamanca, Salamanca, Spain

Francesca Ghirga, Sapienza Università di Roma, Rome, Italy

Fúlvio R. Mendes Gina Negri, Universidade Federal de São Paulo (UNIFESP), São Paulo, Brazil

Giuliano Delle Monache, Sapienza Università di Roma, Rome, Italy

Gordon M. Cragg, National Cancer Institute-Frederick, Frederick, MD, USA

Ilaria D'Acquarica, Sapienza Università di Roma, Rome, Italy

James D. McChesney, Ironstone Separations, Inc., Etta, MS, USA

Jean-Luc Wolfender, University of Geneva and University of Lausanne, Switzerland

Jie Li, Division of Medicinal Chemistry and Pharmacognosy, The Ohio State University, Columbus, OH, USA

Jielu Zhao, Department of Chemistry, Virginia Polytechnic Institute and State University, Blacksburg, VA, USA

Joaquim M. Duarte-Almeida, Universidade Federal de São Paulo (UNIFESP), São Paulo, Brazil

José L. López-Pérez, Department of Pharmaceutical Chemistry, University of Salamanca, Salamanca, Spain

José R. Santin, Chemical-Pharmaceutical Investigations Center (NIQFAR), Universidade do Vale do Itajaí (UNIVALI), Itajaí-SC, Brazil

Jun Qi, Department of Chemistry, Virginia Polytechnic Institute and State University, Blacksburg, VA, USA

Kip Vought, Regulus Pharmaceutical Consulting, Inc., Boulder, CO, USA

Kurt Hostettmann, University of Geneva and University of Lausanne, Switzerland

Laura Svetaz, Universidad Nacional de Rosario, Rosario, Argentina

Leonardo Gobbo-Neto, Departamento de Física e Química, Universidade de São Paulo, (USP), Ribeirão Preto – SP, Brazil

Lidia M. Lima, Faculty of Pharmacy, Universidade Federal do Rio de Janeiro, Rio de Janeiro, RJ, Brazil

Luiz C. Klein-Júnior, Chemical-Pharmaceutical Investigations Center (NIQFAR), Universidade do Vale do Itajaí (UNIVALI), Itajaí-SC, Brazil

Mahabir Gupta, Universidad de Panamá, Panama

Marcela Raimondi, Universidad Nacional de Rosario, Rosario, Argentina

Marco Barba, Sapienza Università di Roma, Rome, Italy

Marcos Derita, Universidad Nacional de Rosario, Rosario, Argentina

María Á. Castro, Department of Pharmaceutical Chemistry, University of Salamanca, Salamanca, Spain

Maximiliano Sortino, Universidad Nacional de Rosario, Rosario, Argentina

Melina Di Liberto, Universidad Nacional de Rosario, Rosario, Argentina

Norberto P. Lopes, Departamento de Física e Química, Universidade de São Paulo, (USP), Ribeirão Preto – SP, Brazil

Paul G. Grothaus, National Cancer Institute-Frederick, Frederick, MD, USA

Pilar Menendez, Catedra de Farmacognosia y Productos Naturales, Facultad de Química, Montevideo, Uruguay

Raymond Cooper, Stratum Nutrition, Novus International, St. Charles, MO, USA

Ricardo Tabach, Universidade Federal de São Paulo (UNIFESP), São Paulo, Brazil

Rivaldo Niero, Universidade do Vale do Itajaí (UNIVALI), Itajaí-SC, Brazil

Rogério Marcos Lenzi, Universidade do Vale do Itajaí (UNIVALI), Itajaí-SC, Brazil

Rosendo A. Yunes, Universidade Federal de Santa Catarina (UFSC), Florianópolis-SC, Brazil

Ruth M. Lucinda da Silva, Universidade do Vale do Itajaí (UNIVALI), Itajaí-SC, Brazil

Sérgio F. de Andrade, Chemical-Pharmaceutical Investigations Center (NIQFAR), Universidade do Vale do Itajaí (UNIVALI), Itajaí-SC, Brazil

Sigrun Chrubasik, Department of Forensic Medicine, University of Freiburg, Freiburg, Germany; and Sydney University, Sydney, Australia

Susana Zacchino, Universidad Nacional de Rosario, Rosario, Argentina

Tania M.B. Bresolin, Universidade do Vale do Itajaí (UNIVALI), Itajaí-SC, Brazil

Valdir Cechinel-Filho, Universidade do Vale do Itajaí (UNIVALI), Itajaí-SC, Brazil

Yulin Ren, Division of Medicinal Chemistry and Pharmacognosy, The Ohio State University, Columbus, OH, USA

PREFACE

This Edition presents new contributions about the secondary metabolites from higher plants, which are of great interest throughout the world. They are important sources of new and promising drugs, and also used as models or lead compounds for obtaining synthetic drugs. The complex and notable variety of structures from several organic classes, and the strategic methods of isolation, structural elucidation, synthesis, biological evaluation, quality control, intellectual property rights, cooperation university-industry, etc., make this an exciting and challenging field for investigators and the pharmaceutical industry. This book addresses these and other subjects related to the field, through the contributions of currently productive and renowned scientists working on bioactive natural products around the world. The topics included in this book have been carefully selected to give readers a broad overview of the importance of the biodiversity as a source of new drugs, as well as the more complex strategies used to discover new drugs from higher plants. This book provides some important, up-to-date and elucidative examples of the above-mentioned subjects, and will be useful for graduate and postgraduate students, researchers and the pharmaceutical industries.

The Editor thanks to all the authors and co-authors for their valuable and relevant cooperation making this book a reality. Sincere thanks also to Dr. Jonathan Rose, Editor of John Wiley & Sons, Inc. for invitation, encouragement and important cooperation. Finally, special thanks to University of Itajaí Valley (UNIVALI) and MSc. Luiz Carlos Klein Junior (Giggio) for his excellent technical assistance.

Valdir Cechinel-Filho
Editor

1

NATURAL PRODUCTS IN DRUG DISCOVERY: RECENT ADVANCES

Gordon M. Cragg, Paul G. Grothaus, and David J. Newman

1.1 INTRODUCTION

Throughout the ages, Nature has provided humans with the means to cater for their basic needs, not the least of which are medicines for the treatment of a wide spectrum of diseases. Plants, in particular, played a significant role forming the basis of sophisticated traditional medicine systems. Records dating from around 2600 BCE document the uses of approximately 1000 plant-derived substances in Mesopotamia. These included oils of *Cedrus* species (cedar) and *Cupressus sempervirens* (cypress), *Glycyrrhiza glabra* (licorice), *Commiphora* species (myrrh), and *Papaver somniferum* (poppy juice), all of which continue to be used today for the treatment of ailments ranging from coughs and colds to parasitic infections and inflammation [1]. Although Egyptian medicine dates from about 2900 BCE, the best-known record is the "Ebers Papyrus," which dates from 1500 BCE and documents over 700 drugs, mostly of plant origin [1]. The Chinese *Materia Medica* has been extensively documented over the centuries [2]; the first record dates from about 1100 BCE (Wu Shi Er Bing Fang, containing 52 prescriptions), and is followed by works such as the Shennong Herbal (~100 BCE; 365 drugs) and the Tang Herbal (659 CE; 850 drugs). Likewise, documentation of the Indian Ayurvedic system dates from before 1000 BCE (Charaka; Sushruta and Samhitas with 341 and 516 drugs, respectively) [3,4].

The Greeks and Romans made substantial contributions to the rational development of the use of herbal drugs in the ancient "Western" world. The Greek physician,

Plant Bioactives and Drug Discovery: Principles, Practice, and Perspectives, Fourth Edition.
Edited by Valdir Cechinel-Filho.

Dioscorides (100 CE), accurately documented the collection, storage, and use of medicinal herbs while traveling with Roman armies throughout the then "known world," while Galen (130–200 CE), a practitioner and teacher of pharmacy and medicine in Rome, is well known for his complex prescriptions and formulae used in compounding drugs. It was, however, the Arabs who preserved much of the Greco-Roman expertise during the Dark and Middle Ages (fifth–twelfth centuries), and they expanded it to include the use of their own resources, together with Chinese and Indian herbs unknown to the Greco-Roman world. A comprehensive review of the history of medicine may be found on the website of the National Library of Medicine (NLM), United States National Institutes of Health (NIH), at http://www.nlm.nih.gov/hmd/collections/archives/index.html.

1.2 THE ROLE OF TRADITIONAL MEDICINE AND PLANTS IN DRUG DISCOVERY

Plant-based systems have continued to play an essential role in health care of many cultures [5,6], and the World Health Organization (WHO) has estimated that approximately 65% of the world's population relies mainly on plant-derived traditional medicines for their primary health care [7]. Plant products also play an important role in the health care systems of the remaining population, mainly residing in "developed" countries [7]. Of 122 compounds identified in a survey of plant-derived pure compounds used as drugs in countries hosting WHO-Traditional Medicine Centers, 80% were found to be used for the same or related ethnomedical purposes, and were derived from only 94 plant species [7]. Relevant examples are given by Fabricant and Farnsworth [8].

Probably the best example of ethnomedicine's role in guiding drug discovery and development is that of the antimalarial drugs, particularly, quinine and artemisinin. The isolation of quinine (Figure 1.1) was reported in 1820 by the French pharmacists, Caventou and Pelletier from the bark of *Cinchona* species (e.g., *Cinchona officinalis*) [9]. The bark, long used by indigenous groups in the Amazon region for the treatment of fevers, was introduced into Europe in the early 1600s for the treatment of malaria, and quinine formed the basis for the synthesis of the commonly used antimalarial drugs, chloroquine (Figure 1.1) and mefloquine, which largely replaced quinine in the mid-twentieth century. As resistance to both these drugs developed in many tropical regions, another plant having a long history of use in traditional Chinese medicine (TCM) for the treatment of fevers, *Artemisia annua* (Qinghaosu), gained prominence [10], and the discovery of artemisinin (Figure 1.1) by Chinese scientists in 1971 provided an exciting new natural product lead compound [11]. Artemisinin analogs, such as artesunate (Figure 1.1), are now used for the treatment of malaria in many countries, and many other analogs of artemisinin have been prepared in attempts to improve its activity and utility [12]. These include totally synthetic molecules with the trioxane moiety included, such as arterolane tosylate (OZ277, Figure 1.1) [13], which is in Phase II trials under Ranbaxy, artemisinin dimers [14], and the amino-artemisinin, artemisone [15].

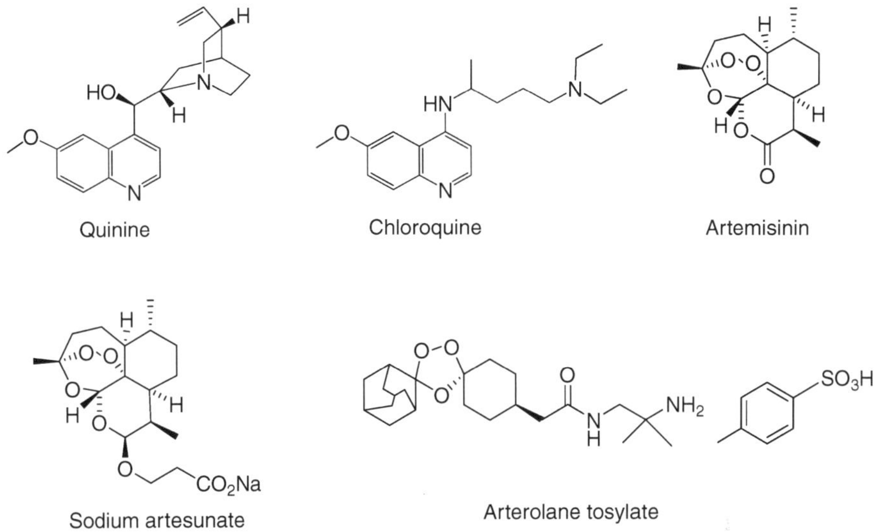

FIGURE 1.1 Natural antimalarial agents and analogs.

Resistance to artemisinin-based drugs is now being observed [16]. In order to counter this development, variations on the basic structure have been launched in combination with other antimalarials (usually variations on the chloroquine structure) such as dihydroartemisinin and piperaquine phosphate (Artekin®); artemether and lumefantrine (Coartem®); artesunate and mefloquine (Artequin®); and artesunate, sulfamethoxypyrazine, and pyrimethamine (Co-Arinate®). Currently there is one other fixed dose combination with an artemisinin derivative in Phase III clinical trials, pyronaridine/artesunate (Pyramax®) [17,18].

While artemisinin and more soluble derivatives have altered the treatment of resistant malaria, the costs of collection of sufficient quantities of the source plants is high, and the overall cost of the drugs may exceed what can be afforded by the countries where the drug is required for general treatment. In an attempt to avoid dependence on wild or even cultivated plant harvesting and thereby reduce costs, the Keasling group, in conjunction with the Gates Foundation and Amyris Pharmaceuticals, has transferred the genes from the producing plant into *Escherichia coli* and also *Saccharomyces cerevisiae*. They have successfully expressed the base terpene (amorpha-4,11-diene) and followed up with modification of the base structure both chemically, and to some extent, biochemically via P450 enzymes [19]. Titers exceeding 25 g/L of amorpha-4,11-diene have been produced by fermentation and are followed by chemical conversion to artemisinin, thereby allowing for the development of a potentially viable process to provide an alternative source of artemisinin [20].

Other significant drugs developed from traditional medicinal plants include: the antihypertensive agent, reserpine, isolated from *Rauwolfia serpentina* used in Ayurvedic medicine for the treatment of snakebite and other ailments [3]; ephedrine, from

Ephedra sinica (Ma Huang), a plant long used in traditional Chinese medicine, and the basis for the synthesis of the antiasthma agents (beta agonists), salbutamol and salmetrol; and the muscle relaxant, tubocurarine, isolated from *Chondrodendron* and *Curarea* species used by indigenous groups in the Amazon as the basis for the arrow poison, curare [9]. Although plants have a long history of use in the treatment of cancer [21], cancer, as a specific disease entity, is likely to be poorly defined in terms of folklore and traditional medicine, and consequently many of the claims for the efficacy of such treatment should be viewed with some skepticism [22]. Of the plant-derived anticancer drugs in clinical use, some of the best known are the so-called vinca alkaloids, vinblastine and vincristine, isolated from the Madagascar periwinkle, *Catharanthus roseus*; etoposide and teniposide which are semisynthetic derivatives of the natural product epipodophyllotoxin; paclitaxel (Taxol®), which occurs along with several key precursors (the baccatins) in the leaves of various *Taxus* species, and the semisynthetic analog, docetaxel (Taxotere®); and topotecan (hycamptamine), irinotecan (CPT-11), 9-amino- and 9-nitro-camptothecin, all semisynthetically derived from camptothecin, isolated from the Chinese ornamental tree, *Camptotheca acuminata.* These agents together with other plant-derived anticancer agents have been reviewed [23,24].

1.3 THE ROLE OF MARINE ORGANISMS IN DRUG DISCOVERY

While marine organisms do not have a significant history of use in traditional medicine, the world's oceans, covering more than 70% of the earth's surface, represent an enormous resource for the discovery of potential chemotherapeutic agents. Of the 33 animal phyla listed by Margulis and Schwartz [25], 32 are represented in aquatic environments, with 15 being exclusively marine, 17 marine and nonmarine (with 5 of these having more than 95% of their species only in marine environments), and only 1, *Onychophora*, being exclusively nonmarine. With the development of reliable scuba diving techniques enabling the routine accessibility of depths close to 40 m, the marine environment has been increasingly explored as a source of novel bioactive agents. The use of remotely operated vehicles (ROVs) permits the performance of selective deepwater collections with minimal environmental damage, but the high cost of ROV operations precludes their extensive use in routine collection operations. The systematic investigation of marine environments as sources of novel biologically active agents only began in earnest in the mid-1970s, and the rapidly increasing pace of these investigations over the recent decades has clearly demonstrated that the marine environment is a rich source of bioactive compounds, many of which belong to totally novel chemical classes not found in terrestrial sources [26].

While the focus of research has been on the discovery of potential new anticancer agents [23], the first marine-derived product to gain approval as a drug was Ziconotide, a non-narcotic analgesic that is currently marketed as Prialt® [27]. This compound is a constituent of combinatorial libraries of several hundred peptides that serve as the venom injected by species of the cone snail genus, *Conus*, to stun their

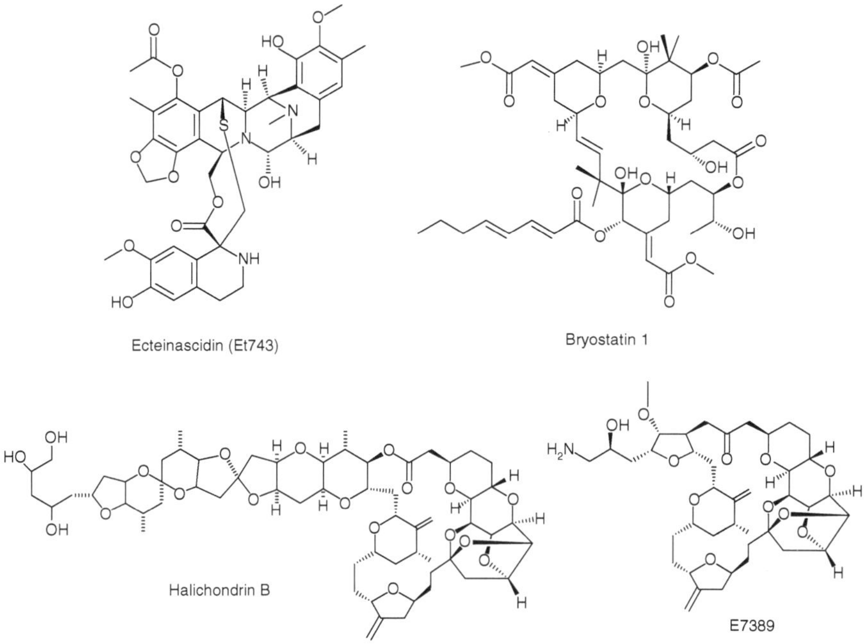

FIGURE 1.2 Some marine-derived bioactive agents.

prey prior to capture [28]. As alluded to above, however, the focus has been on the discovery of potential new anticancer agents. These include the complex alkaloid ecteinascidin 743 (Figure 1.2) isolated from the colonial tunicate *Ecteinascidia turbinata*, which was launched for the treatment of soft tissue sarcomas (STS) in Europe in late 2007 and for the treatment of relapsed ovarian cancer in Europe and the United States in 2009 under the name Yondelis®; it is also in a Phase III trial against ovarian cancer and in Phase II trials for breast, prostate, and pediatric sarcomas. The development of a semisynthetic route from the microbial product cyanosafracin B has solved the issue of compound supply, always a problem with marine-sourced materials [23,24,29]. Further examples of marine-derived anticancer agents are halichondrin B (Figure 1.2) [23,24,30], a complex polyether isolated in very low yield from several sponge sources, and bryostatin 1 (Figure 1.2), another complex macrolide originally isolated by Pettit and his collaborators from the bryozoan, *Bugula neritina* [23,24,31].

Studies on the total synthesis of halichondrin B revealed that the right-hand half of molecule retained all or most of the potency of the parent compound, and large-scale synthesis of the analog E7389 (Eribulin; Figure 1.2) provided adequate supplies for advanced preclinical and clinical development (Section 1.9.1.3). It is currently in advanced clinical trials against a range of cancers, and has shown promising activity in Phase III trials in patients with recurrent or metastatic breast cancer [32]. Bryostatin 1 has excellent antitumor activity, and despite major supply problems, enough cGMP-grade material was isolated from wild collections to provide sufficient material for clinical trials, which indicated that the drug would be most effective in combination therapy; several Phase II trials of this nature are in progress. Synthetic studies have focused on the preparation of simpler analogs possessing comparable or better activity, particularly related to binding to some of the protein kinase C isozymes, which are the main mechanistic target of the bryostatins. The result has been the preparation of compounds, bryologs, with greater potency than bryostatin 1 in *in vitro* cell line assays [24,31]. These agents together with several other promising marine-derived anticancer agents have been reviewed [23,24].

1.4 THE ROLE OF MICROORGANISMS IN DRUG DISCOVERY: AN HISTORICAL PERSPECTIVE

With the serendipitous discovery of penicillin from the filamentous fungus, *Penicillium notatum*, by Fleming in 1929, and the observation of its broad therapeutic use in the 1940s, a new era in medicine, "The Golden Age of Antibiotics," was born, leading to the intensive investigation of Nature as a source of novel bioactive agents [33]. Microorganisms are a prolific source of structurally diverse bioactive metabolites and have led to the development of some of the most important products of the pharmaceutical industry (see Figure 1.3). These include antibacterial agents, such as the penicillins (from *Penicillium* species), cephalosporins (from *Cephalosporium acremonium*), aminoglycosides, tetracyclines, and other polyketides of many structural types (from the *Actinomycetales*); immunosuppressive agents, such as the cyclosporins (from *Trichoderma* and *Tolypocladium* species) and rapamycin (from *Streptomyces* species) (Figure 1.3); cholesterol lowering agents, such as mevastatin (compactin; from *Penicillium* species; Figure 1.3) and lovastatin (from *Aspergillus* species); and anthelmintics and antiparasitic drugs, such as the ivermectins (from *Streptomyces* species) [9].

Microbes are also a source of some of the most important cancer chemotherapeutic agents, the so-called antitumor antibiotics, which include members of the anthracycline [34], bleomycin [35], mitomycin [36], the enediynes [37], and the staurosporines [38], all isolated from various *Streptomyces* species. Clinically useful agents from these families are daunomycin (Figure 1.3) and related agents, doxorubicin, idarubicin, and epirubicin; the glycopeptidic bleomycins A_2 and B_2 (blenoxane®); the mitosanes such as mitomycin C (Figure 1.3); and the enediynes exemplified by the monoclonal antibody-linked calicheamicin conjugate, Mylotarg® (Figure 1.3).

More recent additions to the microbial antitumor armamentarium are the macrocyclic epothilones, such as epothilones A and B (Figure 1.3), isolated from myxobacteria which have a mechanism of action similar to that of Taxol, and which have been extensively studied, both from the biosynthetic and synthetic standpoints [23,24,39]. Currently, there are at least 16 molecular entities in varying stages of testing ranging from biological testing to Phase III clinical trials. Among those in clinical trials (completed, active, or recruiting) as of July, 2010, either as single agents or in combination with other agents (http://www.clinicaltrials.gov/) are ixabepilone

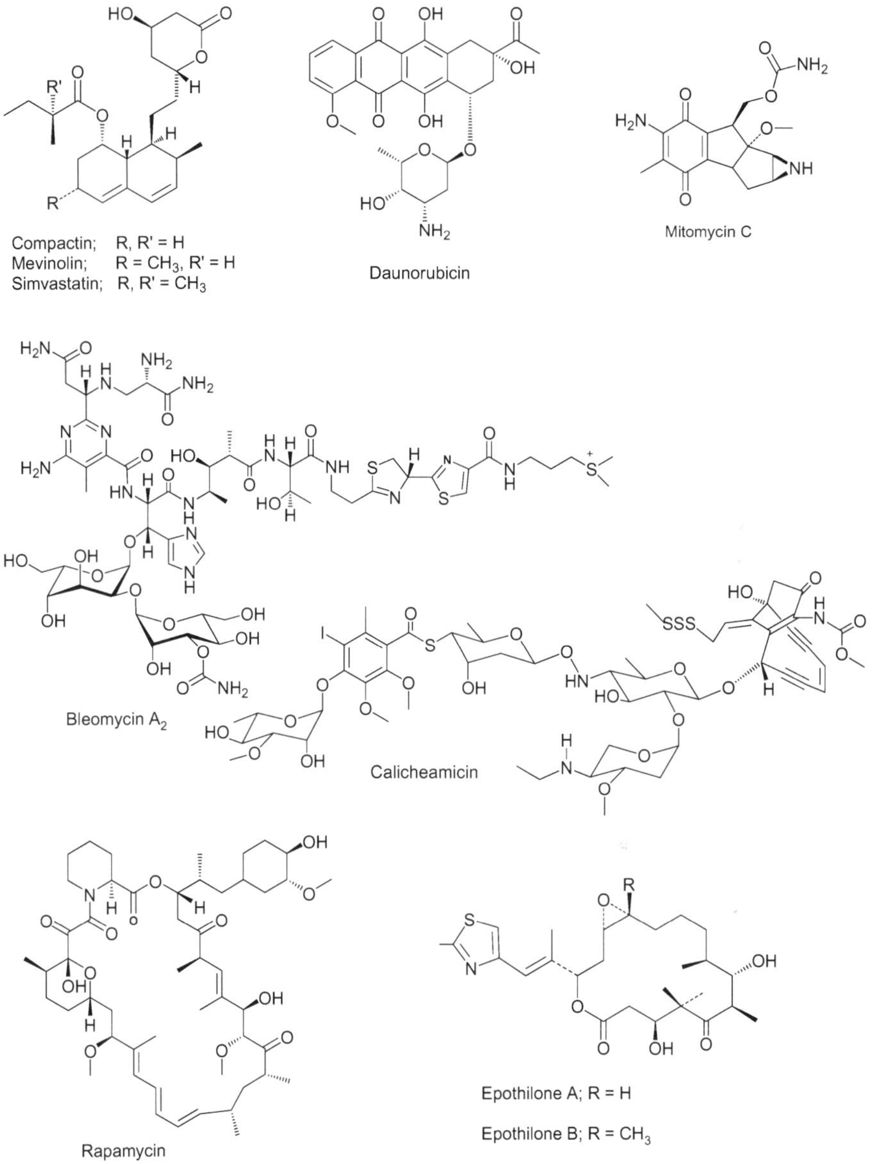

FIGURE 1.3 Microbial-derived drugs.

or 16-azaepothilone B (148 trials), which was approved in October, 2007, by the U.S. Food and Drug Administration for the treatment of aggressive metastatic or locally advanced breast cancer refractory to currently available chemotherapies; epothilone B, patupilone, or EPO-906 (54 trials); epothilone D, 9,10-didehydroepothilone D or KOS-1584 (33 trials); and sagopilone, a fully synthetic analog (14 trials completed or active).

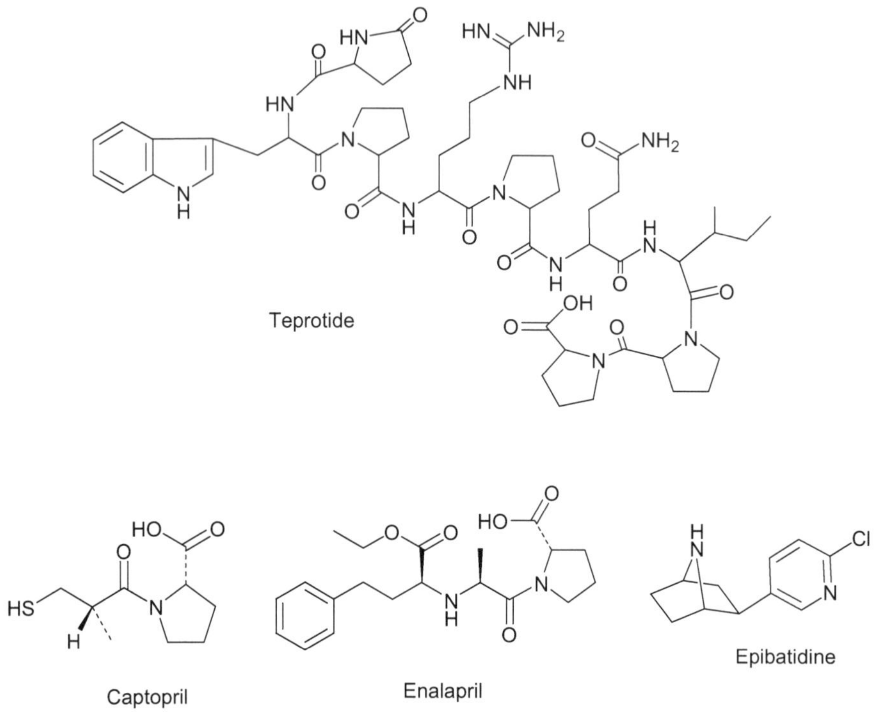

FIGURE 1.4 Leads and drugs from amphibian and reptilian sources.

1.5 OTHER SOURCES

The venom of the pit viper, *Bothrops jararaca*, yielded teprotide which led to the design and synthesis of the ACE inhibitors, captopril and enalapril (Figure 1.4) [9], used in the treatment of cardiovascular disease, while a novel class of painkillers has been developed based on epibatidine (Figure 1.4), isolated from the skin of the poisonous frog, *Epipedobates tricolor* [40]. Another significant discovery was the isolation of exendin-4 from the venom of the Gila monster, *Heloderma suspectum* [41], which led to the development of the polypeptide, exenatide, which was approved by the FDA in April 2005 for the treatment of diabetes mellitus type 2 and is marketed as Byetta.

1.6 THE IMPORTANCE OF NATURAL PRODUCTS IN DRUG DISCOVERY AND DEVELOPMENT

The drugs listed in Sections 1.2–1.5 represent but a small fraction of the total number of bioactive compounds discovered from natural sources, many of which have been used in their original unmodified form, or which have been structurally modified, to yield effective drugs for the treatment of a wide range of diseases. Still more have served as lead compounds that have provided the basis and inspiration for the design and synthesis of a wide range of effective drugs. In a paper published in 2007 [42], Newman and Cragg analyzed the sources of new drugs over the period 01/1981–06/2006, and classified these compounds as N (an unmodified natural product), ND (a modified natural product), S (a synthetic compound with no natural product conception), S*, S*/NM (a synthetic compound with a natural product pharmacophore; /NM indicating competitive inhibition), and S/NM (a synthetic compound showing competitive inhibition of the natural product substrate). A recently updated analysis (October, 2008) using the same coding as above has indicated that, while 67% of the 1024 small molecule, new chemical entities (NCEs) are formally synthetic, 18% correspond to synthetic molecules containing pharmacophores derived directly from natural products classified as S* and S*/NM, and 12% are actually modeled on a natural product inhibitor of the molecular target of interest, or mimic (i.e., competitively inhibit) the endogenous substrate of the active site, such as ATP (S/NM). Thus, only 37% of the 1024 NCEs can be classified as truly synthetic (i.e., devoid of natural inspiration) in origin (S). In certain disease categories such as anti-infectives (anti-bacterial, -fungal, -parasitic, and -viral), 68.3% are naturally derived or inspired (N; ND; S*; S*/NM; S/NM), while in the cancer treatment area 79.8% are in this category, with the figure being 62.9% if the S/NM category is excluded [24].

The brief discussion in Sections 1.1–1.6 above highlights the immense complexity and molecular diversity of natural products, and it should be clear to most observers that very few would have been discovered without the application of natural products chemistry. In addition to structural diversity, however, natural products offer another important feature of considerable interest to those involved in drug development, namely that they often possess highly selective and specific biological activities based on mechanisms of action. Two excellent examples are the HMG-CoA reductase inhibition exhibited by statins such as simvastatin (Figure 1.3), and the tubulin-assembly promotion activity of paclitaxel, neither of which would have been discovered without the natural product leads and investigation of their mechanisms of action. The bioactivity of natural products stems from the hypothesis that essentially all natural products have some receptor-binding activity; the problem is to find to which receptor a given natural product is binding [43]. Experience shows that organisms often provide investigators with complex libraries of unique bioactive constituents, analogous to the libraries of crude synthetic products initially produced by combinatorial chemistry techniques. Thus, the natural products approach can be viewed as complementary to the synthetic approach, each providing access to (initially) different lead structures. Indeed, as discussed below, combinatorial chemistry is an extremely powerful tool for the optimization of an active natural product structure (natural product lead), and the task of the natural products researcher is thus to select those initial lead compounds of pharmacological interest from the "natural combinatorial libraries" produced by extraction of organisms.

1.7 CLASSICAL NATURAL SOURCES: UNTAPPED POTENTIAL

The exceptional complexity and molecular diversity of natural products has been emphasized in earlier sections, but even more remarkable is the fact that the potential of these unique natural resources has barely been explored. Despite the intensive investigation of terrestrial flora, it is estimated that only 6% of the approximately 300,000 species (some estimates are as high as 500,000 species) of higher plants have been systematically investigated pharmacologically, and only some 15% phytochemically [8,44,45]. The potential of the marine environment as a source of novel drugs remains virtually unexplored [46], and until recently, investigations have largely been restricted to tropical and subtropical regions; however, the exploration is being expanded to colder regions including Antarctica [47–49].

The selective and reproducible production of bioactive compounds has been induced through exposure of the roots of hydroponically grown plants to chemical elicitors [50]. In addition, non-natural analogs of natural metabolites can be produced by the feeding of seedlings with derivatives of selected biosynthetic precursors, as illustrated by the production of non-natural terpene indole alkaloids related to the vinca alkaloids through the feeding of seedlings of *C. roseus* with various tryptamine analogs [51].

1.8 THE UNEXPLORED POTENTIAL OF MICROBIAL DIVERSITY

Until fairly recently, the inability to cultivate most naturally occurring microorganisms has severely restricted the study of natural microbial ecosystems, and it has been estimated that less than 1% of microorganisms seen microscopically have been cultivated. Yet, despite this limitation, an impressive number of highly effective microbially derived chemotherapeutic agents has been discovered and developed thus far. Given the observation that "a handful of soil contain billions of microbial organisms" [52], and the assertion that "the workings of the biosphere depend absolutely on the activities of the microbial world" [53], the microbial universe clearly presents a vast untapped resource for drug discovery [54]. In addition, the substantial advances in the understanding of the gene clusters encoding multimodular enzymes involved in the biosynthesis of a host of microbial secondary metabolites, such as polyketide synthases (PKSs) and/or nonribosomal peptide synthetases (NRPSs), has enabled the sequencing and detailed analysis of the genomes of long-studied microbes, such as *Streptomyces avermitilis*. These studies have revealed the presence of additional PKS and NRPS clusters leading to the discovery of novel secondary metabolites not detected in standard fermentation and isolation processes [55]. Such genome mining has been used in the discovery of a novel peptide, coelichelin, from the soil bacterium, *Streptomyces coelicolor* [56], and this concept is further expanded on in the discussion in Section 1.8.3.

1.8.1 Improved Culturing Procedures

Relatively recent developments of procedures for cultivating and identifying microorganisms are aiding microbiologists in their assessment of the earth's full range of microbial diversity [57]. For example, using "nutrient-sparse" media simulating the original natural environment has enabled the massive parallel cultivation of gel-encapsulated single cells (gel microdroplets; GMDs) derived from microbes separated from environmental samples (seawater and soil) [58]. Thus, "the simultaneous and relatively noncompetitive growth of both slow- and fast-growing microorganisms" has been achieved, thereby preventing the overgrowth by fast-growing "microbial weeds," and resulting in the identification of previously undetected species (using 16S rRNA gene sequencing), as well as the culturing and scale-up cultivation of previously uncultivated microbes [58]. The cultivation of uncultivated members of bacteria of the divisions *Acidobacteria* and *Verrucomicrobia* from agricultural soil and from the guts of wood-feeding termites has been reported using "nutrient-sparse" agar media under hypoxic or anoxic conditions for periods over 30 days, with the addition of humic acids or their analogs, as well as quorum-signaling compounds [59]. The need for the presence of small signaling molecules, such as short peptides, in the media has also been shown to be necessary for initiating growth of the otherwise "uncultivable" strain *Psychrobacter* sp. strain MSC33 [60]. Another recent method involves the sandwiching of innocula of selected sediment samples between semipermeable membranes contained in diffusion chambers which are returned to their natural source environments for periods of 4 weeks, thereby allowing free exchange of chemicals by diffusion with the external environment [61]. This method has been adapted to a novel high-throughput technique for the *in situ* parallel cultivation and isolation of previously uncultivated microbes from a range of environments [62]. These developments have been accompanied by progress in the cultivation of marine microbes as reported in a paper on the cultivation of Gram-positive marine microbes [63]. Thus, the potential for discovery of novel bioactive agents is immense.

1.8.2 Extraction of Environmental Samples (the Metagenome)

Despite improvement in culturing techniques, greater than 99% of microscopically observed microbes still defy culture. Extraction of nucleic acids (the metagenome) from environmental samples, however, permits the identification of uncultured microorganisms through the isolation and sequencing of ribosomal RNA or rDNA (genes encoding for rRNA). A recent review covers the significant achievements thus far in this area of research [64], as well as the huge potential for the discovery of novel bioactive compounds from as yet unculturable microbes [65]. Samples from soils and seawater are currently being investigated [66,67], and whole-genome shotgun sequencing of environmental-pooled DNA obtained from water samples collected in the Sargasso Sea off the coast of Bermuda by the Venter group, indicated the presence of at least 1800 genomic species, which included 148 previously unknown bacterial phylotypes [67]. Venter et al. are also examining microbial communities in water samples collected by the *Sorcerer II* Global Ocean Sampling (GOS) expedition, and their data predict more than six million proteins, nearly twice the number of proteins present in current databases, with some of the predicted proteins bearing no similarity to any currently known proteins, and therefore, representing new families [68]. Similar methods are also being applied to the investigation of other habitats, such as the microflora of insects [69] and marine animals [70].

Soil is reported to be the most biodiverse environment on Earth, with estimates of approximately 1000 Gbp of microbial genome sequences per gram of soil compared to the Human Genome project in which 3 Gbp were sequenced, and sequencing projects that target other microbial habitats, such as the Sargasso Sea in which 6 Gbp were sequenced [71]. The soil microbial community thus constitutes a vast resource of genes and pathways having substantial potential for the discovery of novel drug leads and generating useful information for a multitude of other processes. Thus, sequencing of the soil metagenome presents a new and ambitious challenge, which should bring considerable economic and environmental value. With this in mind, a coordinated international effort is being established aiming at combining the skills of the global scientific community to focus on sequencing and annotating the soil metagenome; the TerraGenome International Sequencing Consortium (http://www.terragenome.org/) has been launched to coordinate these efforts [71]. As a start, it is proposed to completely sequence a "reference" soil metagenome from a specific site in the United Kingdom, with the aim of generating information on several biosynthetic processes including drug discovery. Evidence that such an ambitious project is feasible and suitable tools are available is clear from a recently published report by van Elsas et al. on the Metacontrol project [72].

The cloning and understanding of the novel genes discovered through these processes, and the heterologous expression of gene clusters encoding the enzymes involved in biosynthetic pathways in viable host organisms, such as *E. coli*, should permit the production of novel metabolites produced from as yet uncultured microbes. The production of the antibiotic, pantocin A (Figure 1.5), from the bacterium, *Pantoea agglomerans*, is an example of such heterologous expression of genomic DNA [73]. The production of pantocin A by the source microbe grown in liquid culture proved to be impractical due to low titers and the complexity of the mixture of metabolites produced. Expression of a genomic DNA library isolated from *P. agglomerans* in *E. coli*, however, provided access to reasonable quantities of the small molecule antibiotics of interest [73].

1.8.3 Cryptic Clusters in Bacteria and Fungi

As has been mentioned in the introduction to Section 1.8, advances in the understanding of the gene clusters encoding multimodular enzymes involved in the biosynthesis of many microbial secondary metabolites, coupled with advances in the sequencing and detailed analysis of the genomes of long-studied microbes, such as *S. avermitilis*, has led to the discovery of novel secondary metabolites not detected in standard fermentation and isolation processes [55]. The early work on the numbers of such clusters in an individual microbe was mainly performed on the genomes of two very important *Streptomyces* species, *S. avermitilis* (where the number of putative

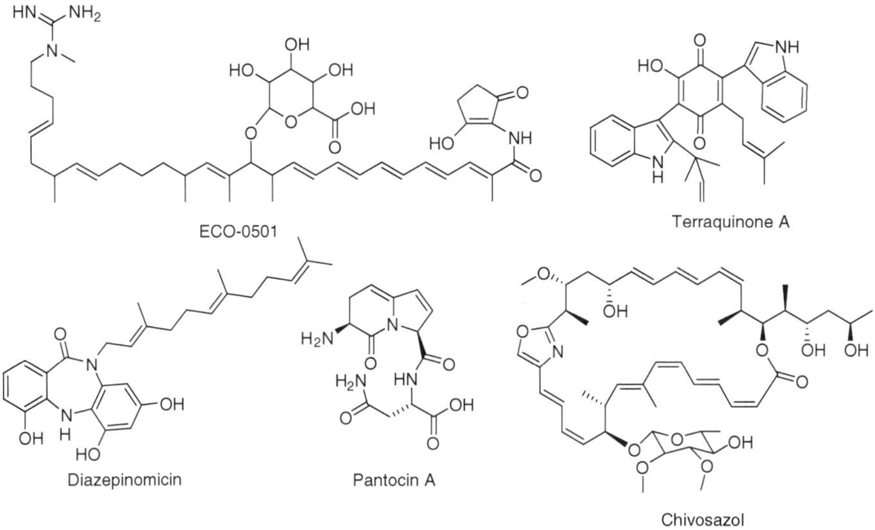

FIGURE 1.5 New compounds from genome mining.

clusters reached into the low 30s) [74,75], and *S. coelicolor* (where numbers are now reaching into the high teens to low 20s) [76]. From this pioneering work, it is now becoming evident that the genomes of the *Streptomycetes* and, by extension, Actinomycetes in general, contain large numbers of previously unrecognized secondary metabolite clusters. Thus, analysis of the genomic sequence of the well-known vancomycin producer, *Amycolatopsis orientalis* (ATCC 43491), led to the prediction of the molecular weight of the novel antibiotic ECO-0501, which following detection by high-performance liquid chromatography-mass spectroscopy (HPLC-MS), led to its isolation (Figure 1.5) [77]. The compound had a very similar biological profile to vancomycin, but it was masked by this compound. Another example illustrating the use of the same technique is the anticancer agent diazepinomicin (Figure 1.5), currently in Phase II clinical trials [78].

Genomic analyses have also been applied to the myxobacteria. The identification of the gene, *ChiR*, controlling the production of the extremely potent antifungal antibiotic, chivosazol (Figure 1.5), has been reported [79]; also discussed in this paper is the major problem of the identification and application of the transcriptional control mechanisms involved in secondary metabolite expression, whether in homologous or heterologous hosts [79]. Further work on the genetics and other aspects secondary metabolite production by myxobacteria have been reviewed by Weissman and Müller [80].

Genomic analysis of the fungus, *Aspergillus nidulans*, suggested the presence of clustered secondary metabolite genes having the potential to generate up to 27 polyketides, 14 nonribosomal peptides, 1 terpene, and 2 indole alkaloids, as well as identifying the potential controller of expression of these clusters [81]; this was demonstrated by expressing terrequinone A (Figure 1.5), a compound not previously reported from this species. Similar predictions can be made for *Aspergillus fumigatus* and *Aspergillus oryzae* based on the analysis of the potential number of secondary metabolite clusters in these fungi [81]. A recent review has expanded the discussion on the control of secondary metabolite production in fungi [82].

1.8.4 Marine Microbes

Deep-ocean sediments are proving to be a valuable source of new actinomycete bacteria that are unique to the marine environment [63,83]. Use of a combination of culture and phylogenetic approaches has led to the description of the first truly marine actinomycete genus named *Salinispora* [63,84,85], and its members are proving to be ubiquitous, being found in concentrations of up to 10^4 per mL in sediments on tropical ocean bottoms and in more shallow waters, as well as appearing on the surfaces of numerous marine plants and animals. On culturing using the appropriate selective isolation techniques, significant antibiotic and cytotoxic activity has been observed, and has resulted in the isolation of a potent cytotoxin, salinosporamide A (Figure 1.6), a very potent proteasome inhibitor ($IC_{50} = 1.3$ nM) [86,87], currently in Phase I clinical trials as NPI-0052. The discovery and development of this compound and other salinosporamides have been reviewed [88]. More recently, the isolation and cultivation of another new actinomycete genus, named *Marinispora*, has been reported, and novel macrolides called marinomycins have been isolated [89]. Marinomycins A–D (Figure 1.6) show potent activity against drug-resistant bacterial pathogens and some melanomas [89]. A recent review discusses the isolation of these and over 60 other antitumor compounds from marine actinomycetes [83]. Recent publications on the novel and diverse chemistry of these new microbial genera include the isolation of potential chemopreventive agents, saliniketals A and B from *Salinispora arenicola* [90], while two new cyclic peptides, thalassospiramides A and B, possessing immunosuppressive activity have been isolated from a new member of the marine alpha-proteobacterium *Thalassospira* [91]. Extensive screening of extracts of actinomycete bacteria isolated and characterized from the sediment samples collected in one of the largest Norwegian fjords, the Trondheim fjord, showed strong antimicrobial activity, clearly demonstrating that actinomycetes from marine sediments in Norwegian fjords can be potential sources for the discovery of novel anti-infective agents [92].

1.8.5 Cyanophytes

Cyanophytes, which are actually prokaryotes often referred as blue-green algae in early papers, are prolific producers of bioactive secondary metabolites. Reviews by Welker and von Döhren [93], Tan [94], and Tidgewell et al. [95], provide overviews of the multiplicity of molecules isolated from either simple extraction of wild harvests or fermentation of purified organisms cultured from collections (unialgal but, especially in the case of the filamentous forms, not necessarily axenic).

The pioneering work of Moore and Patterson in Hawaii on cyanobacteria as sources of potential anticancer agents, yielded one cyanobacterial secondary metabolite, cryptophycin (Figure 1.6), which led to a large synthetic program, with a derivative, cryptophycin 52 (Figure 1.6), reaching Phase II clinical trials in cancer [96]. Currently, cryptophycin 52 is no longer in trials due to toxicity, but the

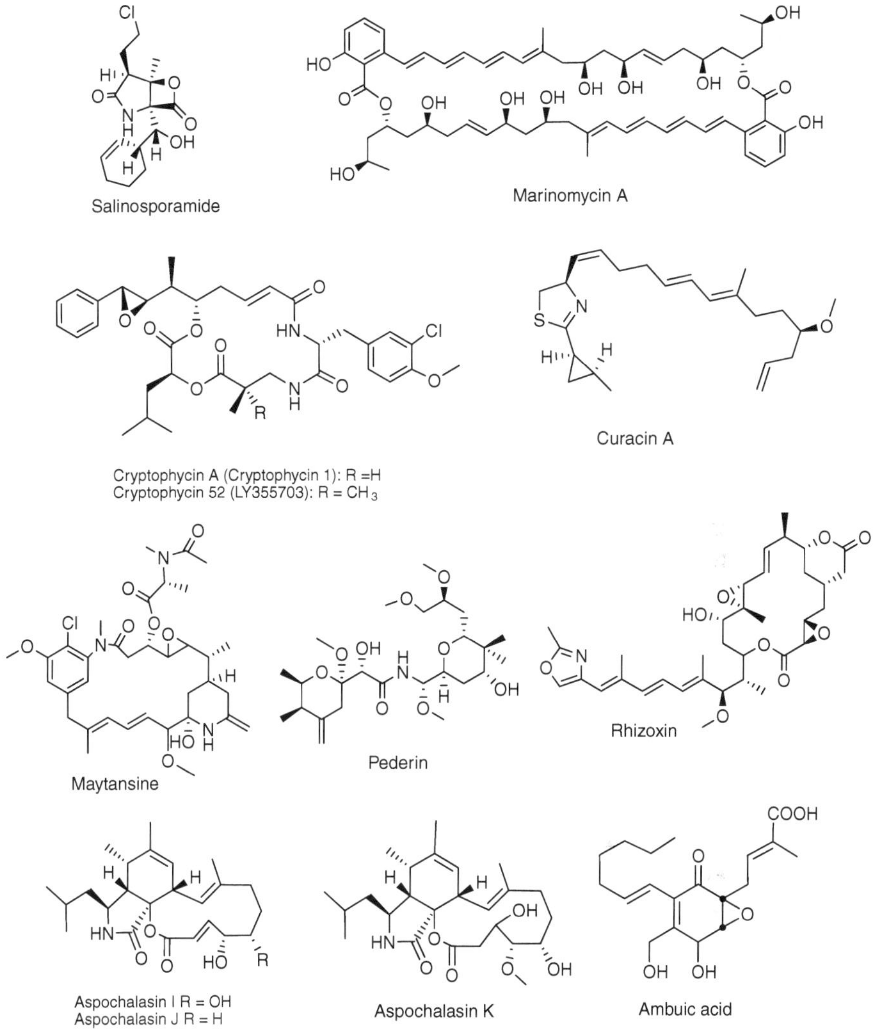

FIGURE 1.6 Bioactive compounds from relatively unexplored microbial sources.

complete biosynthetic pathway of the base molecule has been identified and cloned [97]. Later studies by the Gerwick group yielded curacin A (Figure 1.6) from *Lyngbya majuscula.* Curacin A had excellent *in vitro* cytotoxic activity and is a tubulin interactive agent, but it was effectively insoluble in any formulation that was compatible with *in vivo* testing in animals. The complete biosynthetic cluster was located and cloned by following a series of stable isotope feeding experiments that both identified the precursors and allowed the identification and cloning of the biosynthetic pathway [98]. The actual spatial production of curacin A and other secondary metabolites within *L. majuscula* in the presence of other cyanophytes has been reported, and illustrates the impact of modern instrumentation on such studies [99], and the transcription and regulation of secondary metabolism in this marine cyanobacterium has been characterized [100]. The possibilities of *in vitro* manipulation of such a cluster have recently been reviewed by Walsh's group [101].

1.8.6 Microbial Symbionts

It is now fairly well established that many bioactive compounds isolated from various macroorganisms which can include plants, marine and terrestrial invertebrates, and even fungi, are actually metabolites synthesized by symbiotic bacteria [102–105]. These include the anticancer compounds, the maytansinoids (Figure 1.6), originally isolated from several plant genera of the Celastraceae family [106], and the pederins (Figure 1.6), isolated from beetles of the genera *Paederus* and *Paederidus*, as well as derivatives based on the pederin skeleton from several marine sponges [107–109]. These "pederine-like molecules," now numbering more than 34, are isolated from at least 8 different animal genera, and these symbiotic sources, together with a range of antitumor agents isolated from marine organisms that closely resemble bacterial metabolites, have been discussed in detail in the reviews by Piel [102,105]. In addition to documenting the wide range of structures and potential producers that have so far been identified, these articles [102,105] also elaborate on what can now be done using this information to produce novel but unnatural variations on known potential anticancer agents from these sources. For example, reaction of mycalamide A with the *PedO* gene product has generated the novel biosynthetic hybrid, 18-*O*-methylmycalamide A, which has increased cytotoxicity compared to the parent compound [110].

An interesting example of a complex symbiotic–pathogenic relationship involving a bacterium–fungus–plant interaction has been discovered in the case of rice seedling blight. The toxic metabolite, rhizoxin (Figure 1.6), originally isolated from the contaminating *Rhizopus* fungus, has actually been found to be produced by an endo-symbiotic *Burkholderia* bacterial species [111]. Rhizoxin exhibits potent antitumor activity, but its further development as an anticancer drug has been precluded by toxicity problems. Thus, in addition to offering potentially new avenues for pest control, this unexpected finding has enabled the isolation of significantly higher yields of rhizoxin as well as rhizoxin analogs through the large-scale cultivation of the bacterium independently of the fungal host [112]. This may have significant implications in the development of rhizoxin analogs with improved pharmacological properties.

As illustrated for rhizoxin above, the production of important bioactive agents by symbiotic microbes offers potentially viable solutions to the supply problem which, in particular, often hinder the advanced development of marine-derived agents. Even though most symbionts may remain uncultivated, bacterial production systems might be established by the isolation of biosynthetic genes from marine metagenomes, and expressing them in culturable bacterial hosts [113,114].

1.8.7 Plant Endophytes

As indicated in Section 1.2, plants have been relatively extensively studied as sources of bioactive metabolites, but the role of endophytic microbes that reside in the tissues between living plant cells has only recently started to receive attention. Relationships between endophytes and their host plants may vary from symbiotic to pathogenic, and studies have revealed an interesting realm of novel chemistry [115–117]. The wide range of new bioactive molecules reported include novel wide-spectrum antibiotics, kakadumycins, isolated from an endophytic *Streptomycete* associated with the fern leafed grevillea (*Grevillea pteridifolia*) from the Northern Territory of Australia [118]; ambuic acid (Figure 1.6), an antifungal agent, which recently has been described from several isolates of *Pestalotiopsis microspora* found in many of the world's rainforests [119]; peptide antibiotics, the coronamycins, from a *Streptomyces* species associated with an epiphytic vine (*Monastera* species) found in the Peruvian Amazon [120]; and cytotoxic aspochalasins I, J, and K (Figure 1.6), isolated from endophytes of plants from the southwestern desert regions of the United States [121].

The discovery that various important anticancer agents are produced in small quantities by endophytic fungi isolated from plants is of particular significance. Examples are Taxol from *Taxomyces* [122] and many *Pestalotiopsis* species [123], as well as camptothecin [124,125], podophyllotoxin, an epimer of the precursor to the anticancer drug etoposide [126,127], vinblastine [128], and vincristine [129,130], all produced in relatively small amounts by endophytic fungi isolated from the producing plants. The fact that these compounds have been shown not to be artifacts offers the prospect for their increased production, provided the gene/gene product controlling their production by the relevant endophytes can be identified. Similar discoveries could provide an entry into greatly increased production of other key bioactive natural products.

1.8.8 Extremophiles

Extremophilic microbes (extremophiles) abound in extreme habitats, including deep-sea sediments and vents (see Section 1.8.4). Extremophiles include acidophiles (acidic sulfurous hot springs), alkalophiles (alkaline lakes), halophiles (salt lakes), piezo (baro)- and (hyper)thermophiles (deep-sea sediments and vents) [131–136], and psychrophiles (arctic and antarctic waters, alpine lakes) [137–139]. Until recently, investigations have centered on the isolation of thermophilic and hyper-thermophilic enzymes (extremozymes) [140–143], but there is increasing evidence that these extreme environments are also yielding novel bioactive chemotypes.

Recent reviews highlight the promising range of compounds isolated thus far from extremophilic microbes [144,145]. Abandoned mine-waste disposal sites have yielded unusual acidophiles, which thrive in the acidic, metal-rich waters, polluted environments, which are generally toxic to most prokaryotic and eukaryotic organisms [146]. The novel sesquiterpenoid and polyketide–terpenoid metabolites, berkeleydione and berkeleytrione (Figure 1.7) showing activity against metallopro-teinase-3 and caspase-1, activities relevant to cancer, Huntington's disease and other diseases, have been isolated from *Penicillium* species found in the surface waters of Berkeley Pit Lake in Montana [147–149]. An interesting observation has been the isolation of the diketopiperazine disulfide, glionitrin A (Figure 1.7), from a coculture of a *Sphingomonas* bacterial strain and an *A. fumigatus* fungal strain, both derived from coal mine drainages [150]; the compound was not produced in detectable amounts by either strain grown as a monoculture. Examples of novel compounds

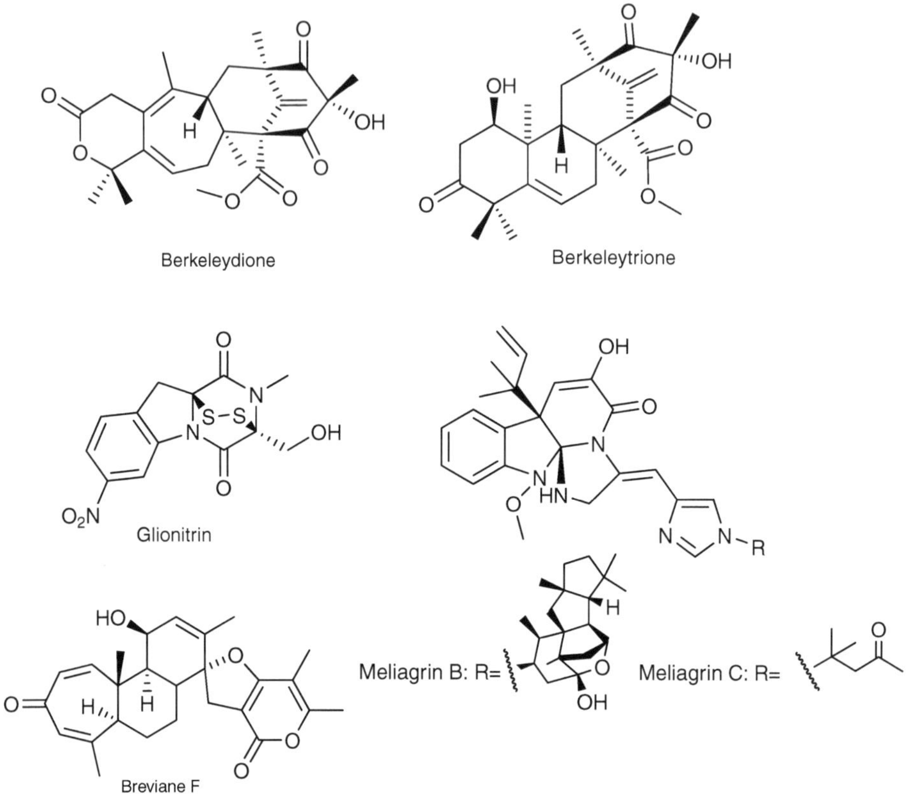

FIGURE 1.7 Novel compounds from extreme environments.

produced by deep-sea fungi (both *Penicillium* species) isolated from sediments collected at depths exceeding 5000 m, are the spiroditerpenoids, brevianes F–H (Figure 1.7) [151], and four new alkaloids, including two new meleagrin analogs, meleagrins B and C (Figure 1.7), two new diketopiperazines, roquefortines F and G, and six new diterpenes, conidiogenones B–G [152].

1.8.9 Combinatorial Biosynthesis

The substantial advances made in the understanding of the role of multifunctional polyketide synthase enzymes in bacterial aromatic polyketide biosynthesis have led to the identification of many such enzymes, together with their encoding genes. These advances have been reported and reviewed in a plethora of articles. Readers are referred to some of the reviews published during the past decade for detailed coverage [153–161], as well as to some reviews specifically covering NRPSs, which are responsible for the biosynthesis of nonribosomal peptides (NRPs) [162–165]. The rapid developments in the analysis of microbial genomes has enabled the identification of a multitude of gene clusters encoding for polyketides, NRPs and hybrid polyketide–NRP metabolites, and have provided the tools for engineering the biosynthesis of novel "non-natural" natural products through gene shuffling, domain deletions, and mutations [155,166,167]. Results of the application of these combinatorial biosynthetic techniques to the production of novel analogs of anticancer agents such as anthracyclines, ansamitocins, epothilones, enediynes, and aminocoumarins have been reviewed by Shen et al. [168].

Some aspects related to the biosynthetic potential of myxobacteria [169], and to the production of novel "unnatural" epothilones [170] have been reviewed. An example illustrating the power of this technique is the efficient scale-up production of epothilone D, the des-epoxy precursor of epothilone B (Figure 1.3). It entered clinical trials as a potential anticancer agent under the code name of KOS-862, but has now been discontinued in favor of a congener, 9,10-didehydroepothilone D [171] known currently as KOS-1584, which is in Phase II clinical trials. The polyketide gene cluster producing epothilone B has been isolated and sequenced from two *Sorangium cellulosum* strains [172,173]. The epoxidation of epothilone D to epothilone B has been shown to be due to the last gene in the cluster, *epo*K, encoding a cytochrome P450, and heterologous expression of the gene cluster minus the *epo*K in *Myxococcus xanthus* has resulted in large-scale production of crystalline epothilone D [174].

1.9 DEVELOPMENT OF DRUGS FROM NATURAL PRODUCTS: A MULTIDISCIPLINARY PROCESS

Limited availability and structural complexity have historically been the major impediments to the development of clinically useful drugs from natural product leads. Quantities of the source biomass are often limited or, in the case of microbial sources, unculturable and the natural products themselves are often produced in trace quantities within the biomass. Advances in genomic mining and the engineering of biosynthetic pathways have revolutionized the discovery of novel natural products. These methods can also be utilized to enable large-scale production of natural products in the native or engineered organisms.

Despite several notable exceptions (e.g., adriamycin or taxanes in the antitumor area), the probability of a directly isolated natural product becoming the actual drug used for the treatment of a given disease in the future is relatively low. Natural molecules, however, can serve as lead compounds for the development of analogs, generated by combinatorial biosynthesis and/or medicinal chemistry, with optimized pharmacological properties. Evolution has selected natural products to bind to biological macromolecules and thus, natural products represent "privileged structures," [175] which are excellent templates for the synthesis of novel, biologically active, molecules. Advances in synthetic strategy and methodology are surmounting the barriers presented by the structural complexity of most natural products. Of course, evaluation and optimization of the structure–activity relationships (SARs) of the derived analogs from all these approaches requires suitable biological assays. Thus, effective natural product–based drug discovery and development requires a truly multidisciplinary, collaborative approach.

1.9.1 Synthesis Based on Natural Products

While natural products often exhibit highly potent and selective bioactivity, they underwent evolutionary selection to serve the needs of their producing organisms not to serve as human therapeutics, and thus have not been fine-tuned to possess the potency, selectivity, and pharmacokinetic properties desired in a clinically useful drug. Optimization to improve physicochemical and pharmacokinetic properties frequently entails modification, removal or introduction of functional groups and stereocenters, or more drastic remodeling of the basic scaffold. Combinatorial biosynthetic methods offer access to unique structural diversity, but this diversity is limited by the available biosynthetic pathways of the host organism. The power of synthetic chemistry can be harnessed to access a greater extent of possible modifications and structural diversity than biosynthetic methods alone.

1.9.1.1 Derivatization and Semisynthesis. Simple functional group transformations, achieved by chemical and/or enzymatic methods, are possibly the simplest approach to optimization of a natural product lead. Large numbers of analogs can be rapidly generated by such semisynthetic approaches; however, the structural diversity of the analogs accessible by derivatization may be limited as many desired transformations cannot be accomplished due to incompatibilities with preexisting functional groups or the lack of a feasible reaction. There are numerous examples of this approach including the taxanes [176,177], camptothecins [178], and combretastatins [179].

Sometimes another, readily available natural product can serve as a starting material for the semisynthesis of a natural product of interest which itself is not readily available from biomass. The development of paclitaxel (Taxol) was severely hampered by the scarcity of *Taxus brevifolia*, whose bark was its original source. The compound supply issue and original commercial production were solved by semisynthesis from 10-deacetylbaccatin III, which is readily available from the needles of various *Taxus* species, a renewable resource. Details of the development of paclitaxel and other taxane derivatives have been comprehensively reviewed [176,177]. Another prominent example is provided by ecteinascidin 743 (Et-743, Yondelis; Figure 1.2). This complex alkaloid was discovered from the rare colonial tunicate *E. turbinata* [180,181]. The issue of compound supply for advanced studies was solved by the development of a semisynthetic route from the readily available microbial product cyanosafracin B [182; Section 1.3]. The discovery and development of ecteinascidin has been comprehensively reviewed [29,183].

1.9.1.2 Total Synthesis. Most of the dramatic advances in the field of synthetic organic chemistry are the result of the challenges posed by the total synthesis of complex natural products [184]. Adequate supply can be a serious limiting factor in the preclinical and clinical development of some naturally derived drugs, and the focus of many top synthetic groups on devising economically feasible synthetic strategies is a very welcome development for both clinicians conducting clinical trials and patient populations. An excellent example is the marine-derived anticancer agent discodermolide, where total synthesis provided sufficient quantities for thorough clinical trials [185]. Unfortunately, these trials have now been terminated due to lack of objective responses and toxicity [186].

1.9.1.3 Diverted Total Synthesis. The process of total synthesis can often lead to the identification of the pharmacophore, the arrangement of steric and electronic features necessary to ensure optimal interaction with a biological target and trigger or block its biological response. This insight, coupled with a synthetic strategy that facilitates introduction of deep-seated structural variation, allows for the "molecular editing" of unnecessary structural complexity. Although the basic strategy had been practiced for many years by both academic and industrial groups, Danishefsky and coworkers have recently formally defined this approach and coined the term "diverted total synthesis" (DTS) [187,188]. DTS involves the synthesis of an advanced intermediate, of lesser complexity than the target natural product, which can be elaborated by different synthetic sequences to yield multiple analogs of varying complexity containing the common pharmacophore, but inaccessible from the natural product itself. Synthesis can be accomplished by both conventional medicinal chemistry or combinatorial chemistry approaches.

The development of Eribulin (E7389; Figure 1.2) from a marine-derived antitumor agent, halichondrin B, is a compelling example of the power of the DTS approach [30; Section 1.3]. Because of its extraordinary antitumor activity, halichondrin B was chosen for preclinical development in 1992. Clinical development was severely impeded due to the limited amounts of compound available from natural sources. Total synthesis studies revealed that the right-hand half of the molecule retained all or most of the potency of the parent compound, and led to the discovery of Eribulin, which is far less structurally complex, is prepared by synthesis, has greater *in vivo* stability, and possesses comparable bioactivity to, and lower toxicity than, halichondrin B [30]. Eribulin is currently in Phase III clinical trials [32] and the U.S. Food and Drug Administration (FDA) granted its manufacturer, Eisai Co., a priority review for use as treatment for advanced breast cancer. Approval for second line use against resistant breast cancer was given by the FDA in November 2011 [189].

In some instances, the original natural product may fail in clinical trials, but totally synthetic analogs continue to be developed. Thus, while clinical trials of the marine-derived anticancer agents, dolastatin 10 (Figure 1.8) and dolastatin 15, have been terminated, synthetic analogs continued in various phase of clinical trials [190]. Auristatin PE (TZT-1027 or soblidotin; Figure 1.8) is the only dolastatin analog still in clinical trials as the base molecule. There are some very interesting modifications that have been made by medicinal chemists in order to deliver this close relative of dolastatin 10 by use of monoclonal antibodies targeted at specific epitopes [191,192]. A significant number of combinations of this base molecule with varying monoclonal antibodies are currently in preclinical to Phase II clinical trials, predominately against hematologic cancers.

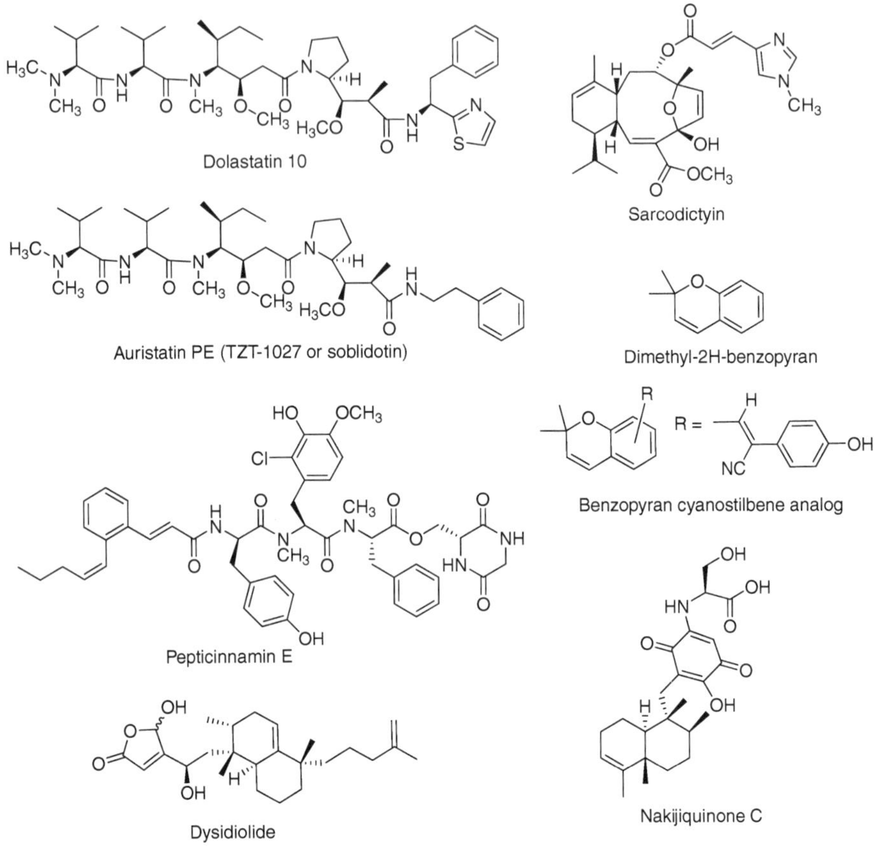

FIGURE 1.8 Synthetic and privileged natural product lead compounds and scaffolds.

Thus, Seattle Genetics has SGN-35 where the antibody is an anti-CD30 linked to auristatin PE in Phase I heading for Phase II [193], CuraGen has the antibody CR011 linked to auristatin PE in Phase II trials for metastatic breast cancer and melanoma [194] and Progenics has PSMA-ADC, a dimeric-specific PSMA antibody also conjugated to auristatin PE in Phase I against prostate cancer [195]. All of these examples are licensed from Seattle Genetics, although the exact linkages between the auristatin molecule and the antibody are subtly different in each case. A combination of the antibody trastuzumab, coupled to the maytansenoid DM1 is currently in Phase III clinical trials under the auspices of Roche, with a filing for approval expected in mid-2012.

1.9.2 Natural Product–Inspired Combinatorial Synthesis

Combinatorial chemistry is a set of techniques developed for the parallel or simultaneous synthesis of large collections of compounds (chemical libraries) for high-throughput screening (HTS) against biological targets. The technology was rapidly embraced within the pharmaceutical industry and used to generate very large libraries of compounds with the expectation that HTS screening of vast numbers of compounds would prove to be more efficient and cost-effective than traditional approaches to drug discovery. These expectations led to the abandonment or de-emphasis of natural products research at many companies. While there are claims that combinatorial chemistry is generating new leads [196], the declining numbers of new NCEs [197] indicate that the use of *de novo* combinatorial chemistry approaches to drug discovery over the past decade have been disappointing, with some of the earlier libraries being described as "poorly designed, impractically large, and structurally simplistic" [196]. As stated in that article, "an initial emphasis on creating mixtures of very large numbers of compounds has largely given way in industry to a more measured approach based on arrays of fewer, well-characterized compounds" with "a particularly strong move toward the synthesis of complex natural product-like compounds—molecules that bear a close structural resemblance to approved natural product-based drugs."

Approaches to the combinatorial synthesis of natural product–inspired libraries can be grouped into three basic categories although the boundaries among these are often indistinct. Combinatorial synthesis starting from known bioactive natural product scaffolds leads to libraries of natural product derivatives displaying appendage and stereochemical diversity. The second strategy, biology-oriented synthesis (BIOS) [198–200] expands on this basic concept by utilizing the structural information from natural products and their protein targets to focus on the most relevant chemical space for a particular target. The third approach, diversity-oriented synthesis (DOS) [201–204], aims to create highly diverse libraries of novel synthetic compounds with complex three-dimensional architectures that resemble natural products. DOS libraries often incorporate skeletal (scaffold) as well as appendage and stereochemical diversity. There is overlap among these approaches as well as with DTS and traditional medicinal chemical approaches to analog development.

1.9.2.1 Combinatorial Synthesis of Natural Product–Derived Libraries. Individual natural products often selectively modulate unrelated targets, a property that led to the recognition that natural product scaffolds are privileged structures as defined by Evans et al. in 1988 [175]; they have the necessary compromise of flexibility and rigidity to present functional groups in a favorable spatial arrangement to bind to biomolecular targets. As such, natural product scaffolds are obvious starting points for the application of combinatorial chemistry to prepare focused libraries of analogs for SAR studies. Nicolaou et al. [205] stated the underlying thesis as follows: "We were particularly intrigued by the possibility that using scaffolds of natural origin, which presumably have undergone evolutionary selection over time, might confer favorable bioactivities and bioavailabilities to library members."

Only a handful of the many recently published reports of the use of natural product scaffolds in combinatorial libraries will be cited to exemplify this approach. The synthesis of a library based on the scaffold of sarcodictyin (Figure 1.8) by Nicolaou et al. is one of the earliest examples [206]. Waldmann et al. prepared a pepticinnamin E

(Figure 1.8) library by solid-phase synthesis [207]. Solid-phase synthesis of combinatorial libraries was used to probe which regions of the epothilone A molecule are important to retention or improvement of activity [208], and combinatorial synthesis of vancomycin dimers yielded compounds with improved activity against drug-resistant bacteria [209]. Wipf et al. prepared some highly modified analogs of the antimitotic natural product curacin A (Figure 1.6), and found a simpler analog which was more potent than curacin A in inhibiting the assembly of tubulin [210]. A particularly versatile scaffold for library synthesis has been 2,2-dimethyl-2H-benzopyran (Figure 1.8); a search of the natural product literature yielded nearly 4000 analogs, with another 8000 structures identified through the inclusion of a slight modification of the search. In one example, application of solid-phase synthetic methods led to the identification and subsequent optimization of benzopyrans with a cyanostilbene substitution (Figure 1.8) that are effective against vancomycin-resistant bacteria [205,211,212].

The synthesis of combinatorial libraries based on natural product scaffolds is now a proven tool for the optimization of the known biological and pharmacokinetic properties of the parent natural product lead. It is also proving to be a potent tool for the discovery of analogs exhibiting biological activities beyond those previously associated with the parent natural product.

1.9.2.2 Biology-Oriented Synthesis of Natural Product–Inspired Libraries. BIOS is a new concept for the design of combinatorial libraries based on natural products developed by Waldmann and coworkers [198–200]. This concept is based on the recognition of fundamental and complementary properties of natural products and their protein targets. Since natural products bind both their biosynthetic enzymes and their target macromolecules, they necessarily populate biologically relevant regions of chemical space. Nature, through coevolution of natural products and their macromolecular targets, which are mainly proteins, has explored only a tiny fraction of the available chemical space. The number of three-dimensional protein folds has been shown to be even more conserved during evolution than the underlying sequences since topologically similar shapes can be formed by different sequences. Estimates of the number of proteins in humans range between 100,000 and 450,000 but the number of topologically different protein folds is actually much lower, with estimates of 600–8000 [213]. The natural product space and the protein structure space explored by Nature have to be highly complementary since they are limited in size and highly conserved. Thus, a natural product which is an inhibitor of a specific protein fold represents a biologically validated starting point for the development of analogs that may inhibit proteins with similar folds, and even allow for the discovery of specificity. These concepts are fundamentally similar to the privileged structure concept [175], but BIOS has the added dimension of using protein-folding patterns as the basis for subsequent screens.

BIOS is based upon the merger of two concepts previously developed by the Waldmann group [200]. The scaffolds of natural products can be mapped in a hierarchical manner to create a scaffold tree, a "structural classification of natural products" (SCONP) [214,215], which allows for logical pathways for the structural simplification of scaffolds. In the second concept, "protein structure similarity clustering" (PSSC), proteins are clustered by three-dimensional shape around the ligand binding sites, regardless of sequence similarity [216–218]. The ligand of any member of a PSSC could be expected to exhibit some degree of complementarity toward other members of the PSSC and thus serve as a starting point for the development of modulators of the other members of the PSSC.

BIOS represents a refinement of combinatorial libraries based on natural product scaffolds by focusing on the most biologically relevant chemical space for the target. Furthermore, it allows the transfer of knowledge about the modulation of a target by a natural product to a whole cluster of structurally related proteins, even when those proteins catalyze mechanistically different reactions.

A combinatorial library inspired by the marine natural product dysidiolide (Figure 1.8) demonstrated the power and potential of the BIOS approach. The authors postulated that the γ-hydroxybutenolide group of dysidiolide was the major determinant of phosphatase activity. Testing of a 147-member library built around this molecule yielded a compound 10-fold more potent ($IC_{50} = 350$ nM) than the parent compound against Cdc25A [219]. In addition, other members of the library were identified with low micromolar activities against the enzymes acetylcholinesterase and 11β-hydroxysteroid dehydrogenase type 1, which fall within the same PSSC as Cdc25A [220].

A second example of the success of BIOS is the discovery of inhibitors of *Tie-2*, insulin-like growth factor 1 receptor (IGF-1R), and vascular endothelial growth factor receptor 2 (VEGFR-2 and -3). Nakijiquinone C (Figure 1.8), isolated from a marine sponge and first reported by Kobayashi et al. [221], in 1995, was shown to be an inhibitor of epidermal growth factor receptor (EGFR), c-ErbB2 and protein kinase C (PKC), in addition to having cytotoxic activity against L1210 and KB cell lines [221]. Testing of a library of 74 compounds, built around the basic nakijiquinone C structure, against a battery of kinases with similar protein domain folds, yielded seven new inhibitors with low micromolar activity *in vitro*, including one VEGFR-2 inhibitor and four inhibitors of Tie-2 kinase, a protein intimately involved in angiogenesis, and for which, at the beginning of the study, no inhibitors were known [222]. The details of the models used, the chemistry leading to the nakijiquinone-based compounds, and the ribbon structures of the kinase domain of the insulin receptor, with the corresponding homology domains of the as yet uncrystallized VEGFR-2 and *Tie-2*, have been fully reviewed [218,223].

1.9.2.3 Diversity-Oriented Synthesis of Natural Product–Like Libraries. The third approach, DOS, is both related to and fundamentally different from combinatorial approaches around natural product scaffolds. Since every member of a combinatorial library is unique, all combinatorial syntheses serve to create diversity and could be classified as DOS. In this review, we restrict the term to DOS from simple starting materials as described by Schreiber et al. in much of their pioneering work in the area.

DOS is based on the arguable premise that regions of chemical space, not defined by natural products or known drugs, may be fertile regions for discovering novel small molecules that modulate biomacromolecules in useful ways, either as probes of function or as drug leads. The previous two approaches, based on known natural product scaffolds, aim to densely populate a specific region of chemical space, which is biologically relevant to a defined target. By contrast, DOS aims to achieve a nonfocused, diverse coverage of chemical space by the efficient and divergent synthesis of large libraries of structurally complex and structurally diverse compounds. Thus, while the molecules' structural complexity can be described as natural product-like, they are often not based on known natural product scaffolds. An in-depth discussion of DOS is outside the scope of this review and readers are referred to the excellent reviews on the subject, its relationship to natural products, and its applications to chemical genetics and drug discovery [202,204,224–230].

The most dramatic example of the power of DOS to generate novel chemical diversity is the 2009 report by Morton et al. of the synthesis of a 96-membered library based on 84 distinct molecular scaffolds [231]. Astonishingly, 65% of the scaffolds in this library are novel. When the skeletal diversity of the library was assessed by Waldmann's hierarchical scheme [214,215], the resulting scaffold tree was very similar to Waldmann's analysis of natural products. Morton et al. have not yet reported biological data for the library but the "natural product–likeness" should allow access to large regions of biologically relevant chemical space.

The synergy of combinatorial chemistry and natural products chemistry holds great potential for the discovery of new molecules populating productive regions of biologically relevant chemical space. Ultimate success will be gauged by the discovery of molecules with novel biological functions [229]. Thus far, the strategy of appending diverse substrates onto natural product scaffolds has performed remarkably well, yielding chemical probes such as secramine [232], uretupamine [233,234], and haptamide B [235].

More time is needed to critically assess the success of more recent innovations such as BIOS and DOS to deliver new small molecule probes and drugs. Ultimately, what is important is not the structural similarity of these new small molecules to natural product leads but whether they possess novel and useful biological functions.

1.10 CONCLUSIONS

While Nature, particularly plants, has long been an indispensable source of medicinal products, the vast untapped resources of the marine and microbial worlds have only recently become more readily accessible. The discovery of an impressive array of novel structure types from these sources has been reported over the past few decades, and the discovery rate and structural diversity of promising bioactive natural product leads are increasing with the rapid advances being made in collection, isolation, and structural elucidation techniques. Added to these advances, the explosion of genetic information over the past decade is leading to the advent of genetic techniques that permit the efficient isolation and expression of microbial biosynthetic cassettes, and the implementation of combinatorial biosynthetic technology and genome mining. A revised paradigm for the development of novel drugs is emerging. It is based on the application of synthetic strategies encompassing DTS, and the elaboration of "privileged structures from Nature," or synthetic compounds modeled on such privileged structures, through use of directed combinatorial synthesis, BIOS, and DOS. The application of these refined combinatorial synthetic methodologies to the preparation and screening of focused, natural product–inspired libraries has great potential to lead to the more efficient development of novel agents and drug entities for the treatment of many disease states. The successful achievement of these goals requires the implementation of close multidisciplinary and international collaboration in both the discovery and development phases.

REFERENCES

[1] Borchardt, J.K. (2002). The beginnings of drug therapy: Ancient Mesopotamian medicine. *Drug News Perspective*, *15*, 187–192.

[2] Huang, K.C., *The Pharmacology of Chinese Herbs*, CRC Press, Boca Raton, FL, 1999.

[3] Kapoor, L.D., *CRC Handbook of Ayurvedic Medicinal Plants*, CRC Press, Boca Raton, FL, 1990.

[4] Dev, S. (1999). Ancient-modern concordance in Ayurvedic plants: Some examples. *Environmental Health Perspectives*, *107*, 783–789.

[5] Johnson, T., *CRC Ethnobotany Desk Reference*, CRC Press, Boca Raton, FL, 1999.

[6] Moerman, D.E., Medicinal Plants of Native America, Vol. 1, University of Michigan Museum of Anthropology, Ann Arbor, MI, 1986.

[7] Farnsworth, N.R., Akerele, R.O., Bingel, A.S., Soejarto, D.D., Guo, Z. (1985). Medicinal plants in therapy. *Bulletin of the World Health Organization*, *63*, 965–981.

[8] Fabricant, D.S., Farnsworth, N.R. (2001). The value of plants used in traditional medicine for drug discovery. *Environmental Health Perspectives*, *109*, 69–75.

[9] Buss, A.D., Waigh, R.D., Natural products as leads for new pharmaceuticals. In: Wolff, M.E. editor. *Burger's Medicinal Chemistry and Drug Discovery. Principles and Practice*, 5th ed., John Wiley & Sons, Inc., New York, 1995, pp. 983–1033.

[10] Wongsrichanalai, C., Pickard, A.L., Wernsdorfer, W.H., Meshnick, S.R. (2002). Epidemiology of drug-resistant malaria. *Lancet Infectious Diseases*, *2*, 209–218.

[11] Klayman, D.L. (1985). Qinghaosu (artemisinin): An antimalarial drug from China. *Science*, *228*, 1049–1055.

[12] O'Neill, P.M., Posner, G.H. (2004). A medicinal chemistry perspective on artemisinin and related endoperoxides. *Journal of Medicinal Chemistry*, *47*, 2945–2964.

[13] Vennerstrom, J.L., Arbe-Barnes, S., Brun, R., Charman, S.A., Chiu, F.C.K., Chollet, J., Dong, Y., Dorn, A., Hunziker, D., Matile, H., McIntosh, K., Padmanilayam, M., Santo Tomas, J., Scheurer, C., Scorneaux, B., Tang, Y., Urwyler, H., Sergio, W., Charman, W.N. (2004). Identification of an antimalarial synthetic trioxolane drug development candidate. *Nature*, *430*, 900–904.

[14] Posner, G.H., Chang, W., Hess, L., Woodard, L., Sinishtaq, S., Usera, A.R., Miao, W., Rosenthal, A.S., Kalinda, A.S., D'Angelo, J.G., Petersen, K.S., Stohler, R., Chollet, J., Santo-Tomas, J., Snyder, C., Rottmann, M., Wittlin, S., Brun, R., Shapiro, T.A. (2008). Malaria-infected mice are cured by oral administration of new artemisinin derivatives. *Journal of Medicinal Chemistry*, *51*, 1035–1042.

[15] Nagelschmitz, J., Voith, B., Wensing, G., Roemer, A., Fugmann, B., Kotecka, B.M., Rieckmann, K.H., Edstein, M.D. (2008). First assessment in humans of the safety, tolerability, pharmacokinetics, and *ex vivo* pharmacodynamic antimalarial activity of the new artemisinin derivative artemisone. *Antimicrobial Agents and Chemotherapy*, *52*, 3085–3091.

[16] Dondorp, A.M., Nosten, F., Yi, P., Das, D., Phyo, A.P., Tarning, J., Lwin, K.M., Ariey, F., Hanpithakpong, W., Lee, S.J., Ringwald, P., Silamut, K., Imwong, M., Chotivanich, K., Lim, P., Herdman, T., An, S.S., Yeung, S., Singhasivanon, P., Day, N.P., Lindegardh, N., Socheat, D., White, N.J. (2009). Artemisinin resistance in *Plasmodium falciparum* malaria. *New England Journal of Medicine*, *361*, 455–467. Erratum in: *New England Journal of Medicine*, *361*, 1714.

[17] Eastman, R.T., Fidock, D.A. (2009). Artemisinin-based combination therapies: Vital tool in efforts to eliminate malaria. *Nature Reviews Microbiology*, *7*, 864–874.

[18] Cui, L., Su, X.Z. (2009). Discovery, mechanisms of action and combination therapy of artemisinin. *Expert Review of Anti-Infective Therapy*, *7*, 999–1013.

[19] Keasling, J.D. (2008). Synthetic biology for synthetic chemistry. *ACS Chemical Biology*, *3*, 64–76.

[20] Tsuruta, H., Paddon, C.J., Eng, D., Lenihan, J.R., Horning, T., Anthony, L.C., Regentin, R., Keasling, J.D., Renninger, N.S., Newman, J.D. (2009). High-level production of amorpha-4,11-diene, a precursor of the antimalarial agent artemisinin, in *Escherichia coli*. *PloS One*, *4*, e4489.

[21] Hartwell, J.L., *Plants Used Against Cancer*, Quarterman, Lawrence, MA, 1982.

[22] Cragg, G.M., Boyd, M.R., Cardellina, J.H., Newman, D.J., Snader, K.M., McCloud, T.G., Ethnobotany and drug discovery: The experience of the US National Cancer Institute. In: Chadwick, D.J., Marsh, J., editors. *Ethnobotany and the Search for New Drugs*, Vol. 185, John Wiley & Sons, Inc., Ciba Foundation Symposium, New York, 1994, pp. 178–196.

[23] Cragg, G.M., Newman, D.J. (2009). Nature: A vital source of leads for anticancer drug development. *Phytochemistry Reviews*, *8*, 313–331.

[24] Cragg, G.M., Grothaus, P.G., Newman, D.J. (2009). Impact of natural products on developing new anti-cancer agents. *Chemical Reviews*, *109*, 3012–3043.

[25] Margulis L, Schwartz KV, *Five Kingdoms: An Illustrated Guide to the Phyla of Life on Earth*, W. H. Freeman & Co., New York, 1988.

[26] Molinski, T.F., Dalisay, D.S., Lievens, S.L., Saludes, J.P. (2009). Drug development from marine natural products. *Nature Reviews Drug Discovery*, *8*, 69–85.

[27] Wallace, M.S. (2006) Ziconotide: A new nonopioid intrathecal analgesic for the treatment of chronic pain. *Expert Review of Neurotherapeutics*, *6*, 1423–1428.

[28] Bulaj, G., Buczek, O., Goodsell, I., Jiminez, E.C., Kranski, J., Nielsen, J.S., Garrett, J.E., Olivera, B.M. (2003). Efficient oxidative folding of conotoxins and the radiation of venomous cone snails. *Proceedings of the National Academy of Sciences of the USA*, *100*, 14562–14568.

[29] Henriquez, R., Faircloth, G., Cuevas, C., Ecteinascidin 743 (ET 743), Yondelis™; Aplidine, and Kahalalide F. In: Cragg, G.M., Kingston, D.G.I., and Newman, D.J. editors. *Anticancer Agents from Natural Products*, CRC Press, Boca Raton, FL, 2005, pp. 215–240.

[30] Yu, M.J., Kishi, Y., Littlefield, B.A., Discovery of E7389, a fully synthetic macrocyclic ketone analog of halichondrin B. In: Cragg, G.M., Kingston, D.G.I., and Newman, D.J.

editors. *Anticancer Agents from Natural Products*, CRC Press, Boca Raton, FL, 2005, pp. 241–265.

[31] Newman, D.J., The bryostatins. In: Cragg, G.M., Kingston, D.G.I., and Newman, D.J. editors. *Anticancer Agents from Natural Products*, CRC Press, 2005, pp. 137–150.

[32] Twelves, C., Loesch, D., Blum, J.L., Vahdat, L.T., Petrakova, K., Chollet, P.J., Akerele, C.E., Seegobin, S., Wanders, J., Cortes, J. (2010). A phase III study (EMBRACE) of eribulin mesylate versus treatment of physician's choice in patients with locally recurrent or metastatic breast cancer previously treated with an anthracycline and a taxane. 2010 ASCO Annual Meeting. *Journal of Clinical Oncology*, *28*, 18s.

[33] Scriabine, A., Discovery and development of major drugs. Currently in use. In: Landau, R., Achilladelis, B., and Scriabine, A.editors. *Pharmaceutical Innovation. Revolutionizing Human Health*, Chemical Heritage Press, Philadelphia, 1999, pp. 148–270.

[34] Arcamone, F.M., Anthracyclines. In: Cragg, G.M., Kingston, D.G.I., and Newman, D.J. editors. *Anticancer Agents from Natural Products*, CRC Press, Boca Raton, FL, 2005, pp. 299–320.

[35] Hecht, S.M., Bleomycin group antitumor agents. In: Cragg, G.M., Kingston, D.G.I., and Newman, D.J. editors. *Anticancer Agents from Natural Products*, CRC Press, Boca Raton, FL, 2005, pp. 357–381.

[36] Remers, W.A., The mitomycins. In: Cragg, G.M., Kingston, D.G.I., and Newman, D.J. editors. *Anticancer Agents from Natural Products*, CRC Press, Boca Raton, FL, 2005, pp. 475–497.

[37] Hamann, P.R., Upeslacis, J., and Borders, D.B., Enediynes. In: Cragg, G.M., Kingston, D.G.I., and Newman, D.J. editors. *Anticancer Agents from Natural Products*, CRC Press, Boca Raton, FL, 2005, pp. 451–474.

[38] Prudhomme, M., Staurosporines and structurally related indolocarbazoles as antitumor agents. In: Cragg, G.M., Kingston, D.G.I., and Newman, D.J.editors. *Anticancer Agents from Natural Products*, CRC Press, Boca Raton, FL, 2005, pp. 499–517.

[39] Hofle, G., Reichenbach, H., Epothilone, a myxobacterial metabolite with promising antitumor activity. In: Cragg, G.M., Kingston, D.G.I., and Newman, D.J. editors. *Anticancer Agents from Natural Products*, CRC Press, Boca Raton, FL, 2005, pp. 413–450.

[40] Daly, J.W., Spande, T.F., Garrafo, H.M. (2005). Alkaloids from amphibian skins. A tabulation of over eight hundred compounds. *Journal of Natural Products*, *68*, 1556–1575.

[41] Eng, J., Kleinman, W.A., Singh, G., Raufman, J.P. (1992). Isolation and characterization of Exendin-4, an Exendin-3 analogue from *Heloderma suspectum* venom. *Journal of Biological Chemistry*, *267*, 7402–7405.

[42] Newman, D.J., Cragg, G.M. (2007). Natural products as sources of new drugs over the last 25 years. *Journal of Natural Products*, *70*, 461–477.

[43] Cragg, G.M., Newman, D.J. Natural product sources of drugs: Plants, microbes, marine organisms and animals. In: Kennewell, P.D., Triggle, D., and Taylor, J. editors. *Comprehensive Medicinal Chemistry II*, Vol. 1 Elsevier, Oxford, 2006, pp. 355–403.

[44] Balandrin, M.F., Kinghorn, A.D., Farnsworth, N.R., Plant-derived natural products in drug discovery and development: An overview. In: Kinghorn, A.D., and Balandrin, M.F. editors. *Human Medicinal Agents from Plants*, Vol. 534 American Chemical Society, Washington, DC, 1993, pp. 2–12.

[45] Raskin, I., Ribnicky, D.M., Komarnytsky, S., Ilic, N., Poulev, A., Borisjuk, N., Brinker, A., Moreno, D.A., Ripoll, C., Yakoby, N., O'Neal, J. M., Cornwell, T., Pastor, I., Fridlander, B. (2002). Plants and human health in the twenty-first century. *Trends in Biotechnology*, *20*, 522–531.

[46] Newman, D.J., Hill, R.T. (2006). New drugs from marine microbes: The tide is turning. *Journal of Industrial Microbiology Biotechnology*, *33*, 539–544.

[47] (a)Perry, N.B., Ettouati, L., Litaudon, M., Blunt, J.W., Munro, M.H.G. (1994). Alkaloids from the antarctic sponge *Kirkpatrickia varialosa.* Part 1: Variolin B, a new antitumour and antiviral compound. *Tetrahedron*, *50*, 3987–3992; (b)Trimurtulu, G., Faulkner, D.J., Perry, N.B., Ettouati, L., Litaudon, M., Blunt, J.W., Munro, M.H.G., Jameson, G.B. (1994). Alkaloids from the antarctic sponge *Kirkpatrickia varialosa.* Part 2: Variolin A and *N*(3′)-methyl tetrahydrovariolin B. *Tetrahedron*, *50*, 3993–4000.

[48] Diyabalanage, T., Amsler, C.D., McClintock, J.B., Baker, B.J. (2006). Palmerolide A, a cytotoxic macrolide from the antarctic tunicate *Synoicum adareanum. Journal of the American Chemical Society*, *128*, 5630–5631.

[49] Conxita Avila, C., Taboada, S., Núñez-Pons, L. (2008). Antarctic marine chemical ecology: What is next? *Marine Ecology*, *29*, 1–71.

[50] Poulev, A., O'Neal, J.M., Logendra, S., Pouleva, R.B., Timeva, V., Garvey, A.S., Gleba, D., Jenkins, I.S., Halpern, B.T., Kneer, R., Cragg, G.M., Raskin, I. (2003). Elicitation, a new window into plant chemodiversity and phytochemical drug discovery. *Journal of Medicinal Chemistry*, *46*, 2542–2547.

[51] McCoy, E., O'Connor, S.E. (2006). Directed biosynthesis of alkaloid analogs in the medicinal plant. *Catharanthus roseus. Journal of the American Chemical Society*, *128*, 14276–14277.

[52] Pace, N.R. (1997). A molecular view of microbial diversity and the biosphere. *Science*, *276*, 734–740.

[53] Madigan, M.T., Martinko, J.M., Parker, J.B., *Biology of Microorganisms*, Prentice-Hall, Upper Saddle River, NJ, 1996.

[54] Rappé, M.S., Giovannoni, S.J. (2003) The uncultured microbial majority. *Annual Review of Microbiology*, *57*, 369–394.

[55] McAlpine, J.B., Bachmann, B.O., Piraee, M., Tremblay, S., Alarco, A.-M., Zazopoulos, E., Farnet, C.M. (2005). Microbial genomics as a guide to drug discovery and structural elucidation: ECO-02301, a novel antifungal agent, as an example. *Journal of Natural Products*, *68*, 493–496.

[56] Lautru, S., Deeth, R.J., Bailey, L., Challis, G.M. (2005). Discovery of a new peptide natural product by *Streptomyces coelicolor* genome mining. *Nature Chemical Biology*, *1*, 265–269.

[57] Park, E.Y. (2004). Recent progress in microbial cultivation techniques. *Advances in Biochemical Engineering/Biotechnology*, *90*, 1–33.

[58] (a)Zengler, K., Toledo, G., Rappe, M., Elkins, J., Mathur, E.J., Short, J.M., Keller, M. (2002). Cultivating the uncultured. *Proceedings of the National Academy of Sciences of the USA*, *99*, 15681–15686; (b) Keller, M., Zengler, K. (2004). Tapping into microbial diversity. *Nature Reviews Microbiology*, *2*, 141–150.

[59] Stevenson, B.S., Eichorst, S.A., Wertz, J.T., Schmidt, T.M., Breznak, J.A. (2004). New strategies for cultivation and detection of previously uncultured microbes. *Applied and Environmental Microbiology*, *70*, 4748–4755.

[60] Nichols, D. Lewis, K., Orjala, J., Mo, S., Ortenberg, R., O'Connor, P., Zhao, C., Vouros, P., Kaeberlein, T., Epstein, S.S. (2008). Short peptide induces an "uncultivable" microorganism to grow *in vitro*. *Applied and Environmental Microbiology*, *74*, 4889–4897.

[61] Bollmann, A., Lewis, K., Epstein, S.S. (2007) Incubation of environmental samples in a diffusion chamber increases the diversity of recovered isolates. *Applied and Environmental Microbiology*, *73*, 6386–6390.

[62] Nichols, D., Cahoon, N., Trakhtenberg, E.M., Pham, L., Mehta, A., Belanger, A., Kanigan, T., Lewis, K., Epstein, S.S. (2010). High-throughput *in situ* cultivation of "uncultivable" microbial species. *Applied and Environmental Microbiology*, *76*, 2445–2450.

[63] Gontang, E.A., Fenical, W., Jensen, P.R. (2007). Phylogenetic diversity of gram-positive bacteria cultured from marine sediments. *Applied and Environmental Microbiology*, *73*, 3272–3282.

[64] Lefevre, F., Robe, P., Jarrin, C., Ginolhac, A., Zago, C., Auriol, D., Vogel, T.M., Simonet, P., Nalin, R. (2008). Drugs from hidden bugs: Their discovery via untapped resources. *Research in Microbiology*, *159*, 153–161.

[65] Wilkinson, B., Micklefield, J. (2007). Mining and engineering natural-product biosynthetic pathways. *Nature Chemical Biology*, *3*, 379–386.

[66] Rondon, M.R., August, P.R., Bettermann, A.D., Brady, S.F., Grossman, T.H., Liles, M.R., Loiacono, K.A., Lynch, B.A., MacNeil, I.A., Minor, C., Tiong, C.L., Gilman, M., Osburne, M.S., Clardy, J., Handelsman, J., Goodman, R.M. (2000). Cloning the soil metagenome: A strategy for accessing the genetic and functional diversity of uncultured microorganisms. *Applied and Environmental Microbiology*, *66*, 2541–2547.

[67] Venter, J.C., Remington, K., Heidelberg, J.F., Halpern, A.L., Rusch, D., Eisen, J.A., Wu, D., Paulsen, I., Nelson, K.E., Nelson, W., Fouts, D.E., Levy, S., Knap, A.H., Lomas, M.W., Nealson, K., White, O., Peterson, J., Hoffman, J., Parsons, R., Baden-Tillson, H., Pfannkoch, C., Rogers, Y.-H.,O., Smith, H.O. (2004). Environmental genome shotgun sequencing of the Sargasso Sea. *Science*, *304*, 66–74.

[68] Yooseph, S., Sutton, G., Rusch, D.B., Halpern, A.L., Williamson, S.J., Remington, K., Eisen, J.A., Heidelberg, K.B., Manning, G., Li, W., Jaroszewski, L., Cieplak, P., Miller, C.S., Li, H., Mashiyama, S.T., Joachimiak, M.P., van Belle, C., Chandonia, J.-M., Soergel, D.A., Zhai, Y., Natarajan, K., Lee, S., Raphael, B.J., Bafna, B., Friedman, R., Brenner, S.E., Godzik, A., Eisenberg, D., Dixon, J.E., Taylor, S.S., Strausberg, R.L., Frazier, M., Craig Venter, J. (2007). The *Sorcerer II* global ocean sampling expedition: Expanding the universe of protein families. *PLoS Biology*, *5*, e16.

[69] Warnecke, F., Luginbühl, P., Ivanova, N., Ghassemian, M., Richardson, T.H., Stege, J.T., Cayouette, M., McHardy, A.C., Djordjevic, G., Aboushadi, N., Sorek, R., Tringe, S.G., Podar, M., Martin, H.G., Kunin, V., Dalevi, D., Madejska, J., Kirton, E., Platt, D., Szeto, E., Salamov, A., Barry, K., Mikhailova, N., Kyrpides, N.C., Matson, E.G., Ottesen, E.A., Zhang, X., Hernández, M., Murillo, C., Acosta, L.G., Rigoutsos, I., Tamayo, G., Green, B.D., Chang, C., Rubin, E.M., Mathur, E.J., Robertson, D.E., Hugenholtz, P., Leadbetter, J.R. (2007). Metagenomic and functional analysis of hindgut microbiota of a wood-feeding higher termite. *Nature*, *450*, 560–565.

[70] Fieseler, L., Hentschel, U., Grozdanov, L., Schirmer, A., Wen, G., Platzer, M., Hrvatin, S., Butzke, D., Zimmermann, K., Piel, J. (2007). Widespread occurrence and genomic context of unusually small polyketide synthase genes in microbial consortia

associated with marine sponges. *Applied and Environmental Microbiology*, *73*, 2144–2155.

[71] Vogel, T.M., Simonet, P., Jansson, J.K., Hirsch, P.R., Tiedje, J.M., van Elsas, J.D., Bailey, M.J., Nalin, R., Philippot, L. (2009). TerraGenome: A consortium for the sequencing of a soil metagenome. *Nature Reviews Microbiology* *7*, 252.

[72] van Elsas, J.D., Costa, R., Jansson, J., Sjöling, S., Bailey, M., Nalin, R., Vogel, T. M. van Overbeek, L. (2008). The metagenomics of disease-suppressive soils—Experiences from the METACONTROL project. *Trends in Biotechnology*, *26*, 591–601.

[73] Jin, M., Liu, L., Wright, S.A.I., Beer, S.V., Clardy, J. (2003). Structural and functional analysis of pantocin A: An antibiotic from *Pantoea agglomerans* discovered by heterologous expression of cloned genes. *Angewandte Chemie International Edition*, *42*, 2898–2901.

[74] Omura, S., Ikeda, H., Ishikawa, J., Hanamoto, A., Takahashi, C., Shinose, M., Takahashi, Y., Horikawa, H., Nakazawa, H., Osonoe, T., Kikuchi, H., Shiba, T., Sakaki, Y., Hattori, M. (2001). Genome sequence of an industrial microorganism *Streptomyces avermitilis*: Deducing the ability of producing secondary metabolites. *Proceedings of the National Academy of Sciences of the U.S.A.*, *98*, 12215–12220.

[75] Ikeda, H., Ishikawa, J., Hanamoto, A., Shinose, M., Kikuchi, H., Shiba, T., Sakaki, Y., Hattori, M., Omura, S. (2003). Complete genome sequence and comparative analysis of the industrial microorganism *Streptomyces avermitilis*. *Nature Biotechnology*, *21*, 526–531.

[76] Bentley, S.D., Chater, K.F., Cerdeño-Tárraga, A.M., Challis, G.L., Thomson, N.R., James, K.D., Harris, D.E., Quail, M.A., Kieser, H., Harper, D., Bateman, A., Brown, S., Chandra, G., Chen, C.W., Collins, M., Cronin, A., Fraser, A., Goble, A., Hidalgo, J., Hornsby, T., Howarth, S., Huang, C.H., Kieser, T., Larke, L., Murphy, L., Oliver, K., O'Neil, S., Rabbinowitsch, E., Rajandream, M.A., Rutherford, K., Rutter, S., Seeger, K., Saunders, D., Sharp, S., Squares, R., Squares, S., Taylor, K., Warren, T., Wietzorrek, A., Woodward, J., Barrell, B.G., Parkhill, J., Hopwood, D.A. (2002). Complete genome sequence of the model actinomycete *Streptomyces coelicolor* A3(2). *Nature*, *417*, 141–147.

[77] Banskota, A.H., McAlpine, J.B., Sørensen, D., Ibrahim, A., Aouidate, M., Piraee, M., Alarco, A.M., Farnet, C.M., Zazopoulos, E. (2006). Genomic analyses lead to novel secondary metabolites. *Journal of Antibiotics*, *59*, 533–542.

[78] Gourdeau, H., McAlpine, J.B., Ranger, M., Simard, B., Berger, F., Beaudry, F., Farnet, C. M., Falardeau, P. (2008). Identification, characterization and potent antitumor activity of ECO-4601, a novel peripheral benzodiazepine receptor ligand. *Cancer Chemotherapy and Pharmacology*, *61*, 911–921.

[79] Rachid, S., Gerth, K., Kochems, I., Müller, R. (2007). Deciphering regulatory mechanisms for secondary metabolite production in the myxobacterium *Sorangium cellulosum* So ce56. *Molecular Microbiology*, *63*, 1783–1796.

[80] Weissman, K.J., Müller, R. (2009). A brief tour of myxobacterial secondary metabolism. *Bioorganic and Medicinal Chemistry*, *17*, 2121–2136.

[81] Bok, J.W., Hoffmeister, D., Maggio-Hall, L.A., Murillo, R., Glasner, J.D., Keller, N.P. (2006). Genomic mining for *Aspergillus* natural products. *Chemical Biology*, *13*, 31–37.

[82] Hoffmeister, D., Keller, N.P. (2007). Natural products of filamentous fungi: Enzymes, genes, and their regulation. *Natural Products Reports*, *24*, 393–416.

[83] Olano, C., Méndez, C., Salas, J.A. (2009) Antitumor compounds from marine actinomycetes. *Marine Drugs*, *7*, 210–248.

[84] Mincer, T.J., Jensen, P.R., Kauffman, C.A., Fenical, W. (2002). Widespread and persistent populations of a major new marine *Actinomycete taxon* in ocean sediments. *Applied and Environmental Microbiology*, *68*, 5005–5011.

[85] Udwary, D.W., Zeigler, L., Asolkar, R.N., Singan, V., Lapidus, A., Fenical, W., Jensen, P.R., Moore, B.S. (2007). Genome sequencing reveals complex secondary metabolome in the marine actinomycete *Salinispora tropica. Proceedings of the National Academy of Sciences of the U.S.A.*, *104*, 10376–10381.

[86] Feling, R.H., Buchanan, G.O., Mincer, T.J., Kauffman, C.A., Jensen, P.R., Fenical, W. (2003) Salinosporamide A: A highly cytotoxic proteasome inhibitor from a novel microbial source, a marine bacterium of the new genus *Salinospora. Angewandte Chemie International Edition*, *42*, 355–357.

[87] Fenical, W., Jensen, P.R., Palladino, M.A., Lam, K.S., Lloyd, G.K., Potts, B.C. (2009). Discovery and development of the anticancer agent salinosporamide A (NPI-0052). *Bioorganic and Medicinal Chemistry*, *17*, 2175–2180.

[88] Potts, B.C., Lam, K.S. (2010). Generating a generation of proteasome inhibitors: From microbial fermentation to total synthesis of salinosporamide A (Marizomib) and other salinosporamides. *Marine Drugs*, *8*, 835–880.

[89] Kwon, H.C., Kauffman, C.A., Jensen, P.R., Fenical, W. (2006) Marinomycins A–D, antitumor–antibiotics of a new structure class from a marine actinomycete of the recently discovered genus "*Marinispora*." *Journal of the American Chemical Society*, *128*, 1622–1632.

[90] Williams, P.G., Asolkar, R.N., Kondratyuk, T., Pezzuto, J.M., Jensen, P.R., Fenical, W. (2007) Saliniketals A and B, bicyclic polyketides from the marine actinomycete *Salinispora arenicola. Journal of Natural Products*, *70*, 83–88.

[91] Oh, D.C., Strangman, W.K., Kauffman, C.A., Jensen, P.R., Fenical, W. (2007). Thalassospiramides A and B, immunosuppressive peptides from the marine bacterium *Thalassospira* sp. *Organic Letters*, *9*, 1525–1528.

[92] Bredholt, H., Fjærvik, E., Johnsen, G., Zotchev, S.B. (2008). Actinomycetes from sediments in the Trondheim Fjord, Norway: Diversity and biological activity. *Marine Drugs*, *6*, 12–24.

[93] Welker, M., von Döhren, H. (2006). Cyanobacterial peptides—Nature's own combinatorial biosynthesis. *FEMS Microbiology Reviews*, *30*, 530–563.

[94] Tan, L.T. (2007) Bioactive natural products from marine cyanobacteria for drug discovery. *Phytochemistry*, *68*, 954–979.

[95] Tidgewell, K., Clark, B.R., Gerwick, W.H., The natural products chemistry of cyanobacteria. In: Mander, L., and Liu, H.-W.editors. *Comprehensive Natural Products Chemistry. Chemistry and Biology*, Vol. 2 Elsevier Ltd., Oxford, 2010, pp. 141–188.

[96] Al-awar, R.S., Sih, C., The isolation, characterization, and development of a novel class of potent antitumor macrocyclic depsipeptides: The cryptophcins. In: Cragg, G.M., Kingston, D.G.I., and Newman, D.J. editors. *Anticancer Agents from Natural Products*, CRC Press, Boca Raton, FL, 2005, pp. 151–170.

[97] Magarvey, N.A., Beck, Z.Q., Golakoti, T., Ding, Y., Huber, U., Hemscheidt, T.K., Abelson, D., Moore, R.E., Sherman, D.H. (2006). Biosynthetic characterization and chemoenzymatic assembly of the cryptophycins. Potent anticancer agents from Nostoc cyanobionts. *ACS Chemical Biology*, *1*, 766–779.

[98] Chang, Z., Sitachitta, N., Rossi, J.V., Roberts, M.A., Flatt, P.M., Jia, J., Sherman, D.H., Gerwick, W.H. (2004). Biosynthetic pathway and gene cluster analysis of curacin A, an anti-tubulin natural product from the tropical marine cyanobacterium *Lyngbya majuscula*. *Journal of Natural Products*, *67*, 1356–1367.

[99] Esquenazi, E., Coates, C., Simmons, L., Gonzalez, D., Gerwick, W.H., Dorrestein, P.C. (2008). Visualizing the spatial distribution of secondary metabolites produced by marine cyanobacteria and sponges via MALDI-TOF imaging. *Molecular BioSystems. Molecular BioSystems*, *4*, 562–570.

[100] Jones, A.C., Gerwick, L., Gonzalez, D., Dorrestein, P.C., Gerwick, W.H. (2009). Transcriptional analysis of the jamaicamide gene cluster from the marine cyanobacterium *Lyngbya majuscula* and identification of possible regulatory proteins. *BMC Microbiology*, *9*, 247.

[101] Sattely, E.S., Fischbach, M.A., Walsh, C.T. (2008). Total biosynthesis: in vitro reconstitution of polyketide and nonribosomal peptide pathways. *Natural Products Reports*, *25*, 757–793.

[102] Piel, J. (2004). Metabolites from symbiotic bacteria. *Natural Products Reports*, *21*, 519–538.

[103] Thomas, T.R.A., Kavlekar, D.P., LokaBharathi, P.A. (2010). Marine drugs from sponge–microbe association—A review. *Marine Drugs*, *8*, 1417–1468.

[104] Penesyan, A., Kjelleberg, S., Egan, S. (2010). Development of novel drugs from marine surface associated microorganisms. *Marine Drugs*, *8*, 438–459.

[105] Piel, J. (2009). Metabolites from symbiotic bacteria. *Natural Products Reports*, *26*, 338–362.

[106] Yu, T.-W., Floss, H.G., Ansamitocins (maytansinoids). In: Cragg, G.M., Kingston, D.G. I., and Newman, D.J.editors. *Anticancer Agents from Natural Products*, Taylor and Francis, Boca Raton, FL, 2005, pp. 321–337.

[107] Piel, J., Butzke, D., Fusetani, N., Hui, D., Platzer, M., Wen, G., Matsunaga, S. (2005). Exploring the chemistry of uncultivated bacterial symbionts: Antitumor polyketides of the pederin family. *Journal of Natural Products*, *68*, 472–479.

[108] Piel, J.; Hofer, I., Hui, D. (2004). Evidence for a symbiosis island involved in horizontal acquisition of pederin biosynthetic capabilities by the bacterial symbiont of *Paederus fuscipes* beetles. *Journal of Bacteriology*, *186*, 1280–1286.

[109] Piel, J., Hui, D., Wen, G., Butzke, D., Platzer, M., Fusetani, N., Matsunaga, S. (2004). Antitumor polyketide biosynthesis by an uncultivated bacterial symbiont of the marine sponge *Theonella swinhoei*. *Proceedings of the National Academy of Sciences of the U.S. A.*, *101*, 16222–16227.

[110] Zimmermann, K., Engeser, M., Blunt, J.W., Munro, M.H., Piel, J. (2009). Pederin-type pathways of uncultivated bacterial symbionts: Analysis of *o*-methyltransferases and generation of a biosynthetic hybrid. *Journal of the American Chemical Society*, *131*, 2780–2781.

[111] Partida-Martinez, L.P., Hertweck, C. (2005). Pathogenic fungus harbours endosymbiotic bacteria for toxin production. *Nature*, *437*, 884–888.

[112] Partida-Martinez, L.P., Hertweck, C. (2007). A gene cluster encoding rhizoxin biosynthesis in "*Burkholderia rhizoxina*," the bacterial endosymbiont of the fungus *Rhizopus microsporus*. *ChemBioChem*, *8*, 41–45.

[113] Uria, A., Piel, J. (2009). Cultivation-independent approaches to investigate the chemistry of marine symbiotic bacteria. *Phytochemistry Reviews*, *8*, 401–414.

[114] Hertweck, C. (2009). Hidden biosynthetic treasures brought to light. *Nature Chemical Biology*, *5*, 450–452.

[115] Gunatilaka, A.A.L. (2006). Natural products from plant-associated microorganisms: Distribution, structural diversity, bioactivity, and implications of their occurrence. *Journal of Natural Products*, *69*, 509–526.

[116] Strobel, G., Daisy, B., Castillo, U., Harper, J. (2004). Natural products from endophytic microorganisms. *Journal of Natural Products*, *67*, 257–268.

[117] Tan, R.X., Zou, W.X. (2001). Endophytes: A rich source of functional metabolites. *Natural Products Reports*, *18*, 448–459.

[118] Castillo, U., Harper, J.K., Strobel, G.A., Sears, J., Alesi, K., Ford, E., Lin, J., Hunter, M., Maranta, M., Ge, H., Yaver, D., Jensen, J.B., Porter, H., Robinson, R., Millar, D., Hess, W.M., Condron, M., Teplow, D. (2003). Kakadumycins, novel antibiotics from *Streptomyces* sp. NRRL 30566, an endophyte of *Grevillea pteridifolia*. *FEMS Microbiology Letters*, *224*, 183–190.

[119] Li, J.Y., Harper, J.K., Grant, D.M., Tombe, B.O., Bashyal, B., Hess, W.M., Strobel, G.A. (2001). Ambuic acid, a highly functionalized cyclohexenone with antifungal activity from *Pestalotiopsis* spp. and *Monochaetia* sp. *Phytochemistry*, *56*, 463–468.

[120] Ezra, D., Castillo, U.F., Strobel, G.A., Hess, W.M., Porter, H., Jensen, J.B., Condron, M.A., Teplow, D.B., Sears, J., Maranta, M., Hunter, M., Weber, B., Yaver, D. (2004). Coronamycins, peptide antibiotics produced by a verticillate *Streptomyces* sp. (MSU-2110) endophytic on *Monstera* sp. *Microbiology*, *150*, 785–793.

[121] Zhou, G.-X., Wijeratne, E.M.K., Bigelow, D., PiersonIII, L.S., VanEtten, H.D., Gunatilaka, A.A.L. (2004). Aspochalasins I, J, and K: Three new cytotoxic cytochalasans of *Aspergillus flavipes* from the rhizosphere of *Ericameria laricifolia* of the Sonoran desert. *Journal of Natural Products*, *67*, 328–332.

[122] Stierle, A., Strobel, G., Stierle, D. (1993). Taxol and taxane production by *Taxomyces andreanae*, an endophytic fungus of Pacific yew. *Science*, *260*, 214–216.

[123] Li, J.-Y., Sidhu, R.S., Bollon, A., Strobel, G.A. (1998). Stimulation of taxol production in liquid cultures of *Pestalotiopsis microspora*. *Mycology Research*, *102*, 461–464.

[124] Puri, S.C., Verma, V., Amna, T., Qazi, G.N., Spiteller, M. (2005). An endophytic fungus from *Nothapodytes foetida* that produces camptothecin. *Journal of Natural Products*, *68*, 1717–1719.

[125] Amna, T., Puri, S.C., Verma, V., Sharma, J.P., Khajuria, R.K., Musarrat, J., Spiteller, M., Qazi, G.N. (2006). Bioreactor studies on the endophytic fungus *Entrophosphora infrequens* for the production of an anticancer alkaloid camptothecin. *Canadian Journal of Microbiology*, *52*, 189–196.

[126] Eyberger, A.L., Dondapati, R., Porter, J.R. (2006). Endophyte fungal isolates from *Podophyllum peltatum* produce podophyllotoxin. *Journal of Natural Products*, *69*, 1121–1124.

[127] Puri, S.C., Nazir, A., Chawla, R., Arora, R., Riyaz-ul-Hasan, S., Amna, T., Ahmed, B., Verma, V., Singh, S., Sagar, R., Sharma, A., Kumar, R., Sharma, R.K., Qazi, G.N. (2006). The endophytic fungus *Trametes hirsuta* as a novel alternative source of podophyllotoxin and related tetralin lignans. *Journal of Biotechnology*, *122*, 494–510.

[128] Guo, B., Li, H., Zhang, L. (1998). Isolation of a fungus producing vinblastine. *Journal of Yunnan University*, *20*, 214–215.

[129] Zhang, L.Q., Guo, B., Li, H., Zeng, S., Shao, H., Gu, S., Wei, R. (2000). Preliminary study on the isolation of an endophytic fungus of *Catharanthus roseus* and its fermentation to produce products of therapeutic value. *Zhong Cao Yao (Chinese Traditional and Herbal Drugs)*, *31*, 805–807.

[130] Yang, X., Zhang, L., Guo, B., Guo, S. (2004). Preliminary study of a vincristine-producing endophytic fungus isolated from leaves of *Catharanthus roseus*. *Zhong Cao Yao (Chinese Traditional and Herbal Drugs)*, *35*, 79–81.

[131] Abe, F., Horikoshi, K. (2001). The biotechnological potential of piezophiles. *Trends in Biotechnology*, *19*, 102–108.

[132] Persidis, A. (1998). *Extremophiles. Nature Biotechnology*, *16*, 593–594.

[133] Rossi, M., Ciaramella, M., Cannio, R., Pisani, F.M., Moracci, M., Bartolucci, S. (2003). Extremophiles 2002. *Journal of Bacteriology*, *185*, 3683–3689.

[134] Jørgensen, B.B., Boetius, A. (2007). Feast and famine—Microbial life in the deep-sea bed. *Nature Reviews Microbiology*, *5*, 770–781.

[135] Pikuta, E.V., Hoover, R.B., Tang, J. (2007). Microbial extremophiles at the limits of life. *Critical Reviews in Microbiology*, *33*, 183–209.

[136] Wagner, I.D., Wiegel, J. (2008). Diversity of thermophilic anaerobes. *Annals of the New York Academy of Sciences*, *1125*, 1–43.

[137] Cavicchioli, R., Siddiqui, K.S., Andrews, D., Sowers, K.R. (2002). Low-temperature extremophiles and their applications. *Current Opinion in Biotechnology*, *13*, 253–261.

[138] Feller, G. (2003). Psychrophilic enzymes: Hot topics in cold adaptation. *Nature Reviews Microbiology*, *1*, 200–208.

[139] Cavicchioli, *R.* (2006). Cold-adapted archaea. *Nature Reviews Microbiology*, *4*, 331–343.

[140] Gomes, J., Steiner, W. (2004). The biocatalytic potential of extremophiles and extremozymes. *Food Technology and Biotechnology*, *42*, 223–235.

[141] Hoyoux, A., Blaise, V., Collins, T., D'Amico, S., Gratia, E., Huston, A.L., Marx, J.C., Sonan, G., Zeng, Y.X., Feller, G., Gerday, C. (2004). Extreme catalysts from low-temperature environments. *Journal of Bioscience and Bioengineering*, *98*, 317–330.

[142] Schiraldi, C., De Rosa, M. (2002). The production of biocatalysts and biomolecules from extremophiles. *Trends in Biotechnology*, *20*, 515–521.

[143] van den Burg, B. (2003). Extremophiles as a source for novel enzymes. *Current Opinion in Microbiology*, *6*, 213–218.

[144] Pettit, R.K. (2010). Culturability and secondary metabolite diversity of extreme microbes: Expanding contribution of deep sea and deep-sea vent microbes to natural product discovery. *Marine Biotechnology*, *13*, 1–11.

[145] Wilson, Z.E., Brimble, M.A. (2009). Molecules derived from the extremes of life. *Natural Products Reports*, *26*, 44–71.

[146] Johnson, D.B., Hallberg, K.B. (2003). The microbiology of acidic mine waters. *Research in Microbiology*, *154*, 466–473.

[147] Stierle, A.A., Stierle, D.B., Kemp, K. (2004). Novel sesquiterpenoid matrix metalloproteinase-3 inhibitors from an acid mine waste extremophile. *Journal of Natural Products*, *67*, 1392–1395.

[148] Stierle, D.B., Stierle, A.A., Hobbs, D., Stokken, J., Clardy, J. (2004). Berkeleydione and berkeleytrione, new bioactive metabolites from an acid mine organism. *Organic Letters*, *6*, 1049–1052.

[149] Stierle, A.A., Stierle, D.B., Patacini, B. (2008). The berkeleyamides, amides from the acid lake fungus *Penicillum rubrum*. *Journal of Natural Products*, *71*, 856–860.

[150] Park, H.B., Kwon, H.C., Lee, C. -H, Yang, H.O. (2009). Glionitrin A, an antibiotic–antitumor metabolite derived from competitive interaction between abandoned mine microbes. *Journal of Natural Products*, *72*, 248–252.

[151] Li Y., Ye D., Chen X., Lu X., Shao Z., Zhang H., Che Y. (2009). Breviane spiroditerpenoids from an extreme-tolerant *Penicillium* sp. isolated from a deep sea sediment sample. *Journal of Natural Products*, *72*, 912–916.

[152] Du, L., Li, D., Zhu, T., Cai, S., Wang, F., Xiao, X., Gu, Q. (2009). New alkaloids and diterpenes from a deep ocean sediment derived fungus *Penicillium* sp. *Tetrahedron*, *65*, 1033–1039.

[153] Khosla, C. (2000). Natural product biosynthesis: A new interface between enzymology and medicine. *Journal of Organic Chemistry*, *65*, 8127–8133.

[154] Staunton, J., Weissman, K.J. (2001). Polyketide biosynthesis: A millennium review. *Natural Products. Reports*, *18*, 380–416.

[155] Walsh, C.T. (2004). Polyketide and nonribosomal peptide antibiotics: Modularity and versatility. *Science*, *303*, 1805–1810.

[156] Van Lanen, S.G., Shen, B. (2006). Microbial genomics for the improvement of natural product discovery. *Current Opinion in Microbiology*, *9*, 252–260.

[157] Abhirup Das, A., Khosla, C. (2009). Biosynthesis of aromatic polyketides in bacteria. *Accounts of Chemical Research*, *42*, 631–639.

[158] Kopp, F., Marahiel, M.A. (2007). Where chemistry meets biology: The chemoenzymatic synthesis of nonribosomal peptides and polyketides. *Current Opinion in Biotechnology*, *18*, 513–520.

[159] Walsh, C.T. (2008). The chemical versatility of natural-product assembly lines. *Accounts of Chemical Research*, *41*, 4–10.

[160] Sattely, E.S., Fischbach, M.A., Walsh, C.T. (2008). Total biosynthesis: *In vitro* reconstitution of polyketide and nonribosomal peptide pathways. *Natural Products Reports*, *25*, 757–793.

[161] Weissman, K.J., Müller, R. (2008). Protein–protein interactions in multienzyme megasynthetases. *ChemBioChem*, *9*, 826–848.

[162] Felnagle, E.A., Jackson, E.E., Yolande A. Chan, Y.A., Angela, M., Podevels, A.M., Andrew, D., Berti, A.D., Matthew, D., McMahon, M.D., Thomas, M.G. (2008). Nonribosomal peptide synthetases involved in the production of medically relevant natural products. *Molecular Pharmaceutics*, *5*, 191–211.

[163] Marahiel, M.A., Essen, L.-O., Nonribosomal peptide synthetases: Mechanistic and structural aspects of essential domains. In: Hopwood, D.A., editor. *Methods in Enzymology. Complex Enzymes in Microbial Natural Product Biosynthesis, Part A: Overview Articles and Peptides*. Elsevier, Inc., Oxford, 2009, pp. 337–351.

[164] Koglin, A., Walsh, C.T. (2009). Highlight: Structural insights into nonribosomal peptide enzymatic assembly lines. *Natural Products Reports*, *26*, 987–1000.

[165] Nolan, E.M., Walsh, C.T. (2009). How nature morphs peptide scaffolds into antibiotics. *ChemBioChem*, *10*, 34–53.

[166] Clardy, J., Walsh, C.T. (2004). Lessons from natural molecules. *Nature*, *432*, 829–837.

[167] Kira J., Weissman, K.J. (2007). Mutasynthesis—Uniting chemistry and genetics for drug discovery. *Trends in Biotechnology*, *25*, 139–142.

[168] Thomas, M.G., Bixby, K.A., Shen, B., Combinatorial biosynthesis of anticancer natural molecules. In: Cragg, G.M., Kingston, D.G.I., and Newman, D.J. editors. *Anticancer Agents from Natural Products*, Taylor & Francis, Boca Raton, FL, 2005, pp. 519–552.

[169] Wenzel, S.C., Müller, R. (2009). The biosynthetic potential of myxobacteria and their impact in drug discovery. *Current Opinion in Drug Discovery and Development*, *12*, 220–230.

[170] Tang, L., Chung, L., Carney, J.R., Starks, C.M., Licari, P., Katz, L. (2005). Generation of new epothilones by genetic engineering of a polyketide synthase in Myxococcus xanthus. *Journal of Antibiotics*, *58*, 178–184.

[171] Chou, T.C., Zhang, X., Zhong, Z.Y., Li, Y., Feng, L., Eng, S., Myles, D.R., Johnson, R., Wu, N., Yin, Y.I., Wilson, R.M., Danishefsky, S.J. (2008). Therapeutic effect against human xenograft tumors in nude mice by the third generation microtubule stabilizing epothilones. *Proceedings of the National Academy of Sciences of the U.S.A.*, *105*, 13157–13162.

[172] Julien, B., Shah, S., Ziemann, R., Goldman, R., Katz, L., Khosla, C. (2000). Isolation and characterization of the epothilone biosynthetic gene cluster from *Sorangium cellulosum*. *Gene*, *249*, 153–160.

[173] Molnar, I., Schupp, T., Ono, M., Zirkle, R.E., Milnamow, M., Nowak-Thompson, B., Engel, N., Toupet, C., Stratmann, A., Cyr, D.D., Gorlach, J., Mayo, J.M., Hu, A., Goff, S., Schmid, J., Ligon, J.M. (2000). The biosynthetic gene cluster for the microtubule-stabilizing agents epothilones A and B from *Sorangium cellulosum* So ce90. *Chemical Biology*, *7*, 97–109.

[174] Lau, J., Frykman, S., Regentin, R., Ou, S., Tsuruta, H., Licari, P. (2002). Optimizing the heterologous production of epothilone D in *Myxococcus xanthus*. *Biotechnology and Bioengineering*, *78*, 280–288.

[175] Evans, B.E., Rittle, K.E., Bock, M.G., DiPardo, R.M., Freidinger, R.M., Whitter, W.L., Lundell, G.F., Veber, D.F., Anderson, P.S., Chang, R.S.L., Lotti, V.J., Cerino, D.J., Chen, T.B., Kling, P.J., Kunkel, K.A., Springer, J.P., Hirshfield, J. (1988). Methods for drug discovery: Development of potent, selective, orally effective cholecystokinin antagonists. *Journal of Medicinal Chemistry*, *31*, 2235–2246.

[176] Kingston, D.G.I., Taxol and its analogs. In: Cragg, G.M., Kingston, D.G.I., and Newman, D.J., editors. *Anticancer Agents from Natural Products*, CRC Press, Boca Raton, FL, 2005, pp. 89–122.

[177] Kingston, D.G.I. (2008). A natural love of natural products. *Journal of Organic Chemistry*, *73*, 3975–3984.

[178] Rahier, N.J., Thomas, C.J., Hecht, S.M., Camptothecin and its analogs. In: Cragg, G.M., Kingston, D.G.I., and Newman, D.J. editors. *Anticancer Agents from Natural Products*, CRC Press, Boca Raton, FL, 2005, pp. 5–22.

[179] Pinney, K.G., Jelinek, C., Edvardsen, K., Chaplin, D.J. Pettit, G.R., The discovery and development of the combretastatins. In: Cragg, G.M., Kingston, D.G.I., and

Newman, D.J.editors. *Anticancer Agents from Natural Products*, CRC Press, Boca Raton, FL, 2005, pp. 23–46.

[180] Rinehart, K.L., Holt, T.G., Fregeau, N.L., Stroh, J.G., Keifer, P.A., Sun, F., Li, L.H., Martin, D.G. (1990). Ecteinascidins 729, 743, 745, 759A and 770: Potent antitumor agents from the Caribbean tunicate *Ecteinascidia* turbinata. *Journal of Organic Chemistry*, *55*, 4512–4515.

[181] Wright, A.E., Forleo, D.A., Gunawardana, G.P., Gunasekera, S.P., Koehn, F.E., McConnell, O.J. (1990). Antitumor tetrahydroisoquinoline alkaloids from the colonial ascidian *Ecteinascidia turbinata*. *Journal of Organic Chemistry*, *55*, 4508–4512.

[182] Cuevas, C., Pérez, M., Martín, M.J., Chicharro, J.L., Fernández-Rivas, C., Flores, M., Francesch, A.M., Gallego, P., Zarzuelo, M., de La Calle, F., García, J., Polanco, C., Rodríguez, I., Manzanares, I. (2000). Synthesis of ecteinascidin ET-743 and phthalascidin Pt-650 from cyanosafracin B. *Organic Letters*, *2*, 2545–2548.

[183] Newman, D.J., Cragg, G.M. (2006). Natural products from marine invertebrates and microbes as modulators of antitumor targets. *Current Drug Targets*, *7*, 279–304.

[184] Nicolaou, K.C., Vourloumis, D., Winssinger, N., Baran, P.S. (2000). The art and science of total synthesis at the dawn of the twenty-first century. *Angewandte Chemie International Edition*, *39*, 44–122.

[185] Mickel, S.J., Niederer, D., Daeffler, R., Osmani, A., Kuesters, E., Schmid, E., Schaer, K., Gamboni, R., Chen, W.C., Loeser, E., Kinder, F.R., Jr., Konigsberger, K., Prasad, K., Ramsey, T.M., Repic, O., Wang, R.M., Florence, G., Lyothier, I., Paterson, I. (2004) Large-scale synthesis of the anti-cancer marine natural product (+)-discodermolide. Part 5: Linkage of fragments C_{1-6} and C_{7-24} and finale. *Organic Process Research and Development*, *8*, 122–130.

[186] Freemantle, M., (2004). Discodermolide. *Chemical and Engineering News*, *82*, 33–35.

[187] Njardarson, J.T., Gaul, C., Shan, D.; Huang, X.Y., Danishefsky, S.J. (2004). Discovery of potent cell migration inhibitors through total synthesis: Lessons from structure–activity studies of (+)-migrastatin. *Journal of the American Chemical Society*, *126*, 1038–1040.

[188] Wilson, R.M., Danishefsky, S.J. (2006). Small molecule natural products in the discovery of therapeutic agents: The synthesis connection. *Journal of Organic Chemistry*, *71*, 8329–8351.

[189] http://www.bloomberg.com/news/2010-06-02/eisai-s-eribulin-breast-cancer-treatment-gets-priority-review-by-u-s-fda.html

[190] Flahive, E., Srirangam, J., The dolastatins. Novel antitumor agents from *Dollabella auricularia*. In: Cragg, G.M., Kingston, D.G.I., and Newman, D.J.editors. *Anticancer Agents from Natural Products*, CRC Press, Boca Raton, FL, 2005, pp. 191–214.

[191] Law, C.-L., Gordon, K.A., Toki, B.E., Yamane, A.K., Hering, M.A., Cerveny, C.G., Petroziello, J.M., Ryan, M.C., Smith, L., Simon, R., Sauter, G., Oflazoglu, E., Doronina, S.O., Meyer, D.L., Francisco, J.A., Carter, P., Senter, P.D., Copland, J.A., Wood, C.G., Wahl, A.F. (2006). Lymphocyte activation antigen CD70 expressed by renal cell carcinoma is a potential therapeutic target for anti-CD70 antibody–drug conjugates. *Cancer Research*, *66*, 2328–2337.

[192] Sutherland, M.S.K., Sanderson, R.J., Gordon, K.A., Andreyka, J., Cerveny, C.G., Yu, C., Lewis, T.S., Meyer, D.L., Zabinski, R.F., Doronina, S.O., Senter, P.D., Law, C.-L., Wahl, A.F. (2006). Lysosomal trafficking and cysteine protease metabolism confer

target-specific cytotoxicity by peptide-linked anti-CD30–auristatin conjugates. *Journal of Biological Chemistry*, *281*, 10540–10547.

[193] Oflazoglu, E., Kissler, K.M., Sievers, E.L., Grewal, I.S., Gerber, H.-P. (2008). Combination of the anti-CD30–auristatin-E antibody–drug conjugate (SGN-35) with chemotherapy improves antitumour activity in Hodgkin lymphoma. *British Journal of Haematology*, *142*, 69–73.

[194] Pollack, V.A., Alvarez, E., Tse, K.F., Torgov, M.Y., Xie, S., Shenoy, S.G., MacDougall, J.R., Arrol, S., Zhong, Z., Gerwein, R.W., Hahne, W.F., Senter, P.D., Jeffers, M.E., Lichtenstein, H.S., LaRochelle, W.J. (2007). Treatment parameters modulating regression of human melanoma xenografts by an antibody-drug conjugate (CR011-vcMMAE) targeting GPNMB. *Cancer Chemotherapy and Pharmacology*, *60*, 423–435.

[195] Ma, D., Hopf, C.E., Malewicz, A.D., Donovan, G.P., Senter, P.D., Goeckeler, W.F., Maddon, P.J., Olson, W.C. (2006). Potent antitumor activity of an auristatin-conjugated, fully human monoclonal antibody to prostate-specific membrane antigen. *Clinical Cancer Research*, *12*, 2591–2596.

[196] Borman, S. (2003). The many faces of combinatorial chemistry. *Chemical and Engineering News*, *81*, 45–56.

[197] Class, S. (2002). Pharma overview. *Chemical and Engineering News*, *80*, 39–49.

[198] Breinbauer, R., Manger, M., Scheck, M., Waldmann, H. (2002). Natural product guided compound library development. *Current Medicinal Chemistry*, *9*, 2129–2145.

[199] Breinbauer, R., Vetter, I.R., Waldmann, H. (2002). From protein domains to drug candidates—Natural products as guiding principles in the design and synthesis of compound libraries. *Angewandte Chemie International Edition*, *41*, 2878–2890.

[200] Kaiser, M., Wetzel, S., Kumar, K., Waldmann, H. (2008). Biology-inspired synthesis of compound libraries. *Cellular and Molecular Life Sciences*, *65*, 1186–1201.

[201] Burke, M.D., Berger, E.M., Schreiber, S.L. (2003). Generating diverse skeletons of small molecules combinatorially. *Science*, *302*, 613–618.

[202] Burke, M.D., Schreiber, S.L. (2004). A planning strategy for diversity-oriented synthesis. *Angewandte Chemie International Edition*, *43*, 46–58.

[203] Burke, M.D., Berger, E.M., Schreiber, S.L. (2004). A synthesis strategy yielding skeletally diverse small molecules combinatorially. *Journal of the American Chemical Society*, *126*, 14095–14104.

[204] Nielsen, T.E., Schreiber, S.L. (2008). Towards the optimal screening collection: A synthesis strategy. *Angewandte Chemie International Edition*, *47*, 48–56.

[205] Nicolaou, K.C., Pfefferkorn, J.A., Roecker, A.J., Cao, G.Q., Barluenga, S., Mitchell, H.J. (2000). Natural product-like combinatorial libraries based on privileged structures. 1. General principles and solid-phase synthesis of benzopyrans. *Journal of the American Chemical Society*, *122*, 9939–9953.

[206] Nicolaou, K.C., Kim, S., Pfefferkorn, J., Xu, J., Ohshima, T., Hosokawa, S., Vourloumis, D., Li, T. (1998). Synthesis and biological activity of sarcodictyins. *Angewandte Chemie International Edition*, *37*, 1418–1421.

[207] Thutewohl, M., Kissau, L., Popkirova, B., Karaguni, I.M., Nowak, T., Bate, M., Kuhlmann, J., Müller, O., Waldmann, H. (2002). Solid-phase synthesis and biological evaluation of a pepticinnamin E library. *Angewandte Chemie International Edition*, *41*, 3616–3620.

[208] Nicolaou, K.C., Roschangar, F., Vourloumis, D. (1998). Chemical biology of epothilones. *Angewandte Chemie International Edition, 37*, 2014–2045.

[209] Nicolaou, K.C., Hughes, R., Cho, S.Y., Winssinger, N., Labischinski, H., Endermann, R. (2001). Synthesis and biological evaluation of vancomycin dimers with potent activity against vancomycin-resistant bacteria: Target-accelerated combinatorial synthesis. *Chemistry—A European Journal, 7*, 3824–3843.

[210] Wipf, P., Reeves, J.T., Balachandran, R., Giuliano, K.A., Hamel, E., Day, B.W. (2000). Synthesis and biological evaluation of a focused mixture library of analogues of the antimitotic marine natural product curacin A. *Journal of the American Chemical Society, 122*, 9391–9395.

[211] Nicolaou, K.C., Pfefferkorn, J.A., Barluenga, S., Mitchell, H.J., Roecker, A.J., Cao, G.Q. (2000). Natural product-like combinatorial libraries based on privileged structures. 3. The "libraries from libraries" principle for diversity enhancement of benzopyran libraries. *Journal of the American Chemical Society, 122*, 9968–9976.

[212] Nicolaou, K.C., Pfefferkorn, J.A., Mitchell, H.J., Roecker, A.J., Barluenga, S., Cao, G.Q., Affleck, R.L., Lillig, J.E. (2000). Natural product-like combinatorial libraries based on privileged structures. 2. Construction of a 10000-membered benzopyran library by directed split-and-pool chemistry using nanokans and optical encoding. *Journal of the American Chemical Society, 122*, 9954–9967.

[213] Koonin, E.V., Wolf, Y.I., Karev, G.P. (2002). The structure of the protein universe and genome evolution. *Nature, 420*, 218–223.

[214] Koch, M.A., Schuffenhauer, A., Scheck, M., Wetzel, S., Casaulta, M., Odermatt, A., Ertl, P.; Waldmann, H. (2005). Charting biologically relevant chemical space: A structural classification of natural products (SCONP). Proceedings of the National Academy of Sciences of the U.S.A., *102*, 17272–17277.

[215] Schuffenhauer, A., Ertl, P., Roggo, S., Wetzel, S., Koch, M.A., Waldmann, H. (2007). The Scaffold Tree—Visualization of the scaffold universe by hierarchical scaffold classification. *Journal of Chemical Information and Modeling, 47*, 47–58.

[216] Dekker, F.J., Koch, M.A., Waldmann, H. (2005). Protein structure similarity clustering (PSSC) and natural product structure as inspiration sources for drug development and chemical genomics. *Current Opinion in Chemical Biology, 9*, 232–239.

[217] Koch, M.A., Breinbauer, R., Waldmann, H. (2003). Protein structure similarity as guiding principle for combinatorial library design. *Biological Chemistry, 384*, 1265–1272.

[218] Koch, M.A., Waldmann, H. (2005). Protein structure similarity clustering and natural product structure as guiding principles in drug discovery. *Drug Discovery Today, 10*, 471–483.

[219] Brohm, D., Metzger, S., Bhargava, A., Müller, O., Lieb, F., Waldmann, H. (2002). Natural products are biologically validated starting points in structural space for compound library development: Solid-phase synthesis of dysidiolide-derived phosphatase inhibitors. *Angewandte Chemie International Edition, 41*, 307–311.

[220] Koch, M.A., Wittenberg, L.O., Basu, S., Jeyaraj, D.A., Gourzoulidou, E., Reinecke, K., Odermatt, A., Waldmann, H. (2004). Compound library development guided by protein structure similarity clustering and natural product structure. *Proceedings of the National Academy of Sciences of the U.S.A., 101*, 16721–16726.

[221] Kobayashi, J., Madono, T., Shigemori, H. (1995). Nakijiquinones C and D, new sesquiterpenoid quinones with a hydroxy amino acid residue from a marine sponge inhibiting c-erbB-2 kinase. *Tetrahedron*, *51*, 10867–10874.

[222] Stahl, P., Kissau, L., Mazitschek, R., Giannis, A., Waldmann, H. (2002). Natural product derived receptor tyrosine kinase inhibitors: Identification of IGF1R, Tie-2, and VEGFR-3 inhibitors. *Angewandte Chemie International Edition*, *41*, 1174–1178.

[223] Kissau, L., Stahl, P., Mazitschek, R., Giannis, A., Waldmann, H. (2003). Development of natural product-derived receptor tyrosine kinase inhibitors based on conservation of protein domain fold. *Journal of Medicinal Chemistry*, *46*, 2917–2931.

[224] Klekota, J., Brauner, E., Roth, F.P., Schreiber, S.L. (2006). Using high-throughput screening data to discriminate compounds with single-target effects from those with side effects. *Journal of Chemical Information and Modeling*, *46*, 1549–1562.

[225] Spring, D.R. (2003). Diversity-oriented synthesis; a challenge for synthetic chemists. *Organic and Biomolecular Chemistry*, *1*, 3867–3870.

[226] Spandl, R.J., Díaz-Gavilán, M., O'Connell, K.M., Thomas, G.L., Spring, D.R. (2008). Diversity-oriented synthesis. *Chemical Record*, *8*, 129–142.

[227] Spandl, R.J., Bender, A., Spring, D.R. (2008). Diversity-oriented synthesis; a spectrum of approaches and results. *Organic and Biomolecular Chemistry*, *6*, 1149–1158.

[228] Peuchmaur, M., Wong, Y.S. (2008). Expanding the chemical space in practice: Diversity-oriented synthesis. *Combinatorial Chemistry and High Throughput Screening*, *11*, 587–601.

[229] Cordier, C., Morton, D., Murrison, S., Nelson, A., O'Leary-Steele, C. (2008). Natural products as an inspiration in the diversity-oriented synthesis of bioactive compound libraries. *Natural Products Reports*, *25*, 719–737.

[230] Walsh, D.P., Chang, Y.T. (2006). Chemical genetics. *Chemical Reviews*, *106*, 2476–2530.

[231] Morton, D., Leach, S., Cordier, C., Warriner, S., Nelson, A. (2009). Synthesis of natural-product-like molecules with over eighty distinct scaffolds. *Angewandte Chemie International Edition*, *48*, 104–109.

[232] Pelish, H.E., Westwood, N.J., Feng, Y., Kirchhausen, T., Shair, M.D. (2001). Use of biomimetic diversity-oriented synthesis to discover galanthamine-like molecules with biological properties beyond those of the natural product. *Journal of the American Chemical Society*, *123*, 6740–6741.

[233] Blackwell, H.E., Pérez, L., Stavenger, R.A., Tallarico, J.A., Cope Eatough, E., Foley, M.A., Schreiber, S.L. (2001). A one-bead, one-stock solution approach to chemical genetics: Part 1. *Chemical Biology*, *8*, 1167–1182.

[234] Clemons, P.A., Koehler, A.N., Wagner, B.K., Sprigings, T.G., Spring, D.R., King, R.W., Schreiber, S.L., Foley, M.A. (2001). A one-bead, one-stock solution approach to chemical genetics: Part 2. *Chemical Biology*, *8*, 1183–1195.

[235] Koehler, A.N., Shamji, A.F., Schreiber, S.L. (2003). Discovery of an inhibitor of a transcription factor using small molecule microarrays and diversity-oriented synthesis. *Journal of the American Chemical Society*, *125*, 8420–8421.

A Second Edition of Anticancer Agents from Natural Products was published in November 2011 and readers may refer to this edition for the latest information discussed in references 29-31; 34-39; 96, 106, 168, 176, 178-9; 190.

2

MODERN APPROACHES IN THE SEARCH FOR NEW ACTIVE COMPOUNDS FROM CRUDE EXTRACTS OF NATURAL SOURCES

EMERSON F. QUEIROZ, KURT HOSTETTMANN, AND JEAN-LUC WOLFENDER

2.1 INTRODUCTION

Nature is a source of a considerable number of natural products (NPs) with an almost infinite variety of structural variations. These molecules often have specific functions, and many of them have biological activities that can be useful to humans. They may provide lead compounds for the development of new drugs or they may be indispensable tools in biomedical research. Although many NPs are substrates for life processes, they can also consist of toxins, hormones, or molecules with other functions. Before the development of modern medicine, natural products were the only therapy available to humans. Nature is still an important reservoir of new molecules with potential therapeutic interest. In the case of plants, only a relatively small percentage of the 350,000 known species have been studied from a phytochemical or pharmacological perspective. Research in pharmacognosy has demonstrated that potent bioactive products can be obtained from plants [1]. The reservoir of NPs can be greatly extended if the entire genre of biodiversity is included in the search, such as microorganism, marine organisms, fungi, and animals.

NPs and their derivatives have been historically viewed as an invaluable source of therapeutic agents. However, in the past decade, research in NPs within the pharmaceutical industry has declined because of a lack of compatibility between complex

Plant Bioactives and Drug Discovery: Principles, Practice, and Perspectives, Fourth Edition.
Edited by Valdir Cechinel-Filho.

natural extract libraries and high-throughput screening. However, recent technological advances that alleviate these issues, coupled with current lead-generation strategies, have led to a renewed interest in NP drug discovery [2]. It has long been recognized that NP structures, including those from sources other than plants, have characteristics of high chemical diversity [3], biochemical specificity, and other molecular properties that make them favorable lead structures for drug discovery, which serve to differentiate them from libraries of synthetic and combinatorial compounds [4]. A series of studies have revealed that NPs have a greater number of chiral centers and increased steric complexity compared to either synthetic drugs or combinatorial libraries [5]. NPs can be viewed as a population of privileged structures selected by evolutionary pressures to interact with a wide variety of proteins and other biological targets for specific purposes; this is a view supported by the fact that NPs have become effective drugs in a wide variety of therapeutic indications. According to Cragg and coworkers, 61% of the 877 new small-molecule chemical compounds registered as drugs worldwide during the period of 1981–2002 are or have been inspired by NPs [4]. In specific therapeutic areas, for instance, anti-infective (e.g., antibacterial, antifungal, antiparasitic, and antiviral) drugs, the percentage of drugs inspired, or derived by a NP has reached 70% [6]. To discover new bioactive compounds from natural sources extracts should be simultaneously submitted to a chemical screening and to various biological or pharmacological targets. The chemical screening or metabolite profiling is aimed at distinguishing between already known compounds (dereplication) and new molecules directly in crude extracts. Thus, the tedious isolation of known compounds can be avoided and a targeted isolation of constituents presenting novel or unusual spectroscopic features can be undertaken. It is important to note that metabolite profiling clearly refers here to the detection and identification of plant metabolites and is different from the metabolic profiling process associated with the detection of metabolized products issued from a given lead compound [7]. Metabolite profiling in crude extracts is not an easy task because NPs display very important structural diversity [8]. For each compound, the order of the atoms and their stereochemical orientations have to be elucidated *de novo* in a complex manner, and the compounds cannot simply be sequenced, as is the case for genes or proteins. Consequently, unlike genomics and proteomics, a single analytical technique capable of profiling all secondary metabolites in crude extract does not currently exist [9]. Thus, to develop innovative strategies for the metabolite profiling of crude extracts, researchers must take advantage of innovative hyphenated techniques created over the last decade.

Various aspects related to the metabolite identification possibilities of such methods will be discussed herein. However, the principle operation of these powerful techniques will not be explained in detail because these can be found in other, more specialized reviews [10–14]. In this chapter, emphasis is placed on the strategy used to select the material and develop efficient dereplications of NPs as well as to identify (online) bioactive constituents based on the combination of LC-PDA, LC-MS, UHPLC-MS, LC-MS/MS, SPE-NMR, and CapNMR techniques applied to crude extract profiling. The role of these techniques in the structural investigation of unstable products or the investigation of compounds that are difficult to isolate at the preparative level will also be highlighted.

2.2 SELECTION OF THE NATURAL MATRICES

When a phytochemical investigation is undertaken in plants with the aim of isolating and identifying active substances, the correct choice of vegetable material must be made. Considering the number of plants that have not yet been studied from both phytochemical and pharmacological viewpoints, this choice can be difficult. The factors to be considered consist of the following: chemotaxonomic criteria, information from traditional medicine, zoopharmacognosy, field observations, and random collection [15]. Recent approaches have also involved reversed pharmacognosy, which entails the reisolation of interesting NP scaffolds with potential activity on new targets based on computational predictions [16,17]. Chemotaxonomy, the science of the classification of plants as a function of the structures of their chemical constituents, can introduce useful elements. Constituents are often specific to a given botanical family, genus or species. If an NP has interesting therapeutic properties, it may be possible to find analogous substances in species closely related to the plants from which it has been isolated. The selection of plants based on data from traditional medicine can also lead to the discovery of promising new molecules. Plants from tropical and subtropical regions are abundant and represent an important reservoir of new molecules with potential therapeutic activities waiting to be discovered. International agreements have been established to prevent these natural resources, which are primarily located in third-world countries, from being excessively exploited by industries from developed countries without equitable benefit sharing. Within the framework of these standards, a special agreement has been established for endangered species to ensure their protection. An example of this trade is given by the case of prostratin.

In 1991, based on traditional medicine in Samoan, the National Cancer Institute (NCI) isolated prostratin from the bark of the *Homalanthus nutans*, a plant belonging to the Euphorbiaceae family used for the treatment of a form of hepatitis. Subsequently, this compound was discovered to hold enormous therapeutic potential as an anti-AIDS drug, and it was patented. An agreement between the Aids Research Alliance (ARA) and the Samoan government was signed. Twenty percent of the profits made from prostratin were negotiated to go directly to Samoa. In 2004, to ensure a plentiful supply of prostratin without the need to directly depend on the mamala's tree bark, the prostratin genes were cloned at the University of California at Berkeley. An agreement was signed establishing that any royalties from the sale of a gene-derived drug would also be equally shared with the people of Samoa [18,19]. According to Nogueira et al., the prostratin case is a rare example of an effective compensation mechanism established between ARA, UC Berkeley, and the Samoan government [20].

Clearly, field observations are very important during plant-collecting expeditions. A species that grows in a hostile environment (e.g., tropical forest), in which there is danger of attack from insects, fungi, bacteria, or virus, will attempt to protect itself by synthesizing insecticidal, fungicidal, antibacterial, or virucidal constituents. An example of field observations was performed by Prof. Lema in Ethiopia. He observed that the berries of *Phytolacca dodecandra* (Phytolaccaceae), which was used as a traditional soap, killed aquatic snails [21]. This observation allowed for the discovery of a new class of molluscicidal compounds, the saponins. Berries from *P. dodecandra* provide a readily available molluscicide to control schistosomiasis, a parasitic disease caused by several species of fluke of the genus *Schistosoma*; the snails represent the intermediate host of the parasite [22].

Random collection also represents an important strategy. Extracts, semipurifed fractions or pure NPs, can be screened on a large scale using high-throughput screening. An example of this strategy is the discovery of taxol, from *Taxus brevifolia* L. (Taxaceae), one of the most important anticancer drugs [23]. Many NPs derived from microorganisms or marine sources have also been discovered using the random collection method.

2.3 RAPID ONLINE IDENTIFICATION AND DEREPLICATION

When searching for active metabolites, biological screening followed by activity-guided fractionation is the standard procedure [24]. Bioassays also serve as a guide during the isolation process. However, the number of available targets is limited. Moreover, bioassays do not always predict clinical efficacy, and the bioassay-guided fractionation strategy frequently leads to the isolation of known metabolites. Therefore, the chemical screening of crude extracts constitutes an efficient and complementary approach that allows for the localized and targeted isolation of new types of constituents with potential activities. This procedure also enables the recognition of known metabolites at the earliest stage of separation, thus avoiding costly and time-consuming isolations of common constituents. The potential of the chemical screening strategy has been considerably increased by the recent development of hyphenated techniques, which are able to provide an efficient separation of metabolites and, concurrently, valuable online and at-line structural information (Figure 2.1). The online process refers to spectroscopic data that can be recorded without stopping the HPLC flow rate in direct hyphenation, the at-line mode refers to spectroscopic information recorded on LC-peaks that have been collected or enriched directly after HPLC separation for which spectroscopic data are acquired independently and usually with longer acquisition time.

2.4 HPLC-HYPHENATED METHODS FOR NATURAL PRODUCT IDENTIFICATION

2.4.1 HPLC Separation of Crude Extracts

HPLC can be considered one of the most versatile techniques for the analysis of NPs and can be used to directly profile complex mixtures, as in the case of extracts without derivatization steps. In the case of plant analysis, these extracts are usually obtained

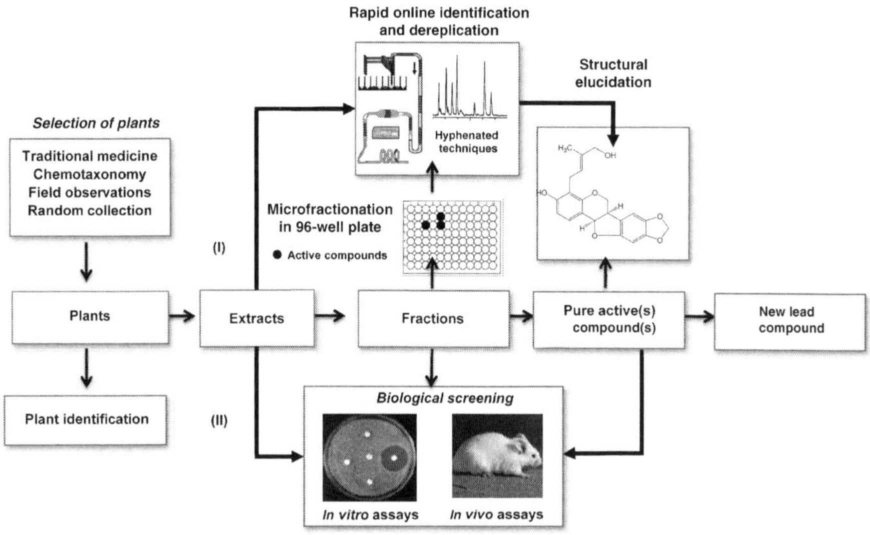

FIGURE 2.1 Discovery procedure of active compound using hyphenated techniques.

by maceration of the dried plant material with nonpolar (CH_2Cl_2) or polar (MeOH) solvents; in some cases, the extracts are obtained by liquid–liquid acid–base extractions (e.g., alkaloids).

For other organisms, different extraction methods have been used, but the final extracts are usually either polar, nonpolar, or cover an intermediate polarity range. In some cases, sample preparation procedures can be used to enrich a given class of metabolites; these typically involve liquid–liquid extraction (LLE) or solid phase extraction (SPE). HPLC has notably evolved over the years, and various columns with different stationary phases have enabled the separation of a very large range of NPs [25]. However, the majority of applications are primarily related to the reversed phase (RP) mode. The latest developments on HPLC include the recent introduction of highly pH-stable phases, sub-2 μm particles [26] monolith and fused core columns. These recent innovations have considerably improved the performances of HPLC systems, in terms of their resolution, speed and reproducibility. Efficiency exceeding 100,000 plates and peak capacities over 900 can be attained by coupling columns [27]. For the separation of crude extracts, either the raw mixtures or samples already treated by simple SPE or LLE can be directly injected into the HPLC system. The separations are primarily performed on C_{18} material with MeCN–H_2O or MeOH–H_2O solvent systems in the gradient elution mode. To improve the separation efficiency, various modifiers that can strongly influence the sensitivity of detection are added to the mobile phase [28]. In multiple hyphenated systems, the presence of several different online detectors (hyphenated systems) can lead to the need for an eluent composition that is compatible with each of the detectors [29].

HPLC is very useful for obtaining information on crude extract composition (metabolite profiling). For this kind of work, two main types of detectors can be defined: simple detectors used to record chromatographic traces for profiling or quantification purposes (e.g., UV, ELSD, and ECD) and detectors for hyphenated systems that generate multidimensional data (chromatographic and spectroscopic) for online identification and dereplication purposes (e.g., PDA, MS, and NMR) [28]. This chapter will focus primarily on the last category. The hyphenated methods are LC-photodiode array detection (PDA) [10], LC-mass spectrometry (LC-MS, LC-MS/MS, or LC-MS^n) [11], and LC-nuclear magnetic resonance (NMR). NMR can be either directly hyphenated to an LC (LC-NMR) [12] at-line with preconcentration methods (e.g., LC-SPE-NMR [13]) or after microfractionation with a microflow LC-NMR (CapNMR) [14].

In an ideal situation, all of these hyphenated techniques should provide similar spectroscopic information to those obtained by the structural analysis of the isolated pure compounds. However, this is not always true for on-flow methods, and restricted structural information is often only available because of instrumental constraints and compromises related to the nature of the HPLC mobile phase. For example, solvent or modifiers with high UV cut-offs in LC-UV-PDA hinder the recording of relevant absorption bands of the analytes or eluent modifiers that can cause ion suppression in LC-MS; in addition, the use of a nondeuterated solvent in LC-NMR may mask the structure informative in a given region of the spectra. These constraints are even worse when multiple detectors are coupled to LC-NMR-MS systems [30] or in more extended hypernated systems; however, the advantage of obtaining all structural information within a single analysis is possible [29].

To avoid compromises that are related to direct hyphenation, new strategies have utilized a rapid at-line coupling to NMR or occasionally to MS; this involves more automation for microfractionation or a SPE preconcentration step prior to the NMR measurements or for peak parking (collection of LC peak followed by infusion in MS) of in-depth MS/MS experiments. Particularly in NMR, these at-line methods provide spectra of a higher quality compared to those required online.

2.4.1.1 UHPLC Analysis on Crude Extracts. One of the greatest advances in the analysis of NPs was the introduction of ultra-high-pressure liquid chromatography (UHPLC). This technique greatly improved the resolution in metabolite profiling studies and also enhanced the throughput when rapid fingerprinting of crude extracts needed to be performed [31].

The introduction of sub-2 μm particle supports have induced a concomitant increase in efficiency and optimal velocity due to improved mass transfer [26]. Therefore, high-throughput separations can be obtained with short column lengths for metabolite fingerprinting, whereas highly efficient separations are achieved with relatively long columns for detailed metabolite profiling [32]. However, the use of such small particles requires dedicated systems that are able to work at pressures up to 1000 bars. Because this approach provides very narrow LC peaks, MS detectors with rapid response times are generally mandatory; in addition, UHPLC-TOF-MS has been recognized as very efficient for profiling studies in metabolomics in recent years [31,33].

Conventional HPLC profiling methods can easily be transferred to UHPLC, which is a major advantage that has sped up the introduction of the technique in many analytical laboratories. UHPLC essentially consists of changing the column dimensions (i.e., length, particle size (dp), and internal diameter (ID)); the equations for geometrical changes usually used for scale-up between the analytical and preparative steps can also be employed to determine the mobile phase flow rate, injection volume, and gradient profile to be used in UHPLC [34–36]. This task can be automated using various calculators available free of charge on supplier or academic websites. By applying such rules, it is possible to maintain identical selectivity between HPLC and UHPLC, provided that the same phase chemistry is used. By selecting the adequate length of the UHPLC column, it is possible to increase the throughput by a factor 9 compared to a conventional HPLC. As an example, if the original separation has been performed on a 150 mm/5 μm column, a 50 mm/1.7 μm stationary phase should be selected in UHPLC to attain equivalent performance; this would be accompanied by a ninefold reduction of analysis time because of the threefold shorter column and three times higher linear velocity [37]. Such ultra-fast separations have been experimentally demonstrated, and analysis times in the range 1–5 min can be expected [38–40]. Conversely, if an identical column length is used for both HPLC and UHPLC, it is hypothetically possible to increase the plate count by a factor of 3 between columns packed with 5 and 1.7 μm particles, and the plate count can reach up to 40,000 with a 150 mm long column packed with 1.7 μm particles. However, it can become difficult to work in optimal flow rate conditions because of the important backpressure generated by long columns packed with sub-2 μm particles. Some separations involving 150 mm or even longer UHPLC columns have been reported in the literature and show very high efficiency [41,42].

The transfer of HPLC to UHPLC methods is illustrated for the profiling of the standardized extract of *Ginkgo biloba*, one of the most important phytopreparations marketed worldwide (Figure 2.2). As expected from theory on such a complex biological matrix, a ninefold reduction was observed in the analysis time by transferring the 60 min HPLC gradient (on a 150 × 4.6 mm ID, 5 μm column) to a short UHPLC gradient (on a 50 × 2.1 mm ID, 1.7 μm UHPLC column), whereas the use of the UHPLC column (150 × 2.1 mm ID, 1.7 μm) with the same gradient time provided a notable increase in resolution. Applications of UHPLC to the profiling of NPs have been recently reviewed [43].

2.4.2 LC-PDA

Photodiode array detection provides UV spectra directly online and is particularly useful for the detection of compounds with characteristic chromophores. For example, polyphenols can be efficiently localized by this method. With these types of compounds, UV spectral libraries can be built and used for dereplication, but these compounds must be analyzed in the same LC conditions because the composition of the mobile phase might slightly affect the UV bands; however, this feature is limited compared to the LC-MS method. The same method can also be used to assess peak purity by recording several spectra across given LC peaks and checking for differences between the spectra. More information on the localization of the hydroxyl groups in polyphenols can be obtained after the postcolumn addition of the UV-shift

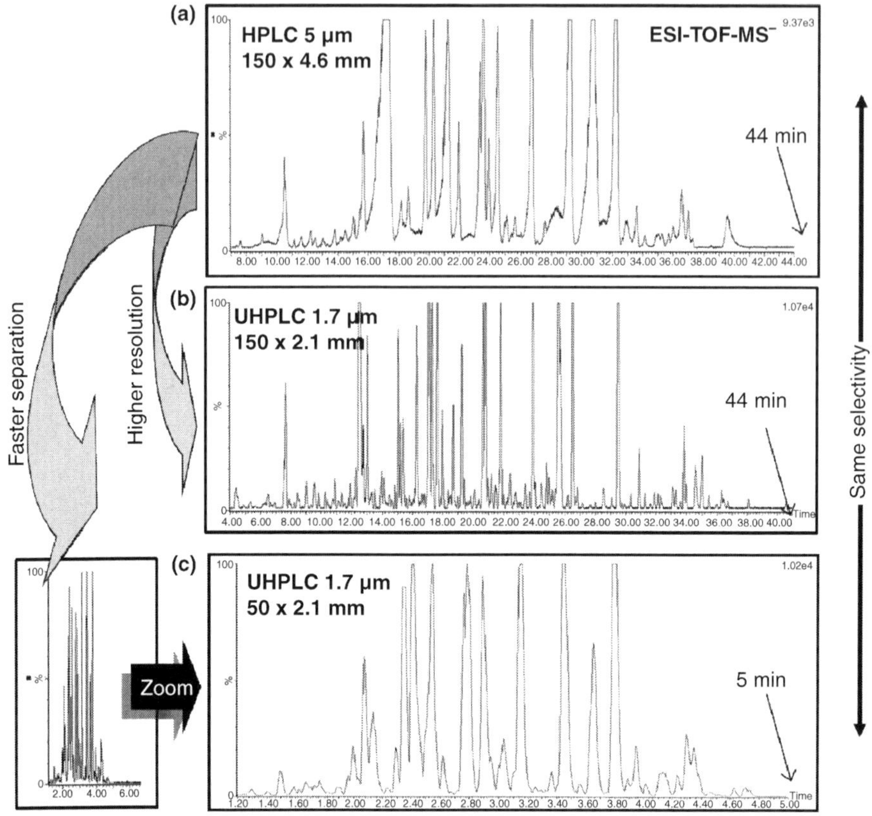

FIGURE 2.2 Comparison of chromatograms of a standardized *Ginkgo biloba* extract with a transfer of methods. (a) Classical HPLC analysis performed on a 150 × 4.6 mm ID, 5 μm column, with a gradient of 5–40% ACN in 60 min at 1 mL/min. (b) HPLC method transferred on a 150 × 2.1 mm ID, 1.7 μm UHPLC column, with the same run time, flow reduced to 0.35 mL/min. In this case, the resolution was notably increased. (c) Geometric transfer of method calculated by HPLC modeling software on a 50 × 2.1 mm ID, 1.7 μm UHPLC column with the same phase chemistry: the gradient time was reduced to 6.76 min (i.e., ninefold reduction), and the flow rate to 0.6 mL/min. Note that the performance should have been the same, but was slightly lower because of the relatively larger dead volumes caused by the ESI probe. Detection: ESI-TOF-MS, *m/z* range 100–1000 in negative mode.

reagents, which are commonly used for the structural characterization of these products in their pure form [44]. Such techniques have been successfully applied (e.g., for the characterization of apple polyphenols and *Trifolium* flavonoids) [45,46].

2.4.3 HPLC-MS

HPLC-MS is an analytical chemistry technique that combines the physical separation capabilities of high-pressure liquid chromatography with the mass analysis capabilities of mass spectrometry. The coupling of HPLC with MS (LC-MS) is extremely powerful in terms of the detection, quantification, and identification of a wide range of NPs. Hyphenation enables bidimensional detection, in which each metabolite is resolved in both chromatographic (retention time) and mass spectrometric (*m/z*) dimensions. In classical LC-MS profiling studies, rather long analysis times are required for the chromatographic separation (typically 120–60 min); however, this gradient time can be considerably reduced by using corresponding methods in UHPLC (see above) [47].

The application of LC-MS in the analysis and characterization of NPs represents a major breakthrough [48]. While expensive, the use of a mass spectrometer as an HPLC detector can provide excellent sensitivity and selectivity for the analysis of NPs in complex biological matrices. Furthermore, MS detection provides important structural information online, such as molecular weight, molecular formula, and diagnostic fragments, which are crucial for the dereplication and rapid online characterizations of NPs. However, the type of structural information provided online is strongly dependent on the interface and ionization method applied, as well as the type of analyzer used.

Historically, the primary difficulty in the coupling between LC and MS was that the high liquid flow from HPLC was not compatible with the high vacuum required for the standard operation of a mass spectrometer. Since the early 1980s, a number of different interfaces have been developed to address this issue and overcome this challenge [11,49]. The overwhelming popularity of LC-MS today is largely because of atmospheric pressure ionization (API) interfaces, including electrospray ionization (ESI) and atmospheric pressure chemical ionization (APCI); most of the profiling studies performed concern these two main ionization methods. For both ESI and APCI, ionization occurs at atmospheric pressure. A combination of high voltage and heat is used to provide the ionization that is needed to produce ions that are measured by the MS system. In ESI, the high voltage field (3–5 kV) produces nebulization of the column effluent, resulting in charged droplets that are directed toward the mass analyzer. These droplets become smaller as they approach the entrance to the mass analyzer. As the droplets get smaller, individual ions emerge in a process referred to as "ion evaporation." These ions are then separated by the MS system. In APCI, heat is used to vaporize the column eluent, and then a corona discharge is used to ionize solvent molecules, which then produce the analyte ions via chemical ionization mechanisms. More recently, a third type of ionization source, termed atmospheric pressure photoionization (APPI), has become available. In APPI, heat is used to vaporize the column eluent (similar to APCI), but the ionization is produced by an ultraviolet (UV) lamp that produces 10 eV photons. Depending on the solvent system used, the 10 eV photons will either ionize the mobile phase solvent or a dopant (e.g., toluene) added to the column effluent; these ions then produce the analyte ions through various ionization mechanisms, including charge or proton transfer [50]. Using these three ionization methods, a large variety of NPs ranging from low molecular weight, nonpolar constituents to high molecular weight polar molecules can be successfully analyzed.

It is important to understand that analyte ionization is largely compound-dependent and is primarily governed by proton affinity. As a rule of thumb for a good approximation, acidic molecules (e.g., carboxylic acids or phenolics) will primarily produce $[M-H]^-$ in negative ionization (NI) mode, whereas bases (e.g., alkaloids and amines) will generate $[M+H]^+$ in positive ionization (PI) mode. Compounds such as glycosides will have a high affinity for salts and will tend to form sodium adducts in PI. ESI (PI) is primarily used for alkaloids, APCI (PI) for pigments and carotenoids, ESI (PI/NI) for triterpene glycosides, and APCI or ESI (PI/NI) for phenolics and flavonoids. APCI is generally more suitable for nonpolar constituents, whereas ESI provides a softer ionization and a good response for polar metabolites. Additionally, APCI is generally recognized to be less prone to matrix effects than ESI [51]. Only a few studies have referred to the use of APPI for NP detection [52].

The performances of APPI, APCI, and ESI were evaluated on 106 standards and 241 proprietary drug candidates, occupying a wide chemical space [53]. The results indicated that the detection rates in PI were 94% for APPI, 84% for APCI, and 84% for ESI. Combining PI and NI mode detection, APPI detected 98% of the compounds, whereas APCI and ESI detected 91%. In APPI, for analytes with low polarity and low proton-affinity, M^+ was the predominant ion species generated, whereas for high polarity and high proton-affinity compounds, the $[M+H]^+$ ion species dominated the spectra. Thus, APPI is regarded by some authors [54] as a more universal detection method for LC-MS than APCI and ESI, but it is still not widely used in NP analysis.

For LC-MS applications, many types of mass spectrometers can be used. Those that are of low resolution (LR), such as the single-quadrupole (Q) mass spectrometers, are the most commonly used and least expensive. Those providing high-resolution (HR) and exact mass capabilities, such as time-of-flight (TOF) instruments, are becoming increasingly popular. TOF-MS provides resolution up to 20,000 and excellent mass accuracy (<5 ppm). They can be operated with a high acquisition frequency over a broad mass range, and they are compatible with rapid chromatographic methods and are free from compromising sensitivity. The high mass accuracy of HR-MS measurements, usually below 5 ppm, and the ratio of intensities of the isotopic patterns, together with the use of rules for heuristic filtering of molecular formula [55] and chemotaxonomic information, allow for an efficient mass-based putative metabolite identification and dereplication. An efficient filtering of putative molecular formulae from a given exact mass is strategic, and it considerably restricts the number of possibilities when searching a NP database, especially when cross-search involving chemotaxonomical information can be performed.

To obtain additional structure information, molecular ions generated by soft ionization methods can be fragmented in tandem mass spectrometers using MS/MS or MS^n experiments. In this respect, triple-quadrupole (QQQ) MS/MS systems are the most commonly used. They are also particularly useful for bioanalytical assays and for specific quantification purposes in complex matrices through multiple reaction monitoring (MRM), providing very specific detection [54]. Ion-trap (IT) mass spectrometers have the unique capability of producing multiple-stage MS/MS (MS^n) data that may be essential for structural elucidation. In addition to these types of mass spectrometers, there are a growing number of additional varieties, including hybrid systems that combine LR and HR analyzers for specific applications [11,50]. In these MS/MS systems, the fragmentation can be performed by selecting specific molecular ions or by performing data-dependant experiments in which the mass spectrometer automatically selects the most abundant ion at a given retention time for fragmentation.

For the online structural determination of NPs, the MS/MS spectra are primarily useful for the partial determination of sugar sequences of various glycosides; for the classical fragmentation of flavonoids or related compounds for the determination of substituent positions on the A or B rings; or for the differentiation of isomers [56–58]. For other studies, the interpretation of the MS/MS spectra often requires the analysis of many related products to extract structurally relevant information. For the analysis of completely unknown constituents, this approach alone cannot record enough information for structural determination. The impact of high-resolution MS/MS considerably improves the interpretation of the MS/MS spectra based on molecular formula assignment of the fragment obtained [50]. This feature will play an increasingly important role in the rapid identification of NPs with the concomitant development of MS/MS databases that are not instrument dependent.

2.4.4 LC-NMR

Due to the lack of efficient commercial databases, especially for LC-MS/MS spectral comparison, the dereplication process often requires additional spectroscopic information to ascertain the identity of known NPs or to partially identify unknown metabolites. In this respect, LC-NMR provides important complementary structural information or, in some cases, a complete structural assignment of NPs, online [59]. The interest of directly coupling HPLC with NMR resides not only in the fact that full structure elucidation and stereochemical information can be obtained (with 2D spectra in the stop-flow mode) but also in the fact that it is a highly nonselective detection technique, which can provide the comprehensive detection of any hydrogen-containing compound present in the HPLC eluate [59].

The coupling of HPLC with NMR began in the early 1980s [60]. This process provided the necessary improvements in probe technology, the construction of high-field magnets and the progresses in pulse field gradients and solvent suppression. LC-NMR emerged in the mid-1990s as a very efficient method for the online identification of NPs in complex matrices [61]. It took nearly 20 years to establish LC-NMR, and this long period can be primarily attributed to the intrinsic low sensitivity of NMR. Currently, the online coupling of HPLC with NMR is still insensitive compared to the LC-MS; however, many strategies that involve various operational modes of hyphenated systems and at-line coupling of the NMR to the HPLC with online sample methods, such as SPE-NMR or tubeless technology (e.g., microflow NMR (CapNMR)) have again boosted the utility of this powerful identification method. LC-NMR has now evolved into comprehensive and very powerful LC-NMR-MS platforms [62,63] used for many applications in drug discovery, including dereplication [64], rapid *de novo* identification of new NPs [65], and biomarker identification in metabolomics [66].

LC-NMR represents an interesting complementary technique to LC-PDA-MS for the online identification of LC peaks. Recent applications have demonstrated its utility in various fields of analytical chemistry [67].

2.4.5 SPE-NMR, Microflow NMR, and NMR with Cryogenized Probes

As shown in Figure 2.3, one possibility for overcoming the limitation of on-flow LC-NMR (Figure 2.3a) is to use preconcentration methods prior to NMR detection. This involves performing NMR detection at-line after the HPLC separation by either trapping the LC-peaks on SPE (LC-SPE-NMR) [13,68,69] or by HPLC microfractionation, drying and reinjection of the concentrated LC-peak in deuterated solvent in a microflow capillary LC-NMR probe (CapNMR) [14,63,70].

LC-SPE-NMR has become very popular for increasing LC-NMR detection. The advantage of this technique is that it is able to hyphenate NMR with online SPE, which is a powerful technique for a reproducible, rapid, and selective preparation. In LC-SPE-NMR, the separated peaks are diluted postcolumn with water and are automatically trapped on SPE [30]. After a drying step with nitrogen to remove all

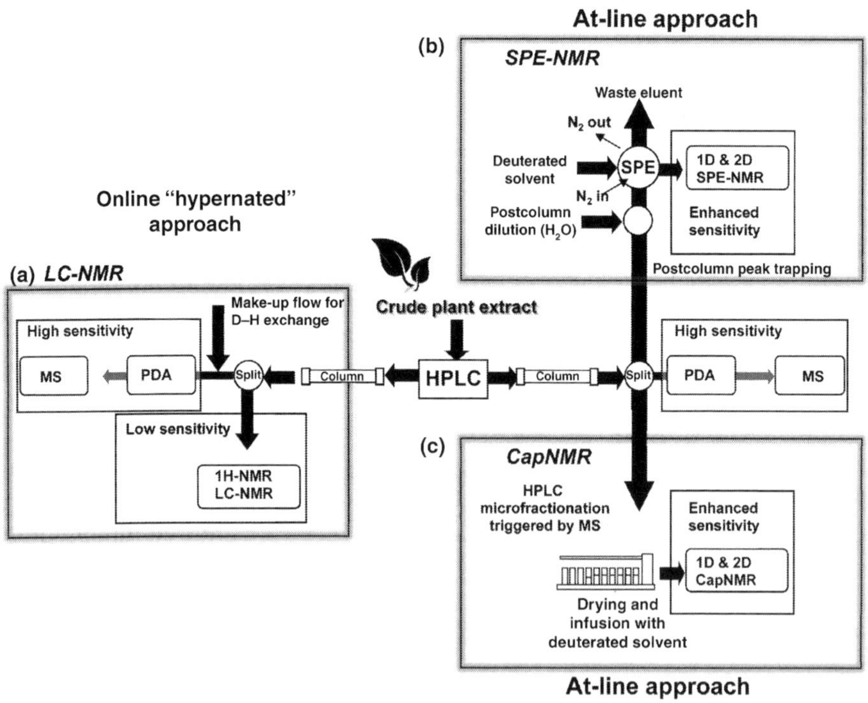

FIGURE 2.3 Summary of the different ways to hyphenate NMR with HPLC either online or at-line. (a) On-flow LC-NMR in a hyphenated system with PDA and MS detection. (b) LC-SPE-NMR with postcolumn addition of H_2O for trapping drying and reinjection in a LC-NMR flow cell. (c) CapNMR detection after HPLC microfractionation. Reprinted (adapted) with permission from [139]. Copyright (2010) Bentham Science Publishers Ltd.

solvents used for chromatographic separation, the analytes can be transferred with the deuterated solvent of choice to the NMR flow-cell probe (Figure 2.3b) [71]. This technique presents the advantage that the separation can be performed with normal protonated solvent and, moreover, multiple trapping of the same analyte from repeated LC injection circumvents problems of low concentration. The operation can be fully automated on state-of-the-art systems, but a good optimization of the trapping conditions is required for optimum performances [69,71]. The use of either semi-preparative or analytical columns with and without multiple trapping parameters has been investigated [72]. It was found that the choice of the optimum separation strategy is primarily related to the polarity of the analyte of interest.

An example of the effect of multiple trapping on sensitivity is illustrated using the analysis of isoflavonoids from *Smirnowia iranica* in Figure 2.4. The ^{1}H-NMR spectrum of glyasperin H, a representative compound that is present in small amounts, is presented in peak 10. Up to six repeated trappings of peak 10 resulted in a linear increase of the signal-to-noise ratio in the NMR spectra (Figure 2.4b), which enabled the recording of informative 2D NMR spectra, such as the ^{1}H–^{1}H COSY spectra (Figure 2.4c) [73].

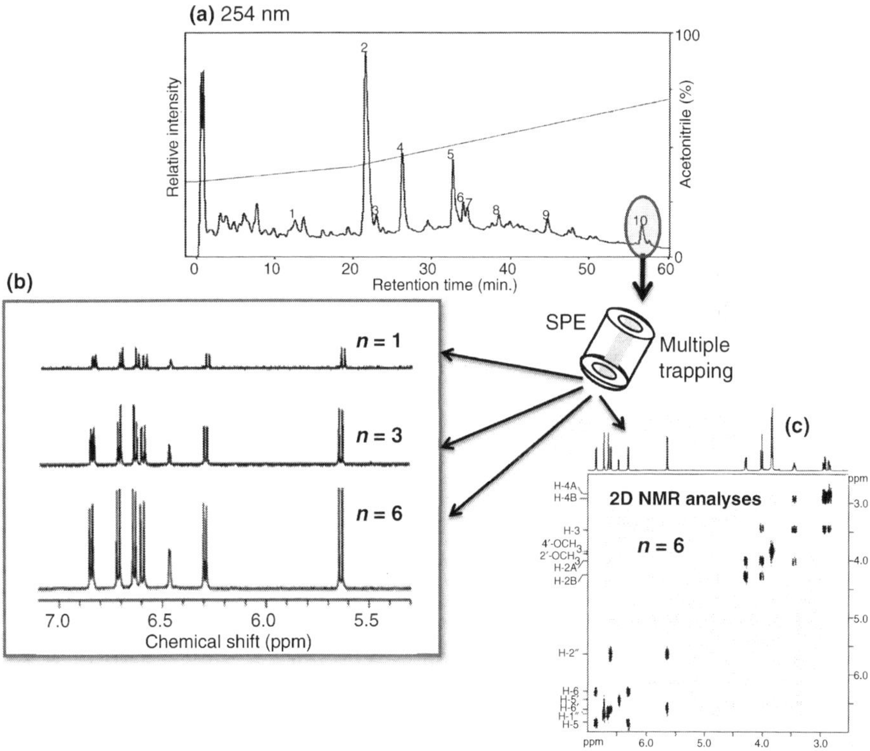

FIGURE 2.4 (a) Chromatogram of the ethanolic extract of *Smirnowia iranica* roots on a C_{18} column, acetonitrile gradient profile in waters at 254 nm. (b) Multiple trapping of peak 10 on LC-SPE-NMR. (c) COSY spectrum with peak 10 after 6 trappings [73]. Reprinted (adapted) with permission from [73]. Copyright (2005) American Chemical Society.

Compared to stop-flow LC-NMR, the increased sensitivity provided by analyte focusing multiple trapping and the advantage of a fully deuterated solvent operation, provide the ability to perform the full set of 2D correlation experiments on the analytes, including HMBC, in a reasonable analysis time. A typical LC-SPE-NMR would require a full set of 2D spectra and approximately 120 μg in a 60 μL cell on a 600 Hz equipped with a room-temperature probe head. For a secondary metabolite (MW < 700 Da), a series of high-quality homonuclear and heteronuclear 2D spectra would need to be recorded overnight. A single ^{1}H-NMR spectra can be obtained with approximately 10 nmol of substances within a reasonable timeframe [74].

The sensitivity of the flow cell can be enhanced by using a cryogenized probe by a factor of roughly 3–4 [68]. For LC-NMR, cryofit inserts for the adaptation of 30 μL probes in a 3 mm ID inverse cryoprobe have been successfully used for the analysis of very limited amounts of plant material [75].

One of the last improvements in flow cell design has been the introduction of reduced-diameter solenoid NMR coils (microflow NMR, capillary NMR CapNMR), which represents a particularly attractive approach to enhance NMR sensitivity for mass-limited samples [14]. Similar to LC-SPE-NMR, microflow NMR represents another alternative for the indirect hyphenation of NMR with HPLC. The same microfractionation procedure described for LC-SPE-NMR can be used for analyzing extracts; however, in this case, the LC peaks are not trapped on SPE cartridges but simply collected into vials or in 96-well plates and usually dried by speed vacuum evaporation. Each sample can be dissolved in a volume slightly exceeding the volume of the flow cell (6.5 μL for a 5 μL flow cell volume), filtered, injected into the probe, and parked in the center of the micro-coil by pushing it with an adjusted push volume (typically 8–10 μL) [76,77]. The disadvantage of at-line work is that more sample handling is needed compared to the on-flow approach. However, NMR spectra of higher quality are obtainable in standard deuterated solvents at low cost, whereas other information, such as UV and MS data, can be obtained online during the microfractionation. Automation for direct infusion of 10 μL samples from a low-volume 384-well plate has recently been demonstrated [78] and has been optimized for injection volumes on the microliter scale for a LC-MS-CapNMR platform using the microdroplet NMR approach [63]. This type of approach is also important for checking the chemical integrity of NPs in a high-throughput pharmaceutical screening program. These at-line measurements also have the considerable advantage of providing spectra recorded in standard deuterated solvents with the highest possible concentration for a given amount of compound. Furthermore, access to ^{13}C-NMR information deduced indirectly from inverse detection (HSQC and HMBC) experiments is feasible, providing detailed structural information for *de novo* structural determination of NPs.

Swertiamarin, a secoiridoid with a molecular weight of 374, provides an example of the type of spectra that can be recorded by microflow NMR; the compound after microfractionation is illustrated in Figure 2.5c and compared to that obtained by repeated injections at different concentrations on an on-flow LC-NMR (Figure 2.5a on-flow plot). The detection limits were approximately 20 μg on-column in the on-flow mode with a standard LC-NMR flow probe (60 μL on a 500 MHz) (Figure 2.5a), whereas a spectrum of good quality was obtained with approximately

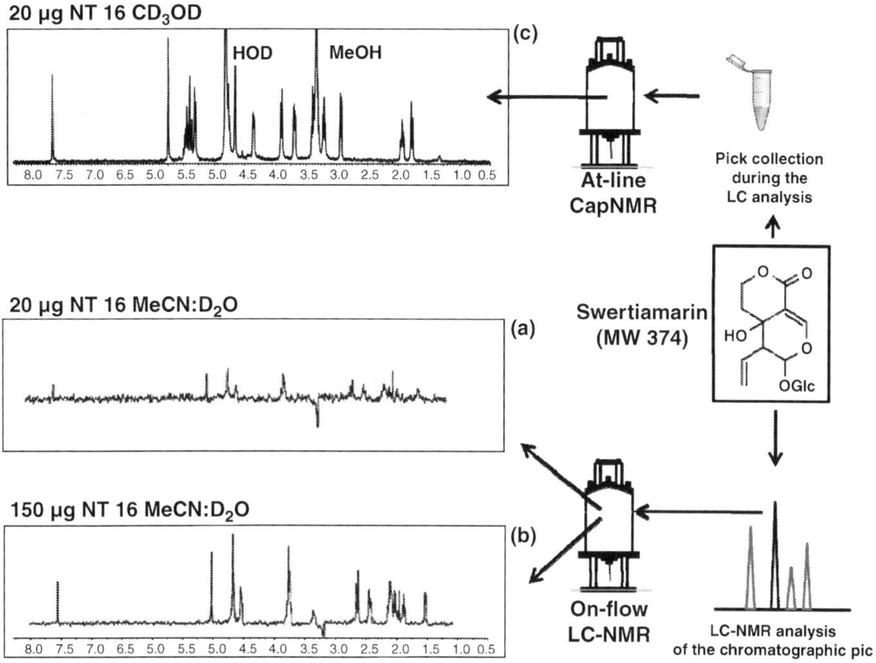

FIGURE 2.5 Estimation of sensitivity and the quality of NMR spectra recorded either by on-flow LC-NMR (500 MHz, flow cell 60 μL active volume) or at-line CapNMR (500 MHz, flow cell 1.5 μL active volume) by the injection of different amount of swertiamarin (2), always with the same number of transients (NT = 16) [79]. Reprinted (adapted) with permission from [79]. Copyright 2006, Informa UK Ltd.

150 μg (Figure 2.5b). If the same amount of compound (20 μg) was injected in a volume corresponding to a microflow NMR CapNMR probe (5 μL on a 500 MHz), a much better S/N ratio would be obtained because of the probe design and the effect of the increase in concentration (Figure 2.5c).

With an automated microflow NMR platform, the interpretable ^{1}H-NMR spectra were obtained from analytes at the 200 ng level in 1 h/well [80]. With a more classical microflow NMR approach, a few tenths of μg (depending on the magnet strength and MW) would typically be needed to obtain a full set of 2D correlations overnight. On a 600 MHz system, COSY NMR spectra have been reported to be acquired with less than 10 μg in approximately 1 h, 30 μg for HSQC in 5 h, and 70 μg for HMBC in approximately 15 h. Direct ^{13}C-NMR data can already be obtained with 200 μg of NPs [80].

2.5 STUDIES ON NATURAL PRODUCTS USING LC-NMR, MICROFLOW NMR, AND SPE-NMR

In crude extract profiling studies, LC-NMR or the other at-line methods have been predominantly used for dereplication studies or for the targeted *de novo* identification of given bioactive metabolites or biomarkers issued from metabolomic studies. Various aspect of the use of hyphenated methods for rapid structural determination have been illustrated. The number of applications for rapid online or at-line identification of NPs is constantly growing. In Table 2.1, a summary of the most recent applications (2006–2010) is presented, including the type of experiments and the probe used. For earlier applications, the reader can refer to other reviews [13,59].

2.5.1 Application of Online and At-Line LC-NMR Methods for Dereplication Studies

As shown in Table 2.1, an important number of applications in the field of NPs have been performed by the LC-SPE-NMR method, and applications of microflow NMR are also rapidly growing. The interest in these at-line methods primarily resides in the quality of the spectra that are generated compared to classical LC-NMR, as discussed previously. In most applications, in addition to recording high-quality 1D-NMR data with the at-line methods, 2D-NMR experiments, such as ^{1}H–^{1}H (COSY, TOCSY) and ^{1}H–^{13}C (HSQC and HMBC), can be performed using a few micrograms of the analyte for in-depth structural investigations.

The first LC-SPE-NMR analysis on NPs was performed on an extract of Greek oregano and led to the successful identification of various flavonoids, rosmarinic acid, and monoterpene carvacrol based on a combination of UV, MS, and NMR spectra [68]. In 2007, the technique was applied in the separation and structural verification of the major known constituents present in the Greek *Hypericum perforatum* extract. The chromatographic separation was performed in a C18 column using a mixture of MeCN–H_2O as a mobile phase. For the online NMR detection, the analytes eluted from the column were trapped one by one in separate SPE cartridges and thereafter transported into the NMR flow-cell. LC-PDA-SPE-NMR and LC-PDA-MS allowed for the characterization of constituents of *H. perforatum*, primarily naphtodianthrones, phloroglucinols, and phenolic acids. Two phloroglucinols (hyperfirin and adhyperfirin) that had been previously reported to be precursors in the biosynthesis of hyperforin and adhyperforin were detected for the first time [116].

This technique has been employed in the identification of alkaloids from *Lobelia siphilitica* and *Hippobroma longiflora* (Campanulaceae) and sesquiterpene lactones from *Distephanus angulifolius* (Asteraceae) [83,84]. Another interesting study has demonstrated its use for the identification of iridoid and caffeoyl glycosides from *Pinguicula lusitanica* (Lentibulariaceae) [89].

As an example, the online identification of antioxidant compounds using DPPH and ABTS radicals from the methanolic extracts of *Tilia europaea*, *Urtica dioica*, *Lonicera periclymenum*, and *H. perforatum* was performed using a combination of LC-MS and LC-UV-SPE-NMR as well as online bioassays [93]. To increase the sensitivity of LC-UV-SPE-NMR, multiple trapping was used, which allowed larger quantities of the compound to accumulate on the corresponding SPE cartridges from three successive LC runs, similarly to what was demonstrated in Figure 2.5 for the analysis of glyasperin H. The trapped compounds were extracted from the corresponding SPE cartridges with 560 µL of acetonitrile-d_3 into the NMR flow-cell.

TABLE 2.1 Recent Applications (2006–2010) of Direct NMR, LC-NMR, LC-SPE-NMR, and CapNMR for Online or At-Line Metabolite Fingerprinting or Profiling

Source	Application	Methodology	Experiment	Instrumentation	References
Metabolite Identification/Dereplication					
Haplophyllum acutifolium	Alkaloids profiling	LC-PDA-MS-SPE-NMR	^{1}H-NMR, COSY, HSQC, HMBC, NOESY	ESI (+); 600 MHz, 30 μL	[81]
Machilus philippinensis	Flavonoids glycosides profiling	LC-MS, HPLC-SPE-NMR	^{1}H-NMR	ESI (+), 400 MHz, 30 μL	[82]
Distephanus angulifolius	Sesquiterpene lactones characterization	LC-MS-SPE-NMR	^{1}H-NMR	ESI-IT-MS (+/–), 600 MHz, 5 mm	[83]
Lobelia siphilitica, Hippobroma longiflora	Alkaloid identification	LC-MS-SPE-NMR	^{1}H-NMR	600 MHz, 5 mm	[84]
Harpagophytum procumbens	Metabolite profiling	LC-MS, HPLC-SPE-NMR	^{1}H-NMR, ^{13}C, COSY, TOCSY, HSQC, HMBC	ESI, APCI, 600 MHz, 30 μL	[85]
Neolitsea sericea var. *aurata*	Flavonoids glycosides profiling	LC-SPE-NMR	^{1}H-NMR	400 MHz, 30 μL	[86]
Trinia glauca and other Apiaceae spp.	Polyacetylenes profiling	LC-SPE-NMR	^{1}H-NMR, ^{13}C-NMR, ^{19}F-NMR	500 MHz, insert 60 μL	[87]
Penicillium roqueforti	Sesquiterpenes identification	LC-MS-SPE-NMR	^{1}H-NMR, ^{13}C-NMR, COSY, TOCSY, NOESY, gHSQC, gHMBC	ESI-IT-MS, 600 MHz, 5 mm	[88]
Pinguicula lusitanica	Iridoid and phenylpropanoid glycosides profiling	LC-MS, LC-SPE-NMR	^{1}H-NMR	ESI-TOF-MS (–), 600 MHz	[89]
N. sericea var. *aurata*	Isoquinoline alkaloids profiling	LC-SPE-NMR	^{1}H-NMR, COSY, NOESY	600 MHz, 30 μL	[90]
Blepharis aspera	Phenylpropanoids profiling	LC-MS, LC-SPE-NMR	^{1}H-NMR	ESI-IT-MS (+), 600 MHz, 30 μL	[91]

(*continued*)

TABLE 2.1 (*Continued*)

Source	Application	Methodology	Experiment	Instrumentation	References
Corydalis solida	Alkaloids profiling	LC-MS^n, LC-SPE-NMR	^{1}H-NMR, COSY, HSQC, HMBC	ESI-IT-MS (+), 600 MHz, insert 30 μL	[75]
Thespesia danis	Gossypol profiling	LC-MS-SPE-NMR	^{1}H-NMR	ESI-IT-MS (+/–), 600 MHz, 30 μL	[92]
Tilia europea, Urtica dioica, Lonicera periclymenum, Hypericum perforatum	Antioxidant metabolite profiling	LC-MS, LC-SPE-NMR	^{1}H-NMR	ESI (–), APCI (+/–)-IT-MS, 400 MHz, 120 μL	[93]
Phyllanthus reticulatus	Flavonoid glycosides profiling	LC-SPE-NMR	^{1}H-NMR	400 MHz, 120 μL	[94]
Schizanthus grahamii	Tropane alkaloids profiling	LC-NMR, stop-flow and LC-SPE-NMR	^{1}H-NMR; ^{1}H-NMR, COSY	600 MHz, insert 60 μL; 600 MHz, 30 μL	[95]
Swietenia macrophylla	Limonoids profiling	LC-NMR, stop-flow	^{1}H-NMR, TOCSY, HSQC, gHMBC	600 MHz, 120 μL	[96]
Erythrophleum fordii	Amides diterpenoids	LC-NMR, stop-flow	^{1}H-NMR, COSY	500 MHz, 60 μL	[97]
Chamomilla recutita	Metabolite profiling	LC-MS/MS, LC-NMR, stop-flow	^{1}H-NMR	ESI-IT-MS (+/–), 400 MHz, 60 μL	[98]
Bixa orellana	Carotenoids profiling	LC-MS, CapHPLC-CapNMR, stop-flow	^{1}H-NMR, COSY	APCI (+), 600 MHz, 2 μL	[99]
Sepedonium chrysospermum	Metabolite profiling	LC-bioactivity profiling, MS, CapNMR	^{1}H-NMR, ^{13}C-NMR, COSY, HSQC, HMBC	ESI-TOF-MS (+), 500 MHz, 1.5 μL	[100]
Psorothamnus arborescens	Isoflavones profiling	LC-MS, CapNMR	^{1}H-NMR, ^{13}C-NMR, COSY, HSQC, HMBC	ESI-TOF-MS (+/–), 600 MHz, 5 μL	[101]

Thapsia garganica	Sesquiterpene lactone and phenylpropanoids profiling	LC-MS, LC-SPE-CapNMR	^{1}H-NMR, COSY, TOCSY, HSQC, HMBC	ESI-IT-MS (+/–), 500 MHz, 1.5 μL	[102]
Muehlenbeckia sp., *Aspergillus* sp.	Metabolite profiling	LC-MS, CapNMR	^{1}H-NMR, COSY, HSQC, HMBC	ESI-TOF-MS (+), 500 MHz, 1.5 μL	[103]
Fischerella ambigua	Metabolite profiling	LC-MS-MS/MS, CapNMR, stop-flow	^{1}H-NMR	ESI-IT-MS (+), 500 MHz, 1.5 μL	[63]
Dactylospongia sp.	Metabolite profiling	LC-NMR on flow, LC-NMR stop flow	^{1}H-NMR, ^{13}C, gCOSY, gHSQC, HMBC	500 MHz, 60 μL	[104]
Belamcanda chinensis	Isoflavonoids profiling	LC-MS, LC-NMR on flow	^{1}H-NMR	ESI-IT-MS (+), 500 MHz, 60 μL	[105]
Nandina domestica	Aporphine alkaloids profiling	LC-NMR, stop-flow	^{1}H-NMR	500 MHz, 60 μL	[106]
Lysidice brevicalyx	Metabolite profiling	LC-MS^{n}, LC-NMR stop flow	^{1}H-NMR, dqfCOSY	ESI-IT-MS (+/–), 500 MHz, 60 μL	[107]
Spider venom	Metabolite profiling	Direct-MS, direct-NMR	^{1}H-NMR, dqfCOSY, NOESY, HMQC, HMBC	ESI-QMS (+/–), 600 MHz	[108]
Sisymbrium officinale	Metabolite profiling	LC-PDA, direct-NMR	^{1}H-NMR, DOSY, TOCSY	600 MHz	[109]
Foliar endophytes of *Picea glauca*	Polyketides profiling	LC-UV-NMR, stop-flow	^{1}H-NMR	600 MHz, LC-NMR insert 60 μL	[110]
Teucrium luteum	Diterpenoids profiling	UHPLC-MS, LC-SPE-NMR	^{1}H-NMR, COSY, HSQC, HMBC	ESI-TOF-MS (+/–), 600 MHz, 30 μL	[111]
Angelica dahurica	Furanocumarins profile	LC-PDA-MS, LC-NMR on flow	^{1}H-NMR	ESI (+), 500 MHz, 60 μL	[112]
Bufo melanostictus	Bufadienolide sulfates profile	LC-PDA-MS^{n}, LC-SPE-NMR	^{1}H-NMR, COSY, HSQC, HMBC	ESI (+), 600 MHz	[113]
Ianthella flabelliformis	Araplysillin derivatives profiling	LC-PDA-MS, LC-SPE-NMR	^{1}H-NMR, COSY, HSQC, HMBC, NOESY	ESI-FT-ICR-MS (–), 600 MHz	[114]

(*continued*)

TABLE 2.1 (*Continued*)

Source	Application	Methodology	Experiment	Instrumentation	References
Malbranchea sulfurea	Malbranpyrroles profiling	LC-PDA-MS, ICD-MS, LC-SPE-NMR	^{1}H-NMR, COSY, HSQC, HMBC, NOESY, TOCSY, ROESY	ESI-MS (+), 500 MHz	[115]
Metabolomics: Quality Assessment					
H. perforatum	Metabolite profiling	LC-MS/MS, LC-SPE-NMR	^{1}H-NMR, COSY, TOCSY	ESI-IT-MS (–), 400 MHz, 120 μL	[116]
Tomato juice, palm oil, mandarin juice	Carotenoids profiling	LC-NMR stop flow	^{1}H-NMR, gCOSY	500 MHz, 60 μL	[117]
Cannabis sativa	Metabolite profiling	Direct-NMR	^{1}H-NMR	500 MHz	[118]
Panax ginseng	Metabolite fingerprinting	Direct-NMR	^{1}H-NMR, *J*-resolved, COSY, HSQC, HMBC	400 MHz	[119]
Solanum lycopersicum	Metabolite profiling	LC-MS, direct-NMR	^{1}H-NMR	ESI-QTOF-MS (–), 500 MHz	[120]
Echinacea purpurea, H. perforatum, Ginkgo biloba, Valeriana officinalis	Metabolite fingerprinting	Direct-NMR, direct-MS	^{1}H-NMR	500 MHz, ESI-qqqMS (+/–)	[121]
Artemisia annua	Metabolite profiling	Direct-NMR	^{1}H-NMR	500 MHz	[122]
P. ginseng	Metabolite profiling	Direct-NMR	^{1}H-NMR, DOSY, HMBC	500 MHz	[123]
Galphimia glauca	Galphimine profiling	Direct-NMR	^{1}H-NMR	500 MHz	[124]
Metabolomics: Stress/Interactions					
Arabidopsis thaliana	Oxylipins profiling	LC-MS, CapNMR	^{1}H-NMR, COSY	APCI-MS (–), 500 MHz, 1.5 μL	[125]
A. thaliana	Jasmonate derivatives profiling	UHPLC-MS, CapNMR	^{1}H-NMR, TOCSY	ESI-TOF-MS (–), 500 MHz, 1.5 μL	[66]
Botryosphaeria obtusa and *Eutypa lata*	Metabolite profiling	UHPLC-MS, CapNMR	^{1}H-NMR, gCOSY, TOCSY, NOESY, gHSQC, gHMBC	ESI-TOF-MS (+/–), 500 MHz, 1.5 μL	[65]

Brassica rapa	Metabolite fingerprinting	Direct-NMR	^{1}H-NMR	400 MHz	[126]
B. rapa	Metabolite fingerprinting	Direct-NMR	^{1}H-NMR, *J*-resolved, COSY, HMBC, HSQC	500 MHz	[127]
B. rapa	Metabolite profiling	Direct-NMR	^{1}H-NMR, *J*-resolved, COSY, HMBC, HSQC	400 MHz	[128]
Silybum marianum	Metabolite fingerprinting	Direct-NMR	^{1}H, COSY, HSQC, HMBC	400 MHz	[129]
Populus tremula	Lignin and polysaccharides profiling	Direct-NMR	2D ^{13}C–^{1}H HSQC	600 MHz, 5 mm	[130]
Kitasatospora	Metabolite profiling	Direct-MS, direct-NMR	^{1}H-NMR	ESI-FTICR-MS, 500 MHz, 5 mm	[131]
Biotransformation by *Corydalis* and *Macleaya* cell	Isoquinoline alkaloids profiling	LC-MS/MS, LC-NMR, stop-flow	^{1}H-NMR	APCI-QQQMS (+), 500 MHz, 60 μL	[132]
Metabolomics: Genotype/Taxonomy					
G. glauca	Metabolite fingerprinting	Direct-NMR	^{1}H-NMR	500 MHz	[124]
Dilatris pillansii, D. corymbosa, Dodonaea viscosa	Metabolite fingerprinting	LC-UV, direct-NMR	^{1}H-NMR	500 MHz, 5 mm	[133]
B. rapa	Metabolite profiling	Direct-NMR	^{1}H-NMR, *J*-resolved	500 MHz	[134]

COSY: correlation spectroscopy; DEPT: distortionless enhancement by polarization transfer; HMBC: heteronuclear multiple-bond correlation; HSQC: heteronuclear single quantum correlation; NOESY: nuclear Overhauser effect spectroscopy; DOSY: diffusion ordered spectroscopy; TOCSY: total correlation spectroscopy; ROESY: rotating frame Overhause effect spectroscopy; CapNMR: capillary NMR; SPE: solid-phase extraction; ESI: electrospray ionization; APCI: atmospheric pressure chemical ionization; Q: quadrupole; QQQ: triple quadrupole; QTOF: quadrupole time of flight; IT: ion trap; FTICR: Fourier transform ion cyclotron resonance; (+): positive mode ionization; (–) negative mode ionization; LC: high-performance liquid chromatography, UHPLC: ultra high performance liquid chromatography; UV: ultra violet; PDA: photodiode array, ICD-MS: intact cell desorption/ionization on silicon mass.

The screening of the four methanolic extracts allowed for the identification of 14 different antioxidants using exclusively chromatographic hyphenated techniques. Structure elucidation revealed that the antioxidant activity of the extracts was primarily attributed to flavonoid glycosides and caffeoylquinic acids [93].

Castro et al. demonstrated that it was possible to identify new *neo*-Clerodane diterpenes of plant origin that are difficult to monitor because of their nonspecific UV/vis absorption [111]. The authors used the LC-MS-SPE-NMR technique for the isolation and characterization of three new *neo*-clerodane diterpenes, 3-hydroxyteucroxylepin, and teuluteumin A and teuluteumin B, from *Teucrium luteum* [111].

LC-SPE-NMR was also applied to analyze the venom of *Bufo melanostictus*, which has been widely used for the treatment of heart failure, sores, pains, and various cancers in China [113]. The MeOH extract of the toad venom from *B. melanostictus* was fractionated by preparative HPLC, and the structures of five new bufadienolide sulfates, as well as one new bufogenin, were rapidly elucidated on the basis of LC-DAD-MS^n and LC-SPE-NMR data [113].

To improve sensitivity, LC-SPE-NMR spectra have also been acquired using cryogenic flow probes. In such probes, the electronic noise is reduced and results in a significant improvement of S/N ratio. For example, the analysis of *Penicillium roqueforti* using LC-SPE-NMR allowed the identification of three eremophilane sesquiterpenes, one of which was described for the first time [88]. The structural determinations were based on complete gHSQC, gHMBC, and NOESY experiments in combination with high-resolution ESI-MS.

Recently, Wu and coworkers described six photosensitive polyketides, malbranpyrroles A–F from the thermophilic fungus *Malbranchea sulfurea*, by using intact cell desorption/ionization on silicon mass (ICD-MS) and LC-SPE-NMR [115]. These two strategies facilitated the search and structural determination of unstable NPs. Moreover, the intact-cell MALDI mass spectrometry analysis indicated that only brown hyphae of *M. sulfurea* can produce malbranpyrroles. The biosynthetic pathway of malbranpyrroles was determined by ^{13}C isotope precursors and amino acid feeding experiments [115].

A significant number of recent applications (Table 2.1) are performed using microflow NMR (CapNMR). For example, the carotenoids of *Bixa orellana* have been studied using a capillary LC-NMR system equipped with a home-built double-resonant solenoidal NMR microprobe. For the unambiguous assignment of natural carotenoid stereoisomers, the authors used a closed-loop LC-NMR technique online. The technique showed important advantages compared to the procedure of the at-line microisolation because this strategy often resulted in carotenoid isomerization and degradation. Using this system, the authors were able to determine the structure of the main bixin stereoisomers [99].

Munro and coworkers used the microflow NMR instrumentation for the identification of active compounds in the mycoparasitic fungus *Sepedonium chrysospermum* [100]. This dereplication approach allowed the rapid recognition of a new peptaibol using only 700 μg of crude *S. chrysospermum* extract. The high-quality 1D and 2D homonuclear ^1H-NMR and ESI-MS/MS data obtained on the isolated sample of chrysaibol (33 μg) were sufficient for the definition of a planar structure, whereas

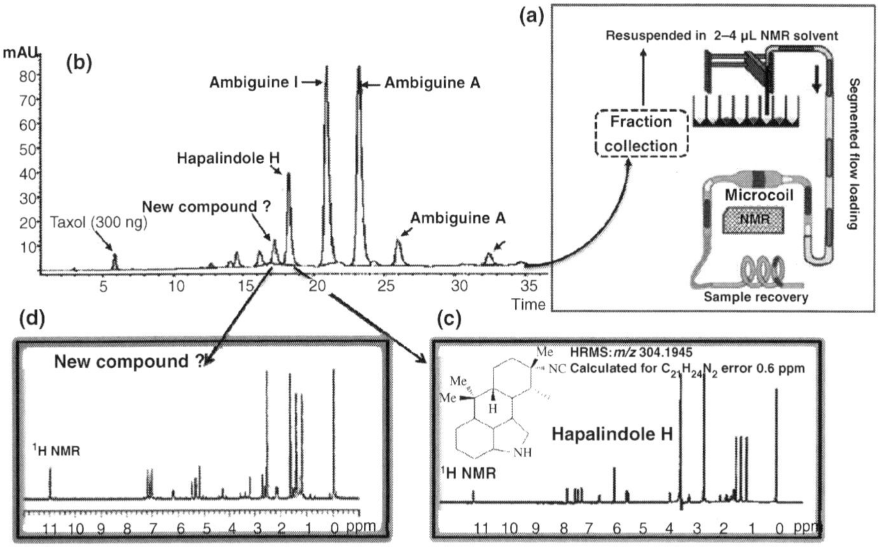

FIGURE 2.6 (a) Schematic diagram of the microdroplet autosampler, as applied to natural product discovery. Microfractions were resuspended in a small volume (2–5 μL) of deuterated solvent and loaded into a microcoil NMR probe, with an observed volume of 1–2 μL, using a microplate automation. Samples were recovered after NMR analysis for additional analyses, archival, or bioassay. (b) UV chromatogram of the separation of a bioactive cyanobacterial extract and analyzed with the LC-MS-NMR system, indicating known and unknown compounds, were found. (c) ^{1}H-NMR spectra of LC peak at 18.3 min identified from the literature as hapalindole H and (d) NMR spectra of the 17.3 min peak not found in the literature or natural product databases. Therefore, it was prioritized for detailed structure studies by scale-up purification [63]. Reprinted (adapted) with permission from [63]. Copyright (2008) American Chemical Society.

the absolute configurations of the amino acid residues were defined by Marfey's method. Chrysaibol was cytotoxic against the P388 murine leukemia cell line and showed notable activity against *Bacillus subtilis*.

The use of the capillary NMR technique on an automated platform has been extensively optimized for sample-limited investigations by Lin et al. that use a microscale LC-MS-NMR platform (Figure 2.6a) to identify active cyanobacterial metabolites [63]. The system combined two innovations in microscale analysis, nanoSplitter LC–MS and microdroplet NMR. The nanoSplitter provides the high sensitivity of nanoelectrospray MS, allowing 98% of the HPLC effluent from a LC column to be collected and concentrated for NMR analysis. Microdroplet NMR is a microfluidic droplet NMR loading method that provides several fold higher sample efficiency than conventional flow injection methods. Interpretable 1D NMR spectra were obtained from analytes at the 200 ng level, in 1 h/well automated NMR data acquisitions. The system offered impressive limits of detection (LOD), at the 50 ng level for NMR, excellent reproducibility (RSD: 1.17%), and sample recovery on the order of 93%. The applicability of this system to the investigation of a bioactive cyanobacterial extract was demonstrated. The platform recognized four known NPs, ambiguine A, I, E, and hapalindole H, from a single 30 μg LC injection of cyanobacterial crude extract LC (Figure 2.6b) and, most impressively, identified one LC-MS peak as a novel bioactive compound (Figure 2.6c). Taxol (300 ng) was spiked into this mixture to give an idea of the sensitivity of the method developed (Figure 2.6b).

Microflow NMR requires the microfractionation of the crude extract and drying of the fractions, usually in 96-well microtiter plates. The same plates can also be used for measuring various bioactivities and thus localizing the bioactive NPs of interest. In this respect, the methanol extract of *Rhynchosia viscosa*, a Tanzanian medicinal plant, was microfractionated from microflow NMR experiments using only 20 mg of crude extract. One-tenth of each fraction was tested in an *in vivo* zebrafish assay directly in the 96-well plate format for bioactivity, revealing four fractions with antiangiogenic activity. Microflow NMR and TOF-MS analysis of these microfractions containing only a few micrograms of compounds enabled the *de novo* structure identification to be performed and an early estimation of antiangiogenic activity measurement in the wells holding the activity. This miniaturized zebrafish-based natural product discovery platform united a rapid isolation procedure with a reliable and sensitive *in vivo* bioassay. It provides a promising means to localize and identify bioactive natural products in complex extracts with minimal amounts of plant material and considerably accelerated bioactivity-guided fractionation procedures [135].

Microflow NMR was also found to be essential for *de novo* identification of trace biomarkers, such as jasmonic acid derivatives in the context of MS-based metabolomic studies [65,66]. These examples illustrate the potential for the sample limited application in both NP discoveries as well as in metabolomics. The methodology can be applied in other fields requiring trace analysis of components of complex mixtures.

Classical LC-NMR profiling is still performed to support dereplication studies. For example, Kang et al. applied LC-NMR in the identification of isoflavonoids from *Belamcanda chinensis* [105]. The LC-NMR data were acquired using a 500 MHz NMR system equipped with an LC-NMR cryoflow probe with a 60 μL flow cell. ^{1}H-NMR spectra of the five main peaks were obtained in the on-flow mode. The WET solvent suppression was used to eliminate the acetonitrile and residual water peak signals. The use of LC-NMR in drug discovery has also been used in the study of limonoids of *Swietenia macrophylla*. Separation of the limonoids was conducted using reversed-phase gradient HPLC elution coupled to an NMR (600 MHz) spectrometer equipped with a 120 μL flow cell in the stopped-flow mode. The structural elucidation of the limonoids was attained using the ^{1}H-NMR, TOCSY, gHSQC, and gHMBC spectra [96]. Christen and coworkers compared two fully automated HPLC-NMR methods for the structural elucidation of four isomeric tropane alkaloids from the stem-bark of an endemic Chilean plant, *Schizanthus grahamii* (Solanaceae). The first approach interfaced a conventional HPLC column to NMR by means of a loop storage unit. After elution with a mobile phase consisting of deuterated water and standard protonated organic solvents, the separated analytes were momentarily stored in a loop cassette and then individually transferred to the NMR flow probe for measurements. The second strategy combined HPLC with parallel ion-trap MS detection and NMR spectroscopy using SPE-NMR. The SPE cartridges were dried under a gentle stream of nitrogen, and analytes were sequentially eluted and directed to a cryogenically cooled flow-probe with deuterated solvents. The structures of four isomeric alkaloids were unambiguously determined by combining NMR assignments with MS data [95].

The chemical profile of the antitumor active extract of the marine sponge *Dactylospongia* sp. was performed using a combination of both at-line (HPLC followed by NMR and MS) and online (on-flow and stop-flow LC-NMR) techniques. The on-flow LC-NMR analysis was employed to initially identify compounds present in the dichloromethane extract, whereas stop-flow LC-NMR experiments were conducted on the major components present, resulting in the partial identification of pentaprenylated *p*-quinol. The complete structure was achieved by 2D-NMR spectra after isolation by preparative HPLC. This study represents one of the few reports describing the application of LC-NMR to chemically profile secondary metabolites from a marine organism [104].

Recently, Kang et al. used LC-NMR-MS and NO inhibitory assays to investigate the anti-inflammatory constituents of extracts from *Angelica dahurica*, a plant used for the treatment of colds, headaches, and toothaches in Korean traditional medicine [112]. Reversed-phase isocratic chromatography was performed using the acetonitrile–water solvent system on a C_{30} column. The identification of the compounds was based on information from ESI-MS and ^{1}H-NMR. NO inhibitory activities for five main fractions of the extract were evaluated; these fractions were identified by LC-NMR-MS as containing furanocoumarins: byakangelicol, oxypeucedanin, imperatorin, phellopterin, and isoimperatorin. The results obtained showed that the anti-inflammatory activities of *A. dahurica* could be linked to imperatorin and phellopterin [112].

2.6 APPLICATION OF DIRECT NMR METHODS FOR CHEMICAL PROFILING OF CRUDE EXTRACTS

In the previous examples discussed, NMR was used to analyze purified NPs either in conjunction with online hyphenation with HPLC or after at-line collection. NMR is also increasingly used to directly analyze complex mixtures without the need for prior chromatographic separation. The method is simple, does not require specific sample preparation, and provides detection of all protons in a quantitative manner. The main drawback is that it lacks sensitivity, and the identification of single compounds in extracts may be hindered by overlapping signals [121]. In conjunction with multivariate data analysis (MVDA), NMR fingerprinting is presently used in NP research, primarily for quality control purposes and in metabolomics.

2.6.1 Application of Direct NMR Methods in Quality Control

HPLC has been widely used for the standardization and for the quality control of plant extracts used as phytopharmaceuticals for the profiling of medicinal herbs. HPLC coupled to UV is robust, inexpensive, and simple; thus, it is widely used for the quality control of herbal extracts [27]. Conversely, to obtain a more comprehensive profile and form a more complete picture of the ratio of all the main constituents, quality control can be performed by direct NMR profiling (Table 2.1) because the method is simple and reproducible. NMR-based metabolic fingerprinting has been applied to diverse fields of plant research. The classification and identification of adulterated plant products are among the major interests [136]. Direct NMR analysis has been applied for the comparison of metabolic profiles and the fingerprinting of a number of plant extracts used in phytotherapy, such as *Echinacea purpurea*, *G. biloba*, *Valeriana officinalis*, and *Panax ginseng* [122,123,137].

NMR fingerprinting was also applied to various hot- and cold-water extracts as well as ethanol/water mixtures (tinctures) of *Cannabis sativa* to better understand how these extracts differ in their metabolic profile. The method utilized the diffusion-edited ^{1}H-NMR (1D DOSY) and ^{1}H-NMR with a suppression of the ethanol and water signals. The authors observed that temperature and polarity of the solvents used for the extraction were two factors that affect the total amount of Δ^9-THC in the extracts and its relative quantity with respect to Δ^9-THC-acid and other metabolites. With this method, it was possible, without an evaporation or separation step, to distinguish between tinctures from different cannabis cultivars [118].

For quality control, hyphenated NMR methods are primarily used for peak identification and not for comparison purposes. In this respect SPE-NMR was used to establish the chemical profile extracts of *H. perforatum* [116]. LC-NMR was also applied in this field for the profiling of carotenoids in foods, such as tomato and mandarin juice as well as palm oil. The analyses were performed in a stop-flow LC-NMR. This method has also been extremely useful for the efficient analysis of carotenoids that are fragile and sensitive to light and air [117].

2.6.2 Application of Direct NMR Methods in Metabolomics

For metabolomics, many MS- and NMR-based methods that also involve hyphenated LC-MS or GC-MS have been used for the comprehensive profiling of extracts [138]. Currently, no LC-NMR or at-line NMR methods have been used for the comparison of crude plant extracts. All studies either involved the use of direct NMR profiling with adequate MVDA or at-line NMR methods, such as microflow NMR, to confirm the identity of biomarkers highlighted by MS-based metabolic approaches. Many studies discussing stress/interaction and genotype/taxonomy have been reported using *Brassica rapa*. For example, the metabolic alterations of *B. rapa* leaves attacked by larvae of the *Plutella xylostella* and *Spodoptera exigua* were investigated by NMR, followed by a MVDA [128]. The highly overlapped ^{1}H-NMR spectra complicated the identification of the metabolites. To overcome the spectral complexity, several 2D NMR techniques were applied. Of those evaluated, *J*-resolved spectroscopy, which affords an additional coupling constant, provided a wide range of structural information for differentiating the metabolites. Based on the *J*-resolved spectra combined with principal component analysis (PCA), the major signals contributing to the discrimination were alanine, threonine, glucose, sucrose, feruloyl malate, sinapoyl malate, and gluconapin. Recently, the metabolomic changes induced by metal ions, such as copper, iron, and manganese, were evaluated in the same plant. ^{1}H-NMR and 2D NMR spectra coupled with PCA and partial least square-discriminant analysis (PLS-DA) were applied. The ^{1}H-NMR analysis followed by the application of chemometric methods revealed a number of metabolic consequences. Glucosinolates and hydroxycinnamic acids conjugated with malates were found to be the discriminating metabolites, as were primary metabolites such as carbohydrates and amino acids [127].

Metabolomics has also proved to be very effective for assessing bioactivity. Takeda et al. used direct ^{1}H-NMR profiling and multivariate data analysis for the investigation of the crude extract of *Galphimia glauca,* a plant popularly employed in Mexico for the treatment of central nervous system disorders [124]. The crude extracts from wild plant populations, which were collected from six different locations, were investigated to differentiate their chemical profile. Pharmacological and phytochemical studies led to the identification of the anxiolytic and sedative compounds consisting of a mixture of nor-secofriedelanes named the galphimine series. PCA of the ^{1}H-NMR spectra of 39 crude extracts revealed clear differences among the populations; two of the populations out of the six studied manifested differences related to the presence of galphimines. This information consistently correlated with the corresponding HPLC analysis and with neuropharmacological activity [124]. The strategy applied in this study opens new possibilities for the correct selections of plant populations with suitable metabolic and pharmacologic profiles for the development of standardized herbal medicines.

2.7 CONCLUSIONS

In the drug discovery process, NPs still represent an important source of new pharmacophores, and the development of efficient dereplication protocols is essential for a rational approach. The complementary application of LC-SPE-NMR and microflow NMR with other LC-hyphenated techniques and with online or at-line bioassays is playing an increasingly important role in natural product research in the search of new lead compounds.

The progresses made in mass spectrometry, in terms of sensitivity and resolution, facilitate the extraction of the molecular formula of each constituents over an important dynamic range, whereas MS/MS experiments generate useful complementary structural information for a wide range of NPs. Following different strategies that also involve the acquisition of PDA spectra, many compounds can also be dereplicated by additional cross-searches based on chemotaxonomic information and searches in spectral databases.

As shown, NMR can be used directly for profiling extracts for rapid fingerprinting, providing an efficient way to rapidly assess the quality of samples or to compare extracts in metabolomic studies. NMR detection complemented with other LC-hyphenated techniques, either online or at-line, is playing an increasingly important role as a strategic tool to support profiling studies. Even if it is relatively insensitive compared to the other analytical detectors, NMR remains the only technique that can unambiguously solve structural identification issues, especially for *de novo* metabolite determination. According to specific needs, NMR can be hyphenated to LC systems in different ways. Online LC-NMR is now being gradually replaced by alternative at-line approaches, such as LC-SPE-NMR or microflow NMR. These latter methods do not require compromises to be made and provide high-quality 1D and 2D NMR, even for the minor constituents of an extract, and they can now be fully automated. However, for labile constituents, online LC-NMR methods remain mandatory and still provide a rapid and efficient means of evaluating the nature and the real proportion of the main constituents in a crude extract. These hyphenated methods have evolved into sophisticated LC-NMR-MS platforms that are becoming essential in many drug discovery studies.

The integration of these sensitive spectroscopic methods hyphenated with HPLC separation and with highly sensitive bioassays can be coupled online or used at-line after microfractionation; they will considerably speed up the drug discovery process and thus contribute to the discovery of new useful drugs from natural sources. The combination of chemoanalytical and biological data as well as automated dereplication will become essential. These aspects will represent the greatest challenge to face for the next few years in this field of research.

ABBREVIATIONS

APCI	atmospheric pressure chemical ionization
API	atmospheric pressure ionization
APPI	atmospheric pressure photoionization
CapNMR	capillary nuclear magnetic resonance
CID	collision-induced dissociation
ELSD	evaporative light-scattering detection
ESI	electrospray ionization
FT	Fourier transform
gHMBC	gradient heteronuclear multiple-quantum correlation (NMR)
gHSQC	gradient heteronuclear single-quantum correlation (NMR)
HPLC	high-performance liquid chromatography
HR	high resolution
LLE	liquid–liquid extraction
LOD	limit of detection
LOQ	limit of quantification
LR	low resolution
MS	mass spectrometry
MS/M	tandem mass spectrometry
MS^n	multiple-stage mass spectrometry
MVDA	multivariate data analysis
MW	molecular weight
NI	negative ion

NMR	nuclear magnetic resonance
NOESY	nuclear Overhauser enhancement spectroscopy (NMR)
NPs	natural products
PCA	principal component analysis
PDA	photodiode array
PI	positive ion
PID	precursor ion discovery
Q	quadrupole
QQQ	triple quadrupole
SPE	solid phase extraction
TOF	time of flight
UHPL	ultra-high-pressure liquid chromatography
UV	ultraviolet
WET	fast solvent suppression (LC-NMR)

REFERENCES

[1] Hostettmann, K., Potterat, O., Wolfender, J.-L. (1998). The potential of higher plants as a source of new drugs. *Chimia*, *52*, 10–17.

[2] Koehn, F.E., Carter, G.T. (2005). The evolving role of natural products in drug discovery. *Nature Reviews Drug Discovery*, *4*, 206–220.

[3] Dobson, C.M. (2004). Chemical space and biology. *Nature*, *432*, 824–828.

[4] Wermuth, C.G., *The Practice of Medicinal Chemistry*, Academic, London, 2003.

[5] Feher, M., Schmidt, J.M. (2003). Property distributions: Differences between drugs, natural products, and molecules from combinatorial chemistry. *Journal of Chemical Information and Modeling*, *43*, 218–227.

[6] Newman, D.J., Cragg, G.M. (2007). Natural products as sources of new drugs over the last 25 years. *Journal of Natural Products*, *70*, 461–477.

[7] Hostettmann, K., Wolfender J.-L., Terreaux, C. (2001). Modern screening techniques for plant extracts. *Pharmaceutical Biology*, *39*, 18–32.

[8] Villas-Boas, S.G., Mas, S., Aakesson, M., Smedsgaard, J., Nielsen, J. (2005). Mass spectrometry in metabolome analysis. *Mass Spectrometry Reviews*, *24*, 613–646.

[9] Dunn, W.B., Ellis, *Di.* (2005). Metabolomics: Current analytical platforms and methodologies. *Trends in Analytical Chemistry*, *24*, 285–294.

[10] Huber, L., George, S.A., *Diode Array Detection in HPLC*, Marcel Dekker, New York, 1993.

[11] Niessen, W.M.A., *Liquid Chromatography—Mass Spectrometry*, Taylor & Francis, Boca Raton, FL, 2006.

[12] Albert, K., *On-Line LC-NMR and Related Techniques*, John Wiley & Sons, Chichester, 2002.

[13] Jaroszewski, J.W. (2005). Hyphenated NMR methods in natural products research, Part 2: HPLC-SPE-NMR and other new trends in NMR hyphenation. *Planta Medica*, *71*, 795–802.

[14] Olson, D.L., Norcross, J.A., O'Neil-Johnson, M., Molitor, P.F., Detlefsen, D.J., Wilson, A.G., Peck, T.L. (2004). Microflow NMR: Concepts and capabilities. *Analytical Chemistry*, *76*, 2966–2974.

[15] Queiroz, E.F., Marston, A., Hostettmann, K., Contemporary methodological approaches in the search for new lead compounds from higher plants. In: Elisabetsky E. and Etkin N.L., editors. *Ethnopharmacology*, Encyclopedia of Life Support Systems (EOLSS), Developed under the Auspices of the UNESCO, Eolss Publishers, Oxford, 2005. Available at http://www.eolss.net.

[16] Blondeau, S., Do, Q.T., Scior, T., Bernard, P., Morin-Allory, L. (2010). Reverse pharmacognosy: Another way to harness the generosity of nature. *Current Pharmaceutical Design*, *16*, 1682–1696.

[17] Rollinger, J.M., Langer, T., Stuppner, H. (2006). Strategies for efficient lead structure discovery from natural products. *Current Medicinal Chemistry*, *13*, 1491–1507.

[18] Gustafson, K.R., Cardellina, J.H., McMahon, J.B., Gulakowski, R.J., Ishitoya, J. Szallasi, Z., Lewin, N.E., Blumberg, P.M., Weislow, O.S., Beutler, J.A. (1992). A nonpromoting phorbol from the Samoan medicinal plant *Homalanthus nutans* inhibits cell killing by HIV-1. *Journal of Medicinal Chemistry*, *35*, 1978–1986.

[19] Wender, P.A., Kee, J.-M., Warrington, J.M. (2008). Practical synthesis of prostratin, DPP and their analogs, adjuvant leads against Latent HIV. *Science*, *320*, 649–652.

[20] Nogueira, R.C., de Cerqueira, H.F., Soares, M.B. (2010). Patenting bioactive molecules from biodiversity: The Brazilian experience. *Expert Opinion on Therapeutic Patents*, *20*, 145–157.

[21] Lema, A. (1965). A preliminary report on the molluscicidal properties of Endod (*Phytolacca dodecandra*). *Ethiopian Medical Journal*, *3*, 187–190.

[22] Tchuenté, L.A. (2006). Lutte contre la schistosomose: Défi et perspectives pour le XXIe siècle. *Bulletin de la Société de Pathologie Exotique*, *5*, 372–376.

[23] Kingston, D.G.I. (2001). Taxol, a molecule for all seasons. *Chemical Communications*, *10*, 867–880.

[24] Pieters, L., Vlietinck, A.J. (2005). Bioguided isolation of pharmacologically active plant components, still a valuable strategy for the finding of new lead compounds? *Journal of Ethnopharmacology*, *100*, 57–60.

[25] Natishan, T.K. (2004). Recent developments of achiral HPLC methods in pharmaceuticals using various detection modes. *Journal of Liquid Chromatography & Related Technologies*, *27*, 1237–1316.

[26] Nguyen, D.T.T., Guillarme, D., Rudaz, S., Veuthey, J.L. (2006). Fast analysis in liquid chromatography using small particle size and high pressure. *Journal of Separation Science*, *29*, 1836–1848.

[27] David, F., Vanhoenacker, G., Tienpont, B., Francois, I., Sandra, P. (2007). Coupling columns and multidimensional configurations to increase peak capacity in liquid chromatography. *LC-GC Europe*, *20*, 154–158.

[28] Wolfender, J.-L. (2009). HPLC in natural product analysis: The detection issue. *Planta Medica*, *75*, 719–734.

[29] Wilson, I.D., Brinkman, U.A.T. (2003). Hyphenation and hypernation. The practice and prospects of multiple hyphenation. *Journal of Chromatography A*, *1000*, 325–356.

[30] Corcoran, O., Spraul, M. (2003). LC-NMR-MS in drug discovery. *Drug Discovery Today*, *8*, 624–631.

[31] Grata, E., Guillarme, D., Glauser, G., Boccard, J., Carrupt, P.A., Veuthey, J.L., Rudaz, S., Wolfender, J.-L. (2009). Metabolite profiling of plant extracts by ultra-high-pressure liquid chromatography at elevated temperature coupled to time-of-flight mass spectrometry. *Journal of Chromatography A*, *1216*, 5660–5668.

[32] Grata, E., Boccard, J., Guillarme, D., Glauser, G., Carrupt, P.A., Farmer, E., Wolfender, J.-L., Rudaz, S. (2008). UPLC–TOF-MS for plant metabolomics: A sequential approach for wound marker analysis in Arabidopsis thaliana. *Journal of Chromatography B*, *871*, 261–270.

[33] Wilson, I.D., Nicholson, J.K., Castro-Perez, J., Granger, J.H., Johnson, K.A., Smith, B.W., Plumb, R.S. (2005). High resolution ultra performance liquid chromatography coupled to oa-TOF mass spectrometry as a tool for differential metabolic pathway profiling in functional genomic studies. *Journal of Proteome Research*, *4*, 591–598.

[34] Guillarme, D., Nguyen, D.T.T., Rudaz, S., Veuthey, J.L. (2007). Method transfer for fast liquid chromatography in pharmaceutical analysis: Application to short columns packed with small particles. Part I. Isocratic separation. *European Journal of Pharmaceutics and Biopharmaceutics*, *66*, 475–482.

[35] Guillarme, D., Nguyen, D.T.T., Rudaz, S., Veuthey, J.L. (2008). Method transfer for fast liquid chromatography in pharmaceutical analysis: Application to short columns packed with small particles. Part II. Gradient separation. *European Journal of Pharmaceutics and Biopharmaceutics*, *68*, 430–440.

[36] Grumbach, E., Wheat, T., Mccabe, D., Diehl, D., Mazzeo, J. (2006). Improving LC separations: Transferring methods from HPLC to UPLC. *LC-GC North America*, *24*, 80–81.

[37] De Villiers, A., Lestremau, F., Szucs, R., Gelebart, S., David, F., Sandra, P. (2006). Evaluation of ultra performance liquid chromatography. Part I. Possibilities and limitations. *Journal of Chromatography A*, *1127*, 60–69.

[38] Novakova, L., Matysova, L., Solich, P. (2006). Advantages of application of UPLC in pharmaceutical analysis. *Talanta*, *68*, 908–918.

[39] Wren, S.A.C., Tchelitcheff, P. (2006). Use of ultra-performance liquid chromatography in pharmaceutical development. *Journal of Chromatography A*, *1119*, 140–146.

[40] Lurie, I.S., Toske, S.G. (2008). Applicability of ultra-performance liquid chromatography-tandem mass spectrometry for heroin profiling. *Journal of Chromatography A*, *1188*, 322–326.

[41] Cabooter, D., Lestremau, F., Lynen, F., Sandra, P., Desmet, G. (2008). Kinetic plot method as a tool to design coupled column systems producing 100,000 theoretical plates in the shortest possible time. *Journal of Chromatography A*, *1212*, 23–34.

[42] Plumb, R.S., Jones, M.D., Rainville, P., Castro-Perez, J.M. (2007). The rapid detection and identification of the impurities of simvastatin using high resolution sub 2 micron particle LC coupled to hybrid quadrupole time of flight MS operating with alternating high-low collision energy. *Journal of Separation Science*, *30*, 2666–2675.

[43] Eugster, P., Guillarme, D. Rudaz, S., Veuthey, J.L., Carrupt, P.A., Wolfender, J.-L. (2011). Ultra high pressure liquid chromatography for crude plant extracts profiling. *Journal of Association of Analytical Communities*, *94*, 51–70.

[44] Markham, K.R., *Techniques of Flavonoid Identification*, Academic Press, London, 1982.

[45] Alonso-Salces, R.M., Queiroz, E.F., Ndjoko, K., Ioset, J.R., Hostettmann, K., Gallom B., Berrueta, L.A., Vicente, F. (2004). On-line characterization of apple polyphenols by high performance liquid chromatography coupled with mass spectrometry and ultraviolet absorbance. *Journal of Chromatography A*, *1046*, 89–100.

[46] Polasek, J., Queiroz, E.F., Hostettmann, K. (2007). On-line identification of *Trifolium* flavonoids by liquid chromatography-UV detection using post-column derivatization and liquid chromatography-mass spectrometry. *Phytochemical Analysis*, *26*, 13–23.

[47] Urbain, A., Marston, A., Marsden-Edwards, E., Hostettmann, K. (2009). Ultra-performance liquid chromatography/time-of-flight mass spectrometry as a chemotaxonomic tool for the analysis of Gentianaceae species. *Phytochemical Analysis*, *20*, 134–138.

[48] Cheng, K.W., Cheng, F., Wang, M., Liquid chromatography-mass spectrometry in natural product research. In: Colegate, S.M. and Molyneux, R.J. editors. *Bioactive Natural Products: Detection, Isolation, and Structural Determination*, 2nd ed., CRC Press, London, 2008, pp. 245–266.

[49] He, X.G. (2000). On-line identification of phytochemical constituents in botanical extracts by combined high-performance liquid chromatographic-diode array detection-mass spectrometric techniques. *Journal of Chromatography A*, *880*, 203–232.

[50] Korfmacher, W.A. (2005). Principles and applications of LC-MS in new drug discovery. *Drug Discovery Today*, *10*, 1357–1367.

[51] Jessome, L.L., Volmer, D.A. (2006). Ion suppression: A major concern in mass spectrometry. *LC-GC North America*, *24*, 83–89.

[52] Syage, J.A., Short, L.C., Cai, S.S. (2008). Atmospheric pressure photoionization—The second source for LC-MS? *LC-GC North America*, *26*, 286–300.

[53] Cai, Y.X., Kingery, D., McConnell, O., Bach, A.C. (2005). Advantages of atmospheric pressure photoionization mass spectrometry in support of drug discovery. *Rapid Communications in Mass Spectrometry*, *19*, 1717–1724.

[54] Pisitkun, T., Hoffert, J.D., Yu, M.J., Knepper, M.A. (2007). Tandem mass spectrometry in physiology. *Physiology*, *22*, 390–340.

[55] Kind, T., Fiehn, O. (2007). Seven Golden Rules for heuristic filtering of molecular formulas obtained by accurate mass spectrometry. *BMC Bioinformatics*, *8*, 105–125.

[56] Waridel, P., Wolfender, J.-L., Ndjoko, K., Hobby, K.R., Major, H.J., Hostettmann, K. (2001). Evaluation of quadrupole time-of-flight tandem mass spectrometry and ion-trap multiple-stage mass spectrometry for the differentiation of C-glycosidic flavonoid isomers. *Journal of Chromatography A*, *926*, 29–41.

[57] Queiroz, E.F. Wolfender, J.-L., Atindehoua, K.K., Traore, D., Hostettmann, K. (2002). On-line identification of the antifungal constituents of *Erythrina vogelii* by liquid chromatography with tandem mass spectrometry, ultraviolet absorbance detection and nuclear magnetic resonance spectrometry combined with liquid chromatographic micro-fractionation. *Journal of Chromatography A*, *974*, 123–134.

[58] Colombo, R., Yariwake, J.H., Queiroz, E.F., Ndjoko, K., Hostettmann, K. (2005). On-line identification of sugarcane (*Saccharum officinarum* L.) methoxyflavones by liquid chromatography–UV detection using post-column derivatization and liquid chromatography–mass spectrometry. *Journal of Chromatography A*, *1082*, 51–59.

[59] Jaroszewski, J.W. (2005). Hyphenated NMR methods in natural products research, Part 1: Direct hyphenation. *Planta Medica*, *71*, 691–700.

[60] Bayer, E., Albert, K., Nieder, M., Grom, E., Wolff, G., Rindlisbacher, M. (1982). On-line coupling of liquid chromatography and high-field nuclear magnetic resonance spectroscopy. *Analytical Chemistry*, *54*, 1747–1750.

[61] Wolfender, J.-L., Ndjoko, K., Hostettmann, K. (2001). The potential of LC-NMR in phytochemical analysis. *Phytochemical Analysis*, *12*, 2–22.

[62] Spraul, M., Freund, A.S., Nast, R.E., Withers, R.S., Maas, W.E., Corcoran, O. (2003). Advancing NMR sensitivity for LC-NMR-MS using a cryoflow probe: Application to the analysis of acetaminophen metabolites in urine. *Analytical Chemistry*, *75*, 1536–1541.

[63] Lin, Y., Schiavo, S., Orjala, J., Vouros, P., Kautz, R. (2008). Microscale LC-MS-NMR platform applied to the identification of active cyanobacterial metabolites. *Analytical Chemistry*, *80*, 8045–8054.

[64] Wolfender, J.-L., Queiroz, E.F., Hostettmann, K., Development and application of LC/NMR techniques to the identification of bioactive natural products. In: Colegate, S. M. and Molyneux, R.J. editors. *Bioactive Natural Products, Detection, Isolation and Structural Determination*, 2nd ed., CRC Press, London, 2008, pp. 143–190.

[65] Glauser, G., Gindro, K., Fringeli, J., de Joffrey, J.-P., Rudaz, S., Wolfender, J.-L. (2009). Differential analysis of mycoalexins in confrontation zones of grapevine fungal pathogens by ultra-high pressure liquid chromatography/time-of-flight mass spectrometry and capillary nuclear magnetic resonance. *Journal of Agricultural and Food Chemistry*, *57*, 1127–1134.

[66] Glauser, G., Guillarme, D., Grata, E., Boccard, J., Thiocone, A., Carrupt, P.A., Veuthey, J.L., Rudaz, S., Wolfender, J.-L. (2008). Optimized liquid chromatography–mass spectrometry approach for the isolation of minor stress biomarkers in plant extracts and their identification by capillary nuclear magnetic resonance. *Journal of Chromatography A*, *1180*, 90–98.

[67] Exarchou, V., Krucker,M., van Beek, T.A., Vervoort, J., Gerothanassis, I.P., Albert, K. (2005). LC-NMR coupling technology: recent advancements and applications in natural products analysis. *Magnetic Resonance in Chemistry*, *43*, 681–687.

[68] Exarchou, V., Godejohann, M., Van Beek, T.A., Gerothanassis, I.P., Vervoort, J. (2003). LC-UV-solid-phase extraction-nmr-ms combined with a cryogenic flow probe and its application to the identification of compounds present in greek oregano. *Analytical Chemistry*, *75*, 6288–6294.

[69] Clarkson, C., Sibum, M., Mensen, R., Jaroszewski, J.W. (2007). Evaluation of on-line solid-phase extraction parameters for hyphenated, high-performance liquid chromatography-solid-phase extraction-nuclear magnetic resonance applications. *Journal of Chromatography A*, *1165*, 1–9.

[70] Webb, A.G. (2005). Microcoil nuclear magnetic resonance spectroscopy. *Journal of Pharmaceutical and Biomedical Analysis*, *38*, 892–903.

[71] Wilson, S.R., Malerod, H., Petersen, D., Simic, N., Bobu, M.M., Rise, F., Lundanes, E., Greibrokk, T. (2006). Controlling LC-SPE-NMR systems. *Journal of Separation Science*, *29*, 582–589.

[72] Miliauskas, G., van Beek, T.A., de Waard, P., Venskutonis, R. P., Sudholter, E.J.R. (2006). Comparison of analytical and semi-preparative columns for high-performance liquid chromatography-solid-phase extraction-nuclear magnetic resonance. *Journal of Chromatography A*, *1112*, 276–284.

[73] Lambert, M., Staerk, D., Hansen, S.H., Sairafianpour, M., Jaroszewski, J.W. (2005). Rapid extract dereplication using HPLC-SPE-NMR: Analysis of isoflavonoids from *Smirnowia iranica*. *Journal of Natural Products*, *68*, 1500–1509.

[74] Seger, C., Sturm, S. (2007). HPLC-SPE-NMR—A novel hyphenation technique. *LC-GC Europe*, *20*, 587–588.

[75] Sturm, S., Seger, C., Godejohann, M., Spraul, M., Stuppner, H. (2007). Conventional sample enrichment strategies combined with high-performance liquid chromatography–solid phase extraction–nuclear magnetic resonance analysis allows analyte identification from a single minuscule Corydalis solida plant tuber. *Journal of Chromatography A*, *1163*, 138–144.

[76] Wolfender, J.-L., Queiroz, E.F., Hostettmann, K. (2005). Phytochemistry in the microgram domain, a LC-NMR perspective. *Magnetic Resonance in Chemistry*, *43*, 697–709.

[77] Hu, J.F., Yoo, H.D., Williams, C.T., Garo, E., Cremin, P.A., Zeng, L., Vervoort, H.C., Lee, C.M., Hart, S.M., Goering, M.G., O'Neil-Johnson, M., Eldridge, G.R. (2005). Miniaturization of the structure elucidation of novel natural products. Two trace antibacterial acylated caprylic alcohol glycosides from Arctostaphylos pumila. *Planta Medica*, *71*, 176–180.

[78] Jansma, A., Chuan, T., Albrecht, R.W., Olson, D.L., Peck, T.L., Geierstanger, B.H. (2005). Automated microflow NMR: Routine analysis of five-microliter samples. *Analytical Chemistry*, *77*, 6509–6515.

[79] Wolfender, J.-L., Queiroz, E.F., Hostettmann, K. (2006). The importance of hyphenated techniques in the discovery of new lead compounds from nature. *Expert Opinion on Drug Discovery*, *1*, 237–260.

[80] Queiroz, E.F., Wolfender, J.-L., Hostettmann, K. (2009). Modern approaches in the search for new lead antiparasitic compounds from higher plants. *Current Drug Targets*, *10*, 202–211.

[81] Staerk, D., Kesting, J.R., Sairafianpour, M., Witt, M., Asili, J., Emami, S.A., Jaroszewski, J.W. (2009). Accelerated dereplication of crude extracts using HPLC-PDA-MS-SPE-NMR: Quinolinone alkaloids of *Haplophyllum acutifolium*. *Phytochemistry*, *70*, 1055–1061.

[82] Lee, S.-S., Lin, H.-C., Chen, C.-K. (2008). Acylated flavonol monorhamnosides, [alpha]-glucosidase inhibitors, from *Machilus philippinensis*. *Phytochemistry*, *69*, 2347–2353.

[83] Pedersen, M.M., Chukwujekwu, J.C., Lategan, C.A., Staden, J.V., Smith, P.J., Staerk, D. (2009). Antimalarial sesquiterpene lactones from *Distephanus angulifolius*. *Phytochemistry*, *70*, 601–607.

[84] Kesting, J.R., Tolderlund, I.-L., Pedersen, A.F., Witt, M., Jaroszewski, J.W., Staerk, D. (2009). Piperidine and tetrahydropyridine alkaloids from *Lobelia siphilitica* and *Hippobroma longiflora*. *Journal of Natural Products*, *72*, 312–315.

[85] Clarkson, C., Staerk, D., Hansen, S.H., Smith, P. J., Jaroszewski, J.W. (2006). Identification of major and minor constituents of *Harpagophytum procumbens* (devil's claw) using HPLC-SPE-NMR and HPLC-ESIMS/APCIMS. *Journal of Natural Products*, *69*, 1280–1288.

[86] Lam, S.H., Chen, C.K., Wang, J.S., Lee, S.S. (2008). Investigation of flavonoid glycosides from *Neolitsea sericea* var. *aurata* via the general method and HPLC-SPE-NMR. *Journal of the Chinese Chemical Society*, *55*, 449–455.

[87] Seger, C., Godejohann, M., Spraul, M., Stuppner, H., Hadacek, F. (2006). Reaction product analysis by high-performance liquid chromatography-solid-phase extraction-

nuclear magnetic resonance—Application to the absolute configuration determination of naturally occurring polyyne alcohols. *Journal of Chromatography A*, *1136*, 82–88.

[88] Sorensen, D., Raditsis, A., Trimble, L.A., Blackwell, B.A., Sumarah, M.W., Miller, J.D. (2007). Isolation and structure elucidation by LC-MS-SPE/NMR: PR toxin- and cuspidatol-related eremophilane sesquiterpenes from *Penicillium roqueforti*. *Journal of Natural Products*, *70*, 121–123.

[89] Grevenstuk, T., van der Hooft, J.J.J., Vervoort, J., de Waard, P., Romano, A. (2009). Iridoid and caffeoyl phenylethanoid glycosides of the endangered carnivorous plant *Pinguicula lusitanica* L. (Lentibulariaceae). *Biochemical Systematics and Ecology*, *37*, 285–289.

[90] Lee, S.S., Lai, Y.C., Chen, C.K., Tseng, L.H., Wang, C.Y. (2007). Characterization of isoquinoline alkaloids from *Neolitsea sericea* var. *aurata* by HPLCSPE-NMR. *Journal of Natural Products*, *70*, 637–642.

[91] Mmatli, E.E., Malerod, H., Wilson, S.R., Abegaz, B., Greibrokk, T., Lundanes, E., Malterud, K.E., Petersen, D., Rise, F. (2007). Identification of major metal complexing compounds in *Blepharis aspera*. *Analytica Chimica Acta*, *597*, 24–31.

[92] Sprogoe, K., Staerk, D., Ziegler, H.L., Jensen, T.H., Holm-Moller, S.B., Jaroszewski, J.W. (2008). Combining HPLC-PDA-MS-SPE-NMR with circular dichroism for complete natural product characterization in crude extracts: Levorotatory gossypol in *Thespesia danis*. *Journal of Natural Products*, *71*, 516–519.

[93] Exarchou, V., Fiamegos, Y.C., van Beek, T.A., Nanos, C., Vervoort, J. (2006). Hyphenated chromatographic techniques for the rapid screening and identification of antioxidants in methanolic extracts of pharmaceutically used plants. *Journal of Chromatography A*, *1112*, 293–302.

[94] Lam, S.H., Wang, C Y., Chen, C.K., Lee, S.S. (2007). Chemical investigation of *Phyllanthus reticulatus* by HPLC-SPE-NMR and conventional methods. *Phytochemical Analysis*, *18*, 251–255.

[95] Bieri, S., Varesio, E., Veuthey, J.-L., Muñoz, O., Tseng, L.-H., Braumann, U., Spraul, M., Christen, P. (2006). Identification of isomeric tropane alkaloids from *Schizanthus grahamii* by HPLC-NMR with loop storage and HPLC-UV-MS/SPE-NMR using a cryogenic flow probe. *Phytochemical Analysis*, *17*, 78–86.

[96] Schefer, A.B., Braumann, U., Tseng, L.-H., Spraul, M., Soares, M.G., Fernandes, J.B., da Silva, M.F.G.F., Vieira, P.C., Ferreira, A.G. (2006). Application of high-performance liquid chromatography-nuclear magnetic resonance coupling to the identification of limonoids from mahogany tree (*Swietenia macrophylla*, Meliaceae) by stopped-flow 1D and 2D NMR spectroscopy. *Journal of Chromatography A*, *1128*, 152–163.

[97] Jing, Q., Ying-Hong, W., Jian-Bei, L., Shi-Shan, Y., Yuan, L., Yun-Bao, L. (2007). Rapid structural determination of new trace cassaine-type diterpenoid amides in fractions from *Erythrophleum fordii* by liquid chromatography-diode array detection/electrospray ionization tandem mass spectrometry and liquid chromatography/nuclear magnetic resonance. *Rapid Communications in Mass Spectrometry*, *21*, 2109–2119.

[98] Weber, B., Herrmann, M., Hartmann, B., Joppe, H., Schmidt, C.O., Bertram, H.J. (2008). HPLC-MS and HPLC/NMR as hyphenated techniques for accelerated characterization of the main constituents in Chamomile (*Chamomilla recutita* [L.] Rauschert). *European Food Research and Technology*, *226*, 755–760.

[99] Rehbein, J., Dietrich, B., Grynbaum, M.D., Hentschel, P., Holtin, K., Kuehnle, M., Schuler, P., Bayer, M., Albert, K. (2007). Characterization of bixin by LC-MS and LC-NMR. *Journal of Separation Science*, *30*, 2382–2390.

[100] Mitova, M.I., Murphy, A.C., Lang, G., Blunt, J.W., Cole, A.L.J., Ellis, G., Munro, M.H.G. (2008). Evolving trends in the dereplication of natural product extracts. 2. The isolation of chrysaibol, an antibiotic peptaibol from a New Zealand sample of the mycoparasitic fungus *Sepedonium chrysospermum*. *Journal of Natural Products*, *71*, 1600–1603.

[101] Hu, J.F., Garo, E., Hough, G.W., Goering, M.G., Yoo, H.D., O'Neil-Johnson, M., Eldridge, G.R. (2007). Isoflavones from *Psorothamnus arborescens* via high-throughput natural product chemistry methods. *Journal of Chemical Research*, *38*, 296–297.

[102] Lambert, M., Wolfender, J.-L., Staerk, D., Christensen, B., Hostettmann, K., Jaroszewski, J.W. (2007). Identification of natural products using HPLC-SPE combined with CapNMR. *Analytical Chemistry*, *79*, 727–735.

[103] Lang, G., Mayhudin, N.A., Mitova, M.I., Sun, L., van der Sar, S., Blunt, J.W., Cole, A.L.J., Ellis, G., Laatsch, H., Munro, M.H.G. (2008). Evolving trends in the dereplication of natural product extracts: New methodology for rapid, small-scale investigation of natural product extracts. *Journal of Natural Products*, *71*, 1595–1599.

[104] Dias, D.A., Urban, S. (2009). Application of HPLC-NMR for the rapid chemical profiling of a southern australian sponge, Dactylospongia sp. *Journal of Separation Science*, *32*, 542–548.

[105] Kang, S.W., Kim, M.C., Kim, C.Y., Jung, S.H., Um, B.H. (2008). The rapid identification of isoflavonoids from *Belamcanda chinensis* by LC-NMR and LC-MS. *Chemical & Pharmaceutical Bulletin*, *56*, 1452–1454.

[106] Iwasa, K., Takahashi, T., Nishiyama, Y., Moriyasu, M., Sugiura, M., Takeuchi, A., Tode, C., Tokuda, H., Takeda, K. (2008). Online structural elucidation of alkaloids and other constituents in crude extracts and cultured cells of *Nandina domestica* by combination of LC-MS/MS, LC-NMR, and LC-CD analyses. *Journal of Natural Products*, *71*, 1376–1385.

[107] Jing, Q., You-cai, H., Jian-bei, L., Ying-hong, W., Jin-lan, Z., Zeper, A., Shi-shan, Y., Yun-bao, L. (2008). Structural characterization of constituents with molecular diversity in fractions from *Lysidice brevicalyx* by liquid chromatography/diode-array detection/electrospray ionization tandem mass spectrometry and liquid chromatography/nuclear magnetic resonance. *Rapid Communications in Mass Spectrometry*, *22*, 755–765.

[108] Schroeder, F.C., Taggi, A.E., Gronquist, M., Malik, R.U., Grant, J.B., Eisner, T., Meinwald, J. (2008). NMR-spectroscopic screening of spider venom reveals sulfated nucleosides as major components for the brown recluse and related species. *Proceedings of the National Academy of Sciences*, *105*, 14283–14287.

[109] Politi, M., Braca, A., Altinier, G., Sosa, S., Ndjoko, K., Wolfender, J.-L., Hostettmann, K., Jimenez-Barbero, J. (2008). Different approaches to study the traditional remedy of "hierba del canto," Sisymbrium officinale (L.) Scop. *Latin American and Caribbean Bulletin of Medicinal and Aromatic Plants*, *7*, 30–37.

[110] Sumarah, M.W., Puniani, E., Blackwell, B.A., Miller, J.D. (2008). Characterization of polyketide metabolites from foliar endophytes of *Picea glauca*. *Journal of Natural Products*, *71*, 1393–1398.

[111] Castro, A., Moco, S., Coll, Vervoort, J. (2010). LC-MS-SPE-NMR for the isolation and characterization of neo-clerodane diterpenoids from *Teucrium luteum* subsp. flavovirens. *Journal of Natural Products*, *73*, 962–965.

[112] Kang, S.W., Kim, C.Y., Song, D.G., Pan, C.H., Cha, K.H., Lee, D.U., Um, B.H. (2010). Rapid identification of furanocoumarins in *Angelica dahurica* using the online LC-

MMR-MS and their nitric oxide inhibitory activity in RAW 264.7 cells. *Phytochemical Analysis*, *21*, 322–327.

[113] Gao, H., Zehl, M., Kaehlig, H., Schneider, P., Stuppner, H., Banuls, L.M.Y, Kiss, R., Kopp, B. (2010). Rapid structural identification of cytotoxic bufadienolide sulfates in toad venom from *Bufo melanostictus* by LC-DAD-MS*n* and LC-SPE-NMR. *Journal of Natural Products*, *73*, 603–608.

[114] Motti, C.A., Freckelton, M.L., Tapiolas, D.M., Willis, R.H. (2009). FTICR-MS and LC-UV/MS-SPE-NMR applications for the rapid dereplication of a crude extract from the sponge *Ianthella flabelliformis*. *Journal of Natural Products*, *72*, 290–294.

[115] Yang, Y.-L., Liao, W.-Y., Liu, W.-Y., Liaw, C.-C., Shen, C.N., Huang, Z.-Y., Wu S.-H. (2009). Discovery of new natural products by intact-cell mass spectrometry and LC-SPE-NMR: malbranpyrroles, novel polyketides from thermophilic fungus *Malbranchea sulfurea*. *Chemistry – A European Journal*, *15*, 11573–11580.

[116] Tatsis, E.C., Boeren, S., Exarchou, V., Troganis, A.N., Vervoort, J., Gerothanassis, I.P. (2007). Identification of the major constituents of *Hypericum perforatum* by LC/SPE/NMR and/or LC-MS. *Phytochemistry*, *68*, 383–393.

[117] Chisato, T., Takashi, M., Makiko, S. (2009). Application of LC-NMR to analysis of carotenoids in foods. *Journal of Separation Science*, *32*, 1–5.

[118] Politi, M., Peschel, W., Wilson, N., Zloh, M., Prieto, J.M., Heinrich, M. (2008). Direct NMR analysis of cannabis water extracts and tinctures and semi-quantitative data on D^9-THC and D^9-THC-acid. *Phytochemistry*, *69*, 562–570.

[119] Yang, S.Y., Kim, H.K., Lefeber, A.W.M., Erkelens, C., Angelova, N., Choi, Y.H., Verpoorte, R. (2006). Application of two-dimensional nuclear magnetic resonance spectroscopy to quality control of ginseng commercial products. *Planta Medica*, *72*, 364–369.

[120] Moco, S., Forshed, J., De Vos, R., Bino, R., Vervoort, J. (2008). Intra- and inter-metabolite correlation spectroscopy of tomato metabolomics data obtained by liquid chromatography-mass spectrometry and nuclear magnetic resonance. *Metabolomics*, *4*, 202–215.

[121] Kim, H.K., Choi, Y.H., Verpoorte, R. (2010). NMR-based metabolomic analysis of plants. *Nature Protocols*, *5*, 536–549.

[122] Van der Kooy, F., Verpoorte, R., Meyer, J.J.M. (2008). Metabolomic quality control of claimed anti-malarial *Artemisia* afra herbal remedy and *A. afra* and *A. annua* plant extracts. *South African Journal of Botany*, *74*, 186–189.

[123] Kang, J., Lee, S., Kang, S., Kwon, H., Park, J., Kwon, S., Park, S. (2008). NMR-based metabolomics approach for the differentiation of ginseng (*Panax ginseng*) roots from different origins. *Archives of Pharmaceutical Research*, *31*, 330–336.

[124] Cardoso-Taketa, A.T., Pereda-Miranda, R., Choi, Y.H., Verpoorte, R., Villarreal, M.L. (2008). Metabolic profiling of the Mexican anxiolytic and sedative plant *Galphimia glauca* using nuclear magnetic resonance spectroscopy and multivariate data analysis. *Planta Medica*, *74*, 1295–1301.

[125] Thiocone, A., Farmer, E.E., Wolfender, J.-L. (2008). Screening for wound-induced oxylipins in *Arabidopsis thaliana* by differential HPLC-APCI/MS profiling of crude leaf extracts and subsequent characterization by capillary-scale NMR. *Phytochemical Analysis*, *19*, 198–205.

[126] Liang, Y.S., Choi, Y.H., Kim, H.K., Linthorst, H.J.M., Verpoorte, R. (2006). Metabolomic analysis of methyl jasmonate treated *Brassica rapa* leaves by 2-dimensional NMR spectroscopy. *Phytochemistry*, *67*, 2503–2511.

[127] Jahangir, M., Abdel-Farid, I.B., Choi, Y.H., Verpoorte, R. (2008). Metal ion-inducing metabolite accumulation in *Brassica rapa*. *Journal of Plant Physiology*, *165*, 1429–1437.

[128] Widarto, H.T., Van der Meijden, E., Lefeber, A.W.M., Erkelens, C., Kim, H.K., Choi, Y.H., Verpoorte, R. (2006). Metabolomic differentiation of *Brassica rapa* following herbivory by different insect instars using two-dimensional nuclear magnetic resonance spectroscopy. *Journal of Chemical Ecology*, *32*, 2417–2428.

[129] Sanchez-Sampedro, A., Kim, H.K., Choi, Y.H., Verpoorte, R., Corchete, P. (2007). Metabolomic alterations in elicitor treated *Silybum marianum* suspension cultures monitored by nuclear magnetic resonance spectroscopy. *Journal of Biotechnology*, *130*, 133–142.

[130] Hedenstrom, M., Wiklund-Lindstrom, S., Oman, T., Lu, F., Gerber, L., Schatz, P., Sundberg, B., Ralph, J. (2009) Identification of lignin and polysaccharide modifications in *Populus* wood by chemometric analysis of 2D NMR spectra from dissolved cell walls. *Molecular Plant*, *2*, 933–942.

[131] Haesler, F., Hagn, A., Frommberger, M., Hertkorn, N., Schmitt-Kopplin, P., Munch, J.C., Schloter, M. (2008). *In vitro* antagonism of an actinobacterial *Kitasatospora* isolate against the plant pathogen *Phytophthora citricola* as elucidated with ultrahigh resolution mass spectrometry. *Journal of Microbiological Methods*, *75*, 188–195.

[132] Cui, W., Iwasa, K., Sugiura, M., Takeuchi, A., Tode, C., Nishiyama, Y., Moriyasu, M., Tokuda, H., Takeda, K. (2007). Biotransformation of phenolic 1-benzyl-*N*-methyltetrahydroisoquinolines in plant cell cultures followed by LC/NMR, LC-MS, and LC/CD. *Journal of Natural Products*, *70*, 1771–1778.

[133] Schneider, B., Holscher, D. (2007). Laser microdissection and cryogenic nuclear magnetic resonance spectroscopy: An alliance for cell type-specific metabolite profiling. *Planta*, *225*, 763–770.

[134] Abdel-Farid, B.B., Kim, H.K., Choi, Y.H., Verpoorte, R. (2007). Metabolic characterization of *Brassica rapa* leaves by NMR spectroscopy. *Journal of Agricultural and Food Chemistry*, *55*, 7936–7943.

[135] Crawford, A., Bohni, N., Maes, J., Kamuhabwa, A., Moshi, M., Esguerra, C., de Witte, P., Wolfender, J.-L. (2010). Zebrafish bioassay-guided microfractionation combined with CapNMR a comprehensive approach for the identification of anti-inflammatory and anti-angiogenic constituents of *Rhynchosia viscosa*. *Planta Medica*, *76*, 1331–1332.

[136] Ko, R.J. (1998). Adulterants in Asian patent medicines. *The New England Journal of Medicine*. *339*, 847.

[137] Politi, M., Zloh, M., Pintado, M.E., Castro, P.M.L., Heinrich, M., Prieto, J.M. (2009). Direct metabolic fingerprinting of commercial herbal tinctures by nuclear magnetic resonance spectroscopy and mass spectrometry. *Phytochemical Analysis*, *20*, 328–334.

[138] Wolfender, J.-L., Glauser, G., Boccard, J., Rudaz, S. (2009). MS-based plant metabolomic approaches for biomarker discovery. *Natural Product Communications*, *4*, 1417–1430.

[139] Wolfender, J.-L., Marti, G., Queiroz, E.F. (2010). Advances in techniques for profiling crude extracts and for the rapid identification of natural products. *Current Organic Chemistry, 14*, 1808–1832.

3

NATURAL PRODUCTS AS LEAD COMPOUNDS IN MEDICINAL CHEMISTRY

Eliezer J. Barreiro, Carlos A. M. Fraga, and Lidia M. Lima

3.1 MEDICINAL CHEMISTRY DEFINITION AND THE IMPORTANCE OF THE LEAD COMPOUND IN DRUG DISCOVERY

The importance of natural products in therapeutics has been generally understood from time immemorial [1]. The amerindians' knowledge of hallucinogenic plants used for their pagan rituals, as well as the aphrodisiac properties of several potions prepared from various vegetable species, have accompanied man for millennia. Throughout the ages, the search for the well-being and for the pleasure has always stimulated man to approach nature, teaching him to make good use of plants and their components. For instance, the possibility of finding new medicines from natural sources is one of the more old man's activities [2]. Currently, represents one of the most mentioned reasons for preserving biodiversity. The use of natural products as medicines found several examples in therapeutics and also has been the source of new original structural patterns that represents an authentic "molecular inspiration" for the design of new drugs by introducing molecular modifications in these authentic natural lead compounds [3]. In this chapter, it will be treated the importance of natural products (NP) as lead-candidate (LC), [4] precursors of new drugs, introducing examples from distinct therapeutic classes. In order to contextualize it some NP, used as medicines, will be include as few examples and by authors choice the more classical ones [5].

Plant Bioactives and Drug Discovery: Principles, Practice, and Perspectives, Fourth Edition.
Edited by Valdir Cechinel-Filho.

Medicinal chemistry is typically interdisciplinary, geared toward streamlining drug action at the molecular level. It was recently defined by the Committee of the International Union of Pure and Applied Chemistry (IUPAC) and published in the *European Journal of Medicinal Chemistry* [6], as follows: "Medicinal Chemistry is a chemistry-based discipline, also involving aspects of biological, medical and pharmaceutical sciences. It is concerned with the invention, discovery, design, identification and preparation of biologically active compounds, the interpretation of their mode of interaction at the molecular level, the construction of their structure–activity relationships, and the study of their metabolism." The interdisciplinarity between chemistry and biology, essential for medicinal chemistry, was brilliantly recognized by Arthur Kornberg, Nobel Laureate in Physiology and Medicine in 1959, when he said at the Congress of the American Association for the Advance of Science, in 1987: "We have the paradox of the two cultures, chemistry and biology, growing further apart even as they discover more common ground. For the chemists, the chemistry of biological systems is either too mundane or too complex ..." [7]. By harmonizing these two sciences, seemingly distinct, are created the necessary and essential conditions to the interaction of medicinal chemists, and pharmacologists, that lead to the discovery or design of new molecular patterns that can be consider as candidates for new drugs. A drug candidate is a lead compound (LC) [8] when it presents pharmacological properties in *in vivo* protocols and when it is expected to bind selectively to the recognition site on the receptor target, to elicit the desired functional response. A lead compound must have adequate bioavailability and biodistribution, which means adequate absorption, distribution, metabolism and elimination (ADME) favorable properties [9], to elicit the desired responses in animals and humans; it must also pass all formal toxicity evaluation in animals. Trials of subacute toxicity involve determining the therapeutic index, and histological investigation of major organs (e.g., liver, kidney, lung, blood, and CNS) prior to preclinical trials being carried out with different members of series congener built. In the rational drug discovery (DD) [10] process, once an LC is discovered, the next step that will occupy the medicinal chemist is optimizing it. For this goal,

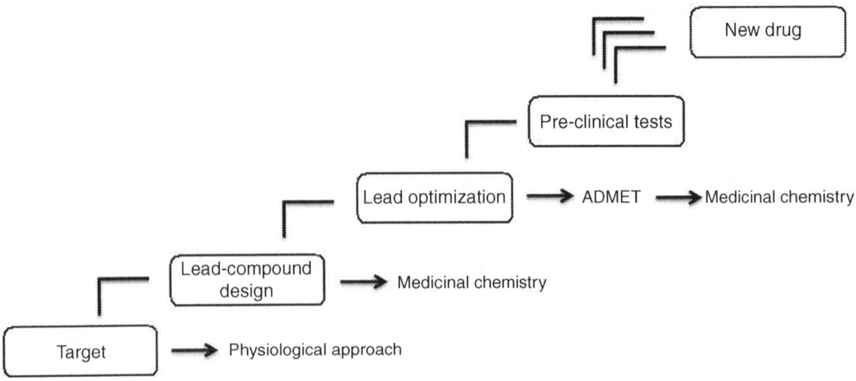

FIGURE 3.1 The lead compound approach in the complex process of rational discovery of new drugs.

several medicinal chemistry molecular modifications strategies as bioisosterism [11] or molecular simplification and others [12] are used; Figure 3.1 illustrates the LC approach of the rational DD process.

It is the first pure compound discovery in a DD research program that presents important properties in *in vivo* assays, indicating to have both pharmacokinetic (PK) and pharmacodynamic (PD) properties representing those of a possible clinical drug candidate. Among all structural properties that one needs to be aware of in the LC, in order to be able to modify it in the molecular level and optimize it, those related to the PK phase, are worth highlighting. Knowledge of your partition coefficient and pK_a, indicating your standard biological stability in different compartments and the transport capacity through membranes, are determining factors in the definition of new modifications to be promoted on its structure. Recognition of functional metabolically vulnerable sites, usually named "soft points," particularly those related to the endothelial hepatic system microsomal crosshairs, dependent on cytochrome P450 (CYP450) [13], where oxidation reactions that are essential to production of various endogenous substances involved in the control of normal body physiology takes place. Therefore, the process of discovering new drugs is complex and obeys a rigid hierarchy of decisions, representing a compelling multidisciplinary challenge as is illustrated, in a summarized form, in Figure 3.2.

Each DD program is unique and setbacks are not rare, making it difficult to accurately predict the duration or costs involved; it is now well accepted that predicted eventual toxicity in the very early stages seems to be important to identify the level of safety of this new drug candidate. Furthermore, modern DD has become a hybrid process, no longer linear as it was considered a few years ago, and efforts are typically made in order to obtain new LCs in shorter times, with fewer expenses [14]. Different modern strategies are employed, usually named hyphen strategies, resulting from the combination of two specific technologies (e.g., HTS-combinatorial chemistry) [15], especially at pharmaceutical industry laboratories or at small high-tech companies. It is necessary to recognize that the

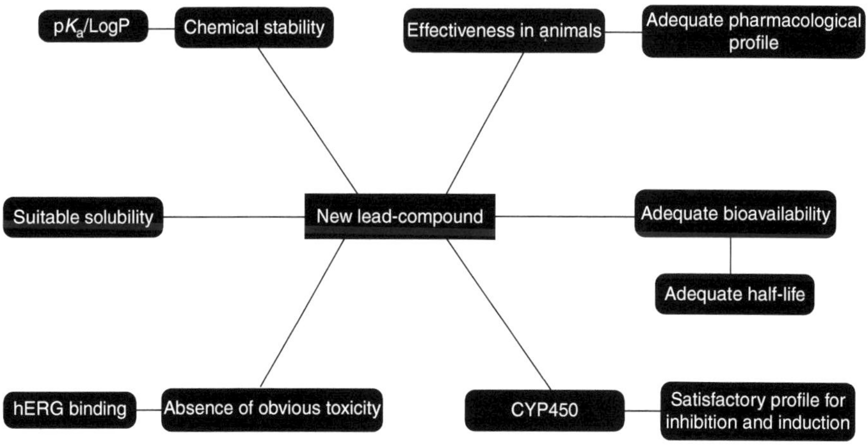

FIGURE 3.2 Structural properties that must be identified in the new lead compound.

DD process represents an important example, maybe even the most important example, of the critical role that the basic research from different health fields of study take on a specific sequence of tasks to – if successfully – culminate in a new drug for the treatment of a human disease. This overall pathway is well structured and starts from the careful selection of the drug target, the identification of a new LC, its structural modification to an optimized, and safe compound when tested in animals, therefore generating an authentic drug candidate for preclinical and clinical testing.

3.2 NATURAL PRODUCTS AS DRUGS [16]

Plants, fungi, insects, marine organisms, and bacteria are important sources of biologically active substances, so much that most of the drugs under clinical trial or usage are either of natural origin or were developed through planned chemical synthesis of natural products [17]. Although there are currently several strategies and methodologies available to synthesize and discover new medicines, the chemistry of natural products is one of the most successful ones throughout history [18]. Many secondary metabolites are special and eminent researchers praise them as valuable raw materials of numerous contemporary medicines [19], stating that the partnership between medicinal and natural products chemists could be strategic for the discovery of innovative medicines [2]. As such, the production of novel medicines, particularly for cancer [20–22], pain [23], malaria [24], heart disease [25a], blood cholesterol control [25b], and several others human disease treatment, represents important examples of this strategy for the discovery of new medicines and will be detailed herein.

3.2.1 Digitalis

The powdered leaves of *Digitalis purpurea* have been used in Western medicine for more than 200 years, with a major active constituent being the cardiac (steroidal) glycoside digitoxin (1) [26]. Although they present a reduced therapeutic index, until today glycosides of *Digitalis* have been employed as cardiotonics, since an adequate substitute has not yet been identified, that is, one that not only shows similar pharmacological effectiveness and potency but also a higher therapeutic index. Digitalis may be considered as "the Decane" of vegetable natural products presenting therapeutic applications [26].

This millenary therapeutically useful class of natural product presents differences because of the nature of the C-17α,β-unsaturated lactone subunit. Cardenolides (1) shows a γ-lactone system, whereas butenolides (2) homologs present a δ-lactone ring [27]. These compounds present the cyclopentane–perhydrophenantrene system, typical of steroid hormones, presenting *cis*-A/B and *cis*-C/D junctions with a different conformational arrangement relative to steroidal hormones, as illustrated in the comparison with progesterone (3), one of the main female hormones.

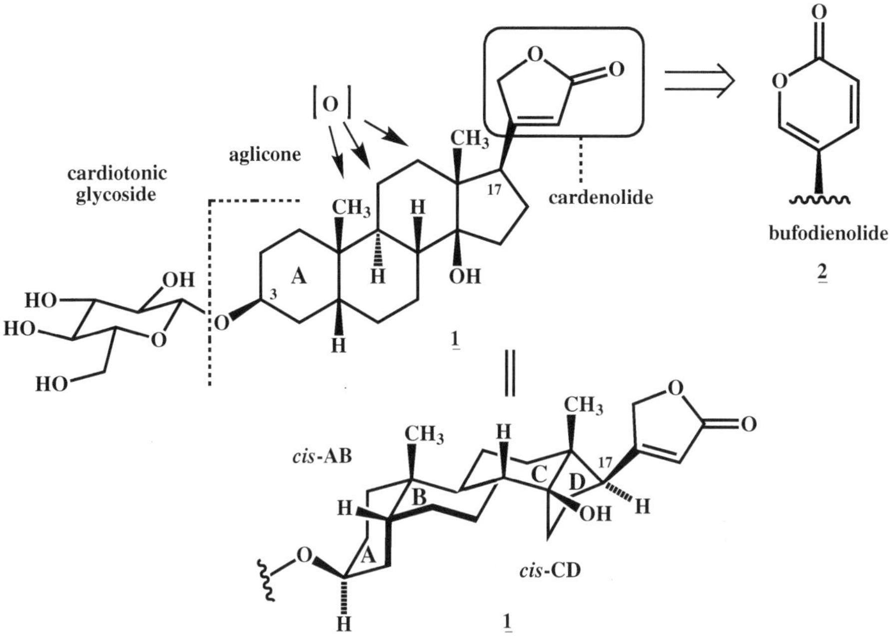

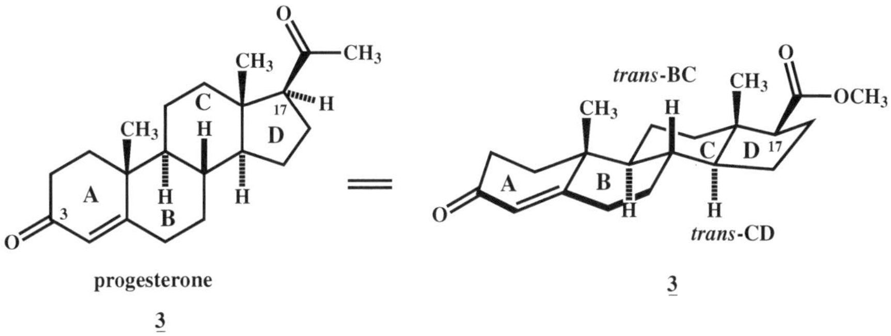

3.2.2 Alkaloids from Plants as Drugs

Alkaloids represent the most important class of bioactive natural products useful to drugs. Distributed by a large number of botanical and others organism, this class of natural compounds represents an important number of drugs or precursors of different therapeutic classes. The pilocarpine (4) [28] found in *Pilocarpus jaborandi*, a Rutaceae widely found in Brazil, is an imidazole alkaloid presenting potent cholinergic properties. It acts on the muscarinic receptors of acetylcholine (mAChR) and is used in ophthalmology for treating glaucoma.

Emetine (5) [29] represents another relevant contribution from the Brazilian florae to therapeutics. It is a tetrahydroisoquinoline alkaloid isolated from *Cephaelis ipecacuanha* and *Camptotheca acuminata* (Rubiaceae) presenting potent amoebicide properties, widely used for treating dysentery. This alkaloid also has important emetic properties, being therefore employed in expectorant preparations. Reserpine (6) [30] is another indole alkaloid, isolated only in 1952 from root of the Indian shrub *Rauwolfia serpentina* with the pentacyclic structure, oxygenated at C-6 of indole moiety, possessing antipsychotic and antihypertensive properties, that has been used for the control of high blood pressure and for the relief of psychotic symptoms. This natural product presents severe side effects, and due the development of better drugs for these purposes, it is rarely used nowadays. Galanthamine (7) [31] is another alkaloid, which was isolated in 1953, from *Galanthus woronowii*, belonging to the Amaryllidaceae family, that presents acetylcholinesterase (AChE) inhibiting properties and was introduced in the therapeutics under the name of Reminyl™ by Janssen Pharmaceutica to control Alzheimer disease (AD).

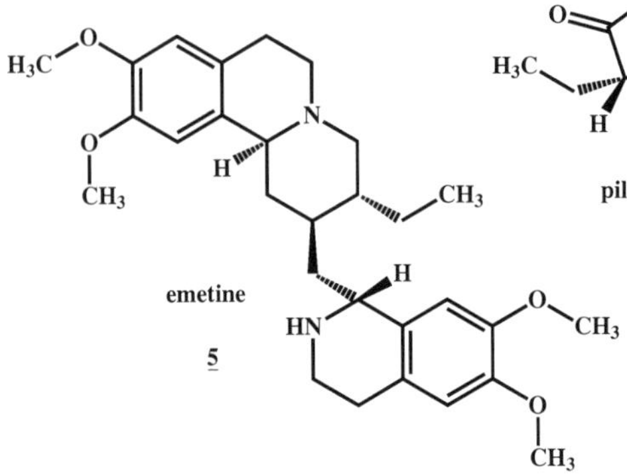

emetine

5

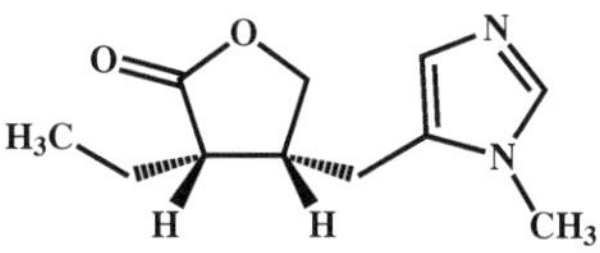

pilocarpine

4

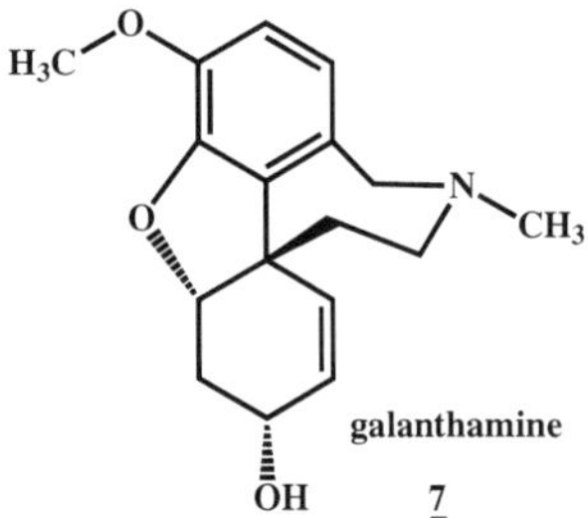

galanthamine

7

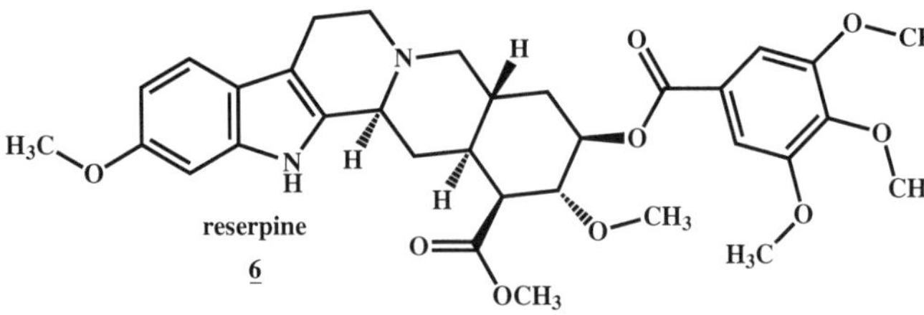

reserpine

6

A large number of other alkaloids compounds could be included here as drug candidates, and readers will find below some representative examples of this class of drug precursors [32].

3.2.3 Penicillin and the Antibiotics Era

Antibiotics, which are real life-saving drugs, have had a unique development process since the discovery of penicillin-G (8). This drug is also one of the most important examples of the role played by chance in the discovery of new drugs. Thanks to a series of fortunate events and to the wit of Sir Alexander Fleming, then Director of the Inoculation Department of St. Mary's Hospital in Paddington, London, the antibiotic action of penicillin-G (8) was recognized around 1930, although it had been discovered in 1896 by French medical student Ernest Duchesne [33]. Probably due to its chemical instability, chemists at the time did not attempt to characterize its structure. There were a few reports in the literature at Fleming's time about the antimicrobial properties of fungi; however, their therapeutic properties were yet to be proven. On September 3, 1928, upon returning from vacation, Fleming had a visitor to his laboratory, when he observed that some culture plates had been forgotten on his counter, instead of being placed in the thermal incubator. These plates were contaminated, showing a reduced growth in the colonies of *Staphylococcus* that had been seeded in the beginning of summer. This fortunate observation led Fleming to treat the contaminated plate with a formaldehyde solution, to preserve it, and store it. The later realization that the contaminant was a fungus of the *Penicillium* genus happened upon consulting with the vaccines department in the same hospital. The observation of the inhibition of the growth of the *Staphylococcus* strains and the recognition of the presence of metabolites of the *Penicillium notatum* fungus with antibacterial properties led Fleming to identify penicillin-G, later having his complete work on the topic published in 1929. Maybe due to the several difficulties in isolating the pure active principle, the therapeutic importance of penicillin was neither recognized by Fleming nor by the members of the Royal Society, who refused, at the time, his application. A few years later, around 1940, the process of isolating penicillin was finally developed by Sir Ernst Chain and Lord Howard Florey, in Oxford, encouraged by the then recent discovery of prontosil, in 1935 [33]. In the following year, with the availability of the isolated substance, the antibacterial properties of penicillin were proven in laboratory animals, showing adverse side effects caused by the presence of impurities. In the same year, Edward Abraham purified penicillin chromatographically and showed its antimicrobial properties through intravenous injection in rats. These results were published and clinical studies with volunteers were started, confirming its instability in the gastric juice and its quick renal elimination. These studies had a decisive contribution to the knowledge of the best form of administration, which allowed usage to begin in February of that year. In the summer of 1943, at the Dyson Perrins laboratories at Oxford University, Sir Robert Robinson et al. elucidated the structure of penicillin-G, making its synthesis feasible.

The production of penicillin through a fermentation route was developed, which allowed for widespread therapeutic usage, approximately 12 years after the identification of its properties in the fungal colonies, validating the old premise by Pasteur: "*La vie empeche la vie*," which created the bases for chemotherapy. The discovery of penicillin-G awarded Sir A. Fleming, Sir Ernst Boris Chain, and Lord Howard Walter Florey the Nobel Prize in Medicine in 1945. Penicillin-G (8) was the first representative of the important class of β-lactam antibiotics, which also includes the cephalosporins (e.g., cefalexin, 9) [34]. Currently this class of chemotherapeutics is among the most sold pharmaceuticals in the world, illustrating its exceptional therapeutic importance.

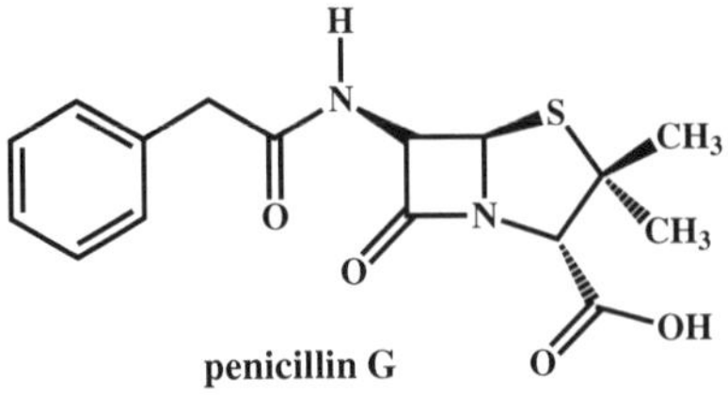

penicillin G

8

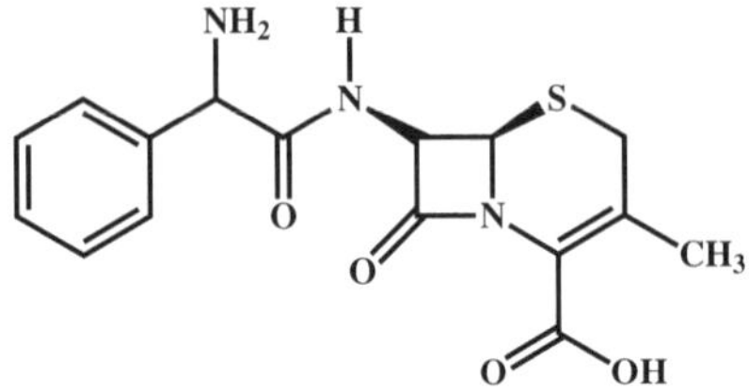

cefalexin

9

The mechanism of action currently accepted for β-lactam antibiotics consists in the inhibition of the *D*-alanine carboxypeptidase (*DD*-Cbase) activity of the microorganism, preventing the insertion of the dipeptide unit *D*-Ala-*D*-Ala (11, Scheme 3.1) in the final stage of the building of its external cellular membrane. The β-lactam unit, which is electrophillically reactive, is essential and is part of forming covalent ligations, probably irreversible ones, with peptides essential to the correct function of the external membrane of microorganisms. The molecular bases of this mechanism were elucidated by Blumberg and Strominger in 1974, who made the structural similarities between penicillin-G (8) and the *D*-alanyl-*D*-alanine (11) termination of the peptideglycan in the membrane of susceptible microorganisms evident [35].

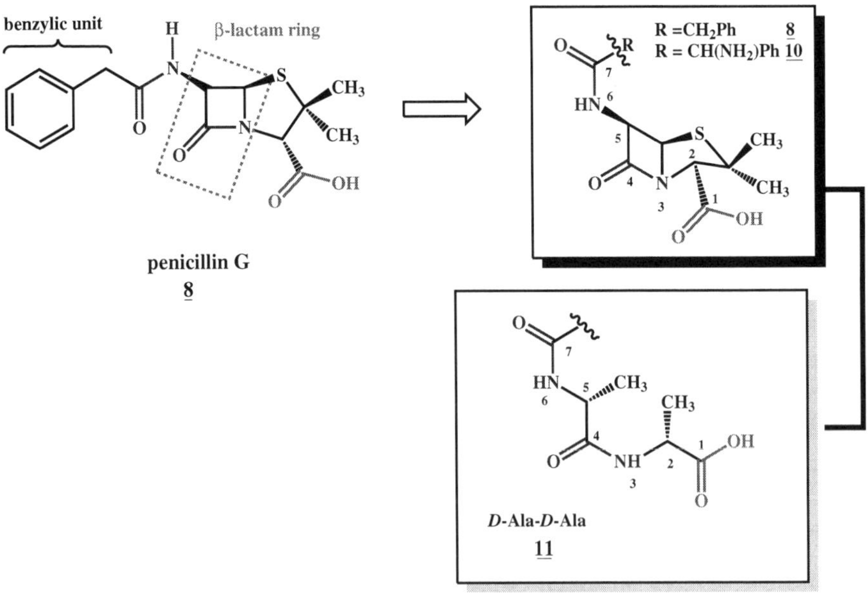

SCHEME 3.1

The structure of penicillin-G (8), containing a masked "dipeptide" unit (C-3/C-8), which includes the β-lactam pharmacophoric group (*PhG*), when compared to the terminal *D*-alanil-*D*-alanine dipeptide unit of the peptideglycan, shows the molecular similarity between them, one that even respects the stereogenic centers of the terminal dipeptide (C-3 and C-6) [35].

Due to its peptide function, first-generation penicillins could not resist the acidity of the intestinal tract, being rendered inactive when taken orally. However, subsequent structural modifications (e.g., amoxilin, 10) or identification of other β-lactam structures in other fungi have led to new generations of β-lactam antibiotics that remain active when taken orally, as illustrated by thienamycin (12) [36], derived from the carbapenem class, and isolated from fungi, where the sulfur atom is no longer part of the thiazolidine nucleus, being replaced with a methylene group, representing an example of natural bioisosterism [11].

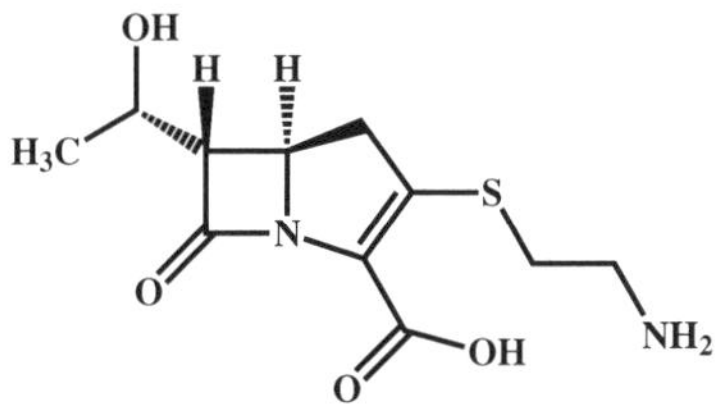

thienamicin

12

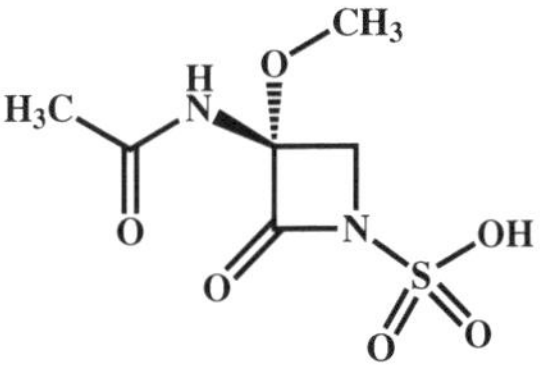

monobactam moiety

13

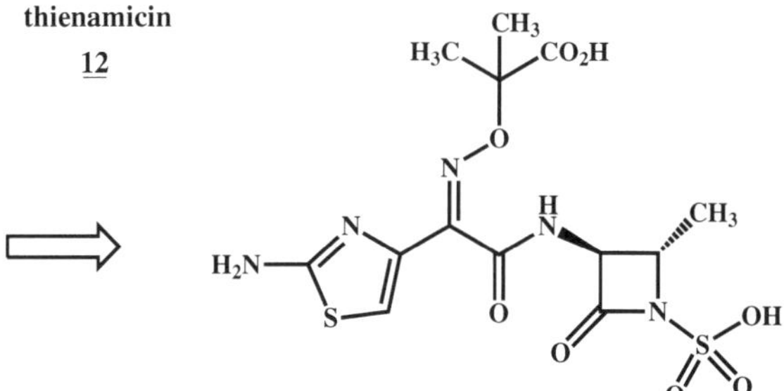

aztreonan

14

The history of penicillin-G (8) is an example of the possibility of "domesticating" the natural, "wild" chemical structure, after identifying the main group(s) responsible for the therapeutic activity, that is, PhG (PhG) [8], so that occasional limitations in the pharmacokinetic stage can be overcome, and new structural characteristics that potentialize its therapeutic efficacy may be introduced. The original structure of penicillin-G (8) has been significantly changed, in a planned way, leading to new generations of powerful synthetic β-lactam antibiotics, of broad action spectrum and active when taken orally, as illustrated by monobactams (13) [37], (14) [37], discovered in 1981, simultaneously in Japan and the United States, within the scope of a planned screening to identify, from natural sources, β-lactam prototypes. This new class of antibiotics, structurally simplified in comparison to the original penicillins, was modified as to introduce the structural characteristics capable of the ensuring activity on strains of microorganisms resistant to penicillin that have the β-lactamase enzyme, a peptidase responsible for the hydrolysis of the β-lactam PhG bioinactivating the antibiotic.

Countless compounds with an azetidine nucleus, made functional by a sulfonic acid group, were obtained, out of which we highlight aztreonam (14) [38] as the first representative of this class to be marketed, active in gram-negative strains that are resistant to penicillin.

Due to the discovery of penicillin, antibiotic therapy began, and it had significant impact in the increase of hope and quality of life for men and women all over the world, and consequently, for the health of humankind. Several other antibiotics, from different chemical classes, discovered since then through random screening, have shown their properties through other action mechanisms, different from those of the β-lactam antibiotics, which are inhibitors of the biosynthesis of the external cellular membrane of microorganisms [39].

3.2.4 Anticancer Drugs

A variety of anticancer agents with different mechanisms of action are natural in origin [40], as the anthracycline derivatives isolated from *Streptomyces* sp., especially daunorubicin (15) and hydroxylated analog doxorubicin (16). Both agents are antineoplastic pharmaceuticals as potent as bleomycin A_2 (17, Blenoxane™), a drug isolated from *Streptomyces werticullus* by Umezawa in 1965, which more recently joined the therapeutic arsenal [41].

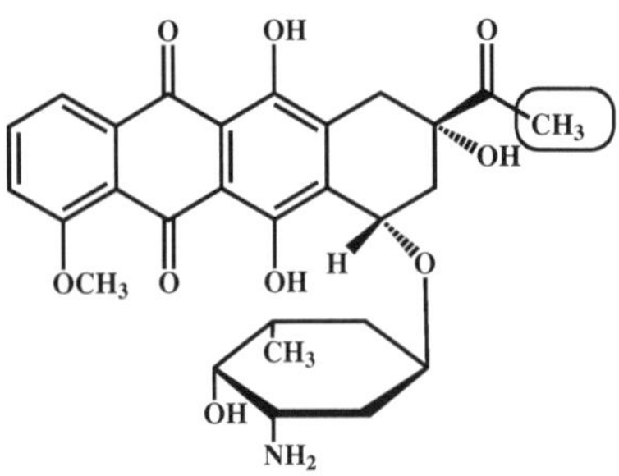

daunorubicin

15

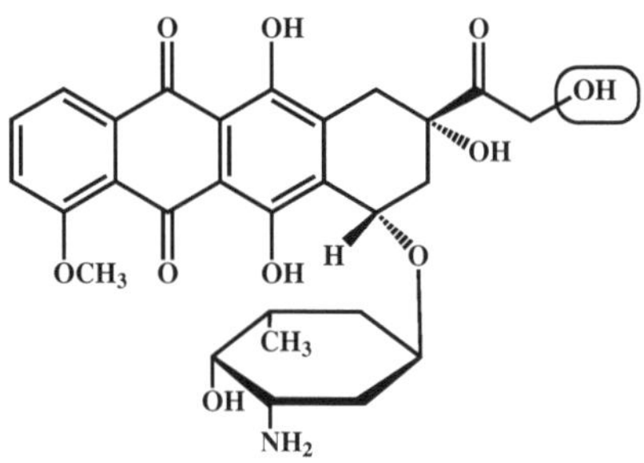

doxorubicin

16

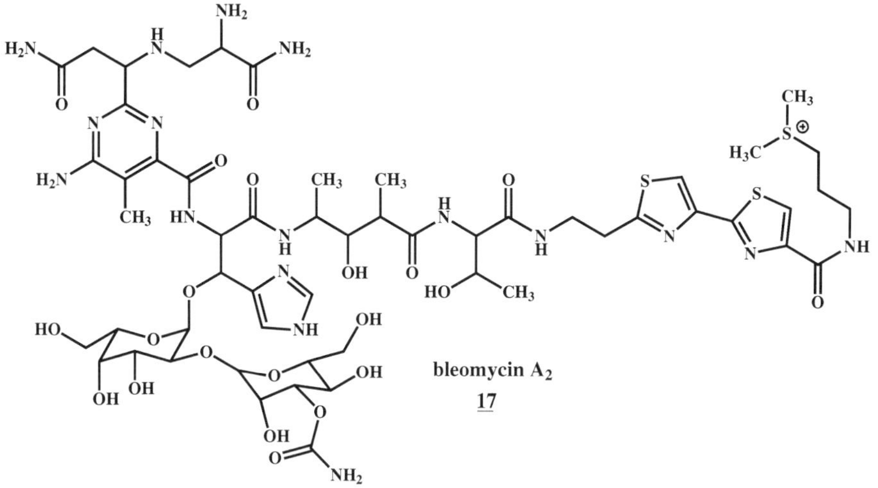

bleomycin A$_2$
<u>17</u>

The mechanism of antineoplastic action of daunorubicin (15) was elucidated, proving that the naphthoquinone unit of this natural product constitutes the main *PhG* to the action of a quinone-oxidase, which leads to the formation of reactive radical species responsible for the irreversible covalent bonds that promote cell growth inhibition [42].

From the plants vincristine (18) and vinblastine (19) [43] – dimeric indole alkaloids – effective pharmaceuticals have been isolated from *Vinca rosea*, widely employed in the treatment of leukemias, especially in children. These pharmaceuticals represent a powerful therapeutic tool for fighting this disease and are produced from natural sources by Eli Lilly Laboratories, which process ca. 8000 kg of flowers of *Vinca* annually to guarantee the necessary quantities for the annual consumption.

The continuous search for agents capable of controlling cancer motivated the National Cancer Institute (NCI) from the United States to promote an extensive research program in order to investigate antitumoral properties in vegetable natural products, launched in the mid-1950s [44]. This program counted on the work of researchers Monroe E. Wall, and Mansukh C. Wani, who supervised the research group from the Laboratory of Natural Products from the *Research Triangle Institute* in North Carolina, USA. These researchers have certainly made the most valuable contribution to cancer chemotherapy by discovering two natural substances introduced in the therapeutics. The first one, named paclitaxel (Taxol™, 20) [45], was isolated from the tree bark of *Taxus brevifolia Nutt.,* and presents a characteristic structure, promoting tubulin polymerization and stabilization of microtubules formed, representing a new pharmacological mechanism of intervention in cellular proliferation. The discovery of this substance and the elucidation of its original structure of mixed biosynthesis, presenting an original tetracyclic terpene skeleton and a side chain containing a residue of rare α-hydroxyaminoacid, as well as its novel mechanism of action, turned paclitaxel into an object of innumerable studies and of a wide range of publications, as illustrated in the book *The Story of Taxol* by Goodman and Walsh [46]. Paclitaxel (20) has inspired innumerable research projects with the aim of discovering a formulation that can be administered orally, since this natural substance is only administered through injection. Furthermore, the discovery of Taxol™ represents a relevant example of the contribution of the basic research to the contemporary therapeutic arsenal. In fact, the discovery of paclitaxel has inspired the discovery of docetaxel (21) [47], a new taxane development by Poitier and Greene in France.

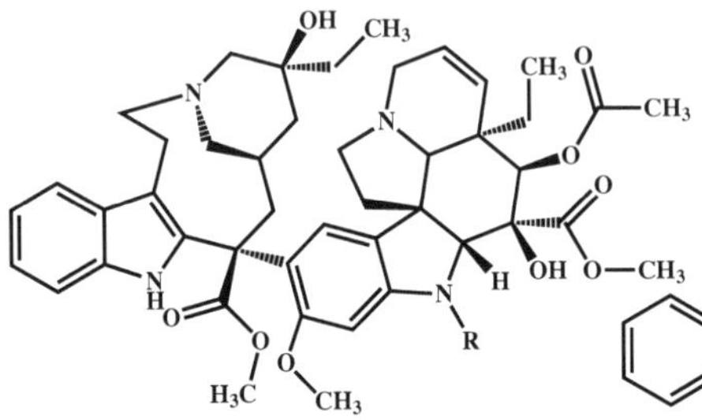

R = H 18 vincristine
R = CHO 19 vimblastine

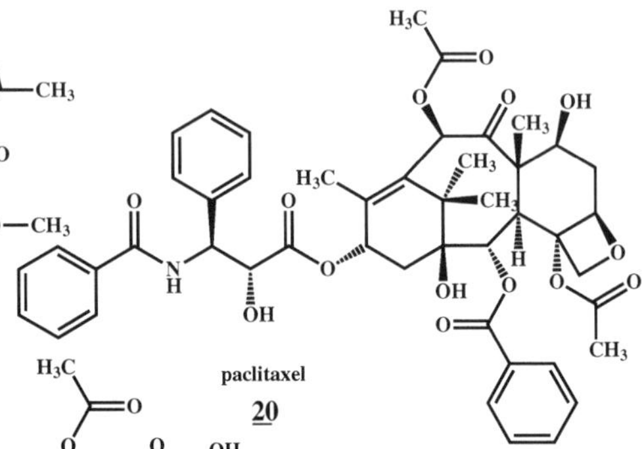

paclitaxel
20

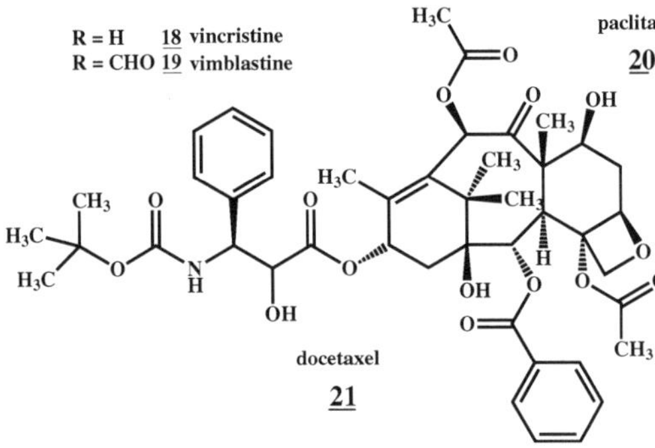

docetaxel
21

Wall and Wani's second contribution is camptothecin (22) [48], isolated from the Chinese tree *C. acuminata*. In contrast to paclitaxel, this alkaloid substance, also of mixed biosynthesis with an end-iridoid represented by the lactone ring, was accidentally discovered in the scope of a research program of the Agricultural Department of the USA. Having started in 1958, this program was originally aimed at discovering new steroidal substances from vegetable extracts capable of representing raw materials suitable for the synthesis of cortisone. In 1960, many of the thousands extracts studied were assayed in adenocarcinomas, resulting in the identification of antitumoral properties of the camptothecin (22), whose structure was elucidated in 1966. Once again, similarly to paclitaxel, the discovery of camptothecin permitted the identification of new mechanisms to control cellular proliferation through the inhibition of the topoisomerase I enzyme, which is therapeutic action target of 22 [49].

Irinotecan (23) [50] is a topoisomerase 1 inhibitor, which prevents DNA from unwinding. Chemically, it is a semisynthetic analog of camptothecin with modifications in the C-6 of the quinoline system, similar to topotecan (24) that has two structural modifications at the same ring [51]. Both synthetic drugs have better PK properties and are mainly used in the treatment of colon cancer, particularly in combination with other chemotherapy agents. Another anticancer drug created from a natural product lead is etoposide (26) [52]. This drug coming from the roots of *Podophyllum hexandrum Royle*, a plant from China, which produces a lignan constituent as podophyllotoxin (PDT, 25), an aryl-tetralin lignan presenting a γ--lactone ring. From this "wild" molecular pattern, inadequate for therapeutic application due to toxicity, structural modifications were then introduced, especially to ring B in C-4, leading to the discovery of new anticancer agents as 26.

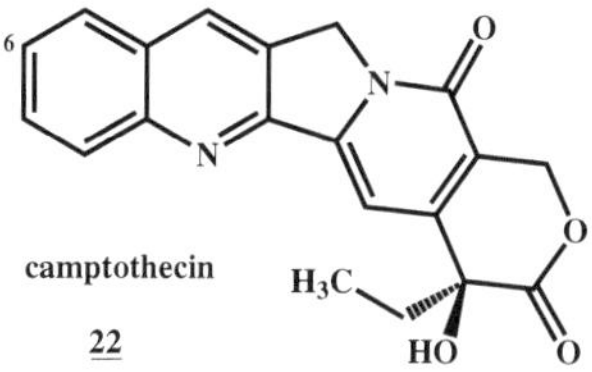

camptothecin

22

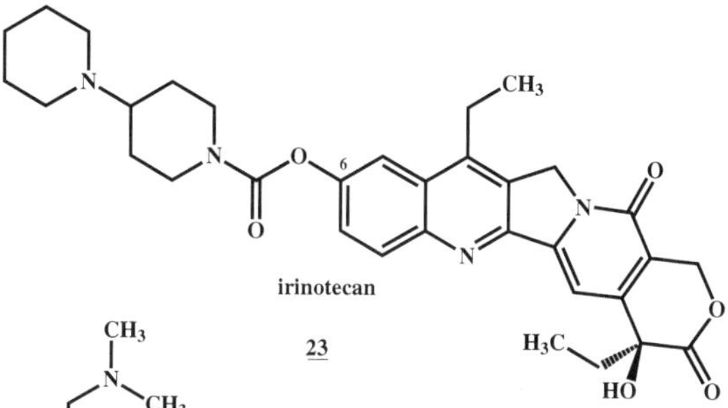

irinotecan

23

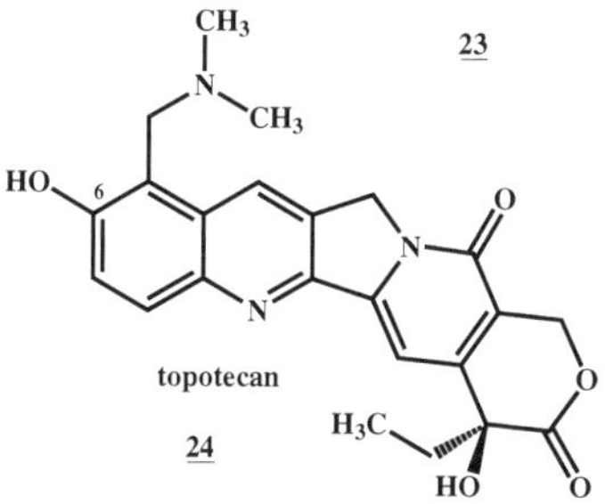

topotecan

24

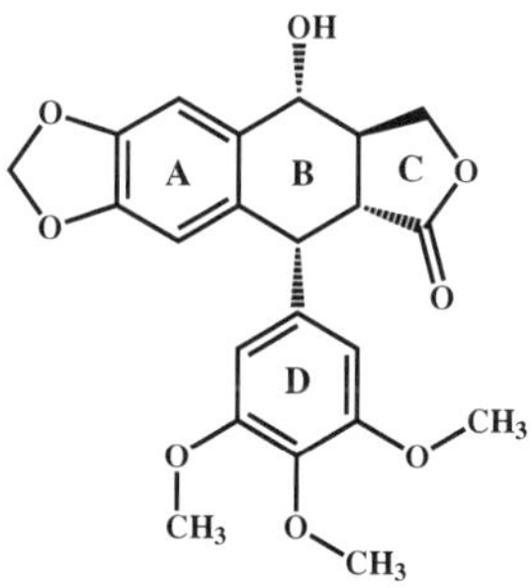

PDT

25

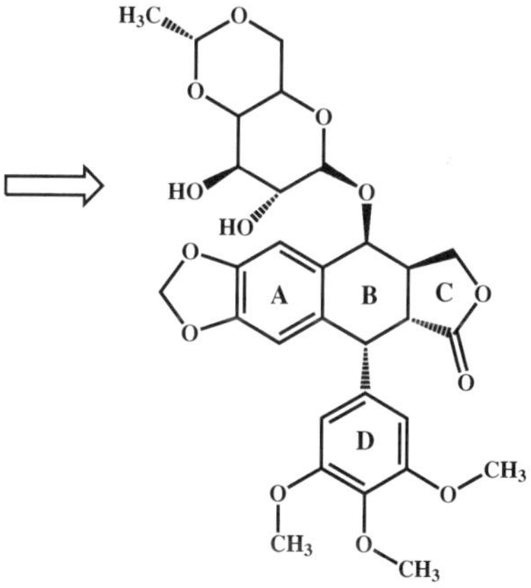

etoposide

26

One of the most therapeutically promising substances of marine origin has a peculiarity in its origin and nomenclature. Ecteinascidin (28) [53] – as it was originally named – also called trabectedin was isolated from *Ecteinascidia turbinata,* an ascidian organism, and described in 1980. The chemical structure was elucidated by Rinehart at the University of Illinois, in 1984 [54], and was initially called ET-743 (28) due to its maximum absorption under UV, and it has a chemotype known as renieramycins in other classes of substances from sponges, such as the one described most recently by Fusetani et al., renieramycin-J (27) [55], which has a bis-tetrahydroquinoline fused system in the pentacyclic structure. The main original characteristic of the ET-743 structure is the presence of a sulfur bridge, a benzo-dioxole ring and a tetrahydroquinoline unit in its multifunctionalized polycyclic structure. This substance is sold by Zeltia and Johnson & Johnson under the brand name Yondelis™ for the treatment of advanced soft tissue sarcoma. It is not naturally abundant, as are many bioactive substances of the marine origin. For this reason, it had its total synthesis described in 2000 [56] by the group of synthetic chemists of the Nobelist Elias James Corey of Harvard University. The company PharmaMar, based in Madrid, Spain, discovered, through the work of researchers Manzanares et al. [57], an elegant hemisynthesis for ET-743, using another natural substance as raw material, one which is more abundant and easier to isolate, cianosafracin B (29), produced in attractive amounts through fermentation with *Pseudomonas fluorescens*. A third total synthesis was described, in 2002, by the Fukuyama group in Japan [58].

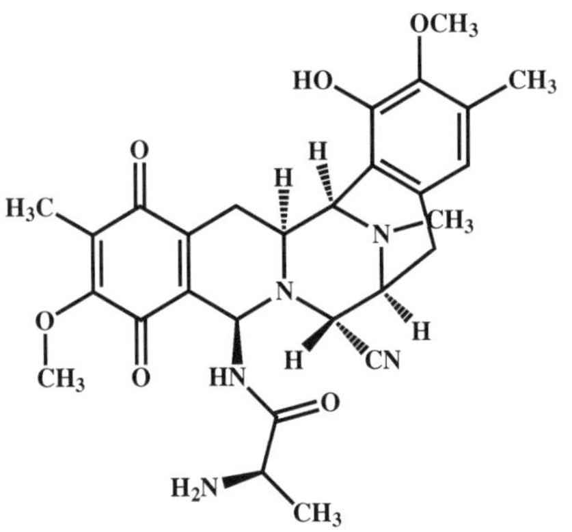

renieramycin-J

27

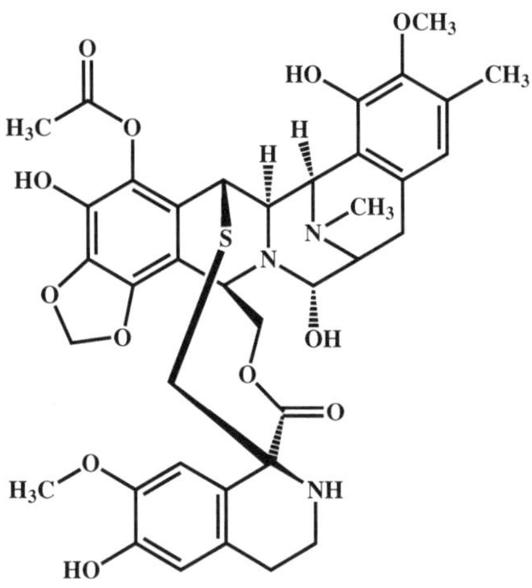

ecteinascidin or ET-743

28

The structural complexity of ET-743 (28) combined with its remarkable biological activities and the lack of availability from natural sources has motivated the study of simplified synthetic or hemisynthetic analogs, such as phthalascidin (Pt 650-2; 30) that has a phthalimide unit, described as a structurally simplified analog with properties similar to those of the original natural product (28) [59].

Initial studies of the structure activity relationship (SAR) of 28 were conducted from the synthesis of countless modified derivates as described by Liu and co-workers, in 2006 [60], which synthesized new analogs, simpler than the natural product by

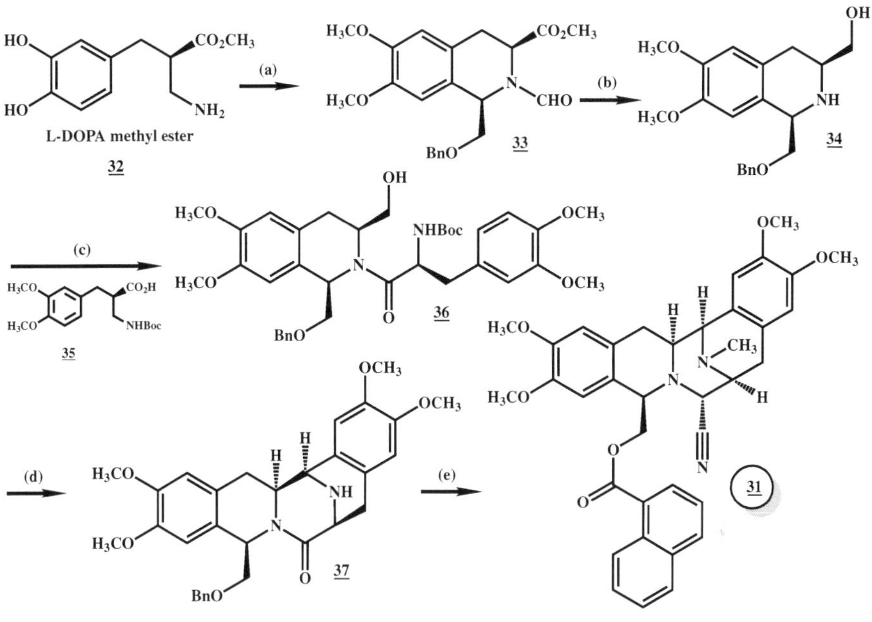

SCHEME 3.2

replacing the benzodioxole moiety of ET-743 by a cathechol dimethylether framework, coming from *L*-DOPA used as synthetic starting material by the route depicted in Scheme 3.2. The authors introduce in the southern part of the natural molecule several different aryl isostere groups [11], leading to the especially active naphthyl analog (31, Scheme 3.3), which presented important antitumoral properties [61].

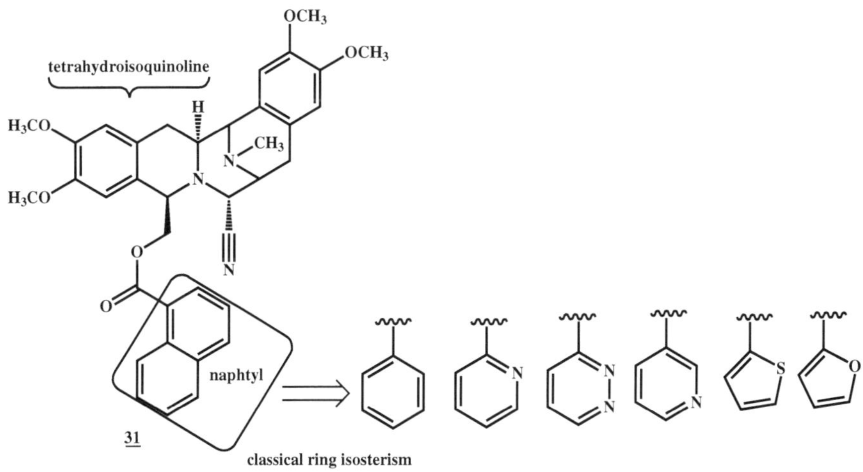

SCHEME 3.3

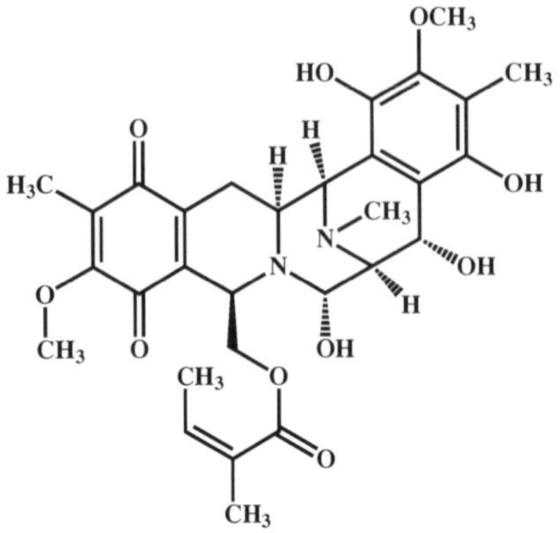

cianosafracin B

29

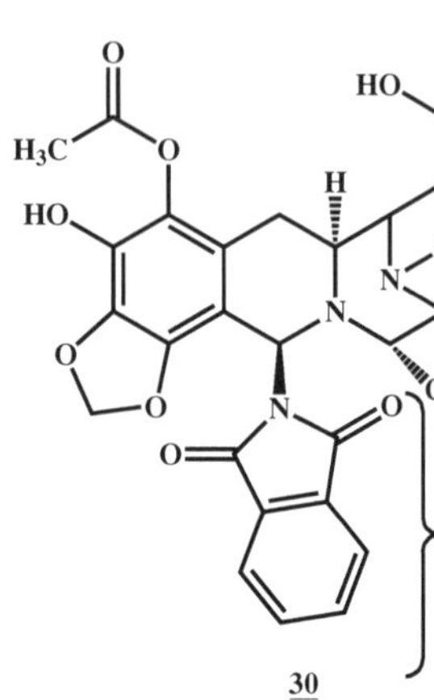

30

3.2.5 The Recent Discovery of Ziconotide as a Novel Analgesic Drug [61–65]

Ziconotide (SNX-111; Prialt ™) [61–65] is a peptide painkiller with 25 amino acids, isolated from the *Conus magus* shellfish (ω-conotoxin), that has been studied for many years, and it is estimated that each species contains up to 200 different biologically active peptides that bind in ionic channels of voltage dependent or activated by external ligands. This analgesic drug is prescribed for the mitigation of severe and chronic pain acting as a nonopioid and non-NSAID analgesic agent blocking the calcium channel of n-type at picomoles concentrations. Ziconotide is thousand times more potent than morphine and acts through distinct mechanisms by changing the sensitivity of the fiber c-type. In contrast to opioid agonists, which are bound to protein G receivers and that lead to a decrease in the number of receptors that continues after treatment and result in its desensitization, treatment with ziconotide does not induce tolerance after chronic use.

3.3 NATURAL PRODUCTS AS LEAD COMPOUND FOR NEW MEDICINES DISCOVERY

As previewed above, different chemical classes of natural products originated a variety of pharmaceuticals belonging to distinct therapeutic categories, as exemplified by the alkaloids: morphine (38) and quinine (41) [1].

3.3.1 From Morphine (38) to Synthetic Hypnoanalgesic Drugs

Isolated in 1804 by German pharmacist Friedrich Wilhelm Adam, morphine (38) inspired the later discovery of 4-phenyl-piperidine [66] as a new class of hypnoanalgesics. This molecular simplification has enabled subsequent molecular optimizations [67] and better understanding of the action mechanism of this important natural and millenary analgesic alkaloid originally isolated from *Papaver somniferum*. The natural alkaloid was the model for medicinal chemists that made several structural modifications from its functional groups. Through the presence of an allyl alcohol in the structure (ring C), numerous chemical degradation products were obtained under acidic conditions, yielding less functionalized compounds that preserve the piperidine ring that presented the analgesic activity. Among these, new simplified analogs were meperidine derivate (39), a component of 4-phenylpiperidine hypnoanalgesic drugs [68].

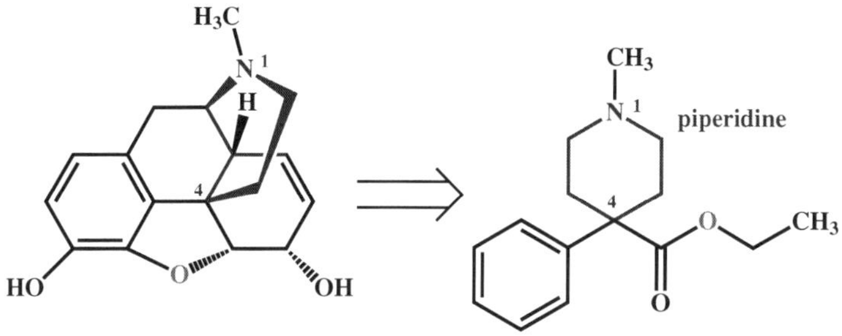

morphine

38

meperidine

39

This drug, discovered in 1939, is considered a pioneer of the 4-phenylpiperidine hypnoanalgesic class, with reduced tolerance-inducing properties when compared to natural morphine. Hydrochloride salt began to be used in therapy in 1947, as a drug to treat severe pain, being used through injection, like morphine. The vulnerability of ester at C-4 of this structure to plasma unspecific esterases leads to the formation of corresponding acid *N*-methyl-4-carboxylic phenylpiperidine (39a) [69], which presents lower lipophilicity ($C \log P = 2.03$) and is not able to permeate the blood–brain barrier (BBB) (Scheme 3.4). Bioformation in plasma of this metabolite reduces the analgesic properties, and new molecular modifications were performed with the intent to prevent or eliminate this metabolic vulnerability. Thus, new ketone derivative (40, Scheme 3.4) was synthesized and introduced in therapy as a potent hypnoanalgesic drug with an important half-life due to its resistance to plasma esterases, representing the second generation of hypnoanalgesic drugs obtained from morphine [70].

3.3.2 Alkaloids as Source of Antimalarial Drugs

Another classic example is coming from indolic alkaloid quinine (41) [71], isolated from *Cinchona officinalis*. This natural product has antimalarial properties and has inspired the discovery of synthetic 4-amino and 8-aminoquinoline derivatives, represented by chloroquine (42) and primaquine (43), respectively. Some years later, in the Walter Reed Institute of US Army, the synthetic antimalarial drug mefloquine

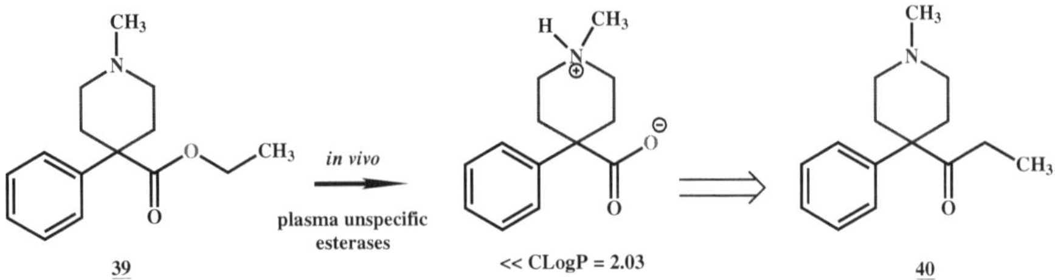

39 39a 40

SCHEME 3.4

(44) was developed, illustrating the molecular simplification strategy in the drug design (Scheme 3.5) [72].

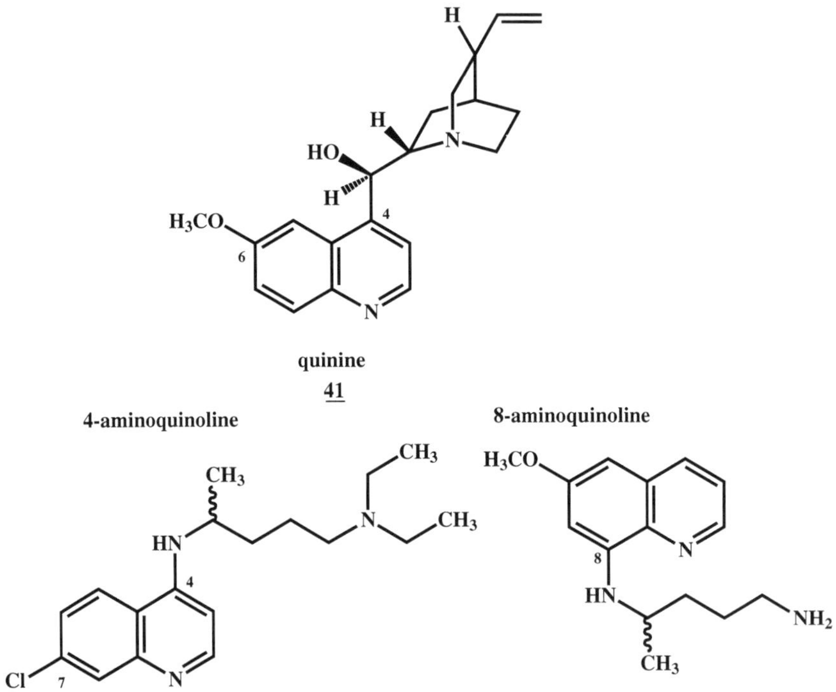

quinine
41

chloroquine
42

primaquine
43

3.3.3 Ganglionary Blockers: The Drug Class of Amazon Natives

Curare, an arrow poison used by Amazon natives, composed of raw, crude extract from the roots and peels of a variety of plants from the region of the Amazon and Orinoco rivers, used for hunting and fishing, presents (+)- tubocurarine (45)- a

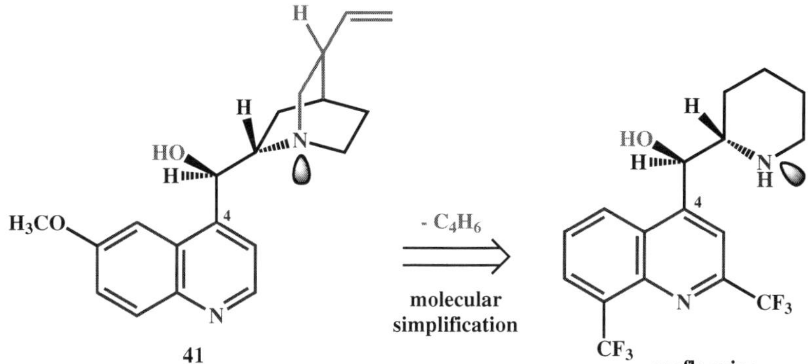

41

mefloquine
44

SCHEME 3.5

bisbenzyltetrahydroquinole alkaloid complex, which has a water-soluble quaternary ammonium salt as its main active principle. This substance, present in up to 4% of the extract of *Chondrodendron tomentosum* (Menispermaceae), is responsible for the paralyzing, lethal properties of curare, especially due to its action on the nicotinic receptors of acetylcholine in the muscles [73].

Tubocurarine (45) inspired the appearance of the therapeutic class of ganglionary blockers, muscle relaxants useful for intestinal surgeries, of which hexamethonium (46), succinylcholine (47), and pancuronium (48) [74] are main examples, which also represent a significant molecular simplification of the lead-natural product. It is interesting to note that curare was studied at Institute Pasteur, Paris, France, in the first medicinal chemistry laboratory, created and directed by Ernest Fourneau.

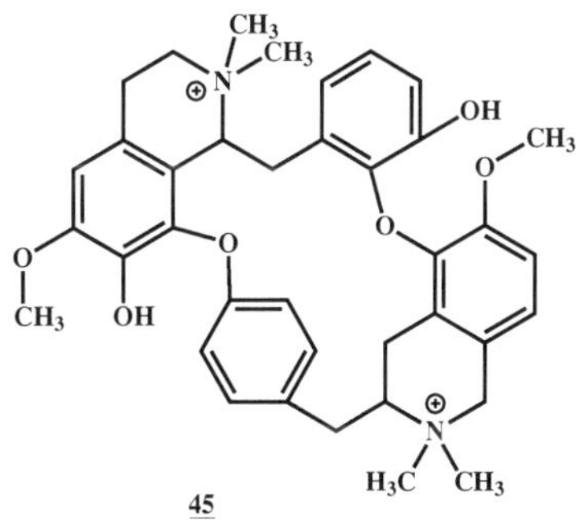

45

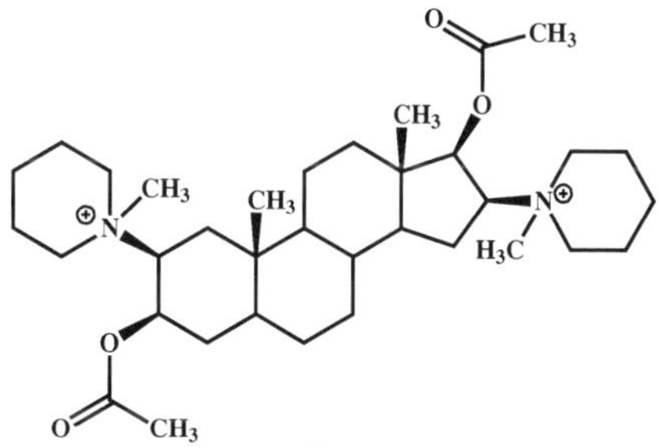

48

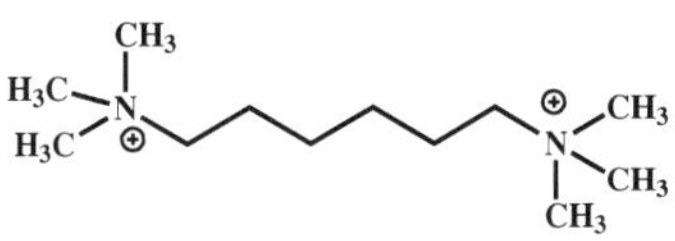

46

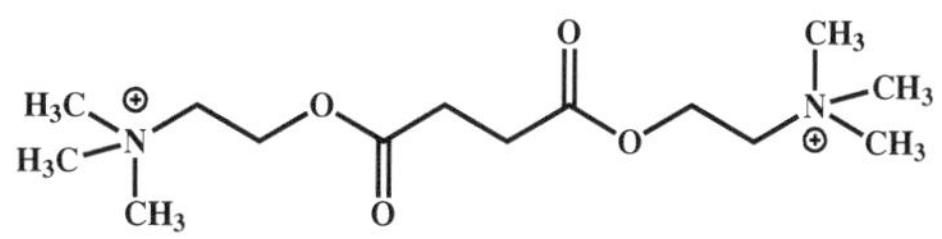

47

3.3.4 Antiviral Drugs from the Sea

Zidovudine (AZT, 49), an important chemotherapeutical resource available for the treatment of acquired immune deficiency syndrome (AIDS), was discovered through properties identified in nucleosides isolated from sea algae [75]. Burroughs-Welcome laboratories (currently Glaxo Smith-Kline, GSK) were developing, at the time, research projects to find new antiviral agents, based on the premise that for the nucleosides to be biologically active; the osidic radical would not necessarily have to be ribose or deoxyribose, as the results achieved by the cytarabine derivate had suggested (Ara-C, 50) a compound with antileukemic properties released by Upjohn (later Pharmacia and currently GSK). Through changes introduced in the osidic subunit, Burroughs-Welcome researchers discovered AZT (49), exploring natural antiviral drugs from marine sources and modified ones, aiming at treatments for the *Herpes* virus in particular. AZT was obtained and assayed, but it did not prove effective against this type of virus. With the discovery of HIV by Gallo and Montagner in the US and France, respectively, this substance was reassessed and it was shown to be active at the viral reverse-transcriptase (RT) enzyme, becoming the first drug used in the chemotherapy of retrovirus infections [76].

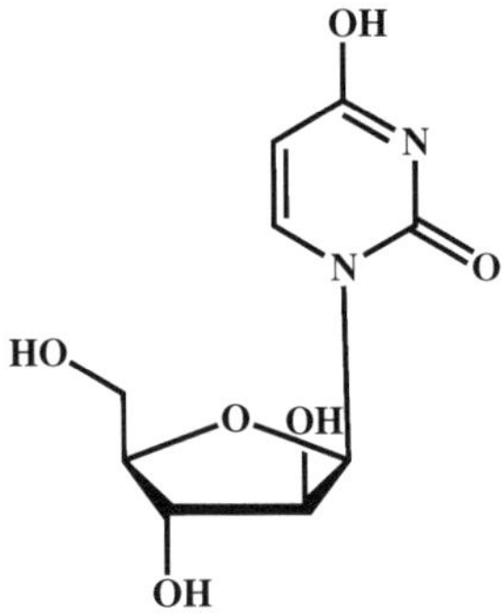

Ara-C

50

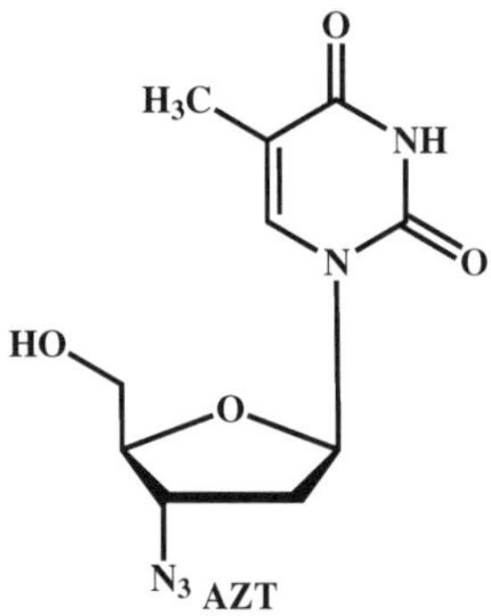

AZT

49

3.3.5 From Prostaglandins Derived from Caribbean Corals to Misoprostol

Plexaura homomalla corals, abundant in the Caribbean sea, have supplied significant amounts of prostaglandin A_2 (PGA_2) (51), which were isolated by Spraggins and Weinheimer, and have become an important raw material for the hemisynthesis of PGE_2 (52) and $PGF_{2\alpha}$ (53) depicted in Scheme 3.6, considered the primary PGs of most physiological importance in mammals [77]. Due to this abundant natural source, hemisynthetic studies have made it feasible to achieve enough primary PGs necessary to the physiological studies of the role of PGs, which have shown the important physiological properties of these powerful cellular mediators. These studies were useful to obtain new modified analogs of different class of PG's with therapeutic

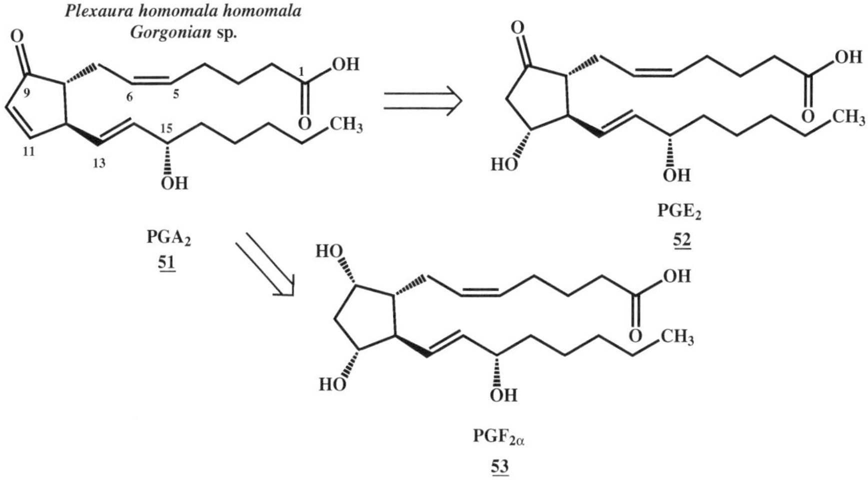

SCHEME 3.6

interest like misoprostol (54; Cytotec™) [78], an important drug for the treatment of gastric ulcers acting as agonist of EP receptors at gastrointestinal treat. Another drug derived from PGs is gemeprost (55) [79], a modified prostanoid derivate, structurally related to PGE_1 with abortifactant properties. It is interesting to note that PGE_2*per se* is used for erectile dysfunction and due to the excessive metabolic lability of the allylic alcohol subunit to prostaglandin dehydrogenase (PGDH), it needs to be administered by intrapenile injection [80].

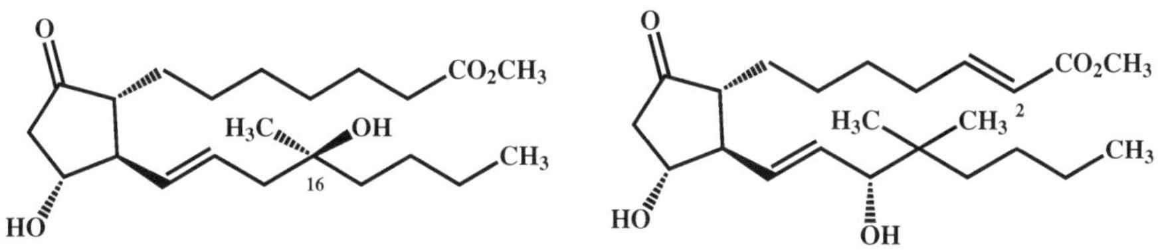

misoprostol 54 gemeprost 55

3.3.6 Antiobesity Drugs Derived from Natural Products

Tetrahydrolipstatin (57; Xenical™) is another important and innovative drug, launched in 1999 by Hoffmann–LaRoche, which originates from modifications to a natural product [81]. This substance resulted from a fungi natural product (*Streptomyces toxytricini NR0619*), lipstatin (56) [82]. Through a simple saturation of both *cis*-double bonds present in the natural prototype, lipstatin had its structure modified originating the tetrahydrogenated named orlistat, usually prescribed for the treatment of obesity as a inhibitor of pancreatic lipases with an IC_{50} of 1,2 μM modifying therefore the intestinal lipid absorption pattern. This drug has in his structure a β-acyl-*N*-formylamine group and a four-member lactonic ring (β-lactone), subunits rarely present in drug molecules. As its main adverse effect, orlistat (57) has the inconvenient of provoking distinct degrees of diarrhea in some patients.

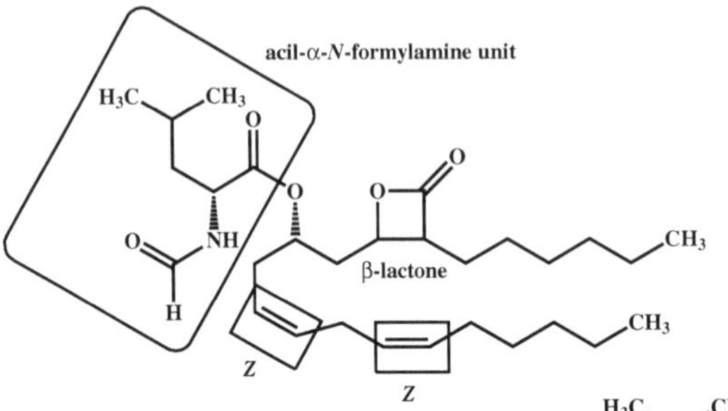

liptastin
56

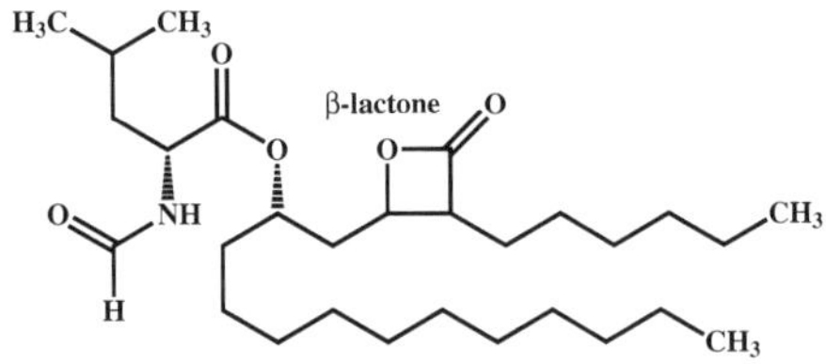

tetrahydroliptastin
57

3.3.7 The NP Inspiration to the Modern Synthetic Anticancer Class

Epothilones (58), (59) [83], identified in fermentation broth of mycobacterium *Sorangium cellulosum,* has demonstrated important antitumoral properties in cellular growth inhibition assays. These properties were later identified as resulting from the same mechanism exhibited by paclitaxel (20), that is, acting in microtubules. These substances proved to be active in models of cellular growth inhibition resistant to paclitaxel, representing an important alternative for the treatment of cancer. Four representatives from these 16-member macrolides illustrated by epothilones-A (58) and B (59) are known and they differ by the presence of a methyl radical in the macrocyclic ring at the bottom of the epoxide in 58, which is absent in 59. Other structural variations identified in this class of macrolides refer to the oxidation degree of the methyl bound to the tiazole system that occurs as hydroxymethylene [84] in some representatives [85].

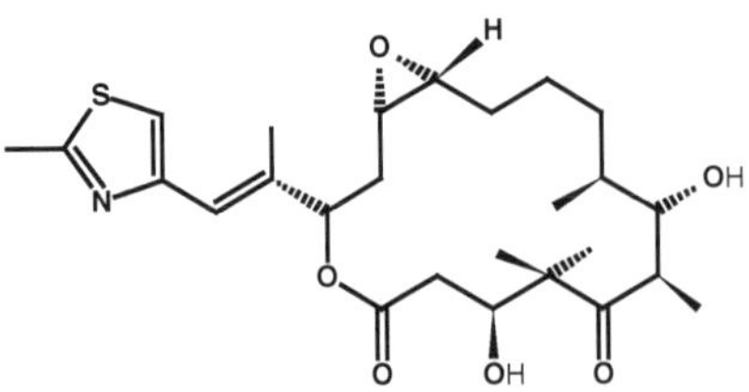

epothilone A
58

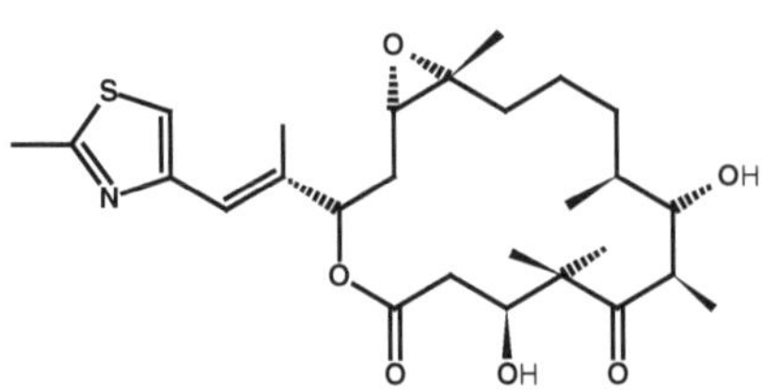

epothilone B
59

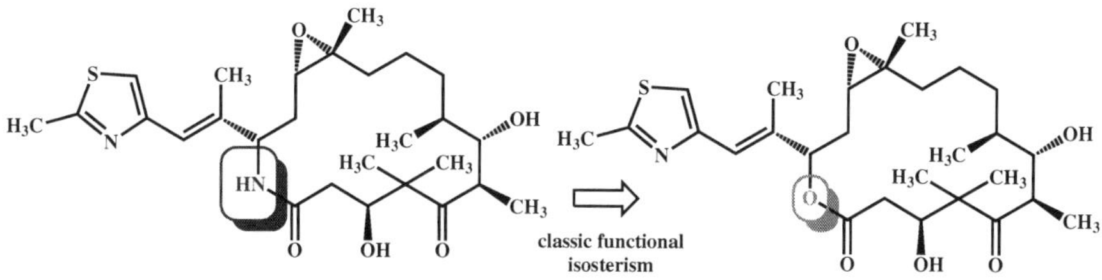

Ixabepilone
60

59

SCHEME 3.7

Ixabepilone, also known as aza-epothilone B (60, BMS-247550, Ixempra™) [86,87] is a semisynthetic cytotoxic macrolide drug with potent anticancer action that was launched in 2007, by Bristol–Myers Squibb (BMS), as the first representative of the epothilone class of natural products in the pharmaceutical market. The 60, which was designed from epothilone B (59) isolated from the bacteria *S. cellulosum* through classic isosteric [11] functional replacement of cyclic oxygen atom of the macrolactone ring by a nitrogen amide function creating the macrolactam ring of 60 (Scheme 3.7). Although the natural product was quite powerful in *in vitro* tests, including acting on taxanes resistant cell lines, it presented only modest activity *in vivo* due to its metabolic liability resulting in adverse PK properties. The replacement of the ester function of epothilone B (59) by grouping the amide group produced ixabepilone (60), which turns more than twenty lines of cancerous cells with $IC_{50} = 1.4 - 34.5$ nM, proving to be a very useful parenteral drug indicated for the treatment of breast cancer.

3.3.8 Discovery of Statins: The Top-Class of Drug in the World Market from Fungi

In 1976, Akira Endo, a Japanese researcher, working for Sankyo Co. Laboratories, isolated polyketide ML236B derivative from fungus *Penicillium citrinum*, naming it compactin, also known as mevastatin (61) [88]. A while later, the same substance was identified in another species of *Penicillium*, that is, *P. brevicompactum* by Brown et al. at Beecham Pharmaceuticals, in England [89]. As it presents the δ-lactone-β-hydroxylate function, which, in its acyclic form, is capable of mimetizing the intermediate involved in the reduction of HMGCo-A (62) promoted by HMG-CoA reductase (HMGCo-AR; Scheme 3.8), compactin (61) was assayed based on this molecular similarity (Scheme 3.9) [90], as a candidate for this enzyme inhibitor. In fact, this natural fungal metabolite presented higher affinity for the enzyme than the natural substract itself, with $K_i = 1.4$ nM, and it was shown to be a competitive inhibitor of HMGCo-AR [91].

The identification of this natural inhibitor motivated several research laboratories to dedicate their efforts to the search of other active derivatives on HMGCo-AR, leading to the discovery of a new class of antilipemic agents – statins – [92] that represent an authentic therapeutic innovation and will be further discussed.

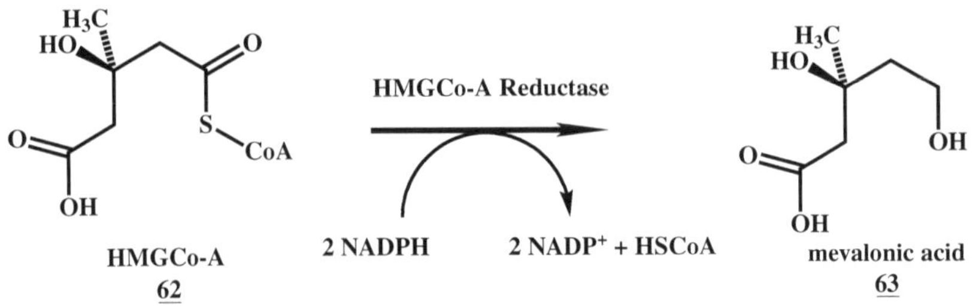

SCHEME 3.8

The relationship between the pathogenic form of blood cholesterol index and the risk of heart accidents is well established, and the employment of cholesterol biosynthesis inhibitors in this treatment will be therapeutically useful [93]. The biosynthesis of cholesterol (64) is a complex process that involves approximately 26 steps and can be interrupted by inhibition of hidroximetilglutaril Co-reductase (HMG-CoA) [94]. This enzyme substrate biocatalyzes reduction for mevalonic acid (63) involving the NADPH (Scheme 3.8). Its inhibition does not lead to toxic effects resulting from the accumulation of its substrate, in contrast with the inhibition of cholesterol biosynthesis in later stages of the biosynthesis pathway, where harmful effects result from the accumulation of desmosterol intermediate (65).

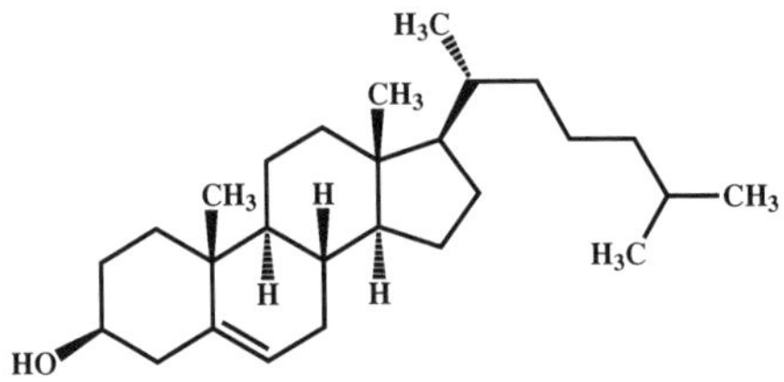

cholesterol

64

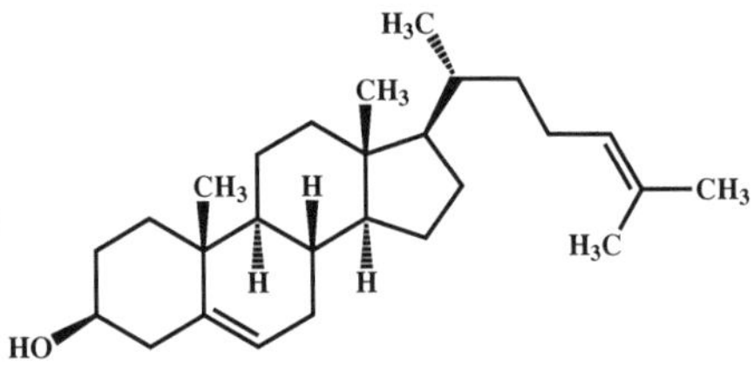

desmosterol

65

Endo results have encouraged several studies in this area and, in 1987, researchers at Merck Co., under leadership of Arthur A. Patchett, identified lovastatin (also known as mevilonin; 66) [95], isolated from *Aspergillus terreus*. This substance has a δ-lactone system that corresponds to the cyclic product formed in reduction of HMG-CoA reductase, being an authentic natural prodrug. In 1987, lovastatin (66) was marketed under the name of Mevacor™. Other analogs were described by the same company as simvastatin (67; Zoccor™), a derivate of 2′-dimethyl mevilonin was introduced in the American pharmaceutical market, in 1992. Simvastatin (67) has seven stereogenic centers, five of them being in the decalinic central system and two additional ones at δ-lactone ring. Considering these structural factors, the lactone ring could be considered as pro-pharmacophore to produce *in vivo* by lactone ring open (Scheme 3.10), the main *PhG* of this class of drugs involved in the molecular recognition of inhibitors by HMG-CoA reductase.

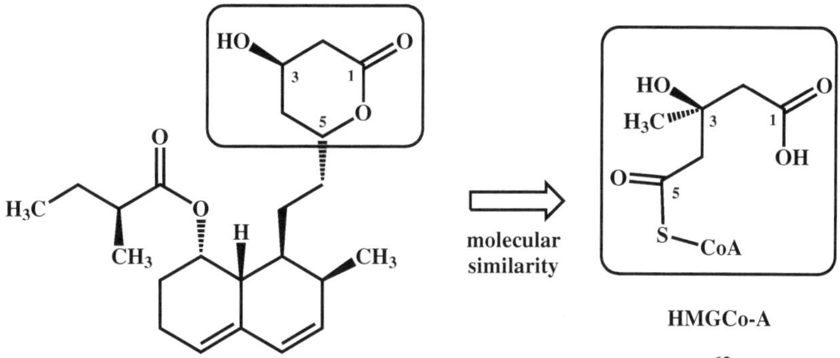

compactin or mevastatin
61

HMGCo-A
62

SCHEME 3.9

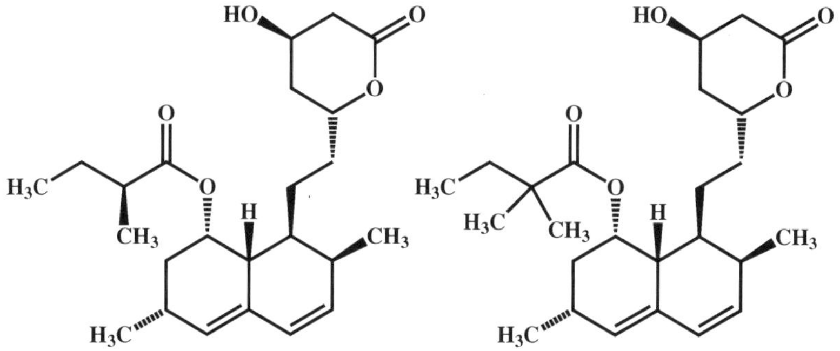

mevilonin or lovastatin
66

simvastatin
67

A second generation of statins, structurally distinct and simplified when compared to the prototype 67, with five fewer stereogenic centers was developed [96]. This next

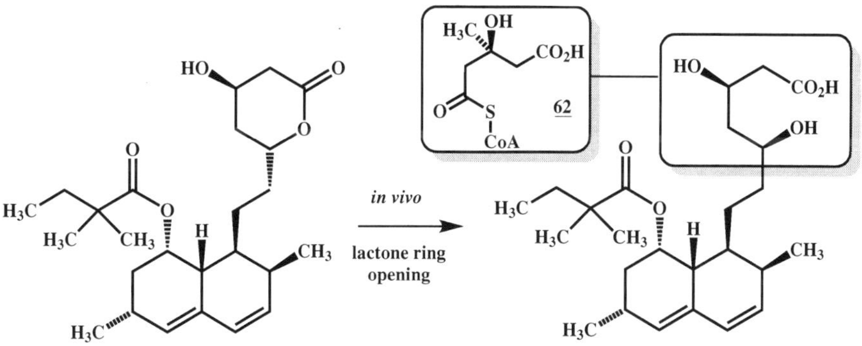

SCHEME 3.10

generation of statins has a diphenylpyridine aromatic scaffold linked to the *PhG* instead of the chiral decalinic central system (Scheme 3.11), and an additional unsaturated "spacer" with an *E*-configuration, in order to restrict the excessive conformational flexibility in the Northern part of the molecule, resulting in the second generation of selective inhibitors of HMG-CoA reductase [97]. Later, Bruce Roth and coworkers at Parke-Davis laboratories drew up a new class of HMG-CoA reductase inhibitors, containing both the cyclic central aromatic system as well as the isosteric 1,2-diphenylpyrrole moiety and the lactone-*PhG* as an open acid, creating atorvastatin, (68) the number one top-selling drug in the world for the past *ca.* 5 years with *ca.* US$ 10bi/year [98].

This class of antilipemic drugs has also rosuvastatin (69) as an important member [99], this drug shows additional advantages when compared to the other members of the class available on the market as, for example, the property of reducing the concentration of plasma lipoproteins of low density, at a lowest dose and more quickly, allowing a better fit to the dosage.

In 2008, Pfizer researchers in Ann Arbor, USA [100], identified antilipemic properties in sodium salts of pyrazolic isosteres (70), where the nitrogen atom of the central ring of terphenylic classical system was replaced by a pyrazole ring. Recently, Nissan Chemical Industries, Ltd. and Kowa Pharmaceuticals, Japan companies discovered pitavastatin (71), which was launched in US in 2009

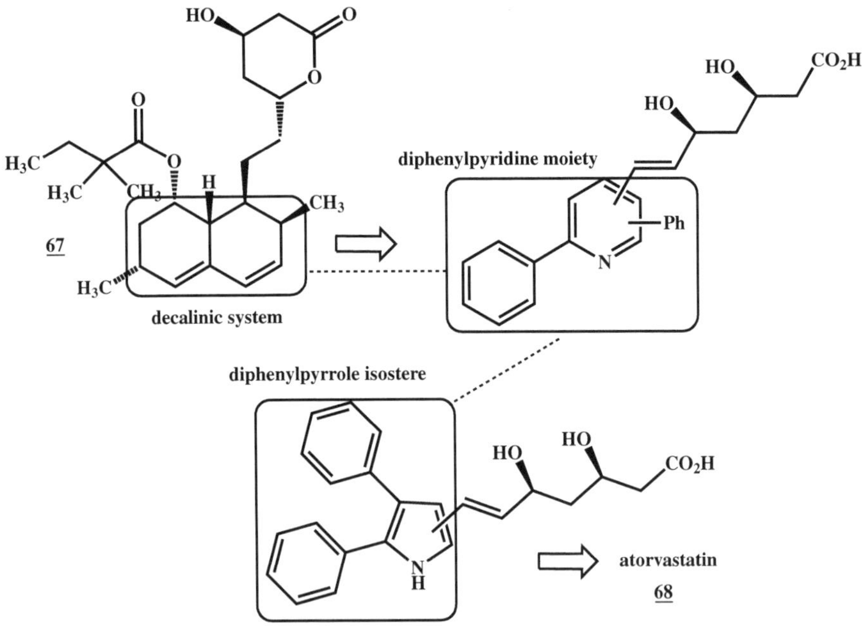

SCHEME 3.11

(Livalo™) represents the most recent statin in the pharmaceutical market. This compound presents a greater bioavailability than all others statins in the market.

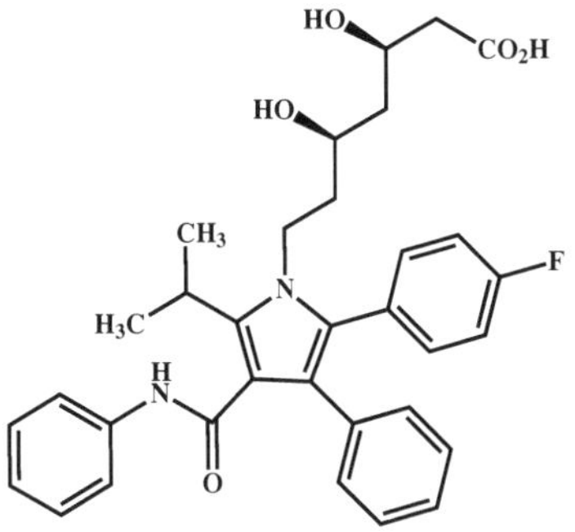

atorvastatin
68

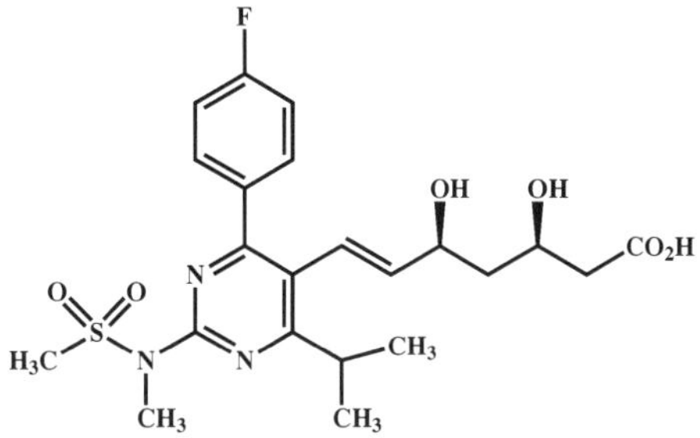

rosuvastatin
69

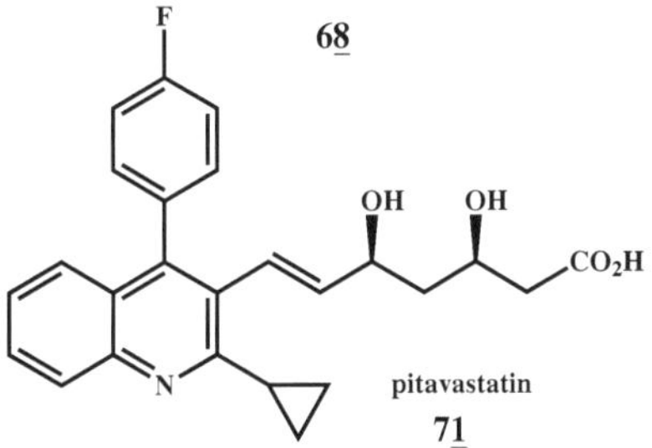

pitavastatin
71

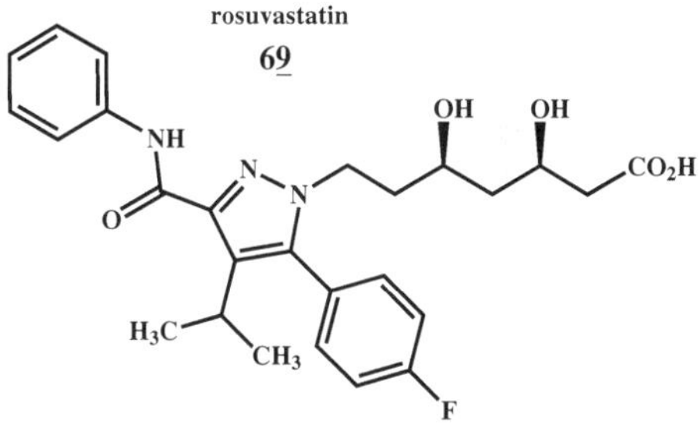

PF-3052334
70

3.4 NATURAL PRODUCTS AS LEAD COMPOUNDS FOR NEW DRUG CANDIDATES

In the 1970s, Chinese scientists isolated artemisinin (72) [101] from the leafy portion of *Artemisia annua* and by the next decade, they had formulated and studied its derivates as antimalarial drugs. Artemisinin (72) is a labile functionalized sesquiterpene lactone endoperoxide, very effective and safe against chloroquine (CQ) sensitive and chloroquine (CQ) resistant strains of *Plasmodium falciparum,* but it has certain limitations such as poor oil and water solubility and a high rate of recrudescence [102]. Hence, a lot of efforts have been made to develop semisynthetic artemisinin derivates [103–105]. Ether derivates of artemisinin have the advantage of better oil solubility and are prepared by treating dihydroartemisinin 73 with an appropriate alcohol in the presence of an acid catalyst [106]. Artemether 74 and arteether 75 are the two most important artemisinin (72) derivates. Both of them show better oil solubility and improved activity and are currently in clinical use.

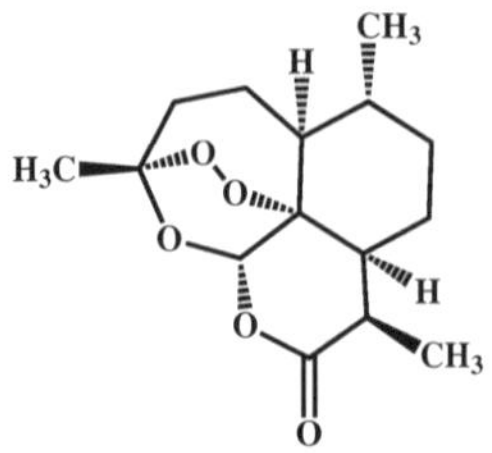

artemisinin

72

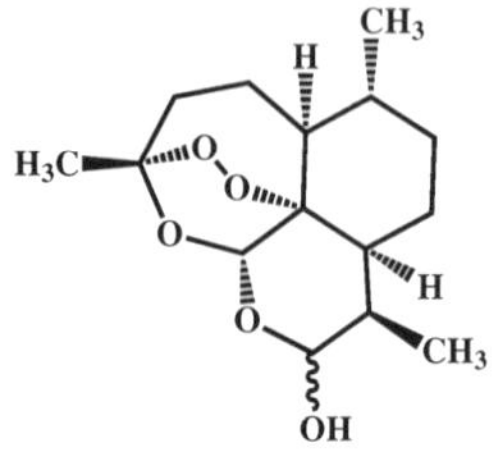

dihydroartemisinin

73

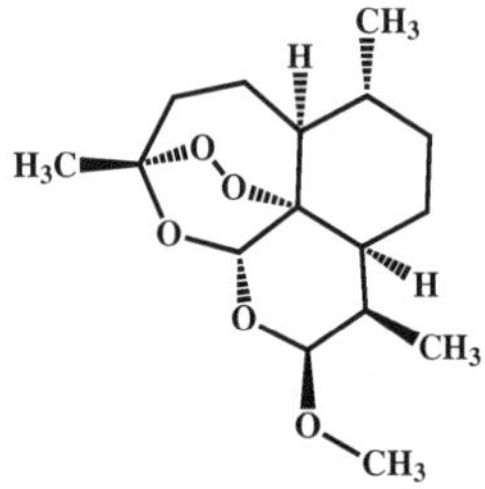

artmether

74

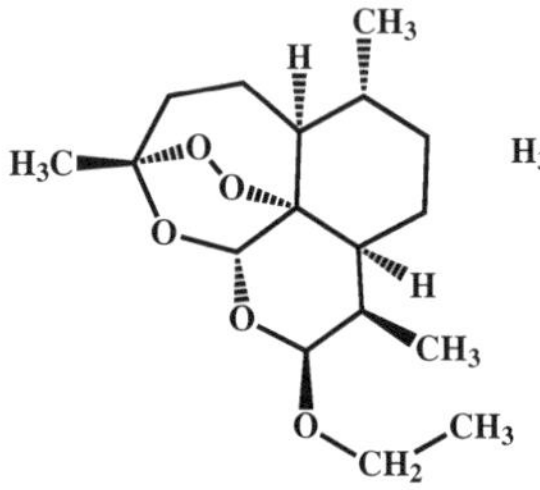

artether

75

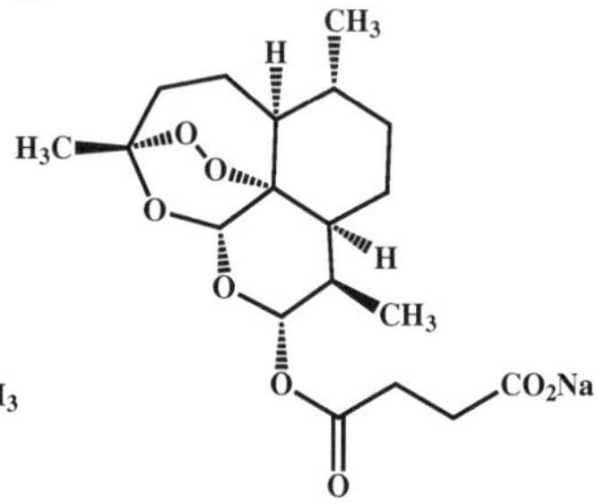

sodium artesunate

76

Charles W. Jefford and co-workers at the University of Geneve, SW, were the first to report the *in vitro* antimalarial activity of simple 1,2,4-trioxanes (A, Scheme 3.12), the structural subunit present in the natural product and considered as *PhG* for the antimalarial activity [107]. Among all the 1,2,4-trioxanes screened, *cis*-fused cyclopentene trioxanes showed the most promising activity profile.

Today, as recommended by the World Health Organization (WHO) and as adopted by most countries where malaria is endemic, artemisinin combination therapy (ACTa)

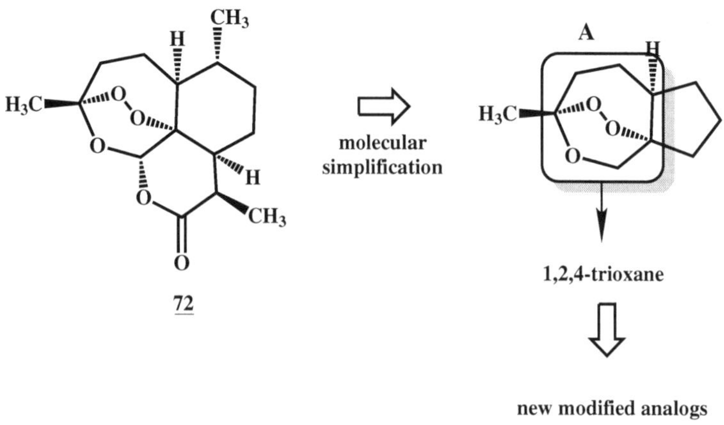

SCHEME 3.12

is popular [108,109]. Typically natural trioxane derivates such as artemether or sodium artesunate (76) are used.

Several research groups have studied the SAR of trioxane derivates as antimalarial agents. Gary H. Posner from Johns Hopkins University in Baltimore, Maryland, USA, has described modified compounds as (77) and (78) with a very important biological profile, active even in resistant *Plasmodium* sp. [110].

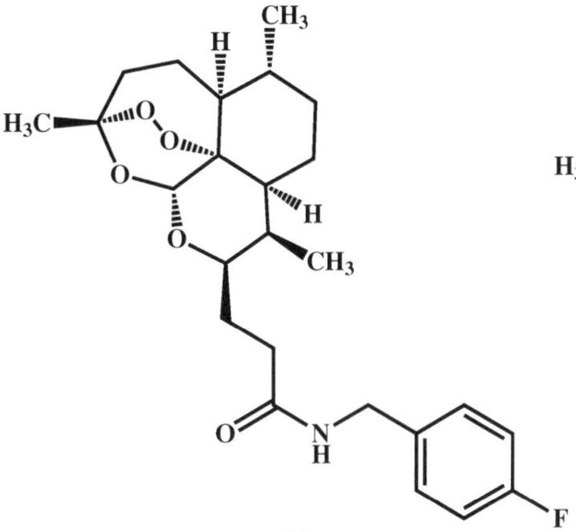

77

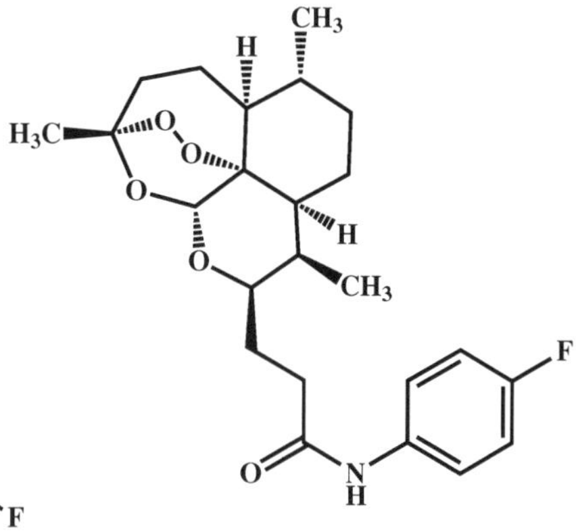

78

The increasing prevalence of multiple drug-resistant strains of *P. falciparum* in most malaria endemic areas has significantly reduced the efficacy of current antimalarial drugs as closed structural related compounds febrifugine (79) and isofebrifugine (80), isolated as the active components against malaria in the Chinese herb Chang Shan (*Dichroa febrifuga* Lour) [111]. In this class of antimalarial drugs, the crucial role of 4-quinazoline ring and 1″-amino group of the piperidine ring in the activity has been elucidated. For instance, new synthetic analogues (81) and (82) have been obtained and have demonstrated an important antimalarial profile [112].

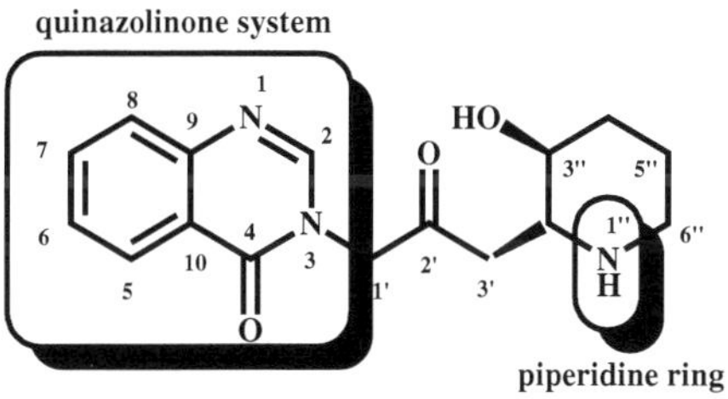

febrifugine

79

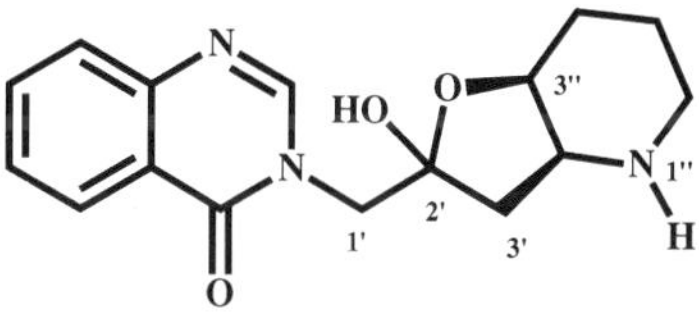

isofebrifugine

80

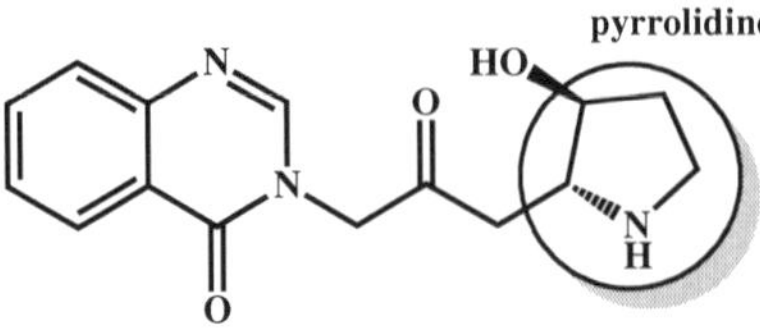

81

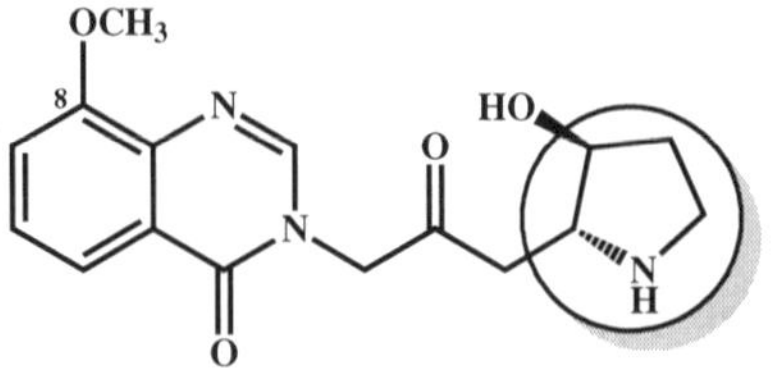

82

Studies of natural biodiversity as a source of pharmaceutically interesting structural patterns have significantly evolved with the introduction of new techniques, aside from those of sophisticated structural elucidation, which allows working with smaller amounts of substances, regardless of their structural complexity. The use of combined biochemical approaches was created, using autobiographical techniques in fine layer affinity chromatography for specific targets, with excellent levels of sensitivity. The use of combinatory techniques [113] to widen the "natural" diversity of certain classes of compounds, and recently, an example of the use of molecular modeling to "anchor" several natural patterns was described by Suh et al. [114] at the Faculty of Pharmacy at the National University of Seoul, South Korea, who described the inhibiting properties of cyclooxigenase-2 (COX-2) in acanthoic acid (83), a sesqueterpene of pimarane class, isolated from medicinal plants commonly used in local folk medicine for the treatment of symptoms of rheumatic conditions. From the molecular modeling studies using COX-2, the authors could optimize the inhibiting properties of the original natural substance, identifying newer and more potent analogs (e.g., 84, IC_{50} COX-2 = 13.2μM) [114].

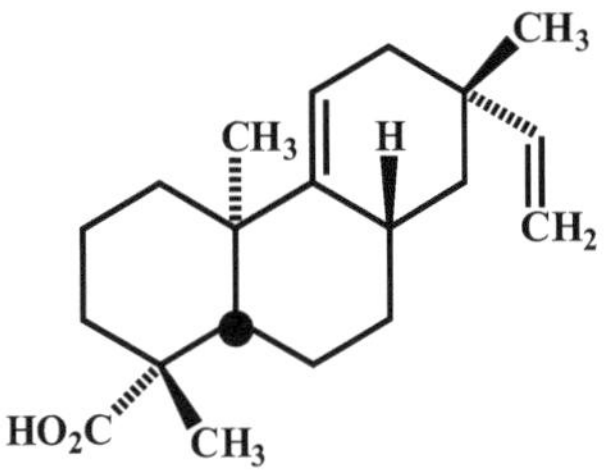

acanthoic acid

83

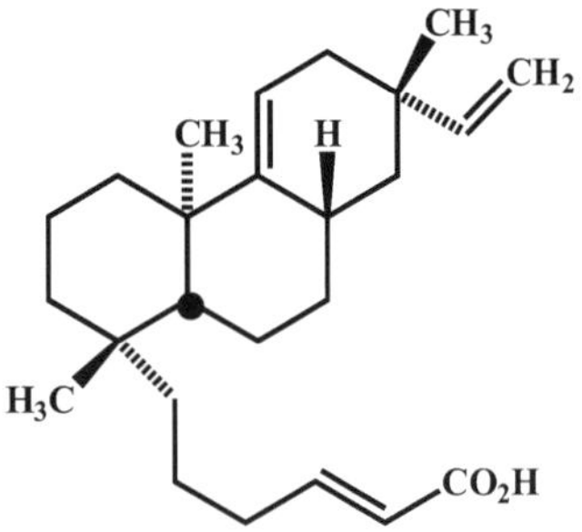

84

ABT-418 (85) [115], a synthetic lead compound candidate for the analgesic nonopioid central action developed by Abbott Laboratories, contains the 5-(3-methyl) isoxazolyl nucleus in its structure, mimicking the pyrimidine ring of natural alkaloid nicotine (86). It presents high affinity for the nicotine receptors of acetylcholine (nACh) (Scheme 3.13). The methyl homologation of this LC at the pyrrolidinic ring

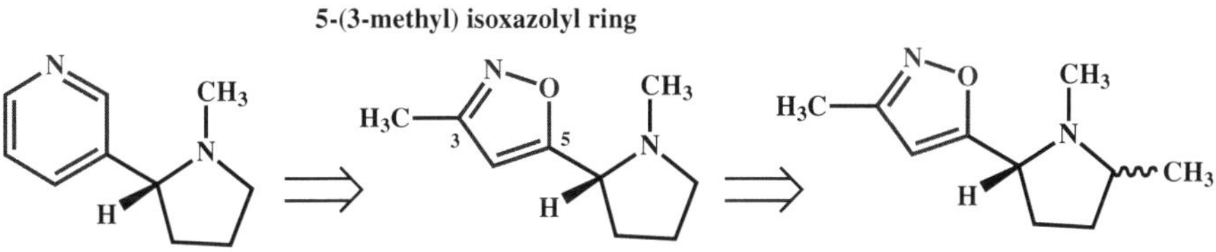

nicotine

86

ABT-418

85

87

SCHEME 3.13

gave 87, which presented an improvement of the analgesic activity and affinity to nAChR. This methylated homolog (87) [116] presented a longer mean lifetime in the biophase, due to the steric protection introduced by the methyl group *vis-à-vis* the oxidative enzymes of the hepatic metabolism.

Huperzine-A (88) [117], an amide alkaloid presenting a characteristic tricyclic structure, was isolated from the creeping weed *Huperzia serrata* (sin. *Lycopodium serratum*) and proved to be a potent inhibitor of the acethylcolinesterase enzyme (AChE). It has been characterized as an authentic natural prototype for the development of substances useful for the treatment of AD [118].

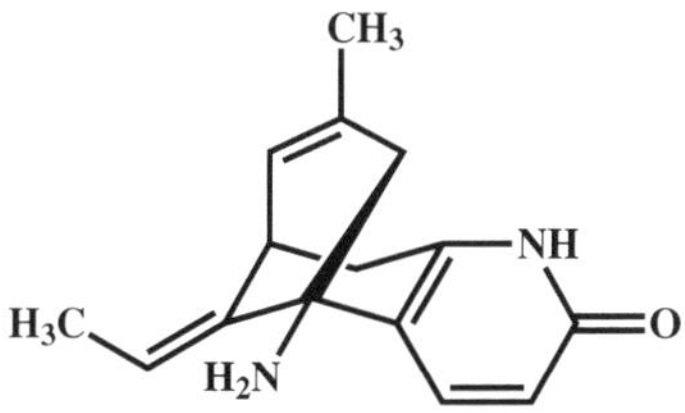

huperzine-A

88

The fanatic search for new molecular patterns exhibiting AChE inhibiting properties has led to the development of bioautographic assays for the enzyme, enabling fast *in vitro* assays of dozens of substances. It has been recently reported by Bolzani et al. [119], within the scope of a systematic study of the vegetable natural resources remaining in the Brazilian Atlantic Forest, a piperidine alkaloid named spectaline (89), isolated from the Leguminous *Cassia spectabilis* (sin. *Senna spectabilis*) [119]. The carefully structural inspection of 89 made evident the molecular similarity to acethylcholine – substract from acetylcholinesterase (AChE) – at level of the cyclic subunit (A), indicating the possibility of obtaining new inhibitors of AChE through adequate structural modifications of this alkaloid, which could represent new LC's for treatment of AD. This approach permitted the discovery of compounds LASSBio-767

(90) [120] and LASSBio-822 (91) [120], which presented potent AChE inhibitory properties [121], with a high selectivity pattern *vis-à-vis* butyrylcholinesterase (BuChE), responsible for some of the side effects of the pharmaceuticals available for treating AD.

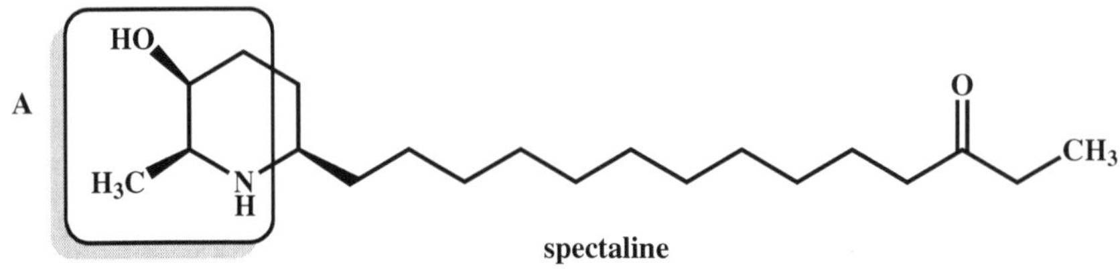

spectaline

89

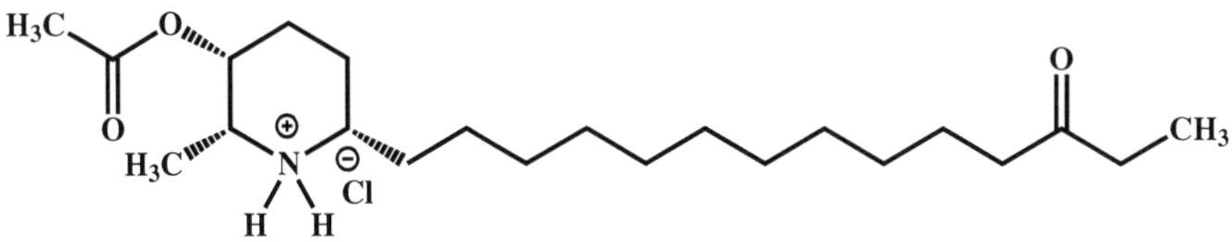

LASSBio-767

90

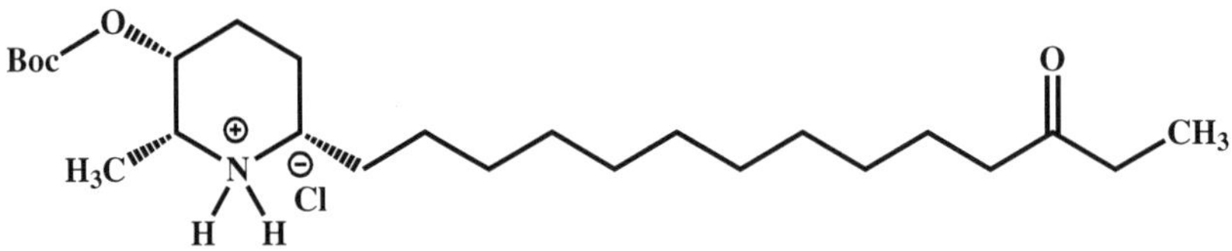

LASSBio-822

91

Flavopiridol (93), also known as L86-8275 or HMR 1275 is a semisynthetic flavonoid derived from rohitukine (92) [122], an alkaloid isolated from a plant indigenous to India. In initial studies, flavopiridol (93) showed potent *in vitro* EGF receptor tyrosine kinase (EGFR) and protein kinase A (PKA) inhibitor (IC_{50}= 21 and 122 μM, respectively) [122]. In addition, flavopiridol (93) has demonstrated potent and specific *in vitro* inhibition of all cyclin-dependent kinases (*cdk*) tested, blocking the cell cycle progression. This natural compound has a *peri*-hydroxy-flavone framework that is structurally similar to the PhG of *tinibes* (94), a synthetic class of kinase inhibitors anticancer drugs [123]. This class of tyrosine kinase inhibitor drugs has as *PhG* (e.g., 94; Scheme 3.14) an 4-arylamino-quinoxaline moiety corresponding to a H-donor (*amino*) and a H-acceptor (N_3) equivalent to *peri*-hydroxy-ketone of flavopiridol (93).

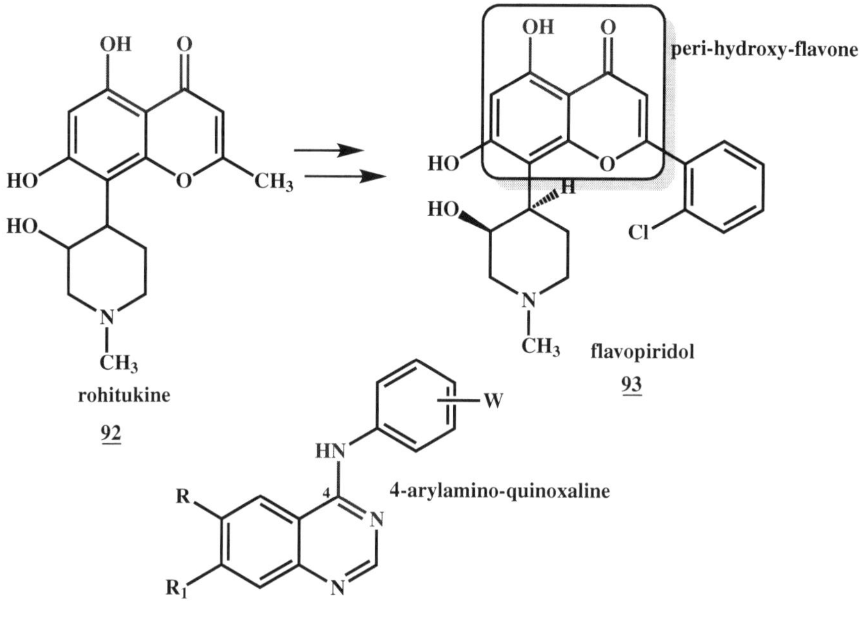

94

SCHEME 3.14

3.5 CONCLUSIONS

Natural products were the only source of drugs known to mankind for a long time [124]. Pharmaceutical sciences having inspired the creation of pharmacognosy, when using natural products as drugs go from a mystical-religious period of time to one of science, and where its precise description is established [125]. Subsequently, with the evolution of chemistry, the isolation of active principles begins, especially on Colleges of Pharmacy in Germany and France, and from that point on, the structural modification studies of natural products of therapeutic interest, aimed at improving their PK and PD as well as the importance of NP as LCs to new medicines.

Currently, the contemporary therapeutic arsenal is mostly composed of synthetic substances (*ca.* 85%), but several of these synthetic drugs were inspired by NP [126] as illustrated by the introduction of acetyl salicylic acid (95; ASA), in 1889, considered to be the first synthetic drug. This compound was discovered by Felix Hoffmann in Bayer laboratories in Germany, from salicin (96), a natural product found in willow (*Spirea* sp.) [127].

A modern example is synthetic selective phosphodiesterase-5 (PDE-5) inhibitors, recommended for the treatment of erectile dysfunction [128] and originally represented by sildenafil (97) [129], started from methyl-xanthines, especially theophylline (98).

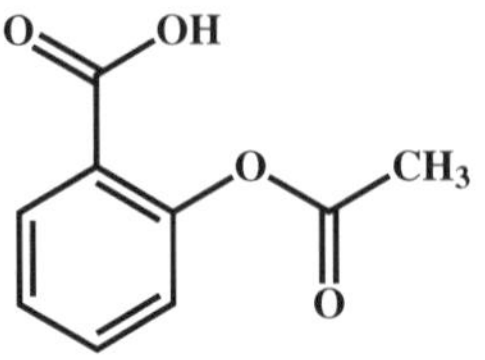

ASA

95

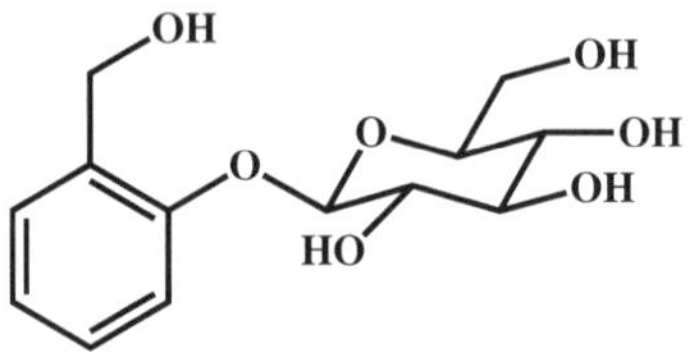

salicin

96

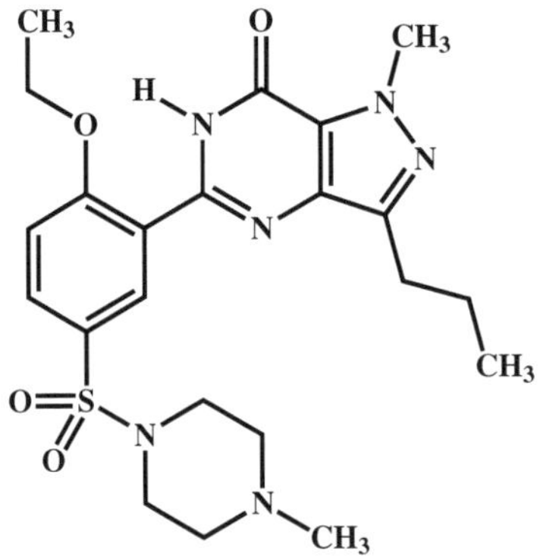

sildenafil

97

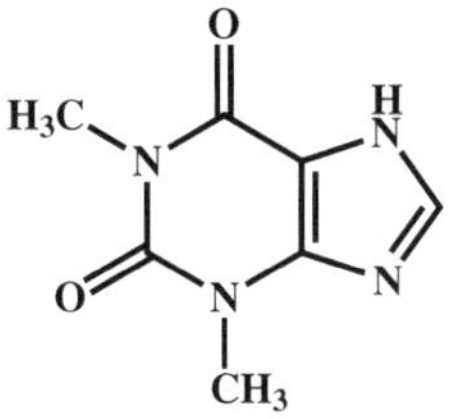

theophylline

98

A definitive example is the class of synthetic regulators of hypertension, such as angiotensin-converting enzyme (ACE) inhibitors, pioneeringly developed by Ondetti, Cushman, and collaborators in the Squibb Labs and represented by captopril (99) [130], which was inspired by work undertaken by Oscar Mauricio Rocha e Silva and Wilson Beraldo and, subsequently, by Sergio Henrique Ferreira working in the pharmacology laboratories of the School of Medicine of São Paulo University at Ribeirão Preto, Brazil, with the poison lance head viper (*Bothrops sp*) that resulted in the discovery of bradykinin [131], creating the foundation for the knowledge of the renin-angiotensin system (RAS) blood pressure control [132]. The introduction of ACE-inhibitors in the treatment of hypertension represents an authentic therapeutic innovation and nowadays any important pharmaceutical company markets a "*pril*" ACE-inhibitor (*inter-alia:* enalapril, 100; lisinopril,101; ramipril, 102; fosinopril, 103) [133].

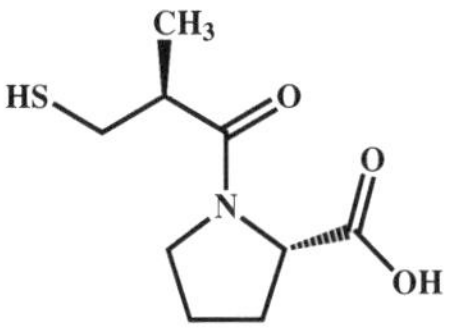

captopril

99

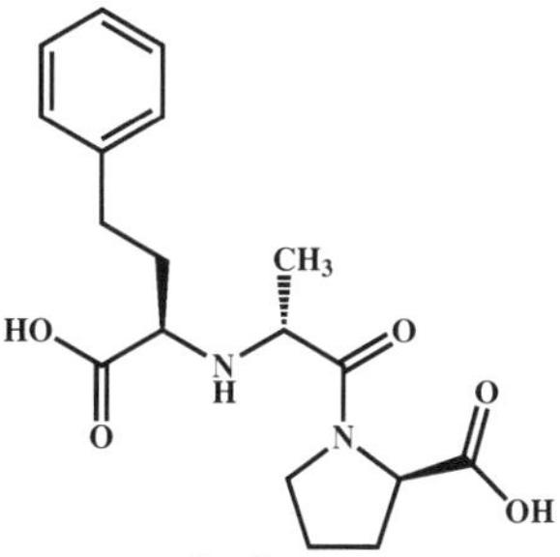

enalapril

100

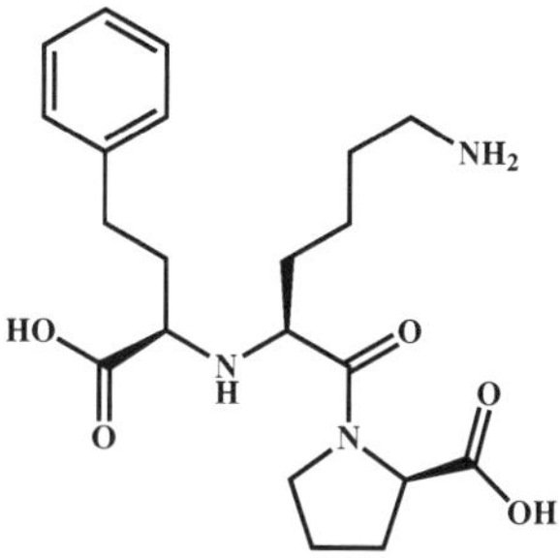

lisinopril

101

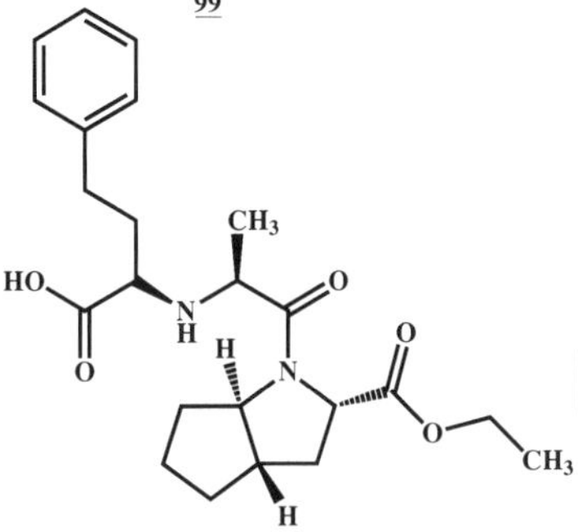

ramipril

102

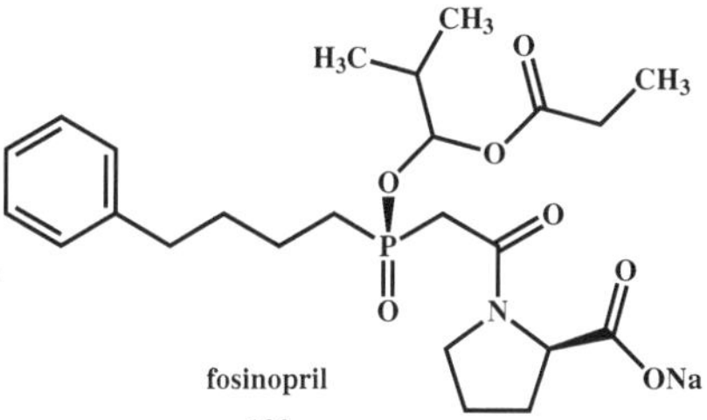

fosinopril

103

ACKNOWLEDGMENTS

The authors hope that the examples included in this chapter, far from exhausting the subject, could well represent the important role of numerous natural products from distinct sources in the drug discovery process, at different time. The authors thank Professor Valdir Cechinel Filho for the kind invitation to write this chapter in this book.

REFERENCES

[1] (a) Barreiro, E.J., Bolzani, V.S. (2009). Biodiversidade: Fonte potencial para a descoberta de fármacos. *Química Nova*, *32*, 679–688; (b) Baker, D.D., Chu, M., Oza, U., Rajgarhia, V. (2007). The value of natural products to future pharmaceutical discovery. *Natural Products Reports*, *24*, 1225–1244; (c) Viegas- Jr., C., Bolzani, V.S., Barreiro, E.J. (2006). Os produtos naturais e a química medicinal moderna. *Química Nova*, *29*, 326–337; (d) Koehn, F.E., Carter, G.T. (2005). The evolving role of natural products in drug discovery. *Nature Reviews Drug Discovery*, *4*, 206–220; (e) Pinto, A.C., Silva, D.H.S., Bolzani, V.S., Lopes, N.P., Epifânio, R.A. (2002). Produtos naturais: Atualidade, desafios e perspectivas. *Química Nova*, *25*, 45–61; (f) Mann, J., *Murder, Magic, and Medicine*, Oxford University Press, Oxford, 2000, pp. 17–29; (g) Shu, Y.Z. (1998). Recent natural products based drug development: A pharmaceutical industry perspective. *Journal of Natural. Products*, *61*, 1053–1071; (h) Verpoorte, R. (1998). Exploration of

nature's chemodiversity: The role of secondary metabolites as leads in drug development. *Drug Discovery Today*, *3*, 232–238; (i) Agosta, W.C. (1997). Medicines and drugs from plants. *Journal of Chemical Education*, *74*, 857–860; (j) Wrigley, K., Chicarelli-Robinson, M.I. (1997). Natural products research and pharmaceuticals in the 1990s. *Annual Reports in Medicinal Chemistry*, *32*, 285–294; (k) Barreiro, E.J. (1990). Produtos naturais de origem vegetal e o desenvolvimento de fármacos. *Química Nova*, *13*, 29–39.

[2] Corey, E.J., Kürti, L., Czakó, B., *Molecules and Medicines*, John Wiley & Sons, New Jersey, 2007.

[3] Dewick, P.M., *Medicinal Natural Products: A Biosynthetic Approach*, John Wiley & Sons, Chichester, UK, 2009.

[4] Wermuth, C.G., Ganellin, C.R., Lindberg, P., Mitscher, L.A. (1998). Glossary of terms used in medicinal chemistry (IUPAC recommendations 1997). *Annual Reports in Medicinal Chemistry*, *33*, 385–395.

[5] (a) Butler, M.S. (2005). Natural products to drugs: Natural product derived compounds in clinical trials. *Natural Products Reports*, *22*, 162–195; (b) Tringali, C., *Bioactive Compounds from Natural Sources: Natural Products as Lead Compounds in Drug Discovery*, CRC Press, London, 2011;. (c) Butler, M.S. (2004). The role of natural product chemistry in drug discovery. *Journal of Natural Products*, *67*, 2141–2153; (d) Daly, J.W. (1998). Thirty years of discovering arthropod alkaloids in amphibian skin. *Journal of Natural Products*, *61*, 162–172; (e) Newman, D.J., Cragg, G.M., Snader, K.M. (2003). Natural products as sources of new drugs over the period 1981-2002. *Journal of Natural Products*, *66*, 1022- 1037; (f) Daly, J.W. (2003). Ernest Guenther award in chemistry of natural products. Amphibian skin: A remarkable source of biologically active arthropod alkaloids. *Journal of Medicinal Chemistry*, *46*, 445–452; (g) Butler, M.S. (2004). The role of natural product chemistry in drug discovery. *Journal of Natural Products*, *67*, 2141–2153.

[6] Wermuth, C.G., Ganellin, C.R., Lindberg, P., Mitscher, L.A. (1998). Glossary of terms used in medicinal chemistry. *Pure and Applied Chemistry*, *70*, 1129–1143.

[7] Kornberg, A. (1987). The two cultures: Chemistry and biology. *Biochemistry*, *26*, 6888–6891.

[8] The IUPAC glossary of terms used in medicinal chemistry states that: "Lead discovery is the process of identifying active new chemical entities, which by subsequent modification may be transformed into a clinically useful drug".

[9] (a) Leeson, P.D., Springthorpe, B. (2007). The influence of drug-like concepts on decision-making in medicinal chemistry. *Nature Reviews Drug Discovery*, *6*, 881–890; (b) Lombardo, F., Gifford, E., Shalaeva, M.Y. (2003). In silico ADME prediction: Data, models, facts and myths. *Mini Reviews in Medicinal Chemistry*, *3*, 861–875.

[10] (a) Cragg, G.M., Newman, D.J., Snader. K.M. (1997) Natural products in drug discovery and development. *Journal of Natural Products*, *60*, 52–60; (b) Drews, J. (2000). Drug discovery: A historical perspective. *Science*, *287*, 1960–1964.

[11] Lima, L.M., Barreiro, E.J. (2005). Bioisosterism: A useful strategy for molecular modification and drug design. *Current Medicinal Chemistry*, *12*, 23–49.

[12] Wermuth, C.G., *The Practice of Medicinal Chemistry*, Academic Press, Amsterdam, 2008.

[13] Fura, A., Shu, Y.Z., Zhu, M., Hanson, R.L., Roongta, V., Humphreys, W.G. (2004). Discovering drugs through biological transformation: Role of pharmacologically active metabolites in drug discovery. *Journal of Medicinal Chemistry*, *47*, 4339–4351.

[14] Newman, D.J. (2008). Natural products as leads to potential drugs. An old process or the new hope for drug discovery? *Journal of Medicinal Chemistry*, *51*, 2589–2599.

[15] Borchardt, J.K. (1999). Combinatorial biosynthesis: Panning for pharmaceutical gold. *Modern Drug Discovery*, *2*, 22–29.

[16] (a) Mann, J., Davidson, R.S., Hobbs, J.B., Banthorpe, D.V., Harborne, J.B., *Natural products: Their chemistry and biological significance*, John Wiley & Sons, New York, 1994;. (b) Newman, D.J., Cragg, G.M. (2004). Marine natural products and related compounds in clinical and advanced preclinical trials. *Journal of Natural Products*, *67*, 1216–1238.

[17] Hale, K.J., *The chemical synthesis of natural products*, Blackwell Publishing, Oxford, 2000.

[18] Herbal Pharmacology in the People's Republic of China, National Academy of Science, Washington, DC, 1975.

[19] (a) Josephson, J. (2000). Tapping tradicional healers' treasures. *Modern Drug Discovery*, *3*, 45–50; (b) Ertl, P., Roggo, S., Schuffenhauer, A. (2008). Natural product-likeness score and its application for prioritization of compound libraries. *Journal of Chemical Information and Modeling*, *48*, 68–74.

[20] Srivastava, V., Negi, A.S., Kumar, J.K., Gupta, M.M., Khanuja, S.P.S. (2005). Plant-based anticancer molecules: A chemical and biological profile of some important leads. *Bioorganic and Medicinal Chemistry*, *13*, 5892–5908.

[21] Butler, M.S. (2004). The role of natural product chemistry in drug discovery. *Journal of Natural Products*, *67*, 2141–2153.

[22] Mukherjee, A.K., Basu, S., Sarkar, N., Ghosh, A.C. (2001). Advances in cancer therapy with plant based natural products. *Current Medicinal Chemistry*, *8*, 1467–1486.

[23] Salim, A.A., Chin, Y.-W., Kinghorn, A.D., Drug discovery from plants. In: Ramawat, K. G., Merillon, J.M. editors. Bioactive Molecules and Medicinal Plants. Springer-Verlag, Heidelberg, 2008, pp. 1–24.

[24] (a) Rasoanaivo, P., Petitjean, A., Ratsimamanga-Urverg, S., Rakoto-Ratsimamanga A. (1992). Medicinal plants used to treat malaria in Madagascar. *Journal of Ethnopharmacology*, *37*, 117–127; (b) Chouteau, F., Ramanitrahasimbola, D., Rasoanaivo, P., Chibale, K. (2005). Exploiting a basic chemosensitizing pharmacophore hypothesis. Part 1: Design, synthesis and biological evaluation of novel bicyclic chemosensitizers against drug-resistant malaria parasites. *Bioorganic and Medicinal Chemistry Letters*, *15*, 3024–3028; (c) Willcox, M.L., Cosentino, M.J., Pink, R., Wayling, S., Bodeker, G. (2001). Natural products for the treatment of tropical diseases. *Trends in Parasitology*, *17*, 58–60.

[25] (a) Campbell, S.F. (2000). Science, art and drug discovery: A personal perspective. *Clinical Science*, *99*, 255–260; (b) Steinberg, D., Gotto- Jr., A.M. (1999). Preventing coronary artery disease by lowering cholesterol levels: Fifty years from bench to bedside. *Journal of the American Medical Association*, *282*, 2043–2050.

[26] Lewis, J.A. (2009). Molecule of month. *Current Topics in Medicinal Chemistry*, *9*, 117.

[27] Fraga, C.A.M., Barreiro, E.J. (1996). Cardiotônicos: Histórico & perspectivas de uma antiga e importante classe de agentes terapêuticos. *Química Nova*, *19*, 182–189.

[28] Fifer, E.K., Drugs affecting cholinergic transmission. In: Lemke, T.L., Williams D.A. editors. *Foye's Principles of Medicinal Chemistry*. 6th ed., Lippincott Williams & Wilkins, Baltimore, 2008, pp. 373–374.

[29] Lee, M.R. (2008). Ipecacuanha: The South American vomiting root. *Journal of the Royal College of Physicians of Edinburgh*, *38*, 355–360.

[30] Williams, D.A., Central and peripheral sympatholytics and vasodilators. In: Lemke, T.L., Williams, D.A. editors. *Foye's Principles of Medicinal Chemistry*. 6th ed. Lippincott Williams & Wilkins, Baltimore, 2008, pp. 782.

[31] Scott, L.J., Goa, K.L. (2000). Galantamine: A review of its use in Alzheimer's disease. *Drugs*, *60*, 1095–1122.

[32] (a) *Inter-alia:* Kuo, Y.C., Kuo, P.L., Hsu, Y.L, Cho, C.Y., Lin, C.C. (2006). Ellipticine induces apoptosis through p53-dependent pathway in human hepatocellular carcinoma HepG2 cells. *Life Science*, *78*, 2550–2557; (b) Mal, D., Senapati, B.K., Pahari, P. (2005). A new synthetic route to indoloquinones: Formal synthesis of ellipticine. *Synlett*, *6*, 994–996; (c) Ishikura, M., Hino, A., Katagiri, N. (2000). An efficient total synthesis of ellipticine. *Heterocycles*, *53*, 11–14; (d) Harada, N., Ozaki, K., Oda, K., Nakanishi, N., Ohashi, M., Hashiyama, T., Tsujihara, K. (1997). Synthesis and antitumor activity of 9-acyloxyellipticines. *Chemical and Pharmaceutical Bulletin*, *45*, 1156–1162.

[33] (a) Chain, E. (1979). The early years of the penicillin discovery. *Trends in Pharmacological Science*, *1*, 6–11; (b) Maurois, A., *La vie de Sir Alexander Fleming*, Hachette, Paris, 1959; (c) Chain, E., Florey, H.W., Gardner, A.D., Heatley, N.G., Jennings, M.A., Orr-Ewing, J., Sandres, A.G. (1940). Penicilin as chemotherapeutic agent. *Lancet*, *2*, 226–228.

[34] Polubiec, A., Weuta, H., Stepka, K., Ktos, K., Strazynska, R., Jorasz, J. (1988). Bacterial-infections of the lower urinary-tract treated with ciprofloxacin or cefalexin – a comparative-study. *Infection*, *16*, S62–S64.

[35] Blumberg, P.M., Strominger, J.L. (1974). Interaction of penicillin with bacterial cell: Penicillin-binding proteins and penicillin-sensitive enzymes. *Bacteriological Reviews*, *38*, 291–335.

[36] Kahan, J.S., Kahan, F.M., Goegelman, R., Currie, S.A., Jackson, M., Stapley, E.O., Miller, T.W., Miller, A.K., Hendlin, D., Mochales, S., Hernandez, S., Woodruff, H.B., Birnbaum, J. (1979). Thienamycin, a new beta-lactam antibiotic: I. Discovery, taxonomy, isolation and physical properties. *Journal of Antibiotics*, *32*, 1–12.

[37] Gringauz, A., Introduction to medicinal chemistry, how drugs act and why, Wiley, New York, 1997, pp. 201–211.

[38] Mitscher, L.A., Lemke, T.A., Gentry, E.J., Antibiotics and antimicrobial agents. In: Lemke, T.L., Williams, D.A. editors. Foye's Principles of Medicinal Chemistry. 6th ed. Lippincott Williams & Wilkins, Baltimore, 2008, p. 1065.

[39] Mitscher, L.A., Lemke, T.A., Gentry, E.J., Antibiotics and antimicrobial agents. In: Lemke, T.L., Williams, D.A. editors. *Foye's Principles of Medicinal Chemistry*. 6th ed. Lippincott Williams & Wilkins, Baltimore, 2008, pp. 1044–1064.

[40] Avendaño, C., Menendez, J.C., *Medicinal chemistry of anticancer drugs*, Elsevier, Amsterdam, 2008, p. 371.

[41] Hecht, S.M. (2000). Bleomycin: New perspectives on the mechanism of action. *Journal of Natural Products*, *63*, 158–168.

[42] Gaudiano, G., Koch, T.H. (1991). Redox chemistry of anthracycline antitumor drugs: A use of capdative radials as tools for its elucidation and control. *Chemical Research in Toxicology*, *4*, 2–16.

[43] Johnson, I.S., Armstrong, J.G., Gorman, M., Burnett-Jr, J.P. (1963). The vinca alkaloids: A new class of oncolytic agents. *Cancer Research*, *23*, 1390–1427.

[44] Cragg, G.M., Newman, D.J. (2004). A tale of two tumor targets: Topoisomerase I and tubulin. The Wall and Wani contribution to cancer chemotherapy. *Journal of Natural Products 67*, 232–244.

[45] (a) Wall, M.E., Taxol. In: Lednicer, D.,editor. *Chronicles of Drug Discovery*. American Chemical Society, Washington DC, 1993, pp. 327–348. (b) Fang, W.S., Liang, X.T. (2005). Recent progress in structure activity relationship and mechanistic studies of Taxol^{TM} analogues. *Mini Reviews in Medicinal Chemistry*, *5*, 1–12.

[46] Goodman, J., Walsh, V., *The story of taxol. Nature and politics in the pursuit of an anti-cancer drug*, Cambridge University Press, Cambridge, 2001.

[47] Georgoulias, V. (2002). Docetaxel (taxotere) in the treatment of non-small cell lung cancer. *Current Medicinal Chemistry*, *9*, 869–877.

[48] (a) Oberlies, N.H., Kroll, D.J. (2004). Camptothecin and taxol: Historic achievements in natural products research. *Journal of Natural Products*, *67*, 129–135; (b) Thomas, C.J., Rahier, N., Hecht, S.M. (2004). Camptothecin: Current perspectives. *Bioorganic and Medicinal Chemistry*, *12*, 1585–1604.

[49] Mitscher, L.A. (2005). Bacterial topoisomerase inhibitors: Quinolone and pyridone antibacterial agents. *Chemical Reviews*, *105*, 559–592.

[50] Cragg, G.M., Newman, D.J., Kingston, D.G.I., *Anticancer agents from natural products*, CRC Press, Boca Raton, FL, 2005.

[51] (a) Hossfeld, D.K. (2000). Irinotecan (campto®) - Clinical overview. *Onkologie*, *23*, 1–38; (b) DeVita Jr., V.T., Chu, E. (2008). A history of cancer chemotherapy. *Cancer Research*, *68*, 8643–8653.

[52] (a) Kingsley, C.D., Teniposide and etoposide. In Perry, M.C. editor.: *The Chemotherapy Source Book*, Lippincott Williams & Wilkins, Philadelphia, 2008, pp. 155–156;(b) Roche, V.F., Cancer and chemotherapy. In: Lemke, T.L., Williams, D.A., editors. *Foye's Principles of Medicinal Chemistry*. 6th ed. Lippincott Williams & Wilkins, Baltimore, 2008, pp. 1086–1087.

[53] Rinehart, K.L. (2000). Antitumor compounds from tunicates. *Medicinal Research Reviews*, *20*, 1–27.

[54] Rinehart, K.L., Holt, T.G., Fregeau, N.L., Stroh, J.G., Keifer, P.A., Sun, F., Li, L.H., Martin, D.G. (1990). *Ecteinascidins 729 743*, *745*, 759A, 759B, and 770: Potent antitumor agents from the Caribbean tunicate *Ecteinascidia turbinate*. *Journal of Organic Chemistry*, *55*, 4512–4515.

[55] Oku, N., Matsunaga, S., Van Soest, R.W.M., Fusetani, N. (2003). Renieramycin J, a highly cytotoxic tetrahydroisoquinoline alkaloid, from a marine sponge *Neopetrosia* sp. *Journal of Natural Products*, *66*, 1136–1139.

[56] (a) Martinez, E.J., Corey, E.J. (2000). A new more efficient and effective process for the synthesis of a key pentacyclic intermediate for production of ecteinascidin and phthalascidin antitumor agents. *Organic Letters*, *2*, 993–996; (b) For a more recent example of ecteinascidin synthesis see: Endo, A., Yanagisawa, A., Abe, M., Toluna, S., Kan, T., Fukuyama, T. (2002). Total synthesis of ecteinascidin 743. *Journal of American Chemical Society*, *124*, 6552–6554.

[57] Cuevas, C., Pérez, M., Martin, M.J., Chicharro, J.L., Fernandez-Rivas, C., Flores, M., Francesca, A., Gallego, P., Zarzuelo, M., deLaCalle, F., Garcia, J., Polanco, C.,

Rodríguez, I., Manzanares, I. (2000). Synthesis of ecteinascidin ET-743 and phthalascidin Pt-650 from cyanosafracin B. *Organic Letters*, *2*, 2545–2548.

[58] Endo, A., Yanagisawa, A., Abe, M., Tohma, S., Kan, T., Fukuyama, T. (2002). Total synthesis of ecteinascidin 743. *Journal of American Chemical Society*, *124*, 6552–6554.

[59] Martinez, E.J., Owa, T., Schreiber, L.S., Corey, E.J. (1999). Phthalascidin, a synthetic antitumor agent with potency and mode of action comparable to ecteinascidin 743. *Proceedings of the National Academy of Sciences USA*, 96, 3496–3951.

[60] Liu, Z.Z., Wang, Y., Tang, Y.F., Chen, S.Z., Shen, X.G. (2006). Synthesis and antitumor activity of simplified ecteinascidin-saframycin analogs. *Bioorganic and Medicinal Chemistry Letters*, *16*, 1282–1285.

[61] Costa-Lotufo, L.V., Wilde, D.V., Jimenez, P.C., Epifanio, R.A. (2009). Organismos marinhos como fonte de novos fármacos: Histórico & perspectivas. *Química Nova*, *32*, 703–716.

[62] Li, J.W.H., Vederas, J.C. (2009). Drug discovery and natural products: End of an era or an endless frontier? *Science*, *325*, 161–165.

[63] Butler, M.S. (2008). Natural products to drugs: Natural product-derived compounds in clinical trials. *Natural Products Reports*, *25*, 475–516.

[64] Harvey, A.L. (2008). Natural products in drug discovery. *Drug Discovery Today*, *13*, 894–901.

[65] Olivera, B.M., Rivier, J., Clark, C., Ramilo, C.A., Corpuz, G.P., Abogadie, F.C., Mena, E.E., Woodward, S.R., Hillyard, D.R., Cruz, L.J. (1990). Diversity of Conus neuropeptides. *Science*, *249*, 257–263.

[66] Zimmerman, D.M., Gidda, J.S., Cantrell, B.E., Schoepp, D.D., Johnson, B.G., Leander, J.D. (1994). Discovery of a potent, peripherally selective trans-3, 4-dimethyl-4-(3-hydroxyphenyl)piperidine opioid antagonist for the treatment of gastrointestinal motility disorders. *Journal of Medicinal Chemistry*, *37*, 2262–2265.

[67] (a) Kutter, E., Herz, A., Teschemacher, H.J., Hess, R. (1970). Structure-activity correlations of morphinelike analgetics based on efficiencies following intravenous and intraventricular application. *Journal of Medicinal Chemistry*, *13*, 801–805; (b) Loew, G.H., Lawson, J.A., Uyeno, E.T., Toll, L., Frenking, G., Polgar, W., Ma, L.Y., Camerman, N., Camerman, A. (1988). Structure-activity studies of morphine fragments. I. 4-Alkyl-4-(m-hydroxy-phenyl)-piperidines. *Molecular Pharmacology*, *34*, 363–376; (c) For recent example of molecular simplifications, see: Magedov, I.V., Manpadi, M., Ogasawara, M.A., Dhawan, A.S., Rogelj, S., VanSlambrouck, S., Steelant, W.F.A., Evdokimov, N.M., Uglinskii, P.Y., Elias, E.M., Knee, E.J., Tongwa, P., Antipin, M.Y., Kornienko, A. (2008). Structural simplification of bioactive natural products with multicomponent synthesis. 2. antiproliferative and antitubulin activities of pyrano [3,2-*c*]pyridones and pyrano[3,2-*c*]quinolones. *Journal of Medicinal Chemistry*, *51*, 2561–2570.

[68] Zimmerman, D.M., Nickander, R., Homg, J.S., Wong, D.T. (1978). New structural concepts for narcotic antagonists defined in a 4-phenylpiperidine series. *Nature*, *279*, 332–334.

[69] Zhang, J., Burnell, J.C., Dumaual, N., Bosron, W.F. (1999). Binding and hydrolysis of meperidine by human liver carboxylesterase hCE-1. *Journal of Pharmacology and Experimental Therapeutics*, *290*, 314–318.

[70] (a) Avison, A.W.D., Morrison, A.L. (1950). Synthetic analgesics. Part VI. The synthesis of ketobemidone. *Journal of Chemical Society*, 1469–1471; (b) Froimowitz, M., Cody, V. (1993). Enantiomeric conformers of the opioid agonist ketobemidone hydrochloride in the crystalline state. *Chirality*, *5*, 560–564; (c) Kristensen, K., Christensen, C.B., Christrup, L.L., Nielsen, L.C. (1996). The mu1 and mu2 opioid receptor binding of ketobemidone, norketobemidone and 3-dimethylamino-1, 1-diphenylbutene. *Pharmacology and Toxicology*, *79*, 103–104.

[71] Rocco, F., *Quinine, Malaria and the Quest for a Cure that Changed the World*, Harper Collins Publishers, New York, 2003.

[72] Lemke, T.A. Antiparasitic agents. In: Lemke, T.L., Williams, D.A. editors. Foye's Principles of Medicinal Chemistry. 6th ed. Lippincott Williams & Wilkins, Baltimore, 2008, p. 1098.

[73] Lee, M.R. (2005). Curare: The South American arrow poison. *Journal of the Royal College of Physicians of Edinburgh*, *35*, 83–92.

[74] Fifer, E.K., Drugs affecting cholinergic neurotransmission. In: Lemke, T.L., Williams, D.A. editors. *Foye's Principles of Medicinal Chemistry*. 6th ed. Lippincott Williams & Wilkins, Baltimore, 2008, pp. 385–389.

[75] Fusetani, N., *Drugs from the Sea*, Karger, Basel, 2000.

[76] (a) Mitsuya, H., Weinhold, K., Furman, P., St. Clair, M., Lehrman, S., Gallo, R., Bolognesi, D., Barry, D., Broder, S. (1985). 3′-Azido-3′-deoxythymidine (BW A509U): An antiviral agent that inhibits the infectivity and cytopathic effect of human T-lymphotropic virus type III/lymphadenopathy-associated virus *in vitro*. *Proceedings of the National Academy of Sciences USA*, 82, 7096–100; (b) Horwitz, J.P., Chua, J., Noel, M. (1964). Nucleosides: V. The monomesylates of 1- (2′-deoxy-B-D- lyxofuranosyl) thymine. *Journal of Organic Chemistry*, *29*, 2076–2078.

[77] Crabbé, P., *Prostaglandin Research*, Academic Press, New York, 1977.

[78] Collins, P.W., *Misoprostol, in Chronicles of Drug Discovery*, American Chemical Society, Washington DC, 1993, pp. 101–124.

[79] Elrefaey, H., Calder, L., Wheatley, D.N. (1994). Cervical priming with prostaglandin E1 analogs, misoprostol and gemeprost. *Lancet*, *343*, 1207–1209.

[80] Maw, G.N. (1999). Pharmacological therapy for the treatment of erectile disfunction. *Annual Reports in Medicinal Chemistry*, *34*, 71–80.

[81] Eisenreich, W., Kupfer, E., Stohler, P., Weber, W., Bacher, A. (2003). Biosynthetic origin of a branched chain analogue of the lipase inhibitor, lipstatin. *Journal of Medicinal Chemistry*, *46*, 4209–4212.

[82] Carriere, F., Renou, C., Ransac, S., Lopez, V., DeCaro, J., Ferrato, F., DeCaro, A., Fleury, A., Sanwald-Ducray, P., Lengsfeld, H., Beglinger, C., Hadvary, P., Verger, R., Laugier, R. (2001). Inhibition of gastrointestinal lipolysis by Orlistat during digestion of test meals in healthy volunteers. *American Journal of Physiology – Gastrointestinal and Liver Physiology*, *281*, G-16–G28.

[83] Hofle, G., Bedorf, N., Steinmetz, H., Schumburg, D., Gerth, K., Reichenbach, H. (1996). Epothilone A and B--novel 16-membered macrolides with cytotoxic activity: Isolation, crystal structure, and conformation in solution. *Angewandte Chemie International Edition*, *35*, 1567–1569.

[84] Gerth, K., Bedorf, N., Höfle, G., Irschik, H., Reichenbach, H. (1996). Epothilons A and B: antifungal and cytotoxic compounds from *Sorangium cellulosum* (Myxobacteri(a).

Production, physico-chemical and biological properties. *Journal of Antibiotics*, *49*, 560–563.

[85] (a) Goodin, S., Kane, M.P., Rubin, E.H. (2004). Epothilones: Mechanism of action and biologic activity. *Journal of Clinical Oncology*, *22*, 2015–2025; (b) Roche, V.F., Cancer and chemotherapy. In: Lemke, T.L., Williams, D.A.,editors. *Foye's Principles of Medicinal Chemistry*. 6th ed. Lippincott Williams & Wilkins, Baltimore, 2008, pp. 1186–1187.

[86] Conlin, A., Fornier, M., Hudis, C., Kar, S., Kirkpatrick, P. (2007). Ixabepilone. *Nature Reviews Drug Discovery*, *6*, 953–954.

[87] (a) Pivot, X., Villanueva, C., Chaigneau, L., Nguyen, T., Demarchi, M., Maurina, T., Stein, U., Borg, C. (2008). Ixabepilone, a novel epothilone analog in the treatment of breast cancer. *Expert Opinion on Investigational Drugs*, *17*, 593–599; (b) Denduluri, N., Swain, S.M. (2008). Ixabepilone for the treatment of solid tumors: A review of clinical data. *Expert Opinion on Investigational Drugs*, *17*, 423–435.

[88] (a) Endo, A., Tsujita, Y., Kuroda, M., Tanzawa, K. (1977). Inhibition of cholesterol synthesis *in vitro* and *in vivo* by ML-236A and ML-236B, competitive inhibitors of 3-hydroxy-3-methylglutaryl-coenzyme A reductase. *European Journal of Biochemistry*, *77*, 31–36; (b) Tsujita, Y., Kuroda, M., Tanzawa, K., Kitano, N., Endo, A. (1979). Hypolipidemic effects in dogs of ML-236B, a competitive inhibitor of 3-hydroxy-3-methylglutaryl coenzyme A reductase. *Atherosclerosis*, *32*, 307–313.

[89] Brown, A.G. (1987). Discovery and development of new beta-lactam antibiotics. *Pure and Applied Chemistry*, *59*, 475–484.

[90] Patterson, D.E., Cramer, R.D., Ferguson, A.M., Clark, R.D., Weinberger, L.E. (1996). Neighborhood behavior: A useful concept for validation of "molecular diversity" descriptors. *Journal of Medicinal Chemistry*, *39*, 3049–3059.

[91] (a) Endo, A. (1992). The discovery and development of HMG-CoA reductase inhibitors, *Journal of Lipid Research*, *33*, 1569–1582; (b) Steinberg, D., Gotto-Jr, A.M. (1999). Preventing coronary artery disease by lowering cholesterol levels: Fifty years from bench to bedside. *Journal of the American Medical Association*, *282*, 2043–2050.

[92] Floto, A., Overington, J., Allen, J. (2004). Pleiotropic effects of statins. *Annual Reports in Medicinal Chemistry*, *39*, 239–258.

[93] Harrold, M., Antihyperlipoproteinemics and inhibitors of cholesterol biosynthesis. In: Lemke, T.L., Williams, D.A. editors. *Foye's Principles of Medicinal Chemistry*. 6th ed. Lippincott Williams & Wilkins, Baltimore, 2008, pp. 802–807.

[94] Templeton, W., *Introduction to the Chemistry of Terpenoids and Steroids*, Butterworths, London, 1969.

[95] Tobert, J.A. (2003). Lovastatin and beyond: The history of the HMG-CoA reductase inhibitors. *Nature Reviews Drug Discovery*, *2*, 517–526.

[96] Patchett, A.A. (2002). Alfred Burger award address in medicinal chemistry. Natural products and design: Interrelated approaches in drug discovery. *Journal of Medicinal Chemistry*, *45*, 5609–5616.

[97] Brown, M.S., Goldstein, J.L.A. (1986). A receptor-mediated pathway for cholesterol homeostasis. *Science*, *232*, 34–47.

[98] (a) Lea, A.P., McTavish, D. (1997). *Atorvastatin. A review of its pharmacology and therapeutic potential in the management of hyperlipidaemias. Drugs*, *53*, 828–847; (b) Sever, P.S., Dahlöf, B., Poulter, N.R., Wedel, H., Beevers, G., Caulfield, M.,

Collins, R., Kjeldsen, S.E., Kristinsson, A., McInnes, G.T., Mehlsen, J., Nieminen, M., O'Brien, E., Ostergren, J. (2003). Prevention of coronary and stroke events with atorvastatin in hypertensive patients who have average or lower-than-average cholesterol concentrations, in the Anglo-Scandinavian Cardiac Outcomes Trial-Lipid Lowering Arm (ASCOT-LLA): A multicentre randomized controlled trial. *Lancet*, *361*, 1149–1158.

[99] (a) McTavish, D., Sorkin, E.M. (1991). *Pravastatin. A review of its pharmacological properties and therapeutic potential in hypercholesterolaemia. Drugs*, *42*, 65–89; (b) Pitt, B., Mancini, G.B., Ellis, S.G., Rosman, H.S., Park, J.S., McGovern, M.E. (1995). Pravastatin limitation of atherosclerosis in the coronary arteries (PLAC I): Reduction in atherosclerosis progression and clinical events. PLAC I investigation. *Journal of the American College of Cardiology*, *26*, 1133–1139; (c) Sacks, F.M., Pfeffer, M.A., Moye, L.A., Rouleau, J.L., Rutherford, J.D., Cole, T.G., Brown, L., Warnica, J.W., Arnold, J.M., Wun, C.C., Braunwald, B.R.E. (1996). The effect of pravastatin on coronary events after myocardial infarction in patients with average cholesterol levels. *New England Journal of Medicine*, *335*, 1001–1009.

[100] Jeffrey A. Pfefferkorn,* Chulho Choi, Scott D. Larsen, Bruce Auerbach, Richard Hutchings, William Park, Valerie Askew, Lisa Dillon, Jeffrey C. Hanselman, Zhiwu Lin, Gina H. Lu, Andrew Robertson, Catherine Sekerke, Melissa S. Harris, Alexander Pavlovsky, Graeme Bainbridge, Nicole Caspers, Mark Kowala, and Bradley D. Tait (2008), Substituted pyrazoles as hepatoselective HMG-CoAreductase inhibitors: discovery of (3R,5R)-7-[2-(4-fluoro-phenyl)-4-isopropyl-5-(4-meyhyl-benzylcarbamoyl)-2H-pyrazol-3-yl]-3,5-dihydroxyheptanoic acid (PF-3052334)as acandidate for the treatment of hypercholesteromia, Journal of Medicinal Chemistry, 51, 31–45.

[101] Vroman, J.A., Alvim-Gaston, M., Avery, M.A. (1999). Current progress in the chemistry, medicinal chemistry and drug design of artemisinin based antimalarials. *Current Pharmaceutical Design*, *5*, 101–138.

[102] (a) Li, J., Zhou, B. (2010). Biological actions of artemisinin: Insights from medicinal chemistry studies. *Molecules*, *15*, 1378–1397; (b) Pandey, A.V., Tekwani, B.L., Singh, R.L., Chauhan, V.S. (1999). Artemisinin, an endoperoxide antimalarial, disrupts the hemoglobin catabolism and heme detoxification systems in malarial parasite. *Journal of Biological Chemistry*, *274*, 19383–19388.

[103] Woodard, L.E., Chang, W., Chen, X., Liu, J.O., Shapiro, T.A., Posner, G.H. (2009). Malaria-infected mice live until at least day 30 after a new monomeric trioxane combined with mefloquine are administered together in a single low oral dose. *Journal of Medicinal Chemistry*, *52*, 7458–7462.

[104] O'Neill, P.M., Posner, G.H. (2004). A medicinal chemistry perspective on artemisinin and related endoperoxides. *Journal of Medicinal Chemistry*, *47*, 2945–2964.

[105] Posner, G.H., Chang, W., Hess, L., Woodard, L., Sinishtaj, S., Usera, A.R., Maio, W., Rosenthal, A.S., Kalinda, A.S., D'Angelo, J.G., Petersen, K.S., Stohler, R., Chollet, J., Santo-Tomas, J., Snyder, C., Rottmann, M., Wittlin, S., Brun, R., Shapiro, T.A. (2008). Malaria-infected mice are cured by oral administration of new artemisinin derivatives. *Journal of Medicinal Chemistry*, *51*, 1035–1042.

[106] Jachak, S.M., Saklani, A. (2007). Challenges and opportunities in drug discovery from plants. *Current Science*, *92*, 1251–1257.

[107] Jefford, C.W., Kohmoto, S., Jaggi, D., Timari, G., Rossier, J.C., Rudaz, M., Barbuzzi, O., Gerard, D., Burger, U., Kamalaprija, P., Mareda, J., Bernardinelli, G. (1995). Synthesis,

structure, and antimalarial activity of some enantiomerically pure, *cis*-fused cyclopenteno-1,2, 4-trioxanes. *Helvetica Chimica Acta*, *78*, 647–662.

[108] Chaturvedi, D., Goswami, A., Saikia, P.P., Barua, N.C., Rao, P.G. (2010). *Artemisinin and its derivatives: A novel class of anti-malarial and anti-cancer agents*. Chemical Society *Reviews*, *39*, 435–454.

[109] Hencken, C.P., Jones-Brando, L., Bord, C., Stohler, R., Mott, B.T., Yolken, R., Posner, G.H., Woodard, L.E. (2010). Thiazole, oxadiazole, and carboxamide derivatives of artemisinin are highly selective and potent inhibitors of *Toxoplasma gondii*. *Journal of Medicinal Chemistry*, *53*, 3594–3601.

[110] Rosenthal, A.S., Chen, X., Liu, J.O., West, D.C., Hergenrother, P.J., Shapiro, T.A., Posner, G.H. (2009). Malaria-infected mice are cured by a single oral dose of new dimeric trioxane sulfones which are also selectively and powerfully cytotoxic to cancer cells. *Journal of Medicinal Chemistry*, *52*, 1198–1203.

[111] (a) Koepfli, J.B., Mead, J.F., Brockman Jr.,J.A. (1947). An alkaloid with high antimalarial activity from Dichroa febrifuga. *Journal of American Chemical Society*, *69*, 1837; (b) Kuehl- Jr., F.A., Spencer, C.F., Folkers, K. (1948). Alkaloids of Dichroa febrifuga Lour. *Journal of American Chemical Society.*, *70*, 2091–2093.

[112] (a) Zhu, S., Zhang, Q., Gudise, C., Wei, L., Smith, E., Zeng, Y. (2009). Synthesis and biological evaluation of febrifugine analogues as potential antimalarial agents. *Bioorganic and Medicinal Chemistry*, *17*, 4496–4502; (b) Jiang, S., Zeng, Q., Gettayacamin, M., Tungtaeng, A., Wannaying, S., Lim, A., Hansukjariya, P., Okunji, C.-O., Zhu, S., Fang, D. (2005). Antimalarial activities and therapeutic properties of febrifugine analogs. *Antimicrobial Agents and Chemotherapy*, *49*, 1169–1176.

[113] Zhang, W., Tang, Y. (2008). Combinatorial biosynthesis of natural products. *Journal of Medicinal Chemistry*, *51*, 2629–2633.

[114] Suh, Y.G., Kim, Y.H., Park, M.H., Choi, Y.H., Lee, H.K., Moon, J.Y., Min, K.H., Shin, D.H., Jung, J.K., Park, O.H., Jeon, R.O., Park, H.S., Kang, S.H. (2001). Pimarane cyclooxygenase 2 (COX-2) inhibitor and its structure–activity relationship. *Bioorganic and Medicinal Chemistry Letters*, *11*, 559–562.

[115] (a) Decker, M.W., Brioni, J.D., Sullivan, J.P., Buckley, M.J., Raszkiewicz, J.L., Kang, C.H., Kim, D.J., Giardina, W., Wasicak, J.T., Radek, R., Williams, M., Arneric, S.P. (1994). (*S*)-3-Methyl-5-(1-methyl-2-pyrrolidinyl)isoxazole (ABT 418): A novel cholinergic ligand with cognition-enhancing and anxiolytic activities: II. *In vitro* characterization. *Journal of Pharmacology and Experimental Therapeutics*, *270*, 319–328; (b) Arneric, S.P., Anderson, D.J., Bannon, A.W., Briggs, C.A., Buccafusco, J.J., Brioni, J.D., Cannon, J.B., Decker, M.W., Donnelly-Roberts, D., Gopalakrishnan, M., Holladay, M.W., Kyncl, J., Marsh, K.C., Pauly, J., Radek, R.J., Rodrigues, A.D., Sullivan, J.P. (1995). Preclinical pharmacology of ABT-418: A prototypical cholinergic channel activator for the potential treatment of Alzheimer's disease. *CNS Drug Reviews*, *1*, 1–26; (c) Holladay, M.W., Dart, M.J., Lynch, J.K. (1997). Neuronal nicotinic acetylcholine receptors as targets for drug discovery. *Journal of Medicinal Chemistry*, *40*, 4169–4194; (d) Huang, X., Zheng, F., Zhan, C.-G. (2008). Modeling differential binding of $\alpha4\beta2$ nicotinic acetylcholine receptor with agonists and antagonists. *Journal of American Chemical Society*, *130*, 16691–16696.

[116] (a) Viegas- Jr., C., Bolzani, V.S., Barreiro, E.J., Fraga, C.A.M. (2005). New anti-alzheimer drugs from biodiversity: The role of the natural acetylcholinesterase

inhibitors. *Mini Reviews in Medicinal Chemistry*, *5*, 915–926; (b) John, V., Latimer, L.H., Tung, J.S., Dappen, M.S. (1997). Alzheimer's disease: Recent advances on the amyloid hypothesis. *Annual Reports in Medicinal Chemistry*, *32*, 11–20; (c) John, V., Lierberburg, I., Thorsett, E.D. (1997) Alzheimer's disease: Current therapeutic approaches. *Annual Reports in Medicinal Chemistry*, *28*, 197–206.

[117] (a) Marco, J.L., Carreiras, M.C. (2003). Recent developments in the synthesis of acetylcholinesterase inhibitors. *Mini Reviews in Medicinal Chemistry*, *3*, 518–24; (b) Kozikowski; P., Tückmantel, W. (1999). Chemistry, pharmacology, and clinical efficacy of the chinese nootropic agent huperzine A. *Accounts of Chemical Research*, *32*, 641–650.

[118] Camps, P., El-Achab, R., Morral, J., Muñoz-Torrero, D., Badia, A., Baños, J.E., Vivas, N.M., Barril, X., Orozco, M., Luque, F.J. (2000). New Tacrine – Huperzine A aybrids (Huprines): Highly potent tight-binding acetylcholinesterase inhibitors of interest for the treatment of Alzheimer's disease. *Journal of Medicinal Chemistry*, *43*, 4657–4666.

[119] (a) Bolzani, V.S., Gunatilaka, A.A.L., Kingston, D.G.I. (1995). Bioactive piperidine alkaloids from *Cassia leptophylla*. *Tetrahedron*, *51*, 5929–5934; (b) Viegas-Jr.; C., deRezende, A., Silva, D.H.S., Castro-Gâmboa, I., Bolzani, V.S., Barreiro, E.J., Miranda, A.L.P., Alexandre-Moreira, M.S., Young, M.C.M. (2006). Aspectos químicos, biológicos e etnofarmacológicos do gênero *Cassia*. *Química Nova*, *29*, 1279–1286.

[120] Silva, D.H.S., Viegas-Jr, C., Santos, L.A., Castro-Gamboa, I., Cavalheiro, A.J., Bolzani, V.S., Castro, N.G., Pivatto, M., Young, M.C.M., Rocha, M.S., Fraga, C.A.M., Barreiro, E.J. (2010). Spectaline, cassine and semi-synthetic analogues as potential candidate drugs for the treatment of Alzheimer disease. *Revista Virtual de Química*, *2*, 38–46.

[121] Viegas Jr., C., Bolzani, V.S., Pimentel, L.S.B., Castro, N.G., Cabral, R.F., Costa, R.S., Floyd, C., Rocha, M.S., Young, M.C.M., Barreiro, E.J., Fraga, C.A.M. (2005). New selective acetylcholinesterase inhibitors designed from natural piperidine alkaloids. *Bioorganic and Medicinal Chemistry*, *13*, 4184–4190.

[122] Sedlacek, H.H., Czech, J., Naik, R., Kaur, G., Worland, P., Losiewicz, M., Parker, B., Carlson, B., Smith, A., Senderowicz, A., Sausville, E. (1996). Flavopiridol (L86-8275, NSC-649890), a new kinase inhibitor for tumor therapy. *International Journal of Oncology*, *9*, 1143–1168.

[123] Senderowicz, A.M. (1999). Flavopiridol: The first cyclin-dependent kinase inhibitor in human clinical trials. *Investigational New Drugs*, *17*, 313–320.

[124] Sneader, W., *Drug Discovery: The evolution of modern medicine*, John Wiley & Sons, Chichester, 1985.

[125] Heinrich, M., Barnes, J., Gibbons, S., Williamson, E., *Fundamentals of Pharmacognosy and Phytotherapy*, Churchill Livingstone, Edinburgh, 2004.

[126] (a) Zhang, L., Demain, A.L., Natural products and drug discovery. In: Zhang, L., Demain, A.L.,editors. *Natural Products, Drug Discovery and Therapeutic Medicine*. Humana Press, New Jersey, 2005, pp. 3–32; (b) Grabley, S., Thiericke, R., The impact of natural products on drug discovery. In: Grabley, S., Thiericke, R.,editors. *Drug Discovery from Nature*. Springer-Verlag, Berlin, 1999, pp. 3–36.

[127] (a) Sneader, W. (2000). The discovery of aspirin: A reappraisal. *British Medical Journal*, *23*, 1591–1594; (b) Metcalf, E., *Aspirin's History and Future, in Aspirin, the Miracle Drug*, Penguin Group, New York, 2005, pp. 3–14.

[128] Combs, D.W. (1997). Male contraception. *Annual Reports in Medicinal Chemistry*, *32*, 191–200.

[129] (a) Merkel, L. (1993) Zaprinast – A cGMP-selective phosphodiesterase inhibitor, *Cardiovascular Drug Reviews*. *11*, 501–515; (b) Langtry, H.D., Markham, A. (1999). Sildenafil: A review of its use in erectile dysfunction. *Drugs*, *57*, 967–989; (c) Ghofrani, H. A, Osterloh, I.H., Grimminger, F. (2006). Sildenafil: From angina to erectile dysfunction to pulmonary hypertension and beyond. *Nature Reviews Drug Discovery*, *5*, 689–693.

[130] Ondetti, M.A., Cushman, D.W., Rubin, B., Captopril.In: Bindra, J., Lednicer, D. editors. *Chronicles of Drug Discovery*. John Wiley & Sons, New York, 1983, pp. 1–31.

[131] Beraldo, W.T. (1992). Mauricio Rocha e Silva and the discovery of bradykinin. *Agents and Actions Supplement*, *36*, 134–139.

[132] Harrold, M., Angiotensin-converting enzyme inhibitors. In: Lemke, T.L., Williams, D.A. editors. *Foye's Principles of Medicinal Chemistry*. 6th ed. Lippincott Williams & Wilkins, Baltimore, 2008, pp. 738–752.

[133] DeLima, D.P. (1999). Synthesis of angiotensin-converting enzyme (ACE) inhibitors: An important class of antihypertensive drugs. *Química Nova*, *22*, 375–381.

4

THE IMPORTANCE OF STRUCTURAL MANIPULATION OF NATURAL COMPOUNDS IN DRUG DISCOVERY AND DEVELOPMENT

ARTURO SAN FELICIANO, MARÍA Á. CASTRO, JOSÉ L. LÓPEZ-PÉREZ, AND ESTHER DEL OLMO

4.1 INTRODUCTION

Through millions of years, Nature has provided material resources for the development and survival of living beings, particularly the most simple and essential substances such as oxygen and water and many other substances fundamental for life in the planet. In origin, most substances are generated by the plant kingdom and daily serve to microorganisms, animals, and humans as food, as well as tools, instruments, or means for maintaining and improving their living, health, and welfare. However, it must be taken into account that Nature evolves by itself, energetically driven, and that the welfare or benefit of humans or any other living species are beyond its own goals. Thus, the possibility of the existence of a hypothetical "intelligence" or "intention" of service to living beings must be absolutely excluded. Changes, large or small, physical, geological, or chemical that spontaneously lead and support the planet and the materials and substances that Nature puts at our disposal cannot be considered to have been generated to feed the man or to alleviate disease and suffering, or as services for any other living being. Thus, living beings are obliged to discover and evaluate by themselves, either instinctively or rationally, the usefulness of whatever Nature offers. Nature offers all its plants and organisms but, *a priori*,

Plant Bioactives and Drug Discovery: Principles, Practice, and Perspectives, Fourth Edition.
Edited by Valdir Cechinel-Filho.

none of them can be qualified as alimentary or medicinal. Similarly, plants provide all their components but, in a first instance, none of them can be qualified as a nutrient or a bioactive principle. To add the term medicinal to a particular plant species it is necessary to validate not only its activity, but also its *in vivo* therapeutic usefulness. Characterizing a substance as a natural active principle becomes necessary to isolate and purify it, identify its structure, and evaluate and confirm its proposed activity through the appropriate confirmatory tests. From another perspective, bioactive principles extracted from medicinal plants were certainly not biosynthesized with the aim of building any therapeutic tool that could interact with a defined molecular target to fight a particular disease of humans or animals. Consequently, the isolation of an ideal drug-compound from a medicinal plant is highly improbable. Indeed, most characterized natural bioactive principles, even those most potent and efficacious against disease, have original structures that merit, or even need, some changes to improve their pharmacological, toxicological, and therapeutic profiles.

The first and probably the most interesting example of bioactivity chemomodulation is that of the major opium alkaloid morphine. Knowing that this natural and highly potent narcotic-analgesic drug possessed severe adverse effects, once its structure was established by Robinson in 1925, more than 50 different series of molecules, with hundreds of derivatives and analogs, were synthesized and evaluated, aiming to remove its pernicious drug addiction and the respiratory depression effect, which could lead to causing death in the case of overdosing. Structural changes and structure–activity studies initiated by Eddy, May, and Mossetig [1,2], (Figure 4.1) included from modification, introduction, or removal of simple functional groups to fair skeletal simplifications (levorphanol, meperidine, etc.) or to increased structural complexities (etorphine, buprenorphine, etc.).

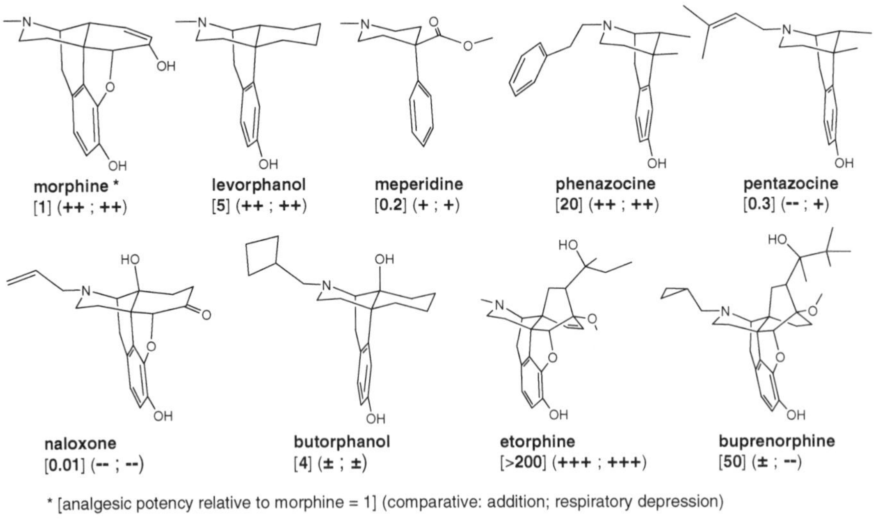

FIGURE 4.1 Simplified 3D structures and pharmaco-toxicological profile of some morphine-related drugs.

As can be seen in the Figure 4.1, the aforementioned adverse effects were corrected in many new morphine-related drugs, while the analgesic potency was enhanced up to several orders of magnitude. Additionally, a number of partial or total antagonists (pentazocine, naloxone, buprenorphine, etc), useful to treat opioid overdose or addiction, were also discovered. Interestingly many of these developments were performed, much before the discoveries of opioid receptors by the mid-1960s [3] and of the role of endorphins as endogenous opioids by the mid-1970s [4]. More interestingly, it should be noted that a simple structural change, as the substitution of the *N*-methyl group by an allyl, an isopentenyl or a cycloalkylmethyl group, served to change the agonist/antagonist character of morphine-type opioids.

In the side opposite to that of bioactive natural compounds, in a considerably much greater number, hundreds of thousands of inactive substances obtained from Nature along the centuries can be found. Many of them are scarce and display complex structures, thus, limiting the possibilities of their study focusing on a potential application. Some others can be accessible in reasonable amounts by direct extraction from their original natural sources and, finally, others are more abundant, commercially available, and cheap, like those produced as residual or by-products in large amounts by agrochemical, cosmetic, and other industries dealing with natural products. The two last groups of substances can be taken into consideration as potential starting materials for designing and synthesizing new bioactive molecules.

To illustrate the aspect of chemoinduction of bioactivity, the example of the use of gallic acid, a cheap product from tannin industries, in the synthesis of bioactive compounds, could serve adequately for the purposes of this chapter. From the pharmacological point of view, gallic acid could be considered inactive, though displays astringent, antiseptic, antioxidant, and other properties associated with its phenolic and carboxylic functions. However, the level of incorporation of the "gallic fragment," either as galloyl, trihydroxybenzyl, pyrogallol or, most commonly, as the respective methyl ethers, is high, and can be quantified through manual or automatic searching in drug libraries. Interestingly, The Merck Index provides a number of well- or less-known drugs, their structures and *names* (Figure 4.2) associated with their respective *activity* or *use*. Some examples, alphabetically ordered: amoproxam, butobendine, capobenic acid, *(antiarrhythmics)*, cinepazet *(antianginal)*, cinepazide,

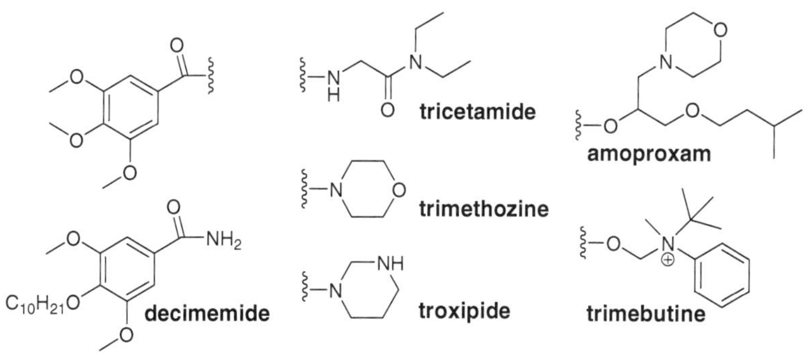

FIGURE 4.2 Structures of some selected gallamide ester and gallate ester drugs.

dilazep *(vasodilators)*, decimemide *(anticonvulsant)*, dimethophrine *(hypertensive)*, ... tricetamide, trimethozine *(sedatives)*, trimazosin *(antihypertensive)*, trimebutine *(antispasmodic)*, trimethobenzamide *(antiemetic)*, trimethoprim *(antibacterial)*, trimetrexate *(antineoplastic)*, trithiozin *(antiulcer)*, among others.

An important detail to be noted when looking at the structures of the drugs included in Figure 4.2 is that the "gallic fragment" constitutes the main skeletal basis or a representative part of the drug molecules, though most often it could not be considered as the pharmacophoric part of any drug. A discussion on why this fragment, so common in polyphenolic and aromatic alkaloidal natural compounds, can also play an important role in the structure of many useful drugs, lies beyond the aims of this chapter. However, some considerations about its relevance appear in the section dedicated to chemomodulation of cyclolignans.

Modern examples of how chemomodulation and chemoinduction of bioactivity of natural products may contribute to the development of new clinically useful drugs can be found among those related to secondary metabolites that have been approved and marketed, or have attained clinical trials phases during the last decades. Aiming to demonstrate that simple structural changes can modify substantially the pharmacological profile of a drug compound, some examples are shown in Figure 4.3.

The chemoinduction of the Na^{+}-channel blocking activity was achieved through simple sulfamoylation/ketalization of the common sugar D-fructose, thus leading to topiramate, an adjunct drug for anticonvulsant therapy, approved and introduced in UK by 1995 [5,6].

1-Deoxynojirimicin is an iminosugar found in mulberry leaves (*Morus sp*, Moraceae), having a structure close to that of 2-deoxyglucose and displaying the alpha-glucosidase inhibitory activity. Together with some derivatives was considered for development as a potential anti-HIV agent. Its chemomodulation, however, led to the simple hydroxyalkyl derivative miglitol, which was approved (Germany, 1998) [7] for treatment of type-2 diabetes, acting through delaying intestinal

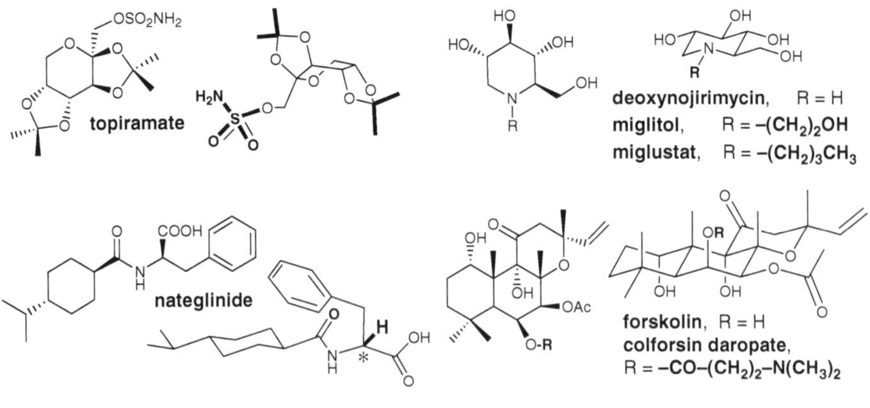

FIGURE 4.3 Some examples of chemomodulation and chemoinduction of bioactivity in natural compounds. Structural changes introduced on the natural compounds are indicated by bolded atoms and thick bonds in the 3D structures.

carbohydrate absorption. The close analog miglustat was introduced (UK, 2003) [8] for the treatment of Gaucher disease, which is caused by a deficiency in the activity of lysosomal glucocerebrosidase [9].

Nateglinide is another marketed (Japan, 1999) antidiabetic agent with structure of *N*-menthanoylated *D*-phenylalanine and insulinotropic activity. The presence of the *p*-menthanoic fragment seems to determine a lesser hypoglycemic effect with respect to other "glinide" and sulfonylurea antidiabetics [10].

Forskolin (coleonol) is a manoyl oxide derivative isolated from roots of *Coleus forskohlii* (Synonym: *Plectranthus barbatus*, Lamiaceae), a multivalent Ayurvedic medicinal species recommended for heart disease, epilepsy, spasmodic digestive and urinary pains, and insomnia [11]. Forskolin displayed recognized the vasodilating and positive inotropic activities due to stimulation of the adenylate cyclase function with increase of cAMP production. Transformed into the soluble hydrochloride salt of its 6β-OH 3-dimethylaminopropionate ester, colforsin daropate [12], attained the official approval (Japan, 1999) [13] as a coronary vasodilator with cardiotonic properties.

Many other examples of chemomodulation of bioactivity, toxicity, and ADME properties of natural products and analogs can be found in a number of published series [14,15], books [16–18], chapters [19–25], and reviews [26,27], focusing on this matter.

As an example of the most recent research on chemoinduction of bioactivity, Zhao et al. have reported how to induce anticancer properties in the old and well-known antibiotic novobiocin [28]. Through chemical manipulation of the three main fragments of the natural compound, central coumarin system, sugar moiety, and amide side-chain, they attained the transformation of this DNA gyrase inhibitor into selective inhibitors of Hsp90, a molecular chaperone modulating several oncogenic pathways, thus obtaining new lead anticancer compounds (Figure 4.4).

Once shown and illustrated with examples the current expectatives related to the incorporation of natural compounds and their derivatives and analogs to the clinical practice, it would be desirable to analyze the strategies and tactics that the chemist could apply in each particular case. The definition of strategies to be applied and the changes to be made on the starting molecule (*what to do?*), for improving its pharmacological profile, correcting an adverse effect, inducing on it a certain type of activity, or another activity different from that already displayed, can be complex. This work usually not only requires deep molecular chemical considerations, but also

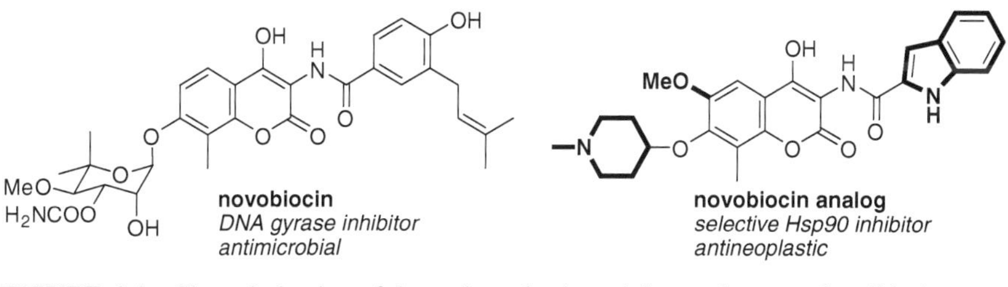

FIGURE 4.4 Chemoinduction of the antineoplastic activity on the natural antibiotic novobiocin. Structural changes inducing the new activity highlighted in thick.

others of pharmacological and toxicological nature, that generally involve the participation of a multidisciplinary team of specialists. Moreover, the definition of the structural manipulation tactics to be applied (*how to do it?*) is based on the researcher ability to define the most appropriate chemical schemes and to efficaciously carry out the planned transformations.

In general, from the pure chemical point of view, the work of chemomodulation or chemoinduction of bioactivity pursues the creation of molecular diversity through the preparation of families of structurally related compounds, aiming to have included the optimum within those synthesized and actually evaluated. Both types of research are conditioned, primarily, by the availability of the natural starting compound, which can be commercially available or directly obtainable from the plant material or can be readily synthesized if its structure is not too complex.

A second key aspect to take into account in chemomodulation is the degree of structural definition in relation with the activity/toxicity to be modulated. It is actually different to modulate the activity or the therapeutic use of compounds for which a specific target is unknown (fragment-based or ligand-based design), that to modulate the physicochemical interaction of a compound with its target, whose chemical structure, and perhaps also that of the interaction complex, is sufficiently defined (structure-based design).

In the following sections, we present two examples of our own research on natural compounds focused to improve or induce bioactivity. The first is related to the chemomodulation of the antineoplastic selectivity and the immunosuppressive activity of the natural lignan podophyllotoxin whose antimitotic mechanism of action and specific target into the cell are well established. The second is focused to the chemoinduction of antiadrenergic (vasodilating) and antiparasitic activities, starting from phytochemical studies related to the dihydrostilbenoid isonotholaenic acid, the major component of nonpolar extracts from an Andean fern.

4.2 CHEMOMODULATION OF PODOPHYLLOTOXIN CYCLOLIGNANS

The lignans constitute a large family of natural compounds widely distributed in the plant kingdom. Biogenetically they are formed by the coupling of two phenylpropane (C_6–C_3) units and they are classified into different subclasses, depending on the junction between both units and the association to other type of metabolites [29]. Thus, they have been grouped into simple lignans, cyclolignans, neolignans, mixed lignanoids, etc. In all these groups it is possible to find bioactive compounds that have been and can be used as leaders for the development of new drugs. In fact, the lignans present in medicinal plants are often recognized as the responsible principles for their therapeutic uses as folk remedies, such as cathartic, anthelmintic, wart removers, vesicant agents, and in the treatment of rheumatoid arthritis or gastric ulcer, among other ailments [30].

Within those mentioned groups, that of cyclolignans results particularly interesting. In this group, the C_6–C_3 units are joined by two C–C bonds making a new ring

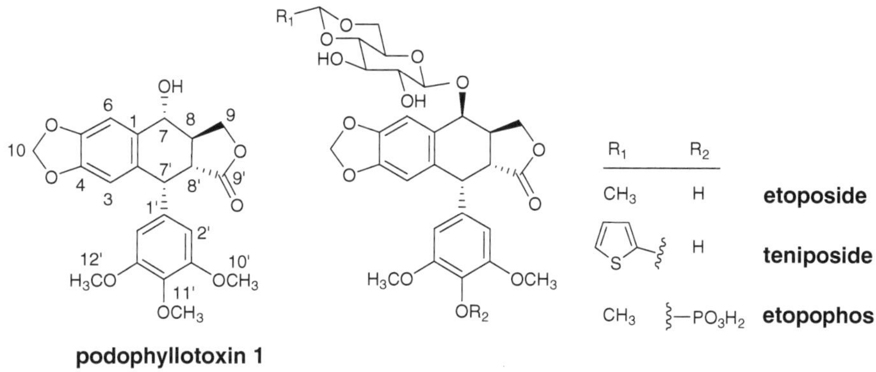

FIGURE 4.5 Podophyllotoxin and its main related antineoplastic drugs.

with three to four stereocenters and important stereochemistry-bioactivity implications. Podophyllotoxin (**1**, Figure 4.5), the major component of *Podophyllum* sp. resins, is probably the main representative element of this group [31,32]. It is clinically used for the treatment of venereal warts caused by papilloma virus [33], and several other properties have been described for it, its derivatives and analogs such as antitumor, cathartic, antirheumatic, antiasthmatic, insecticidal, antitrypanosomal, etc. [34–37]. This variety of activities and potential uses reported for a single substance clearly suggests a lack of selectivity, giving the alert on its predictable toxicity and probable adverse effects, which should be corrected, through the appropriate chemomodulation, to obtain better and less toxic drugs that could attain the approval. Among the potentialities of **1**, the anticancer activity was the most attractive and several clinical trials had been undertaken by the middle of the last century. Trials were soon abandoned, due to severe gastrointestinal toxicity and other adverse effects, but podophyllotoxin then became a good starting material for the synthesis of new antineoplastics with improved pharmacological profiles [30]. Hundreds of derivatives were prepared and evaluated and two semisynthetic derivatives, etoposide and teniposide and, more recently, a soluble prodrug of etoposide, etopophos (Figure 4.5), were approved for clinical use as anticancer agents [38]. They are currently included in treatment protocols for testicular carcinoma, small-cell lung carcinoma, lymphomas, leukaemias, Kaposi's sarcoma, and other malignancies.

Surprisingly, these semisynthetic derivatives and the parent compound **1** showed different mechanisms of action [30,32a]. Podophyllotoxin inhibits tubulin polymerization through interaction at the colchicine binding site. If tubulin does not polymerize, the microtubules and the spindle are not formed. Cells begin the division but chromosomes are unable to separate and cell division is arrested in the G2/M cell-cycle stage. Cells remain with their chromosomes joined together until they disintegrate several hours later.

Etoposide and related cyclolignans do not inhibit tubulin polymerization, but induce a premitotic blockade in the cell-cycle S stage by binding to DNA topoisomerase II (Topo-II), an enzyme required for the unwinding of DNA during replication. Topo-II is assumed to form a transient covalent DNA-protein link, the cleavage complex, which allows one double strand to pass through a temporary break into another double strand. Etoposide binds to and stabilizes the cleavage complex preventing repair of the double-strand breaks.

The main, though simple, chemical modifications that convert podophyllotoxin from a tubulin polymerization inhibitor into a DNA Topo-II inhibitor [29,32] are: demethylation at C-4′, epimerization at C-7, glycosylation of the 7β-OH group and acetalization of the 4-OH and 6-OH groups of the glucopyranose unit by aldehydes. Currently, these derivatives are widely used as anticancer drugs, but they still have problems such as myelosuppression, poor bioavailability, drug resistance, or carcinogenic induction. Because of these problems, the aryltetralin lignans are still the subject of extensive research looking for new drugs with improved pharmaco-toxicological profiles. With this aim, many analogs of etoposide have been prepared by replacement of the C-7 sugar moiety with a nonsugar bulky substituent linked to the lignan through *O*-, *S*- or *N*-linkage [31,39]. In general, ethers, esters, and thioethers of 4′-demethylepipodophyllotoxin show the lower activity as compared with the *N*-linked analogs, while a bulkier substitution at C-7 usually enhances the anti-Topo-II activity and helps to overcome drug resistance. Several derivatives have shown improved profiles in comparison with etoposide and some of them, such as NPF, GL-331, TOP-53, GP-11, or F11782 (tafluposide, a dual Topo-I and Topo-II inhibitor) have undergone anticancer clinical trials, while CPH-82 has been approved for antiarthritic treatment (Figure 4.6) [31,40,41].

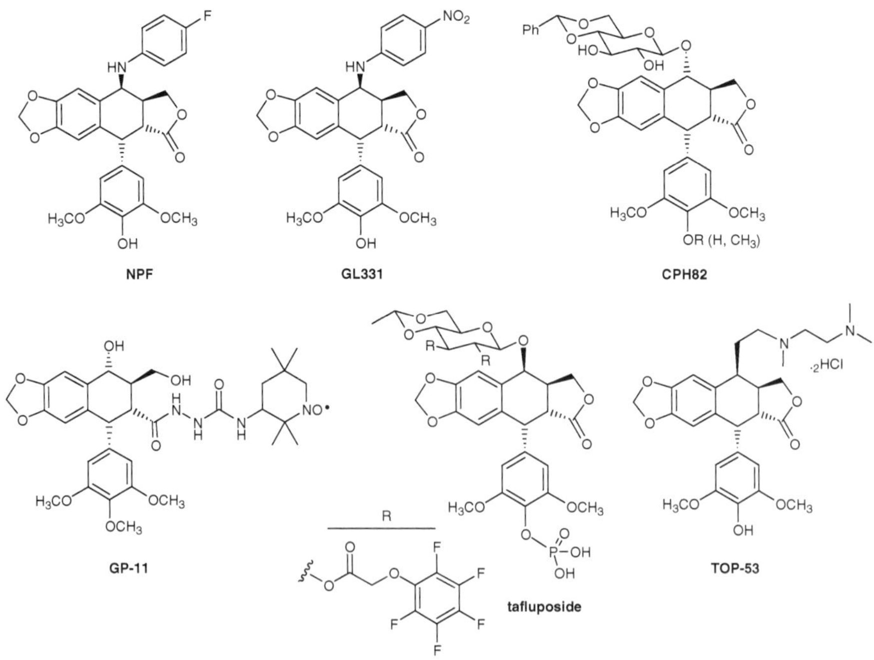

FIGURE 4.6 Other main antineoplastic and antiarthritic podophyllotoxin derivatives.

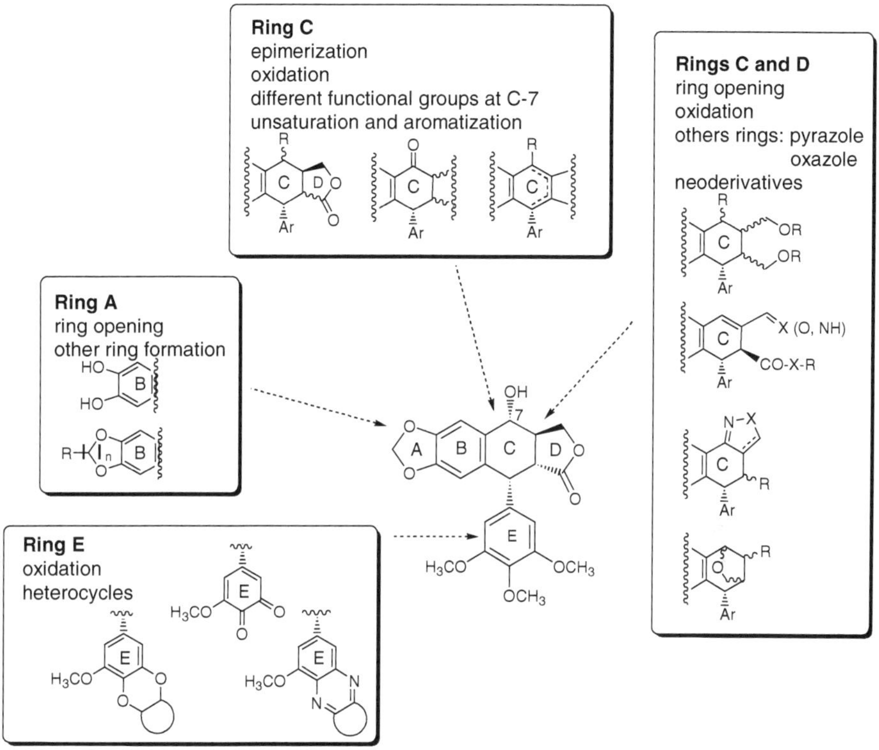

FIGURE 4.7 Main areas and structural changes involved in podophyllotoxin chemomodulation.

Our research group has been interested in the chemomodulation of podophyllotoxin for several years and nearly every ring or function in the cyclolignan skeleton has been modified by us [40,42–45]. Some of the main changes performed are summarized in Figure 4.7, taking into account that changes could have been implemented in different parts of the molecule simultaneously.

Most of those derivatives represented in Figure 4.7 resulted several times less potent than podophyllotoxin; however, it is worth to note that some cyclolignans lacking the lactone ring showed selective cytotoxicity against certain solid-tumor cells, in comparison with normal or leukemic cells and, more interestingly, some other less-cytotoxic compounds displayed a relevant immunosuppressant activity that compared favorably with that of cyclosporine A.

4.2.1 Chemomodulation of the Immunosuppressive Activity of Podolignans

Most chemotherapeutics have a number of undesirable side effects that sometimes include the decreasing of natural defenses and immune responses of the treated patients. This is practically sure in the case of cancer treatment due to the more aggressive nature of the available therapeutic agents. The immunosuppressive effect of drugs can be considered only as an adverse side effect to be avoided or corrected, or on the other hand, can be put to good use in the reduction or suppression of the immune response of patients to avoid the rejection of transplanted organs. At the mid-90s when we started this study, there were only a few references [46,47] related to the potential utility of lignans as immunosuppressant, but there was not report of a direct experimental evaluation of this property in cyclolignans. So, we decided to explore this aspect in podolignans, considering that substantial improvements of the immunosuppressive (IMS) index of these compounds would be attained better through decreasing their cytotoxicity, rather than through the increase of their immunosuppressive potency. Indeed, a proposal of what structural changes should be implemented to enhance the immunosuppressive aspects of lignans would be, even today, totally speculative, while structural aspects supporting the antineoplastic cytotoxicity of podolignans seemed clear at that time. Thus, we looked at the structural facts being the most relevant for the cytotoxicity of podolignans, the trans-fused γ-lactone and the dioxole ring, as the targets to be modified.

Several less-cytotoxic cyclolignans, either lacking the lactone ring or the methylenedioxy group, were selected and submitted to evaluation in the mixed lymphocyte reaction (MLR) *in vitro* test [48], displaying relevant results of immunosuppressant activity for compounds **2**, **3,** and **4** (Table 4.1) in comparison with those observed for cyclosporine A (CyA).

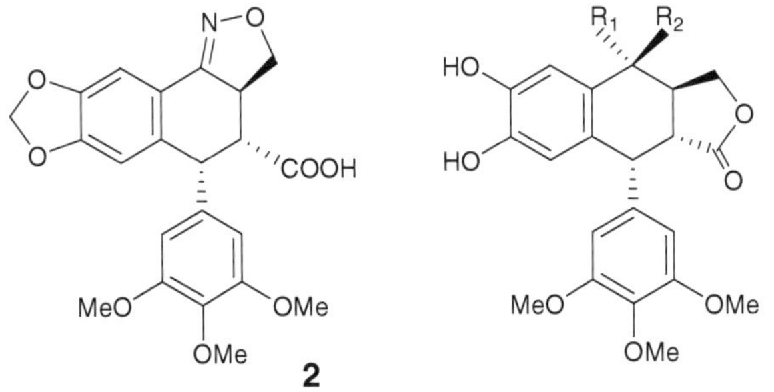

	R_1	R_2
3	OH	H
4	H	OH

TABLE 4.1 ***In vitro*** **Immunosuppressive (IMS) and Cytotoxicity Data for Compounds A–C**

Compound	MLR[a]	LcV[b]	nCIMSI[c]
2	<0.009	>30.0	>3333
3	0.00563	>27.5	>4885
4	0.0000015	>35.0	>23333333
CyA	0.0518	2.75	53

[a] Mixed lymphocyte reaction, IMS activity (IC_{50}, μg/mL).
[b] Lymphocyte viability, cytotoxicity (CC_{50} μg/mL).
[c] Noncytotoxic IMS index (ratio LcV/MLR). Eight different dilutions were tested for each compound per duplicate (concentration ranged from 0.5 to 0.0015 μg/mL).

These promising compounds were then submitted to the *in vivo* experiments graft versus host reaction and skin grafting. The results showed that prevented splenomegaly, in comparison with control animals and also promoted tolerance to grafting, being the first report of an experimental evidence of this property for cyclolignans [49].

4.2.2 Chemomodulation of Antineoplastic Potency and Selectivity of Podolignans

The two mechanisms of action described for podophyllotoxin-related cyclolignans are well recognized; however, the molecular interaction of etoposide in the ternary complex with DNA and Topo-II is not so clear. Eich et al. [50] proposed a molecular mechanism, in which the free phenol at C-4′ would interact with DNA, while Topo-II would attack the lactone carbonyl at C-9′, suggesting that these lignans could be phosphorylated at the C4′-OH by a phosphate group of DNA and successive or simultaneously would acylate any Topo-II nucleophile, through the lactone carbonyl. However, after a conformational study of etoposide, we felt that this mechanism looked improbable. The two bulky groups of etoposide, the modified sugar on the upper part and the pendant trimethoxyphenyl at the lower part, have almost free rotation, and thus it seemed clear that the approach of the large Topo-II to the C-9′ carbonyl of the lignans would be too hindered. On the other hand, the electronically deficient C-9 methylene looks a more accessible position, which could be attacked from the rear side by a nucleophilic function of Topo-II, which would covalently result bonded to the lignan. The chemical process would be energetically favored by the aperture of the *trans*-fused and much strained lactone ring. On this basis, we considered that if C-9 could be an actual point of interaction, *trans*-fused cyclolignanolides could also act as alkylating agents, which would be enough significant in order to support the existence of a third, previously unmentioned, mechanism of action for podolignans. The actual reaction with target biomolecules would largely justify their toxicity and adverse effects. To further validate this idea, we confirmed chemically that weak nucleophiles were able to open the strained *trans*-lactone ring of some podolignans [51].

The ease with which the C-9 methylene reacts, prompted us to modify the electrophilic character at that position, and we prepared derivatives with a carbonyl function at C-9, by opening the lactone ring under basic conditions, formation of the methyl ester, and oxidation under Swern conditions [52]. Among those derivatives, the podophyllic aldehyde **5** showed an interesting selectivity *in vitro*, being about 20 times more potent against HT-29 colon carcinoma cells, than against P-388 leukaemia cells or normal CV-1 monkey kidney cells, while retaining IC_{50} cytotoxicity values similar to those of the nonselective podophyllotoxin (**1**). This fact, added to the change of configuration at C-8′ would serve to reconsider the previously recognized presence of the *trans*-fused γ-lactone ring as necessary for the activity of podolignans.

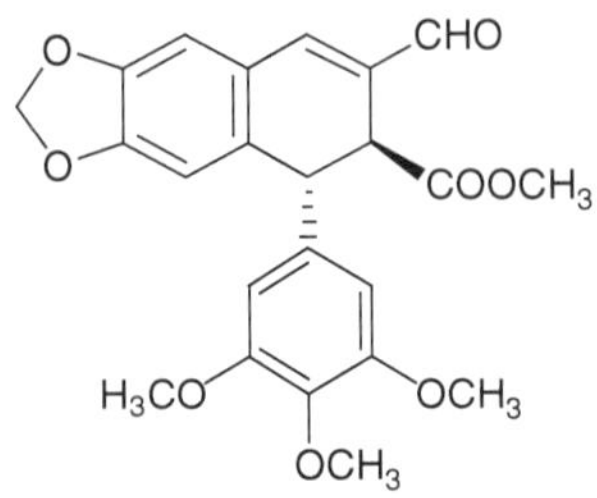

Podophyllic aldehyde (**5**)

Evaluation of **5** at the NCI (USA) on the complete panel of neoplastic cells confirmed the selectivity for colon carcinoma cells (Log TGI = −7.0) and in lesser degree for CNS and breast cancers (Log TGI = −6.1 and −6.3, respectively).

Due to the retention of the cytotoxicity levels after such a substantial change of structure and molecular geometry, with the presence of a β-axially-oriented substituent at C-8, cell cycle and tubulin binding assays were considered necessary to confirm the assumed mechanism of action. Experimentally, the aldehyde **5** displayed its antimitotic action through disorganization of the microtubule network, in a way similar to podophyllotoxin [53], also blocked the cycle of neoplastic cells at the G2/M stage and was able to induce delayed apoptosis of cells, characterized by caspase-3 activation, with a long lag between microtubule disassembly and the onset of apoptosis [42].

In addition, conformational molecular modeling calculations and docking of **5** in tubulin were performed. The graphical results, in comparison with those of podophyllotoxin (**1**), and in agreement with the experimental facts mentioned earlier, are shown in Figure 4.8. Nevertheless, the initial attempted docking of the main conformer of **5** with AUTODOCK [54] was unsuccessful, probably due to steric incompatibility of the β-axially extended methoxycarbonyl group at C-8′. However, actual docking could be achieved for two other less-stable conformers of **5**, though with energies 30–40 kcal/mole higher than those observed for **1**. As can be seen in Figure 4.8c and d the C–O bond of the methoxyl group of the ester function adopted a spatial disposition almost parallel with respect to the four-rings moiety in both conformers (Figure 4.8d), differing substantially from the rather orthogonal arrangement adopted by the methoxyl group in the most stable conformer (not shown). Such conformations and their global molecular orientations permitted to both less-stable conformers to attain the poses permitting to avoid some contacts with the protein, thus allowing their docking.

With these results in hand, the aldehyde **5** seemed to be a good candidate for further chemomodulation and several transformations were carried out to modify the electrophilic character at C-9, in order to analyze the influence of such position on the selectivity observed for **5**. The modifications considered were: changes in the degree of oxidation of both C-9 and C-9′; formation of vinylogues; reactions with nucleophiles or synthesis of purinyl and other heterocyclic analogs. Most of the compounds prepared were evaluated against neoplastic cells and many of them lost

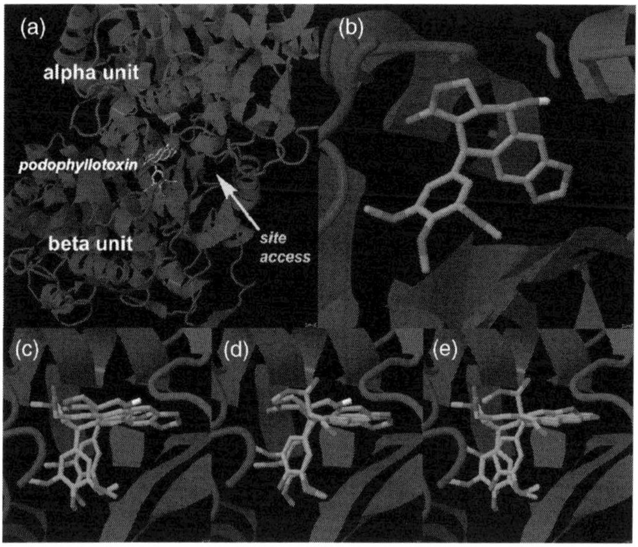

FIGURE 4.8 Docking of podophyllotoxin and two conformers of podophyllic aldehyde 5 in tubulin. (a) Overview of the α and β units of tubulin and podophyllotoxin **1** at the colchicine site, with indication of the probable route of access for docking. (b) Magnification of **1** docked in the active site. (c) Comparison of docked **1** and conformer-1 of aldehyde **5**. (d) Comparison of docked **1** and conformer-2 of aldehyde **5**. (e) Comparison of both conformers of aldehyde **5** docked in tubulin.

the selectivity and/or decreased the potency with respect to the parent compound [42]. However, those derivatives resulted from the reaction of the aldehyde function with nitrogen nucleophiles, such as hydrazines or amines, maintained the cytotoxicity and even improved the selectivity of the parent aldehyde. Particularly interesting were the imine derivatives (compounds **6–9**, Figure 4.9). All of them showed IC_{50} values below the μM level and some of them were selective against HT-29 carcinoma in comparison with other cell lines tested (P-388, A-549, and MEL-28). The ethylimine **6** was some 500 times more potent against HT-29 than against P-388, thus the selectivity increased by a factor of 25, compared to the lead compound, the aldehyde **5**. These studies confirmed the unexpected fact that the γ-lactone ring is not an essential feature for cytotoxicity, whereas the presence of an electrophilic function at C-7/C-9 seems important for the antineoplastic potency and selectivity of this type of cyclolignans [43].

Another objective of the research was to study the influence of the nature and size of the group at C-9′ on the cytotoxicity and selectivity. Hence, several podophyllic aldehyde derivatives, having longer and structurally varied ester chains at C-9′ and also some bioisosteric amides were prepared and evaluated, together with some imine derivatives at C-9 [45]. In general, those compounds containing an ester function at C-9′ (compounds **10–14**, Figure 4.9) were more potent than the corresponding amides and the combination of an aldehyde or an imine function at C-9 with an ester at C-9′,

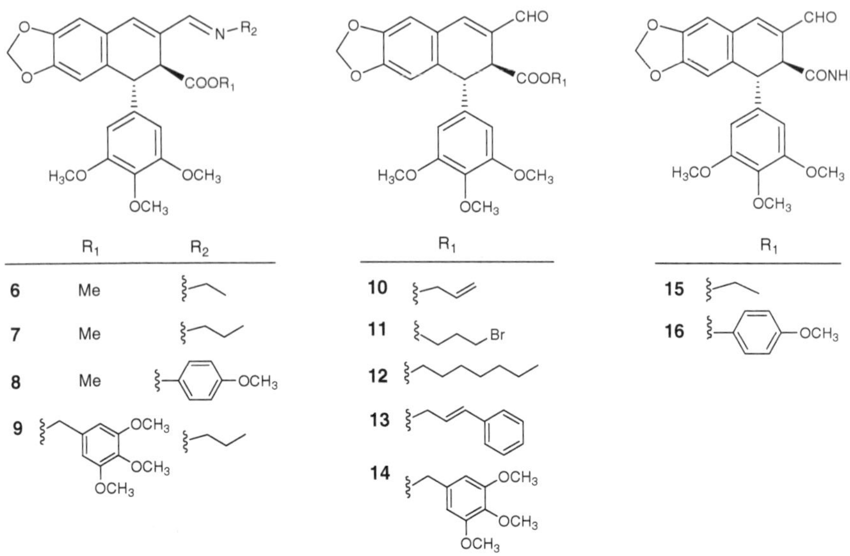

FIGURE 4.9 Selected derivatives of podophyllic aldehyde 5.

was found to be the best arrangement for potency and selectivity against HT-29 colon carcinoma. The transformation of the imine function into the corresponding more stable amine analogs was also performed but, the chemical stabilization had a negative influence on cytotoxicity. The results of this part of the research confirmed that the C-9 position of these cyclolignans plays an important role in their activity and that the presence of an electrophilic function at C-9 (aldehyde or imine), is the most important feature for their potency and selectivity.

In a more recent study on derivatives of aldehyde **5**, compounds **11**, **12**, **15**, and **16** (Figure 4.9), while showing not too high antineoplastic cytotoxicity in comparison with other related members of these series, were able to promote apoptotic cell death without any significant cell-cycle arrest [44]. These potency and mechanistic differences, being displayed by compounds with similar structures, could be considered as an experimental proof supporting the existence of a third mechanism of action for podophyllotoxin-related lignans [43].

4.3 CHEMOINDUCTION OF BIOACTIVITY ON DIHYDROSTILBENOIDS

Natural stilbenoids have a wide distribution in the plant kingdom and display different types of the biological activity [55]. *Trans*-stilbenoids, with resveratrol (Figure 4.10) as representative of the class, have been reported as antitumor, antileukemic, antioxidant, antiinflammatory, antiaging, cardioprotective, and neuroprotective [56–58]. *Cis*-stilbenoids also display interesting biological properties, as combretastatin A4, with potent antimitotic activity due to inhibition of tubulin

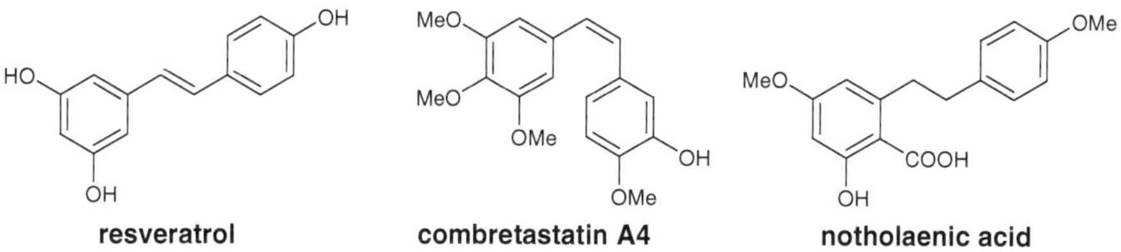

FIGURE 4.10 Representative elements of *trans*-, *cis*- and dihydro-stilbenoids.

polymerization [59]. Dihydrostilbenoids are less frequent in plants, though also have been reported as antioxidant [60], antibacterial [61], and antifungal [62].

The Andean American fern *Notholaena nivea* var. *nivea* is used in the Peruvian traditional medicine as a tea for the treatment of diabetes. Wollenweber et al. [63] studied different *Notholaena* spp. and concluded that dihydrostilbenoids are very common in the genus, serving as chemotaxonomic markers, with particular reference to those regioisomers notholaenic and isonotholaenic acids (Figures 4.10 and 4.11).

Some time ago we were interested in the bioguided studies of *N. nivea* var. *nivea*. First, the reported antidiabetic activity was corroborated in dose-response mode for aqueous, methanolic, and dichloromethane extracts, by testing on normal and diabetic Wistar male rats [64]. Then, bioguided fractionation of the dichloromethane extract allowed associating the activity with the flavonoid fraction and the known flavonoids quercetin and acacetin were experimentally confirmed as the antihyperglucemic components. During the phytochemical work along with flavonoids, triterpenoids, and other metabolites, the dihydrostilbenoid isonotholaenic acid (**17**, Figure 4.11), devoid of antidiabetic or any other known activity, was found as the major component (4.5%) of the dichloromethane extract. With several grams in hands, we wanted to transform **17** into bioactive compounds based on its structural parenthesque with either the antimitotic combretastatins or some antimicrobial phenolic acids and with other phenolic compounds with the vasorelaxant activity previously discovered by our group. We also planned to incorporate some *typically active fragments* to the dihydrostilbenoid skeleton in order to induce a particular type of bioactivity. Consequently, with the previous support of the corresponding literature searches, several expected potentialities for **17** were experimentally checked through the opportune assays.

Firstly, we found literature reports of antiadrenergic properties for hydroxybenzoic acids and some of their derivatives, gallic acid [65], and aspirin [66], as for other

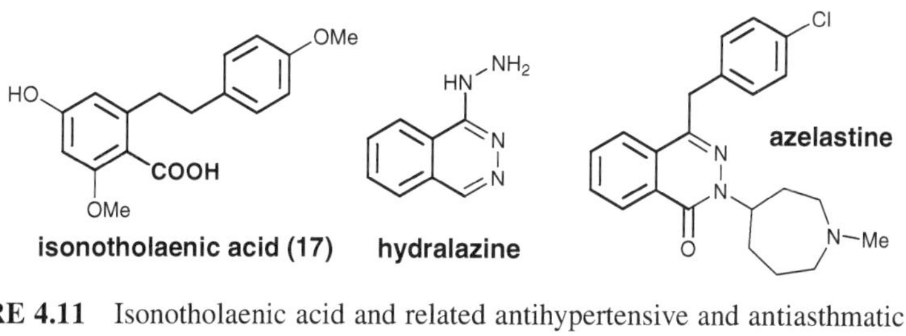

FIGURE 4.11 Isonotholaenic acid and related antihypertensive and antiasthmatic drugs.

phenolic compounds [67–69]. Additionally, heterocyclic compounds containing the stilbene-carbonyl skeletal basis, such as the benzylphthalazinones, were reported to display important antihypertensive and antiasthmatic effects, and such type of compounds also demonstrated to exert inhibitory activity on phosphodiesterases and on platelet aggregation [70].

Vascular tone control is one of the principal keys for hypertension treatment and control [71]. Hydralazine, a phthalazine drug launched in the 1950s, represented one of the first antihypertensive agents, being considered as a lead for the design of directly acting vasodilators [72]. Several families of compounds structurally based on the phthalazinone system were then synthesized and tested with remarkable antihypertensive and vasorelaxant results [73,74]. Azelastine (Figure 4.11), is a phthalazinone-stilbenoid drug useful in the treatment of asthma due to its potent bronchodilator activity [75,76].

Recent reports could also be found on the leishmanicidal properties of natural and synthetic stilbene-related compounds as piceatannol [77], resveratrol [78], benzofurans [79], combretastatins [80], diamidines [81], and tamoxifene [82]. Some of those compounds found in the literature also showed antiplasmodial [83–88] or antitrypanosomal [89,90] activities.

The isonotholaenic acid was then subjected to different bioassays to be evaluated as antiadrenergic (vasorelaxant) and antiparasitic, with poor to moderated results in all of the cases. Compound **17** (at 10^{-4} M) induced a 34% relaxation in rat aorta rings precontracted with 10^{-5} M phenylephrine (PE); it also displayed total grown inhibition on *Leishmania spp* promastigotes at a high concentration (50 μg/mL), and a better antiparasitic behavior on *Plasmodium falciparum* F32 ($IC_{50} = 16$ μg/mL), while was practically inactive against *Trypanosoma cruzi* epimastigotes. In order to obtain active or more active compounds, some derivatives of **17** were prepared, but only several amides and particularly its piperidide attained an improvement in the vasorelaxant and antiparasitic properties. Then, we decide to design and develop totally synthetic procedures to prepare several types of compounds that, while retaining the stilbenecarbonyl fragment of **17**, would incorporate certain pharmacophoric functions or fragments (amide, hydrazide, hydrazone, amidine, azoline, pyrimidine, and others), similar to those commonly present in vasorelaxant and/or antiparasitic drugs. The synthetic scheme should be at the same time simple, efficient, and versatile, allowing us to obtain a wide variety of structural families and substances. This way, we got not only a number of new molecules of more than twelve different skeletons, but also certain lead compounds in the activities considered, to be subsequently improved.

The selected chemical procedure was based on the synthesis of the benzalphthalides (**I**), which would result in versatile intermediates, that on treatment with different reagents would provide a wide variety of structural families, including several with fused-heterocyclic systems. A short resume of the structural possibilities is shown in Scheme 4.1.

The procedure initially applied to obtain benzalphthalides **I**, the Nokihara method [91], involved high temperature and neat reagents. We made an improvement by adapting a Dean–Stark apparatus and adding toluene to the reaction. In other

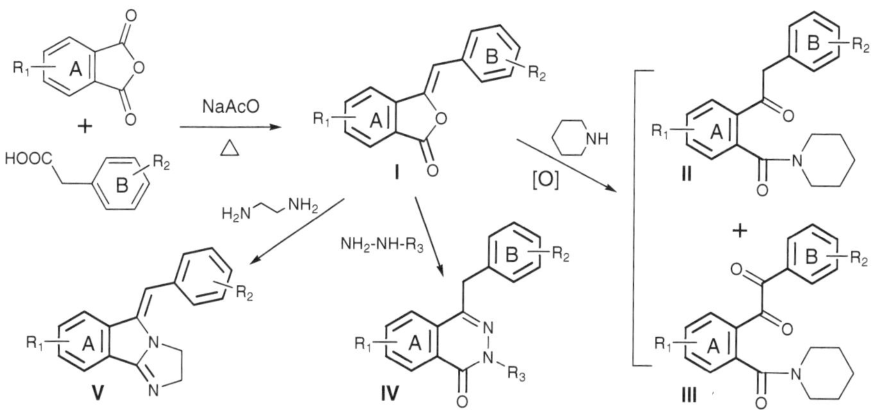

SCHEME 4.1 General procedure to obtain some representative series of stilbenoids (II and III) and hetero-stilbenoids (**I**, **IV**, and **V**) related to isonotholaenic acid.

experiments based on the Lácová microwave method [92], the anhydride and the phenyl acetic acid were fused with a "heating jacket," then KAcO was added and the round-bottom flask containing the mixture, placed into a beaker filled with silica gel. This way yields were improved and the reaction times shortened.

The first families of compounds included only ring B modifications with electron-donating or withdrawing groups at different positions. The compounds were tested on rat aorta rings precontracted with phenylephrine (PE), and in antiparasitic assays against *Leishmania* spp, *T. cruzi* and *P. falciparum*. Significant results were found for the dihydrostilbenoid **IIIa** and the phthalazinone **IVa** as vasorelaxant: 66% and 64% rat aorta relaxation, at 10^{-5} μM, respectively, with a 20-fold increase of potency relative to that of **17**, approximately. Good results were also observed for the imidazoisoindole **Vc** as antiparasitic, with potencies close to or higher than those found for the corresponding reference drugs (Figure 4.12).

For **Vc** we found values of $IC_{50} = 13.9$ μM against *Leishmania* spp. promastigotes (pentamidine: 3.7 μM); of $IC_{50} = 33.6$ and 7.9 μM against epimastigotes and trypomastigotes of *T. cruzi*, respectively, (benznidazol: 28.5 μM, gentian violet: 245 μM); and of $IC_{50} = 0.1$ μM against *P. falciparum*, F-32 strain (chloroquine: 0.04 μM). Taking these results into account, compound **IVa** was selected as the lead compound for further chemomodulation to enhance its vasorelaxant activity and **Vc** for improvement of the antiparasitic activities, by introducing other structural modifications, first on ring B and then on ring A.

4.3.1 Chemomodulation of the Vasorelaxant Activity of Phthalazinone IVa

In the second phase of the research, to ascertain the influence of substitution on the nitrogen atom at position *N*-2, we combined twelve modifications of ring B [$R_3 = H$; 2-Cl; 4-OMe; 4-SMe; 4-Cl; 2,4-Cl_2; 3,4-$(OMe)_2$; 3,4-Cl_2; 3,4-OCH_2O; 3,4,5-$(OMe)_3$ and the change of the phenyl group to the bicyclic α-naphthyl and β-naphthyl groups] with the presence or the absence of a methyl group at *N*-2 ($R_3 = H$, Me; Figure 4.12).

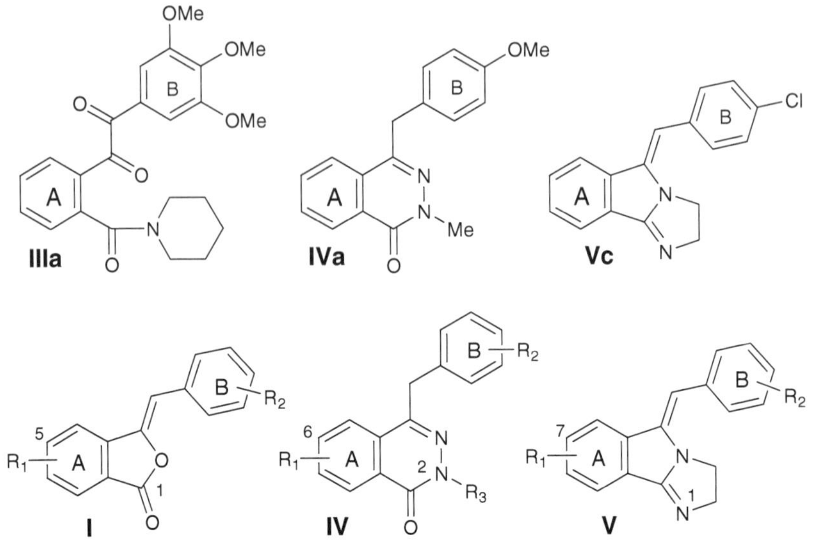

FIGURE 4.12 Representative compounds and structural types of antiparasitic heterostilbenoids.

From the results of the respective bioevaluation we deduced that alkylation at position *N*-2 increased substantially the vasorelaxant activity, while the new substituents incorporated on ring B failed to increase the potency. Thus, among other similarly potent vasorelaxants, compound **IVb** ($R_1 = H$, $R_2 = 4\text{-Cl}$, $R_3 = Me$), which showed the IC_{50} value of 4.3×10^{-7} M, on aorta rings pre-contracted with 10^{-5} M PE, emerged as a greatly improved lead compound. The next step was the introduction of other alkyl/aryl (ethyl, allyl, *n*-butyl, *t*-butyl, phenyl, 4-Br-phenyl, and 4-NO_2-phenyl) groups at *N*-2, but none of these groups contributed positively to the activity.

The third phase of this research focused the evaluation of the influence of the substituents R_1 attached to ring A on the vasorelaxant activity. Changes at this part of the molecule were initially based on the use as starting materials of those commercially available substituted phthalic anhydrides (R_1: 4,7-Cl_2; 5,6-Cl_2; 4-NO_2; 5-Me; 5-COOH), which led to the corresponding phthalazinones [R_1: 5,8-Cl_2; 6,7-Cl_2; 5-NO_2; 6(7)-Me; 6(7)-COOH], respectively. In the cases of starting with the nonsymmetric 5-methyl or 5-carboxyl derivatives of phthalic anhydride, it should be noted that the synthesis would ordinarily led to 1:1 mixtures of 5- and 6-methylbenzalphthalides and, correspondingly, to 6- and 7-methylphthalazinones. In this phase were evaluated 33 additional phthalazinones with structures combining the above-mentioned substituents at ring A with several of those most relevant arrangements for ring B and, in general, methylated at *N*-2. For these phthalazinones another assay of inhibition of PE-induced contraction was performed, including the previous incubation of the aorta rings with phthalazinones at 10^{-5} M. The preincubation with phthalazinones significantly prevented the ulterior contraction as shown in Table 4.2,

TABLE 4.2 Inhibition Data for PE-Induced Contraction on Rat Aorta Rings Preincubated with Selected Phthalazinones (10^{-5}M)

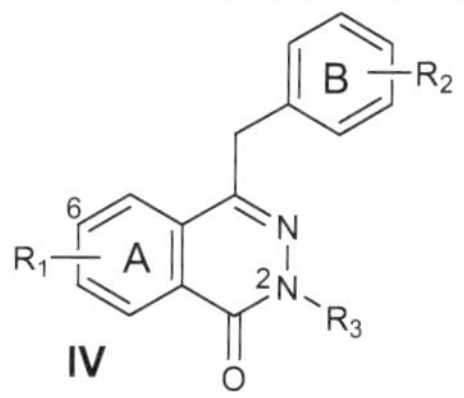

Compound	R_1	R_2	R_3	E_{max}[a] ± S.E.M[b]
IVa	H	4-OMe	Me	~85[c]
IVb	H	4-Cl	Me	62.2 ± 8.3
IVc	6,7-Cl_2	3,4-Cl_2	Me	60.6 ± 10.2
IVd	6,7-Cl_2	4-Cl	Me	59.9 ± 10.6
IVe	5-NO_2	4-Cl	H	58.5 ± 7.4
IVf	6(7)-COOH	4-Cl	Me	57.4 ± 13.0
IVg	6(7)-NO_2	4-Cl	Me	57.2 ± 10.0
IVh	6(7)-Me	2-Cl	Me	54.9 ± 10.4
IVi	6(7)-NO_2	4-OMe	Me	43.7 ± 4.3
IVj	6(7)-Me	4-OMe	Me	39.6 ± 7.6
IVj′	6-Me	4-OMe	Me	64.2 ± 5.0
IVk	6(7)-Me	3,4-OCH_2O–	Me	33.9 ± 7.4
IVm	6(7)-Me	4-Cl	Me	29.3 ± 5.3

[a] Maximal contractile response, expressed as % of that attained in absence of phthalazinone.
[b] Standard Error of the Mean of 4–8 separate experiments.
[c] Estimated from IC_{50} relaxation data.

where it can be seen that many compounds gave better results than **IVb** in inhibiting the PE-induced contraction.

According to these data compounds **IVm** and **IVk** revealed their high ability to prevent contraction by up to more than 2/3 of the maximum achieved by PE in their absence. Other observations, based also on experimental data not included in the table, are the following: substitutions on ring A improve, in general, the vasodilating activity of phthalazinones, with apparent independence of the nature of the substituent (**IVc-m** vs. **IVb**). Comparison between data for the 6-methyl and 7-methyl derivatives **IVj** and **IVj′** suggests a probable preference of position C-7 for the activity. Related to the type of substituents on ring A, the potency decreases in the order Me > NO_2 ≈ COOH within those substituents actually present in the assayed compounds, and with respect to those attached to ring B, the order of decrease is 4-Cl > 3,4-OCH_2O- ≈ 4-OMe > 2-Cl and others. Interestingly, the nitroderivative **IVe** was the only compound not methylated at *N*-2 within those classified as more potent inhibitors than the previous lead compound **IVb**. Indeed, its corresponding methylated analog (not included in Table 4.2), attained only to decrease the Emax value to 74.9 ± 4.3 % of the PE control.

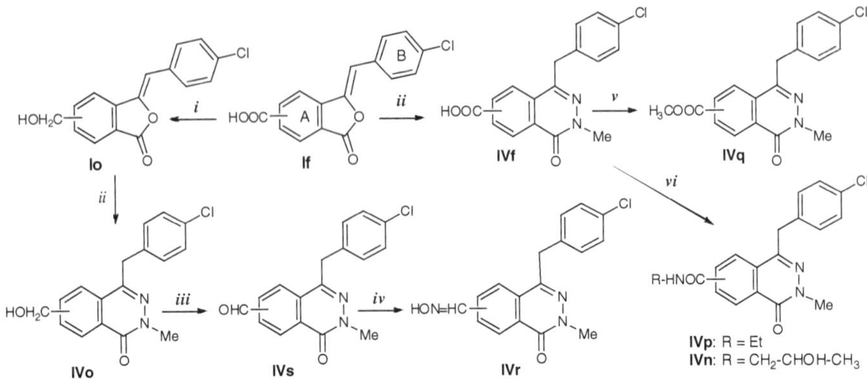

Reagents and conditions:
i) Et_3N, EtOCOCl, THF, -15ºC, 1h; then $NaBH_4$, H_2O, 0ºC, 30 min. *ii*) NH_2-NH-CH_3, silica, MW, 45 min.
iii) $(CO)_2Cl_2$/DMSO, Et_3N, CH_2Cl_2. *iv*) NH_2OH.HCl, pyridine, EtOH. *v*) CH_2N_2/ether. *vi*) R-NH_2, DCC/HOBT, THF.

SCHEME 4.2 Phthalides and phthalazinones with special modifications on ring A (alphabetically named and ordered according to the respective potencies shown in Table 4.3).

The next modifications to be made, while keeping the 4-chloro substituent on ring B and the methyl group at *N*-2, were focused to transforming the substituents at position 6(7) on ring A. Those needed reactions for these changes were applied as convenient, either on the intermediate substituted phthalides or on the resulting phthalazinones as shown in Scheme 4.2. The 6(7)-carboxyphthalazinone **IVf** was selected as the starting material for the last chemomodulation phase.

Data of aorta contraction prevention for the new compounds are presented in Table 4.3, along with those for the unsubstituted phthalazinone **IVb** and the most potent compound of the series previously evaluated **IVm**, included for comparison. As can be seen compounds **IVo–q** ranged within the levels observed for compounds **IVc–j**, being more potent than **IVb** but lesser than **IVm**. However, significant improvement of the inhibitory activity was observed for the aldehyde **IVs** and its oxime **IVr**, that resulted those most potent contraction inhibitors evaluated to the date.

These results and the own structure of the aldehyde **IVs** could have been considered as new starting points for further chemomodulation aiming to attain the optimal vasorelaxant structure. Nevertheless, it was considered that working ahead on this subject, based on indirect design and without information about the mechanism of action or the molecular target for these vasorelaxant phthalazinones, most probably would have not led to further substantial progress in this research.

4.3.2 Mechanistic and Structure–Activity Relationships Studies on the Vasorelaxant Activity of Phthalazinones

Initially, studies on the possible α-agonistic/antagonistic and β-agonistic/antagonistic nature of the effects of phthalazinones on aorta rings were carried out. Unsubstituted phthalazinones at *N*-2 produced a competitive α-antagonism related to phenylephrine (PE) response. The best compound with this behavior was the

TABLE 4.3 Inhibition Data for PE (10^{-5}M)-Induced Contraction on Rat Aorta Rings Preincubated with Several Ring-A Functionalized Phthalazinones

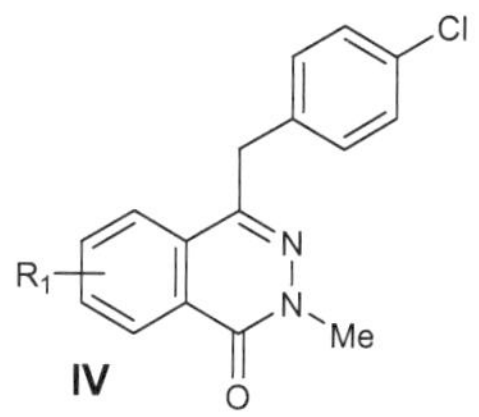

IV

Compound	R_1 [a]	E_{max} % [b] ± S.E.M [c]
IVn	–CONHCH$_2$CH(OH)Me	64.0 ± 9.4
IVb	H	**62.2 ± 8.3**
IVf	–COOH	57.4 ± 13.0
IVo	–CH$_2$OH	37.6 ± 13.4
IVp	–CONHEt	36.3 ± 13.0
IVq	–COOMe	29.6 ± 5.6
IVm	–Me	**29.3 ± 5.3**
IVr	–CH=NOH	**20.5 ± 4.3**
IVs	–CHO	**16.9 ± 2.3**

[a] Substituent at position 6 or 7.
[b] Maximal contractile response, expressed as % of that attained in the absence of phthalazinone.
[c] Standard error of the mean of 4–8 separate experiments.

6(7)-nitroderivative **IVg**, with a $pA_2 = 6.00 \pm 0.1$ M, (Figure 4.12). Nevertheless, many other phthalazinones showed a noncompetitive antagonism. The best substitution for ring B corresponded to 4-Cl, associated with that of the carbaldehyde group at the 6(7) position for ring A, as represented by the most potent inhibitor **IVs**, which decreased the E_{max} to 16.9%, through a noncompetitive mechanism (Figure 4.13).

From the attempt to correlate the vasorelaxant activity of phthalazinones with the electronic and lipophilic parameters of the substituents attached to the aromatic rings, not good SAR conclusions were attained. However, the sterimol parameter ***L*** [93], representing the longitudinal size of the different substituents on ring A, could be considered as an adequate correlation factor for the vasorelaxant effect. Within the series of compounds with carbonyl-related substituents, the found or estimated ***L*** values seemed to correlate well with the respective E_{max} values and, inversely, with the inhibitory efficacy (Table 4.4). Compounds with saturated substituents on ring A, **IVm** and **IVo**, have also been included in the table, to compare their values with those of the similarly sized **IVs** and **IVr**, for showing the convenient presence of an electrophilic function conjugated with the aromatic system of ring A, and also to denote the similar behavior between the activity and size for these saturated substituents.

Within all the series of phthalazinones tested, **IVs** exerted the greatest vasorelaxant effect and, therefore, studies on the mechanism of action were performed with this compound. For this compound the relaxing effect on PE prestimulated rat aorta rings was not completely reversible; however, after comparison with a negative control

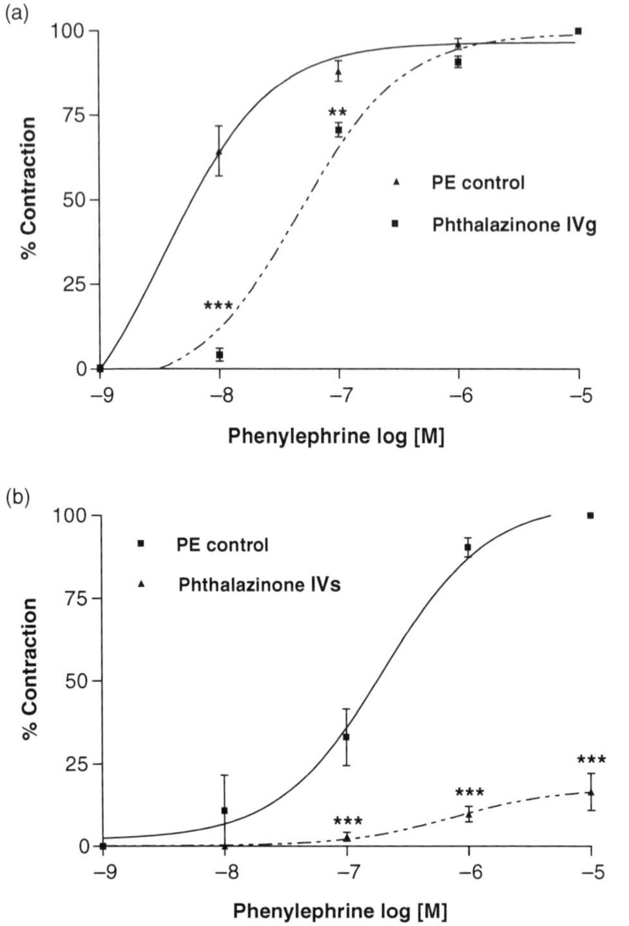

FIGURE 4.13 Inhibition of PE-induced contraction of rat aorta rings by phthalazinones **IVg** (a) and **IVs** (b), showing their competitive and non-competitive mode of action, respectively.

TABLE 4.4 Correlation of E_{max} Contraction Values with Structural and Electronic Parameters for Ring-A Substituents of Phthalazinones

Compound	Ring A [6(7) substituent]	σ_I	F	R	π	L	E_{max} (%)	Inh. (%)
IVs	–CHO	0.25	0.31	0.13	−0.65	**3.53**	16.9	**84.1**
IVr	–CH=NOH	0.27	0.25	−0.13	−0.38	**3.94**	20.5	**79.5**
IVq	$–COOCH_3$	0.34	0.33	0.15	−0.01	**4.73**	29.6	**70.4**
IVp	$–CONHCH_2CH_3$	0.27	0.14	0.24	−0.73	**∼5.95**[a]	36.3	**63.7**
IVn	$–CONHCH_2CHOHCH_3$	–	–	–	–	**∼6.97**[a]	64.0	**36.0**
IVm	$–CH_3$	−0.04	−0.04	−0.13	0.56	**2.87**	29.3	**70.7**
IVo	$–CH_2OH$	0.05	0.00	0.00	−1.03	**3.97**	34.6	**65.4**

[a] Values estimated from comparison with substituents of similar extended size (–COOEt and *n*-Pent, respectively) [92]. (–) values not available.

experiment, the effect resulted not due to fatigue of the smooth muscle. Moreover, experiments performed with KCl-stimulated rat aorta rings showed that **IVs** did not have any relaxing effect and simultaneously they confirmed that this compound did not have any toxic effect on the aorta rings (data not shown).

The inhibitory effect on human platelet aggregation was also examined through experiments on platelets isolated from human rich plasma (PRP), and induced to aggregate with norepinephrine NE. In the presence of **IVs** a certain inhibitory effect was observed, although it was not as remarkable as the vasorelaxation. This fact could confirm the hypothesis that this compound had some affinity for α-adrenergic receptors (data not given).

Further experiments were carried out in the presence of other contractile agents to analyze the vasorelaxant effect of **IVs**. According to the results found, this phthalazinone is not a specific blocker of α-adrenergic receptors, because it also relaxed rat aorta rings precontracted with $PGF_{2\alpha}$ [94] or with the calcium ionophore A-23187 [95]. Furthermore, **IVs** also inhibited the platelet aggregation induced with A-23187 (data not given).

Taking into account all these results, the intracellular changes of calcium in platelets loaded with the calcium-sensitive dye Fura-2-AM [96], and the $^{45}Ca^{2+}$ uptake/release in saponin-permeabilized human platelets [97] were examined. Fura 2-AM experiments showed that **IVs** reduced the intracellular $[Ca^{2+}]$ level if thrombin was used to increase the cytoplasmic calcium level. On the other hand, **IVs** did not change $[Ca^{2+}]$ if the calcium ionophore ionomycin was present. Consequently, this phthalazinone would not have any effect on the entry of calcium ions from the extracellular medium (data not given). Experiments with thapsigargin, that depletes calcium from the stores and blocks its uptake [98], confirmed this hypothesis by observing that the results in the presence of **IVs** were the same as those in the control experiment (data not given).

Finally, the effect of **IVs** on IP_3 receptors was evaluated through the analysis of $^{45}Ca^{2+}$ uptake/release in saponin-permeabilized human platelets [95]. According to the results shown in Figure 4.14, this compound inhibited calcium release from the intracellular stores by blocking the IP_3 receptors.

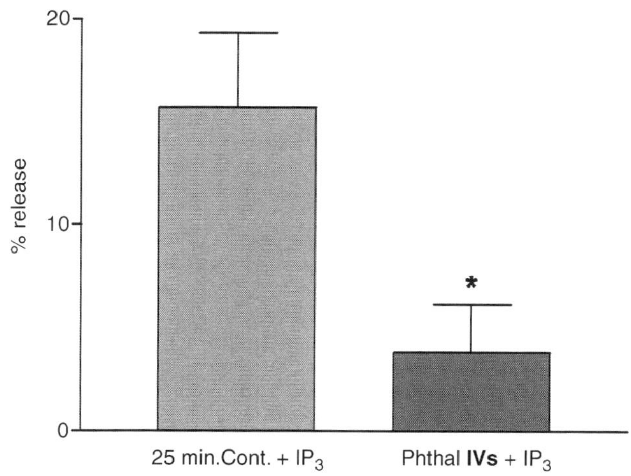

FIGURE 4.14 Effect of the phthalazinone **IVs** on the intracellular release of $^{45}Ca^{2+}$.

In summary, benzylphthalazinones methylated at *N*-2, having a 4-chloro substituent on ring B and a small-sized conjugated substitution at position 6(7) on ring A, specially a carbaldehyde group as in compound **IVs**, display important vasorelaxant and platelet aggregation effects, which appear to be mediated by IP_3 blockade, resulting in a decrease on intracellular calcium release. Further studies are needed to evaluate whether this compound could also have an indirect effect on IP_3 levels, through phospholipase C inhibition.

4.4 CHEMOINDUCTION AND CHEMOMODULATION OF THE ANTIPARASITIC ACTIVITY OF STILBENOIDS

As has been stated above, the natural compound isonotholaenic acid (**17**) resulted relatively active against *Plasmodium*, less active against *Leishmania* and practically inactive against *Trypanosoma*. After slight improvement of the activities through chemomodulation of the original structure, several compounds, representative of diverse types of heterostilbenoids were synthesized and tested, to define the imidazo [2,1-*a*]isoindole system as a new antiparasitic scaffold and compound **Vc** as the main lead compound to be optimized for attaining new antiplasmodial drugs [99]. Along with the new compounds of type **V**, representative elements of the intermediate benzalphthalide (**I**) and of the phthalazinone series (**IV**) (Scheme 4.1), were also submitted to evaluation against the mentioned parasites and the results observed are presented in the following sections.

4.4.1 Antileishmanial Activity

A selection of compounds of the series **I**, **IV,** and **V** were tested *in vitro* against promastigotes of *Leishmania amazonensis* (PH8), *Leishmania braziliensis* (M 2903), and *Leishmania donovani* (PP75), using pentamidine as the reference drug [100].

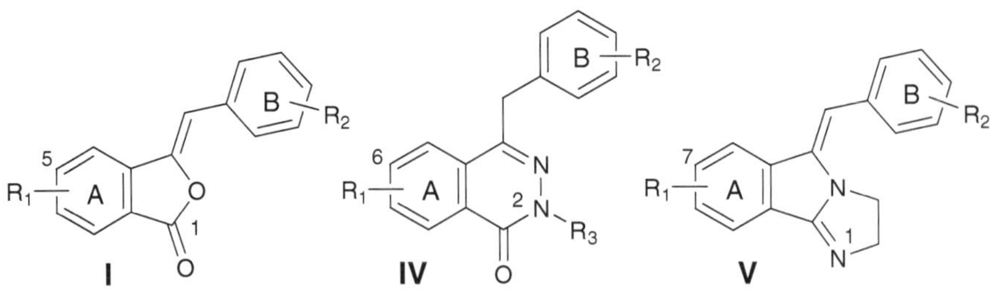

Related to the intermediate series of benzalphthalides (**I**) the combination of 4-NO_2 or 5(6)-Me substituents on ring A, with 4-Cl; 2,4-Cl_2 or 3,4-Cl_2 on ring B, led to compounds 2–4 times more potent than pentamidine. Similar potency results were obtained for compounds of the phthalazinone series (**IV**) equally substituted at the same (ring B) or equivalent (ring A) positions.

In case of the series of imidazoisoindoles (**V**), those compounds without substitution at ring A and with 2-Cl, 4-SMe, 2,4-Cl_2, 3,4-Cl_2 at ring B or α- or β-naphthyl groups instead the B phenyl fragment, behaved similar to **Vc** (4-Cl-phenyl), with antileishmanial IC_{50} values in the range 11–14 μM. In the case of compounds with combined substitutions on rings A and B and with those groups mentioned for series **I** and **IV** the antileishmanial potency raised up to 6.5-fold that of pentamidine in the case of the polychlorinated derivative with 7,8-dichloro substitution on ring A and 3,4-dichloro on ring B.

4.4.2 Trypanocidal Activity

The *in vitro* trypanocidal activity on cultured epimastigote and trypomastigote forms of *T. cruzi* was evaluated. Benznidazol ($IC_{50} = 28.5$ μM) was used as the reference drug for epimastigotes and gentian violet ($IC_{50} = 245$ μM) for trypomastigotes [101]. As in the case of *Leishmania* bioassays, in the absence of a substituent on ring A, only compounds of series **V** gave good results. Compounds with 4-Cl, 2-Cl, 4-SMe, 2,4-Cl_2, 3,4-Cl_2, or α-naphthyl groups on ring B gave results similar to that of benznidazol against epimastigotes and 25–50 times more potent than gentian violet against trypomastigotes. For compounds with substituents on rings A and B, the benzalphthalides **I**, with 5(6)-Me/4-Cl (ring A/ring B, respectively) and 4-NO_2/3,4-Cl_2 were nearly twice as potent as benznidazol. The phthalazinones **IV**, with substituents on both rings, 6(7)-Me/4-Cl; 6(7)-Me/4-SMe; 6,7-Cl_2/4-Cl and 6,7-Cl_2/2,4-Cl_2, were two to four times more potent than benznidazol. Also better trypanocidal results were found for the imidazoisoindole derivatives **V**, particularly for the tetrachlorinated derivative [7(8)-Cl_2/3,4-Cl_2 (rings A/B)], that displayed trypanocidal activities more than 6.3/2.4 times higher than those the reference drugs against epimastigotes/trypomastigotes, respectively.

4.4.3 Antimalarial Activity

Biological assays against *P. falciparum*, Chloroquine (CQ) sensible (F32) and resistant (D2) strains, were performed *in vitro*, with using CQ as the reference drug ($IC_{50} = 0.04$ μM on F32; $IC_{50} = 0.13$ μM on D2). Compounds of series **I–IV** showed no remarkable interest. However, preliminary evaluation of the imidazoisoindole series **V** gave results that suggested the continuation of the research. Twelve different substitution patterns (R_2: **a-m**, Figures 4.12 and 4.15) on ring B, in compounds without substitution ($R_1 = H$) on ring A were evaluated. Those methylsulfanyl (**Ve**) and β-naphthyl (**Vm**) derivatives resulted to be 6.9 and 8.5 times more potent than CQ on the F32 strain, respectively. No further improvement of the antiplasmodial effect was observed for compounds incorporating either the 7(8)-Me or the 7,8-Cl_2 substituents on ring A (Figure 4.15), though several compounds resulted almost as potent as CQ in the *in vitro* assays.

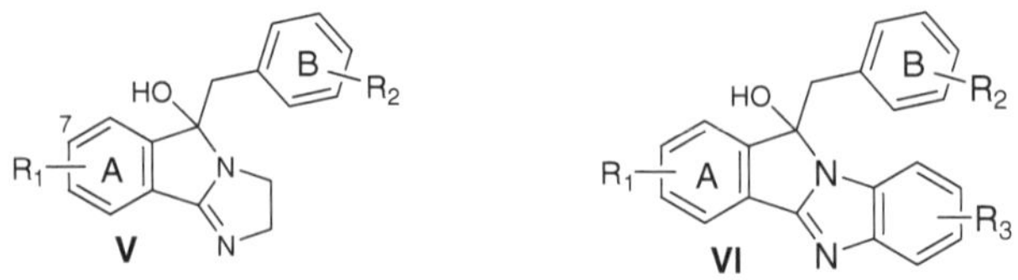

R_1 = H; 7(8)-Me; 7,8-Cl_2.
R_2 (R_1=H), **a:** H; **b:** 2-Cl; **c:** 4-Cl; **d:** 4-OMe; **e:** 4-SMe; **f:** 2,4-Cl_2; **g:** 3,4-Cl_2; **h:** 3,4(-OCH_2O-); **i:** 3,4-$(OMe)_2$; **j:** 3,4,5-$(OMe)_3$; **k:** 1-naphthyl; **m:** 2-naphthyl.

R_1 = H.
R_2 = 4-SMe; 2-naphthyl
R_3 = 2(3)-Me; 2(3)-Cl; 2,3-Cl_2.

FIGURE 4.15 Compounds of series **V** and **VI** evaluated *in vitro* against *Plasmodium falciparum*.

Physicochemical studies based on a molecular topology approach and focused to design more active compounds, predicted that the increase of size in the imidazoline region of series **V** should lead to potent antimalarial compounds; thus, the *typically antiparasitic* benzimidazole system was integrated with the imidazoline fragment in several tetracyclic compounds with a larger structural base, the isoindole [2,1-*a*] benzimidazole system **VI**. However, none of those compounds synthesized and assayed, represented in Figure 4.15, resulted better than compounds **Ve** and **Vm**.

4.4.4 Studies on the Mechanism of Antiplasmodial Activity and *In Vivo* Assays

To investigate the mechanism by which these compounds could act, the potential inhibition of haematin polymerization was evaluated [102]. The most potent inhibitor **Ve** attained only the 4% of the value observed for CQ, thus, suggesting another mechanism of action for these compounds.

In vivo antimalarial assays with some representative elements, within those not substituted on ring A, of series **IV** and **V** were performed on mice infected with *Plasmodium berghei*, at doses of 50 and 25 mg/kg.

The best result of those phthalazinone derivatives (**IV**) was observed for 2-methyl-4-naphth-2-ylmethylphthalazinone that attained a 39% reduction of *Plasmodium* parasitaemia levels, at the dose of 50 mg/kg. Working with imidazoisoindole derivatives **V**, compound **Vm** was revealed as the best antimalarial, attaining to inhibit the parasitaemia by 87%, at the dose of 25 mg/kg, with respect to control mice. Additionally, compounds **Ve** and **Vh** were less efficacious, attaining inhibition values of 84% and 53% of parasitaemia, respectively, at the dose of 50 mg/kg.

In summary, chemoinduction and subsequent chemomodulation of antiparasitic activities in these heterostilbenoids have led to a couple of new antimalarial lead compounds, with further expectancy for the results of other pending *in vivo* assays.

4.5 CONCLUSIONS

Along the sections and through the examples included in this chapter we have summarily described some contributions of Medicinal and Synthetic Chemistry to the improvement of natural drugs or bioactive substances and have also showed how to induce a desired type of activity on inactive molecules. We hope to have contributed in some extent to the understanding of the fundamental roles of chemistry in the bioactivity-guided and mechanism-guided structural modulation, oriented to the discovery, improvement, and development of better therapeutic agents from substances of the natural origin.

ACKNOWLEDGMENTS

Financial support from grants MCT: CTQ2008-02899; JCyL: SA028A10-2; FIS: PI-060118, RICET-ISCIII (RD06-0021-0022 and 0018), for those projects involved in this chapter is gratefully acknowledged. Authors are indebted to Dr. Carmen Cuevas (PharmaMar SA. Tres Cantos, Spain); Dr. Faustino Mollinedo (Centro de Investigación del Cáncer, Salamanca, Spain); Dr. Cinthia Boselli (Universitá degli Studi di Pisa, Italy); Dr. Albert Ferro (University College, London, UK); Dr. Alberto Giménez (Universidad Mayor de San Andrés, La Paz, Bolivia), Dr. Antonio R. Martínez (Universidad Complutense de Madrid, Spain), Dr. Dolores González-Pacanowska (Instituto de Parasitología y Bioquímica "López Neyra". Granada, Spain) and all those members of their teams collaborating in the research resumed in this chapter.

REFERENCES

[1] Eddy, N.B., May, E.L., Mossetig, E. (1952). Chemistry and pharmacology of the methadols and acetylmethadols. *Journal of Organic Chemistry*, *17*, 321–326.

[2] Braenden, O.J., Eddy, N.B., Halbach, H. (1955). Synthetic substances with morphine-like effect relationship between chemical structure and analgesic action. *Bulletin of The World Health Organisation*, *13*, 937–998.

[3] Martin, W.R. (1967). Opioid antagonists. *Pharmacological Reviews*, *19*, 463–521, and references therein.

[4] Hughes, J., Smith, T., Kosterlitz, H., Fothergill, L., Morgan, B., Morris, H. (1975). Identification of two related pentapeptides from the brain with potent opiate agonist activity. *Nature*, *258*, 577–580.

[5] Shank, R.P., Gardocki, J.F., Vaught, J.L., Davis, C.B., Schupsky, J.J., Raffa, R.B., Dodgson, S.J., Nortey, S.O., Mayanoff, B.E. (1994). Topiramate: Preclinical evaluation of structurally novel anticonvulsant. *Epilepsia*, *35*, 450–460.

[6] Rogawski, M.A., Porter, R.J. (1990). Antiepiletic drugs – pharmacological mechanisms and clinical efficacy with consideration of promising developmental stage compounds. *Pharmacological Reviews*, *42*, 223–286.

[7] Gaudilliere, B. (1999). To market, to market-1998. *Annual Reports in Medicinal Chemistry*, *34*, 327–354.

[8] Hedge, S., Carter, J. (2004). To market, to market-2003. *Annual Reports in Medicinal Chemistry*, *39*, 279.

[9] Cox, T.M., Aerts, J.M, Andria, G., Beck, M., Belmatoug, N., Bembi, B., Chertkoff, R., Vom Dahl, S., Elstein, D., Erikson, A, Giralt, M., Heitner, R., Hollak, C., Hrebicek, M., Lewis, S., Mehta, A., Pastores, G.M., Rolfs, A., Miranda, M.C., Zimran, A. (2003).

The role of the iminosugar *N*-butyldeoxynojirimycin (miglustat) in the management of type I (non-neuronopathic) Gaucher disease: A position statement. *Journal of Inherited Metabolic Disease*, *26*, 513–26.

[10] Tack, C.J., Smits P. (1999). New drugs for diabetes. *Netherlands Journal of Medicine*, *55*, 209–211.

[11] Dubey, M.P., Srimal, R.C., Nityanand, S. Dhawan, B.N. (1981). Pharmacological studies on coleonol, a hypotensive diterpene from Coleus forskohlii. *Journal of Ethnopharmacology*, *3*, 1–13.

[12] Yoneyama, M., Sugiyama, A., Satoh, Y., Takahara, A., Nakamura, Y., Hashimoto, K. (2002). Cardiovascular and adenylate cyclase stimulating effects of colforsin daropate, a water-soluble forskolin derivative, compared with those of isoproterenol, dopamine and dobutamine. *Circulation Journal*, *66*, 1150–1154.

[13] Gaudilliere, B., Berna, P. (2000). To market, to market-1999. *Annual Reports in Medicinal Chemistry*, *35*, 331–356.

[14] Rahman, A., series editor. *Studies in Natural Products Chemistry*, Vols. 21–35, Bioactive Natural Products, Elsevier, Amsterdam, 2000–2008.

[15] Gupta, R.R., Maes, B.U.W., series editors. *Topics in Heterocyclic Chemistry. Bioactive Heterocycles I–VII*, Springer, Berlin, 2006–2009.

[16] Liang, X.-T., Fang, W.-S., editors. *Medicinal Chemistry of Bioactive Natural Products*, John Wiley & Sons, Inc., New York, 2006.

[17] Zhang, L., Demain, A.L., editors., *Natural Products: Drug Discovery and Therapeutic Medicine*, Humana Press Inc., Totowa, 2005.

[18] Ramawat, K.G., Merillon, J.-M., editors. *Bioactive Molecules and Medicinal Plants*, Springer, Berlin, 2008.

[19] Hong, J., Chen, S.-H., Case studies in natural-product optimization. In: Zhang, L., Demain, A.L., editors. *Natural Products: Drug Discovery and Therapeutic Medicine*. Humana Press, Totowa, NJ, 2005, pp. 169–196.

[20] Salim, A.A., Chin, Y.-W., Kinghorn, A.D. Drug discovery from plants. In: Ramawat, K.G., Merillon, J.M., editors. *Bioactive Molecules and Medicinal Plants*. Springer, Berlin, 2008, pp. 1–24.

[21] Wang, G., Tang, W., Bidigare, R.R. Terpenoids as therapeutic drugs and pharmaceutical agents. In: Zhang, L., Demain, A.L., editors. *Natural Products: Drug Discovery and Therapeutic Medicine*. Humana Press, Totowa, NJ, 2005, pp. 197–227.

[22] Bahar, M., Deng, Y., Fletcher, J.N., Kinghorn, A.D. Plant-derived natural products in Drug discovery and development: An overview. In: Ikan, R., editor. *Selected Topics in the Chemistry of Natural Products*. World Scientific Publishing Co. Pte. Ltd., Singapore, 2007, pp. 11–48.

[23] Majetich, G. Recent trends in the use of natural products and their derivatives as potential pharmaceutical agents. In: Cutler, S.J. and Cutler, H.G., editors. *Biologically Active Natural Products: Pharmaceuticals*. CRC Press, Boca Raton, FL, 2000, pp. 61–71.

[24] Lee, K.-H. Highlights of research on plant-derived natural products and their analogs with antitumor, anti-HIV, and antifungal activity. In: Cutler, S.J. and Cutler, H.G., editors. *Biologically Active Natural Products: Pharmaceuticals*. CRC Press, Boca Raton, FL, 2000, pp. 82–103.

[25] Houghton, P.J., Howes M.-J. Natural products and related compounds of realized and potential use in treating neurodegenerative disease. In: Ikan, R., editor. *Selected Topics in*

the Chemistry of Natural Products. World Scientific Publioshing Co. Pte. Ltd., Singapore, 2007, pp. 377–426.

[26] Newman, D.J., Cragg, G.M. (2007). Natural products as sources of new drugs over the last 25 years. *Journal of Natural Products*, *70*, 461–477.

[27] Butler, M.S. (2005, 2008). Natural products to drugs: Natural product derived compounds in clinical trials. *Natural Products Report*, *22*, 162–195; *25*, 475–516.

[28] Zhao, H., Donnelly, A.C., Kusuma, B.R., Brandt, G.E.L., Brown, D., Rajewski, R.A., Vielhauer, G., Holzbeierlein, J., Cohen, M.S., Blagg, B.S.J. (2011). Engineering an antibiotic to fight cancer: Optimization of the novobiocin scaffold to produce anti-proliferative agents. *Journal of Medicinal Chemistry*, *54*, 3839–3853.

[29] Moss, G.P. (2000). Nomenclature of lignans and neolignans. *Pure and Applied Chemistry*, *72*, 1493–1523.

[30] Ayres, D.C., Loike, J.D., *Lignans: Chemical, Biological and Clinical Properties*. Cambridge University Press, London, 1990.

[31] Liu, Y.Q., Yang, L., Tian, X. (2007). Podophyllotoxin: Current perspectives. *Current Bioactive Compounds*, *3*, 37–66.

[32] (a) Xu, H., Lu, M., Tian, X. (2009). A review on hemisynhesis, biosynthesis, biological activities, mode of action, and structure–activity relationship of podophyllotoxins: 2003–2007. *Current Medicinal Chemistry 16*, 327–349. (b) Lee, K.H., Xiao, Z. Podophyllotoxins and analogs. In: Cragg, G., Kingston, D.G.I. and Newman, D.J., editors. *Anticancer Agents from Natural Products*. CRC Press, Boca Raton, FL, 2005, pp. 71–87.

[33] (a) Saitoh, T., Kuramochi, K., Imai, T., Takata, K., Takehara, M., Kobayashi, S., Sakaguchia, K., Sugawara, S. (2008). Podophyllotoxin directly binds a hinge domain in E2 of HPV and inhibits an E2/E7 interaction *in vitro*. *Bioorganic and Medicinal Chemistry*, *16*, 5815–5825; (b) Bernard, H.U. (2004). Established and potential strategies against papillomavirus infections. *Journal of Antimicrobial Chemotherapy*, *53*, 137–139; (c) Scheinfeld, N., Lehman, S.D. (2006). An evidence based review of medical and surgical treatments of genital warts, *Dermatology Online Journal*, *12*, 5.

[34] (a) Carlstrom, K., Hedin, P.J., Jonsson, L., Lerndal, T., Lien, J., Weitoft, T., Axelson, M. (2000). Endocrine effects of the podophyllotoxin derivative drug CPH82 (Reumacon®) in patients with rheumatoid arthritis. *Scandinavian Journal of Rheumatology*, *29*, 89–94; (b) Rantapaa, D.S., Landberg, G., Roos, G., Norberg, B. (1994). Cell cycle effects of the anti-rheumatic agent CPH82. *British Journal of Rheumatology*, *33*, 327–331.

[35] (a) Jin, M., Lee, E., Yang, J.H., Lu, Y., Kang, S.G., Chang, Y.C., Lee, S.H., Suh, S.J., Kim, C.H., Chang, H. (2010). Deoxypodophyllotoxin inhibits the expression of intra-cellular adhesion molecule-1 induced by tumor necrosis factor-alpha in murine lung epithelial cells. *Biological and Pharmaceutical Bulletin*, *33*, 1–5; (b) Lin, C.X., Lee, E., Jin, M.H., Yook, J., Quan, Z., Ha, K., Moon, T.C., Kim, M.J., Kim, K.J., Lee, S.H., Chang, H.W. (2006). Deoxypodophyllotoxin (DPT) inhibits eosinophil recruitment into the airway and Th2 cytokine expression in an OVA induced lung inflammation. *Planta Medica*, *72*, 786–791.

[36] (a) Frackenpohl, J., Adelt, I., Antonicek, H., Arnold, C., Behrmann, P., Blaha, N., Bohmer, J., Gutbrod, O., Hanke, R., Hohmann, S., Van Houtdreve, M., Losel, P., Malsam, O., Melchers, M., Neufert, V., Peschel, E., Reckmann, U., Schenke, T., Thiesen, H.P., Velten, R., Vogelsang, K., Weiss, H.C. (2009). Insecticidal

heterolignans-tubuline polymerization inhibitors with activity against chewing pests. *Bioorganic and Medicinal Chemistry*, *17*, 4160–3184; (b) Liu, Y.-Q., Liu, Y., Xiao, H., Gao, R., Tian, X. (2008). Synthesis and insecticidal activities of novel derivatives of podophyllotoxin: Part XII. *Pesticide Biochemistry and Physiology*, *91*, 116–121; (c) Xu, H., Xiao, X. (2009). Natural products-based insecticidal agents 4. Semisynthesis and insecticidal activity of novel esters of 2-chloropodophyllotoxin against *Mythimna separate* Walker *in vivo*. *Bioorganic and Medicinal Chemistry Letters*, *19*, 5415–5418.

[37] Uddin, M.J., Smithson, D.C., Brown, K.M., Crews, B.C., Connelly, M., Zhu, F., Marnett, L.J., Guy, R.K. (2010). Podophyllotoxin analogues active versus *Trypanosoma brucei*. *Bioorganic and Medicinal Chemistry Letters*, *20*, 1787–1791.

[38] (b) Lee, K.-H., Zhiyan, X. (2004). Lignans in treatment of cancer and other diseases. *Phytochemical Reviews*, *2*, 341–362; (b) Bohlin, L., Rosen, B. (1996). Podophyllotoxin derivatives: Drug discovery and development. *Drug Discovery Today*, *1*, 343–351.

[39] Kamal, A., Kumar, B.A., Arifuddin, M., Dastidar, S.G. (2003). Synthesis of 4β-amido and 4β-sulphonamido analogues of podophyllotoxin as potential antitumour agents. *Bioorganic and Medicinal Chemistry*, *11*, 5135–5142.

[40] Gordaliza, M., García, P.A., Miguel del Corral, J.M., Castro, M.A., Gómez-Zurita, M.A. (2004). Podophyllotoxin: Distribution, sources, applications and new cytotoxic derivatives. *Toxicon*, *44*, 441–459.

[41] Kluza, J., Mazinghein, R., Irwin, H., Hartley, J.A., Bailly, C. (2006). Relationships between DNA strand breakage and apoptotic pregression upon treatment of HL-60 leukemia cells with tafluposide or etoposide. *Anticancer Drugs*, *17*, 155–164.

[42] Castro, M.A., Miguel del Corral, J.M., Gordaliza, M., Gómez-Zurita, M.A., García, P.A., San Feliciano, A. (2003). Chemoinduction of cytotoxic selectivity in podophyllotoxin-related lignans. *Phytochemical Reviews*, *2*, 219–233.

[43] Castro, M.A., Miguel del Corral, J.M., Gordaliza, M., García, P.A., Gómez-Zurita, M.A., García-Grávalos, M.D., Iglesia-Vicente, J., Gajate, C., An, F., Mollinedo, F., San Feliciano, A. (2004). Synthesis and biological evaluation of new selective cytotoxic cyclolignans derived from podophyllotoxin. *Journal of Medicinal Chemistry*, *47*, 1214–1222.

[44] Castro, M.A., Miguel del Corral, J.M., Gordaliza, M., García, P.A., Gómez-Zurita, M.A., San Feliciano, A. (2007). Synthesis and cytotoxic evaluation of C-9 oxidized podophyllotoxin derivatives. *Bioorganic and Medicinal Chemistry*, *15*, 1670–1678.

[45] Castro, M.A., Miguel del Corral, J.M., García, P.A., Rojo, M.V., Iglesia-Vicente, J., Mollinedo, F., Cuevas, C., San Feliciano, A. (2010) Synthesis and biological evaluation of new podophyllic aldehyde derivatives with cytotoxic and apoptosis-inducing activities. *Journal of Medicinal Chemistry*, *53*, 983–993.

[46] Hirano, T., Wakasugi, A., Oohara, M., Oka, K., Sashida, Y. (1991). Suppression of mitogen-induced proliferation of human peripheral blood lymphocytes by plant lignans. *Planta Medica*, *57*, 331–334.

[47] Rantapaa, D.S., Lofvenbert, E., Norberg, B. (1993). Effect of CPH82 in rheumatoid arthritis. Accumulation of bone marrow cells in mitosis and clinical response. *Clinical Rheumatology*, *12*, 541–542.

[48] Gordaliza, M., Faircloth, G.T., Castro, M.A., Miguel del Corral, J.M., López-Vázquez, M.L., San Feliciano, A. (1996). Immunosuppressive cyclolignans. *Journal of Medicinal Chemistry*, *39*, 2865–2868.

[49] Gordaliza, M., Castro, M.A., Miguel del Corral, J.M., López-Vázquez, M.L., San Feliciano, A., Faircloth, G.T. (1997). *In vivo* immunosuppressive activity of some cyclolignans. *Bioorganic and Medicinal Chemistry Letters*, *7*, 2781–2786.

[50] Eich, E., Schulz, J., Kaloga, M., Merz, H., Schoder, H.C., Muller, W.E.G. (1991). Interference of epipodophyllotoxins and natural lignanolides with topoisomerase: II. A proposed molecular mechanism. *Planta Medica*, *57*, A7–A8.

[51] Gordaliza, M., Miguel del Corral, J.M., Castro, M.A., López-Vázquez, M.L., San Feliciano, A., García-Grávalos, M.D., Carpy, A. (1995). Synthesis and evaluation of pyrazolignans. A new class of cytotoxic agents. *Bioorganic and Medicinal Chemistry*, *3*, 1203–1210.

[52] Gordaliza, M., Castro, M.A., Miguel del Corral, J.M., López-Vázquez, M.L., García, P.A., San Feliciano, A., García-Grávalos, M.D., Broughton, H. (1997). Preparation and cytotoxicity of podophyllotoxin derivatives lacking the lactone ring. *Tetrahedron*, *53*, 15743–15760.

[53] García-Rocha, M., García-Grávalos, M. D., Ávila, J. (1996). Characterisation of antimitotic products from marine organisms that disorganise the microtubule network: Ecteinascidin 743, isohomohalichondrin-B and LL-15. *British Journal of Cancer*, *73*, 875–883.

[54] Huey, R., Morris, G.M., Olson, A.J., Goodsell, D.S. (2007). A semiempirical free energy force field with charge-based desolvation. *Journal of Computational Chemistry*, *28*, 1145–1152.

[55] Garcia, D., Marrero, J.J., Ravelo, A.G. Stilbenoids: Chemistry and bioactivities. In: Rahman, A., editor. *Studies in Natural Products Chemistry*. Elsevier, Amsterdam, 2008, pp. 453–646.

[56] Kimura, Y. Natural stilbenes and analogues as antineoplastic agents. In: Rahman, A., editor. *Studies in Natural Products Chemistry*. Elsevier, Amsterdam, 2008, pp. 77–127.

[57] Geahlen, R.L., Mc Laughlin, J.L. (1989). Piceatannol (3,4,3′,5′-tetrahydroxy-trans-stilbene) is a naturally occurring protein-tyrosine kinase inhibitor. *Biochemical and Biophysical Research Communications*, *165*, 241–245.

[58] Thakkar, K., Geahlen, R.L., Cushman, M. (1993). Synthesis and protein-tyrosine kinase inhibitory activity of polyhydroxylated stilbene analogs of piceatannol. *Journal of Medicinal Chemistry*, *36*, 2950–2955.

[59] Hamze, A., Giraud, A., Messaoudi, S., Provot, O., Peyrat, J.F., Bignon, J., Liu, J.M., Wdzieczak-Bakala, J., Thoret, S., Dubois, J., Brion, J.D., Alami, M. (2009). Synthesis, biological evaluation of 1,1-diarylethylenes as a novel class of antimitotic agents. *ChemMedChem*. *11*, 1912–1924.

[60] Wang, Y., Edrada-Ebel, R., Tsevegsuren, N., Sendker, J., Braun, M., Wray, V., Lin, W., Proksch. P. (2009). Dihydrostilbene derivatives from the Mongolian medicinal plant *Scorzonera radiata*. *Journal of Natural Products*, *72*, 671–675.

[61] Lin, L.G., Yang, X.Z., Tan g, C.P., Ke, C.Q., Zhang, J.B., Ye, Y. (2008). Antibacterial stilbenoids from the roots of *Stemona tuberosa*. *Phytochemistry*, *69*, 457–463.

[62] Kostecki, K., Engelmeier, D., Pacher, T., Hofer, O., Vajrodaya, S., Greger, H. (2004). Inhibition of leukotriene biosynthesis by stilbenoids from *Stemona* species. *Phytochemistry*, *65*, 99–106.

[63] Wollenweber, E., Doerr, M., Waton H., Bonvin, J.F. (1993). Flavonoid aglycones and a dihydrostilbene from the frond exudate of *Notholaena nivea*. *Phytochemistry*, *33*, 611–612.

[64] Jay, M., Wollenweber, E., Bonvin, J.F. (1979). Novel 2′-substituted flavonols from the farinose exudate of *Notholaena affinis*. *Phytochemistry*, *18*, 153–154.

[65] Prince, P.S.M., Priscilla, H., Devika, P.T. (2009). Gallic acid prevents lysosomal damage in isoproterenol induced cardiotoxicity in Wistar rats. *European Journal of Pharmacology*, *615*, 139–143.

[66] (a) von Känel, R., Kudielka, B.M., Helfricht, S., Metzenthin, P., Preckel, D., Haeberli, A., Cung, T., Fischer, J.E. (2008). The effects of aspirin and nonselective beta blockade on the acute prothrombotic response to psychosocial stress in apparently healthy subjects. *Journal of Cardiovascular Pharmacology*, *51*, 231–238; (b) von Känel, R., Kudielka, B.M., Helfricht, S., Metzenthin, P., Preckel, D., Haeberli, A., Cung, T., Fischer, J.E. (2008). Effects of aspirin and propranolol on the acute psychological stress response in factor VIII coagulant activity: A randomized, double-blind, placebo-controlled experimental study. *Blood Coagulation and Fibrinolysis*, *19*, 75–81; (c) Grisar, J.M., Claxton, G.P., Bare, T.M., Dage, R.C., Cheng, H.C., Woodward, J.K. (1981). Salicylamide derivatives related to medroxalol with alpha- and beta-adrenergic antagonist and antihypertension activity. *Journal of Medicinal Chemistry*, *24*, 327–336.

[67] Deliorman, D., Caliş, I., Ergun, F., Doğan, B.S., Buharalioğlu, C.K., Kanzik, I. (2000). Studies on the vascular effects of the fractions and phenolic compounds isolated from *Viscum album* ssp. *album*. *Journal of Ethnopharmacology*, *72*, 323–329.

[68] Chen, I.J., Yeh, J.L., Liou, S.J., Shen, A.Y. (1994). Guaiacoxypropanolamine derivatives of capsaicin: A new family of beta-adrenoceptor blockers with intrinsic cardiotonic properties. *Journal of Medicinal Chemistry*, *37*, 938–943.

[69] Jen, T., Frazee, J.S., Kaiser, C., Colella, D.F., Wardell, J.R. Jr. (1977). Adrenergic agents. 6. Synthesis and potential beta-adrenergic agonist activity of some meta-substituted p-hydroxyphenylethanolamines related to salbutamol. *Journal of Medicinal Chemistry*, *20*, 1029–1035.

[70] Demirayak, S., Karaburun, A.C., Beis, R. (2004). Some pyrrole substituted aryl pyridazinone and phthalazinone derivatives and their antihypertensive activities. *European Journal of Medicinal Chemistry*, *39*, 1089–1095.

[71] Rang, H.P., Ritter, J.M., Moore, P.K., *Pharmacology*, Elsevier, Churchill Livingstone, 2003, pp. 298–305.

[72] Flórez, J., Armijo J.A., Mediavilla A., *Farmacología Humana*, Masson S.A., Barcelona, 2000, pp. 671–683.

[73] Cherkez, S., Herzig, J., Yellin, H. (1986). Synthesis, saludiuretic, and antihypertensive activity of 6,7-disubstituted 1(2*H*) and 3,4-dihydro-1(2*H*)-phthalazinones. *Journal of Medicinal Chemistry*, *29*, 947–959.

[74] Yassin, F.A., Bayoumy B.E., El-Farargy A.F. (1990). Synthesis of 4,5,6,7-tetraphenyl-8-(substituted)-3(2H)-phthalazinone derivatives likely to posses antihypertensive activity. *Bulletin of the Korean Chemical Society*, *11*, 7–10.

[75] McTavish, D., Sorkin, E.M. (1989). Azelastine: A review of its pharmacodynamic and pharmacokinetic properties, and therapeutic potential. *Drugs*, *38*, 778–800.

[76] Lee, H., Ohya, Y., Doupnik, C.A., Sperelakis, N. (1990). Effects of azelastine on contraction of guinea pig tracheal smooth muscle. *European Journal of Pharmacology*, *187*, 67–74.

[77] Duarte, N., Kayser, O., Abreu, P., Ferreira, M.J. (2008). Antileishmanial activity of piceatannol isolated from *Euphorbia lagascae* seeds. *Phytotherapy Research*, *22*, 455–457.

[78] Kedzierski, L., Curtis, J.M., Kaminska, M., Jodynis-Liebert, J., Murias, M. (2007). *In vitro* antileishmanial activity of resveratrol and its hydroxylated analogues against *Leishmania major* promastigotes and amastigotes. *Parasitology Research*, *102*, 91–97.

[79] Getti, G.T., Aslam, S.N., Humber, D.P., Stevenson, P.C., Cheke, R.A. (2006). The effect of cicerfuran, an arylbenzofuran from *Cicer bijugum*, and related benzofurans and stilbenes on *Leishmania aethiopica, L. tropica* and *L. major*. *Planta Medica*, *72*, 907–911.

[80] Del Rey, B., Ramos, A.C., Caballero, E., Inchaustti, A., Yaluff, G., Medarde, M., Rojas de Arias, A., San Feliciano, A. (1999). Leishmanicidal activity of combretastatin analogues and heteroanalogues. *Bioorganic Medicinal Chemistry Letters*, *9*, 2711–2714.

[81] Reguera, R., Balaña Fouce, R., Cubria, J.C., Alvarez Bujidos, M.L., Ordoñez, D. (1994). Putrescine uptake inhibition by aromatic diamidines in *Leishmania infantum* promastigotes. *Biochemical Pharmacology*, *47*, 1859–1866.

[82] Miguel, D.C., Zauli-Nascimento, R.C., Yokoyama-Yasunaka, J.K., Katz, S., Barbiéri, C.L., Uliana, S.R. (2008). Tamoxifen as a potential antileishmanial agent: Efficacy in the treatment of *Leishmania braziliensis* and *Leishmania chagasi* infections. *Journal of Antimicrobial Chemotherapy*, *63*, 365–368.

[83] Park, W.H., Lee, S.J., Moon, H.I. (2008). Antimalarial activity of a new stilbene glycoside from *Parthenocissus tricuspidata* in mice. *Antimicrobial Agents and Chemotherapy*, *52*, 3451–3453.

[84] Moon, H.I., Sim, J. (2008). Antimalarial activity in mice of resveratrol derivative from *Pleuropterus ciliinervis*. *Annals of Tropical Medicine and Parasitology*, *102*, 447–450.

[85] Kasinathan, R.S., Föller, M., Koka, S., Huber, S.M., Lang, F. (2007). Inhibition of eryptosis and intraerythrocytic growth of *Plasmodium falciparum* by flufenamic acid. *Naunyn Schmiedeberg's Archives of Pharmacology*, *374*, 255–264.

[86] Staines, H.M., Dee, B.C., Shen, M.R., Ellory, J.C. (2004). The effect of mefloquine and volume-regulated anion channel inhibitors on induced transport in *Plasmodium falciparum*-infected human red blood cells. *Blood Cells, Molecules and Diseases*, *32*, 344–348.

[87] Duker-Eshun, G., Jaroszewski, J.W., Asomaning, W.ARRCR.R. Oppong-Boachie, F., Brøgger Christensen, S. (2004). Antiplasmodial constituents of *Cajanus cajan*. *Phytotherapy Research*, *18*, 128–130.

[88] Mishra, N.C., Sharma, M., Sharma, A. (1999). Inhibitory effect of piceatannol, a protein tyrosine kinase inhibitor, on asexual maturation of *Plasmodium falciparum*. *Indian Journal of Experimental Biology*, *37*, 418–420.

[89] Bernardes, C.F., Meyer-Fernandes, J.R., Saad-Nehme, J., Vannier-Santos, M.A., Peres-Sampaio, C.E., Vercesi, A.E. (2000). Effects of 4,4′-diisothyocyanatostilbene-2,2′-disulfonic acid on *Trypanosoma cruzi* proliferation and Ca($^{2+}$) homeostasis. *International Journal of Biochemistry and Cell Biology*, *32*, 519–527.

[90] Dreyfuss, G., Loiseau, P.M., Lachâtre, G., Pénicaut, B., Nicolas, J.A., Craciunescu, D.G. (1995). *Trypanosoma brucei brucei*: Antitrypanosomal evaluation of stilbamidinium hexachloroiridiate on the murine CNS model and iridium serum kinetics in infected sheep. *Tropical Medicine and Parasitology*, *46*, 41–44.

[91] Noda, M., Yamaguchi, M., Ando, E., Takeda, K., Nokihara, K. (1994). Synthesis of 5-(((R,S)-5-((9-fluorenylmethoxycarbonyl)amino)-10,11dihydrodibenzo[a,d] cyclohepten-

2-yl)oxy)valeric acid (CHA) and 5-(((R,S)-5-((9-fluorenylmethoxycarbonyl)amino)dibenzo[a,d]-cyclohepten-2-yl)oxy)valeric acid (CHE) handles for the solid-phase synthesis of C-terminal peptide amides under mild conditions. *Journal of Organic Chemistry*, *59*, 7968–7975.

[92] Lácová, M., Chovancová, J., Veverková, E., Toma, S. (1996). Microwaves-assisted Gabriel synthesis of phthalides. *Tetrahedron*, *52*, 14995–15006.

[93] Silipo, C., Vittoria, A., Three-dimensional structure of drugs. In: Hansch, C., Sammes, P.G., Taylor, J.B., and Ramsden, C.A., editors. *Comprehensive Medicinal Chemistry*. Pergamon Press, Oxford, 1990, pp. 153–204.

[94] Heaslip, R.J., Sickels, B.D. (1989). Evidence that prostaglandins can contract the rat aorta via a novel protein kinase C-dependent mechanism. *Journal of Pharmacology and Experimental Therapeutics*, *250*, 44–51.

[95] Rassheed, H., Tirmizi, A.H., Salahuddin, F., Rizbi, N.B., Arshad, M., Farooq, S.Z., Saeed, S.A. (2004). Calcium signaling in human platelet aggregation mediated by platelet activating factor and calcium ionophore, A23187. *Journal of Biological Sciences*, *4*, 117–121.

[96] Grynkieewicz, G., Poenie, M., Tsien, R.Y. (1985). A new generation of Ca^{2+} indicators with greatly improved fluorescence properties. *Journal of Biological Chemistry*, *260*, 3440–3450.

[97] Authi, K.S., Bokkala, S., Patel, Y., Kakkar, V.V., Munkonge, F. (1993). Ca^{2+} release from platelet intracellular stores by thapsigargin and 2,5-di-(*t*-butyl)-1,4-benzohydroquinone: Relationship to Ca^{2+} pools and relevance in platelet activation. *Biochemical Journal*, *294*, 119–126.

[98] Thastrup, O., Cullen, P.J., Drobak, B.K., Hanley, M.R., Dawson, A.P. (1990). Thapsigargin, a tumor promoter, discharges intracellular Ca^{2+} stores by specific inhibition of the endoplasmic reticulum Ca^{2+}-ATP. *Proceedings of the Natural Academy of Sciences USA*, *87*, 2466–2470.

[99] Del Olmo, E., García-Armas, M., Ybarra, M.I., López, J.L., Oporto, P., Giménez, A., Deharo, E., San Feliciano, A. (2003). The imidazo[2,1-*a*]isoindole system: A new skeletal basis for antiplasmodial compounds. *Bioorganic and Medicinal Chemistry Letters*, *13*, 2769–2772.

[100] Del Olmo, E., García-Armas, M., López-Pérez, J.L., Muñoz, V., Deharo, E., San Feliciano, A. (2001). Leishmanicidal activity of some stilbenoids and related heterocyclic compounds. *Bioorganic and Medicinal Chemistry Letters*, *11*, 2123–2126.

[101] Del Olmo, E., García-Armas, M., López-Pérez, J.L., Ruiz, G., Vargas, F., Giménez, A., Deharo, E., San Feliciano, A. (2001). Anti-trypanosoma activity of some natural stilbenoids and synthetic related heterocyclic compounds. *Bioorganic and Medicinal Chemistry Letters*, *11*, 2755–2757.

[102] Deharo, E., García, R.N., Oporto, P., Giménez, A., Sauvain, M., Jullian, V., Ginsburg, H. (2002). A non-radiolabelled ferriprotoporphyrin IX biomineralisation inhibition test for the high throughput screening of antimalarial compounds. *Experimental Parasitology*, *100*, 252–256.

5

THE ACTION OF PLANTS AND THEIR CONSTITUENTS ON THE CENTRAL NERVOUS SYSTEM

Fúlvio R. Mendes, Giuseppina Negri, Joaquim M. Duarte-Almeida, Ricardo Tabach, and Elisaldo A. Carlini

5.1 INTRODUCTION

Psychoactive substances (including plants) are those that alter some aspects of the mind, including behavior, mood, cognition, and well being. According to Chalout [1], psychoactive drugs can be classified into three categories: depressants, which decrease the activity of the central nervous system (CNS); stimulants, which increase this activity; and mind altering, drugs that disrupt the normal activity of the brain. They are also referred as psycholeptics, psychoanaleptics, and psychodysleptics, respectively [2]. Other drugs, such as nootropics and neuroprotective agents cannot be easily defined within this classification. This chapter will discuss the main plants and their active principles acting as CNS depressants, such as neuroleptics, analgesics, anticonvulsants, anxiolytics, and hypnotics; CNS stimulants (psychostimulants, antidepressants, and adaptogens) and substances that disturb the CNS (hallucinogens). It also will discuss the main plants and active principles with neuroprotective or related activities, such as antioxidant and anticholinesterasic action, and possible natural alternatives that can be used to help prevent drug dependence.

Before addressing each of these effects, and the plants used for these purposes, it is opportune to remember some of the traditional species used in the development of pharmacology, that is, those that have made a decisive contribution to the development of medicine and our understanding of the mechanisms of many diseases.

Plant Bioactives and Drug Discovery: Principles, Practice, and Perspectives, Fourth Edition.
Edited by Valdir Cechinel-Filho.

One such plant species is *Atropa belladonna* L. The name Atropa or Atropus is linked to one of the fates of Greek mythology, who was said to be the one who cuts the thread of life of the mortals, in reference to the toxic effects of the plant. Its popular name "bella donna" comes from Rome, where women would use the juice from the fruits, which has mydriatic properties, to induce pupil dilation and thereby enhance their beauty. The alkaloid atropine, the active principle, which has anticholinergic activity, has been isolated from *A. belladonna* and is still used today in ophthalmology, to dilate the pupils. Other tropane alkaloids present in this species are hyoscyamine, scopolamine, and apoatropine [3].

Physostigmine is an alkaloid isolated from *Physostigma venenosum* Balf. f., a plant that was traditionally used in poisoning rituals in Africa, to help find people accused of crimes [4]. It is a powerful reversible acetylcholinesterase inhibitor, but its clinical use is limited because it has a short half-life and requires multiple daily doses [5].

Rauvolfia serpentina is used in traditional medicine in India to treat snake bites, mental diseases, and epilepsy. This plant contains alkaloids of the yohimbine (reserpine and rescinnamine), heteroyohimbine (ajmalicine), and indole (ajmaline) types. Reserpine acts by decreasing the catecholamine and serotonin levels in the central nervous system [6] and was one of the first drugs to be used for the treatment of schizophrenia. The understanding of the mechanism of action of reserpine has contributed a great deal to the understanding of the physiology of schizophrenia and depression.

Finally, another plant that has made a huge contribution to the advance of medicine is *Papaver somniferum* L., thanks to the narcotic effects of the opium and its tincture, which subsequently led to the isolation of morphine. The alkaloid morphine continues to be one of the most important analgesic drugs used in clinical practice today [7].

Today, various active principles isolated from plants or their derivatives are used in the manufacture of medicines with different purposes. But in many cases, the whole plant, or its standardized extract, are used successfully [8], as this chapter will seek to show. It is clear that plants and their active principles continue to have great importance for human health.

5.2 PLANTS WITH CNS DEPRESSANT ACTIVITY

Substances with CNS depressant (psycholeptic) effects are classified according to their pharmacological effects such as hypnotics/sedatives, anxiolytics, analgesics, neuroleptics, and anticonvulsants. As we shall now see, most plants with depressant action on the CNS are used as anxiolytic or hypnotic.

Since ancient times, many drugs obtained from various plant sources, such as *Matricaria chamomilla*, *Passiflora incarnata, Valeriana officinalis*, *Melissa officinalis*, *Mentha piperita,* and *P. somniferum* have been used to induce sleep, calm states of excessive agitation, and decrease anxiety. Following this line of reasoning, we see that popular culture is rich in plants that have the effect of "calming the nerves" or "helping you get to sleep." Exploration of these uses, through methods of pure and applied chemistry, has resulted in well-known active principles, which are highly

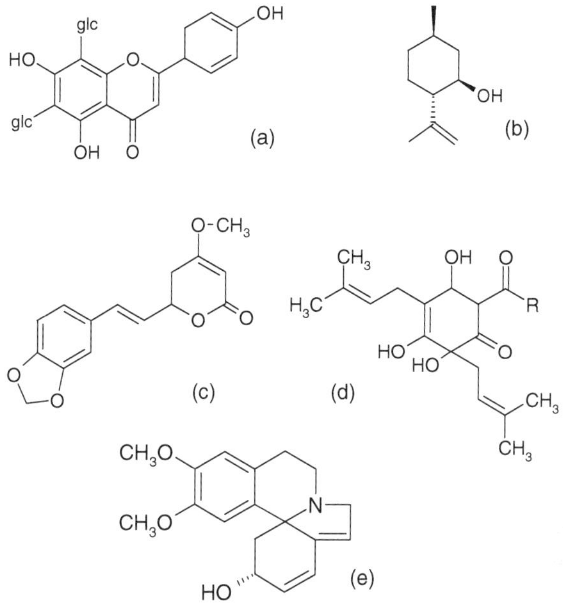

FIGURE 5.1 Chemical structure of some active principles with CNS depressant activity: (a) vicenin-2, (b) isopulegol, (c) 6-methysticin, (d) humulones, and (e) erythravine. Glc = glucopyranose.

regarded and well documented in the literature. Flavonoids and essential oils are the main chemical classes with CNS depressant, and particular, analgesic, and anxiolytic effects, but alkaloids also present similar effects. Some of these active principles are shown in Figure 5.1, and will be mentioned during the course of this chapter.

Ernst [9], in a revision work, evaluated the results of various plants in terms of their anxiolytic effects. According to the author, only the use of kava kava appears to bring consistent results in terms of anxiolytic effect, comparable to benzodiazepines, but further studies are needed to confirm this. The author concludes that there are no well-documented studies using medicinal plants as an alternative to conventional anxiolytics [9]. Some of these main species used as anxiolytics will be addressed below.

5.2.1 Kava-kava: *Piper methysticum* Forst (Piperaceae)

The rhizomes of the Piperacecae kava-kava or kava (*Piper methysticum*) are used in the preparation of a drink used in social and ceremonial rituals on some islands of the South Pacific. Herbal remedies or dietary supplements based on kava-kava are sold as anxiolytics in many countries. The active principles of kava-kava are kavalactones (methysticin, dihydromethysticin) (Figure 5.1) and chalcones (flavokavains), as well as small amounts of essential oil [10] and piperidine alkaloids [11]. Kavalactones (also known as kavapyrones) are powerful inhibitors of many enzymes of the cytochrome system P_{450} (CYP), which suggests the high potential of pharmacokinetic interactions with drugs that are metabolized by the same enzymes. For this reason, kava-kava should be used with caution [12,13].

Despite this drawback, Sarris and Kavanagh [14] affirm that the risk/benefit ratio of kava is highly favorable, with good clinical effectiveness and low risk of adverse reactions, particularly bearing in mind that conventional pharmacological treatment with benzodiazepines has high potential to cause dependence. Kava is a well-established hypnotic drug, with a rapid onset of effect, adequate duration of action, and minimal morning after effects. Clinical studies demonstrate that kava-kava is very effective for the treatment of anxiety disturbances, and for enhancing the quality of sleep [15]. Gastpar and Klimm [16] highlight the advantages of this plant over benzodiazepines, due to the fact that it does not cause dependence and is well tolerated by patients.

However, there are reports in the scientific literature of cases of hepatitis and hepatic failure caused by chronic use of kava, or even the appearance of ataxia resulting from chronic use [17], which led to the banning of this preparation in many countries [18]. Nevertheless, subsequent studies revealed that the majority of these cases were not directly related to the ingestion of this plant. Prior to these reports of toxicity related to the use of kava-kava, it was one of the ten most popular drugs in Europe and North America [13], and the adverse effects were not found previously in the traditional drinks prepared with the infusion with water, unlike the commercial products extracted with organic solvents.

5.2.2 Valerian: *Valeriana officinalis* L. (Valerianaceae)

Valerian (*V. officinalis*) has been used for therapeutic purposes since Greek and Roman times, and is widely distributed in Europe and Asia. Its roots are indicated mainly for their sedative and anticonvulsant effects. Among other uses, its sedative action is used to treat various forms of nervous disorder, nonsevere states of hysteria, anxiety, and neurasthenia. A series of nonglycosilated iridoids called valepotriates have been proposed as active constituents [19]. On the other hand, valerenic acid (a sesquiterpene) is generally used in the standardization of herbal products.

Despite its well-known sedative properties, little is known about its mechanism of action. Neurobiological research has begun to show that the plant, with its active valerenic acid, interacts with the GABA-ergic system ($GABA_A$), a mechanism of action similar to that of benzodiazepine drugs [20]. In animal experiments, valerenic acid, and valerenol, common constituents of valerian, showed tranquilizing and/or sedative activity, through the binding site on $GABA_A$ receptors [21]. Other studies indicate that the constituents of valerian are also partial agonists of different subtypes of serotonergic receptors [22]. The flavanone glycoside 2S (−) hesperidin and the flavone glycoside linarin, isolated from the species *V. officinalis* and *Valeriana wallichii* DC., also present sedative effect [23].

5.2.3 Chamomile: *Matricaria chamomilla* L. (Asteraceae)

Dried flowers of *M. chamomilla* are widely used for their sedative and spasmolytic effects. The flavonoids apigenin, quercetin, patuletin, luteolin, and their glucosides were identified in *Matricaria recutita* L. and *M. chamomilla*, both of which are well known for their anxiolytic properties [24].

Apigenin, together with kaempferol and chrysin, acts as monoaminoxidase (MAO) A and B inhibitors. Some studies show that MAO inhibitor substances may have anxiolytic effect [25] and therefore chamomile flavonoids with this activity would contribute to their anxiolytic effect. Baicalin, another chamomile flavonoid, demonstrated a subtype of agonist selectivity against the $GABA_A$ receptor, without producing amnesia or affecting motor coordination in rats.

5.2.4 Balm, Lemon Balm: *Melissa officinalis* L. (Lamiaceae)

Balm (*M. officinalis*) is a plant that is widely used for its sedative properties [26] and is also used as an anxiolytic. *M. officinalis* contains rosmarinic acid, triterpenoid oleanolic acid, and ursolic acid in high contents, which inhibit GABA transaminase activity and increase GABA levels in the brain [27,28]. This may be related to its depressant effect on the CNS. In high doses, *M. officinalis* can present peripheral analgesic activity [26], while its essential oils have antioxidant and antitumor activity [29].

Clinical trials with *M. officinalis* extract show that its use in healthy volunteers reduces stress induced in a laboratory, as well as altering the mood and increasing calmness, although it can cause impairment in cognitive functions [30]. Kennedy et al. [31] evaluate the possible anxiolytic effect of this plant in association with *V. officinalis*, but the results are not conclusive. New studies are needed to determine whether there is any benefit of this combination.

5.2.5 Hop: *Humulus lupulus* L. (Moraceae)

Hop is used in traditional folk medicine as a bitter depurative and aromatic tonic, while the flowers are used for their calming effects, and also as hypnotics [32,33]. *H. lupulus* is also used as a flavoring in beer.

Most of the phytoconstituents of hop have already been described. Its composition includes volatile oils, around 1% of humulene, myrcene, caryophyllene, farnesene, and bitter resinous principles (15–25%) known as bitter acids, which are the main substances responsible for its hypnotic-sedative effect [32]. Chalcones (xanthohumol and isoxanthohumol), condensed tannins (2–4%), phenolic acids (ferulic and chlorogenic acids), flavonoids glycosydes (kaempferol, quercetin, astragalin, and rutin), amino acids, and estrogenic substances were also found [34].

In general, two major α and β-acids are found in hop. The two series comprise, in fact, three constituents that differ in relation to the nature of the side chain [35]; these are α-acids with three major analogs (cohumulone, humulone, and adhumulone) (Figure 5.1), and β-acids also with three major analogs (colupulone, lupulone, and adlupulone) with a six-membered ring structure [36]. The main phenolic compounds found in hydroethanol extract of *H. lupulus* with anxiolytic properties were characterized as hulupinic acid, cohulupone, two oxidized hop α-bitter acids (principal constituents) – one an oxidized cohumulinone and the other an oxidized humulinone derivative – a procyanidin dimer B, and flavonoids rutin and kaempferol-7-O-rutinoside [37].

Some pharmacological studies were conducted to determine the action of *H. lupulus* on the CNS, seeking to find support for its traditional uses, such as its hypnotic action, and characterize the phytoconstituents responsible for this action [32,38]. Following this line of reasoning, Bravo et al. [39] had already observed that the volatile oils of hop cause a more intense reduction on motor activity, compared with the hydroalcohol or aqueous extracts of this plant. The anxiolytic activity of *H. lupulus* is attributed to the β-acids, reinforcing the fact that the bitter acids are the main active principles responsible for their effects [40].

5.2.6 Tilia: *Tilia cordata* Mill. and *Tilia americana* var *mexicana* (Tiliaceae)

The flowers and stems of *Tilia cordata*, commonly known as tilia or lime tree, are used in traditional folk medicine for their mild calming properties and for insomnia. Although this plant has been known by man for several centuries, there have been few phytochemical and pharmacological studies on this species.

Phytochemical studies have demonstrated that *Tilia* species possess hydrocarbons, esters, aliphatic acids, terpenoids, myricetin, quercetin and kaempferol derivatives, phenolic compounds, condensed tannins, and a coumarin scopoletin [41,42]. *Tilia americana* var *mexicana*, which is endemic to Mexico, has shown a very promising sedative and anxiolytic effect. According to Perez-Ortega et al. [43], different extracts obtained from the flowers of this plant present not only sedative effect, but also anxiolytic effects, through interaction with GABAergic receptors, confirming its popular indication.

The use of the plant proved to be well tolerated, and did not reveal any toxicity. Its anxiolytic activity has been attributed to the flavonoids tiliroside, quercetin, quercitrin, kaempferol, and their glycosides [25] and other quercetin and kaempferol derivatives [43]. The analgesic activity of *T. americana* var. *mexicana*, in turn, is attributed to quercetin, probably exerting this activity through the interaction with 5-HT_{1A} receptors [44].

5.2.7 Bushy Lippia or Falsa Melissa: *Lippia alba* Mill (Verbenaceae)

Another plant with leaves and aerial parts are widely used as a sedative is *Lippia alba*. Essential oils, iridoids, and flavonoids were found in this plant [45]. The anticonvulsant activities of different species of *Lippia* have been attributed to phenylpropanoids, flavonoids, and/or inositols, and also to the volatile terpenoids (beta-myrcene, citral, limonene, and carvone) [46]. Citral, myrcene, and limonene potentiated pentobarbital-induced sleeping time in mice, which was more intense in the presence of citral [47]. Studies indicate that essential oils of *L. alba* present not only a possible anxiolytic effect [48], but also an anticonvulsant effect [49], as was observed for the alpha-terpineol, a monoterpenoid alcohol [50].

The mechanism of action for the anticonvulsant activity of alpha, beta-epoxy-carvone, a monocyclic monoterpene, is not attributed to direct activation of the benzodiazepinic $GABA_A$ receptor, but may be associated with a reduction in isolated nerve excitability, possibly involving a voltage-gated Na^+ channel blockade [51].

5.2.8 Lemon Grass: *Cymbopogon citratus* (DC.) Stapf (Poaceae)

Tea obtained from leaves of *Cymbopogon citratus* is used for its anxiolytic, hypnotic, and anticonvulsant properties in Brazilian folk medicine. These activities have been attributed to essential oils [52]. Tannins, phenolic acids (caffeic and *p*-coumaric acid derivatives), and flavone glycosides (apigenin and luteolin derivatives) were also found in the leaves of this plant [53]. Isoorientin, isoscoparin, isoorientin 2″-O-rhaminoside, and orientin were isolated and identified by spectroscopic methods in the methanol, hydroalcohol, infusion, and decoction extracts of *C. citratus*, and demonstrated high antioxidant power [54].

Blanco et al. [52] observed an anxiolytic effect using the essential oils of the *C. citratus* with high citral content, not reported in previous studies [55]. Blanco et al. [52] attempt to explain these results by suggesting that some active principles present in the essential oils (citral and others) may have been lost during the preparation of "abafado" (a traditionally used tea, which was reproduced in the studies with negative results mentioned above). The discrepancy between the results could also be explained by the use of different plant chemotypes [56]. The essential oil from *C. citratus*, rich in citral, produced significant antinociceptive effect in mice [56].

5.2.9 Erythrina: *Erythrina mulungu* Mart. ex Benth. and *Erythrina velutina* Wild (Leguminosae)

Among the plants with a CNS depressant effect are those belonging to the *Erythrina* genus. This is a well-known genus, not only in Brazil but also in various other parts of the world. The species *E. mulungu* and *E. velutina* are traditionally used by Brazilian communities to treat insomnia and other disorders of the CNS. Some central effects of these species, such as their anticonvulsant, anxiolytic, and analgesic effects, have already been demonstrated. The erythrinian alkaloids, such as, alpha-hydroxy-erysotrine, erythravine (Figure 5.1), and 11 alpha-hydroxy-erythravine, isolated from the flowers of *Erythrina mulungu,* exert an anxiolytic effect [57].

According to Vasconcelos et al. [58], the anticonvulsant effect of hydroalcohol extract of *E. mulungu* and *E. velutina* was more evident in the strychnine-induced model than in the pentylenotetrazol-induced model, indicating a mechanism of action through glycine. According to these authors, although the hydroalcohol extracts of *E. velutina* and *E. mulungu* did not alter the results in the elevated plus maze test, therefore presenting no anxiolytic effect, the data obtained by them revealed a depressant action on the CNS, since the animals treated with this plant presented a reduction in locomotor activity, rearing and grooming in the open field test. Onusic et al. [59], on the other hand, observed an anxiolytic effect in rats, following acute administration of a hydroalcoholic extract of *E. mulungu.*

5.2.10 *Passiflora genus* (Passifloraceae)

The well-known *Passiflora* genus includes approximately 500 species [60]. The leaves of various species of *Passiflora* are used in traditional medicine for their calming, sedative properties. The most studied species of the genus is *P. incarnata* L., which is included in most Pharmacopeias. The leaves of *Passiflora alata* Curtis are also used in the pharmaceutical industry for their anxiolytic properties [61].

The flavones homoorientin, orientin, vitexin, and isovitexin were identified as principal constituents in a commercial sample of *P. incarnata* [62]. Another chemical constituent that appears to play an important role in the sedative effect of *P. incarnata* is chrysin [63]. Isovitexin was found to be the principal constituent of a methanol extract of *Passiflora actinia* Hooker that showed selective anxiolytic activity in the elevated plus-maze and open field tests [64]. Flavone derivatives can act as positive modulators of the gabaergic $GABA_A$ receptors. However, luteolin (3′,4′,5,7-tetrahydroxyflavone) is a very common flavone in plants, which has been reported as being devoid of specific affinity for the benzodiazepine binding site, but with the considerable anxiolytic activity, indicating that it probably acts by another mechanism [65].

Passiflora edulis Sims, a species cultivated mainly for its fruits, is also used to treat or prevent disorders of the CNS, such as insomnia. Animals that received extracts of leaves of *P. edulis* evaluated in the elevated plus maze presented expected behavior after the use of an anxiolytic [66]. Phytochemical analysis of this species showed the predominance of the flavones C-glycosydes, isoorientin, vicenin-2 (Figure 5.1), spinosin, and 6,8-di-*C*-glycosylchrysin [67]. In another study, luteolin-7-*O*-[2-rhamnosylglucoside] isolated from *P. edulis* showed anxiolytic activity, without compromising motor activity [68]. Dhawan et al. [69] observed that although the methanol extract of *P. edulis* presents an anxiolytic effect, it is less potent than the effect of *P. incarnata.*

5.2.11 Essential Oils

The inhalation of essential oils leads to their absorption into the bloodstream, as has been demonstrated in animal and biochemical models, exerting measurable psychological effects. Aromatherapy is used worldwide to relieve chronic pain, anxiety, insomnia, depression, and stress-related problems, and to treat some cognitive disorders.

Popular anxiolytic oils include lavender (*Lavandula angustifolia*), rose (*Rosa damascena*), orange (*Citrus sinensis*), bergamot (*Citrus aurantium*), lemon (*Citrus limon*), sandalwood (*Santalum album*), clary sage (*Salvia sclarea*), Roman chamomile (*Anthemis nobilis*), rose-scented geranium (*Pelargonium* spp.), and the compound linalool, among others [70,71]. Isopulegol (Figure 5.1), present in the *Corymbia citrodora, Eucalyptus citrodora, Zanthoxylum schinifolium*, among other aromatic plants, presented a depressant effect on the CNS [72].

The essential oil of *Rosmarinus officinalis* L. (Lamiaceae), popularly named rosemary, a common household plant grown in many parts of the world, including

Brazil, showed the peripheral antinociceptive activity [73]. Ballard et al. [74] observed that aromatherapy carried out with essential oils of *M. officinalis* was very effective in reducing the agitation normally observed in patients with severe dementia, with a consequent improvement in the quality of life of these people.

In a study carried out by Leite et al. [75], exposure of rats to essential oil of Citrus caused an anxiolytic effect in these animals in the elevated plus maze. It has been demonstrated that exposure to the essential oil of orange can contribute to the treatment of psychiatric disorders, as well as reduce psychological stress, without the adverse effects commonly seen with conventional medicines [76]. Lemon oil presented a strong antistress effect in the elevated plus maze, forced swimming, and open field tests in mice. The antistress effect of lemon oil was significantly blocked by pretreatment with flumazenil, benzodiazepine receptor antagonist, or apomorphine, a nonselective DA receptor agonist [77]. Citral, limonene, and myrcene presented a sedative, relaxing effect [47]. In high doses, they potentiated sleep time induced by pentobarbital in mice, which was more intense in the presence of citral. However, none of them showed anxiolytic effect, but rather, a slight anxiogenic effect at higher doses [47].

5.2.12 Other Plants with Depressant Action on the Central Nervous System

Besides the plants presented in this work, there is an enormous number of species popularly used as tranquilizers, analgesics, and sleep inducers, but these have not yet been fully studied. For example, various preclinical trials suggest antinociceptive effects for chemical extracts and constituents of several plants [7,78]. On the other hand, few species have been clinically evaluated to determine their efficacy as analgesic agents. Although the clinical use of crude extract of plants with analgesic effects has been somewhat rare, several active principles with analgesic activity have been found, and subsequently isolated. The most well known of these is morphine, obtained from the poppy (*P. somniferum*), a plant mentioned at the beginning of this chapter, which also has codeine and thebaine as ingredients with the analgesic activity. From the barks of *Salix alba* (and other species), saliciline (a phenolic glicosyde) was isolated, from which salicylate was obtained, to finally obtain acetylsalicylic acid, possibly the most frequently used medicine all over the world.

A similar scenario is seen with plants for which experiments on animals suggest a neuroleptic effect [79], but which have not yet been submitted to clinical trials on humans. Reserpine, a compound extracted from *R. serpentina*, has already been used as an antipsychotic agent. However, as far as we know, there is no phytotherapic currently available with clinically proven antipsychotic effect.

5.3 PLANTS WITH THE CNS STIMULANT ACTIVITY

Psychostimulant or psychoanaleptic drugs are those that stimulate the activity of the central nervous system. CNS stimulants are rarely used for therapeutic purposes (to treat pathological disturbances or conditions), but are used mainly to increase the state of alert and attention.

Various plants are used around the world, for their stimulant effects. Some of the most well-known plants with this property are coffee (*Coffea arabica*), cocoa (*Erythroxylum coca*), mate (*Ilex paraguariensis*), and tea (*Camellia sinensis*), among others [19]. These plants are rich in constituents with CNS stimulant action, particularly the alkaloid caffeine (1,3,7-trimethylxanthine), considered the most commonly used psychoactive substance in the world. Caffeine and other methylxanthines act as adenosine antagonists, and in high concentrations, inhibit phosphodiesterases, a group of enzymes that metabolize AMPc. Other active principles with plant-derived psychostimulant action are structurally similar to amphetamines, as is the case with cathinone (from *Catha edulis*) and ephedrine (from *Ephedra sinica*). Plants rich in these chemical constituents are often used in weight loss diets, due to their anorexic effect, which also occurs with amphetamine. Another substance with powerful stimulant action on the CNS is cocaine, but unlike those mentioned earlier; it is considered illegal when used as a recreational drug, particularly because of its potential to cause dependence. The main active principles with stimulant action on the CNS are shown in Figure 5.2.

5.3.1 Coffee: *Coffea arabica* L. (Rubiaceae)

Coffee is one of the most popular drinks in the world. It is prepared by infusing the roasted, ground seeds of the coffee tree, a plant that originates in Ethiopia and is cultivated on a large scale in some countries of Africa and in particular, in South America. The two major coffee species consumed worldwide are *Coffea arabica* L. (Arabica) and *Coffea canephora* Pierre ex A. Froehner (Robusta). Arabica coffee is considered to be superior to Robusta due to its organoleptic properties. The main group of phenolic compounds present in green coffee beans are chlorogenic acids, which are esters of trans cinnamic acids and quinic acid. Several classes of chlorogenic acids have been distinguished in green coffee beans, including

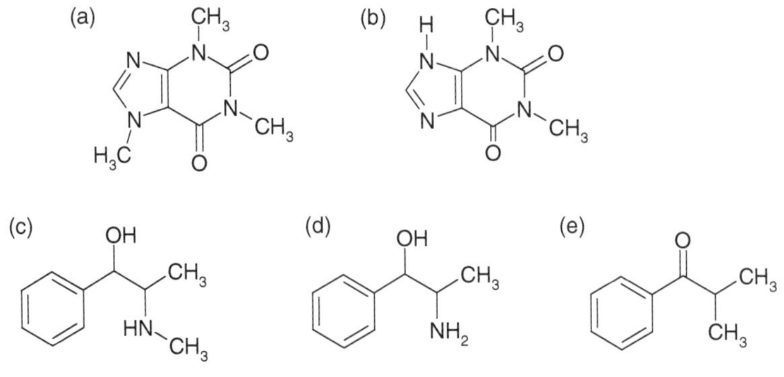

FIGURE 5.2 Chemical structure of some active principles with CNS stimulant activity: (a) caffeine, (b) theophylline, (c) ephedrine, (d) norpseudoephedrine, and (e) cathinone.

caffeoylquinic acids, feruloylquinic acids, dimethoxycinnamoylquinic acid, dicaffeoylquinic acids, and diferuloylquinic acids derivatives [80]. Of the methylxanthines, caffeine, theobromine, and theophylline (Figure 5.2) have also been found in coffee. Caffeine is a major alkaloid in green coffee beans [81] and is the main constituent responsible for the stimulant effect of coffee. The various effects attributed to caffeine include its capacity to reduce tiredness [82], increase attention, and improve psychomotor performance [83].

In the past, coffee was considered harmful to health, but in recent years its reputation has improved, as a healthy drink capable of decreasing the occurrence of certain diseases. Epidemiological studies indicate that regular consumption of coffee can decrease the incidence of heart attack and coronary disease [84] and alcoholic cirrhosis [85]. Various studies indicate that coffee has a neuroprotective effect against Parkinson's disease [86,87], like other plants rich in caffeine (as will be mentioned later in this chapter).

5.3.2 Tea Plant: *Camellia sinensis* L. Kuntze (Theaceae)

C. sinensis is a native shrub of Southern Asia, which is widely cultivated in India and China. Along with coffee, tea infusion brewed from *C. sinensis* is one of the most popular beverages. The leaves of the plant are used in the preparation of green tea and black tea. The two types differ in relation to the state of the leaf when it is collected, and the drying process. Recently, white tea has also become popular, for which only the fresh, immature leaves of the plant are used. All the types of tea are drunk as a stimulant, and also for their antioxidant properties, which are particularly powerful in green tea.

The leaves of the tea plant contain caffeine and, in lesser quantities, theobromine and theophylline as components with stimulating properties [88]. However, the main constituents of *C. sinensis* are phenolic compounds [89], which are responsible for the marked antioxidant effect of tea. Among these components are catechins and phenolic acids, flavonoids, and coumarins [88]. Preparations containing *C. sinensis* are also used in weight loss diets, though their effectiveness in these cases is questionable.

5.3.3 Guarana: *Paullinia cupana* Kunth (Sapindaceae)

Guarana is a native tree of the Amazon rainforest, and is now cultivated in other regions. The guarana seeds were used as a tonic and stimulant by the Amazon Indians, even before the discovery of Brazil. Guarana is used in the form of a powder (ground seeds) or pastes and sticks. It is also present in the composition of various soft drinks and herbal remedies. In traditional folk medicine, guarana is used for its stimulant and energy-enhancing properties, to combat fatigue and mental exhaustion, and it is also acclaimed for its rejuvenating and aphrodisiac properties [90]. Industrial or semi-industrialized products containing guarana are popular among students and professionals, to help them stay awake during the night. However, its excessive use can cause tachycardia, irritation, and insomnia. Guarana is also used in weight loss, probably due to the fact that it stimulates the metabolism and suppresses the appetite.

The purported stimulant properties of guarana have generally been assumed to reflect the presence of caffeine. However, the psychoactive properties of guarana may also be attributable to the relatively high content of other potentially psychoactive components, including saponins and tannins [91]. The plant also has phenolic compounds such as catechin, epicatechin, and proanthocyanidins, which are believed to be responsible for its antioxidant action. Catechins are believed to contribute to weight loss and increase energy expenditure, together with caffeine [92].

A study carried out with young volunteers indicates an increase in cognitive ability in parameters such as attention, working memory, and accuracy, after treatment with *Paullinia cupana*, *Panax ginseng* or with a combination of the two plants [93]. Another study by the same group shows that a multivitamin supplement containing guarana decreased mental fatigue in the volunteers, following mental agility tests [94].

5.3.4 Mate: *Ilex paraguariensis* St. Hil. (Aquifoliaceae)

Mate is a native shrub of South America, cultivated in Paraguay, Argentina and the South of Brazil. The infusion of its dried leaves is used as a tonic and stimulant, to combat fatigue and to improve the cognitive processes. In Brazil, mate is traditionally prepared in the form of a "chimarrão" (sipped hot from a gourd) by the inhabitants of the south of Brazil, and in the form of "tereré" (cold tea) by the inhabitants of the Central West region of the country.

Ilex paraguariensis is metabolically characterized as having higher amounts of xanthines and phenolics, including phenylpropanoids and flavonoids [95]. Many of the positive effects of mate on humans, including its stimulant effect, are related to the presence of methylxanthines (caffeine, theobromine, and theophylline) and phenolic acids [96]. The latter constituents also appear to be the active principles responsible for its marked antioxidant action. This plant is known for its high content of compounds of the hydroxycinnamic acid group, particularly caffeic acid derivatives [97]. Caffeoylquinic acids, dicaffeoylquinic acids, feruloylquinic acids, diferuloyl quinic acids, caffeoyl-p-coumaroylquinic acids, and caffeoyl-sinapoylquinic acids derivatives were isolated from mate. Furthermore, caffeoylshikimates, dicaffeoylshikimates, tricaffeoylshikimate, and feruloylshikimate have also been detected in this plant [98].

5.3.5 Ma Huang: *Ephedra sinica* Stapf (Ephedraceae)

Ma Huang (or Ephedra) is a shrub that originates in China, where it is used by millions of people to relieve respiratory problems. The aerial parts of the plant contain ephedrine alkaloid as the main constituent (Figure 5.2) and smaller quantities of norephedrine, pseudoephedrine, norpseudoephedrine (also called cathine), and methylpseudoephedrine [99,100], which are responsible for its bronchodilator effect. The phenylpropylamine alkaloids (ephedrine alkaloids) of Ma Huang act on the catecholaminergic system, increase the release of noradrenalin and inhibit its reuptake. The alkaloids of Ma Huang cross the hematoencephalic barrier, producing psychostimulant effects, as well as characteristic peripheral effects [101,102]. Ephedrine is considered less potent than adrenalin, but has a long-lasting effect. For this reason, some ephedrine-based drugs are used for cases of acute asthma and in chronic cases that require maintenance medication. They are also usefully for oral use as nasal decongestants in cases of allergic rhinitis. Ephedra is similar to anorexic medications used in the treatment of obesity [103].

Ephedrine was initially introduced as an antiasthmatic drug, but the CNS stimulatory effects of this sympathomimetic amine have been a problem in therapy, since the use of ephedrine causes insomnia, restlessness, and anxiety and is also limited by its cardiovascular effects [99]. It is reported that ephedrine may even induce toxic psychosis.

5.3.6 Khat: *Catha edulis* Forsk (Celastraceae)

Khat refers to the young leaves and twigs of an evergreen shrub, *C. edulis*, found mainly in Africa. Its leaves are used as a general stimulant in some African and Arabic countries. They are used in various countries as appetite suppressants, as part of weight loss diets, although this use is not approved by the World Health Organization. The effects of khat and its constituent alkaloids on appetite have been previously documented [104].

The khat plant contains psychoactive alkaloids with psychostimulant properties, and has been used for centuries as a recreational and religious drug, particularly in Africa and the Middle East [105]. The main active principle of the fresh leaves is cathinone [106] (Figure 5.2), a phenylethylamine that acts as an indirect sympathomimetic agent. Cathinone, like amphetamine, acts by releasing catecholamines from presynaptic storage sites and subsequently inhibiting their uptake. The leaves of khat also contain norephredrine and norpseudoephedrine (cathine), which have similar effects to cathinone, and act in the same way as cathinone metabolites in the human body [106]. Terpenoids, flavonoids, sterols, glycosides, and tannins were also found in the leaves of *C. edulis* [106], but these do not appear to be involved in the stimulant effect of the plant. Abusive use of khat leaves, particularly as an appetite suppressant, can lead to hypertension, aggressiveness, and neurological disturbances, as occurs with amphetamine analog psychostimulants.

5.3.7 Cola nut: *Cola nitida* (Vent.) Schott et Endl. or *Cola acuminata* (Beauv.) Schott et Endl. (Sterculiaceae)

Another African plant with stimulant active principles is cola. The nuts or seeds of the plant are used in the preparation of stimulant extracts and tinctures, as energizers, and also for the drinks industry. The traditional preparations generally use the fresh seeds consumed as a general tonic, to combat physical and mental fatigue.

The cola seeds contain caffeine, theobromine, catechin, epicatechin, and procyanidins [107], which are the constituents responsible for the stimulating, energizing effects.

5.3.8 Coca: *Erythroxylum coca* Lam. (Erythroxylaceae)

The coca plant is native to South America, occurring mainly in the Andes and Amazon regions of Columbia and Bolivia. Its traditional use is in the form of tea (infusion) with the leaves of the shrub, which are sometimes also chewed, to increase energy and combat fatigue and altitude sickness (an effect of the rarefied air). A fascinating description of the effects produced by coca leaves was mentioned by Garcilaso de la Vega: coca satisfies the hungry, gives new strength to the weary and exhausted and makes the unhappy forget their sorrows [108]. Although its traditional use is legal, the coca leaves are the raw material for the extraction of cocaine, which is used as a recreational drug and is a public health problem throughout the world.

Cocaine is an alkaloid tropane with potent inhibitory action on noradrenaline and dopamine reuptake. It also acts by blocking the Na^+ channels, which is related to its local anesthetic effect. Besides cocaine, which is, without doubt, the active principle responsible for the psychotropic effects of coca, the plant also contains flavonols [109], the majority of which occur as 3-*O*-monoglycosides with glucose, galactose, arabinose, xylose, and rhamnose, and as 3-*O*-diglycosides with various combinations of those sugars. Kaempferol, quercetin, and ombuin are the main aglycones of coca flavonols [110].

In the illicit use of the drug, all kinds of preparations are used, from the more crude ones, such as coca paste "merla" (an intermediary product of refining); coca base "crack," which volatizes at around 95°C and is smoked in pipes; to cocaine (chloridrate), in the form of free salt (powder), which is inhaled or injected.

5.4 PLANTS USED AS ANTIDEPRESSANTS

Although various studies with plant extracts have shown similar responses to those of antidepressant drugs in biochemical and animal models, few extracts have been evaluated in clinical trials. *Hypericum perforatum* is the main medicinal plant used in the treatment of depression.

5.4.1 St. John's wort: *Hypericum perforatum* L. (Guttiferae)

H. perforatum is an herbaceous plant that is widely distributed in Europe, Asia, and North America. In the Middle Ages, it was believed that if this plant was picked on the morning of Saint John's day (24th June), when it was in flower, it would have the power to "chase away evil spirits." Hence, the popular name by which the plant is known today throughout Europe: St. John's wort. On the other hand, the attribution "perforatum" in the botanical nomenclature comes from the fact that the leaves of this plant, when observed against the light, have reddish tips that look like small perforations, caused by the presence of hypericin.

The aerial parts of the plant, when in flower, are used in the manufacture of antidepressant herbal remedies. Despite reports that the plant was used in traditional medicine to treat states of depression, this indication gained notoriety only after its effects were scientifically proven. The first studies, in the 1980s, investigated the inhibitory action of *H. perforatum* on monoamino oxidase (MAO) and were based on the fact that other species of the same family (Guttiferae) also presented this activity [111]. The antidepressant effect of the plant was initially attributed to hypericin (naphthodianthrone), but it has now been demonstrated that other constituents have a more determining role in its pharmacological effect, especially the prenylated phloroglucinol hyperforin [112,113]. Studies indicate that besides the inhibition of MAO, *H. perforatum* also inhibits the reuptake of monoamines, among other auxiliary mechanisms. The lipophilic constituent hyperforin (2–5% of the extract) demonstrated, similarly to chemical antidepressants, a significant effect on the synaptosomal uptake inhibition of several neurotransmitters in *in vitro* assays [114,115]. St. John's wort also contains chemical constituents such as flavonoids, biflavones, xanthones, and tannins [116]. Although hyperforin continues to be considered the main active principle, some recent studies suggest the participation of flavonoids in the antidepressant effect of *H. perforatum*. Of these, isorhamnetin was the flavanol that presented the highest antidepressant activity in animal models [117].

Clinical trials suggest that extracts of *H. perforatum* are effective for treating mild to moderate depression, showing superior activity to that of the placebo, and similar to standard drugs [118,119]. However, other works show that extracts of *H. perforatum* interfere with the P_{450} cytochrome system, modifying the bioavailability of various drugs. This fact led to a greater control of the sale and prescription of herbal remedies based on *H. perforatum*, but these are still the most commonly sold herbal remedies worldwide [116].

5.4.2 Other Plants with Antidepressant Potential

Unlike *H. perforatum*, other plants that reputed to have antidepressant effects have been only superficially evaluated. Extracts of the species *Apocynum venetum*, *Areca catechu*, *Cissampelos sympodialis*, *Bacopa monniera*, *Ocimum sanctum*, and *Withania somnifera* have shown significant antidepressant effects in animal models [79]. It is interesting to observe that the latter three fit into the category of adaptogenic plants, as will be discussed in the following section.

5.5 ADAPTOGENIC PLANTS

Adaptogens are plant extracts that can allow an organism to counteract adverse physical, chemical, and biological stressors by generating nonspecific resistance. They are known to increase the availability of energy, reduce stress response, increase endurance, and increase mental alertness.

Their mechanisms of action are still not well known, but it is clear that they act by different mechanisms. The regulation of the neuroendocrine system, through the modulation of the hypothalamic-pituitary-adrenal system, plays a key role in controlling the response to stress [120,121]. Many adaptogens act as antioxidants and

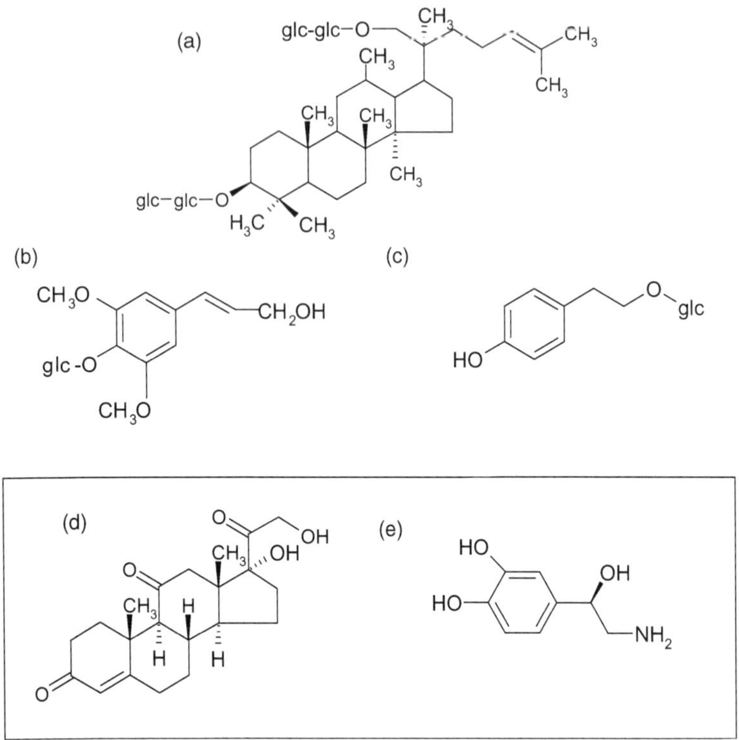

FIGURE 5.3 Chemical structure of some adaptogens compounds: (a) ginsenoside Rb_1, (b) eleutheroside B (syringin), (c) salidroside. The chart shows the analogues endogenous compounds: (d) cortisol, (e) noradrenaline. Glc = glucopyranose.

immunomodulators, among other nonspecific actions. Some chemical constituents of adaptogenic plants are structurally similar to hormones and other endogenous mediators (Figure 5.3), which is probably related to the modulator action of these plants on the homeostasis of the organism. For more details on the mechanisms of action of adaptogens see Panossian and Wikman [121].

5.5.1 Ginseng: *Panax ginseng* C. A. Meyer (Araliaceae)

P. ginseng is a perennial herb native to China and Korea and has been used as an herbal remedy in Eastern Asia for thousands of years. Its herbal root had been broadly used as a functional food in the forms of boiled extract, powder, tea, tablets, and capsules, and it is one of the most famous Chinese herbs used in traditional oriental medicine for the prevention and treatment of age-related disorders, because is believed to promote health and longevity. The name ginseng refers to its reputation as a plant capable of curing all sicknesses: panax (*pan* = all, *axos* = cure) ginseng (*jen* = plant, *sheng* = man; because of the human shape of the root). Thus, *P. ginseng* means "man-shaped plant that cures all."

The main bioactive constituents of ginseng drugs are considered to be triterpene saponins (dammarane-type), generally referred to as ginsenosides. More than 80 ginsenosides have been isolated and most of them possess four types of aglycones, protopanaxadiol, protopanaxatriol, ocotillol, and oleanolic acid [122]. Pharmacologic studies showed that the bioactivities of ginsenosides vary depending on the type of aglycone and the sugar moieties.

The pharmacological effects of ginseng have been demonstrated in the CNS and cardiovascular and immune systems [123]. There have been many studies on the effects of ginseng on the memory, physical resistance, and stress in animals and humans [124,125]. Its beneficial actions on cognition and stress appear to occur by separate mechanisms, the first mainly through an increase in cholinergic neurotransmission (although other systems also participate) and the second mainly through the modulation of the hypothalamic-pituitary-adrenal system [123]. This latter effect may be due to the structural similarity of the ginsenosides present in ginseng, and the steroidal hormones, which include cortisol (Figure 5.3).

5.5.2 Siberian Ginseng: *Eleutherococcus senticosus* (Rupr. and Maxim.) Maxim. (Araliaceae)

Known as Siberian ginseng, *Eleutherococcus senticosus* (syn. *Acanthopanax senticosus*) is a plant that belongs to the ginseng family, discovered by chemotaxonomy through a Soviet government program to investigate the medicinal effects of other plants of the Araliaceae family capable of increasing physical resistance [126]. Extracts of the Siberian ginseng root were used in the preparation of astronauts, soldiers, and athletes of the Soviet Union during the cold war, for training purposes and to increase their physical performance [127].

E. senticosus is rich in phenolic or coumarin glycosides known as *Eleutherosides*, many of which are structurally similar to catecholamines [126] (Figure 5.3). In the roots of *E. senticosus*, a complex mixture of phenylpropane derivatives (syringin, caffeic acid, coniferyl aldehyde, and caffeic acid ethyl ester), sterols (β-sitosterol and daucosterol), sesamin, isofraxidin, hederasaponin, and sugar polymers were found [126].

Extracts of *E. senticosus* have a similar profile of action to *P. ginseng*. Positive results were obtained in tests evaluating the physical capacity and immunostimulant effect [128,129], despite the existence of some studies with negative results.

5.5.3 Damiana: *Turnera diffusa* Willd. var. *aphrodisiaca* (Ward.) Urb. (Turneraceae)

Turnera diffusa (syn. *T. aphrodisiaca*) is an aromatic plant popularly known as damiana that grows wild in the subtropical regions of America and Africa. The leaves of damiana have been used traditionally as a stimulant, aphrodisiac, tonic, diuretic, nerve tonic, and laxative, the leaf infusion being a traditional remedy for diseases related to the gastrointestinal and respiratory systems, and for the treatment of gonorrhea in Latin American societies [130].

Alkaloids, cyanogenic glycosides, flavonoids, and volatile oil constitute major classes of phytoconstituents of the *Turnera* genus [131]. Phytochemical reports on *T. diffusa* indicate the presence of tetraphyllin B (a cyanoglycoside), gonzalitosin I (a flavonoid), arbutin (a phenolic glycoside), damianin, hexacosanol, a volatile oil containing α-pinene, β-pinene, *p*-cymene and 1,8-cineole, and β-sitosterol [130]. Flavonoids, such as, luteolin, apigenin and their derivatives, as well as terpenoids were also found in this plant [132,133].

5.5.4 Other Adaptogens

Other plants have been studied for their adaptogenic properties, particularly not only in the Asian countries, but also in Europe and occasionally in the United States. Those with more promising results include: *Bacopa monniera*, *Schizandra chinensis*, *Rhodiola rosea*, and *Withania somnifera*, among others.

The fruits of *S. chinensis* (Turcz.) Baill (Schizandrae) have been used in traditional medicine as a tonic and antiaging drug. The biological active compounds in the fruits of *S. chinensis* are lignans with dibenzo[*a,c*]cyclooctene moieties derivatives [134].

Bacopa monniera Linn. (Scrophulariaceae), a species that is part of Ayurvedic medicine, popularly known as Brahmi, is used as a nervine tonic to promote mental health and improve memory and intellect [135]. The saponins bacoside A and B were isolated from *B. monniera*. Studies indicate antioxidant and neuroprotective effects of its extract [136].

Rose root (*Rhodiola rosea* L., Crassulaceae) grows in rocky, inhospitable regions of Asia and Europe, particularly in the Arctic countries. Its biological effects of stimulating the nervous system, decreasing depression, enhancing work performance, eliminating fatigue, and preventing high-altitude sickness, have been attributed to salidroside (rhodioloside) (Figure 5.3), rosavins, and p-tyrosol [137].

Another adaptogenic plant that has been well described and studied is Indian ginseng (*Withania somnifera* Dunal, Solanaceae), known locally as ashwagandha. The withanolides found in Indian ginseng roots are structurally similar to the ginsenosides present in the plant *P. ginseng*. Two glycowithanolides (sitoindoside IX or sitoindoside X) showed the antistress activity, and increased learning acquisition and memory retention in both young and old rats [138].

5.6 PLANTS USED TO TREAT NEURODEGENERATIVE DISEASES

Schultes [139] reports various plants used by the Indians of the Amazon and Colombian Basins to treat several symptoms characteristic of the ageing process, such as weakness, shaking (trembling), forgetfulness, and dementia. The different uses are generally specific to certain symptoms, suggesting that the species are used by the Indians not only to treating ageing itself, but possibly also to treat neurodegenerative diseases, such as Parkinson's and Alzheimer's diseases. However, these plants have not yet been fully studied. Next, some of the principal species and chemical components with potential use in the treatment of Alzheimer's and Parkinson's diseases will be presented.

5.6.1 Ginkgo: *Ginkgo biloba* L. (Ginkgoaceae)

Ginkgo is an oriental tree that can live for more than a 1000 years. Its genus goes back 200 million years ago, and it is therefore considered a living fossil [140]. The part of the plant used for medicinal purposes is the leaves, which resemble the two hemispheres of the brain, hence its name biloba (bilobed). Ginkgo is used to improve the memory, particularly in elderly people or those with cognitive impairment, patients with Alzheimer's disease, or in cases of cerebrovascular insufficiency, and is also used to treat circulatory disturbances, tinnitus and vertigo.

Ginkgo biloba leaves contain flavonoids, phenolic acids, diterpene lactones (ginkgolides), and the sesquiterpenoid bilobalide [140,141]. Ginkgolides and bilobalide are considered the main active principles of the plant, responsible for their beneficial effects on the memory. The ginkgo flavonoids include flavonol glycosides, biflavones, proanthocyanidins, alkylphenols, and polyprenols [142].

Although the literature does not classify *G. biloba* as an adaptogenic plant, its various effects and the set of actions are characteristics of this type of substance. Ginkgo has notable antioxidant action [143], decreases the blood viscosity [144], and increases blood flow to the brain [145]. Its effects on improving learning and memory have been studied in laboratory animals, with very favorable results. In relation to the clinical trials, there is still controversy as to its real benefits on cognition. These benefits have been observed in the elderly and in patients with Alzheimer's disease [146,147]. On the other hand, some studies do not indicate any improvement in the cognitive processes of patients treated with *G. biloba*, including a recent trial involving 3069 participants [148]. It is important to bear in mind that although the *G. biloba* extract is considered to be safe, the plant is not recommended for individuals undergoing treatment with anticoagulants. In addition, Leistner and Drewke [149] report the presence of some toxic and allergenic components in the plant, such as ginkgotoxin.

5.6.2 Dong Guai, Chinese Angelica: *Angelica sinensis* (Oliv.) Diels (Apiaceae)

The roots of *Angelica sinensis*, also known in Chinese as Dong Guai or Dang Gui, have been used for more than 2000 years in China to treat various ailments. Previous phytochemical studies have resulted in the isolation and identification of a variety of chemical constituents, including phthalides, polysaccharides, lipids, polyacetylenes, aromatic compounds, and terpenes [150]. The alcoholic extract of the roots of *A. sinensis* reduced the toxicity induced by the administration of the peptide β-amyloid in neuronal cell culture (an *in vitro* model used to estimate the neuroprotective effect) [151]. In a subsequent study using bioassay guided by isolation and identification of the possible active principles, Z-ligustilide,

11-angeloylsenkyunolide F, coniferyl ferulate and ferulic acid were indicated as the main constituents responsible for the neuroprotector effect of *A. sinensis* [152].

5.6.3 Gotu kola: *Centella asiatica* (L.) Urban (Apiaceae)

Centella asiatica (Gotu Kola) is a small herbaceous plant native to Asia and Australia, which grows along ditches and in low, wet areas. The leaves and other parts of the plant are used in cooking, and for various medicinal purposes.

Asiatic acid isolated from *C. asiatica*, as well withanolide A (from *Withania somnifera*), positively modulate multiple targets associated with β-amyloid pathways and thus, may be beneficial in attenuating β-amyloid levels in the Alzheimer's disease brain, both by decreasing β-amyloid production and by increasing β-amyloid degradation [153].

Polyphenols and triterpenes from *C. asiatica* are useful xenobiotic detoxifying agents that decrease lipid peroxidation and enhance brain antioxidants. They may offer useful support in the treatment of Parkinson's disease, by acting as a neuro-protective antioxidant preventing the neuronal brain damage that is associated with this disease [154].

5.6.4 Muirapuama: *Ptychopetalum olacoides* Benth. (Olacaceae)

Ptychopetalum olacoides, a small tree that is native to the Amazon region of Brazil, where it is known as "muirapuama" or "marapuama," is used to treat chronic degenerative conditions of the nervous systems. Roots of *P. olacoides* are prepared in alcoholic infusions to treat "nerve weakness" by Amazonian Indians. Nerve weakness can be described as a syndrome with several symptoms, including lassitude, depression, sexual impotence, and tremors [155].

Clerodane-type diterpenoids found in this plant, showed nerve growth factor potentiating activity on PC-12 cells (*in vitro* assay) in a dose-dependent manner, suggesting that they may be useful candidates for developing drugs for the treatment of neurodegenerative diseases such as Alzheimer's disease [156]. According to da Silva et al. [157], the mechanisms of action of muirapuama include anticholinesterase, dopaminergic, and serotonergic activity.

5.6.5 The Cowhage, Velvet Bean: *Mucuna pruriens* (L.) DC. (Fabaceae)

Mucuna pruriens, a legume extensively used in Ayurveda medicine to treat Parkinson's disease, is reputed to provide benefits against this disease, without inducing dyskinesias [158]. Levodopa (L-dopa) was isolated from the seeds of *M. pruriens* in 1937, but the scientific interest in this species and other plants rich in L-dopa was developed later [159].

The plant has been shown to be more effective compared with synthetic levodopa in an animal model of Parkinson's disease [160]. Katzenschlager et al. [159] suggest that *M. pruriens* might possess advantages over conventional L-dopa preparations in the long term management of Parkinson's disease. However, the plant must be used with caution, as the seeds contain high concentrations of L-dopa, as well as volatile compounds, such as benzoic acid 2-hydroxy methyl ester [161].

5.6.6 Caffeine and Other Adenosinergic Antagonists as Neuroprotective Agents

Various works have shown that caffeine and other adenosinergic antagonists may play an important neuroprotective role, particularly in Parkinson's disease [87,162,163]. There is evidence that the A_{2a} receptors are present in the striatum associated with the D_2 dopaminergic receptors, which it negatively modulates [164]. The administration of A_{2a} agonists decreases the affinity of dopamine by the D_2 receptor. Thus, the final effect of the adenosinergic antagonists is the potentiation of the dopaminergic transmission.

Studies with extracts of plants rich in caffeine, such as mate and black tea indicate a pro-dopaminergic or neuroprotective effect of these extracts in models of Parkinson's disease [165,166].

5.6.7 Antioxidants and Anticholinesterasics of Natural Origin

In recent years considerable attention has been devoted to medicinal plants that possess a high-antioxidant property and reduce free radicals that produce tissue damage. The brain is especially susceptible to oxidative stress produced during normal oxygen metabolism or induced by exogenous sources. Oxidative stress can increase lipid and protein damages and it is associated with some neurodegenerative diseases, such as Alzheimer's disease, Parkinson's disease, and amyotrophic lateral sclerosis [167].

Among the enzymatic reactive oxygen species scavengers are superoxide dismutase, catalase, and glutathione peroxidase, the main antioxidant enzymes. There are also radical scavengers from exogenous source, such as ascorbate (vitamin C), tocopherol (vitamin E), carotenoids and phenolics compounds (flavonoids and tannins) [143]. Moderate consumption of red wine containing polyphenols (proanthocyanidin and resveratrol) reduces the incidence of certain age-related neurological disorders [168]. Volatile compounds, which are found in many plants, as well as in foods and beverages, also showed antioxidant activities [143,169].

The brain in Alzheimer's disease is under increased oxidative stress and this may play an important role in the pathogenesis of neuron degeneration in this disorder. Today acetylcholinesterase inhibitors are the most important compounds for the treatment of the Alzheimer's disease symptoms, increasing the neurotransmitter acetylcholine levels at cerebral cortex synapses, as is the case with the synthetic drugs tacrine, donepezil, and rivastigmine. However, there are various natural active principles with anticholinesterasic activity that have already been tested in the treatment of Alzheimer's disease. Rivastigmine itself is a semisynthetic drug, derived from physostigmine, an alkaloid isolated from *P. venenosum*, mentioned at the beginning of this chapter.

The chemical class with the highest number and diversity of substances with the anticholinesterasic activity is alkaloids (Figure 5.4), particularly indolic

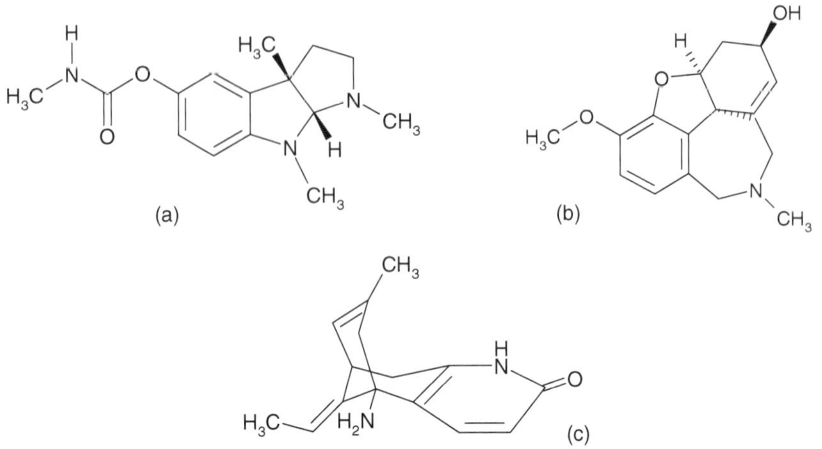

FIGURE 5.4 Examples of alkaloids with anticholinesterasic activity: (a) physostigmine, (b) galantamine, and (c) huperzine A.

alkaloids [5,170]. Galantamine, a reversible acetylcholinesterase inhibitor, has been used to treat various forms of dementia. This phenanthrene alkaloid was isolated, for the first time, from *Galanthus* spp., practically *G. woronowii* and *G. nivalis*, of the Amaryllidaceae family, but it is also obtained from *Narcissus* spp. and *Leucojum* spp., particularly *L. aestivium*. Today, it can also be obtained through organic synthesis. Galantamine is one of the approved treatments for Alzheimer's disease. When administered orally, it has a half-life of 6 h, and its metabolism produces four compounds, one of which is more active than galantamine itself [79].

Huperzine A, a quinolizidine alkaloid isolated from *Huperzia serrata*, besides inhibiting acetylcholinesterase, protects the neuronal cells against the protein β-amyloid, offering a good alternative for the treatment of Alzheimer's disease [171]. This inhibitor can also be used to treat dementias of the cerebrovascular type, which occur after ischemia and are the result of neuronal damage. Huperzine A crosses the hematoencephalic barrier very easily, and is also used as a prophylactic agent against organophosphate poisoning [172].

Some terpenoids are also acetylcholinesterase inhibitors, such as the essential oils 1,8-cineol and α-pinene [173]. Flavonoids, such as linariin and isolariin, also showed anticholinesterase activity. Various active principles have been reported as acetylcholinesterase inhibitors, and continue to have good prospects for the treatment of Alzheimer's disease.

5.7 PLANTS WITH THE MIND-ALTERING ACTIVITY

Mind-altering drugs are a group of substances that qualitatively modify the activity of the brain. In other words, they interfere in or distort its function, causing the individual to perceive the world in an altered way, as though in a dream-like state. Also, a loss of the sense of time, space and emotional state occurs. Historically, these substances have been referred to by a variety of names: psychodysleptics, psychotomimetics, psychotics, hallucinogens, psychogenics, entheogenics, and entactogenics. However, according to Schultes and Hofmann [174], none of these terms fully describes the effects caused by these substances.

A large part of these substances are of natural origin, discovered by primitive populations, who considered them sacred. Plants are the main sources of these substances, by means of their metabolic routes, their main function being to defend the plant against pests [175].

According to Schultes and Hofmann [174], 64 known species of mind-altering drugs are native to the Americas and only six to Europe. In all cases, there is no precise information about the mechanisms of action by which the substances responsible for their pharmacological action act on the human physiological system.

Mind-altering drugs owe their activity to a limited number of psychoactive substances. The altered state normally disappear when the active principles are metabolized and excreted by the body. The active principles generally possess one of three chemical moieties: phenylethylamine (mescaline), indole (tryptamines, β-carbolines) or the anticholinergic tropane esters (atropine, scopolamine) [8]. The interference in serotonergic neurotransmission is normally related to the hallucinogenic effect of these drugs, acting as direct or indirect agonists. However, it is important to remember that psychotomimetic chemistry is as diverse as the plants that produce these substances [175].

The majority of mind-altering drugs of natural origin are alkaloids. A small number is associated with the presence of glycosides, terpenoids, essential oils, and other organic constituents that may be responsible for this action. Below, we list some of the plants more commonly used for this purpose.

5.7.1 Nutmeg: *Myristica fragrans* Houtt (Myristicaceae)

The commercial exploitation of spices from tropical Asia to Europe between the 9th and 15th centuries led to the introduction of narcotic, aphrodisiac, and medicinal substances. One of these products was nutmeg, native of Indonesia, which is so precious that it was also sold on the black market. Mind-altering properties were only associated with it in 1576, when Mattias de L'Obel (Lobélius) described, in his book *Plantarvm: Sev Stirpivm Historia* (Figure 5.5), the case of a pregnant woman who ate 10 to 12 nutmegs and became totally inebriated. Rumphius (1741–1755) describes an incident on the Banda Isles, where two soldiers who slept under a nutmeg tree (*M. fragans*) woke up the next morning completely inebriated, similar to the case described by Lobélius [174].

Today, nutmeg is used as a stimulant, antidiarrhoeal, carminative, stomach tonic, and aphrodisiac. However, it is also considered a low-cost substitute for some recreational drugs, such as marijuana (*Cannabis sativa*). Recently, some cases of nutmeg ingestion have been described, particularly by adolescents, all seeking a low-cost way of achieving a state of euphoria [176]. The plant contains volatile oils, fat, starch, and mucilage. The fixed oils contain myristicin, trimyristin, and myristic

FIGURE 5.5 Cover of the book "Plantarvm," by Lobélius, published in 1576.

acid, while the volatile oils comprise a mixture of terpenes and alkenylbenzene derivatives [177].

The neuropharmacological action of nutmeg is still not clearly defined. The conversion of myristicin to analogous and amphetamine substances, proposed by Shulgin [178], is now disputed. The antidepressant and anticonvulsant actions of nutmeg-based extracts appear to be measured by the adrenergic, dopaminergic, and serotoninergic receptors [179].

5.7.2 Mandrake: *Mandragora officinarum* L. (Solanaceae)

The word Mandrake is of Greek origin, and means "harmful to cattle." Native to South-Eastern Europe, it grows wild in a region extending from Portugal to Greece, and its fruits, leaves and in particular, the root – which are shaped like a man – are used. It is attributed with having aphrodisiac, analgesic, sedative, and anesthetic activities. There are reports of its external use for ulcers and inflammations [174,180]. In Europe, it was used as a stupefacient, being one of the ingredients commonly used by the witches of the Middle Ages in their preparations. According to legend, the cries of the plant can cause those who pluck out its roots to go mad. Tropane alkaloids, such as atropine, hyoscyamine, scopolamine, and scopoline are found in the roots and leaves, with psychoactive and anticholinergic properties. Also, scopoletin (coumarin), sitosterol, sugars, and starch are found in the roots. The fruits contain traces of alkaloids. The presence of aromatic compounds, as well as the high content of sulfurated compounds (over 7%), contributes to the dense, fruity, provocative aroma of the fruits [174,180].

5.7.3 Marihuana, Hemp: *Cannabis sativa* L. (Cannabaceae)

Marihuana is a native plant of Central Asia, and is widespread on all the continents, but its use as a recreational drug (i.e., Cannabis) is prohibited in various countries. Recently, movements have been carried out to decriminalize its recreational use, and considerable progress has been obtained in the discovery and application of its medicinal properties [181,182].

Man's use of marihuana has existed since the discovery of agriculture in the old world, probably more than 10,000 years ago. Marihuana is one of the world's oldest crops. From it, we obtain fiber, used to produce fabrics and rope; oil and seeds, used as food; and therapeutic properties for the treatment of a wide range of sicknesses [182,183]. The chemical composition of marihuana is very complex, but the main constituents are cannabinoids, such as delta-9-tetrahydrocannabinol (Δ^9-THC), canabidiol; and terpenoids, such as caryophyllene oxide. This latter substance has been widely used as a reference for police sniffer dogs in drug detection [184]. Cannabidiol itself does not seem to produce significant effects, but it has a potentiating role in the effects of Δ^9-THC [185].

Cannabinoids bind to the cannabinoid receptors of anandamide: CB_1, located in the CNS and CB_2, present in the cells of the immune system. The psychoactive effects of Δ^9-THC are mediated by the activation of the CB_1 receptors G protein-coupled receptors [186,187].

5.7.4 Salvia: *Salvia divinorum* Eplin et Játiva-M. (Lamiaceae)

This plant is endemic to the Mazatec region of the Sierra Madre Oriental in the Mexican State of Oaxaca. The plant was used by the Aztecs in entheogenic rituals, to promote visionary states of conscience. The recreational use of salvia is a subject of much debate, though it is sold without restriction; its main users are adolescents, who use it as a substitute for cannabis [188]. *S. divinorum* is prepared in various ways, for use as a psychotropic agent. The leaves and the fruits are used dried or fresh, which in turn, are inhaled in the form of cigarettes or incenses, or ingested, when fresh, after chewing, or in the form of an infusion, juice or tincture [189].

The leaves contain the neoclerodane diterpenes salvinorin A (Figure 5.6) and salvinorin B [190]. Studies with salvinorin A, in humans and animals, indicate that its psychoactive effects and potency are similar to those of mescaline [191]. Unlike traditional psychodelics, salvinorin A does not interact with the serotonin 5-HT_{2A} receptors, but probably induces its psychotropic effects through the activation of the kappa-opioid receptor [192,193].

5.7.5 Peyote: *Lophophora williamsii* [Lem.] Coulter (Cactaceae)

Peyote is a hallucinogenic cactus that is associated with the production of physical, visual, and perceptual alterations [194]. This cactus occurs in desert areas, from Central Mexico to the State of Texas in the USA. The ritual use of Peyote goes back to pre-Columbian times. The Aztecs and other Mexican Indians were very familiar with this cactus, and its effects [183]. However, it is also used to treat infections, arrhythmias, arthritis, asthma, colds, intestinal disorders, and diabetes, among other ailments [175].

Peyote contains isoquinoline alkaloids and β-phenethylamine derivatives, all of which are active, but the most important is mescaline (3,4,5-trimethoxy-β-phenethylamine), first identified by Heffter in 1896 [195]. The physiological and psychological effects of mescaline are similar to those of lysergic acid diethylamide (LSD): both are sympathomimetics, radically altering the perception of self and of the reality, increasing suggestibility, and intensifying the emotions. Compared with LSD, mescaline is described as producing more sensual and perceptual effects, with less alteration of thought and sense of self [196].

5.7.6 Jurema: *Mimosa tenuiflora* [Willd.] Poir. (Fabaceae)

This plant is used in traditional rituals for its festive, reflective, therapeutic nature, particularly among the Indians or those who have revived this indigenous science in the North and Northeast regions of Brazil. The plant is consumed as a drink, or smoked in a pipe. The drink is prepared by maceration of the roots in water, and is known as jurema brew or "ajucá." The roots and dried leaves of Jurema are also smoked [197].

The substance responsible for the psychoactive effect is a tryptamine alkaloid called *N*,*N*,-dimethyltryptamine (DMT) (Figure 5.6). In addition to this alkaloid, the following compounds have also been isolated from jurema: 5-hydroxytryptamine, two chalcones (kukulkan A and kukulkan B), three steroids (campesterol-3-*O*-beta-D-glucopyranosyl, sitgmaterol-3-*O*-beta-D-glucopyranosyl, and beta-sitosterol-3-*O*-beta-D-glucopyranosyl), three saponins (mimonoside A, B, and C) and five 2-phenoxychromones (tenuiflorin A, B, and C, along with 6-demethoxycapillarisin and 6-demethoxy-4′-*O*-methylcapillarisin) [198].

The mind-altering effect caused by jurema is similar to that of LSD. Although it is known that the action of DMT, when administered orally, is inhibited by the action of monoamine oxidase (MAO), this does not occur with the jurema brew. One hypothesis raised by some researchers is that β-carbolines (MAO inhibitors) may be formed from the basic tryptamines such as 5-hydroxytryptamine (serotonine) [199,200].

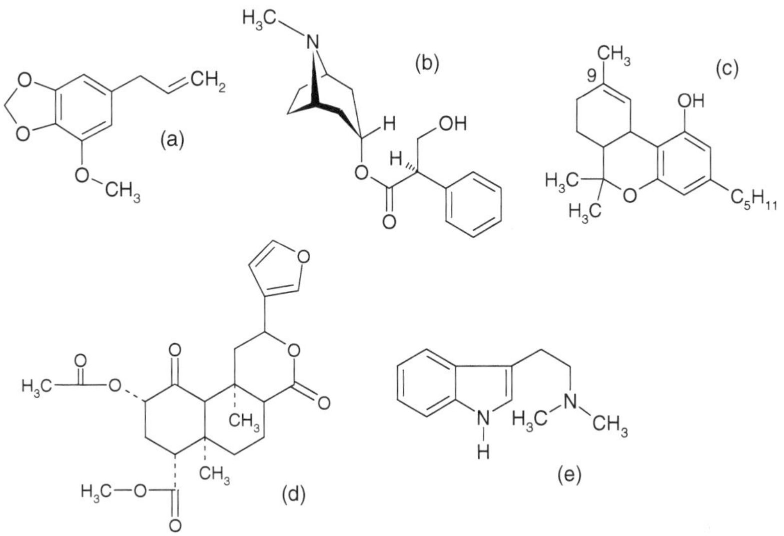

FIGURE 5.6 Chemical structure of some mind-altering substances: (a) myristicin, (b) atropine, (c) Δ^9 tetrahydrocannabinnol, (d) salvinorin A, and (e) *N,N*-dimethyltryptamine.

5.7.7 Ayahuasca

Ayahuasca is a word of indigenous origin (aya = soul, spirit, and waska = rope, vine) described as vine of the spirits, rope of the dead, or vine of the dead [197]. The tea, known as Ayahuasca, Hoasca, Caapi, Daime or Yajé, consists of two plants found in the Amazon rainforest: the jungle vine *Banisteriopsis caapi* (Spruce ex Griseb.) Morton (Malpighiaceae) and the leaves of *Psichotria viridis* R. & Rau. (Rubiaceae).

The main components of the tea are beta-carboline alkaloids (harmine, harmaline, and tetrahydroharmine), which come from the jungle vine *B. Caapi*, and DMT (dimethyltryptamine and *N*-methyltryptamine) from the leaves of *P. viridis*. The hallucinogenic action, responsible for the visionary effect, is attributed to the DMT, which is inactivated by the MAO. However, unlike Jurema, whose inhibitors of this enzyme are not known, it is known that in ayahuasca, the harmine alkaloids and their derivatives are responsible for the inhibitory effects of MAO type A [201,202]. This is a very interesting example of synergism, found among compounds present in the combination of different plant species [203].

5.7.8 Other Mind-Altering Drugs

In general, mind-altering drugs have been used since early times, and have always been related to poisons and "magic" potions, used in the practices of witchcraft or to kill enemies. In Europe of the Middle Ages and the Renaissance, the most commonly used were *A. belladonna*, *Hyosciamus niger*, and *Mandragora officinarum* [197]. Several Solanaceae and species of *Datura* genus also are related as potential hallucinogens.

Disturbing substances are not restricted to plants. Other organisms also produce chemical compounds with this feature, like psylocibine (*O*-phosphoryl-4-hydroxy-*N,N*-dimethyltryptamine), obtained from the mushroom *Psylocibu cubensis* (or *P. mexicana*, among other species) and bufotenine (*N,N*-dimethyl-5-tryptamine), obtained from the *Bufo alvarius* toad. Today, many synthetic mind-altering drugs are used by young people as recreational drugs, or as a means of escaping from the reality, and constitute a major social problem.

5.8 PLANTS USED AGAINST DRUG DEPENDENCE

Few plants have been studied to date, in relation to their therapeutic potential for the treatment of patients with drug dependence. A survey carried out of 76 books found 11 native Brazilian plants used in traditional medicine to treat alcohol abuse or dependence on other drugs [204]; however, there have been no clinical trials to determine the efficacy of these treatments. Some authors suggest that certain hallucinogenic substances have therapeutic potential in the treatment of alcoholism and other chemical dependencies, as well as certain psychiatric disorders [205]. In Brazil, leaders and followers of the so-called ayahuasquero religions report that the consumption of Ayahuasca tea helps people who are dependent on alcohol and other drugs to be free of their dependence [206].

Among the plants for which some studies indicate a potential antialcohol effect are *H. perforatum*, *Tabernanthe iboga,* and *Pueraria lobata* [79]. Studies with extracts of *H. perforatum* demonstrate a reduction in ethanol ingestion by alcohol-preferring rats [207]. Some authors also suggest that the adaptogenic plants *Withania somnifera* and *P. ginseng* have potential to reduce drug dependence or withdrawal [208,209]. However, there have been few studies in this area, and so far, there has been no clinical proof of its effectiveness.

5.9 CONCLUSIONS

Psychoactive drugs have been used by humans from time immemorial. As we have seen in this chapter, many of these plants contribute to the development of drugs for the treatment of disorders of the CNS, with some being used in the form of standardized extracts, some as raw extracts, and others in traditional forms, such as teas [8]. In this chapter, we have addressed only the most well-known and commonly studied medicinal species, as well as their chemical composition and active principles responsible for their effects. Table 5.1 shows a summary of this information for the most established species. For more information about this subject, we suggest the following reviews: Zhang, Kumar and the ESCOP and WHO monographys [79,210–212].

The information on the popular use of medicinal plants has always helped in research on new drugs, guiding researchers on the effects to be investigated, thereby considerably increasing the chances of success. It is also possible to see that certain

TABLE 5.1 Principal plants affecting the central nervous system and the active principles involved in their activities

Group/Scientific Name	Popular Name	Parts Used	Pharmacology Activity	Main Active Principles
CNS depressants				
Cymbopogon citratus	Lemon grass	Leaves	Anxiolytic, analgesic, hypnotic and anticonvulsant	Essential oils, phenolic acids
Erythrina mulungu and *Erythrina velutina*	Erythrina	Several parts	Anxiolytic, anticonvulsant	Erythrinian alkaloids
Humulus lupulus	Hop	Flowers	Hypnotic, sedative, anxiolytic	α and β-bitter acids and phenolic compounds
Lippia alba	Bush lippia, Falsa melissa	Leaves or aerial parts	Sedative, anticonvulsant	Terpenoids
Matricaria chamomilla	Chamomille	Flowers	Sedative, spasmolytic	Flavonoids
Melissa officinalis	Balm, Lemon balm	Aerial parts	Analgesic, anxiolytic	Triterpenoids, rosmarinic acid
Passiflora spp	Passiflora	Leaves	Anxiolytic	Flavones
Piper methysticum	Kava-kava	Rhizomes	Hypnotic, anxiolytic	Kavalactones and chalcones
Tilia cordata	Tilia	Flowers and stems	Anxiolytic, sedative	Flavonoids
Valeriana officinalis	Valerian	Roots	Anxiolytic, anticonvulsant	Valepotriates and valerenic acids
CNS stimulants				
Camellia sinensis	Tea plant	Leaves	Stimulant, antioxidant	Caffeine and phenolic acids
Catha edulis	Khat	Leaves and twigs	Stimulant, anoretic	Cathinone and ephedrine alkaloids
Coffea arabica	Coffee	Seeds	Stimulant	Caffeine and other methylxanthines
Cola nitida or *Cola acuminata*	Cola nut	Nuts	Stimulant	Caffeine, theobromine
Eleutherococcus senticosus	Siberian ginseng	Roots	Adaptogen	Eleutherosides
Ephedra sinica	Ma Huang	aerial parts	Stimulant	Ephedrine alkaloids (phenylethylamines)
Erythroxylum coca	Coca	Leaves	Stimulant	Cocaine
Hypericum perforatum	St. John's worth	Aerial parts	Antidepressant	Hyperforin, xanthones and hypericin

(*Continued*)

TABLE 5.1 (*Continued*)

Group/Scientific Name	Popular Name	Parts Used	Pharmacology Activity	Main Active Principles
Ilex paraguariensis	Mate	Leaves	Stimulant, antioxidant	Methylxanthines and chlorogenic acids
Panax ginseng	Ginseng	Roots	Adaptogen	Ginsenosides
Paullinia cupana	Guarana	Seeds	Tonic, stimulant	Caffeine and tannins
Turnera diffusa	Damiana	Aerial parts	Adaptogen	Alkaloids and flavonoids
Mind-altering plants				
Banisteriopsis caapi and *Psichotria viridis*	Ayahuasca	Stem + leaves	Hallucinogen	Beta-carboline alkaloids and dimetiltriptamine
Cannabis sativa	Marihuana, Hemp	Leaves and flowers	Hallucinogen	Cannabinoids delta-9-tetrahydrocannabinol – (THC)
Lophophora williamsii	Peyote	Aerial parts	Hallucinogen	Alkaloids isoquinoline and β-phenethylamine (mescaline)
Mandragora officinarum	Mandragora	Root, fruits	Hallucinogen	Tropane alkaloids
Mimosa tenuiflora	Jurema	Roots	Hallucinogen	Tryptamine alkaloids
Myristica fragrans	Nutmeg	Nut	Hallucinogen	Trimyristin and myristic acid
Salvia divinorum	Salvia	Leaves and fruits	Hallucinogen	Diterpenes neo-clerodane (salvinorin A and salvinorin B)
Neuroprotector plants				
Angelica sinensis	Dong Guai, Chinese angelica	Roots	Neuroprotector	Phthalides
Centella asiatica	Gotu kola	Leaves	Neuroprotector	Polyphenols, triterpenes, asiatic acid
Ginkgo biloba	Ginkgo	Leaves	Memory enhancer	Ginkgolides and bilobalide
Mucuna pruriens	The cowhage, velvet bean	Endocarp	Anti-Parkinson	L-dopa
Ptychopetalum olacoides	Marapuama	Roots	Neuroprotector	Clerodane-type diterpenoids

biological activities are more common for some phytochemical classes (or chemical groups), which has also helped in the selection of species for scientific investigation. In this chapter, we see that flavonoids and essential oils are the chemical classes that are more closely related to CNS depressant effects, with analgesic, anxiolytic, and hypnotic action. Alkaloids are the main substances not only responsible for the hallucinogenic effects of the plants, but are also involved in many other effects. They act by depressing the CNS, as is the case with morphine, or by as stimulating the CNS, as in the case of caffeine, cathinone, and cocaine. Alkaloids are also the main phytochemical class of anticholinesterasics of natural origin. Tannins, saponins and phenolic compounds, in general, also present various biological activities, including a remarkable antioxidant effect.

However, it is important to emphasize that nature is an inestimable source of diverse substances of chemical classes and structures, with different biological activities on the human organism. Thus, the identification of active constituents isolated from medicinal plants continues to be an important tool for the development of new drugs, particularly those with new mechanisms of action.

ACKNOWLEDGMENTS

The authors thank CEBRID (Centro Brasileiro de Informação sobre Drogas Psicotrópicas) and AFIP (Associação Fundo Incentivo a Psicofarmacologia) for the financial support.

REFERENCES

[1] Chalout, L. (1971). Une nouvelle classification des drogues genesomanogènes. *Toxicomanies*, *4*, 371–375.

[2] Deley, G., Deniker, P., *Methodes Chimiotherapique en Psichiatrie*. Mosson, Paris, 1961.

[3] Griffin, W.J., Lin, G.D. (2000). Chemotaxonomy and geographical distribution of tropane alkaloids. *Phytochemistry*, *53*, 623–637.

[4] Zhao, B., Moochhala, S.M., Tham, S.Y. (2004). Biologically active components of *Physostigma venenosum*. *Journal of Chromatography B – Analytical Technologies in the Biomedical and Life Sciences*, *812*, 183–192.

[5] Mukherjee, P.K., Kumar, V., Mal, M., Houghton, P.J. (2007). Acetylcholinesterase inhibitors from plants. *Phytomedicine*, *14*, 289–300.

[6] Srivastava, A., Tripathi, A.K., Pandey, R., Verma, R.K., Gupta, M.M. (2006). Quantitative determination of reserpine, ajmaline, and ajmalicine in *Rauvolfia serpentina* by reversed-phase high-performance liquid chromatography. *Journal of Chromatographic Science*, *44*, 557–560.

[7] Calixto, J.B., Beirith, A., Ferreira, J., Santos, A.R.S., Cechinel-Filho, V., Yunes, R.A. (2000). Naturally occurring antinociceptive substances from plants. *Phytotherapy Research*, *14*, 401–418.

[8] Carlini, E.A. (2003). Plants and the central nervous system. *Pharmacology, Biochemistry and Behavior*, *75*, 501–512.

[9] Ernst, E. (2006). Herbal remedies for anxiety – a systematic review of controlled clinical trials. *Phytomedicine*, *13*, 205–208.

[10] Matthias, A., Blanchfield, J.T., Penman, K.G., Bone, K.M., Toth, I., Lehmann, R.P. (2007). Permeability studies of kavalactones using a Caco-2 cell monolayer model. *Journal of Clinical Pharmacy and Therapeutics*, *32*, 233–239.

[11] Dragull, K., Yoshida, W.Y., Tang, C.S. (2003). Piperidine alkaloids from *Piper methysticum*. *Phytochemistry*, *63*, 193–198.

[12] Zou, L., Henderson, G.L., Harkey, M.R., Sakay, Y., Li, A. (2004). Effects of Kava (kava-kava, 'Awa, yangona, *Piper methysticum*) on c-DNA-expressed cytochrome P450 enzymes and human cryopreserved hepatocytes. *Phytomedicine*, *11*, 285–294.

[13] Singh, Y.N. (2005). Potential for interaction of kava and St. John's wort with drugs. *Journal of Ethnopharmacology*, *100*, 108–113.

[14] Sarris, J., Kavanagh, D.J. (2009). Kava and St. John's wort: Current evidence for use in mood and anxiety disorders. *The Journal of Alternative and Complementary Medicine*, *15*, 827–836.

[15] Capasso, A., Sorrentino, L. (2005). Pharmacological studies on the sedative and hypnotic effect of Kava kava and Passiflora extracts combination. *Phytomedicine*, *12*, 39–45.

[16] Gastpar, M., Klimm, H.D. (2003). Treatment of anxiety, tension and restlessness states with Kava special extract WS 1490 in general practice: A randomized placebo-controlled double-blind multicenter trial. *Phytomedicine*, *10*, 631–639.

[17] Perez, J., Holmes, J.F. (2005). Altered mental status and ataxia secondary to acute Kava ingestion. *The Journal of Emergency Medicine*, *28*, 49–51.

[18] Wheatley, D. (2005). Medicinal plants for insomnia: A review of their pharmacology, efficacy and tolerability. *Journal of Psychopharmacology*, *19*, 414–421.

[19] Gyllenhaal, C., Merritt, S.L., Peterson, S.D., Block, K.I. Gochenour, T. (2000). Efficacy and safety of herbal stimulants and sedatives in sleep disorders. *Sleep Medicine Reviews*, *4*, 229–251.

[20] Murphy, K., Kubin, Z.J., Shepherd, J.N., Ettinger, R.H. (2010). *Valeriana officinalis* root extracts have potent anxiolytic effects in laboratory rats. *Phytomedicine*, *17*, 674–678.

[21] Benke, D., Barberis, A., Kopp, S., Karl-Heinz, A., Schubiger, M., Vogt, K.E., Rudolph, U., Möhler, H. (2009). $GABA_A$ receptors as *in vivo* substrate for the anxiolytic action of valerenic acid, a major constituent of valerian root extracts. *Neuropharmacology*, *26*, 174–181.

[22] Dietz, B.M., Mahady, G.B., Pauli, G.F., Fransworth, N.R. (2005). Valerian extract and valerenic acid are partial agonists of the $5\text{-}HT_{5a}$ receptor *in vitro*. *Molecular Brain Research*, *138*, 191–197.

[23] Fernández, S., Wasowski, C., Paladini, A.C., Marder, M. (2004). Sedative and sleep-enhancing properties of linarin, a flavonoid-isolated from *Valeriana officinalis*. *Pharmacology, Biochemistry and Behavior*, *77*, 399–404.

[24] McKay, D.L., Blumberg, J.B. (2006). A review of the bioactivity and potential health benefits of chamomile tea (*Matricaria recutita* L.). *Phytotherapy Research*, *20*, 519–530.

[25] Herrera-Ruiz, M., Roman-Ramos, R., Zamilpa, A., Tortoriello, J., Jimenez-Ferrer, J.E. (2008). Flavonoids from *Tilia americana* with anxiolytic activity in plus-maze test. *Journal of Ethnopharmacology*, *118*, 312–317.

[26] Soulimani, R., Fleurentin, J., Mortier, F., Misslin, R., Derrieu, G., Pelt, J.M. (1997). Neurotropic action of the hydroalcoholic extract of *Melissa officinalis* in the mouse. *Planta Medica*, *57*, 105–109.

[27] Awad, R., Ahmed, F., Boubornnais-spear, N., Mullally, M., Ta, C.A., Tang, A., Merali, Z., Maquin, P., Caal, F., Cal, V., Poveda, L., Vindas, P.S., Trudeau, V.L., Arnason, J.T. (2009). Ethnopharmacology of Q'eqchi'Maya antiepileptic and anxiolytic plants: Effects of GABAergic system. *Journal of Ethnopharmacology*, *125*, 257–264.

[28] Ibarra, A., Feuillere, N., Roller, M., Lesburgere, E., Beracochea D. (2010). Effects of chronic administration of *Melissa officinalis* L. extract on anxiety-like reactivity and circadian and exploratory activities in mice. *Phytomedicine*, *17*, 397–403.

[29] Sousa, A.C., Alviano, D.S., Blank, A.F., Alves, P.B.; Alviana, C.S., Gattaz, C. R. (2004). *Melissa officinalis* L. essential oil: Antitumoral and antioxidant activities. *Journal of Pharmacy and Pharmacology*, *56*, 677–681.

[30] Kennedy, D.O., Little, W., Scholey, A.B. (2004). Attenuation of laboratory induced stress in humans following acute administration of *Melissa officinalis* (Lemon Balm). *Psychosomatic Medicine*, *66*, 607–613.

[31] Kennedy, D.O., Little, W., Haskell, C.F., Scholey, E.B. (2006). Anxiolytic effects of a combinations of *Melissa officinalis* and *Valeriana officinalis* during laboratory induced stress. *Phytotherapy Research*, *20*, 96–102.

[32] Zanoli, P., Rivasi, M., Zavatti, M., Brusiani, F., Baraldi, M. (2005). New insight in the neuropharmacological activity of *Humulus lupulus* L. *Journal of Ethnopharmacology*, *102*, 102–106.

[33] Zanoli, P., Zavatti, M. (2008). Pharmacognostic and pharmacological profile of *Humulus lupulus* L. *Journal of Ethnopharmacology*, *116*, 383–396.

[34] Milligan, S.R., Kalita, J.C., Heyerick, A., Rong, H., De Cooman, L., De Keukeleire, D. (1999). Identification of a potent phytoestrogen in hops (*Humulus lupulus* L.) and beer. *The Journal of Clinical Endocrinology and Metabolism*, *83*, 2249–2252.

[35] De Keukeleire, J., Ooms, G., Heyerick, A., Roldan-Ruiz, I., Van Bockstaele, E., De Keukeleire, D. (2003). Formation and accumulation of α-acids, β-acids, desmethylxanthohumol, and xanthohumol during flowering of Hops (*Humulus lupulus* L.). *Journal of Agricultural and Food Chemistry*, *51*, 4436–4441.

[36] De Keukeleire, J., Ooms, G., Heyerick, A., Roldan-Ruiz, I., Van Bockstaele, E., De Keukeleire, J., Janssens, I., Heyerick, A., Ghekiere, G., Cambie, J. Roldan-Ruiz, I., Van Bockstaele, E., De Keukeleire, D. (2007). Relevance of organic farming and effect of climatological conditions on the formation of alpha-acids, beta-acids, desmethylxanthohumol, and xanthohumol in hop (*Humulus lupulus* L.). *Journal of Agricultural and Food Chemistry*, *51*, 4436–4441.

[37] Negri, G., Santi, D., Tabach, R. (2010). Bitter acids from hydroethanolic extracts of *Humulus lupulus* L. used as anxiolytic. *Revista Brasileira de Farmacognosia*, *20*, 850–859.

[38] Lee, K.M., Jung, J.S., Song, D.K., Kräuter, M., Kim, Y.H. (1993). Effects of *Humulus lupulus* extract on the central nervous system in mice. *Planta Medica*, *59*, A691.

[39] Bravo, L., Cabo, A., Fraile., Jimenez, J., Villar, A. (1974). Estudio farmacodinamico del lupulo (*Humulus lupulus* L.). Accion tranqüilizante. *Bollettino Chimico Farmaceutico*, *113*, 310–315.

[40] Schiller, H., Forster, A., Vonhoff, C., Hegger, M., Biller, A., Winterhoff, H. (2006). Sedating effects of *Humulus lupulus* L. extracts. *Phytomedicine*, *13*, 535–541.

[41] Galand, N., Pothier, J., Mason, V., Viel, C. (1999). Separation of flavonoids, coumarins and anthocyanins in plant extracts by overpressured layer chromatography. *Pharmazie*, *54*, 468–471.

[42] Arcos, M.L.B., Cremaschi, G., Werner, S., Coussio, J., Ferraro, G., Anesini, C. (2006). *Tilia cordata* Mill. extracts and scopoletin (isolated compound): Differential cell growth effects on lymphocytes. *Phytotherapy Research*, *20*, 34–40.

[43] Pérez-Ortega, G., Fefer-Guevara, P., Chávez, M., Herrera, J., Martínez, A., Martínez, A.L., González-Trujano, M.E. (2008). Sedative and anxiolytic efficacy of *Tilia americana* var. *mexicana* inflorescences used traditionally by communities of state of Michoacan, Mexico. *Journal of Ethnopharmacology*, *116*, 461–468.

[44] Martinez, A.L., Gonzalez-Trujano, M.E., Aguirre-Hernandez, E., Moreno, J., Soto-Hernandez, M., Lopez-Munoz, F.J. (2009). Antinociceptive activity of *Tilia americana* var. *mexicana* inflorescences and quercetin in the formalin test and in an arthritic pain model in rats. *Neuropharmacology*, *56*, 564–571.

[45] Hennebelle, T., Sahpaz, S., Joseph, H., Bailleul, F. (2008). Ethnopharmacology of *Lippia alba*. *Journal of Ethnopharmacology*, *116*, 211–222.

[46] Neto, A.C., Netto, J.C., Pereira, P.S., Pereira, A.M.S., Taleb-Contini, S.H., Franca, S.C., Marques, M.O.M., Beleboni, R.O. (2009). The role of polar phytocomplexes on anticonvulsant effects of leaf extracts of *Lippia alba* (Mill.) NE Brown chemotypes. *Journal of Pharmacy and Pharmacology*, *61*, 933–939.

[47] Vale, T.G., Furtado, E.C., Santos, J.G., Viana, G.S.B. (2002). Central effects of citral, myrcene and limonene, constituents of essential oil chemotypes from *Lippia alba* (Mill.) NE Brown. *Phytomedicine*, *9*, 709–714.

[48] Vale, T.G., Matos, F.J.A., de Lima, T.CM., Viana, G.S.B. (1999). Behavioral effects of essential oils from *Lippia alba* (Mill.) N.E. Brown chemotypes. *Journal of Ethnopharmacology*, *67*, 127–133.

[49] Viana, G.S., Vale, T.G., Silva, C.M., Matos, F.J. (2000). Anticonvulsant activity of essential oils and active principles from chemotypes of *Lippia alba* (mill) N.E. Brown. *Biological & Pharmaceutical Bulletin*, *23*, 1314–1317.

[50] de Sousa, D.P., Quintans-Junior, L., Almeida, R.N. (2007). Evaluation of the anticonvulsant activity of alpha-terpineol. *Pharmaceutical Biology*, *45*, 69–70.

[51] Almeida, R.N., de Sousa, D.P., Nobrega, F.F.D., Claudino, F.D., Araujo, D.A.M., Leite, J.R., Mattei, R. (2008). Anticonvulsant effect of a natural compound alpha, beta-epoxy-carvone and its action on the nerve excitability. *Neuroscience Letters*, *443*, 51–55.

[52] Blanco, M.M., Costa, C.A.R.A., Freire, A.O., Santos, J.G. Jr, Costa, M. (2009). Neurobehavioral effect of essential oil of *Cymbopogon citratus* in mice. *Phytomedicine*, *16*, 265–270.

[53] Figueirinha, A., Paranhos, A., Perez-Alonso, J.J. (2008). *Cymbopogon citratus* leaves: Characterization of flavonoids by HPLC-PDA-ESI/MS/MS and an approach to their potential as a source of bioactive polyphenols. *Food Chemistry*, *110*, 718–728.

[54] Cheel, J., Theodoluz, C., Rodriguez, J., Schmeda-Hirschmann, G. (2005). Free radical scavengers and antioxidants from Lemongrass (*Cymbopogon citratus* (DC) Stapf). *Journal of Agricultural and Food Chemistry*, *53*, 2511–2517.

[55] Carlini, E.A., Contar, J.D.P., Silva-Filho, A.R., Silveira-Filho, N.G., Frochtengarten, M.L., Bueno, O.F.A. (1986). Pharmacology of lemongrass (*Cymbopogon citratus* Stapf):. I. Effects of teas prepared from the leaves on laboratory animals. *Journal of Ethnopharmacology*, *17*, 37–64.

[56] Viana, G.S.B., Vale, T.G., Pinho, R.S.N., Matos, F.J.A. (2000). Antinociceptive effect of the essential oil from *Cymbopogon citratus* in mice. *Journal of Ethnopharmacology*, *70*, 323–327.

[57] Flausino, O.A., Pereira, A.M., Bolzani, V.S., Nunes-De-Souza, R.L. (2007). Effects of erythrinian alkaloids isolated from *Erythrina mulungu* (Papilionaceae) in mice submitted to animal models of anxiety. *Biological and Pharmaceutical Bulletin*, *30*, 375–378.

[58] Vasconcelos, S.M.M., Lima, N.M., Sales, G.T.M., Cunha, G.M.A., Aguiar, L.M.V., Silveira, E.R., Rodrigues, A.C.P., Macedo, D.S., Fonteles, M.M.F., Sousa, F.C.F., Viana, G.S.B. (2007). Anticonvulsivant activity of hydroalcoholic extracts from *Erythrina velutina* and *Erythrina mulungu*. *Journal of Ethnopharmacology*, *110*, 271–274.

[59] Onusic, G.M., Nogueira, R.L., Pereira, M.S., Viana, M.B. (2002). Effects of acute treatment with a water-alcohol extract from *Erythrina mulungu* on anxiety-related responses in rats. *Brazilian Journal of Medical Biological Research*, *35*, 473–477.

[60] Dhawan, K., Dhawan, S., Sharma, A. (2004). Passiflora: A review update. *Journal of Ethnopharmacology*, *94*, 1–23.

[61] Provensi, G., Noel, F., Lopes, D.V.S., Fenner, R., Betti, A.H., de Costa, F., Morais, E.C., Gosmann, G., Rates, S.M.K. (2008). Participation of GABA-benzodiazepine receptor complex in the anxiolytic effect of *Passiflora alata* Curtis (Passifloraceae). *Latin American Journal of Pharmacy*, *27*, 845–851.

[62] Grundmann, O., Wang, J., Mcgregor, G.P., Butterweck, V. (2008). Anxiolytic activity of a phytochemically characterized *Passiflora incarnata* extract is mediated via the GABAergic system. *Planta Medica*, *74*, 1769–1773.

[63] Speroni, E., Billi, R. (1996). Sedative effects of crude extract of *Passiflora incarnate* after oral administration. *Phytotherapy Research*, *10*, S92–S94.

[64] dos Santos, K.S., Kurtz, S.M.T.F., Muller, S.D., Biavatti, M.W., de Oliveira, R.M.M.W., Santos, C.A.D. (2006). Sedative and anxiolytic effects of methanolic extract from the leaves of *Passiflora actinia*. *Brazilian Archives of Biology and Technology*, *49*, 565–573.

[65] Coleta, M., Campos, M.G., Cotrim, M.D., Lima, T.C., Cunha, A.P. (2008) Assessment of luteolin (3′,4′,5,7-tetrahydroxyflavone) neuropharmacological activity. *Behavioural Brain Research*, *189*, 75–82.

[66] Petry, R.D., Reginatto, F.H., De-Paris, F., Gosmann, G., Salgueiro, J.B., Quevedo, J., Kapczinski, F., Gonzales-Ortega, G., Schenkel, E.P. (2001). Comparative pharmacological study of hydroethanol extract of *Passiflora edulis* leaves. *Phytotherapy Research*, *15*, 162–164.

[67] Sena, L.M., Zucolotto, S.M., Reginatto, F.H., Schenkel, E.P., De Lima, T.C.M. (2009). Neuropharmacological activity of the pericarp of *Passiflora edulis* flavicarpa Degener: Putative involvement of C-glycosylflavonoids. *Experimental Biology and Medicine*, *234*, 967–975.

[68] Coleta, M., Batista, M.T., Campos, M.G., Carvalho, R., Cotrim, M.D., Lima, T.C., Cunha, A.P. (2006). Neuropharmacological evaluation of the putative anxiolytic effects

of *Passiflora edulis* Sims, its sub-fractions and flavonoid constituents. *Phytotherapy Research*, *20*, 1067–1073.

[69] Dhawan, K., Kumar, S., Sharma, A. (2001). Anxiolytic activity of aerial and underground parts of *P. incarnata*. *Fitoterapia*, *72*, 922–926.

[70] Setzer, W.N. (2009). Essential oils and anxiolytic aromatherapy. *Natural Product Communications*, *4*, 1305–1316.

[71] de Almeida, E.R., Rafael, K.R.D., Couto, G.B.L., Ishigami, A.B.M. (2009). Anxiolytic and anticonvulsant effects on mice of flavonoids, linalool, and alpha-tocopherol presents in the extract of leaves of *Cissus sicyoides* L. (Vitaceae). *Journal of Biomedicine and Biotechnology*, *2009*, 1–6.

[72] Silva, M.I.G., Neto, M.R.D., Neto, P.F.T., Moura, B.A., do Amaral, J.F., de Sousa, D.P., Vasconcelos, S.M.M., de Sousa, F.C.F. (2007). Central nervous system activity of acute administration of isopulegol in mice. *Pharmacology, Biochemistry and Behavior*, *88*, 141–147.

[73] Takaki, I., Bersani-Amado, L.E., Vendruscolo, A., Sartoretto, S.M., Diniz, S.P., Bersani-Amado, C.A., Cuman, RK.N. (2008). Anti-inflammatory and antinociceptive effects of *Rosmarinus officinalis* L. essential oil in experimental animal models. *Journal of Medicinal Food*, *11*, 741–746.

[74] Ballard, C.G., Psych, M.R.C., O'Brien, J.T., Reichelt, K., Perry, E. (2002). Aromatherapy as a safe and effective treatment for the management of agitation in severe dementia: The results of a double-blind, placebo-controlled trial of Melissa. *The Journal of Clinical Psychiatry*, *63*, 553–558.

[75] Leite, M.P., Fassin-Júnior, J., Baziloni, E.M.F., Almeida, R.N., Mattei, R., Leite, J.R. (2008). Behavioral effects of essential oil of *Citrus aurantium* L. inhalation in rats. *Revista Brasileira de Farmacognosia*, *18*, 661–663.

[76] Perry, N., Perry, E. (2006). Aromatherapy in the management of psychiatric disorders. *CNS Drugs*, *20*, 257–280.

[77] Komiya, M., Takeuchi, T., Harada, E. (2006). Lemon oil vapor causes an anti-stress effect via modulating the 5-HT and DA activities in mice. *Behavioural Brain Research*, *172*, 240–249.

[78] Almeida, R.N., Navarro, D.S., Barbosa-Filho, J.M. (2001). Plants with central analgesic activity. *Phytomedicine*, *8*, 310–322.

[79] Zhang, Z.J. (2004). Therapeutic effects of herbal extracts and constituents in animal models of psychiatric disorders. *Life Sciences*, *75*, 1659–1699.

[80] Alonso-Salces, R.M., Serra, F., Reniero, F., Heberger, K. (2009). Botanical and geographical characterization of green coffee (*Coffea arabica* and *Coffea canephora*): Chemometric evaluation of phenolic and methylxanthine contents. *Journal of Agricultural and Food Chemistry*, *57*, 4224–4235.

[81] Farah, A., Monteiro, M.C., Calado, V., Franca, A.S., Trugo, L.C. (2006). Correlation between cup quality and chemical attributes of Brazilian coffee. *Food Chemistry*, *98*, 373–380.

[82] Davis, J.M., Zhao, Z., Stock, H.S., Mehl, K.A., Buggy, J., Hand, G.A. (2003). Central nervous system effects of caffeine and adenosine on fatigue. *American Journal of Physiology. Regulatory, Integrative and Comparative Physiology*, *284*, R399–R404.

[83] Hameleers, P.A.H.M., van Boxtel, M.P.J., Hogervost, E., Riedel, W.J., Houx, P.J., Buntinx, F., Jolles, J. (2000). Habitual caffeine consumption and its relation to memory,

attention, planning capacity and psychomotor performance across multiple age groups. *Human Psychopharmacology*, *15*, 573–581.

[84] Woodward, M., Tusntall-Pedoe, H. (1999). Coffee and tea consumption in the Scottish heart health study follow up: Conflicting relations with coronary risk factors, coronary disease, and all cause mortality. *Journal of Epidemiology and Community Health*, *53*, 481–487.

[85] Klatsky, A.L., Morton, C., Udaltsova, N., Friedman, G.D. (2006). Coffee, cirrhosis, and transaminase enzymes. *Archives of Internal Medicine*, *166*, 1190–1195.

[86] Hu, G., Bidel, S., Jousilahti, P., Antikainen, R., Tuomilehto, J. (2007). Coffee and tea consumption and the risk of Parkinson's disease. *Movement Disorders*, *22*, 2242–2248.

[87] Costa, J., Lunet, N., Santos, C., Santos, J., Vaz-Carneiro, A. (2010). Caffeine exposure and the risk of Parkinson's disease: A systematic review and meta-analysis of observational studies. *Journal of Alzheimer's Disease*, *20*, S221–S238.

[88] Komes, D., Horzic, D., Belscak, A., Ganic, K.K., Vulic, I. (2010). Green tea preparation and its influence on the content of bioactive compounds. *Food Research International*, *43*, 167–176.

[89] Spacil, Z., Novakova, L., Solich, P. (2010). Comparison of positive and negative ion detection of tea catechins using tandem mass spectrometry and ultra high performance liquid chromatography. *Food Chemistry*, *123*, 535–541.

[90] Mendes, F.R., Carlini, E.A. (2007). Brazilian plants as possible adaptogens: An ethnopharmacological survey of books edited in Brazil. *Journal of Ethnopharmacology*, *109*, 493–500.

[91] Espínola, E.B., Dias, R.F., Mattei, R., Carlini, E.A. (1997). Pharmacological activity of guarana (*Paullinia cupana* Mart.) in laboratory animals. *Journal of Ethnopharmacology*, *55*, 223–229.

[92] Angelo, P.C.S.A., Nunes-Silva, C.G., Brigido, M.M., Azevedo, J.S.N., Assunção, E.N., Sousa, A.R.B., Patrício, F.J., Rego, M.M., Peixoto, J.C., Oliveira, W.P., Freitas, D.V., Almeida, E.R., Viana, A.M., Souza, A.F., Andrade, E.V., Acosta, P.O., Batista, J.S., Walter, M.E., Leomil, L., Anjos, D.A., Coimbra, R.C., Barbosa, M.H., Honda, E., Pereira, S.S., Silva, A., Pereira, J.O., Silva, M.L., Marins, M., Holanda, F.J., Abreu, R.M., Pando, S.C., Gonçalves, J.F., Carvalho, M.L., Leal-Mesquita, E.R., Silveira, M.A., Batista, W.C., Atroch, A.L., França, S.C., Porto, J.I., Schneider, M.P., Astolfi-Filho, S. (2008). Guarana (*Paullinia cupana* var. sorbilis), an anciently consumed stimulant from the Amazon rain forest: The seeded-fruit transcriptome. *Plant Cell Reports*, *27*, 117–124.

[93] Kennedy, D.O., Haskell, C.F., Wesnes, K.A., Scholey, A.B. (2004). Improved cognitive performance in human volunteers following administration of guarana. *Pharmacology, Biochemistry and Behavior*, *79*, 401–411.

[94] Kennedy, D.O., Haskell, C.F., Robertson, B., Reay, J., Brewster-Maund, C., Luedemann, J., Maggini, S., Ruf, M., Zangara, A., Scholey, A.B. (2008). Improved cognitive performance and mental fatigue following a multi-vitamin and mineral supplement with added guarana (*Paullinia cupana*). *Appetite*, *50*, 506–513.

[95] Kim, H.K., Saifullah, Khan, S., Wilson, E.G., Kricun, S.D.P., Meissner, A., Goraler, S., Deelder, A.M., Choi, Y.H., Verpoorte, R. (2010). Metabolic classification of South American *Ilex* species by NMR-based metabolomics. *Phytochemistry*, *71*, 773–784.

[96] Pagliosa, C.M., Vieira, M.A., Podesta, R., Maraschin, M., Zeni, A.L.B., Amante, E.R., Amboni, R.D.M.C. (2010). Methylxanthines, phenolic composition, and antioxidant

activity of bark from residues from mate tree harvesting (*Ilex paraguariensis* A. St. Hil.). *Food Chemistry*, *122*, 173–178.

[97] Marques, V., Farah, A. (2009). Chlorogenic acids and related compounds in medicinal plants and infusions. *Food Chemistry*, *113*, 1370–1376.

[98] Jaiswal, R., Sovdat, T., Vivan, F., Kuhnert, N. (2010). Profiling and characterization by LC-MSn of the chlorogenic acids and hydroxycinnamoylshikimate esters in mate (*Ilex paraguariensis*). *Journal of Agricultural and Food Chemistry*, *58*, 5471–5484.

[99] Kalix, P. (1991). The pharmacology of psychoactive alkaloids from ephedra and catha. *Journal of Ethnopharmacology*, *32*, 201–208.

[100] Krizevski, R., Bar, E., Shalit, O., Sitrit, Y., Ben-Shabat, S., Lewinsohn, E. (2010). Composition and stereochemistry of ephedrine alkaloids accumulation in *Ephedra sinica* Stapf. *Phytochemistry*, *71*, 895–903.

[101] Abourashed, E.A., El-Alfy, A.T., Khan, I.A., Walker, L. (2003). Ephedra in perspective – a current review. *Phytotherapy Research*, *17*, 703–712.

[102] Chen, W.L., Tsai, T.H., Yang, C.C.H., Kuo, T.B.J. (2010). Effects of ephedra on autonomic nervous modulation in healthy young adults. *Journal of Ethnopharmacology*, *130*, 563–568.

[103] Fleming, R.M. (2008). Safety of ephedra and related anorexic medications. *Expert Opinion on Drug Safety*, *7*, 749–759.

[104] Zelger, J.L., Carlini, E.A. (1980). Anorexigenic effects of two amines obtained from *Catha edulis* Forsk. (Khat) in rats. *Pharmacology, Biochemistry and Behavior*, *12*, 701–705.

[105] Manghi, R.A., Broers, B., Khan, R., Benguettat, D., Khazaal, Y., Zullino, D.F. (2009). Khat use: Lifestyle or addiction? *Journal of Psychoactive Drugs*, *41*, 1–10.

[106] Feyissa, A.M., Kelly, J.P. (2008). A review of the neuropharmacological properties of khat. *Progress in Neuro-psychopharmacology and Biological Psychiatry*, *32*, 1147–1166.

[107] Atawodi, S.E.O., Pfundstein, B., Haubner, R., Spiegelhalder, B., Bartsch, H., Owen, R.W. (2007). Content of polyphenolic compounds in the Nigerian stimulants *Cola nitida* ssp *alba*, *Cola nitida* ssp *rubra* A. Chev, and *Cola acuminata* Schott & Endl and their antioxidant capacity. *Journal of Agricultural and Food Chemistry*, *55*, 9824–9828.

[108] Grinspoon, L., Bakalar, J.B. (1981). Coca and cocaine as medicines: An historical review. *Journal of Ethnopharmacology*, *3*, 149–159.

[109] Hegnauer, R. (1981). Chemotaxonomy of *Erythroxylum* (including some ethnobotanical notes on old world species). *Journal of Ethnopharmacology*, *3*, 279–292.

[110] Gonzalez-Guevara, J.L., Herman, V.C., Gonzalez-Garcia, K.L., Payo-Hill, A.L., Gonzalez-Lavaut, J.A., Molina-Torres, J., Prieto-Gonzalez, S. (2006). Flavonoid glycosides from Cuban *Erythroxylum* species. *Biochemical Systematic and, Ecology*, *34*, 539–542.

[111] Suzuki, O., Katsumata, Y., Oya, M., Bladt, S., Wagner, W. (1984). Inhibition of monoamine oxidases by hypericin. *Planta Medica*, *50*, 272–274.

[112] Jakovljevic, V., Popovic, M., Mimica-Dukic, N., Sabo, A., Gvozdenovic, L. (2000). Pharmacodynamic study of *Hypericum perforatum* L. *Phytomedicine*, *7*, 449–453.

[113] Butterweck, V., Christoffel, V., Nahrstedt, A., Petereit, F., Spengler, B., Winterhoff, H. (2003). Step by step removal of hyperforin and hypericin: Activity profile of different Hypericum preparations in behavioral models. *Life Sciences*, *73*, 627–639.

[114] Chatterjee, S.S., Bhattacharya, S.K., Wonnemann, M., Singer, A., Muller, W.E. (1998). Hyperforin as a possible antidepressant component of Hypericum extracts. *Life Sciences*, *63*, 499–510.

[115] Schulte-Lobbert, S., Holoubek, G., Muller, W.E., Schubert-Zsilavecz, M., Wurglics, M. (2004). Comparison of the synaptosomal uptake inhibition of serotonin by St. John's wort products. *Journal of Pharmacy and Pharmacology*, *56*, 813–818.

[116] Nahrstedt, A., Butterweck, V. (2010). Lessons learned from herbal medicinal products: The example of St. John's wort. *Journal of Natural Products*, *73*, 1015–1021.

[117] Paulke, A., Noldner, M., Schubert-Zslavecz, M., Wurglics, M. (2008). St. John's wort flavonolds and their metabolites show antidepressant activity and accumulate in brain after multiple oral doses. *Pharmazie*, *63*, 296–302.

[118] Ernest, E. (1995). St. John's wort, an antidepressant? A systematic, criteria-based review. *Phytomedicine*, *2*, 67–71.

[119] Gaster, B., Holroyd, J. (2000). St. John's wort for depression – a systematic review. *Archives of Internal Medicine*, *160*, 152–156.

[120] Wagner, H., Nörr, H., Winterhoff, H. (1994). Plant adaptogens. *Phytomedicine*, *1*, 63–76.

[121] Panossian, A., Wikman, G. (2009). Evidence-based efficacy of adaptogens in fatigue, and molecular mechanisms related to their stress-protective activity. *Current Clinical Pharmacology*, *4*, 198–219.

[122] Zhu, S., Zou, K., Cai, S.Q., Meselhy, M.R., Komatsu, K. (2004). Simultaneous determination of triterpene saponins in ginseng drugs by high-performance liquid chromatography. *Chemical and Pharmaceutical Bulletin*, *52*, 995–998.

[123] Kennedy, D.O., Scholey, A.B. (2003). Ginseng: Potential for the enhancement of cognitive performance and mood. *Pharmacology, Biochemistry and Behavior*, *75*, 687–700.

[124] Bahrke, M.S., Morgan, W.R. (2000). Evaluation of ergogenic properties of ginseng: An update. *Sports Medicine*, *29*, 113–133.

[125] Kiefer, D., Pantuso, T. (2003). *Panax ginseng*. *American Family Physician*, *68*, 1539–1542.

[126] Davydov, M., Krikorian, A.D. (2000). *Eleutherococcus senticosus* (Rupr. & Maxim.) Maxim. (Araliaceae) as an adaptogen: A closer look. *Journal of Ethnopharmacology*, *72*, 345–393.

[127] Fulder, S. (1980). The drug that builds Russians. *New Scientists*, *21*, 576–579.

[128] Asano, K., Takahashi, T., Miyashita, M., Matsuzaka, A., Muramatsu, S., Kuboyama, M., Kugo, H., Imai, J. (1986). Effect of *Eleutheroccocus senticosus* extract on human physical working capacity. *Planta Medica*, *52*, 175–177.

[129] Szolomicki, S., Samochowiec, L., Wójcicki, J., Drozdzik, M. (2000). The influence of active components of *Eleutherococcus senticosus* on cellular defence and physical fitness in man. *Phytotherapy Research*, *14*, 30–35.

[130] Kumar, S., Sharma, A. (2005). Anti-anxiety activity studies on homoeopathic formulations of *Turnera aphrodisiaca* Ward. *Evidence-Based Complementary and Alternative Medicine*, *2*, 117–119.

[131] Kumar, S., Taneja, R., Sharma, A. (2005). The genus *Turnera*: A review update. *Pharmaceutical Biology*, *43*, 383–391.

[132] Zhao, J.P., Pawar, R.S., Ali, Z., Khan, I.A. (2007). Phytochemical Investigation of *Turnera diffusa*. *Journal of Natural Products*, *70*, 289–292.

[133] Zhao, J.P., Dasmahapatra, A.K., Khan, S.I., Khan, I.A. (2008). Anti-aromatase activity of the constituents from damiana (*Turnera diffusa*). *Journal of Ethnopharmacology*, *120*, 387–393.

[134] Hancke, J.L., Burgos, R.A., Ahumada, F.A. (1999). *Schizandra chinensis* (Turcz.) Baill. *Fitoterapia*, *70*, 451–471.

[135] Bhattacharya, S.K., Ghosal, S. (1998). Anxiolytic activity of a standardized extract of *Bacopa monniera*: An experimental study. *Phytomedicine*, *5*, 77–82.

[136] Dhanasekaran, M., Tharakan, B., Holcomb, L.A., Hitt, A.R., Young, K.A., Manyam, B.V. (2007). Neuroprotective mechanisms of Ayurvedic antidementia botanical *Bacopa monniera*. *Phytotherapy Research*, *21*, 965–969.

[137] Khanum, F., Bawa, A.S., Singh, B. (2005). *Rhodiola rosea*: A versatile adaptogen. *Comprehensive Reviews in Food Science and Food Safety*, *4*, 55–62.

[138] Kulkarni, S.K., Dhir, A. (2008). *Withania somnifera*: An Indian ginseng. *Progress in Neuro-psychopharmacology and Biological Psychiatry*, *32*, 1093–1105.

[139] Schultes, R.E. (1993). Plants in treating senile dementia in the northwest Amazon. *Journal of Ethnopharmacology*, *38*, 129–135.

[140] Singh, B., Kaur, P., Gopichand, Singh, R.D., Ahuja, P.S. (2008). Biology and chemistry of *Ginkgo biloba*. *Fitoterapia*, *79*, 401–418.

[141] Lin, L.Z., Chen, P., Ozcan, M., Harnly, J.M. (2008). Chromatographic profiles and identification of new phenolic components of *Ginkgo biloba* leaves and selected products. *Journal of Agricultural and Food Chemistry*, *56*, 6671–6679.

[142] van Beek, T.A., Montoro, P. (2009). Chemical analysis and quality control of *Ginkgo biloba* leaves, extracts, and phytopharmaceuticals. *Journal of Chromatography A*, *1216*, 2002–2032.

[143] Packer, L., Hiramatsu, M., Yoshikawa, T., *Antioxidant Food Supplements in Human Health*, Academic Press, San Diego, 1999.

[144] Santos, R.F., Galduróz, J.C.F., Balbieri, A., Castiglioni, M.L.V., Ytaya, L.Y., Bueno, O.F.A. (2003). Cognitive performance, SPECT, and blood viscosity in elderly non-demented people using *Ginkgo biloba*. *Pharmacopsychiatry*, *36*, 127–133.

[145] Krieglstein, J., Beck, T., Seibert, A. (1986). Influence of an extract of *Ginkgo biloba* on cerebral blood flow and metabolism. *Life Sciences*, *39*, 2327–2334.

[146] Kanowski, S., Herrmann, W.M., Stephan, K., Wierich, W., Hörr, R. (1996). Proof of efficacy of the *Ginkgo biloba* special extract EGb 761 in outpatients suffering from mild to moderate primary degenerative dementia of the Alzheimer type or multi-infarct dementia. *Pharmacopsychiatry*, *29*, 47–56.

[147] Mix, J.A., Crews, W.D. Jr, (2002). A double-blind, placebo-controlled, randomized trial of *Ginkgo biloba* extract EGb 761 in a sample of cognitively intact older adults: Neuropsychological findings. *Human Psychopharmacology: Clinical and Experimental*, *17*, 267–277.

[148] Snitz, B.E., O'Meara, E.S., Carlson, M.C., Arnold, A.M., Ives, D.G., Rapp, S.R., Saxton, J., Lopez, O.L., Dunn, L.O., Sink, K.M., De Kosky, S.T. (2009). *Ginkgo biloba* for preventing cognitive decline in older adults: A randomized trial. *The Journal of the American Medical Association*, *302*, 2663–2670.

[149] Leistner, E., Drewke, C. (2010). *Ginkgo biloba* and ginkgotoxin. *Journal of Natural Products*, *73*, 86–92.

[150] Lu, J.L., Zhao, J., Duan, J.A., Yan, H., Tang, Y.P., Zhang, L.B. (2009). Quality evaluation of *Angelica sinensis* by simultaneous determination of ten compounds using LC-PDA. *Chromatographia*, *70*, 455–465.

[151] Huang, S.H., Lin, C.M., Chiang, B.H. (2008). Protective effects of *Angelica sinensis* extract on amyloid beta-peptide-induced neurotoxicity. *Phytomedicine*, *15*, 710–721.

[152] Ho, C.C., Kumaran, A., Hwang, L.S. (2009). Bio-assay guided isolation and identification of anti-Alzheimer active compounds from the root of *Angelica sinensis*. *Food Chemistry*, *114*, 246–252.

[153] Patil, S.P., Maki, S., Khedkar, S.A., Rigby, A.C., Chan, C. (2010). Withanolide A and Asiatic acid modulate multiple targets associated with amyloid-beta precursor protein processing and amyloid-beta protein clearance. *Journal of Natural Products*, *73*, 1196–1202.

[154] Haleagrahara, N., Ponnusamy, K. (2010). Neuroprotective effect of *Centella asiatica* extract (CAE) on experimentally induced Parkinsonism in aged Sprague–Dawley rats. *Journal of Toxicological Sciences*, *35*, 41–47.

[155] Siqueira, I.R., Lara, D.R., Gaieski, F., Silva, F.S., Nunes, D.S., Elisabetsky, E. (1998). Psychopharmacological properties of *Ptychopetalum olacoides* (Olacaceae). *Pharmaceutical Biology*, *36*, 327–334.

[156] Tang, W.X., Kubo, M., Harada, K., Hioki, H., Fukuyama, Y. (2009). Novel NGF-potentiating diterpenoids from a Brazilian medicinal plant, *Ptychopetalum olacoides*. *Bioorganic and Medicinal Chemistry Letters*, *19*, 882–886.

[157] da Silva, A.L., Ferreira, J.G., Martins, B.D., Oliveira, S., Mai, N., Nunes, D.S., Elisabetsky, E. (2008). Serotonin receptors contribute to the promnesic effects of *P. olacoides* (marapuama). *Physiology Behavior*, *95*, 88–92.

[158] Lieu, C.A., Kunselman, A.R., Manyam, B.V., Venkiteswaran, K., Subramanian, T. (2010). A water extract of *Mucuna pruriens* provides long-term amelioration of Parkinsonism with reduced risk for dyskinesias. *Parkinsonism and Related Disorders*, *16*, 458–465.

[159] Katzenschlager, R., Evans, A., Manson, A., Patsalos, P.N., Ratnaraj, N., Watt, H., Timmermann, L., van der Giessen, R., Lees, A.J. (2004). *Mucuna pruriens* in Parkinson's disease: A double blind clinical and pharmacological study. *Journal of Neurology Neurosurgery and Psychiatry*, *75*, 1672–1677.

[160] Manyam, B.V., Dhanasekaran, M., Hare, T.A. (2004). Neuroprotective effects of the antiparkinson drug *Mucuna pruriens*. *Phytotherapy Research*, *18*, 706–712.

[161] Mwatseteza, J.F., Torto, N. (2010). Profiling volatile compounds from *Mucuna* beans by solid phase microextraction and gas chromatography-high resolution time of flight mass spectrometry. *Food Chemistry*, *119*, 386–390.

[162] Xu, K., Bastia, E., Schwarzschild, M. (2005). Therapeutic potential of adenosine $A_{(2A)}$ receptor antagonists in Parkinson's disease. *Pharmacology and Therapeutics*, *105*, 267–310.

[163] Pinna, A., Wardas, J., Simola, N., Morelli, M. (2005). New therapies for the treatment of Parkinson's disease: Adenosine A_{2A} receptor antagonists. *Life Sciences*, *77*, 3259–3267.

[164] Ferré, S., Fredholm, B.B., Morelli, M., Popoli, P., Fuxe, K. (1997). Adenosine–dopamine receptor interaction as an integrative mechanism in the basal ganglia. *Trends in Neurosciences*, *20*, 482–487.

[165] Chaturvedi, R.K., Shukla, S., Seth, K., Chauhan, S., Sinha, C., Shukla, Y., Agrawal, A.K. (2006). Neuroprotective and neurorescue effect of black tea extract in 6-hydroxydopamine-lesioned rat model of Parkinson's disease. *Neurobiology of Disease*, *22*, 421–432.

[166] Milioli, E.M., Cologni, P., Santos, C.C., Marcos, T.D., Yunes, V.M., Fernandes, M.S., Schoenfelder, T., Costa-Campos, L. (2007). Effect of acute administration of hydroalcohol extract of *Ilex paraguaiensis* St Hilaire (Aquifoliaceae) in animal models of Parkinson's disease. *Phytotherapy Research*, *21*, 771–776.

[167] Cui, K., Luo, X., Xu, K., ven Murthy, M.R. (2004). Role of oxidative stress in neurodegeneration: Recent developments in assay methods for oxidative stress and nutraceutical antioxidants. *Progress in Neuro-psychopharmacology and Biological Psychiatry*, *5*, 771–799.

[168] Bastianetto, S., Quirion, R. (2002). Natural extracts as possible protective agents of brain aging. *Neurobiological Aging*, *23*, 891–897.

[169] Dani, C., Oliboni, L.S., Agostini, F., Funchal, C., Serafini, L., Henriques, J.A., Salvador, M. (2010). Phenolic content of grapevine leaves (*Vitis labrusca* var. Bordo) and its neuroprotective effect against peroxide damage. *Toxicology in Vitro*, *24*, 148–153.

[170] Kwon, Y.E., Park, J.Y., No, K.T., Shin, J.H., Lee, S.K., Eun, J.S., Yang, J.H., Shin, T.Y., Kim, D.K., Chae, B.S., Leem, J.Y., Kim, Z.H. (2007). Synthesis, *in vitro* assay, and molecular modeling of new piperidine derivatives having dual inhibitory potency against acetylcholinesterase and $A\beta_{1-42}$ aggregation for Alzheimer's disease therapeutics. *Bioorganic and Medicinal Chemistry*, *15*, 6596–6607.

[171] Adams, M., Gmünder, F., Hamburger, M. (2007). Plants traditionally used in age related brain disorders – a survey of ethnobotanical literature. *Journal of Ethnopharmacology*, *113*, 363–381.

[172] Bai, D. (2007). Development of huperzine A and B for treatment of Alzheimer's disease. *Pure and Applied Chemistry*, *79*, 469–479.

[173] Dohi, S., Terasaki, M., Makino, M. (2009). Acetylcholinesterase inhibitory activity and chemical composition of commercial essential oils. *Journal of Agricultural and Food Chemistry*, *57*, 4313–4318.

[174] Schultes, R.E., Hofmann, A., Plant of the Gods: Origins of hallucinogenic use, EMB-Service for publishers Lucerne, Lucerne, 1989.

[175] Rätsch, C., *The Encyclopedia of Psychoactive Plants*. Park Street Press, Rochester, 2005.

[176] Demetriedes, A.K., Wallman, P.D., McGuiness, A., Gavalas, M.C. (2005). Low cost, high risk: Accidental nutmeg intoxication. *Emergency medicine journal*, *22*, 223–225.

[177] Truitt, E.B. Jr, (1967). The pharmacology of myristicin and nutmeg. *Psychopharmacology Bulletin*, *4*, 14.

[178] Shulgin, A.T. (1966). Possible implication of myristicin as a psychotropic substance. *Nature*, *210*, 380–384.

[179] Olajide, O.A., Ajayi, F.F., Ekhelar, A.I., Awe, S.O., Makinde, J.M., Alada, A.R.A. (1999). Biological effects of *Myristica fragrans* (nutmeg) extract. *Phytotherapy Research*, *13*, 344–345.

[180] Hanus, L.O., Rezanka, T., Spizek, J., Dembitsky, V.M. (2005). Substances isolated from *Mandragora* species. *Phytochemistry*, *66*, 2408–2417.

[181] Carlini, E.A. (2004). The good and the bad effects of (−) trans-delta-9-tetrahydrocannabinol (delta 9-THC) on humans. *Toxicon*, *44*, 461–467.

[182] Mechoulan, R., Lander, N., Srebnik, M., Zamir, I., Breuner, A., Shalita, B., Dickstein, S., Carlini, E.A., Leite, J.R., Edery, H., Porath, G., Recent advances in the use of cannabinoids as therapeutic agents. In: Agurell, S., Dewey, W.L., Willette, R.E., editors. *The Cannabinoids: Chemical, Pharmacologic, and Therapeutic Aspects*. Academic Press Inc., New York, 1984, pp. 777–792.

[183] Schultes, R.E. (1969). Hallucinogens of plant origin. *Science*, *163*, 245–254.

[184] Martin, L.D., Smith, D., Farmilo, C.G. (1961). Essential oil from fresh *Cannabis sativa* and its use in identification. *Nature*, *191*, 774–776.

[185] Zuardi, A.W., Teixeira, N.A., Karniol, I.C. (1984). Pharmacological interaction of the effects of delta 9-trans-tetrahydrocannabinol and cannabidiol on serum corticosterone levels in rats. *Archives Internationales de Pharmacodynamie et de Thérapie*, *269*, 12–19.

[186] Howlett, A.C., Barth, F., Bonner, T.I., Cabral, G., Casellas, P., Devane, W.A., Felder, C.C., Herkenham, M., Mackie, K., Martin, B.R., Mechoulam, R., Pertwee, R.G. (2002). Classification of cannabinoid receptors. *Pharmacology Review*, *54*, 161–202.

[187] Pertwee, R.G. (2006). The pharmacology of cannabinoid receptors and their ligands: An overview. *International Journal of Obesity*, *30*, S13–S18.

[188] Griffin, O.H., Miller, B.L., Khey, D.N. (2008). Legally high? Legal considerations of *Salvia divinorum*. *Journal of Psychoactive Drugs*. *40*, 183–192.

[189] Siebert, D.J. (1994). *Salvia divinorum* and salvinorin A: New pharmacologic findings. *Journal of Ethnopharmacology*, *43*, 53–56.

[190] Shirota, O., Nagamatsu, K., Sekita, S. (2006). Neo-clerodane diterpenes from the hallucinogenic Sage *Salvia divinorum*. *Journal Natural of Products*, *69*, 1782–1786.

[191] Valdés, L.J. (1994). *Salvia divinorum* and the unique diterpene hallucinogen, salvinorin (divinorin) A. *Journal of Psychoactive Drugs*, *26*, 277–283.

[192] Roth, B.L., Baner, K., Westkaemper, R., Siebert, D., Rice, K.C., Steinberg, S., Ernsberger, P., Rothman, R.B. (2002). Salvinorin A: A potent naturally occurring nonnitrogenous kappa opioid selective agonist. *Proceedings of the National Academy of Sciences of the United States of America*, *99*, 11934–11939.

[193] Butelman, E.R., Rus, S., Pisinzano, T.E., Kreek, M.J. (2010). The discriminative effects of the kappa-opioid hallucinogen salvinorin A in nonhuman primates: Dissociation from classic hallucinogen effects. *Psychopharmacology*, *210*, 253–262.

[194] Hermle, L., Funfgeld, M., Oepen, G., Botsch, H., Borchardt, D., Gouzoulis, E., Fehrenbach, R.A., Spitzer, M. (1992). Mescaline-induced psychopathological, neuropsychological, and neurometabolic effects in normal subjects – experimental psychosis as a tool for psychiactric research. *Biological Psychiatry*, *32*, 976–991.

[195] Heffter, A. (1896). Ueber Cacteenalkaloide. *Beichte der Deutschen Chemischen Gesellschaft*, *29*, 216–227.

[196] Grinspoon, L., Bakalar, J.B., *Psychedelic Drugs Reconsidered*. The Lindesmith Center, New York, 1997.

[197] Martinez, S.T., Almeida, M.R.P., Pinto, A.C. (2009). Natural hallucinogens: A flight from Medieval Europe to Brazil. *Química Nova*, *32*, 2501–2507.

[198] Souza, R.S.O., Albuquerque, U.P., Monteiro, J.M., Amorim, E.L.C. (2008). Jurema-preta (*Mimosa tenuiflora* [Willd.] Poir.): A review of its traditional use, phytochemistry and pharmacology. *Brazilian Archives of Biology and Technology*, *51*, 937–947.

[199] Kim, H., Sablin, S.O., Ramsay, R.R. (1997). Inhibition of monoamine oxidase A by β-carboline derivatives. *Arquives of Biochemistry and Biophysics*, *337*, 137–142.

[200] Halberstadt, A.L., Buell, M.R., Masten, V. l., Risbrough, V.B., Geyer, M.A. (2008). Modification of the effects of 5-methoxy-*N,N*-dimethyltryptamine on exploratory behavior in rats by monoamine oxidase inhibitors. *Psychopharmacology*, *201*, 55–66.

[201] Udenfriend, S., Witkop, B., Redfield, B. and Weissbach, H. (1958). Studies with the reversible inhibitors of monamine oxidase: Harmaline and related compounds. *Biochemical Pharmacology*, *1*, 160–165.

[202] Buckholtz, N.S., and Boggan, W.O. (1977). Monoamine oxidase inhibition in brain and liver produced by beta-carbolines: Structure–activity relationships and substrate specificity. *Biochemistry Pharmacology*, *26*, 1991–1996.

[203] Gilbert, B., Alves, L. (2003). Synergy in plant medicines. *Current Medicinal Chemistry*, *10*, 13–20.

[204] Carlini, E.A., Rodrigues, E., Mendes, F.R., Tabach, R., Gianfratti, B. (2006). Treatment of drug dependence with Brazilian herbal medicines. *Revista Brasileira de Farmacognosia 16*, 690–695.

[205] Nichols, D.E. (2004). Hallucinogens. *Pharmacology and Therapeutics*, *10*, 131–181.

[206] Grob, C.S., McKenna, D.J., Callaway, J.C., Brito, G.S., Neves, E.S., Oberlaender, G., Saide, O.L., Labigalini, E., Tacla, C., Miranda, C.T., Strassman, R.J., Boone, K.B. (1996). Human psychopharmacology of hoasca, a plant hallucinogen used in ritual context in Brazil. *The Journal of Nervous and Mental Disease*, *184*, 86–94.

[207] Rezvani, A.H., Overstreet, D.H., Yang, Y., Clarkr, E. Jr, (1999). Attenuation of alcohol intake by extract of *Hypericum perforatum* (St. John's worth) in two different strains of alcohol-preferring rats. *Alcohol and Alcoholism*, *34*, 699–705.

[208] Kim, H.S., Jang, C.G., Lee, M.K. (1990). Antinarcotic effects of the standardized ginseng extract G115 on morphine. *Planta Medica*, *56*, 158–163.

[209] Kulkarni, S.K., Ninan, I. (1997). Inhibition of morphine tolerance and dependence by *Withania somnifera* in mice. *Journal of Ethnopharmacology*, *57*, 213–217.

[210] Kumar, K. (2006). Potential medicinal plants for CNS disorders: An overview. *Phytotherapy Research*, *20*, 1023–1035.

[211] ESCOP, ESCOP monographs. The Scientific foundation for herbal medicinal products. 2nd ed. Argyle House, ESCOP/Thieme. 2003.

[212] WHO, WHO Monographys on selected medicinal plants. World Health Organization publications, Geneva, Volume 1 (1999); Volume 2 (2004); Volume 3 (2007).

6

THE ROLE OF NATURAL PRODUCTS IN DISCOVERY OF NEW ANTI-INFECTIVE AGENTS WITH EMPHASIS ON ANTIFUNGAL COMPOUNDS

Maximiliano Sortino, Marcos Derita, Laura Svetaz, Marcela Raimondi, Melina Di Liberto, Elisa Petenatti, Mahabir Gupta, and Susana Zacchino

6.1 INFECTIOUS DISEASES AND AVAILABLE ANTIMICROBIAL AGENTS

Infectious diseases have had an enormous impact throughout the centuries representing 25% of the disease burden in the adult population, partly as a result of the HIV/AIDS epidemic [1]. At present, they represent a significant cause of morbidity and mortality worldwide, remaining the leading cause of pediatric mortality in most developing countries [2]. Unfortunately, the arsenal of available antimicrobial agents for the treatment of infectious diseases is insufficient to protect people over the long term. The main reasons for this poor prognosis are the unavoidable antimicrobial resistance, the new emerging infections and the re-emergence of old ones [3]. Regarding the latter issue, new outbreaks of the same disease appear from time to time in different parts of the world. That is the case of cholera, which was the cause of death of 13% of the population in Cairo in 1831, and has reappeared in Haiti in October of 2010, having killed at least 142 people in rural central areas of the country and leaving more than 1000 people infected [4–6].

Plant Bioactives and Drug Discovery: Principles, Practice, and Perspectives, Fourth Edition.
Edited by Valdir Cechinel-Filho.

6.2 FUNGAL INFECTIONS AND AVAILABLE ANTIFUNGAL AGENTS

6.2.1 Fungal Infections

Among the different microbes that affect the quality of life of the entire population, fungi have emerged over the past two decades as major causes of human infections, especially among immunocompromised hosts. They produce serious invasive mycoses in individuals submitted to organ transplants or antineoplastic chemotherapy, those with the acquired immunodeficiency syndrome (AIDS), extremely aged persons and patients in intensive care units, among others [7–9]. They also produce superficial fungal infections (those involving the skin and mucosal surfaces) not only in immunocompromised hosts but in healthy individuals, including children of low-developed nations that receive deficient sanitary attention and education, highly diminishing the quality of their lives [10,11].

6.2.2 Available Antifungal Drugs

Although it appears to be a big armamentarium of antifungal drugs in clinical use, in fact only a modest number of drugs, derived from six antifungal classes, are available [8]. Figure 6.1 shows a fungal cell and the main sites of action of the antifungal drugs in clinical use such as cell wall, membrane, and intermediary metabolism.

Polyenes, such as amphotericin B (AmB) and nystatin, bind to ergosterol, the major lipid component of the fungal cell membrane. They form pores leading to the leakage of vital components and, therefore, the death of fungal cell. Allylamines and thiocarbamates inhibit squalene epoxidase and, in turn, azoles control lanosterol 14α-demethylase, both enzymes involved in the biosynthesis of ergosterol. 5-Fluorocytosine interferes with DNA and RNA synthesis and griseofulvin works by disrupting

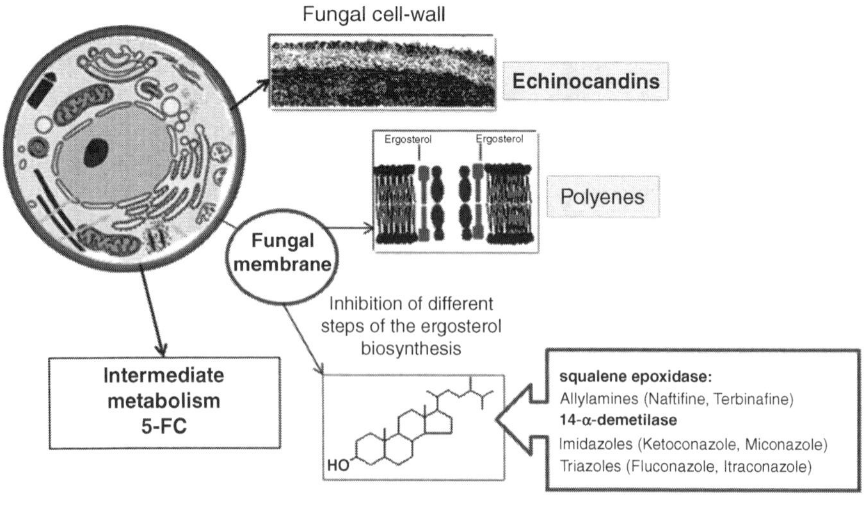

FIGURE 6.1 Schematic representation of a fungal cell and the sites of action of the antifungal agents in clinical use. 5-FC: 5-fluorocytosine.

the spindle microtubule function of the fungal cell. In turn, echinocandins target the fungal cell wall complex (1,3)β-glucan synthase and make the cell vulnerable to osmotic lysis [8,12]. This mode of action is highly appreciated since cell-wall is absent in mammalian hosts [13]; therefore, drugs acting through this mechanism would not cause toxicity in humans.

Among the antifungal drugs in clinical use, nystatin **1**, AmB **2**, griseofulvin **3**, caspofungin **4**, anidulafungin **5**, and micafungin **6** are natural or natural products derived. Polyenes **1** and **2** were isolated from *Streptomyces* spp. The first one, a large macrolide possessing six *E*-double bonds, was isolated from *Streptomyces noursei* in 1950 by the Division of Laboratories and Research of the New York State Department of Health [14]. In turn, polyene **2**, which differs from **1** only in an extra C=C double bond between the 27th and 28th carbons, was originally extracted from cultures of *Streptomyces nodosus* in 1955 at the Squibb Institute for Medical Research [15]. Its name originates from its amphoteric properties.

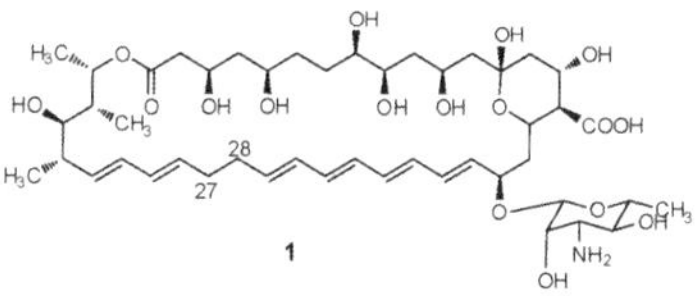
1

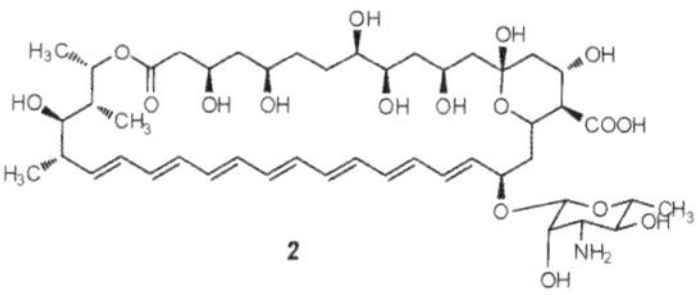
2

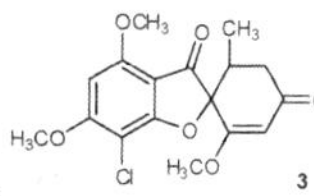
3

Griseofulvin **3**, a tricyclic spiro-diketone with two chiral centers in its structure [16], was first isolated from *Penicillium griseofulvum* and its name originates from its source.

Among the five antifungal drugs that have recently gained approval (the azoles voriconazole (Pfizer, 2002), posaconazole (Schering, 2006), and the echinocandins caspofungin acetate **4** (Merck & Co., Inc.) micafungin sodium **6** (Fujisawa, Astellas Pharmaceuticals, 2002) and anidulafungin **5** (Vicuron and Pfizer, 2006), the last three are natural products derived. The lipohexapeptides **4** and **5** are semisynthetic compounds derived from pneumocandin B **7**, which was isolated from *Glarea lozoyensis*, a fungus found in water [17]. In turn, micafungin **6** is derived from the lipopeptide FR901379 (**8**), which was isolated from the fungus *Coleophoma empetri*, a plant pathogen associated with postharvest fruit rot in cranberries [18].

As noted earlier, all natural antifungal drugs in clinical use are obtained from microbial sources.

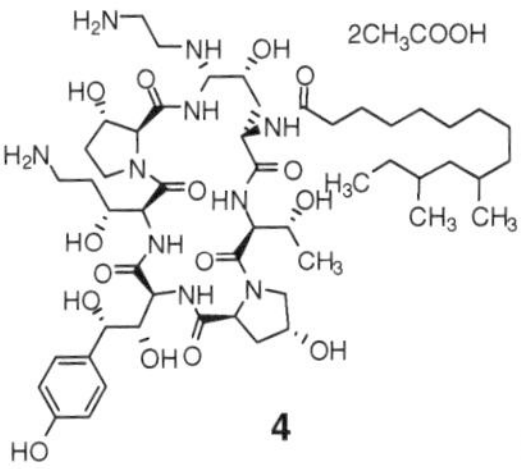
4

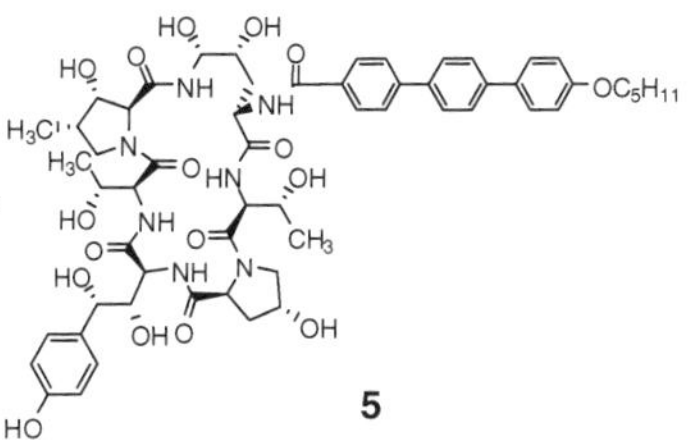
5

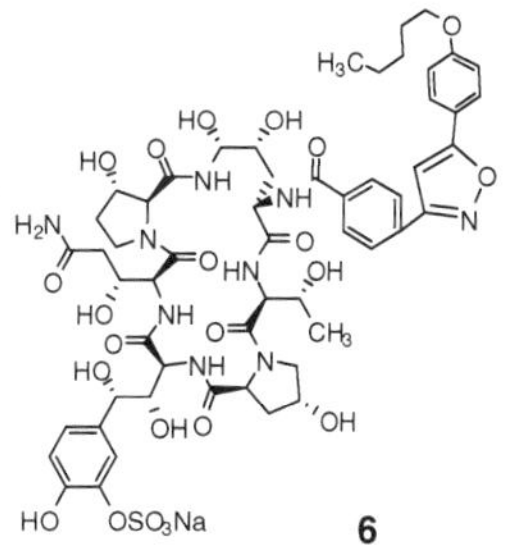
6

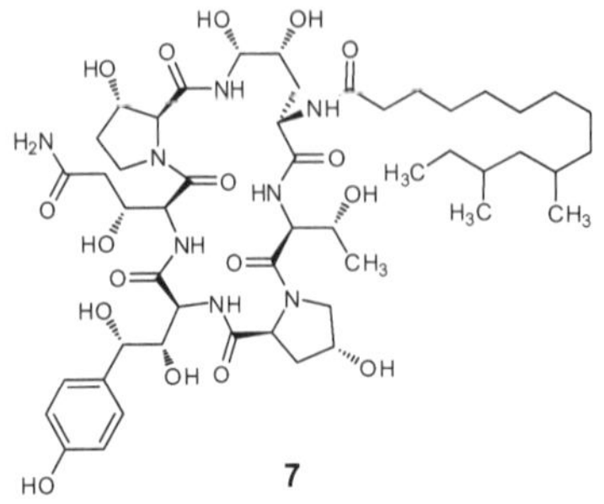

7

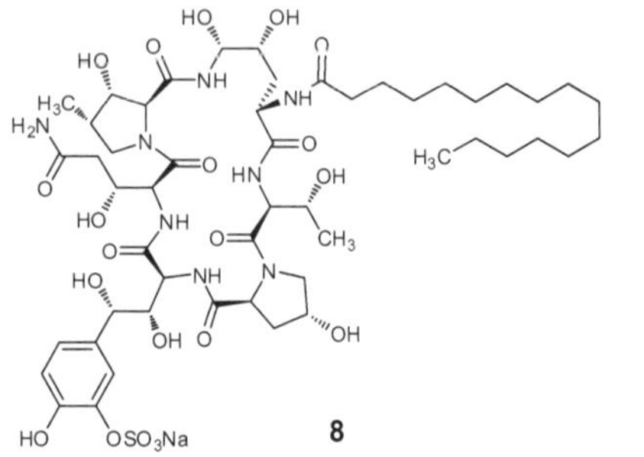

8

6.3 THE NEED OF NEW ANTIFUNGAL AGENTS

All antifungal drugs available for the treatment of fungal infections are not fully effective [18] and possess a certain degree of toxicity. In addition, fungi develop a rapid resistance against them because of their large-scale use [18,19]. Although physicians have shifted to combination therapies, the wonder drug to prevent fungal infections and avoid resistance, has not appeared yet.

Researchers have worked hard in the last three decades to mitigate these issues intending to find novel structures, from natural and synthetic (*sensu lato*) sources [6]. Of course, each laboratory is not prepared for working with all possible sources, since it is necessary to have specialized handling and equipment. This explains the fact that each researcher has focused his work on a single source that he can easily access and handle.

6.3.1 Organisms Recently Investigated as Sources for Antifungal Compounds

Chemists are always looking for new sources of biological diversity in the search of novel bioactive molecules [20]. An analysis of the papers published in *Journal of Natural Products* in the last decade revealed that among the different types of organisms investigated for antifungal properties, 41% were plants, 34%, fungi, 20%, marine organisms, and 5% bacteria (Figure 6.2).

It can be observed that bacteria have been poorly investigated (5%) for antifungal properties, a surprising fact considering that the biosynthetic potential of bacteria encompasses the main pool of genetic diversity in the planet [21]. Most bacteria live in microbial communities where the members are mutually dependent on each other interacting with them through small molecules. Less than 1% of the bacteria on Earth are known, and probably less than 0.1% has ever been cultured, which suggests that there are still many natural products to be discovered from bacteria. Finally, the wealth of bacterial genomic data, now available, shows that there are surely many biosynthetic gene clusters in bacteria for which no associated small molecule was isolated [22].

6.3.2 Plants as Source of Antifungal Metabolites

6.3.2.1 Evolution of the Number of Published Papers in this Area. Among the different possible sources of antifungal compounds, plants maintain a great interest

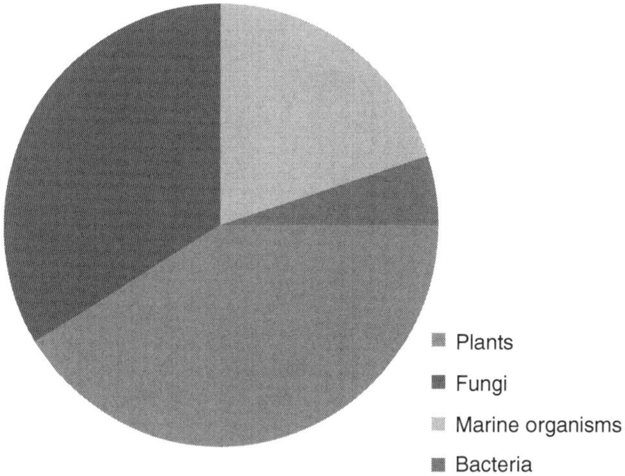

FIGURE 6.2 Types of organisms (%) studied for antifungal properties and published in *J. Nat. Prod.* in the last decade (2001–2010) (*Source*: unpublished personal analysis by the authors, 2010) (*See the color version of this figure in Color Plates section*).

because they provide unlimited opportunities for the isolation of new antifungal compounds due to their unmatched availability of chemical diversity [23,24].

A revision of the papers published in two representative Journals devoted to natural products (*Journal of Ethnopharmacology* and *Planta Medica)* in the last three decades, showed that 221 papers on antifungal screening were published in *Journal of Ethnopharmacology* in that period, increasing from 16 in 1981–1990, 66 in 1991–2000 up to 139 in 2001–2010 (Figure 6.3). Regarding *Planta Medica*, 151 papers on antifungal plants were published in the last three decades, 24 in the first 10 years (1981–1990), 43 in the second decade (1991–2000), and 84 in the third one (2001–2010) (*Source*: authors' unpublished personal analysis, 2010).

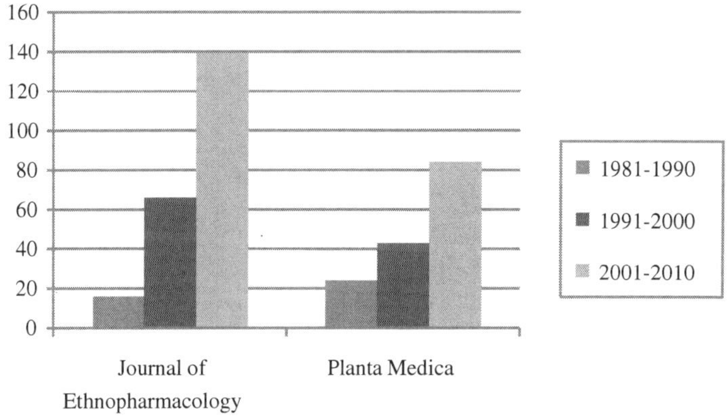

FIGURE 6.3 Number of papers on antifungal properties of plants published in *Journal of Ethnopharmacology* and *Planta Medica* in the last three decades (1981–2010) (*See the color version of this figure in Color Plates section*).

This analysis demonstrates the increasing interest of researchers from Academia in plants as a source of antifungal compounds. In contrast, most big pharmas have abandoned antifungal drug discovery programs based on natural products, including plant biodiversity [25].

6.3.2.2 Selection of Plant Species for the Detection of Antifungal Compounds. The first important concern within a program of discovery of new plants with antifungal properties is the selection of species to be submitted to biological evaluation.

According to a recent review of Ríos and Recio [26], a wide range of criteria were followed for selecting plants to be submitted to antimicrobial studies. Some researchers focused on one plant family; others, on plants growing in a specific region or country; other groups on ethnopharmacologic uses or on inhibitors of one or more clinically important fungal species with a *random* selection of plants.

6.3.2.2.1 Selection According to Plant Family: Analysis of Data from NAPRALERT. With the aim of determining if the selection based on plant family could help in the identification of antifungal compounds, we analyzed the antifungal studies registered in three recent decades in the NAPRALERT database (1975–2005). The plant species studied for antifungal properties were ca. 2000. They belong to 145 botanical families, of which 14 (≈10 %) possess the highest number (811) of studied species (≈40 %). The main families and species are as follows: Asteraceae, 127; Fabaceae, 114; Lamiaceae, 91; Euphorbiaceae, 76; Ranunculaceae, 60; Myrtaceae, 59; Liliaceae, 53; Apiaceae, 52; Rosaceae, 43; Rutaceae, 36; Poaceae, 32; Brassicaceae, 24; Polygonaceae, 23; Solanaceae, 21 (Table 6.1).

TABLE 6.1 The 14 Most Studied Families for Antifungal Properties: ***Source:*** **NAPRALERT database, period 1975–2000**

Family	Total Number of Species within the Botanical Family	Analyzed Entities	Number of Actives	%
Asteraceae	22,700	127	25	20
Fabaceae	18,000	114	26	23
Lamiaceae	6700	91	31	35
Euphorbiaceae	8100	76	5	6.5
Ranunculaceae	2400	60	6	10
Myrtaceae	4600	59	16	27
Liliaceae	4900	53	9	17
Apiaceae	3500	52	14	27
Rosaceae	2800	43	9	21
Rutaceae	1800	36	14	38
Poaceae	9500	32	18	56
Brassicaceae	3200	24	9	37.5
Polygonaceae	1100	23	3	13
Solanaceae	2900	21	12	57

It is clear from Table 6.1 that the families studied for antifungal properties are the most frequent within the flora of the whole world; so, we can assume that the selection of species can be best correlated with the abundance of species and their facile accessibility, than to any other cause for selection.

In addition, results showed that there was not a positive correlation between the number of species studied in each family and the frequency of the actives.

6.3.2.2.2 Selection According to a Certain Region or Country. There are many published papers that delimit the study of antifungal plants to a certain region or country. For example, some papers report studies from Latin American countries [27], such as Uruguay [28], Paraguay [29], Panamá [30], Argentina [31], Brazil [32], México [33], Guatemala [34]; others analyze plants from other regions or countries such as South Africa [35], Canada [36], etc. It appears from a first glance that this kind of selection does not assure better results but, unfortunately, there is not any study establishing a comparison of results from different regions. This task would not be easy because many variables are involved in each region such as biodiversity, accessibility, research possibilities, and many others.

6.3.2.2.3 Selection Based on Ethnopharmacological Information. Two main approaches are commonly used for the selection of plants: the ethnomedical information or the *random* selection (although a third approach was recommended by Mitscher [37]: to test plants previously reported to be active against bacteria). Ríos and Recio [26] suggested to avoid the *random* approach and select plants following an ethnopharmacologic perspective since it appeared to enhance the probability of success in new drug-finding efforts.

Some studies performed by different research groups in different countries, gave support to Rios and Recio suggestion [34,36,38–40]. But, unfortunately, these important antifungal studies were performed with the agar diffusion method which, as it is well known, is a qualitative nonstandardized test that does not give reliable results [41].

Within a multidisciplinary collaborative OAS project carried out during the period 2001–2004 with the participation of research centers in Argentina, Bolivia, Brazil, Colombia, Costa Rica, Guatemala, and Panama, plant extracts from these seven countries were screened for antifungal activities with the microbroth dilution method following the guidelines of CLSI [42], against a same panel of fungi, in a single laboratory.

Results showed [27] that a significant higher chance of detecting plants with the antifungal activity against at least one fungus was found within the group of plants that had reports of ethnopharmacologic uses related to fungal infections (PAU group, 40 %) than within the *random* group (PNAU group, 20.5%, $p < 0.01$). When the type of fungi tested was taken into consideration, a similar higher chance in the PAU group respective of PNAU group (39.3% vs. 20%) was found when plants were tested against dermatophytes ($p < 0.01$) but not against yeasts or *Aspergillus* spp. ($p > 0.05$) (Figure 6.4). Within the detected antifungal plants from both groups, plants of the PAU group tended to display better activities (lower MICs) against dermatophytes

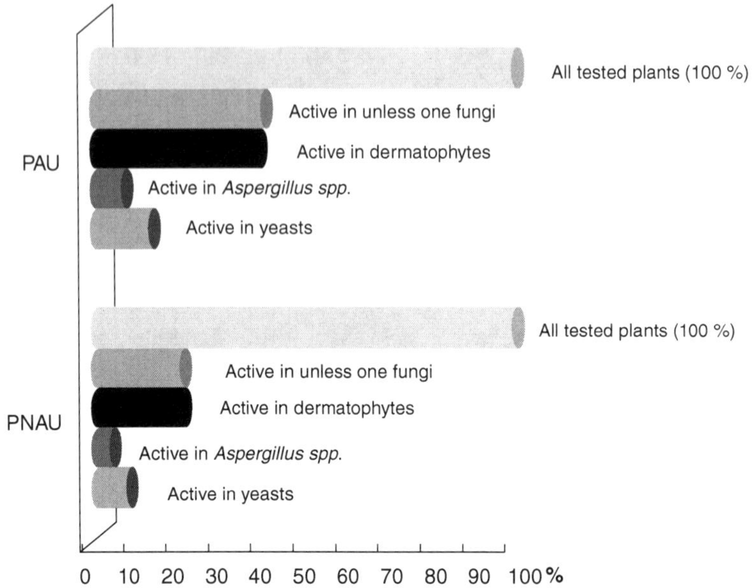

FIGURE 6.4 Percentage of active extracts against: at least one fungal species; yeasts; *Aspergillus* spp.; dermatophytes; within each of both groups of plants: the PAU (plants with ethnopharmacological uses related to mycoses and the PNAU (plants without ethnopharmacological uses related to mycoses).

than those of the PNAU group ($p < 0.05$). This same trend was not observed against yeasts or *Aspergillus* spp (Figure 6.5). So, our findings demonstrated that the ethnopharmacologic knowledge is useful in guiding the discovery of antifungal Latin American plants mainly for infections in which the pathological expression is

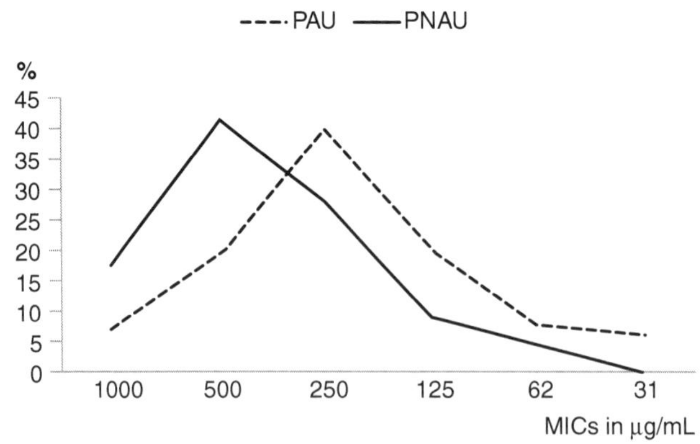

FIGURE 6.5 Percentage of all MICs $\leq$ 1000 μg/mL (considered 100%) acting against dermatophytes within each of both groups of plants: PAU (plants with ethnopharmacological uses related to mycoses) and PNAU (plants without ethnopharmacological uses related to mycoses).

obvious (superficial infections produced by dermatophytes) whose evolution can be clearly followed by traditional healers.

6.3.2.3 Antifungal Compounds Isolated from Natural Sources. Two important reviews on antifungal compounds isolated from natural sources have been published. The review by Cowan, published in 1999 [43], divided the phytochemicals into several categories comprising: phenolics and polyphenols, terpenoids and essentials oils, alkaloids, lectins and polypeptides, polyacetylenes, and other compounds. In a more recent review, Abad et al. [44] examined the studies performed from 1995 to 2006 toward the discovery of antifungal compounds of plant origin. The information was organized into both, crude extracts and natural compounds. The latter were divided into: components of essential oils; terpenoids; saponins; phenolic compounds; alkaloids and peptides; and proteins.

6.3.2.3.1 Alkaloids. Diverse alkaloid structures with antifungal properties have been isolated from plants. For example, a recent paper reported [45] that the copyrine alkaloid 8-hydroxysampangine **8** isolated from *Cleistopholis patens* displayed good activities against *C. albicans* and clinical isolates of non-*albicans Candida* spp. such as *C. glabrata*, *C. guillermondii*, *C. kefyr*, *C. parapsilopsis* and *C. tropicalis* with the agar diffusion method. This same species had been investigated previously (1987 and 1990): in the first of these works, the aporphynoid alkaloids eupolaridine **9** and onychine **10** were isolated from its root bark demonstrating a strong activity against three strains of *C. albicans* with very low MICs [46]. In further paper, the copyrine alkaloid 3-methoxy sampangine **11** was isolated from the same root bark showing high activities against *C. albicans*, *Aspergillus fumigatus* and *Cryptococcus neoformans* [47].

From *Porcelia macrocarpa*, two azanthraquinone alkaloids (**12**, **13**) demonstrated strong antifungal activities against the phytopathogenic fungi *Cladosporium cladosporioides* and *Cladosporium. sphaerospermum* by bioautography [48]. β-carboline alkaloids (**14–17**) from *Peganum harmala* were recently evaluated alone and in combination against *Aspergillus niger* and *C. albicans* [49]. The same species and its alkaloids have been previously evaluated as antimicrobials [50].

A new aporphine alkaloid **18**, along with **19** and 18 berberine-type alkaloids, were isolated from *Fibraurea recisa* stem showing moderate activities against *C. albicans* and non-*albicans Candida* spp. in diffusion and dilution tests [51].

The quaternary alkaloid berberine **20** was isolated by bioassay-guided fractionation from *Berberis heterophylla* [52], showing the antifungal activity against a panel of yeasts and filamentous fungi. It was also isolated from *Mahonia aquifolium*, showing activity against 20 strains of *Candida* spp. [53].

Two alkaloids, D-calycanthine **21** and L-folicanthine **22**, were isolated from the active methanol extract of the seeds of *Chimonanthus praecox* [54], showing significant inhibitory activities against the plant pathogenic fungi *Exserohilum turcicum*, *Bipolaris maydis*, *Alternaria solani*, *Sclerotinia sclerotiorum*, and *Fusarium oxysporum*. Among them, *B. maydis* and *S. sclerotiorum* were found to be the most susceptible species to **21** and **22,** respectively.

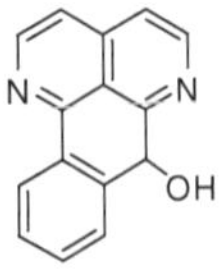
8

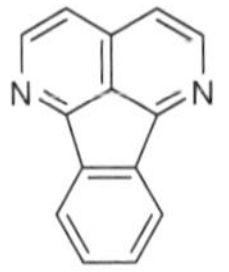
9

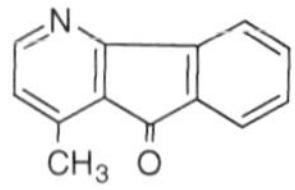
10

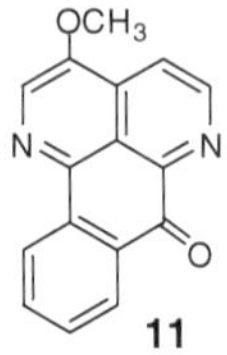
11

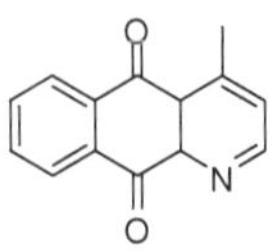
12

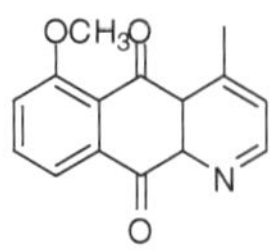
13

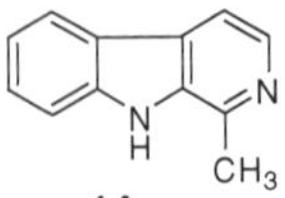
14

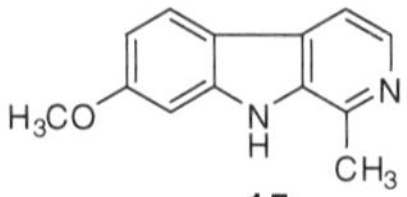
15

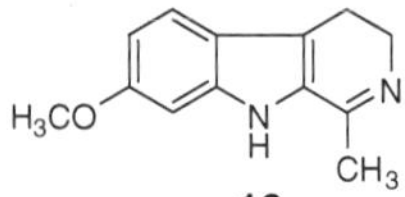
16

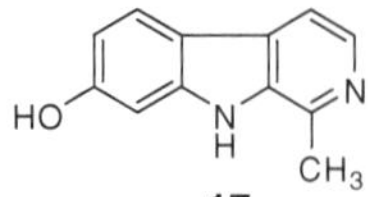
17

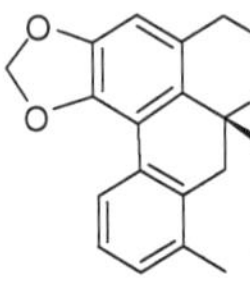
18

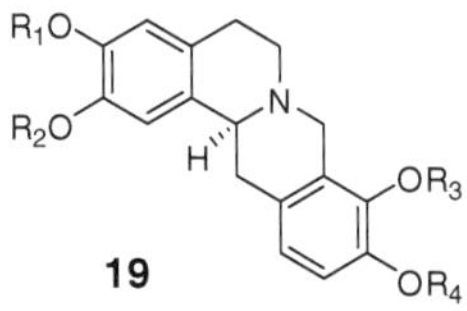

19

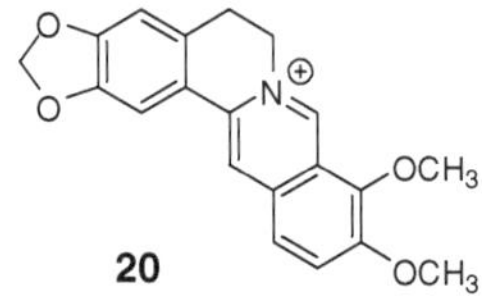
20

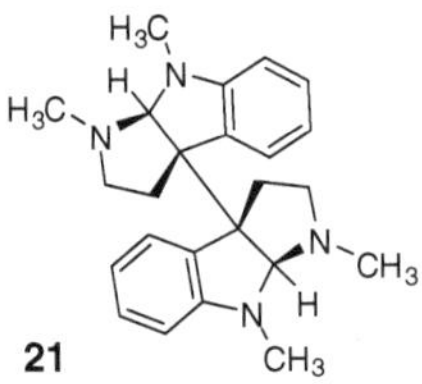
21

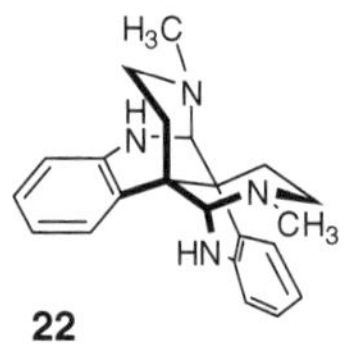
22

The tomato steroidal glycoalkaloid tomatine **23** has antifungal activity toward yeasts, and this activity was associated with membrane permeabilization [55]. Removal of a single sugar from its tetrasaccharide chain resulted in a substantial reduction in the antimicrobial activity. Surprisingly, the complete loss of sugars led to an enhanced antifungal activity.

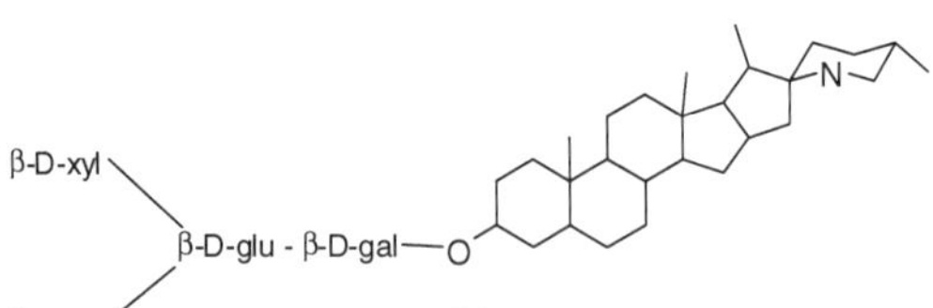

23

6.3.2.3.2 Essential Oils and Their Components as Antifungal Substances. The antifungal activity of essential oils from different families has been recently reviewed [56], Lamiaceae and Asteraceae being the most studied families. Nevertheless, reports on the antifungal activity of oils from Verbenaceae, Lauraceae, Myrtaceae, Rutaceae, and Cupressaceae and other botanical families such as

Liliaceae, Apiaceae, Taxodiaceae, Chenopodiaceae, Theaceae, Geraniaceae, Umbelliferae, etc, have also been found in the literature [57–59].

Although most of these studies do not identify the components responsible for the activity, some monoterpenes or aromatic compounds, isolated from natural essential oils, demonstrated high antifungal activities.

The monoterpene linalool **24** from *Lavandula angustifolia* showed good antifungal activity against 50 clinical isolates of *C. albicans.* Instead, fenchone **25**, limonene **26** and myrtenol **27** appeared to be the most effective components of *Lavandula stoechas* against *Rhizoctonia solani* [57]. In turn, the monoterpene aldehyde citral **28**, extracted from *Cymbopogon citratus* showed antifungal activities against *Rhizopus stolonifer* and *Penicillium italicum* [58].

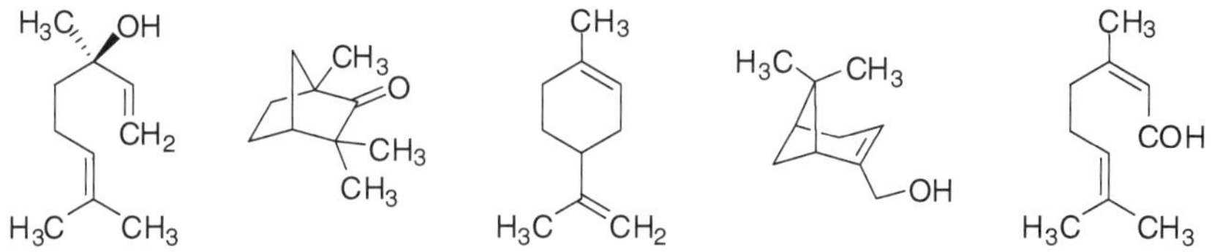

24 25 26 27 28

Regarding aromatic components of essential oils, an interesting quantitative structure-antifungal activity relationship study was performed with 22 oxygenated aromatic compounds [59]. Their antifungal activity was tested against two wood-decaying fungi, the white-rot *Trametes versicolor*, which mainly metabolizes lignin, and the brown-rot *Coniophora puteana*, which digests cellulose in plant cell walls. MIC was used to generate a schematic tree that established an order of the structures of the essential oil components, considering both, the nature, and the position of the substituents in their aromatic rings in a QSAR analysis. Results for the 22 essential oil components revealed that thymol **29**, carvacrol **30**, cuminaldehyde **31**, *trans*-anethole **32**, and methylchavicol **33** were the most active compounds, while hydroquinone **34**, cresol **35**, guaiacol **36**, and coniferaldehyde **37** displayed the lowest activity against *T. versicolor*. Similarly, **29–32** exhibited the greatest inhibitory action against *C. puteana*.

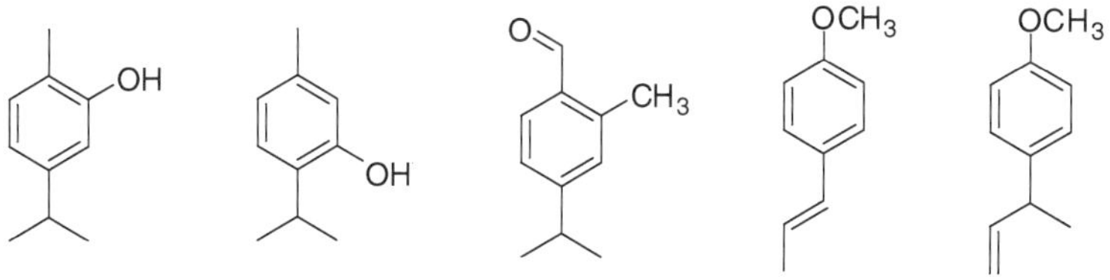

29 30 31 32 33

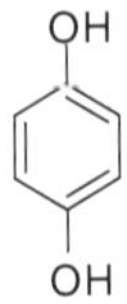

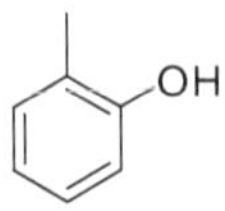

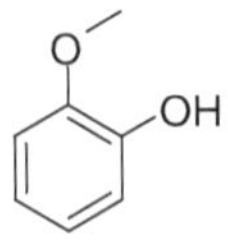

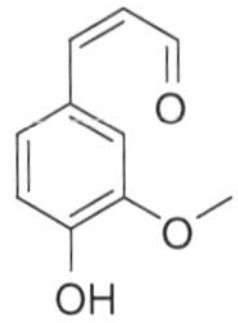

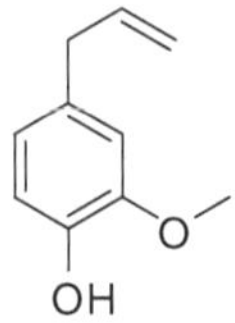

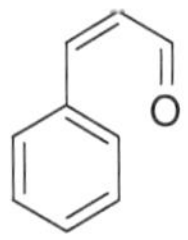

34 **35** **36** **37** **38** **39**

It is important to take into account that aromatic phenols **29, 30**, and eugenol **38** and other essential oil components such as **31–33** and cinnamaldehyde **39** are constantly being investigated for their value in food preservation and as natural antimicrobial agents in human and veterinary medicine [60–65].

6.3.2.3.3 Other Terpenoids. Among sesquiterpenes, polygodial **40**, isolated from *Polygonum hydropiper*, *Warburgia ugandensis* and *Warburgia stuhlmannii*, was reported as a strong antifungal sesquiterpene aldehyde, possessing broad antifungal spectrum and potency, and fungicide properties [66]. In further papers [67,68], **40** and 1β-*p*-methoxy cinnamoyl polygodial **41** demonstrated to be one of the compounds responsible for the antifungal activity of *Polygonum punctatum* or *Drimys brasiliensis* and, in a recent paper, we could demonstrate that polygodial was responsible for the antifungal activity of *P. acuminatum* [69].

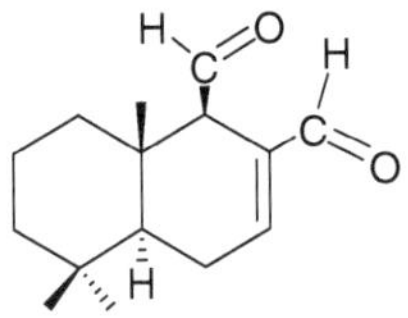

40

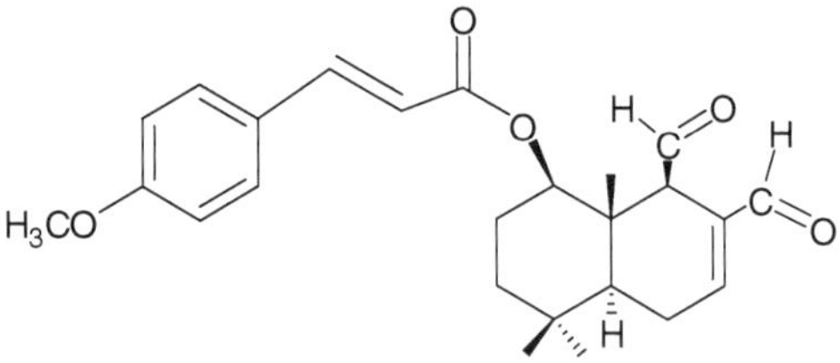

41

In addition, sesquiterpene lactones have demonstrated interesting antifungal properties and their structure–activity relationships were analyzed [70]. Among recent published papers, the antifungal sesquiterpene lactones vernolide **42** and vernodanol **43** have been isolated from *Vernonia amygdalina* [71] and the sesquiterpene 6-cinnamoyl-1-hydroxyeudesm-4-en-3-one (**44**) was identified as the antifungal agent from the roots of *Vernonanthura tweedieana* [72]. Seven minor sesquiterpene lactones from *Centaurea pullata*, showed strong antifungal activities [73].

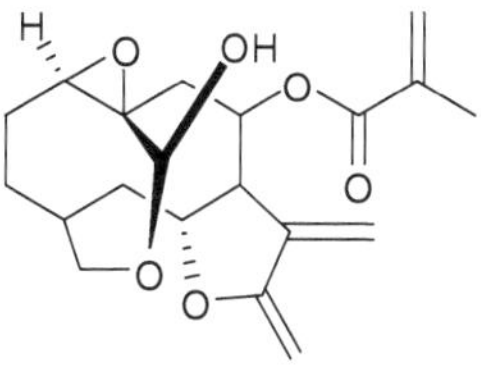

42

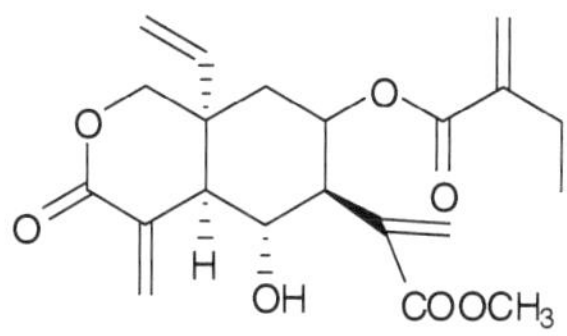

43

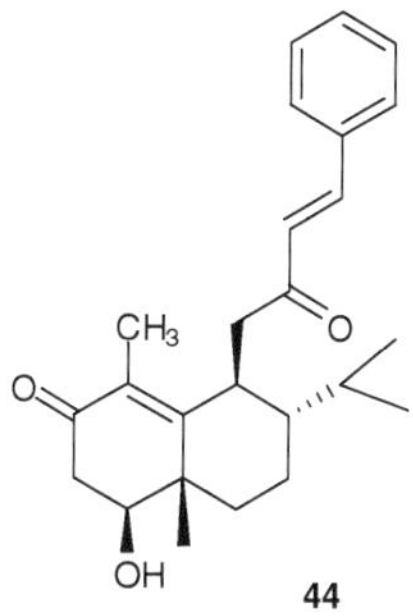

44

6.3.2.3.4 Saponins. The classic view that saponins make poor antifungal drug candidates have been changed in recent years [74]. Regarding triterpenoid saponins, their distribution and antifungal activities were recently reviewed [75].

Some structure relationships of triterpenoid saponins with hederagenin or oleanolic acids as aglycones have shown that the inhibition of fungal growth is influenced by the number and kind of sugar residues, although a higher activity was not observed with an increase in the length of the sugar chain [76]. Two recent examples are in accordance with that observation. Phytolaccoside E **45** (3-*O*-β-D-glucopyranosyl-(1 → 4)-β-D-xylopiranosyl-phytolaccagenin) isolated from *Phytolacca tetramera* possessed lower antifungal activities than Phytolaccoside B **46** (3-*O*-β-D-xylopiranosylphytolaccagenin), which possesses only one sugar in its structure [77]. In turn, phytolaccagenin **47** showed better antifungal activity than **46** [78].

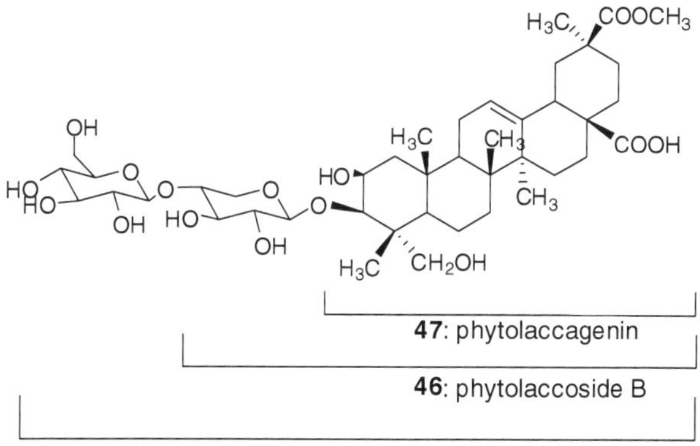

47: phytolaccagenin

46: phytolaccoside B

45: phytolaccoside E

Regarding steroidal saponins from plant origin, particularly those from Dioscoreaceae and Solanaceae families have shown antifungal properties [74,79–81].

Twenty-two spirostanol steroidal saponins from natural sources possessing hecogenin **48**, diosgenin **49**, or chlorogenin **50** with 1–5 sugars into their structures were evaluated for antifungal properties in order to find structure–activity relationships. Results showed that, in contrast to triterpenoid saponins, the sugar chain in each saponin is necessary and plays a key role in its antifungal activity [74].

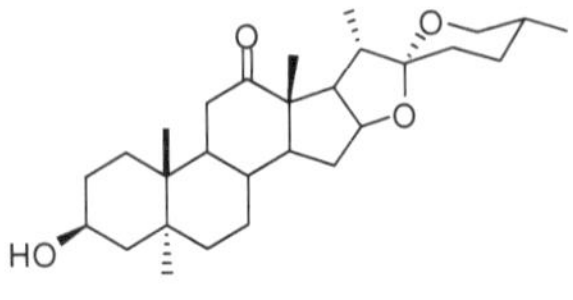

48

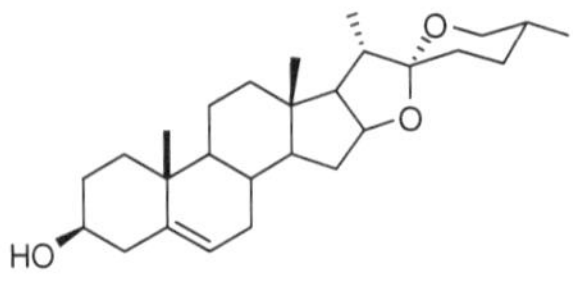

49

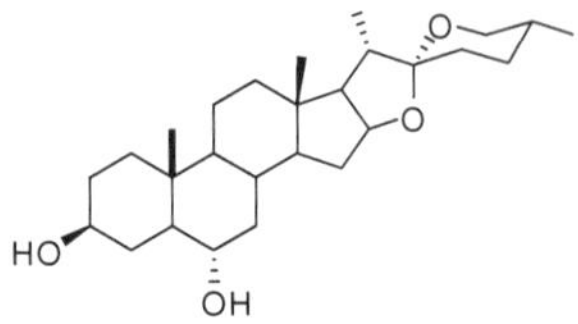

50

6.3.2.3.5 Other Type of Compounds. Many other type of antifungal compounds such as lignans, neolignans, amides, quinones, peptides, proteins, xanthones, flavonoides (chalcones, flavanones), coumarins, polyphenols, polyacetylenes, and others, have been isolated from higher plants.

As some examples of antifungal natural isolated compounds, ellagic acid **51** was isolated from *Lafoensia pacari* a plant from Mato Grosso, Brazil [82]; prenylated coumarins represented by **52** were obtained from *Baccharis darwinii* collected in the Argentinean Patagonia [83]; two sesquiterpenes (**53**, **54**), and a dihydroflavanone (**55**) were isolated from *Heterothalamus alienus* [84]; a new depside (**56**) was found in the lichen *Protousnea poeppigii* [85]; a chalcone (**57**) was isolated from *Zuccagnia punctata* and of propolis formed from it [86]; a series of 8.*O*.4′-neolignans represented by **58**, were isolated from *Myristica fragrans* [87], among many others.

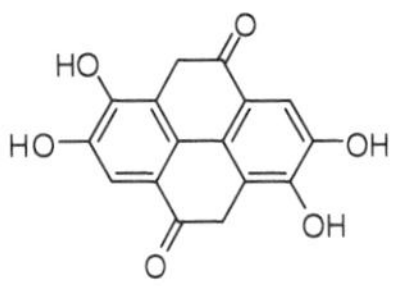

51

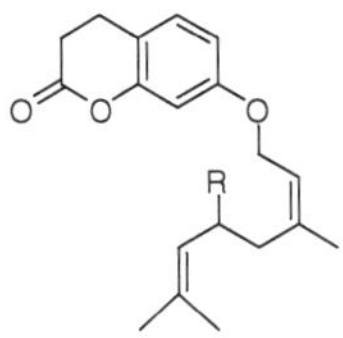

52

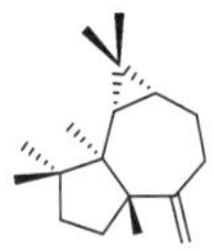

53

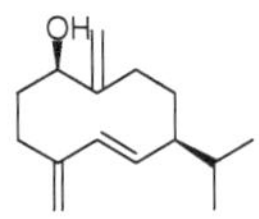

54

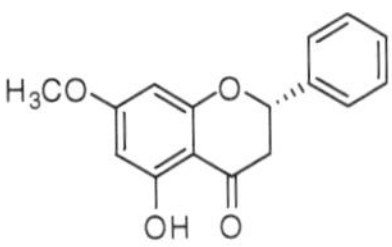

55

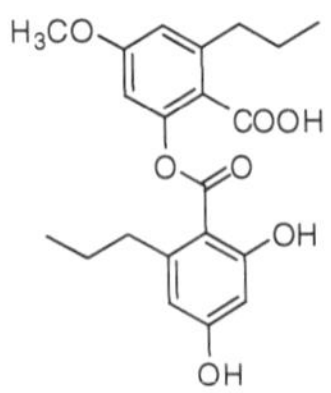

56

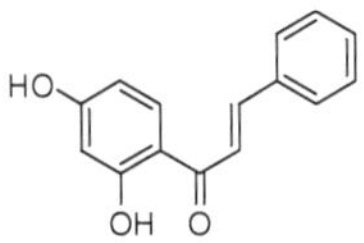

57

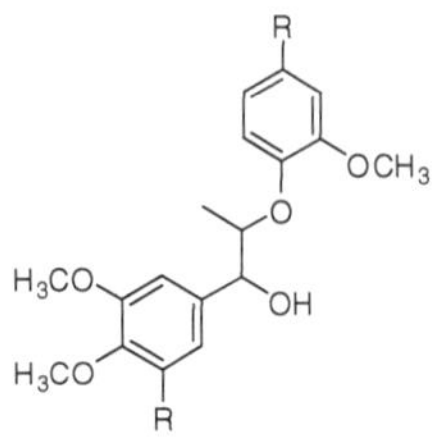

58

6.3.3 Microorganisms as Source of Antifungal Metabolites

6.3.3.1 Antifungal Compounds from Microorganisms of Terrestrial Environments. Vicente et al. [88] reported that 20% of the different families and orders of Basidiomycetes possessed some kind of antifungal activity. The main antifungal compounds were lipopeptides (echinocandins **59–61**, aculeacin **62**, mulundocandin **63** from *Aspergillus* spp., papulacandins **64–66** from *Papularia sphaerosperma*, fusacandin **67** from species of *Fusarium* genus) or acidic triterpenoids (enfumafungin **68** from *Hormonema* spp. and arundifungin **69** from different fungi.

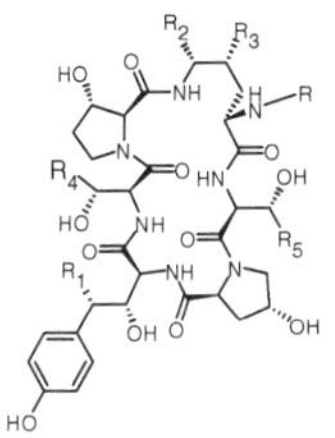

Name	R	R_1	R_2	R_3	R_4	R_5
59: Echinocandin B	linoleoyl	OH	OH	OH	CH_3	CH_3
60: Echinocandin C	linoleoyl	H	OH	OH	CH_3	CH_3
61: Echinocandin D	linoleoyl	H	H	H	CH_3	CH_3
62: Aculeacin	palmitoyl	OH	OH	OH	CH_3	CH_3
63: Mulundocandin	12-methylmyristoyl	OH	OH	OH	CH_2CONH_2	H

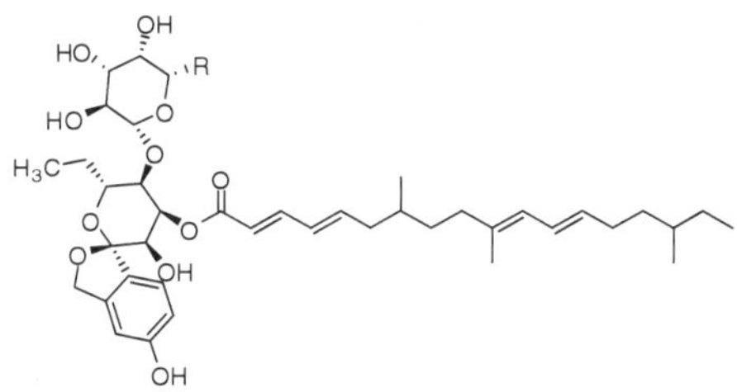

64 papulacandin A: R=

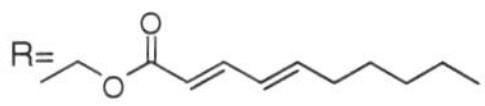

65 papulacandin B: R=

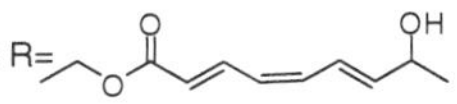

66 papulacandin C: R=

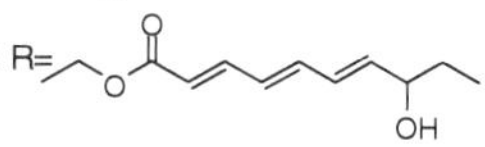

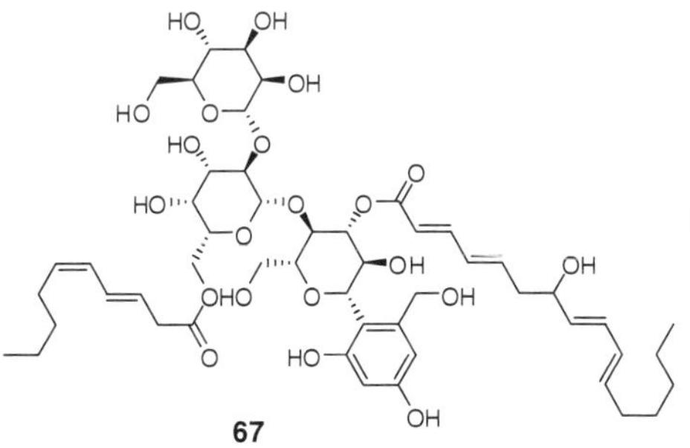

67

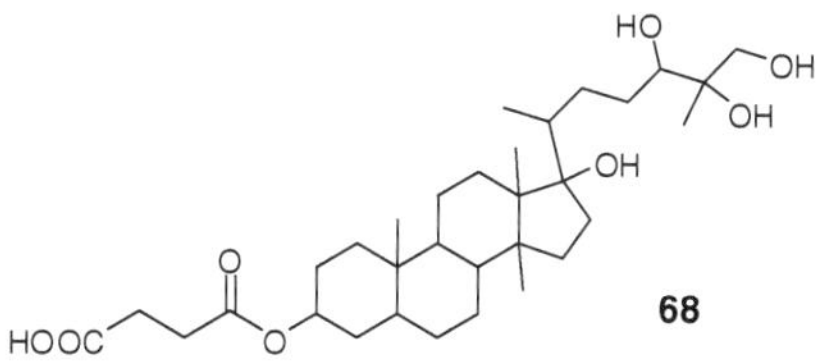

68

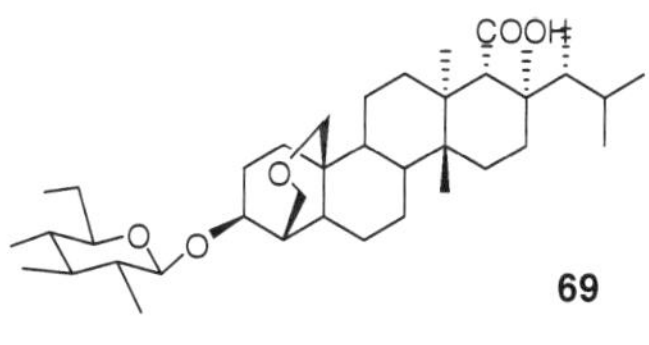

69

6.3.3.2 From Endophytic Fungi. Endophytes, microorganisms that reside in the tissues of living plants have been scarcely studied as potential sources for medicines.

Each individual plant of the many that exist on the earth is host to one or more endophytes [89].

Endophytes in general produce peculiar secondary metabolites with antifungal properties. For example the antifungal peptide cryptocandin **70** and the tetramic acid cryptocin **71** were isolated from *Cryptosporiopsis quercina*, a fungus commonly associated with hardwood species in Europe [89].

From *Pestalotiopsis microspora*, a common rainforest endophyte and from endophytes of *Taxus brevifolia*, the antifungal ambuic acid **72**, two pyrones (pestalopyrone **73** and hydroxypestalopyrone **74**), and the aromatic β-glucoside pestaloside **75** were isolated. *Pestalotiopsis jester*, an endophytic fungus found in the Sepik river area of Papua New Guinea produces jesterone **76** and hydroxyjesterone **77** both exhibiting antifungal activities against a variety of pathogenic fungi [89].

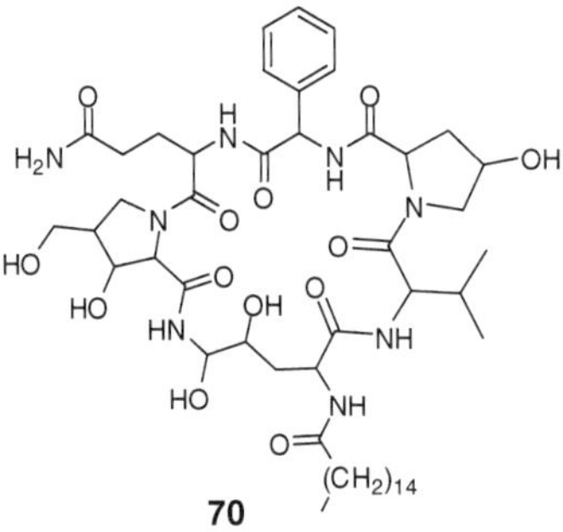

70

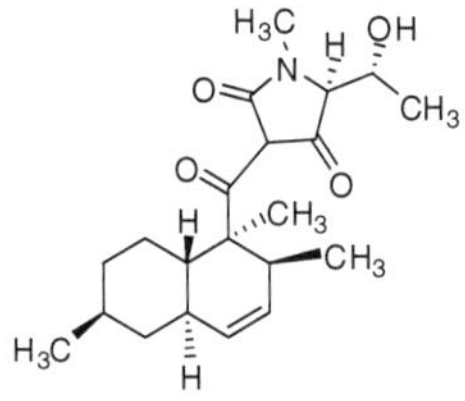

71

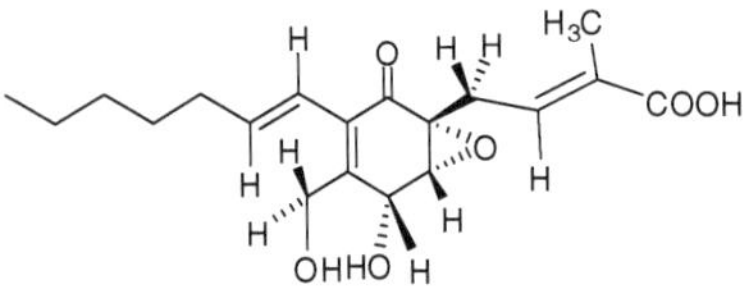

72

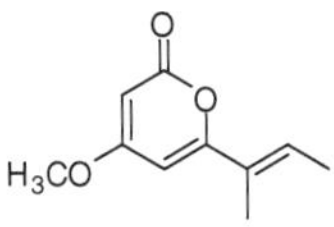

73

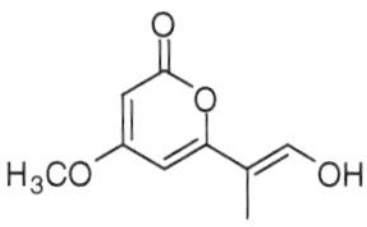

74

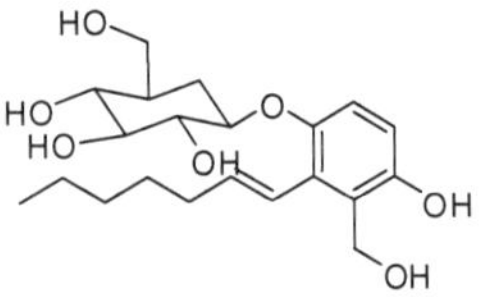

75

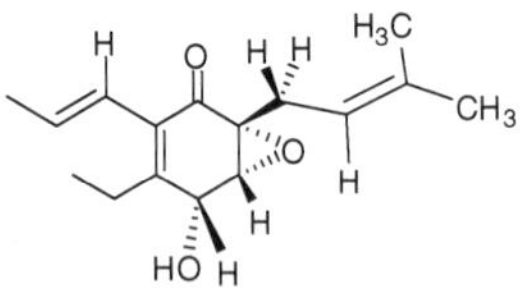

76

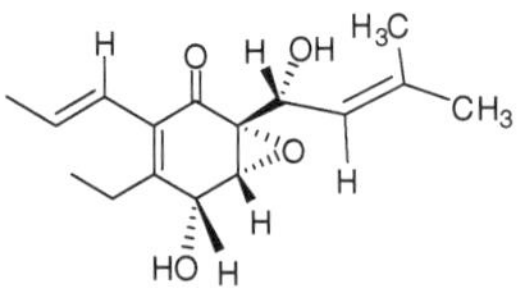

77

6.3.4 Marine Organisms as Sources of Antifungal Metabolites

An analysis of the papers published in *Journal of Natural Products* in the last decade, reveals that among the marine organisms investigated for antifungal properties, 65% were sponges, 17% algae and 6% each, starfishes, corals and sea-cucumbers (Figure 6.6).

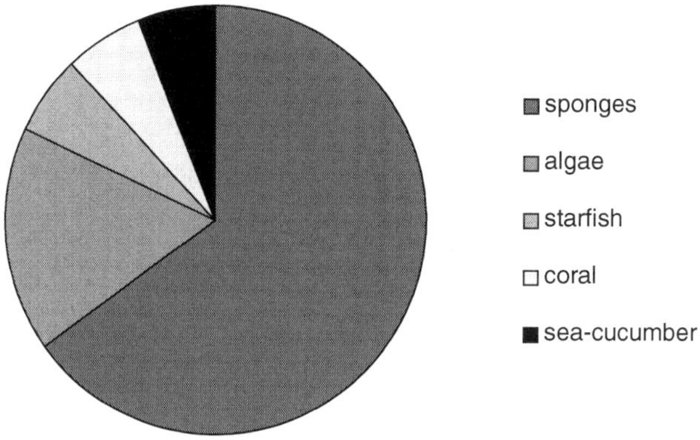

FIGURE 6.6 Percentages of the different marine organisms investigated for antifungal properties in the last decade published in *Journal of Natural Products* (*See the color version of this figure in Color Plates section*).

In addition, in a recent review, Mayer et al. [90] found a substantial increase in the amount of publications on antifungal marine natural products during 2005–2006, with respect to the period 1998–2004.

Among the reported structures, a phenol **78** from the marine sponge *Dysidea herbacea* [91], and the macrocyclic polyether spongistatin **79** from *Hyrtios erecta* [92] showed to possess significant activity against fungal pathogens. Studies on mechanisms of action were performed on them.

Additionally, other marine chemicals with significant antifungal activity were reported. They were the lipopeptide hassallidin A **80**, the polyketide latrunculin **81**, and the fatty acid majusculoic acid **82** [90].

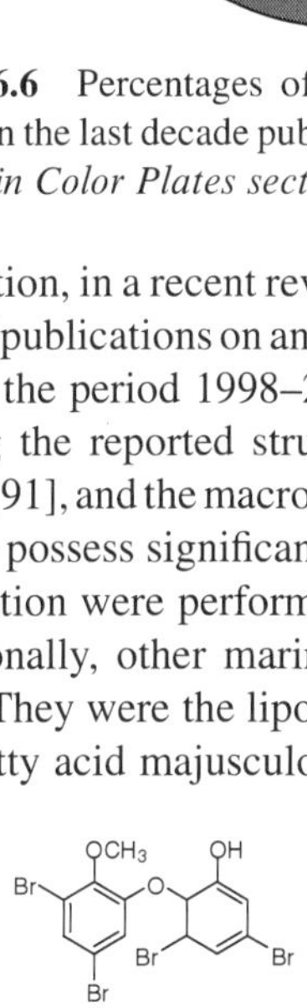

78

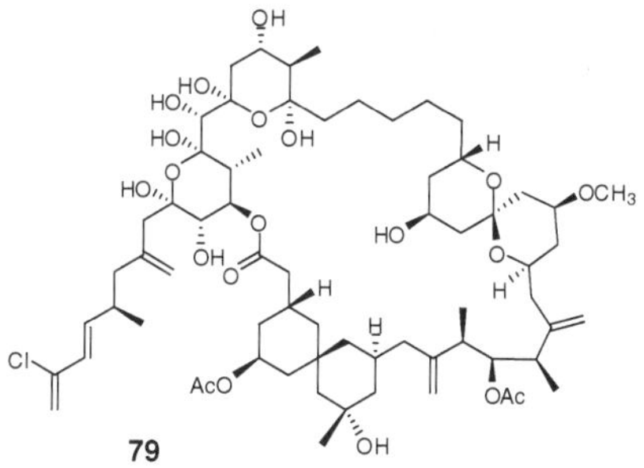

79

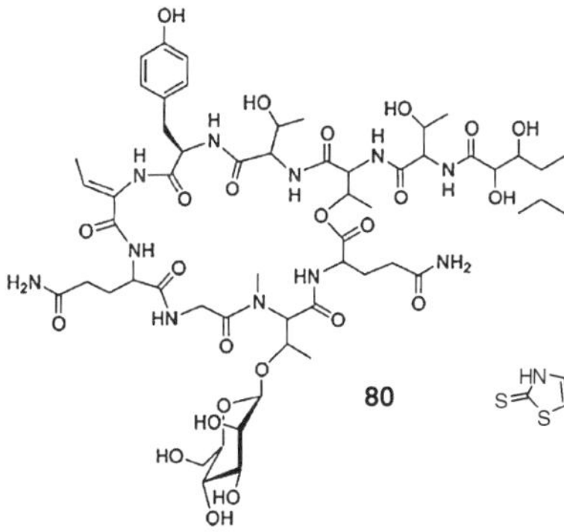

80

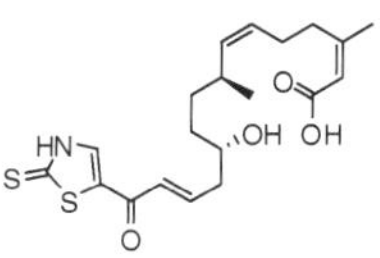

81

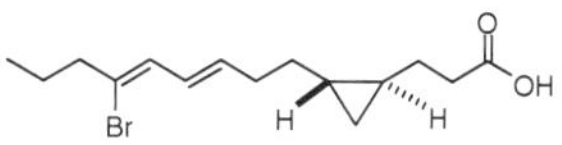

82

More recently, complex triterpenoid antifungal saponins (**83–86**) have been isolated from the sea cucumbers *Actinopyga lecanora* [93], *Holothuria axiloga* [94], *Neothyone gibbosa* [95], and *Bohadschia marmorata* [96] respectively, among others [97].

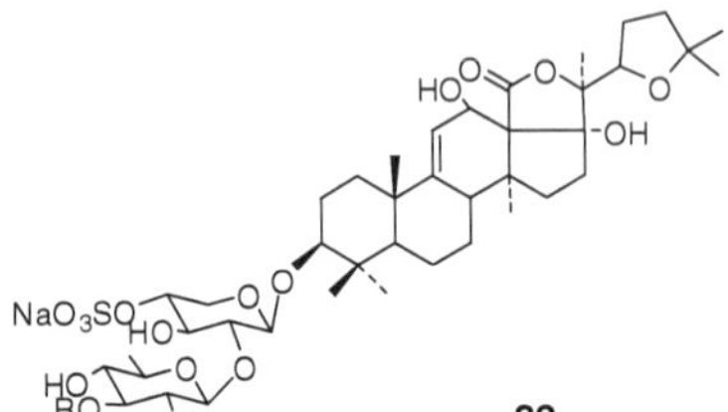

83

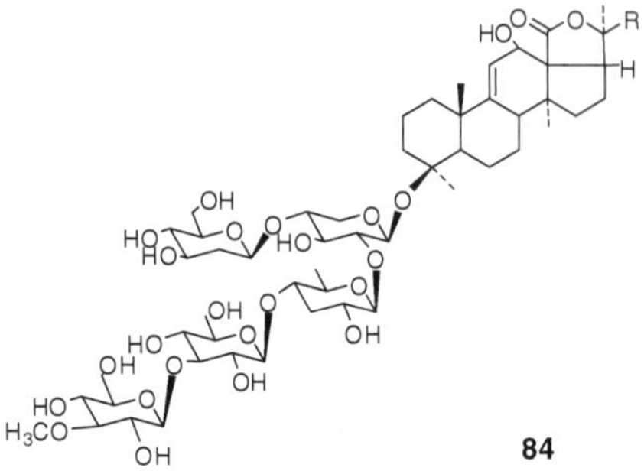

84

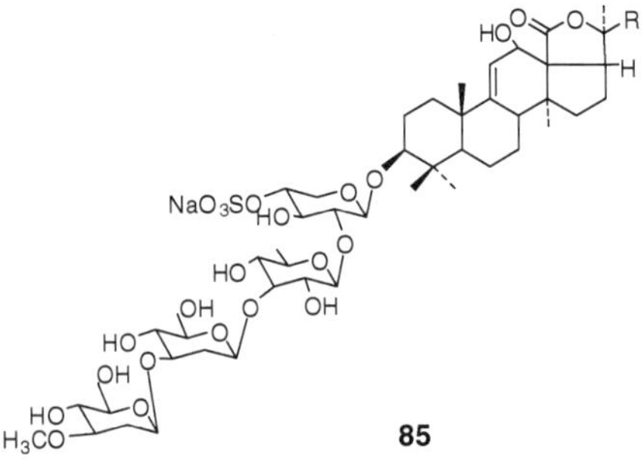

85

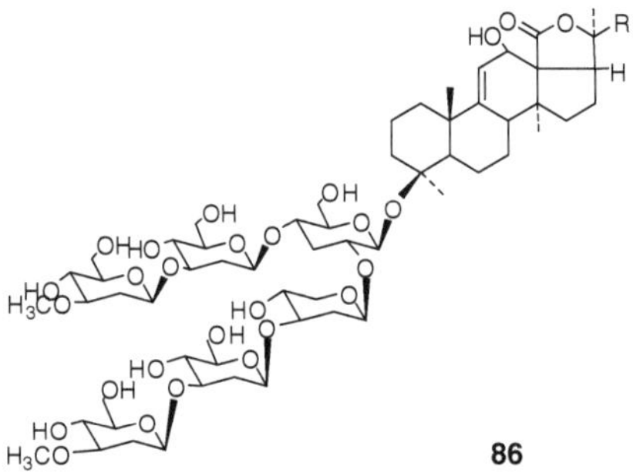

86

A very recent review [98] reports the antifungal peptides discovered in different marine invertebrates. Discoderimine A **87**, isolated from the sponge *Discodermia kilensis* as well as a number of congeners isolated from various sponges, showed anticandidal activity. Nonribosomal antifungal peptides have also been discovered from sponges, nudibranches and tunicates and are analyzed in-depth in the review by Fusetani [99].

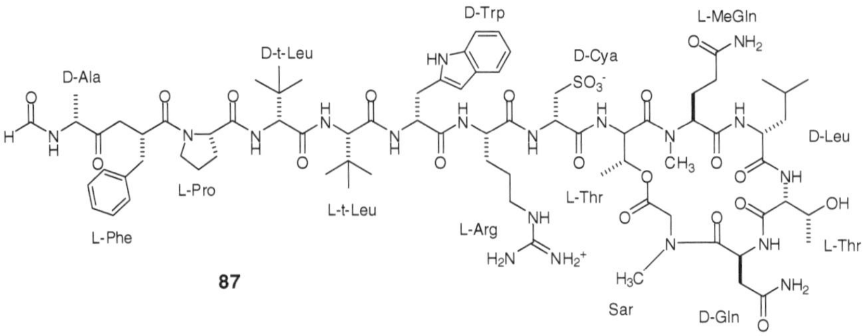

87

Marine microorganisms constitute another interesting source of antifungal compounds from marine environments. Liu et al. [97] reported the libraries existing from microorganisms of marine environment and the possibilities that they offer for the discovery of new microbial drugs.

As an example, two antifungal lipopeptides (**88**, **89**) have been isolated from *Bacillus amyloliquefaciens* SH B10 from deep sea sediment.

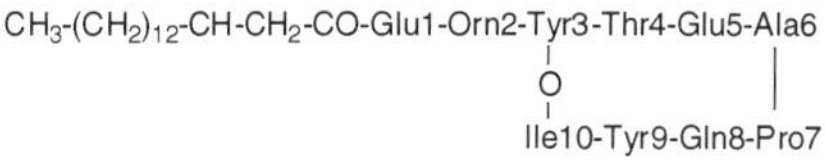

88

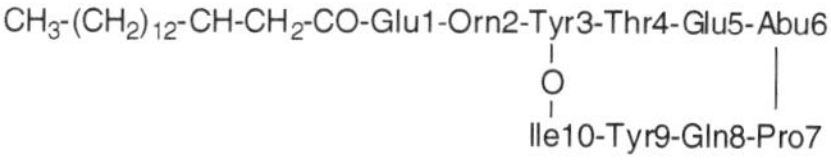

89

6.4 FROM ANTIFUNGAL COMPOUNDS TO ANTIFUNGAL DRUGS: SOME CONSIDERATIONS

The amount and type of compounds discussed above can be considered the tip of an iceberg compared to the whole literature dealing with antifungal compounds of natural origin. The question that arises is: if so many antifungal compounds have been discovered from natural sources, why is it so complicated to bring them to the market? [88,99].

Some of the reasons for this failure are that the discovered compounds are not potent enough or do not show the spectrum of action that make them attractive for further research. Other reason is that all compounds are not selective enough, also affecting mammalian cells, thus being undesirable hits. Other cause is that many compounds possess promising *in vitro* but weak *in vivo* activities, probably due to poor absorption, inactivation by serum binding, high clearance rates, and others.

So, how can one identify and select a lead compound for further study? According to Barrett [99], apart from a strong *in vitro* activity, the following aspects are important to take into account when selecting an antifungal lead: Is the compound structurally novel? Are the mechanisms of action original or potentially useful? Is a clinical proof of concept possible? Are chemical modifications of the structure possible?

As a consequence, new strategies are permanently designed with the aim of detecting drugable hits.

6.5 OTHER STRATEGIES BASED ON NON-TARGETED ASSAYS

6.5.1 Screening of Extracts or Natural Products in Combination with Other Compounds

Synergy-based research on natural products has been gaining approval in recent years. It is based on the premise that natural products can be more effective at a reduced dosage if low dosages of other synergistic compound are introduced simultaneously. This synergistic drug discovery approach will be consonant with a systems biology framework and complementary to the target-based approach. Synergy research has been recently reviewed by Wagner et al. in two comprehensive papers [100,101].

6.5.1.1 Measurement of Synergy. There are some *in vitro* tests to experimentally measure combination effect [102]. The most common and simple assay is the chequerboard method that consist in a two-dimensional array of serial concentrations (around their MICs) of the components of the combination, that is used as the basis for calculating the fractional inhibitory concentration index (FICI) which, according to its value, allows to predict the type of interaction of both compounds.

To calculate the FICI of each combination, fractional inhibitory concentration (FIC) of each component of the combination must be previously calculated. FIC is obtained from the relation between MIC of a compound in combination and alone. Then FICI is the sum of the FICs of both components of the mixture (Scheme 6.1).

If FIC_A and FIC_B are calculated for a series of two compounds A and B in different ratios, then an isobologram can be constructed using the two values as coordinates to express the FICI of each combination of the two compounds. If no synergism exists, the coordinate points form a straight line, but if synergism exists, the points occur in a concave shape. In contrast, if the point give a convex curve, antagonism exists [103]. Different authors interpret the FICI in different ways. According to Berembaum [104] a $FICI \leq 1$ is indicative of synergism, a $FICI = 1$ can be interpreted as no interaction (named additivism or indifference) and a $FICI > 1$, is indicative of antagonism. Instead, another authors suggested $FICI \leq 0.5$ for synergism, $FICI > 0.5$–4 indifference and >4, antagonism [105,106].

6.5.1.2 Different Combinations that have been Tested in the Search for Antifungal Drugs. Different combinations of natural products, such as components of an extract, mixtures of an extract, and a natural product, or combination of an extract or a natural product with a commercial antifungal drug have been tested in the last years within antifungal synergy research [107–110].

6.5.1.2.1 Combination of some components of the same extract or with Natural Products from Different Sources. Some examples showed the synergistic effects between two components of a same extract. For example, scopoletin, a hydroxycoumarin isolated from *Melia azedarach* showed antifungal synergistic effects with vanillin, 4-hydroxy-3-methoxycinnamaldehyde and pinoresinol, which have been previously isolated from the same species as antifungal compounds alone [107,108].

From each combination (A + B) →

$$FIC_A = \frac{MIC_{(A+B)}}{MIC_{(A\ alone)}}$$

$$FIC_B = \frac{MIC_{(A+B)}}{MIC_{(B\ alone)}}$$

→ $FICI = FIC_A + FIC_B$

SCHEME 6.1 Calculations of fractional inhibitory concentration (FIC) of each component on a given mixture and the fractional inhibitory concentration index (FICI) for the mixture.

Another approach is the effect of combination of natural products from different sources. In fact, components of the distillate of *Licaria puchuri-major*, such as anethole, safrole, methyleugenol, eugenol, limonene, and geraniol were combined with polygodial which has been, in turn, isolated from *D. winterii*, *P. hydropiper* or *P. punctatum* [109]. Results showed that the antifungal activity of polygodial was significantly increased when combined with phenolic derived-compounds that do not possess free phenolic groups.

A recent paper by Abdel-Rahman et al. [111] reported the results of twenty different two-components combinations of compounds isolated from *Moringa peregrina* against four fungal species of *Candida* genus (*C. albicans*, *C. krusei*, *C. parapsilopsis* and *C. tropicalis*) and of *S. cerevisiae*. Results showed that among the 100 combinations tested (20 combinations against five fungi), 18 synergistic combinations were found: 7 against each, *C. albicans, C. krusei, C. tropicalis* and *S. cerevisiae* and two against *C. parapsilopsis*.

6.5.1.2.2 Combination of Natural Samples with an Antifungal Drug. There have been interesting findings when mixtures of natural extracts and commercial antifungal drugs were tested, subject that has been reviewed by Hemiswarya et al. [112].

So, essential oils of *Allium sativum* [113] and *Agastache rugosa* [114] and the latex of *Euphorbia characias* [115] were found to act synergistically with ketoconazole, but each one against a different fungal species. In turn, *Santolina chamaecyparissus* [115] acts against *C. albicans* in combination with clotrimazole, and anethole against the same fungus with miconazole and AmB [117].

Several other recent reports describe the action of extracts and secondary metabolites in combination with antifungal agents [118–121].

An interesting report of Quan et al. [122] accounted that *in vitro* synergism was found between natural berberine chloride and fluconazole when 40 strains of resistant *C. albicans* were tested. In this paper, synergism was observed in 32 out of the 40 isolates (80%) while indifference was observed in 8 out of the 40 isolates.

6.5.1.3 High-Throughput Synergy Screening Method. Considering that the *in vitro* screening procedures for the search of combination activities using the chequerboard method is a time-consuming process, some efforts to speed-up the process have been intended, which led to a High Throughput Synergy Screening (HTSS) assay, that identifies metabolites as combination agents with antifungal drugs. The HTSS method was developed by Zhang et al. [123] for the detection of synergistic antifungal microbial extracts with ketoconazole (KTZ). Specifically, a microbial library of 20,000 extracts was screened for hits that synergize the effect of a low dosage of ketoconazole (KTZ) that alone shows little detectable fungicidal activity. Extracts screened in this HTS manner, gave 0.1% hit rate.

Fractionation of the most active extracts revealed beauvericin as the most potent component, because it dramatically synergized the KTZ activity against diverse fungal pathogens by the checkerboard assay.

To perform the assay, 96-wells microplates were used, according to the following scheme of Figure 6.7: (1) antifungal drug at an inhibitory concentration (higher than

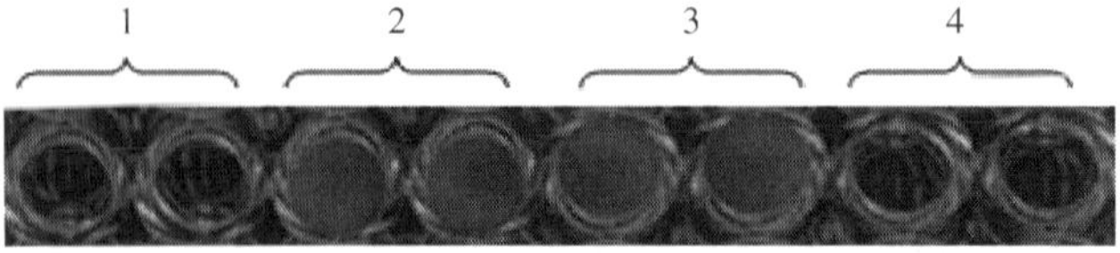

FIGURE 6.7 Eight of the 96-wells of a microtiter plate schematizing the high-throughput synergy screening method (HTSS) with plant extracts. (1) Antifungal drug at a concentration higher that its MIC (inhibition of the fungus is observed); (2) antifungal drug at a concentration lower that its MIC (no inhibition is observed); (3) plant extract at a concentration < MIC (no inhibition); (4) combination of (2) + (3) (inhibition). In the example, the expected result is shown in (4), were noninhibitory concentrations of each, the fungal drug and the extract, produce inhibition of the fungal growth [122].

MIC); (2) antifungal drug at a non-inhibitory concentration (<MIC); (3) plant extract at a non-inhibitory concentration (<MIC); 4) a combination of (2) and (3). A fungal inoculum was added to all wells [42]. Figure 6.7 (4) shows the expected results where the combination of both, the antifungal drug plus the plant extract (each at concentrations lower than their own MICs) inhibit the tested fungal strain, although each component of the mixture did not inhibit the fungus *per se* at those concentrations.

6.6 STRATEGIES BASED ON TARGETED ASSAYS FOR THE DISCOVERY OF ANTIFUNGAL COMPOUNDS

There is, at present, a great interest in detecting the mode of action of new, and previously described, antifungal compounds [6].

Classical targets for fungi have included the fungal plasma membrane (synthesis of ergosterol, phospholipids, and sphingolipids; H^+-ATPases; efflux pumps; channel formation by peptides), DNA and protein synthesis (DNA topoisomerases, DNA nucleases; RNA polymerase; elongation factors 2 and 3; myristoylation of proteins), the intermediary metabolism (nucleic acid, amino acid, and polyamine syntheses), virulence factors and, importantly enough, the fungal cell-wall (synthesis, cross-linking and degradation of glucan, chitin, and mannoproteins) [124]. There are many other targets that have been explored, such as sulfur fixation pathway [125], apoptotic cell-death responses in yeasts [126], and others.

Some recent reports dealing with the study on mechanisms of action of natural molecules are detailed below.

6.6.1 Natural Products Inhibitors of Fungal Cell-Wall Assembly or Synthesis

Lipopeptides isolated from *Pochonia bulbinosa* [127]; (+)-isoepoxydon, brefeldin A and ophiobolin A, isolated from fungi of the genera *Hemicarpenteles* and *Bipolaris* [128], glycolipids from Convolvulaceous species [129] and enfumafungin from an endophytic fungus of *Hormonema* genus are examples of inhibitors of (1,3)β-D-glucan synthase from natural sources.

In turn, catechins isolated from the stem bark of *Taxus cuspidata* [130]; ursolic acid from *Crataegus pinnatifida* [131] and tannins from *Euphorbia pekinensis* [132] are examples of inhibitors of chitin synthase [133]. Phytolaccoside B isolated from *Phytolacca tetramera* caused hyphal malformations due mainly to an enhancement of chitin synthase with concomitant accumulation of chitin [134].

6.6.2 Natural Products Inhibitors of the Fungal Cell Membrane

Steroidal saponins from *Solanum chrysotrichum* produced a loss of the cytoplasmic membrane continuity and, therefore, cellular death [135].

Iturins, peptides isolated from *Bacillus subtillus*, exert its antifungal activity through the formation of pores in the fungal cytoplasmic membrane [136,137].

Other mechanisms have also been investigated: azaphilones isolated from *Penicillium solitum* inhibited fungal, but not human, fatty acid synthase [138]. Plagiochin E, an antifungal active macrocycle, isolated from liverwort *Marchantia polymorpha*, induced apoptosis in *C. albicans* through a metacaspase-dependent apoptotic pathway [139]. Corilagin and chebulagic acid, isolated from the methanol extract of *Erodium stephanianum*, demonstrated to inhibit the topoisomerase I-dependent relaxation of plasmid DNA, mechanism different from that of camptothecin, a compound isolated from *Camptotheca acuminata*, which stabilizes the cleavage complex topoisomerase-DNA [140]. In turn, eupolaridine from species of the genus *Eupomatia*, demonstrated stabilization of the cleavage complex formed by fungal and mammalian type-I isomerase with DNA. The fungal topoisomerase I shows a greater response to eupolaridine than the human topoisomerase I, demonstrating that natural products have specificity for the fungal enzyme [141]. Polygodial, isolated from *P. punctatum*, *P. acuminatum*, *D. winterii* and other species, inhibited mitochondrial ATP synthesis, a few explored mechanism of action [129].

6.6.3 Natural Products Inhibitors of Virulence Factors

The secreted aspartic proteases (SAP) of *C. albicans* have been shown to be a major virulence factor in *Candida* infections. Therefore, the inhibition of SAP has been proposed as a new approach in the treatment of candidiasis. The bioassay-guided fractionation of the active chloroform fraction using the SAP assay led to the isolation of four new xanthones from the stem bark of *Tovomita krukovii* [142].

6.6.4 Discrimination of Modes of Action of Antifungal Substances by Use of Metabolic Foot Printing

To test each antifungal compound in the different targeted assays is a very labor- and time-consuming process.

Since one of the keys for the development of new antifungal drugs is the identification of its main target of action, some efforts have been made to detect the mechanism of antifungal action by using an innovative methodology. Allen et al. [143] have developed a technique where thc metabolic fingerprints of cells or tissues exposed to substances with known targets were analyzed using pattern recognition methods. In fact, *S. cerevisiae* cultures were exposed at first to antifungal drugs with different modes of action such as sterol biosynthesis, respiratory chain, or aminoacid synthesis. The metabolic "footprints" of cultures were analyzed by direct injection mass spectrometry. Principal Components Analysis (PCA), Discrimination Function Analysis (DFA), and Hierarchiral Cluster Analysis (HCA), carried out by using programs written in MATLAB (Mathworks, Cambridge, United Kingdom) allowed to determine that the greatest separation was between respiratory and nonrespiratory inhibitors and within the nonrespiratory branch, the sterol biosynthesis inhibitors formed one branch while the AA biosynthesis inhibitor and uncoupler formed another. The metabolomic fingerprinting of the same strains treated with other metabolites are proposed to be used for discriminating their modes of action.

6.6.5 New Targets Based on Genomics

The principles of genomic-based antifungal drug discovery begin when the complete DNA sequence of a fungal species has been determined. New and important data were obtained from the complete sequentiation of the genomes of *C. albicans* [144], *A. fumigatus* [145], *C. neoformans* [146] and *S. cerevisiae* [147]. Then, the sequence of fungal genes are compared with those of the human genome and fungal genes with a low degree of homology to human genes are listed as potentially fungus-specific targets of interest [22]. This approach can be extended to select genes that allow the finding of potential broad-spectrum antifungals by looking for those that are similar across several fungal species.

Once the list of genes of interest is compiled, the list is reduced for several reasons that include the availability of a HTS assay targeting the selected genes, which was reviewed by Spaltman et al. [148]. After the design of the HTS assay, new natural sources or existing natural products must be screened with this assay. Two review articles [149,150] explain the way in which genomic information can be translated into drug discovery.

Although genomics open a big avenue for the discovery of antifungal drugs, it is important to take into account that the rate of discovery of new antifungal agents has not perceptibly increased since the mid-1990s, when the first partial fungal genome sequence become available [22].

At present, most of the pharmaceutical companies are already using the information obtained from genomics and also from proteomics to find new targets for antifungal compounds [151]. Among them, Genome Therapeutics (www.genome-corp.com) is a leader in the commercialization of genomics-based drug discovery. This company has elaborated the PathoGenome™ database, which is one of the most comprehensive commercial sources of microbial genomic information available, including *Candida* species. Together with its strategic partners, genome therapeutics is using genomic information to develop a new generation of genomics-based antifungal compounds.

Elitra Pharmaceuticals is another pharmaceutical company that began to shift the traditional paradigm of screening targets singly, toward massive parallel screening of all possible drug targets directly in the pathogen of choice. To develop these technologies, Elitra established collaborations with several pharmaceutical companies [152]. They developed a rapid, large-scale target validation method, termed GRACE™ (Gene Replacement and Conditional Expression). According to Foulkes [154], in 2002 they have identified over 900 essential genes in *C. albicans* that encode antifungal drug targets and, in collaboration with celera genomics (Rockville, MD, USA), they completed the sequence of the second major human fungal pathogen, *A. fumigatus*, and were extending their functional genomics platform to this pathogen.

This enterprise hopes to collaborate with pharmaceutical companies that have potent compounds but that are unable to develop them in the absence of a known target.

6.7 CONCLUSION

The need of antifungal agents has grown up in recent years in response to the challenge of invasive mycoses. Great efforts have been made to detect new compounds with antifungal properties. Most scientists use non-targeted assays and almost all type of structures have demonstrated antifungal properties. Although many natural compounds with antifungal properties are permanently described, only very few ones have demonstrated to be good candidates for further research.

With the perception that traditional whole-fungus antifungal discovery had become unproductive, there has been a shift toward new strategies. One of them is the combination of compounds. Another one is the target-based search. Some of them base their work in genomics, identifying essential genes for the growth of fungi and afterwards, use these genes as targets to identify new antifungal compounds with a selective well-defined mode of action. The goal is the rapid target identification of the mechanism of action of highly active antifungal compounds.

All methodologies have their advantages and drawbacks. Clearly, a combination of technologies, old and new, will provide the optimum platform for antifungal discovery [22].

REFERENCES

[1] World Health Statistics 2009, World Health Organization, Geneva. Available at http://www.who.int/healthinfo/statistics.

[2] Mahady, G., Huang, Y., Doyle, B., Locklear, T., Natural products as antibacterial agents. In: Ur-Rahman, A. editor. *Studies in Natural Products Chemistry*,Vol. 35. Elsevier Science Publishers, Amsterdam, 2008, pp. 423–444.

[3] Brown, E., Wright, G. (2005). New targets and screening approaches in antimicrobial drug discovery. *Chemical Reviews*, *105*, 759–774.

[4] Detels, R., Emerging infectious diseases. In: Holland, W., Olsen J, Florey C. editors. *The Development of Modern Epidemiology*. Oxford University Press, Oxford, 2008, pp. 134–145.

[5] Heymann D., Emerging and re-emerging infections. In: Detels, R., Beaglehole, R., Lansang, M., Gulliford, M. editors. *Oxford Textbook of Public Health*. Oxford University Press, Oxford, 2009, pp. 1264–1276.

[6] Available at www.who.int/csr/don/2010_10_26

[7] Petrikkos, G., Skiada, A. (2007). Recent advances in antifungal chemotherapy, *International Journal of Antimicrobial Agents*, *30*, 108–117.

[8] Mathew, B., Nath, M. (2009). Recent approaches to antifungal therapy for invasive mycoses. *ChemMedChem*, *4*, 310–323.

[9] Monk, B., Goffeau, A. (2008). Outwitting multidrug resistance to antifungals. *Science*, *321*, 367–369.

[10] Weitzman, I., Summerbell, R. (1995). The dermatophytes. *Clinical Microbiology Reviews*, *8*, 240–259.

[11] Kontoyiannis, D., Mantadakis, E., Samonis, G. (2003). Systemic mycoses in the immunocompromised host: An update in antifungal therapy. *Journal of Hospital Infection*, *53*, 243–258.

[12] Freixa, B., Vila, R., Vargas, L., Lozano, N., Adzet, T., Cañigueral, S. (1998). Screening for antifungal activity on nineteen Latinoamerican plants. *Phytotherapy Research*, *10*, 427–431.

[13] Leonard, W., Belyk, K., Conlon, D., Bender, D., Di Michele, L., Liu, J., Hughes, D. (2007). Synthesis of the antifungal β-1,3-glucan synthase inhibitor CANCIDAS (Caspofungin acetate) from Pneumocandin B. *Journal of Organic Chemistry*, *72*, 2335–2343.

[14] Steinberg, T., Wright, E., Oura, M. (1955). A new antifungal antibiotic, amphotericin B. *Antibiotics Annual*, *3*, 566–573.

[15] Gold, W., Stout, H., Pagano, J., Donovic, R. (1955). Amphotericins A and B, antifungal antibiotics produced by a streptomycete: I. *In vitro* studies. *Antibiotics Annual*, *3*, 579–586.

[16] Brian, P., Curtis, P., Hemming, H. (1949). A substance causing abnormal developmet of fungal hyphae produced by *Penicillium janczewskii* Zal: III. Identity of 'curling factor' with griseofulvin. *Transactions of the British Mycological Society*, *32*, 30–33.

[17] Peláez, F., Cabello, A., Platas, G., Diez, M., González del Val, A., Martán, I., Vicente, F., Bills, G., Giacobbe, R., Schwartz, R., Onishi, J., Meinz, M., Abbruzzo, G., Flattery, A., Kong, L., Kurtz, M. (2000). The discovery of enfumafungin, a novel antifungal compound produced by an endophytic *Hormonema* spp biological activity and taxonomy of the producing organisms. *Systematics and Applied Microbiology*, *23*, 333–343.

[18] Espinel-Ingroff, A. (2009). Novel antifungal agents, targets or therapeutic strategies for the treatment of invasive fungal diseases: A review of the literature (2005–2009). *Revista Iberoamericana de Micología*, *26*, 15–22.

[19] Mukherjee, P., Leidich, S., Isham, N., Leitner, I., Ryder, N., Ghannoum, M. (2003). Clinical *Trichophyton rubrum* strain exhibiting primary resistance to terbinafine. *Antimicrobial Agents and Chemotherapy*, *47*, 82–86.

[20] Tulp, M., Bohlin, L. (2004). Unconventional natural sources for future drug discovery. *Drug Discovery Today*, *9*, 450–458.

[21] Harvey, A. (2008). Natural products in drug discovery. *Drug Discovery Today*, *13*, 894–890.

[22] Odds, F. (2005). Genomics, molecular targets and the discovery of antifungal drugs. *Revista Iberoamericana de Micología*, *22*, 229–237.

[23] Cos, P., Vlietink, A., Vanden Berghe, D., Maes, L. (2006). Anti-infective potential of natural products: how to develop a stronger *in vitro* "proof-of-concept". *Journal of Ethnopharmacology*, *106*, 290–302.

[24] Maregesi, S., Pieters, L., Ngassapa, O., Apers, S., Vigerhoets, R., Cos, P., Vanden Berghe, D., Vlietink, A. (2008). Screening of some Tanzanian medicinal plants from Bunda district for antibacterial, antifungal and antiviral activities. *Journal of Ethnopharmacology*, *119*, 58–66.

[25] Rouhi M. (2003). Rediscovering natural products. *Chemistry and Engineering News*, *81*, 77–91.

[26] Ríos, J., Recio, M. (2005). Medicinal plants and antimicrobial activity. *Journal of Ethnopharmacology*, *100*, 80–84.

[27] Svetaz, L., Zuljan, F., Derita, M., Petenatti, E., Tamayo, G., Cáceres, A., Cechinel-Filho, V., Giménez, A., Pinzón, R., Zacchino, S., Gupta, M. (2010). Value of the ethnomedical information for the discovery of plants with antifungal properties: A survey among seven Latin American countries. *Journal of Ethnopharmacology*, *127*, 137–158.

[28] Alonso Paz, E., Cerdeiras, M., Fernández, J., Ferreira, F., Moyna, P., Soubes, M., Vázquez, A., Vero, S., Zunino, L. (1995). Screening of Uruguayan medicinal plants for antimicrobial activity. *Journal of Ethnopharmacology*, *45*, 67–70.

[29] Portillo, A., Vila, R., Freixa, B., Adzet. T., Cañigueral, S. (2001). Antifungal activity of Paraguayan plants used in traditional medicine. *Journal of Ethnopharmacology*, *76*, 93–98.

[30] Vila, R., Santana, A., Pérez-Roser, R., Valderrama, A., Castelli, M., Mendonca, S., Zacchino, S., Gupta, M.P., Cañigueral, S. (2010). Composition and biologícal activity of the essentials oil from leaves of *Plinia cerrocampanensis*, a new source of α-bisabolol. *Bioresourse Technology*, *10*, 2510–2514.

[31] Muñoz, J., Cruañes, A., Vivot, E., Cruañes, M., Ruiz, S., Ruiz, C., Salinas, A., Santecchia, C., López, S., Gattuso, S., Zacchino, S. (1998). *In vitro* antifungal evaluation and studies on mode of action of eight selected species from the Argentinian Flora. *Phytomedicine*, *585*, 389–395.

[32] Fenner, R., Sortino, M., Kuze Rates, S., Dall'Agnol, R., Ferraz, A., Bernardi, A., Albring, A., Nör, C., von Poser, G., Schapoval, E., Zacchino, S. (2005). Antifungal activity of some Brazilian *Hypericum* spp. *Phytomedicine*, *12*, 236–240.

[33] Alanís-Garza, B., González-González, G., Salazar-Aranda, R., Waksman de Torres, N., Rivas-Galindo, V. (2007). Screening of antifungal activity of plants, from the northeast of Mexico. *Journal of Ethnopharmacology*, *114*, 468–471.

[34] Cáceres, A., López, B., Giron, M., Logemann, H. (1991). Plants used in Guatemala for the treatment of dermatophytic infections: I. Screening for antimycotic activity of 44 plant extracts. *Journal of Ethnopharmacology*, *31*, 263–276.

[35] Motsei, M., Lindsey, K., van Staden, J., Jager, A. (2003). Screening of traditionally used South African plants for antifungal activity against *Candida albicans*. *Journal of Ethnopharmacology*, *86*, 235–241.

[36] McCutcheon, A., Ellis, S., Hancock, R., Towers, G. (1994). Antifungal screening of medicinal plants of British Columbian native peoples. *Journal of Ethnopharmacology*, *44*, 157–169.

[37] Mistcher, L., Leu, R., Bathala, M., Wu, W., Beal, J. (1972). Antimicrobial agents from higher plants: I. Introduction, rationale and methodology. *Lloydia*, *35*, 157–166.

[38] Coelho de Souza, G., Haas, A., von Poser, G., Schapoval, E., Elisabetsky, E. (2004). Ethnopharmacological studies of antimicrobial remedies in the south of Brazil. *Journal of Ethnopharmacology*, *90*, 135–143.

[39] Jones, N., Arnason, J., Abou-Zaid, M., Akpagana, K., Sanchez-Vindas, P., Smith, M. (2000). Antifungal activity of extracts from medicinal plants used by First Nations Peoples of eastern Canada. *Journal of Ethnopharmacology*, *73*, 191–198.

[40] Cruz, M., Santos, P., Barbosa, A. Jr., de Mélo, D., Alviano, C., Antoniolli, A., Alviano, D., Trindade, R. (2007). Antifungal activity of Brazilian medicinal plants involved in popular treatments of mycoses. *Journal of Ethnopharmacology*, *111*, 409–412.

[41] Braga, F., Bouzada, M., Fabri, R., Matus, M., Moreira, F., Scio, E., Coimbra, E. (2007). Antileishmanial and antifungal activity of plants used in traditional medicine in Brazil. *Journal of Ethnopharmacology*, *111*, 396–402.

[42] CLSI, Clinical and Laboratory Standards Institute, formerly National Committee for Clinical and Laboratory Standards (NCCLS). Reference methods for broth dilution antifungal susceptibility testing of yeasts and filamentous fungi; Approved Standards. document M27-A2 (2002) and A3 (2008), M38 A (2002) and M38-A2 (2008), Wayne, PA.

[43] Cowan, M. (1999). Plant products as antimicrobial agents. *Clinical Microbiology Reviews*, *12*, 564–582.

[44] Abad, M., Ansuategui, M., Bermejo, P. (2007). Active antifungal substances from natural sources, *ARKIVOC*, *7*, 116–145.

[45] Akendengué, B., Champy, P., Nzamba, J., Roblot, F., F., Loiseau, P., Bories, C. (2009). Antifungal and anthelmintic activities of *Cleistopholis patens* (Annonaceae). *Planta Medica*, *75*, 1143–1145.

[46] Hufford, C., Liu, S., Clark, A. (1987). Anticandidal activity of eupolaridine and onychine alkaloids from *Cleistopholis patens*. *Journal Natural Products*, *50*, 961–964.

[47] Liu, S., Oguntimen, B., Hufford, C., Clark, A. (1990). 3-methoxysampangine, a novel antifungal copyrine alkaloid from *Cleistopholis patens*. *Antimicrobial Agents and Chemotherapy*, *34*, 529–533.

[48] Lago, J. Chaves, M., Cruz, M., Agripino, D., Young, M. (2007). Evaluation of antifungal and DNA-damaging activities of alkaloids from branches of *Porcelia macrocarpa*. *Planta Medica*, *73*, 292–295.

[49] Nenaah, G. (2010). Antibacterial and antifungal activities of (beta)-carboline alkaloids of *Peganum harmala* L. seeds and their combination effects. *Fitoterapia*, *81*, 779–782.

[50] Al-Shamma, A., Drake, S., Flynn, D., Mitscher, L., Park, Y., Rao, G., Simpson, A., Swayze, J., Veysoglu, T., Wu, S. (1981). Antimicrobial agents from higher plants. Antimicrobial agents from *Peganum harmala* seeds. *Journal of Natural Products*, *44*, 745–747.

[51] Rao, G., Zhang, S., Wang, H., Li, Z., Gao, S., Xu, G. (2009). Antifungal alkaloids from the fresh rattan stem of *Fibraurea recisa* Pierre. *Journal of Ethnopharmacology*, *123*, 1–5.

[52] Freile, M., Giannini, F., Sortino, M., Zamora, M., Juarez, A., Zacchino, S., Enriz, R. (2006). Antifungal activity of aqueous extracts and of berberine isolated from *Berberis heterophylla*. *Acta Farmaceutica Bonaerense*, *25*, 83–88.

[53] Slobodníková, L., KoSt'álová, D., Labudová, D., Kotulová, D., Kettmann, V. (2004). Antimicrobial activity of *Mahonia aquifolium* crude extract and its major isolated alkaloids. *Phytotherapy Research*, *18*, 674–676.

[54] Zhang, J., Gao, J., Xu, T., Zhang, X., Ma, Y., Jarussophon, S., Konishi, Y. Antifungal activity of alkaloids from the seeds of *Chimonanthus praecox*. *Chemistry and Biodiversity*, *6*, 838–845.

[55] Simons, V., Morrissey, J., Latijnhouwers, M., Csukai, M. (2006). Dual effects of plant steroidal alkaloids on *Saccharomyces cerevisiae*. *Antimicrobial Agents and Chemotherapy*, *50*, 2732–2740.

[56] Saikia, D., Khanuja, S., Kahol, A., Gupta, S., Kumar, S. (2001). Comparative antifungal activity of essential oils and constituents from three distinct genotypes of *Cymbopogon* spp. *Current Science*, *80*, 1264–1266.

[57] D'Auria, F., Tecca, M., Strippoli, V., Salvatore, G., Battinelli L, Mazzanti G. (2005). Antifungal activity of *Lavandula angustifolia* essential oil against *Candida albicans* yeast and mycelial form. *Medical Mycology*, *43*, 391–396.

[58] Saddiq, A., Khayyat, S. (2010). Chemical and antimicrobial studies of monoterpene: Citral. *Pesticide Biochemistry and Physiology*, *98*, 89–93.

[59] Voda, K., Boh, B., Vrtačnik, M. (2004). A quantitative structure–antifungal activity relationship study of oxygenated aromatic essential oil compounds using data structuring and PLS regression analysis. *Journal of Molecular Modeling*, *10*, 76–84.

[60] Ettayebi, K., El Yamani, J., Rossi-Hassani, B. (2000). Synergistic effects of nisin and thymol on antimicrobial activities in *Listeria monocytogenes* and *Bacillus subtilis*. *FEMS Microbiology Letters*, *183*, 191–195.

[61] Ultee, A., Kets, E., Smid, E. (1999). The phenolic hydroxyl group of carvacrol is essential for action against the food-borne pathogen *Bacillus cereus*. *Applied Environment Microbiology*, *65*, 4606–4610.

[62] Helander, I., Alakomi, H., Latva-Kala, K., Mattila-Sandholm, T., Pol, I., Smid, E., Gorris, L., von Wright, A. (1998). Characterization of the action of selected essential oil components on Gram-negative bacteria. *Journal of Agricultural and Food Chemistry*, *46*, 3590–3595.

[63] Kim, J., Marshall, M., Wei, C. (1995). Antibacterial activity of some essential oil components against five foodborne pathogens. *Journal of Agricultural and Food Chemistry*, *43*, 2839–2845.

[64] Zemek, J., Valent, M., Pódová, M., Košíková, B., Joniak, D. (1987). Antimicrobial properties of aromatic compounds of plant origin. *Folia Microbiologica*, *32*, 421–425.

[65] Boonchird C., Flegel T. (1982). *In vitro* antifungal activity of eugenol and vanillin against *Candida albicans* and *Cryptococcus neoformans*. *Canadian Journal of Microbiology*, *28*, 1235–1241.

[66] Lee, S., Lee, J., Lunde, C., Kubo, I. (1999). *In vitro* antifungal susceptibilities of *Candida albicans* and other fungal pathogens to polygodial, a sesquiterpene dialdehyde. *Planta Medica*, *65*, 204–208.

[67] Almeida T., Lacerda Ribeiro, F., Kloos, H., Zani C. (2001). Polygodial, the fungitoxic component from the Brazilian medicinal plant *Polygonum punctatum*, *Memórias do Instituto Oswaldo Cruz*, *96*, 831–833.

[68] Malheiros, A., Cechinel Filho, V., Schmitt, C., Yunes, R., Escalante, A., Svetaz, L., Zacchino, S., Delle Monache F. (2005). Antifungal activity of drimane sesquiterpenes

from *Drimys brasiliensis* using bioassay-guided fractionation. *Journal of Pharmacy and Pharmaceutical Sciences*, *8*, 335–339.

[69] Derita, M., Leiva M., Zacchino, S. (2009). Influence of plant part, season of collection and content of the main active constituent, on the antifungal properties of *Polygonum acuminatum* Kunth. *Journal of Ethnopharmacology*, *124*, 377–383.

[70] Giesbrecht, A., Davino, S., Nassis, C., Young, M., Lopes, J., Rodrigues, D., Vichnewski, Nasi, A., Dias, D., López, J. (1990). Antimicrobial activity of sesquiterpene lactones. *Química Nova*, *13*, 312–314.

[71] Erasto, P., Grierson, D., Afolayan, A. (2006). Bioactive sesquiterpene lactones from the leaves of *Vernonia amygdalina*. *Journal of Ethnopharmacology*, *106*, 117–120.

[72] Portillo, A., Vila, R., Freixa, B., Ferro, E., Parella, T., Casanova, J., Cañigueral, S. (2005). Antifungal sesquiterpene from the root of *Vernonanthura tweedieana*. *Journal of Ethnopharmacology*, *106*, 49–52.

[73] Djeddi, S., Karioti, A., Sokovic, M., Stojkovic, D., Seridi, R., Skaltsa, H. (2007). Minor sesquiterpene lactones from *Centaurea pullata* and their antimicrobial activity. *Journal of Natural Products*, *70*, 1796–1799.

[74] Yang, C., Zhang, Y., Jacob, M., Khan, S., Zhang, Y., Li., X. (2006). Antifungal activity of C-27 steroidal saponins. *Antimicrobial Agents and Chemotherapy*, *50*, 1710–1714.

[75] Sparg, S., Light, M., van Staden, J. (2004). Biological activities and distribution of plant saponins. *Journal of Ethnopharmacology*, *94*, 219–243.

[76] Favel, A., Steinmetz, M., Regli, P., Vidal-Ollivier, E., Elias, R., Balansard, G. (1994). *In vitro* antifungal activity of triterpenoid saponins. *Planta Medica*, *60*, 50–53.

[77] Escalante, A., Santecchia, C., López, S., Gattuso, M., Gutiérrez, A., Delle Monache, F., González Sierra, M., Zacchino, S. (2002). Isolation of antifungal saponins from *Phytolacca tetramera*, an Argentinean species in critic risk. *Journal of Ethnopharmacology*, *82*, 29–34.

[78] Di Liberto, M., Svetaz, L., Furlán, R., Zacchino, S., Delporte, C., Novoa, M., Asencio, M., Cassels, B. (2010). Antifungal activity of saponin-rich extracts of *Phytolacca dioica* L. and of the sapogenins obtained through hydrolysis. *Natural Products Communications*, *5*, 1013–1018.

[79] Sautour, M., Mitaine-Offer, A.-C., Miyamoto, T., Dongmo, A., Lacaille-Dubois, M. (2004). Antifungal steroid saponins from *Dioscorea cayanensis*. *Planta Medica*, *70*, 90–92.

[80] He, X., Moceck, U., Floss, H., Cáceres, A., Girón, L., Buckley, H., Cooney, G., Manns, J., Wilson, B. (1994). An antifungal compound from *Solanum nigrescens*. *Journal of Ethnopharmacology*, *43*, 173–177.

[81] Zamilpa, A., Tortoriello, J., Navarro, V., Delgado, G., Alvarez, L. (2002). Five new steroidal saponins from *Solanum chrysotrichum* leaves and their antimycotic activity. *Journal of Natural Products*, *65*, 1815–1819.

[82] Silva, I., Raimondi, M., Zacchino, S., Cechinel-Filho, V., Noldin, V., Rao, V., Silva, R., Martins, D. (2010). Evaluation of the antifungal activity and mode of action of *Lafoensia pacari* St. Hill. (Lythraceae) stem bark extracts, fractions and ellagic acid. *Brazilian Journal of Pharmacognosy*, *20*, 422–428.

[83] Kurdelas, R., Tapia, A., Lima, B., Feresin, G., González Sierra, M., Rodríguez, M., Zacchino, S., Enriz, R., Freile, M. (2010). Antifungal activity of extracts and prenylated

coumarins isolated from *Baccharis darwinii* Hook & Arn. (Asteraceae), *Molecules*, *15*, 4898–4907.

[84] Pacciaroni, A., Gette, M., Derita, M., Ariza Espinar, L., Gil, R., Zacchino, S., Silva, G. (2008). Antifungal activity of *Heterothalamus alienus* metabolites. *Phytotherapy Research*, *22*, 524–528.

[85] Schmeda-Hirschmann, G., Tapia, A., Feresin, G., Zacchino, S., Sortino, M., Pertino, M., Rojas, A. (2008). A new antifungal and antiprotozoal depside from the Andean lichen *Protousnea poepigii*. *Phytotherapy Research*, *22*, 349–355.

[86] Agüero, M., González Sierra, M., Lima, B., Svetaz, L., Sánchez, M., Zacchino, S., Feresin, G., Schmeda-Hirschmann, G., Palermo, J., Wunderlin, D., Tapia, A. (2010). Argentinean propolis from *Zuccagnia punctata* Cav. exudates: Phytochemical characterization and antifungical activity. *Journal of Agricultural and Food Chemistry*, *58*, 194–201.

[87] Zacchino, S., Rodríguez, G., Pezzenati, G., Orellana, G., Enriz, R., González Sierra, M. (1997). *In vitro* evaluation of antifungal properties of neolignans 8.O.4′. *Journal of Natural Products*, *60*, 659–662.

[88] Vicente, F., Basilio, A., Cabello, A., Peláez, F. (2003). Microbial natural products as a source of antifungals. *Clinical Microbology and Infectious Diseases*, *9*, 15–32.

[89] Strobel, G, Daisy, B., Castillo, Harper, J. (2004). Natural products from endophytic microorganisms. *Journal of Natural Products*, *67*, 257–268.

[90] Mayer, A. Rodríguez, A., Berlinck, R., Hamann, M. (2009). Marine pharmacology in 2005–6: Marine compounds with anthelmintic, antibacterial, anticoagulant, antifungal, anti-inflammatory, antimalarial, antiprotozoal, antituberculosis, and antiviral activities; affecting the cardiovascular, immune and nervous systems, and other miscellaneous mechanisms of action. *Biochimica Biophysica Acta*, *1790*, 283–308.

[91] Sionov, E., Roth, D., Sandovsky-Losica, H., Kashman, Y., Rudi, A., Chill, A., Berdicevsky, I., Segal, E. (2005). Antifungal effect and possible mode of activity of a compound from the marine sponge *Dysidea herbacea*. *Journal of Infection*, *50*, 453–460.

[92] Pettit, R., McAllister, S., Pettit, G., Herald, C., Johnson, M., Cichacz, Z. (1998). A broad-spectrum antifungal from the marine sponge *Hyrtios erecta*. *International Journal of Antimicrobial Agents*, *9*, 147–152.

[93] Kumar, R., Chaturvedi, A., Shukla, P., Lakshmi, V. (2007). Antifungal activity in triterpene glycosides from the sea cucumber *Actinopyga lecanora*. *Bioorganic and Medicinal Chemistry Letters*, *17*, 4387–4391.

[94] Yuan, W., Yi, Y., Tang, R., Wang, Z., Sun, G., Xue, M., Zhang, W., Tang, H. (2009). Antifungal triterpene glycosides from the sea cucumber *Holothuria (Microthele) axiloga*. *Planta Medica*, *75*, 647–653.

[95] Encarnación, R., Murillo, I., Christophersen, C., Chan-Bacab, M., García, R., Zacchino, S. (2006) Leishmanicidal, antifungal, and cytotoxic activity of triterpenoid glycosides isolated from the sea cucumber *Neothyone gibbosa*. *Natural Products Communications*, *1*, 541–547.

[96] Yuan, W., Yi, Y., Tang, H., Liu, B., Wang, Z., Sun, G., Zhang, W., Li, L. (2009). Antifungal triterpene glycosides from the sea cucumber *Bohadschia marmorata*. *Planta Medica*, *75*, 168–173.

[97] Liu, X., Ashforth, E., Ren, B., Song, F., Dai, H., Liu, M., Wang, J., Xie, Q., Zhang, L. (2010). Bioprospecting microbial natural products libraries from the marine environment for drug discovery. *Journal of Antibiotics*, *63*, 415–422.

[98] Fusetani, N. (2010). Antifungal peptides in marine invertebrates. *Invertebrate Survival Journal*, *7*, 53–66.

[99] Barrett, D. (2002). From natural products to clinically useful antifungals. *Biochimica Biophysica Acta*, *1587*, 224–233.

[100] Wagner, H., Ulrich-Merzenich, G. (2009). Synergy research. Approaching a new generation of phytopharmaceuticals. *Phytomedicine*, *16*, 97–110.

[101] Ulrich-Merzenich, G., Wagner, H., Panek, D., Zeitler, H., Wagner, H., Vetter, H. (2009). New perspectives for synergy research with the "omic"-technologies. *Phytomedicine*, *16*, 495–508.

[102] Odds, F. (2003). Synergy, antagonism and what the chequerboard puts between them. *Journal of Antimicrobial Chemotherapy*, *52*, 1–1.

[103] Houghton, P., Synergy and polyvalence: Paradigms to explain the activity of herbal products. In: Houghton, P., Mukherjee, P. editors. *Evaluation of Herbal Medicine Products*. Pharmaceutical Press, London, 2009, pp. 85–94.

[104] Berembaum, M. (1989). What is synergy? *Pharmacological Reviews*, *41*, 93–141.

[105] Biavatti, M. (2009). Synergy, a new paradigm, for pharmacotherapy. *Brazilian Journal of Pharmaceutical Sciences*, *45*, 371–378.

[106] Borisy, A., Elliot, P., Hurst, M., Lehar, M., Price, R., Serbedzija, G., Zimmermann, G., Foley, M., Stockwell, B., Keith, C. (2003). Systematic discovery of multicomponent therapeutics. *Proceedings of the National Academy of Sciences*, 100, 7977–7982.

[107] Carpinella, M., Giorda, L., Palacios, S. (2003). Antifungal effects of different extracts from *Melia azedarach* L. on phytopathogenic fungi and their isolated active components. *Journal of Agricultural and Food Chemistry*, *51*, 2506–2511.

[108] Carpinella, M., Ferrayoli, C., Palacios, S. (2005). Antifungal synergistic effect of scopoletin, a hydroxycoumarin isolated from *Melia azedarach* L. fruits. *Journal of Agricultural and Food Chemistry*, *53*, 2922–2927.

[109] Himejima, M., Kubo, I. (1992). Antimicrobial agents from *Licaria puchuri-major* and their synergistic effect with polygodial. *Journal of Natural Products*, *55*, 620–625.

[110] Kubo, I., Muroi, H., Himejima, M. (1993). Combination effects of antifungal nagilactones against *C. albicans* and two other fungi with penylpropanoids. *Journal of Natural Products*, *56*, 220–226.

[111] Abdel-Rahman, M., Hegazy, A., Mohsen Sayed, A., Kabiel, H., El-Alfy, T., El-Komy, S. (2010). Study on combined antimicrobial activity of some biologically active constituents from wild *Moringa peregrina* Forssk. *Journal of Yeast and Fungal Research*, *1*, 15–24.

[112] Hemiswarya, S., Kruthiventi, A., Doble, M. (2008). Synergism between natural products and antibiotics against infectious diseases. *Phytomedicine*, *15*, 639–652.

[113] Pyun, M., Shin, S. (2006). Antifungal effects of the volatile oils from *Allium* plants against *Trichophyton* species and synergism of the oils with ketoconazole. *Phytomedicine*, *13*, 394–400.

[114] Shin, S., Kang, C. (2003). Antifungal activity of the essential oils of *Agastache rugosa* Kuntze and its synergism with ketoconazole. *Letters in Applied Microbiology*, *36*, 111–115.

[115] Giordani, R., Trebaux, J., Masi, M., Regli, P. (2001). Enhanced antifungal activity of ketoconazole by *Euphorbia characias* latex against *Candida albicans*. *Journal of Ethnopharmacology*, *78*, 1–5.

[116] Suresh, B., Dhanaraj, S., Elangosgiram, K., Chinaswammy, K. (1997). Anticandidal activity of *Santolina chamaecyparissus* volatile oil. *Journal of Ethnopharmacology*, *55*, 151–159.

[117] Lee, S., Kin, C. (1999). Selective combination effect of anethole to antifungal activities of miconazole and amphotericin B. *Yakhak Hoeji*, *43*, 228–232.

[118] Hirasawa, M., Takada, K. (2004). Multiple effects of green tea catechin on the antifungal activity of antimycotics against *Candida albicans*. *Journal of Antimicrobial Chemotherapy*, *53*, 225–229.

[119] Ogita, A., Tanaka, T. (2009). Enhancement of the fungicidal activity of Amphotericin B by allicin: Effects on intracellular ergosterol trafficking. *Planta Medica*, *75*, 222–226.

[120] Mahboubi, M., Ghazian Bidgoli, F. (2010). *In vitro* synergistic efficacy of combination of amphotericin B with *Myrtus communis* essential oil against clinical isolates of *Candida albicans*. *Phytomedicine*, *17*, 771–774.

[121] Giordani, R., Gachon, C., Buc, J., Regli, P., Jacob, J. (1999). Antifungal action of *Hevea brasiliensis* latex. Its effect in combination with fluconazole on *Candida albicans* growth. *Mycoses*, *42*, 465–474.

[122] Quan, H., Cao, Y.-Y., Xu, Z., Zhao, J., Gao, P., Qin, X., Jiang, Y. (2006). Potent *in vitro* synergism of fluconazole and berberine chloride against clinical isolates of *Candida albicans* resistant to fluconazole. *Antimicrobial Agents and Chemotherapy*, *50*, 1096–1099.

[123] Zhang, L., Yan, K., Zhang, Y., Huang, R., Bian, J., Zheng, Ch., Sun, H., Chen, Z., Sun, N., An, R., Min, F., Zhao, W., Zhuo, Y., You, J., Song, Y., Yu, Z., Liu, Z., Yang, K., Gao, H., Dai, H., Zhang, X., Wang, J., Fu, C., Pei, G., Liu, J., Zhang, S., Goodfellow, M., Jiang, Y., Kuai, J., Zhou, G., Chen, X. (2007). High throughput synergy screening identifies microbial metabolites and antiviral activities. *Proceedings of the National Academy of Sciences*, 104, 4606–4611.

[124] Groll, A., De Lucca, A., Walsh, T. (1998). Emerging targets for the development of novel antifungal therapeutics. *Trends in Microbiology*, *6*, 117–124.

[125] Aoki, Y., Kondoh, M., Nakamura, M., Fujii, T., Yamazaki, T., Shimada, H., Arisawa, M. (1994). A new methionine antagonist that has antifungal activity: Mode of action. *Journal of Antibiotics*, *47*, 909–916.

[126] Wu, X., Chang, W., Cheng, A., Sun, L., Lou, H. (2010). Plagiochin E, an antifungal active macrocyclic bis, induced apoptosis in *Candida albicans* through a metacaspase-dependent apoptotic pathway. *Biochimica Biophysica Acta*, *1800*, 439–447.

[127] Koehn, F., Kirsch, D., Feng, X., Janso, J., Young, M. (2008). A cell wall-active lipopeptide from the fungus *Pochonia bulbillosa*. *Journal of Natural Products*, *71*, 2045–2048.

[128] Fukushima, Y., Sakagami, Y., Marumo, S. (1993). β-glucan biosynthesis inhibitors isolated from fungi as hyphal malformation inducer. *Bioorganic and Medicinal Chemistry Letters*, *3*, 1219–1222.

[129] Castelli, M., Cortés, J., Escalante, A., Bah, M., Pereda-Miranda, R., Ribas, J., Zacchino, S. (2002). *In vitro* inhibition of 1,3-β-glucan synthase by glycolipids from Convolvulaceous species. *Planta Medica*, *68*, 739–742.

[130] Kim, S., Hwang, E., Nam., J., Son, K., Bok, S., Kim, H., Kwan, B. (1999). Inhibition of chitin synthase II by catechins from stem bark of *Taxus cuspidata*. *Planta Medica*, *65*, 97–98.

[131] Jeong., T., Hwang, E., Lee, H., Lee, E., Kim, Y., Min, B., Bae, K., Bok, S., Kim, S. (1999). Chitin synthase II inhibitory activity of ursolic acid, isolated from *Crataegus pinnatifida*. *Planta Medica*, *65*, 261–263.

[132] Hwang, E., Ahn, B., Lee, H., Kim, Y., Lee, K., Bok, S., Kim, Y., Kim, S. (2001). Inhibitory activity for chitin synthase II from *Saccharomyces cerevisiae* by tannins and related compounds. *Planta Medica*, *67*, 501–504.

[133] Ruiz, J., San Blas, G. (2003). Chitin synthesis as a target for antifungal drugs. *Current Drugs Target - Infections Disorders*, *3*, 77–91.

[134] Escalante, A., Gattuso, M., Pérez, P. Zacchino, S. (2008). Evidence for the mechanism of action of antifungal Phytolaccoside B isolated from *Phytolacca tetramera* Hauman. *Journal of Natural Products*, *71*, 1720–1725.

[135] López-Villegas, E., Herrera-Arellano, A., Martínez-Rivera, M., Álvarez, L., Cano-Napauseno, M., Marquina, S., Rodríguez-Tovar, A., Tortoriello, J. (2009). Ultraestructural changes on clinical isolates of *Trichophyton rubrum*, *T. mentagrophytes* and *M. gypseum* caused by *Solanum chysotrichum* saponin SC-2. *Planta Medica*, *75*, 1517–1520.

[136] Maget, R., Peypoux, F. (1994). Iturins, a special class of pore-forming lipopeptides: Biological and a physicochemical properties. *Toxicology*, *87*, 151–174.

[137] Latoud, C., Peypoux, F., Michel, G. (1987). Action of iturin A, an antifungal antibiotic from *Bacillus subtilis*, on the yeast *Saccharomyces cerevisiae*: Modifications of membrane permeability and lipid composition. *Journal of Antibiotics*, *11*, 1588–1595.

[138] Laakso, J., Raulli, R., McElhaney, G., Actor, P., Underiner, T., Hotovec, B., Mocek, U., Cihlar, R., Broedel, S. (2003). CT2108A and B: New fatty acids synthase inhibitors as antifungal agents. *Journal of Natural Products*, *66*, 1041–1046.

[139] Hecht, S., Berry, D., Mackenzie, L., Busby, R., Nasuti, C. (1992). A strategy for identifying novel, mechanistically unique inhibitors of topoisomerase I. *Journal of Natural Products*, *55*, 401–413.

[140] Fostel, J., Montgomery, D., Shen, L. (1992). Characterization of DNA topoisomerase I from *Candida albicans* as a target for drug discovery. *Antimicrobial Agents and Chemotherapy*, *36*, 2131–2138.

[141] Zhang, Z., ElSohly, H., Jacob, M., Pasco, D., Walker, L., Clark, A. (2002). Secreted aspartic proteases from *Tovomita krukovii*. *Planta Medica*, *68*, 49–54.

[142] Allen, J., Davey, H., Broadhust, D., Rowland, J., Oliver, S., Kell, D. (2004). Discrimination of modes of action of antifungal substances by use of metabolic footprinting. *Applied and Environmental Microbiology*, *70*, 6157–6165.

[143] Scherer, S., Gene discovery and comparative genomics: Progress and prospects. In: Calderone R.,editor. *Candida and Candidiasis*.American Society of Microbiology, Washington DC, 2002, pp. 259–265.

[144] Denning, D., Anderson, M., Turner, G., Latge, J., Bennett, J. (2002). Sequencing the *Aspergillus fumigatus* genome. *The Lancet Infectious Diseases*, *2*, 251–253.

[145] Schein, J., Tangen, K., Chiu, R., Shin, H., Lengeler, K., McDonald, W., Bosdet, I., Heitman, J., Jones, S., Marra, M., Kronstad, J. (2002). Physical maps for genome

analysis of 163 serotype A and D strains of the fungal pathogen *Cryptococcus neoformans*. *Genome Research*, *12*, 1445–1453.

[146] Goffeau, A., Coster, F., Foury, F., Jonniaux, J.-L., Purnelle, B., Talla, E., Tettelin, H., Van Dyck, L. (1997). The yeast genome directory. *Nature*, *387*, 1–105.

[147] Spaltman, F., Blunck, M., Ziegelbauer, K. (1999). Computer-aided target selection -prioritizing targets for antifungal drug discovery. *Drug Discovery Today*, *4*, 17–26.

[148] Anderson, A. (2003). The process of structure-based drug design. *Chemistry and Biology*, *10*, 787–797.

[149] Stahura, F., Bajorath, J. (2001). Bio-and chemo-informatics beyond data management: crucial challenges and future opportunities. *Drug Discovery Today*, *7*, S41–S47.

[150] De Backer, M., Van Dijck, P., Luyten W. (2002). Functional genomics approaches for the identification and validation of antifungal drug targets. *American Journal of Pharmacogenomics*, *2*, 113–127.

[151] Gordon Foulkes, J. (2002). Elitra Pharmaceuticals: New paradigms for antimicrobial drug discovery. *Drug Discovery Today*, *7*, S12–S15.

7

ANTIULCER AGENTS FROM HIGHER PLANTS

LUIZ C. KLEIN-JÚNIOR, JOSÉ R. SANTIN, AND SÉRGIO F. DE ANDRADE

7.1 INTRODUCTION

The gastrointestinal tract has its own conformation, with a large surface area that permits high absorption, and glands that produce digestive juices, thereby enabling digestion and absorption of nutrients, in order to maintain the functioning of the organism. One of the segments of the gastrointestinal tract is the stomach. This organ is formed by several layers of smooth muscle and is lined by a mucous membrane. It is anatomically divided into four regions: cardia, fundus, body, and antrum [1,2].

The stomach surface contains mucus-secreting cells, which cover the entire surface; peptic cells, which secrete large amounts of pepsinogen; parietal cells, which secrete hydrochloric acid and intrinsic factor; G cells, which secrete gastrin; and chromaffin cells, which secrete histamine [3,4].

The stomach has a series of protective barriers, for its primordial function, digestion of proteins, and protection against self-digestion and injuries. The mucosal layer plays a key role, with four levels of protection. The first level is the factors secreted into the lumen (surface phospholipids, bicarbonate, mucus, and prostaglandins); the second level of protection is the epithelium itself; the third level is related to the joint action of the afferent sensory nerve and the microcirculation mucosa and submucosa; and the fourth level of defense is the mucosal immune system, including mast cells and macrophages [5,6]. Thus, disorders involving the secretory and protector functions may be related to the pathogenesis of peptic ulcer [4].

Plant Bioactives and Drug Discovery: Principles, Practice, and Perspectives, Fourth Edition.
Edited by Valdir Cechinel-Filho.

Peptic ulcer is a chronic lesion, usually solitary, occurring anywhere in the gastrointestinal tract that is exposed to the aggressive action of gastric juice acid. This lesion is more common in the stomach or duodenum, leading to necrosis of the mucosal surface, and also of the muscle layer [7,8].

The usual cause of ulcer formation is an imbalance between the gastrointestinal defenses (mucus barrier and bicarbonate secretion) and aggressive factors (mainly rate of secretion of gastric juice) [3], associated with exogenous predisposing factors related to lifestyle such as stress, smoking, alcohol consumption, nutritional deficiencies, ingestion of nonsteroidal anti-inflammatory drugs (NSAIDs) and *Helicobacter pylori* [9–11].

Epidemiological data show that gastric ulcer is a disease that becomes increasingly common in the world population [10,12]. About 10% of the population in industrialized countries is affected in some life phase, and on the American continent alone, more of two million adults suffer from this condition. In relation to sex distribution, this disorder affects approximately 11% to 20% of men and 8–11% of women. Gastric ulcers tend to be a chronic disease, which can persist for up to 20 years, being characterized by repeated episodes of healing and re-exacerbation [13].

For many centuries, gastric ulcers have been treated using antacids to neutralize gastric secretions, and dietary restrictions and surgical procedures were common. In the 1970s, a revolution in the treatment of gastric ulcers occurred with the advent of the histamine (H_2) receptor antagonists (cimetidine and ranitidine), which act directly on the parietal cells, causing a decrease in the secretion of gastric acid [14]. In the 1980s, another class of drugs for the treatment of ulcers was launched on the market, which act by inhibition of the proton pump (H^+/K^+ ATPase), omeprazole being the first drug of this class. The aim of antiulcer therapy is to reduce pain, promote healing, and prevent re-exacerbation of the ulcers. Drug therapy seeks to re-establish the balance between offensive and defensive factors of the mucosa, that is, to protect the mucosal surface from the action of gastric juice [15].

The drugs currently used to treat this condition are the histamine (H_2) receptor antagonists (cimetidine, ranitidine, nizatidine, and famotidine), proton pump inhibitors H^+/K^+ATPase (omeprazole, lansoprazole, pantoprazole, and esomeprazole), antacids (magnesium hydroxide, magnesium trisilicate, aluminum hydroxide gel, and sodium bicarbonate), anticholinergics, and drugs that protect the mucosa (sucralfate, bismuth chelate, and misoprostol), and the combination of antibiotics (clarithromycin, amoxicillin, and tetracycline), in cases of ulceration caused by infection by *H. pylori* [4]. In rare cases, surgical procedures, such as vagotomy, are still indicated [16].

Despite extensive knowledge of the pathophysiology of peptic ulcers, and the fact that pharmaceutical companies produce a wide range of drugs to treat this condition, it is observed that (1) peptic ulcer is a pathology with a high incidence worldwide, (2) is a chronic disease that requires an endoscopic intervention therapy, often relapsing and requiring surgical intervention, causing financial troubles, both for patients and for the health system, (3) there is no product on the pharmaceutical market that is 100%

effective, and most of the drugs available have adverse effects, and (4) plants are a source of new therapeutic constituent with potentially activities, which is an interesting strategy for drug discovery [17]. Thus, the search for new, more effective and less toxic drugs are very relevant. Herbal and natural products are increasingly assuming a key role in this sense, since many extracts and/or isolated compounds have shown promising results in the treatment and/or prevention of gastric ulcers. The main plants used in the treatment of gastric ulcers, and some isolated substances, are described below.

7.2 MEDICINAL PLANTS WITH ANTIULCER ACTIVITY

A wide variety of medicinal plants is used by the populations of several countries, for the treatment of disorders of the digestive system, especially those related to gastric ulcer. For this reason, scientific research has been conducted, to investigate and confirm the pharmacological properties of some species.

7.2.1 Maytenus Genus

A classic example of a medicinal plant used by the people of South America, particularly Brazil, for the treatment of gastric ulcers, is the genus *Maytenus*. This genus belongs to the *Celastraceae* family and comprises about 200 species, approximately 40% of which occur in Brazilian territory, distributed from north to south [18]. Preparations obtained from the leaves of *Maytenus ilicifolia* are widely used in Brazil, where the plantis popularly known by different names, such as congorça, erva cancerosa, espinho-de-deus, quebrachillo, sombra-de-toro, concorosa, and finally the most commonly used name by the local population, "espinheira santa" or "espinheira." However, other species of this genus, such as *Maytenus aquifolium*, *Maytenus Robusta*, and *Maytenus truncate*, are also known by the same names and are also used in folk medicine to treat gastric disorders [19–22].

The biological properties of the extracts of leaves of *Maytenus*, as well as compounds isolated from these extracts, have been widely studied by several research groups using different experimental models *in vivo* and *in vitro*.

In order to confirm the antiulcer action of two species of *Maytenus*, a study was conducted with an extract obtained by extraction with hot water containing equal parts of *M. aquifolium* and *M. ilicifolia*, and evaluated in the model of acute stress-induced ulcer and ulcer induced by NSAID. This study showed significant results for the extract, when compared to the control group. In the same study, a significant increase in pH was also observed, comparable to the group treated with the drug cimetidine. Thus, the data corroborate with the popular use of *Maytenus* for treatment of gastric ulcers [23].

The dry extract of *M. ilicifolia* in three different doses (140, 280, and 420 mg/kg, i.p.) in the ulcer model induced by cold restraint in rats, was able to increase the volume and gastric pH, revealing similar results to the previous one. The dry extract of

M. aquifolium obtained by spray-drier was evaluated in rats in two doses (250 and 500 mg/kg) using different models: ulcer induced by ethanol acidified with hydrochloric acid, by indomethacin and by stress. The extract showed activity in all the experiments, and at all the doses evaluated [24,25].

A study to investigate possible mechanisms of action of aqueous extract of dried leaves of *M. ilicifolia* on gastric secretion was performed using isolated frog gastric mucosa incubated in an ussing chamber. The drug cimetidine was used as control. The results of this study suggest that the extract of *M. ilicifolia* reduces gastric acid secretion in isolated frog gastric mucosa, by acting in a manner similar to the drug cimetidine, antagonizing the histamine H_2 receptor [26].

The hexane and ethyl acetate *M. ilicifolia* extracts were evaluated in the model of ulcer induced by restraint and cold stress using a single dose of 320 mg/kg. Both extracts showed a decrease in the index of gastric lesions, compared with the control group, besides increasing the volume and pH of gastric secretion [27].

To assess the gastroprotective properties and mechanisms involved in gastroprotection of a flavonoid-rich fraction obtained from the aqueous extract of leaves of *M. ilicifolia*, the models of acute ulcer induced by ethanol and by indomethacin and chronic ulcer induced by acetic acid were used. In all the assays, intraperitoneal administration of flavonoid-rich fraction had the capacity to protect the mucosa in the ulcerogenic agents used. The fraction, when subjected to pylorus ligature test, presented a potent reduction in the release of gastric acid, accompanied by a decreasing of the nitric oxide release in the gastric secretion, 4 h after pylorus ligation. Inhibition of gastric acid secretion *in vivo* was correlated with *in vitro* inhibition of the rabbit gastric H^+/K^+-ATPase activity. In conclusion, the authors propose that the flavonoid-rich fraction of *M. ilicifolia* protects animals from gastric lesions with high potency through the inhibition of gastric acid secretion [28].

In order to ascertain which compounds are related to the antisecretory activity in rats, the lyophilized aqueous extract of *M. ilicifolia* (140 mg/kg) and fractions (14 mg/kg) were tested to verify the volume of gastric secretion and pH. By using HPLC, it was possible to verify that only the fractions containing the compounds tri and tetra-flavonoid glycosides mauritianin (**1**) and tetra-glycoside kaempferol (**2**) were able to significantly increase the volume and pH of gastric secretion, indicating that these compounds may play an important role in the gastroprotective effect of the extract obtained from the leaves of *M. ilicifolia* [29].

Studies using arabinogalactan polysaccharide that is released from the infusion of the leaves of *M. ilicifolia* significantly inhibited the formation of lesions induced by administration acute ethanol, showing a significant antiulcer effect [30].

Moreover, in another study, the polygalacturonic acid obtained from the leaves of *M. ilicifolia* using aqueous extraction was evaluated in the ethanol-induced ulcer model in Wistar rats, using doses of 1, 10, and 100 mg/kg given orally. The results showed that the acid causes a dose-dependent reduction in lesion formation. The mechanism suggested for the gastroprotective activity of polysaccharides was their ability to produce a protective layer on the gastric epithelium, and also their capacity to decrease acid and pepsin secretion and increase mucus synthesis. Thus, the results show that these polysaccharides may contribute to the antiulcer property of *M. ilicifolia* [31].

In some studies, researchers attribute the gastroprotective activity of *M. ilicifolia* the presence of two triterpenes, friedelin **(3)** and 3β-friedelinol **(4)**. To check the veracity of this information, a mixture of triterpenes obtained from the leaves extract of *M. ilicifolia* were subjected to the NSAID-induced ulcer test in rats. Through this experiment, it was observed that the two triterpenes isolated from the extract of *M. ilicifolia* do not have antiulcer activity [32].

M. ilicifolia extracts were submitted to a double-blind clinical trial in order to assess their effectiveness in treating patients with dyspepsia. The study included 23 patients, 13 of whom received treatment with capsules containing 200 mg of lyophilized infusion of *M. ilicifolia*, administered twice daily for a period of 28 days. The other 10 patients received placebo capsules (brown sugar) twice a day, during the same period. The group that received the treatment containing *M. ilicifolia* showed an improvement compared to the placebo, particularly in the symptoms related to heartburn. In another study involving 20 patients with peptic ulcer, 10 patients were treated with capsules containing 200 mg of lyophilized infusion of *M. ilicifolia* twice daily for 28 days, and the other 10 patients received capsules containing brown sugar (placebo) twice a day, during the same period. In this trial, there was no significant difference between the groups. The authors of the study point out that the analysis of the results was affected by the high dropout rate of patients. Nevertheless, the patients treated with *M. ilicifolia* showed clinical and endoscopic improvement compared with those who received the placebo [33,34].

M. ilicifolia is the main species of the genus *Maytenus*, and is the most scientifically studied one and most commonly used for the treatment of gastric ulcers by the population. However, this species is in the process of extinction due to indiscriminate use [35]. Thus, taking into account the chemical constitution and similar indications for popular use, many other species of *Maytenus* were tested for the antiulcer activity.

One of these species is *M. robusta*, a plant that is well adapted in southern Brazil. The extract of *M. robusta* was assessed using different doses (50, 250, and 500 mg/kg) and submitted to different experimental models (acute ulcer induced by ethanol, indomethacin and stress-induced ulcers). It was observed, in these experiments, that the extract significantly reduced the lesion index at all doses evaluated. In addition, in the pylorus ligature model, a reduction in volume and acidity of gastric juice was observed, as well as a significant increase in pH, thus presenting similar results to those obtained in studies with *M. ilicifolia* [36].

In another study, experiments were performed with hexane, chloroform, ethyl acetate, and aqueous fractions obtained from the leaves of *M. robust*, in order to assess the gastroprotective action against ethanol-induced ulcers in rats using a dose of 150 mg/kg. All fractions showed significant reduction in lesion score, total area damaged, and percentage of lesion when compared with the control group. Due to the significant results obtained from the ethyl acetate fraction, it was subjected to column chromatography, enabling the isolation of 3,15-dioxo-21α-hydroxyfriedelan (**5**). This triterpene was evaluated in the ethanol/HCl-induced ulcer model in mice. Triterpene, as well as omeprazole, both at a dose of 30 mg/kg, significantly decreased all the parameters when compared to the control group ($p < 0.05$). Thus, the authors suggest that the antiulcer effect observed in the extract and fractions may be partly attributable to this compound [22].

The leaves of *M. aquifolium* are commonly used in Brazilian folk medicine to treat gastric ulcers and dyspepsia. This species was also tested with the aim of investigating its gastroprotective activity. Hydromethanol extract (1000 mg/kg, p.o.) was administered in mice submitted to the ulcer induced by ethanol and indomethacin/bethanechol. The extract significantly reduced all the parameters in the ethanol-induced ulcer model, when compared to the control group; however, in the model of ulcer induced by indomethacin/bethanechol, contradictorily, the extract increased the ulcer effects of indomethacin/bethanechol. A study with infusion of the leaves of *M. aquifolium* was also performed, showing significant results. Fractionation of the infusion yielded two flavonoid tetrasaccharides: quercetin 3-*O*-α-*L*-rhamnopyranosyl(1 $\rightarrow$ 6)-*O*-[β-D-glucopyranosyl(1 $\rightarrow$ 3)-*O*-α-L-rhamnopyranosyl(1 $\rightarrow$ 2)-*O*-β-D-galactopyranoside and kaempferol 3-*O*-α-L-rhamnopyranosyl(1 $\rightarrow$ 6)-*O*-β-*D*-glucopyranosyl(1 $\rightarrow$ 3)-*O*-α-L-rhamnopyranosyl (1 $\rightarrow$ 2)-*O*-β-D-galactopyranoside [37].

Another species used in Brazilian traditional medicine to treat gastric ulcers is *M. rigida*. The crude ethanol extract of the leaves of *M. rigida* was evaluated in animal models to verify its antiulcer activity. The extract was tested at doses of 250, 500, and 750 mg/kg in the model of acute ethanol-induced ulcer in rats. The extract, at all the doses evaluated had significant effects, thus supporting the popular applications *M. rigida* in the treatment of gastric ulcers [38].

M. truncata Reiss (Celastraceae), also known as "espinheira-santa," is another species of *Maytenus* widely distributed in Brazilian territory. The leaves are used in traditional medicine in the form of tea to treat ulcers. Crude extracts of *M. truncata* prepared with ethyl acetate and methanol were tested in rats, to evaluate their gastroprotective activity at doses of 120 and 240 mg/kg. Treatment with both extracts reduced the severity of lesions induced by cold restraint. The extracts also caused an increase in gastric pH. Thus, the study corroborates the popular use of this species in the treatment of gastric lesions [20].

The gastroprotective activity of extracts from the leaves of *M. truncata*, prepared with chloroform, ethyl acetate, and methanol was also evaluated. The ethyl acetate and methanol extracts, when administered to rats at a dose of 120 and 240 mg/kg, showed a significant decrease in the severity of injuries [21].

The gastroprotective effect of *M. obtusifolia* is also described. The extracts obtained from the leaves, using doses of 62.5, 125, 250, and 500 mg/kg, significantly reduced the ulcerative lesion index compared with the negative control [39].

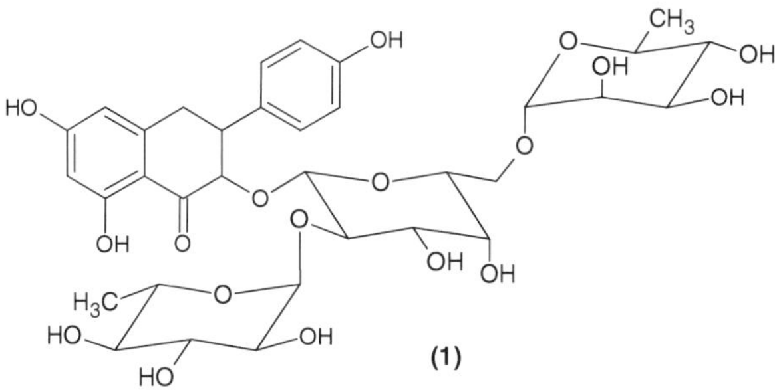
OH
HO
OH
O
O
O
O
OH
OH
H3C
HO
HO
OH
CH3
OH
OH
OH
O
O

(1)

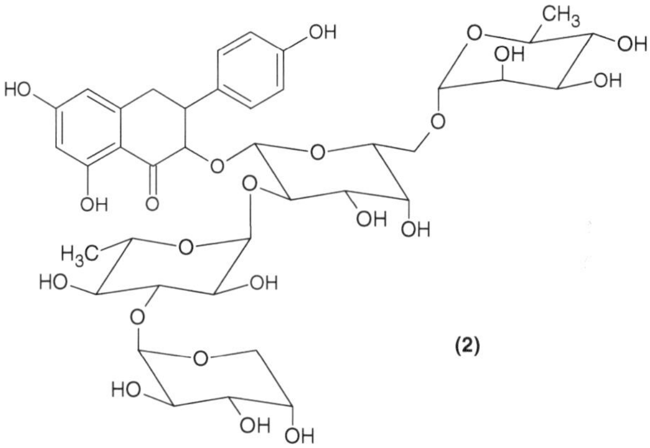
OH
HO
OH
O
CH3
OH
OH
OH
H3C
HO
OH
HO
OH
OH
OH

(2)

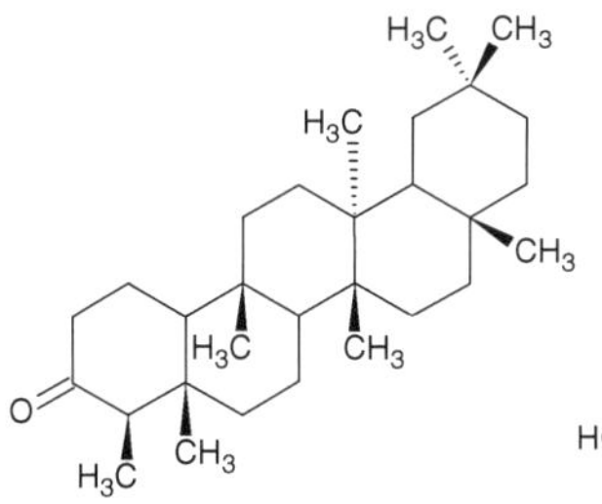
H3C
CH3
H3C
CH3
H3C
CH3
O
H3C
CH3

(3)

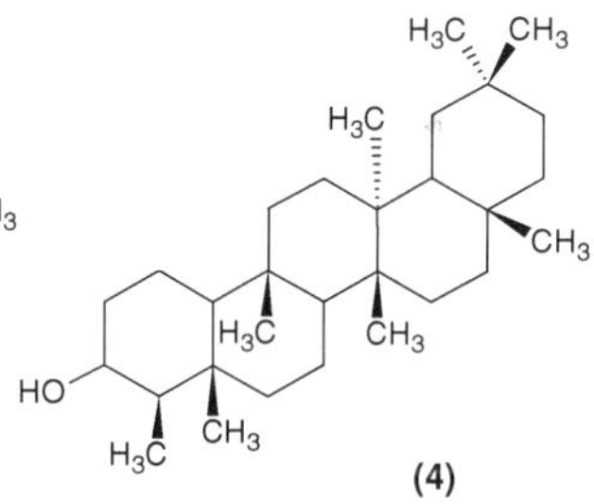
H3C
CH3
H3C
CH3
H3C
CH3
HO
H3C
CH3

(4)

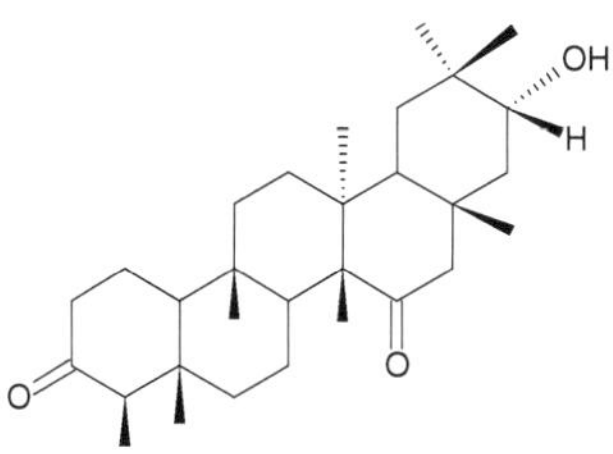
OH
H
O
O

(5)

7.2.2 Glycyrrhiza Genus

Preparations obtained from roots of *Glycyrrhiza glabra*, belonging to the family Leguminosae, popularly known as licorice, sweet root, yast-madhu, kanzohand gan cao, have been used extensively for centuries, to treat gastric ulcers. This species, originally from the Mediterranean, Middle East, and Asia, is widely cultivated in southern Russia and throughout Europe. The genus *Glycyrrhiza* consists of about 30 species, the main ones being *G. glabra, G. uralensis, G. inflata, G. aspera, G. korshinskyi,* and *G. eurycarpa* [40].

The main compound found in licorice is the glycyrrhizic acid, which corresponds to 10% to 25% of the licorice extract. The triterpene saponin is responsible for sweet taste of licorice, being roughly 50 times sweeter than sugar. This chemical compound is known for its anti-inflammatory and antiulcer properties [41–43].

It has been demonstrated in studies that the compounds obtained from licorice have the ability to increase the concentration of prostaglandins in the digestive system, thus stimulating the secretion of mucus, an important factor for gastric protection. Other studies also show that these compounds exhibit antipepsin effect and a cytoprotective role on the epithelial cells lining the stomach. Combined, these effects can lead to scar formation and healing of gastric ulcers [43].

In one study, the compound glycyrrhizin significantly inhibited the formation of lesions in rats submitted to the models of ethanol-induced ulcer and chronic ulcer induced by acetic acid, corroborating with the popular use of licorice in the treatment of gastric ulcers [44].

However, the compound glycyrrhizin can cause side effects because it may act by inhibiting the 11 β-hydroxysteroid enzyme responsible for inactivating cortisol. As a result of continuous use and exposure to high levels of glycyrrhizin, a hypermineralocorticoid-like effect can be detected in both animals and humans. These effects are reversible after withdrawal of licorice or glycyrrhizin [42]. For this reason, a free fraction of this compound was prepared and studied [45]. The free fraction, called FM100 by the author, showed significant inhibition of secretion of gastric juice when compared to the controls [44]. From FM 100, Saitoh et al. [46], isolated new isoflavones and chalcones, which could contribute to the antiulcer activity of licorice.

Simple derivatives of glycyrrhetinic acid (a genin), such as carbenoxolone **(6)** (an ester derivative) and deglycyrrhizinated licorice, have been widely used to treat gastric ulcers. Carbenoxolone was developed in the early 1960s in London, becoming the most widely used derivative of licorice for the treatment of gastric lesions. Carbenoxolone was the first drug capable of accelerating the healing process of gastric tissue through a different mechanism of inhibition of gastric secretion [41,47].

Several studies have been carried out to identify the mechanism of action of this compound. Studies have shown that carbenoxolone is able to stimulate mucus production, promote cell proliferation, and increase PGE_2 synthesis. More recent studies suggest that nitric oxide contributes somehow to the protective effects of carbenoxolone. However, due to the emergence of new effective therapies and many side effects, the use of carbenoxolone had limited use, because continued use leads to sodium retention and hypokalemia, which may lead the patient to a state of hypertension, edema, and heart failure [41].

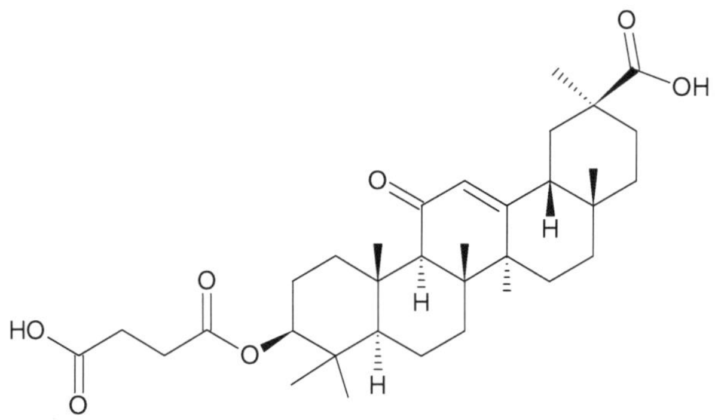

(6)

7.2.3 Plants Investigated in our Laboratory

Our research group has also studied some plants that are used for the treatment of gastric ulcers. One of the species studied was *Achyrocline satureioides* (Asteraceae), popularly known as "marcela" or "macela," which is widely used by the population of southern Brazil and in other countries, including Argentina and Uruguay. The studies conducted with hydroalcohol extract obtained 70% of the inflorescences of *A. satureoides*. The extract was evaluated in different experimental models, at the doses of 100, 250, and 500 mg/kg (p.o.). In models of acute ulcer induced by indomethacin and ethanol, the extract showed a reduction in all parameters when compared to the control group ($p < 0.05$). The extract was also able to increase the production of gastric mucus, without changing parameters such as pH and concentration of H^+ [48].

Another plant studied was *Brassica oleracea* L. var. *acephala* DC (kale), popularly known in Brazil as "couve." The juice from the leaves of *B. oleracea* is used extensively by the Brazilian population to treat gastric disorders such as ulcers. Our group studied the action of 50% hydroalcoholic extract obtained from the leaves of *B. oleracea* (25, 50 and 100 mg/kg, p.o.) in different experimental models. The extract at doses of 50 and 100 mg/kg showed gastroprotective activity in models of acute ulcer induced by ethanol/HCl and indomethacin, when compared to the control. Also, the extract was able to increase mucus production and increase gastric pH, decreasing the concentration of H^+ ions in the pylorus ligature model.

Polygala cyparissias St. Hilaire & Moquin (Polygalaceae), distributed in Argentina, Uruguay and especially in the southern coast of Brazil, is popularly known as "avenca-da-praia," "pinheiro-da-praia" and "gelol," This species is rich in xanthones, a class of secondary metabolites related to the antiulcer activity.

The study was performed using acetone extract and chloroform and methanol fractions, which were evaluated using the models of ulcer induced by ethanol/HCl and indomethacin in mice. The acetone extract and the methanol fraction, at doses of 50, 125, and 250 mg/kg, led to a significant decrease in all the parameters; however, the chloroform fraction showed no significant results [11].

Other species studied by our research group were essential oil from *Syzygium aromaticum*, an herb known as clove, which is popularly used in the treatment of dental pain, respiratory problems, inflammation and gastric problems. In this study, the essential oil, at all the doses evaluated (50, 100, and 250 mg/kg), reduced the severity of lesions induced by ethanol and indomethacin in rats. The oil showed no significant results in the pylorus ligature model. However, it did show an increase in mucus production. Studies to verify the involvement of nitric oxide and endogenous endogenous sulfhydryl in gastroprotection were also conducted, but the results were not significant. Thus, it is suggested that the gastroprotective activity of the essential oil is based on its action of stimulating the synthesis of mucus [49].

"Maromba" or "marroio" are the popular names of *Marrubium vulgare* (Horehound), a plant used to treat inflammation, respiratory and gastrointestinal diseases. The methanol extract obtained from the leaves of *M. vulgare* caused a significant reduction in the rate of injuries across all models evaluated (ethanol and indomethacin) at doses of 25, 50, and 100 mg/kg. The extract also caused an increase in pH and a decrease in the concentration of H^+, also increasing mucus production. With regard to the participation of nitric oxide and endogenous sulfhydryls, the results demonstrated that the gastroprotection induced by the extract is partially related to the activities of the endogenous sulfhydryls and nitric oxide. The data suggested that the effectiveness of the extract is based on its ability to stimulate mucus synthesis, an important gastroprotective factor.

In another study, our group evaluated the gastroprotective activity of the alkaloidal extract of *Galipea longiflora* using different experimental models. The extract using doses of 50, 125, and 250 mg/kg reduced the lesion score, total area damaged and the percentage of lesions in ulcer models induced by ethanol/HCl, indomethacin and stress, when compared to the controls. Also, the extract in the model of pylorus ligature reduced the volume of gastric secretion and increased the pH. There are indications that the mechanisms involved in the antiulcer activity are related to the decrease in gastric secretion [50].

A native plant of Brazil, *Baccharis dracunculifolia* DC (Asteraceae), commonly known as "Alecrim do campo" is used in folk medicine to treat inflammations, liver disorders, and gastric ulcers. The essential oil obtained from aerial parts of *B. dracunculifolia* was evaluated in models of ulcer induced by ethanol, stress, and indomethacin at doses of 50, 250, and 500 mg/kg. In these tests, the essential oil significantly reduced all the parameters. In the gastric pylorus ligature, the essential oil increased pH and decreased gastric volume [51].

These are just some examples of plants studied to treat ulcers. They demonstrate that medicinal plant extracts are promising sources of phytomedicines and bioactive compounds of medicinal interest.

7.3 SECONDARY METABOLITES AS A SOURCE OF ANTI-ULCER DRUG LEADS

Ulcer therapy has progressed from vagotomy to proton pump inhibitors [52]. However, several drugs employed for the treatment of ulcer are not completely effective and produce many adverse effects [53]. Therefore, there is a growing interest in alternative therapies and the use of natural products derived from plants [54].

Studies involving antiulcer activity of secondary metabolites usually focus on their capacity to increase the mucosal protective factors, or to reduce the harmful impact of aggressive factors [41]. The endogenous gastroprotective factors include the prostaglandins, mucus secretion, antioxidant defenses, and nitric oxide. On the other hand, the aggressive factors include gastric pH, inflammation, *Helicobacter pylori* infection and reactive oxygen species (ROS).

7.3.1 Effect on Endogenous Gastroprotective Factors

7.3.1.1 Prostaglandins Content. Prostaglandins play an important role in gastroprotective activity. They inhibit gastric acid secretion, accelerate ulcer healing via angiogenesis, enhance epithelial repair, stimulate the secretion of mucus and bicarbonate, and increase the resistance of the epithelial cells to potential damage by cytotoxins [6,55,56].

Several natural compounds had demonstrated action over the synthesis of prostaglandins: boswellic acid [57], 1′*S*-1′-acetoxychavicol acetate [58], and isopulegol [59]. Ferruginol **(7)**, an abietane diterpene isolated from *Prumnopitys andina* was evaluated in the model of HCl/ethanol-induced gastric lesions in mice, demonstrating a gastroprotective effect at the dose of 25 mg/kg (p.o.) and reducing the appearance of lesions by 60%. In the acetic acid-induced subacute gastric ulcer model, it was observed that **(7)**, at the doses of 25 and 50 mg/kg (p.o.), showed a curative ratio of 36.0% and 92.5%, respectively. These activities seems to be strongly related to the capacity of **(7)** to increase PGE_2 content on AGS cell cultures, a factor that is probably responsible for its ability to stimulate the growth of MRC-5 fibroblasts and AGS cells, improving the repair of the tissue lesion [60].

7.3.1.2 Gastric Mucus Secretion. Gastric mucus is an important protective factor, consisting of a gel formed by water and phospholipids. It acts as an antioxidant agent, and is also responsible for the maintenance of a near-neutral pH at the mucosal surface. The protective action of the mucus barrier depends on the gel structure and on the thickness of the layer covering the surface [6,61].

Barbatusin [62], aparisthman [63], and eugenol [49] are some of the compounds that act by increasing mucus secretion. Limonene **(8)** (245 mg/kg, p.o.), a monoterpene present in the essential oil of *Citrus aurantium*, provided effective gastroprotection when evaluated in ethanol- and NSAID-induced gastric ulcer (99% in both models). Given that NSAIDs diminish the secretion of mucus, and that ethanol has the capacity to dissolve the constituent gastric mucus, the activity of **(8)** on mucus secretion was evaluated. It was observed that the terpene was able to augment the amount of mucus adhering to the gastric mucosa, justifying the previously observed gastroprotective action [64].

7.3.1.3 Endogenous Antioxidant Defenses. ROS play a critical role in the development of gastric lesions induced by stress, ethanol and NSAIDs. ROS are responsible for lipid peroxidation in membranes, a decrease in the membrane permeability, and the activities of enzymes and receptors, and may cause cancer-inducing mutation. Therefore, intracellular antioxidants, like superoxide dismutase (SOD), glutathione peroxidase, catalase (CAT), and glutathione (GSH) are very important for gastric protection [65].

There are some natural products that have been studied, and have demonstrated this activity in preclinical models *in vitro* and *in vivo*, such as pennogenin 3-*O*-α-L-rhamnopyranosyl(1 $\rightarrow$ 2)-[α-L-arabinofuranosyl(1 $\rightarrow$ 4)]-β-D-glucopyranoside [66], 1,8-cineol [67] and kaempferol 3-*O*-1$''$-β-D-glucopyranosyl-3-*O*-1$'''$-β-D-glucopyranoside [68]. Another metabolite that has already been studied for its capacity to enhance the antioxidant defenses of the organism is the mixture of anacardic acids **(9)** obtained from *Anacardium occidentale*. It markedly reduced the gastric damage induced by ethanol at doses of 10, 30, and 100 mg/kg (p.o.) (25.3%, 63.6%, and 82.8% of inhibition, respectively). The main mechanism responsible for this gastroprotective action is related to the prevention of GSH depletion and an increase in the activities of SOD and CAT, reflecting their antioxidant potential [69].

7.3.1.4 Oxinitrergic System. Nitric oxide is involved in gastric acid secretion and the modulation of the gastric mucosal integrity, together with endogenous prostaglandins. It is produced via NO-synthase and appears to be one of the major factors involved in the regulation of gastric blood flow and gastric microcirculation [70].

Momordin Ic [71], lupeol [72], and 2-phenylquinoline [50] are some natural products whose gastroprotective activity is dependent on nitric oxide. Encecanescin **(10)**, a chromene isolated from *Eupatorium aschenbornianum*, was evaluated at doses of 3, 10, 30, and 100 mg/kg (p.o.), demonstrating a maximal gastroprotective effect of 68.6%. Through pretreatment with *N*-nitro-L-arginine methyl ester (L-NAME), a nitric oxide synthase inhibitor, the participation of nitric oxide was evaluated in the mechanism of gastroprotective action. It was shown that in the presence of L-NAME, the effect of **(10)** is inhibited. Thus, nitric oxide is involved in the gastroprotection elicited by the chromene [73].

7.3.1.5 Other Mechanisms Involving Endogenous Factors. Despite great advances in the understanding of ulcers, their etiology has not been completely elucidated [74]. Besides the mechanisms described above, there are other pathways involved in ulcerogenesis. One of them is related to the participation of the vanilloid receptors (TRPV1). Capsaicin-sensitive sensory neurons, through TRPV1, are involved in gastroenteric motility regulation, acid secretion, increase of the gastric blood flow due to the action of calcitonin gene-related peptide and stimulation of bicarbonate and mucus secretion [75]. Polygodial **(11)**, a sesquiterpene obtained from the leaves of *Tasmannia lanceolata*, significantly inhibited mucosal lesions induced by ethanol (0.025–0.5 mg/kg, p.o.), HCl (0.2–1.0 mg/kg, p.o.), aspirin (0.2–1.0 mg/kg, p.o.), and indomethacin

(0.2–2.0 mg/kg, p.o.). In the ethanol-induced ulcer model with pretreatment with ruthenium red, a capsaicin antagonist, a reduction was observed in the protection afforded by **(11)**. This suggests that its gastroprotective action is mediated by interactions with capsaicin-sensitive sensory neurons via TRPV1 [76].

Another mechanism of action involves the opening of K_{ATP} channels, which appears to be related to a variety of physiological functions of the stomach, like gastric blood flow regulation and stomach contractility. Moreover, endogenous prostaglandins, which as activators of these channels, are responsible, at least in part, for its gastroprotective activity [59]. It was shown that quebrachitol **(12)**, obtained from *Magonia glabrata*, produced inhibitory effect against ethanol-induced ulcer of 69%, 64%, and 53% (12.5, 25, and 50 mg/kg, respectively; p.o.). When evaluated with a pretreatment with glibenclamide, a K_{ATP} channel blocker, the protective effect was significantly antagonized. This demonstrates that the activity of **(12)** is dependent of K_{ATP} channels [77].

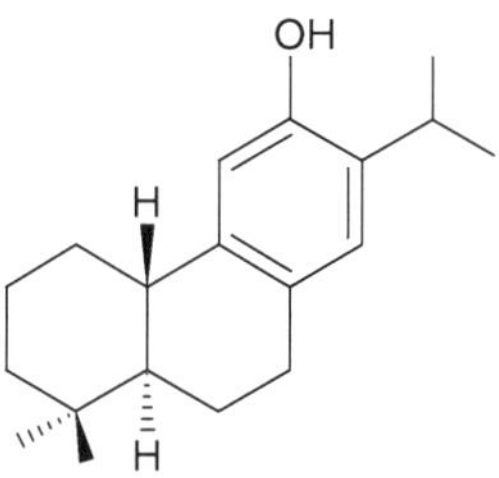

(7)

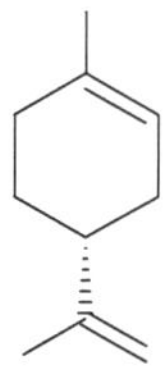

(8)

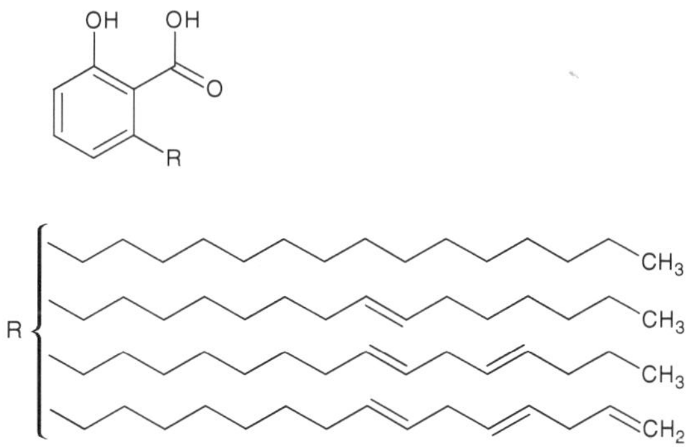

(9)

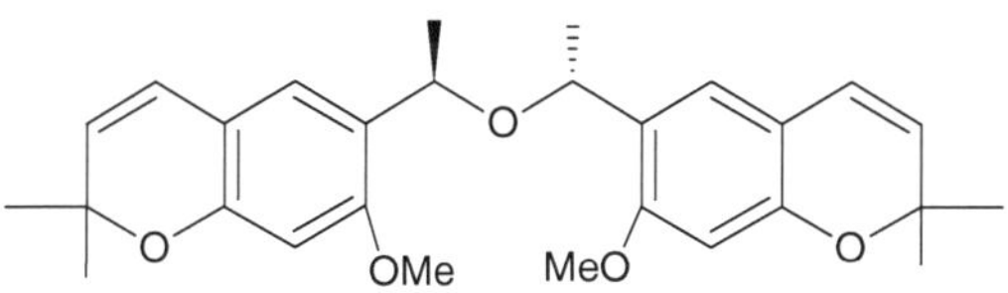

(10)

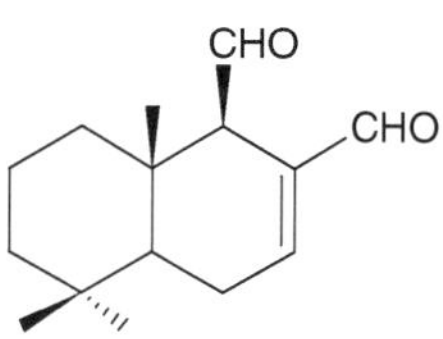

(11)

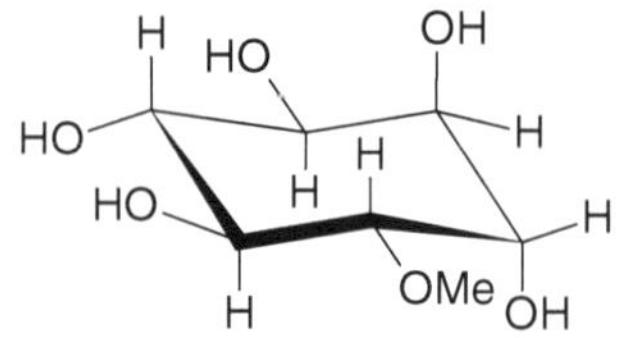

(12)

7.3.2 Protective Activity Against Aggressive Factors

7.3.2.1 ***Gastric pH.*** It is well known that gastric luminal acid play a critical role in ulcer formation. In fact, ulcer healing is significantly accelerated by antacids [78]. Acid secretion is regulated by several receptors, such as histamine H_2, muscarinic M_3 and gastrin CCK_2, located in the parietal cells. However, the ultimate step for acid secretion is H^+/K^+-ATPase (proton pump), which catalyzes the transport of H^+ at the expenses of ATP hydrolysis [79].

One metabolite that demonstrated action on gastric pH was the verbascoside **(13)**. This was isolated from *Tectona grandis*, and was able to reduce the ulcer index by decreasing the total acidity of the gastric juice. It was then checked for *in vitro* H^+/K^+-ATPase in isolated gastric microsomes from rat stomach, where it inhibited activity with an IC_{50} of 60.98 μg/ml (evaluated at doses ranging from 10 to 100 μg/ml) [79]. Others compounds that had already demonstrated this action over gastric pH are cinnamic acid [10], neohesperidin [78], and bisabolangelone [80].

7.3.2.2 ***Inflammation.*** Inflammation in the stomach can influence the protective ability of the mucosa to resist injury induced by aggressive factors, as well as its capacity to repair the injured tissue. Several chemical mediators are involved with the inflammatory response, like nitric oxide (generated through the inducible NOS), prostaglandins (generated by COX-2), leukotrienes, platelet-activating factor, neuropeptides, and cytokines. Usually, the inflammation is self-limiting; however, when the factor that initiated the inflammation persists, it can leads to an excessive tissue injury, becoming an ulcer [81].

Many molecules demonstrated antiulcer potential via the anti-inflammatory activity: diffractaic acid [82], wogonin [83], and genistein [84]. Curcumin **(14)**, a bioactive constituent from *Curcuma longa*, inhibited the ulcer formation in the doses of 20, 40, and 80 mg/kg (p.o.). This activity may be due to the capacity of **(14)** to inhibit IL-6 secretion. IL-6 is a proinflammatory cytokine that is responsible for the activation of neutrophils, lymphocytes and monocytes, initiating the oxidative bursts with the secretion of toxic metabolites and lysosomal enzymes, responsible for local tissue damage in ulcer disease [85].

7.3.2.3 ***Helicobacter pylori Infection.*** *H. pylori* is a microaerophilic, gram-negative, spiral-shaped flagellated bacterium that lives in the stomach and duodenum [86]. It is an important causal factor in the pathogenesis of gastritis and ulcer disease, inducing gastric inflammation, oxidative stress, DNA damage, apoptosis of epithelial cells, and induction of cell cycle deregulation [87]. Some abnormalities are linked with *H. pylori* infection, including increased basal and stimulated acid output, reduced inhibitory effect of somatostatin on gastrin release, and defective inhibition of acid secretion in response to antral distension [56]. Several secondary metabolites have been studied for the anti-*H. pylori* activity, such as licoisoflavone B [88], isorhamnetin [86], and luteolin [89].

7.3.2.4 ***Reactive Oxygen Species.*** Despite the protective activity against aggressive factors described earlier, there are other mechanisms that may be related to this issue. It has been shown that oxygen-derived mechanisms are involved in the pathogenesis of acute gastric mucosal lesions and in delaying chronic gastric ulcer healing. For this reason, besides compounds that can improve endogenous antioxidant defenses, the search for exogenous antioxidants it is of great importance. Yesilada et al. [90], through an activity-guided fractionation, isolated five free radical scavenging flavonoid glycosides from the flowers of *Spartium junceum*, a plant used for the treatment of ulcers in Turkish folk medicine. Luteolin 4′-β-glucoside **(15)** and quercetin 5-methylether-3-β-glucoside **(16)** were the most active (22.59 and 19.08 U/ml, respectively), as demonstrated through the electron spin resonance and spin trapping method.

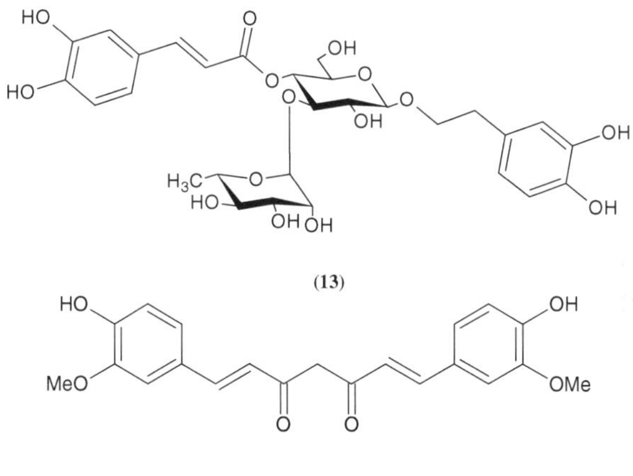

(13)

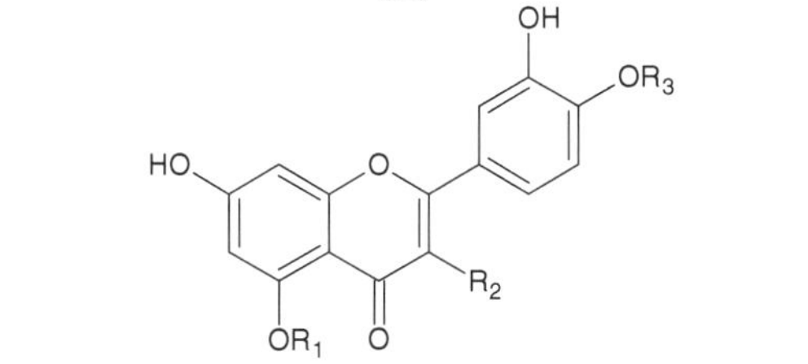

(14)

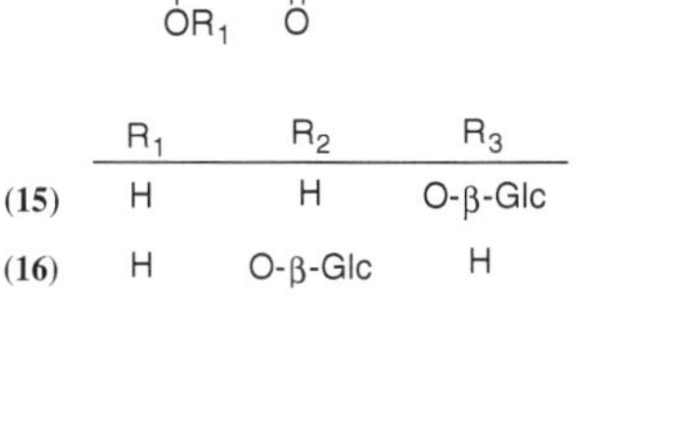

	R_1	R_2	R_3
(15)	H	H	O-β-Glc
(16)	H	O-β-Glc	H

7.4 CONCLUSIONS

One of the important drugs with antiulcer activity is the terpene carbenoxolone. But, like other antiulcer drugs, it is not completely effective, and demonstrates several side-effects. Thus, there is clearly a need for new and more effective antiulcer agents. There are several studies reporting the gastroprotective activity of different plant extracts and secondary metabolites. However, most of these studies are preliminary, with no delineation of the mechanism of action, adverse effects, toxic properties, and clinical trials. It was demonstrated, through this chapter, that there are many possible targets and several compounds that can be used as models for more potent and effective synthetic derivatives, possibly enabling the development of a new antiulcer drug obtained from plants. Thus, with interest for both academics and pharmaceutical companies, the efforts will certainly yield great rewards in the near future.

REFERENCES

[1] Merchant, J.L. (2007). Tales from the crypts: regulatory peptides and cytokines in gastrointestinal homeostasis and disease. *The Journal of Clinical Investigation, 117*, 6–12.

[2] Schubert, M.L., Peura, D.A. (2008). Control of gastric acid secretion in health and disease. *Gastroenterology, 134*, 1842–1860.

[3] Guyton, A.C., Hall, J.E., *Fisiologia Médica*, Guanabara Koogan, Rio de Janeiro, Brazil, 2002.

[4] Rang, H.P., Dale, M.M., Ritter, J.M., *Farmacologia*. Guanabara Koogan, Rio de Janeiro, Brazil, 2001.

[5] Wallace, J.L., Miller, M.J.S. (2000). Nitric oxide in mucosal defense: A little goes a long way. *Gastroenterology, 119*, 512–520.

[6] Tulassay, Z., Herszényi, L. (2010). Gastric mucosal defense and cytoprotection. *Best Practice and Research Clinical Gastroenterology, 24*, 99–108.

[7] Halter, F., Schmassmann, A., Tarnawski, A. (1995). Review article: Healing of experimental gastric ulcers. Interference by gastric acid. Digestive Disease Science, *40*, 2481–2486.

[8] Robbins, S.L., *Patologia Estrutural e Funcional*. Guanabara Koogan, Rio de Janeiro, Brazil, 2000.

[9] Belaiche, J., Burette, A., De Vos, M., Louis, E., Huybrechts, M., Deltenre, M. (2002). Study Group of NSAID-GI complications. Observational survey of NSAID-related upper gastrointestinal adverse events in Belgium. *Acta Gastroenterologic Belgian, 65*, 65–73.

[10] Barros, M.P., Lemos, M., Maistro, E.L., Leite, M.F., Sousa, J.P.B., Bastos, J.K., Andrade, S.F. (2008). Evaluation of antiulcer activity of the main phenolic acids found in Brazilian Green Propolis. *Journal of Ethnopharmacology, 120*, 372–377.

[11] Klein-Júnior, L.C., Gandolfi, R.B., Santin, J.R., Lemos, M., Cechinel-Filho, V., Andrade, S.F. (2010). Antiulcerogenic activity of extract, fractions, and some compounds obtained from *Polygala cyparissias* St. Hillaire & Moquin (Polygalaceae). *Naunyn-Schmiedeberg's Archives of Pharmacology, 381*, 121–126.

[12] Klopell, F.C., Lemos, M., Sousa, J.P.B., Comunello, E., Maistro, E.L., Bastos, J.K., Andrade, S.F. (2007). Nerolidol, an antiulcer constituent from the essential oil of *Baccharis dracunculifolia* DC (Asteraceae). *Zeistschrift fur Naturforschung, 62*c, 537–542.

[13] Naik, Y., Jayaram, S., Nayaka, M.A.H., Lakshman, Dharmesh, S.M. (2007). Gastroprotective effect of swallow root (*Decalepis hamiltonii*) extract: Possible involvemente of $H^{+}K^{+}$ATPase inhibition and antioxidative mechanism. *Journal of Ethnopharmacology, 112*, 173–179.

[14] Yuhong, Y., Ireneusz, T.P., Richard, H.H. (2006). Peptic ulcer disease today. *Nature Clinical Practice Gastroenterology and Hepatology, 3*, 80–89.

[15] Aihara, T., Nakamura, E., Amagase, K., Tomita, K., Fujishita, T. (2003). Pharmacological control of gastric acid secretion for the treatment of acid-related peptic disease: past, present, and future. *Pharmacology and Therapeutics, 98*, 109–127.

[16] Souza-Brito, A.R.M. (1997). Ethnopharmacology and sustainable development of new plant-derived drugs. *Ciência e Cultura, 49*, 402–406.

[17] Batista, L.M. Atividade antiulcerogênica de extratos e frações obtidos dos escapos das espécies *Syngonanthus bisulcatus* Rul e *Syngonanthus arthrotrichus* Silveira em modelos animais. 2003, 97p. Tese de Doutorado.Inst. Biologia – UNICAMP.

[18] Howard RA, *Flora of the Lesser Antilles: Leeward and Windward Islands – Dicotyledoneae*. Harvard University, Massachusetts, 1989.

[19] Gonzalez, F.G., Portela, T.Y., Stipp, E.J., Di Stasi, L.C. (2001). Antiulcerogenic and analgesic effects of *Maytenus aquifolium, Sorocea bomplandii* and *Zolernia ilicifolia*. *Journal of Ethnopharmacology, 77*, 41–47.

[20] Silva, J.L., Silva, R.P., Jorge, R.M., Silva, G.D., Vieira-Filho, S.A., Fonseca, A.P.N.D., Tagliati, C.A. (2005). Avaliação da atividade antiulcerogênica da *Maytenus truncata* Reiss (Celastraceae). *Brazilian Journal of Pharmacognosy, 15*, 30–35.

[21] Fonseca, A.P.N.D., Silva, G.D.F., Carvalho, J.J., Salazar, G.D.C.M., Duarte, L.P., Silva, R.P., Jorge, R.M., Tagliati, C.A., Zani, C.L., Alves, T.M.A., Peres, V., Vieira Filho, S.A. (2007). Estudo fitoquímico do decocto das folhas de *Maytenus truncata* Reiss e avaliação das atividades antinociceptiva, antiedematogênica e antiulcerogênica de extratos e decocto. *Quimica Nova, 30*, 842–847.

[22] Andrade, S.F., Comunello, E., Noldin, V.F., Monache, F.D., Cechinel-Filho, V., Niero, R. (2008). Antiulcerogenic activity of fractions and 3,15-dioxo-21α-hydroxyfriedelane isolated from *Maytenu srobusta* (Celastraceae). *Archives of Pharmcal Research, 31*, 41–46.

[23] Souza-Formigoni, M.L.O., Oliveira, M.G., Silveira-Filho, N.G., Braz, S., Carlini, E.A. (1991). Antiulcerogenic effects of two *Maytenus* species in laboratory animals. *Journal of Ethnopharmacology, 34*, 21–27.

[24] Oliveira, M.L., Duarte, L.P., Silva, G.D.F., Filho, S.A., Knupp, V.F., Alves, F.G. (2007). 3-Oxo-12alpha-hydroxyfriedelane from *Maytenus gonoclada:* Structure elucidation by (1)H and (13)C chemical shift assignments and 2D-NMR spectroscopy. *Magnetic Resonance in Chemistry, 45*, 895–898.

[25] Bersani-Amado, C.A., Massao, L.B., Baggio, S.R., Johanson, L., Albiero, A.L., Kimura, E. (2000). Antiulcer effectiveness of *Maytenus aquifolium* spray dried extract. *Phytotherapy Research, 14*, 543–545.

[26] Ferreira, P.M., Oliveira, C.N., Oliveira, A.B., Lopes, M.J., Alzamora, F., Vieira, M.A. (2004). A lyophilized aqueous extract of *Maytenus ilicifolia* leaves inhibits histamine-mediated acid secretion in isolated frog gastric mucosa. *Planta Medica, 219*, 319–324.

[27] Jorge, R.M., Leite, J.P.V., Oliveira, A.B., Tagliati, C.A. (2004). Evaluation of antinociceptive, anti-inflammatory and antiulcerogenic activities of *Maytenus ilicifolia*. *Journal of Ethnopharmacology*, *94*, 93–100.

[28] Baggio, C.H., Freitas, C.S., Otofuji, G.M., Cipriani, T.R., Souza, L.M., Sassaki, G.L., Iacomini, M., Marques, M.C., Mesia-Vela, S. (2007). Flavonoid-rich fraction of *Maytenus ilicifolia* Mart. ex. Reiss protects the gastric mucosa of rodents through inhibition of both H^+,K^+-ATPase activity and formation of nitric oxide. *Journal of Ethnopharmacology*, *113*, 433–440.

[29] Leite, J.P.V., Braga, F.C., Romussi, G., Persoli, R.M., Tabach, R., Carlini, E.A., Oliveira, A.B. (2010). Constituents from *Maytenus ilicifolia* leaves and bioguided fractionation for gastroprotective activity. *Journal of Brazilian Chemical Society*, *21*, 248–254.

[30] Cipriani, T.R., Mellinger, C.G., de Souza, L.M., Baggio, C.H., Freitas, C.S., Marques, M.C., Gorin, P.A., Sassaki, G.L., Iacomini, M. (2006). A polysaccharide from a tea (infusion) of *Maytenus ilicifolia* leaves with anti-ulcer protective effects. *Journal of Natural Products*, *69*, 1018–1021.

[31] Cipriani, T.R., Mellinger, C.G., Souza, L.M., Baggio, C.H., Freitas, C.S., Marques, M.C.A., Gorin, P.A.J., Sassaki, G.L., Iacomini, M. (2009). Polygalacturonic acid: Another anti-ulcer polysaccharide from the medicinal plant *Maytenus ilicifolia*. *Carbohydrate Polymers*, *78*, 361–363.

[32] Queiroga, C.L., Silva, G.F., Dias, P.C., Possenti, A., Carvalho, J.E. (2000). Evaluation of the antiulcerogenic activity of friedelan-3β-ol and friedelin isolated from *Maytenus ilicifolia* (Celastraceae). *Journal of Ethnopharmacology*, *72*, 465–468.

[33] Geocze, S., Vilela, M.P., Chaves, B.D.R., Ferrari, A.P. (1988). Tratamento de pacientes portadores com dispepsia alta ou úlcera péptica com preparações de espinheira-santa (*Maytenus ilicifolia*). *Publicação CEME*, *2*, 75–87.

[34] Melo, S.F., Soares, S.F., Costa, R.F., Silva, C.R., Oliveira, M.B. N, Bezerra, R.J.A.C., Caldeira-de-Araújo, A., Bernardo-Filho, M. (2001). Effect of the *Cymbopogon citratus, Maytenus ilicifolia* and *Baccharis genistelloides* extracts against the stannous chloride oxidative damage in *Escherichia coli*. *Mutation Research*, *496*, 33–38.

[35] Niero, R., Moser, R., Busato, A.C.B., Yunes, R.A., Reis, A., Cechinel-Filho, V. (2001). A comparative chemical study of *Maytenus ilicifolia* Mart. Reiss and *Maytenus robusta* Reiss (Celastraceae). *Zeitschrift fur Naturforschung*, *56c*, 158–161.

[36] Andrade, S.F., Lemos, M., Comunello, E., Noldin, V.F., Cechinel-Filho, V., Niero, R. (2007). Evaluation of the antiulcerogenic activity of *Maytenus robusta* (Celastraceae) in different experimental ulcer models. *Journal of Ethnopharmacology*, *113*, 252–257.

[37] Sousa, D.P., Almeida, R.N. (2005). Neuroleptic-like properties of the chloroform extract of *Maytenus obtusifolia* Mart. roots. *Biological and Pharmaceutical Bulletin*, *28*, 224–225.

[38] Carlini, E.A., Frochtengarten, M.L., Toxicologia clínica (Fase I) da espinheira-santa (*Maytenus ilicifolia*). In: Central de Medicamentos (Brasil). *Estudo de ação antiúlcera gástrica de plantas brasileiras (Maytenus ilicifolia "espinheira-santa" e outras)*. Central de Medicamentos (Brasil), 1988, pp. 67–73.

[39] Mota, K.S.L., Pita, J.C.L.R., Estevam, E.C., Medeiros, V.M., Tavares, J.F., Agra, M.F., Diniz, M.F.F.M., Silva, M.S., Batista, L.M. (2008). Evaluation of the toxicity and antiulcerogenic activity of the ethanol extract of *Maytenus obtusifolia* Mart. leaves. *Brazilian Journal of Pharmacognosy*, *18*, 441–446.

[40] Asl, M.N., Hosseinzadeh, H. (2008). Review of pharmacological effects of *Glycyrrhiza* sp. and its bioactive compounds. *Phytotherapy Research*, *22*, 709–724.

[41] Borrelli, F., Izzo, A.A. (2000). The plant kingdom as a source of anti-ulcer remedies. *Phytotherapy Research*, *14*, 581–591.

[42] Isbrucker, R.A., Burdock, G.A. (2006). Risk and safety assessment on the consumption of licorice root (*Glycyrrhiza* sp.), its extract and powder as a food ingredient, with emphasis on the pharmacology and toxicology of glycyrrhizin. *Regulatory Toxicology and Pharmacology*, *46*, 167–192.

[43] Aly, A.M., Al-Alousi, L., Salem, H.A. (2005). Licorice: A possible anti-inflammatory and anti-ulcer drug. *AAPS Pharmaceutical Sciences and Technology*, *6*, E74–E82.

[44] Takagi, K., Okabe, S., Saziki, R. (1969). A new method for the production of chronic gastric ulcer in rats and the effect of several drugs on its healing. *Japan Journal Pharmacology*, *19*, 418–422.

[45] Takagi, K., Ishii, Y. (1967). Peptic ulcer inhibiting properties of a new fraction from licorice root (FM100): I. Experimental peptic ulcer and general pharmacology. *Arzneimittel Forschung*, *17*, 1544–1548.

[46] Saitoh, T., Kinoshita, T., Shibata, S. (1976). Flavonols of licorice root. *Chemical and Pharmaceutical Bulletin*, *24*, 1242–1245.

[47] Baron, J.H. (1977). Effect of carbenoxolone sodium on human gastric acid secretion. *Gut*, *18*, 721–722.

[48] Santin, J.R., Lemos, M., Klein-Júnior, L.C., Niero, R., de Andrade S.F. (2010). Antiulcer effects of *Achyrocline satureoides*(Lam.) DC (Asteraceae) (Marcela), a folk medicine plant, in different experimental models. *Journal of Ethnopharmacology*, *130*, 334–339.

[49] Santin, J.R., Lemos, M., Klein-Júnior, L.C., Machado, I.D., Costa, P., Oliveira, A.P., Tilia, C., Souza, J.P., Sousa, J.P.B., Bastos, J.K., Andrade, S.F. (2011). Gastroprotective activity of essential oil of the *Syzygium aromaticum* and its major component eugenol in different animal models. *Naunyn-Schmiedeberg's Archives of Pharmacology*, *383*, 149–158.

[50] Zanatta, F., Gandolfi, R.B., Lemos, M., Ticona, J.C., Gimenez, A., Clasen, B.K., Cechinel-Filho, V., Andrade, S.F. (2009). Gastroprotective activity of alkaloid extract and 2-phenylquinoline obtained from the bark of *Galipea longiflora* Krause (Rutaceae). *Chemical-Biological Interactions*, *180*, 312–317.

[51] Massignani, J.J., Lemos, M., Maistro, E.L., Schaphauser, H.P., Jorge, R.F., Sousa, J.P., Bastos, J.K., de Andrade, S.F. (2009). Antiulcerogenic activity of the essential oil of *Baccharis dracunculifolia* on different experimental models in rats. *Phytotherapy Research*, *23*, 1355–1360.

[52] Wallace, J.L., Granger, D.N. (1996). The cellular and molecular basis of gastric mucosal defense. *The Journal of the Federation of American Societies for Experimental Biology*, *10*, 731–740.

[53] Falcão, H.S., Mariath, I.R., Diniz, M.F.F.M., Batista, L.M., Barbosa-Filho, J.M. (2008). Plants of the American continent with antiulcer activity. *Phytomedicine*, *15*, 132–146.

[54] Schmeda-Hirschmann, G., Yesilada, E. (2005). Traditional medicine and gastroprotective crude drugs. *Journal of Ethnopharmacology*, *100*, 61–66.

[55] Arakawa, T., Higuchi, K., Fukuda, T., Fujiwara, Y., Kobayashi, K., Kuroki, T. (1998). Prostaglandins in the stomach: An update. *Journal of Clinical Gastroenterology*, *27*, S1–S11.

[56] Chan, F.K.L., Leung, W.K. (2002). Peptic-ulcer disease. *The Lancet*, *360*, 933–941.

[57] Singh, S., Khajuria, A., Taneja, S.C., Khajuria, R.K., Singh, J., Johri, R.K., Qazi, G.N. (2008). The gastric ulcer protective effect of boswellic acids, a leukotriene inhibitor from *Boswellia serrata,* in rats. *Phytomedicine*, *15*, 408–415.

[58] Matsuda, H., Pongpiriyadacha, Y., Morikawa, T., Ochi, M., Yoshikawa, M. (2003). Gastroprotective effects of phenylpropanoids from the rhizomes of *Alpinia galangal* in rats: Structural requirements and mode of action. *European Journal of Pharmacology*, *471*, 59–67.

[59] Silva, M.I.G., Moura B.A., Aquino-Neto, M.R., Tomé, A.R., Rocha, N.F.M., Carvalho, A.M.R., Macêdo, D.S., Vasconcelos, S.M.M., Sousa, D.P., Viana, G.S.B., Sousa, F.C.F. (2009). Gastroprotective activity of isopulegol on experimentally induced gastric lesions in mice: Investigation of possible mechanisms of action. *Naunyn-Schmiedeberg's Archives of Pharmacology*, *380*, 233–245.

[60] Rodríguez, J.A., Theoduloz, C., Yáñez, T., Becerra, J., Schmeda-Hirschmann, G. (2006). Gastroprotective and ulcer healing effect of ferruginol in mice and rats: Assessment of its mechanism of action using in vitro models. *Life Sciences*, *78*, 2503–2509.

[61] Hiruma-Lima, C.A., Santos, L.C., Kushima, H., Pellizzon, C.H., Silveira, G.G., Vasconcelos, P.C.P., Vilegas, W., Souza Brito, A.R.M. (2006). *Qualea grandiflora,* a Brazilian "Cerrado" medicinal plant presents an important antiulcer activity. *Journal of Ethnopharmacology*, *104*, 207–214.

[62] Rodrigues, P.A., Morais, S.M., Souza, C.M., Silva, A.R.A., Andrade, G.M., Silva, M.G.V., Albuquerque, R.L., Rao, V.S., Santos, F.A. (2010). Gastroprotective effect of barbatusin and 3-beta-hydroxy-3-deoxybarbatusin, quinonoid diterpenes isolated from *Plectranthus grandis,* in ethanol-induced gastric lesions in mice. *Journal of Ethnopharmacology*, *127*, 725–730.

[63] Hiruma-Lima, C.A., Gracioso, J.S., Toma, W., Almeida, A.B., Paula, A.C.B., Brasil, D.S.B., Muller, A.H., Souza Brito, A.R.M. (2001). Gastroprotective effect of aparisthman, a diterpene isolated from *Aparisthmium cordatum,* on experimental gastric ulcer models in rats and mice. *Phytomedicine*, *8*, 94–100.

[64] Moraes, T.M., Kushima, H., Moleiro, F.C., Santos, R.C., Rocha, L.R.M., Marques, M.O., Vilegas, W., Hiruma-Lima, C.A. (2009). Effects of limonene and essential oil from *Citrus aurantium* on gastric mucosa: Role of prostaglandins and gastric muçus secretion. *Chemico-Biological Interactions*, *180*, 499–505.

[65] Odabasoglu, F., Cakir, A., Suleyman, H., Aslan, A., Bayir, Y., Halici, M., Kazaz, C. (2006). Gastroprotective and antioxidant effects of usnic acid on indomethacin-induced gastric ulcer in rats. *Journal of Ethnopharmacology*, *103*, 59–65.

[66] Matsuda, H., Pongpiriyadacha, Y., Morikawa, T., Kishi, A., Kataoka, S., Yoshikawa, M. (2003). Protective effects of steroid saponins from *Paris polyphylla* var. *yunnanensis* on ethanol- or indomethacin-induced gastric mucosal lesions in rats: Structural requirement for activity and mode of action. *Bioorganic and Medicinal Chemistry Letters*, *13*, 1101–1106.

[67] Santos, F.A., Rao, V.S.N. (2001). 1,8-Cineol, a food flavoring agent, prevents ethanol-induced gastric injury in rats. *Digestive Diseases and Sciences*, *46*, 331–337.

[68] Yesilada, E., Gurbuz, I. (2010). Evaluation of the antiulcerogenic activity profile of a flavonoldiglucoside from *Equisetum palustre* L. *Journal of Ethnopharmacology*, *131*, 17–21.

[69] Morais, T.C., Pinto, N.B., Carvalho, K.M.M.B., Rios, J.B., Ricardo, N.M.P.S., Trevisan, M.T.S., Rao, V.S., Santos, F.A. (2010). Protective effect of anacardic acids from cashew (*Anacardium occidentale*) on ethanol-induced gastric damage in mice. *Chemico-Biological Interactions*, *183*, 264–269.

[70] Wallace, J.L. (2006). Nitric oxide, aspirin-triggered lipoxins and NO-aspirin in gastric protection. *Inflammation Allergy Drug Targets*, *5*, 133–137.

[71] Matsuda, H., Li, Y., Yoshikawa, M. (1999). Roles of capsaicin-sensitive sensory nerves, endogenous nitric oxide, sulfhydrils, and prostaglandins in gastroprotection by momordinIc, an oleanolic acid oligoglycoside, on ethanol-induced gastric mucosal lesions in rats. *Life Sciences*, *65*, 27–32.

[72] Lira, S.R.S., Rao, V.S., Carvalho, A.C.S., Guedes, M.M., Morais, T.C., Souza, A.L., Trevisan, M.T.S., Lima, A.F., Chaves, M.H., Santos, F.A. (2009). Gastroprotective effect of lupeol on ethanol-induced gastric damage and the underlying mechanism. *Inflammopharmacology*, *17*, 221–228.

[73] Sánchez-Mendoza, M.E., Reyes-Trejo, B., Sánchez-Gómez, P., Rodríguez-Silverio, J., Castillo-Henkel, C., Cuevas-Cervantes, H., Arrieta, J. (2010). Bioassay-guided isolation of an anti-ulcer chromene from *Eupatorium aschenbornianum*: Role of nitric oxide, prostaglandins and sulfydryls. *Fitoterapia*, *81*, 66–71.

[74] Ramakrishman, K., Salinas, R.C. (2007). Peptic ulcer disease. *American Family Physician*, *76*, 1005–1012.

[75] Peng, J., Li, Y. The vanilloid receptor TRPV1: Role in cardiovascular and gastrointestinal protection. *European Journal of Pharmacology*, *627*, 1–7.

[76] Pongpiriyadacha, Y., Matsuda, H., Morikawa, T., Asao, Y., Yoshikawa, M. (2003). Protective effects of polygodial on gastric mucosal lesions induced by necrotizing agents in rats and the possible mechanisms of action. *Biological and Pharmaceutical Bulletin*, *26*, 651–657.

[77] Olinda, T.M., Lemos, T.L.G., Machado, L.L., Rao, V.S., Santos, F.A. (2008). Quebrachitol-induced gastroprotection against acute gastric lesions: Role of prostaglandins, nitric oxide and K^+_{ATP} channels. *Phytomedicine*, *15*, 327–333.

[78] Lee, J., Lee, S., Kim, Y.S., Jeong, C.S. (2009). Protective effects of neohesperidin and poncirin isolated from the fruits of *Poncirus trifoliate* on potential gastric disease. *Phytotherapy Research*, *23*, 1748–1753.

[79] Singh, N., Shukla, N., Singh, P., Sharma, R., Rajendran, S.M., Maurya, R., Palit, G. Verbascoside isolated from *Tectona grandis* mediates gastric protection in rats via inhibiting proton pump activity. *Fitoterapia*, *81*, 755–761.

[80] Wang, J., Zhu, L., Zou, K., Cheng, F., Dan, F., Guo, Z., Cai, Z., Yang, J. (2009). The anti-ulcer activities of bisabolangelone from *Angelica polymorpha*. *Journal of Ethnopharmacology*, *123*, 343–346.

[81] Wallace, J.L., Ma., L. (2001). Inflammatory mediators in gastrointestinal defense and injury. *Experimental Biology and Medicine*, *226*, 1003–1015.

[82] Bayir, Y., Odabasoglu, F., Cakir, A., Aslan, A., Suleyman, H., Halici, M., Kazaz, C. (2006). The inhibition of gastric mucosal lesion, oxidative stress and neutrophil-infiltration in rats by the lichen constituent diffractaic acid. *Phytomedicine*, *13*, 584–590.

[83] Park, S., Hahm, K., Oh, T., Jin, J., Choue, R. (2004). Preventive effect of the flavonoid, wogonin, against ethanol-induced gastric mucosal damage in rats. *Digestive Diseases and Sciences*, *49*, 384–394.

[84] Takekawa, S., Matsui, T., Arakawa, Y. (2006). The protective effect of the soybean polyphenol genistein against stress-induced gastric mucosal lesions in rats, and its hormonal mechanisms. *Journal of Nutritional Science and Vitaminology*, *52*, 274–280.

[85] Tuorkey, M., Karolin, R. (2009). Anti-ulcer activity of curcumin on experimental gastric ulcer in rats and its effect on oxidative stress/antioxidant, IL-6 and enzyme activities. *Biomedical and Environmental Sciences*, *22*, 488–495.

[86] Ustün, O., Ozçelik, B., Akyön, Y., Abbasoglu, U., Yesilada, E. (2006). Flavonoids with anti-*Helicobacter pylori* activity from *Cistuslaurifolius* leaves. *Journal of Ethnopharmacology*, *108*, 457–461.

[87] Lee, K., Yeo, M., Choue, J., Jin, J., Park, S.J., Cheong, J., Lee, K.J., Kim, J., Hahm, K. (2004). Protective mechanism of epigallocatechin-3-gallate against *Helicobacter pylori*-induced gastric epithelial cytotoxicity via the blockage of TLR-4 signalling. *Helicobacter*, *9*, 632–642.

[88] Fukai, T., Marumo, A., Kaitou, K., Kanda, T., Terada, S., Nomura, T. (2002). Anti-*Helicobacter pylori* flavonoids from licorice extract. *Life Sciences*, *71*, 1449–1463.

[89] Chung, J.G., Hsia, T.C., Kuo, H.M., Li, Y.C., Lee, Y.M., Lin, S.S., Hung, C.F. (2001). Inhibitory actions of luteolin on the growth and arylamine *N*-acetyltransferase activity in strains of *Helicobacter pylori* from ulcer patients. *Toxicology in Vitro*, *15*, 191–198.

[90] Yesilada, E., Tsuchiya, K., Takaishi, Y., Kawazoe, K. (2000). Isolation and characterization of free radical scavenging flavonoid glycosides from the flowers of *Spartium junceum* by activity-guided fractionation. *Journal of Ethnopharmacology*, *73*, 471–478.

8

RECENT PROGRESS IN THE CHEMISTRY AND BIOLOGY OF PACLITAXEL (TAXOL™) AND RELATED TAXANES

JUN QI, JIELU ZHAO, AND DAVID G. I. KINGSTON

8.1 INTRODUCTION

The introduction of paclitaxel (originally known as taxol) into clinical practice in the early 1990s for the treatment of ovarian cancer and later for breast cancer made a significant improvement in the chemotherapy of these diseases. A measure of the impact that paclitaxel and its analog docetaxel have made on cancer chemotherapy is illustrated by a statement made by an oncologist known to one of the authors, that he divided cancer chemotherapy into two eras: "the pre-taxol era" and "the post-taxol era."

Because of the importance of paclitaxel in cancer chemotherapy, numerous review articles and book chapters have been written on its chemistry, its pharmacology, and its medical applications [1–7]. This chapter will thus focus on recent advances in the chemistry and biology of paclitaxel and related compounds, with an emphasis on new taxanes, new chemistry, new approaches to drug delivery, and new analogs in clinical development. A short historical introduction sets the stage for the remainder of the chapter.

Paclitaxel (**1.1**) and docetaxel (**1.2**) (Figure 8.1) are microtubule-interactive mitotic inhibitors that are widely used in cancer chemotherapy. Paclitaxel was isolated from the bark of the Pacific Yew tree, *Taxus brevifolia,* by Wall and Wani at the Research Triangle Institute in 1967 [8], with publication of its chemical

Plant Bioactives and Drug Discovery: Principles, Practice, and Perspectives, Fourth Edition.
Edited by Valdir Cechinel-Filho.

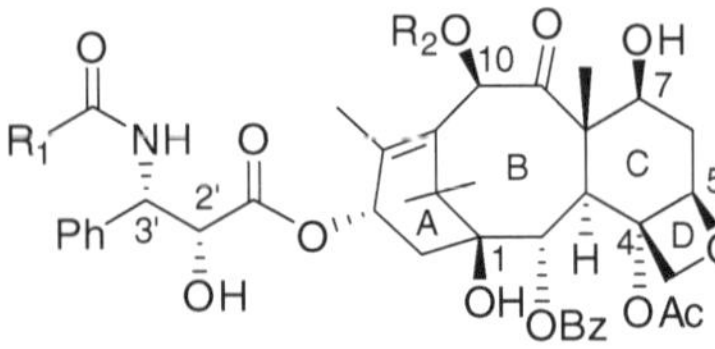

1.1 R_1 = Ph, R_2 = Ac
1.2 R_1 = *t*-BuO, R_2 = H

FIGURE 8.1 Paclitaxel and docetaxel.

structure as well as some biological test results in 1971 [9]. Paclitaxel was originally recognized as a cytotoxic and antileukemic agent with excellent activity, but it had poor solubility in water and was only available in low yield from its natural source (in 0.02% yield from Pacific Yew bark), which lowered interest in developing it as a drug. But the situation changed after further biological evaluations at the National Cancer Institute (NCI) revealed its activity against various human solid tumor xenografts in nude mice in 1977 [10]. Interest in paclitaxel was then significantly increased by the discovery by Horwitz and her coworkers of its interesting mechanism of action in 1979 [11]. They found that paclitaxel bound to β-tubulin, promoted tubulin polymerization, and stabilized microtubules from disassembly. This was an unknown mechanism at that time and was a key discovery in the development of paclitaxel.

Paclitaxel (**1.1**) was formulated for clinical trials as an emulsion in ethanol plus Cremophor EL, and the initial clinical trials encountered problems due to allergic reactions to the Cremophor EL emulsifier. Fortunately, these problems were overcome by premedication and by lengthening the administration from bolus injection to a 24-h infusion, and the first report of the clinical activity of paclitaxel against ovarian cancer appeared in 1989 [12] and against breast cancer in 1991 [13]. Paclitaxel was eventually licensed to Bristol–Myers Squibb and was approved for treatment of ovarian cancer by the U.S. Food and Drug Administration (FDA) in 1992, and for treatment of breast cancer in 1994. Paclitaxel is also used to treat patients with lung cancer, head and neck cancer, as well as advanced forms of Kaposi's sarcoma. Its annual sales peaked in 2000, reaching 1.6 billion dollars.

Docetaxel (**1.2**) is a semisynthetic analog of paclitaxel that was discovered as part of the Potier group's studies of the semisynthesis of paclitaxel by the esterification of 10-deacetyl baccatin III, which is extracted from the needles of the renewable and readily available European yew tree, *Taxus baccata* [14,15]. Docetaxel differs from paclitaxel at two positions in the structure; an *N*-*t*-butoxycarbonyl group replaced the *N*-benzoyl group in paclitaxel at the C3′ position on the side chain; in addition, docetaxel has a free hydroxyl group at the C10 position instead of the acetate ester in paclitaxel, which results in better solubility in water than paclitaxel [16]. Docetaxel has the same mechanism of action as paclitaxel involving promoting tubulin assembly, stabilizing microtubules in the cell and arresting the cell cycle, ultimately leading to apoptosis. It is approximately twice as potent as paclitaxel as a cytotoxic antimicrotubule agent [17]. Docetaxel is currently a well-established antimitotic chemotherapy medication used for the treatment of breast, ovarian, prostate, and non small-cell lung cancer, as well as locally advanced or metastatic breast or non small-cell lung cancer and metastatic castration resistant prostate cancer. It was marketed worldwide by Sanofi–Aventis under the name Taxotere in 2004.The annual sales were approximately 2.5 billion dollars in 2007.

Paclitaxel and docetaxel are two of the most successful anticancer drugs in the past two decades, and the discovery of their unique mechanism of action opened a door to the development of other drugs that act by a similar mechanism, including the recently approved ixabepilone [18]. In the past decade, extensive studies have been focused on these two drugs. Numerous taxane analogs continue to be isolated from natural sources or synthesized and then biologically evaluated to help understand the mechanism of action and structure–activity relationships of paclitaxel and to improve its therapeutic properties.

8.2 NEW CHEMISTRY OF PACLITAXEL

8.2.1 New Taxanes

The natural occurrence of taxanes has been documented in several reviews [19,20], but work has continued during the past ten years on the isolation and identification of bioactive taxanes from natural sources, and these studies have led to the isolation of numerous new members of the taxane family, some of which have promising activities.

Shen and coworkers isolated a series of new taxane diterpenoids from *Taxus mairei* and *Taxus sumatrana* (Taxaceae). In 2000, they isolated the new abeotaxane, taxumairone A (**2.1.1**) (Figure 8.2) from the seeds of *Taxus mairei* [21]. This compound exhibited significant cytotoxicity against human colon (COLO205) and epidermoid carcinoma (KB) cells with ED_{50} values of 0.1 and 1.84 μg/mL, respectively. The same group isolated the three new taxane diterpenoids, taxumairols U-W (**2.1.2–2.1.4**) (Figure 8.2), from extracts of the stem bark of Formosan *Taxus mairei* [22]. Taxumairols U and V exhibited significant activity against human hepatoma tumor cells, while taxumairol W was inactive. The two new taxoids, taxumairol Q (**2.1.5**) and 13-*O*-acetyl wallifoliol (**2.1.6**) (Figure 8.3), were later isolated from the leaves and twigs of *Taxus sumatrana* [23]. Their cytotoxicities were tested *in vitro* against oral epidermoid carcinoma (KB) cells and human liver

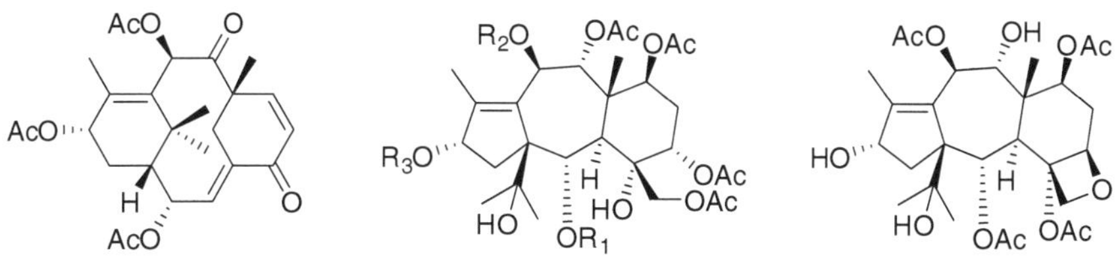

2.1.1

2.1.2 $R_1 = R_2 = H$, $R_3 = Ac$
2.1.3 $R_1 = R_2 = R_3 = H$

2.1.4

FIGURE 8.2 Taxanes **2.1.1–2.1.4**.

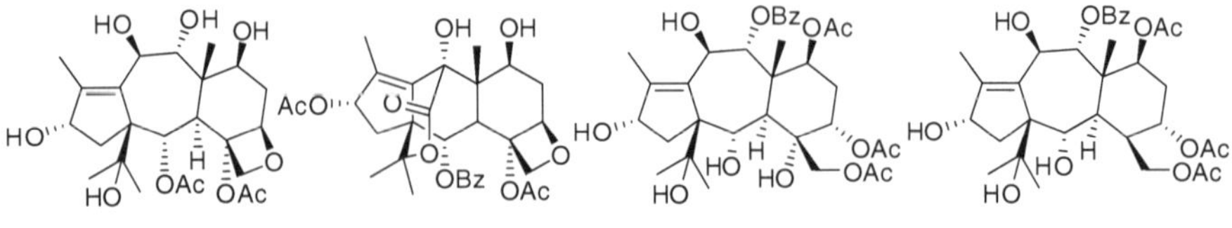

2.1.5 2.1.6 2.1.7 2.1.8

FIGURE 8.3 Taxumairol Q, 13-*O*-acetyl wallifoliol and tasumatrols E and F.

carcinoma Hepa 59T/VGH cells, but they exhibited much lower cytotoxicity than paclitaxel, and only 13-*O*-acetyl wallifoliol was weakly active in KB cells.

Tasumatrols E (**2.1.7**) and F (**2.1.8**) are rearranged taxane diterpenoids with an opened C4–C5 oxetane ring [24]. They belong to the taxchin type of taxoid, which may be biogenetically derived from taxoids of the C-4(20) double bond (Figure 8.3) [25]. These isolated taxoids were tested *in vitro* for cytotoxicities against human A549, NCI-H226, A549, and PC-3 tumor cells. Tasumatrol E (**2.1.7**) showed more promising activity than paclitaxel in all tumor cells, but tasumatrol F (**2.1.8**) was more active than paclitaxel only in A498 and PC-3 tumor cells [24].

Tasumatrols H–L (**2.1.9–2.1.13**) were isolated in 2005. Tasumatrols H–J (**2.1.9–2.1.11**) are rearranged taxane diterpenoids with lactone moieties between C1 and C9 (Figure 8.4). Tasumatrol K (**2.1.12**) is a taxane diterpenoid with a side chain connected at C14. Tasumatrol L (**2.1.13**) is the first 21-carbon taxane ester to be isolated from a natural source. Tasumatrol I (**2.1.10**) and K (**2.1.12**) have significant activities against human liver carcinoma (Hepa59T/VGH), human large cell

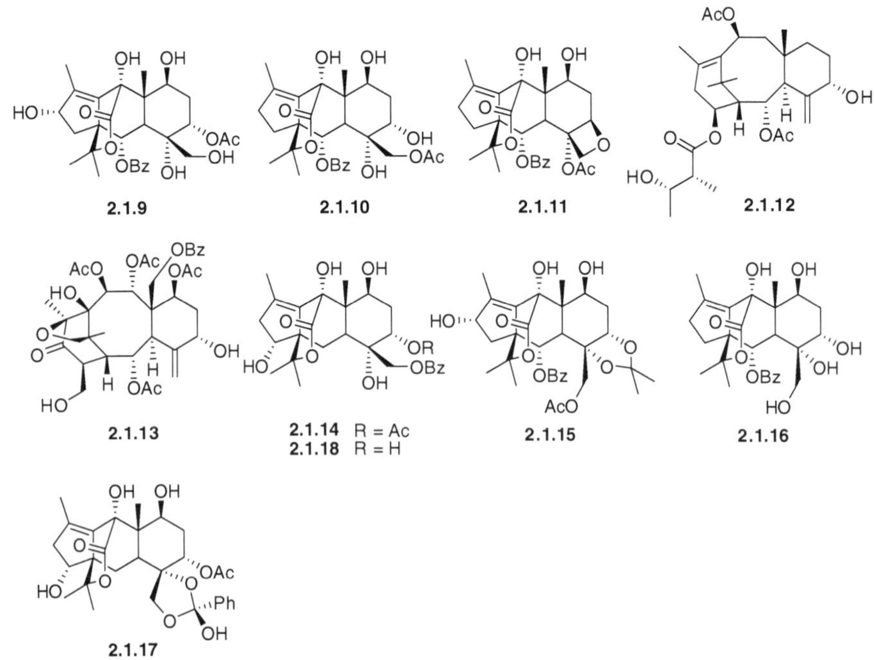

FIGURE 8.4 Tasumatrols H-L and tasumatrols P-T.

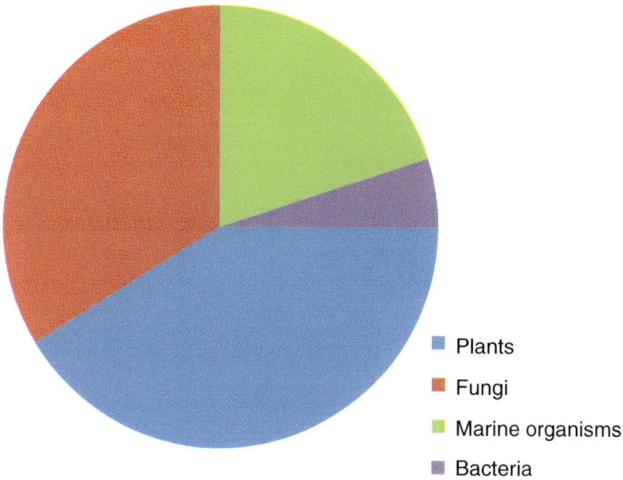

FIGURE 6.2 Types of organisms (%) studied for antifungal properties and published in *J. Nat. Prod.* in the last decade (2001–2010) (*Source*: unpublished personal analysis by the authors, 2010).

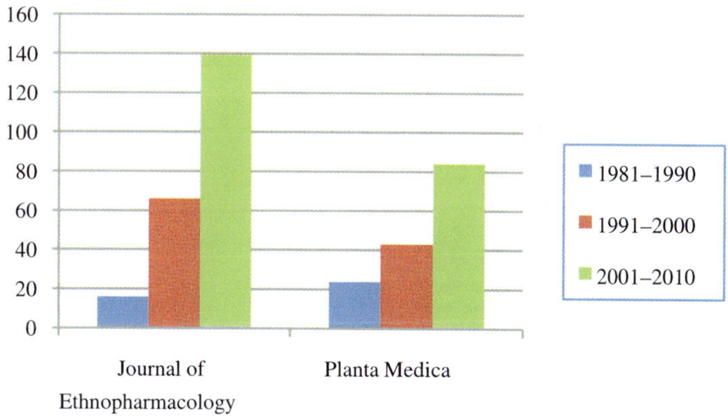

FIGURE 6.3 Number of papers on antifungal properties of plants published in *Journal of Ethnopharmacology* and *Planta Medica* in the last three decades (1981–2010).

Plant Bioactives and Drug Discovery: Principles, Practice, and Perspectives, Fourth Edition.
Edited by Valdir Cechinel-Filho.

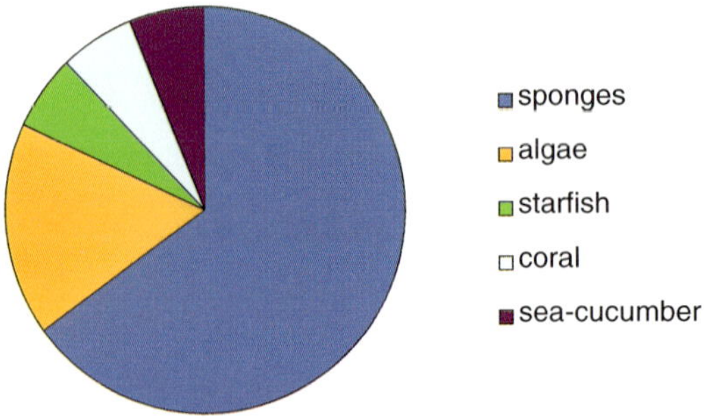

FIGURE 6.6 Percentages of the different marine organisms investigated for antifungal properties in the last decade published in *Journal of Natural Products*.

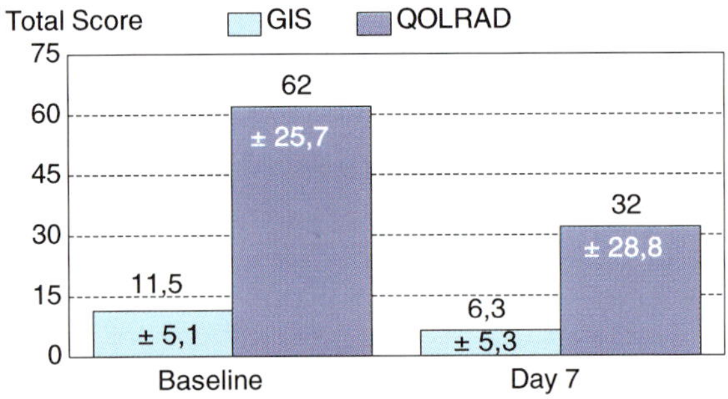

FIGURE 11.2 The total GIS and QOLRAD scores before and after 7 days potato juice therapy, mean ± SD. The absolute values decreased significantly (both $p < 0.001$).

FIGURE 12.1 Adult plants of *Virola surinamensis* (a), fruits details (b), and Waiãpi tribe (c).

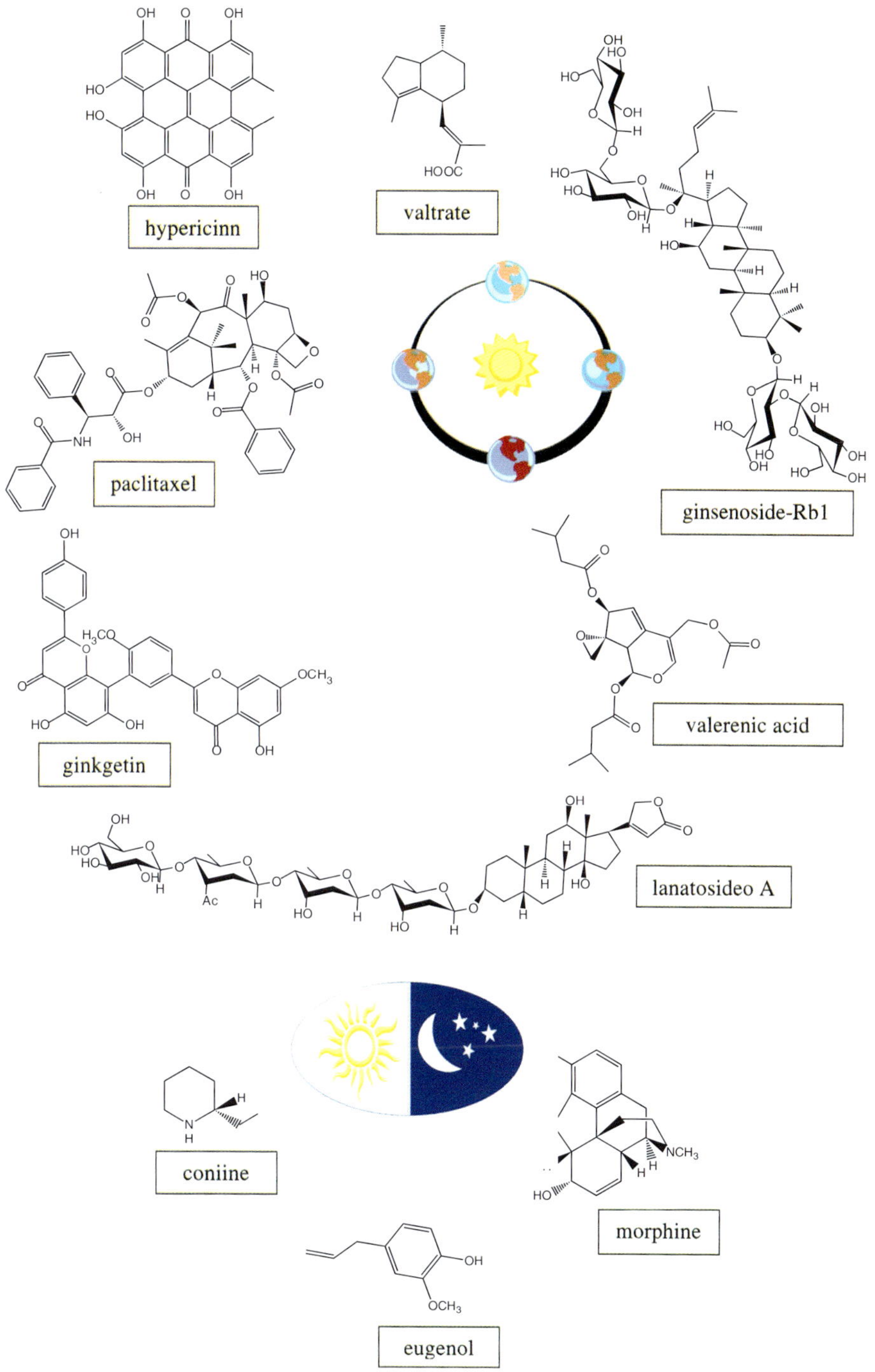

FIGURE 12.6 Examples of secondary metabolites that have their content affected during the year and altered along the day.

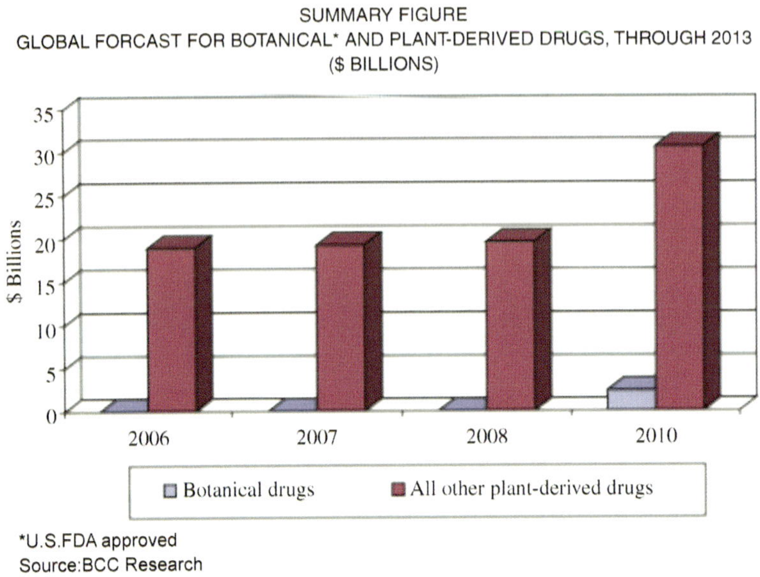

FIGURE 15.1 Global forecast for botanical* and plant-derived drugs, through 2013 ($ billions). Courtesy of BCC Research, Wellesley, MA.

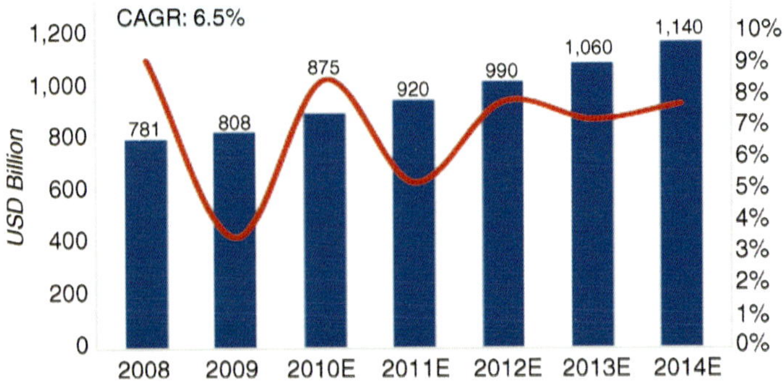

FIGURE 16.1 Global pharmaceutical market estimated. ■ Total market ■ Growth rate (*Source*: adapted from Ref. 19).

carcinoma of the lungs (NCI), human cervical epitheloid carcinoma (Hela), human colon adenocarcinoma (DLD-1), and human medulloblastoma (Med) cell lines [26].

Tasumatrols P–T (**2.1.14–2.1.18**) (Figure 8.4) were isolated from the leaves and twigs of *Taxus sumatrana* in 2007 [27]. These compounds possess an 11(15 → 1), 11 (10 → 9)-diabeotaxane skeleton. Tasumatrols P (**2.1.14**), S (**2.1.17**), and T (**2.1.18**) have an α-hydroxy group at C14, and tasumatrol Q (**2.1.15**) was a natural 4,5-acetonide derivative. Unusually, tasumatrol S (**2.1.17**) contained a spiro-connected 2-hydroxy-2-phenyl-1,3-dioxolane ring. Only tasumatrol P (**2.1.14**) exhibited mild cytotoxicity toward Hela and Daoy cancer cells with ED_{50} values of 22.7 and 22.4 μg/mL, respectively.

The tasumatrols U–Z (**2.1.19–2.1.24**) (Figure 8.5) were isolated from the leaves and twigs of *Taxus sumatrana* in 2008 [28]. Tasumatrol V (**2.1.20**) and Y (**2.1.23**) have a rare five-membered lactone ring between C8 and C9. Their cytotoxicities were investigated *in vitro* against human liver carcinoma (Hepa2), colon adenocarcinoma (WiDr), and cervical epitheloid carcinoma (Hela) cells. Only tasumatrol Y (**2.1.23**) was active against human Hepa2 tumor cells with an ED_{50} value of 3.5 μg/mL; the other taxoids were inactive in all tumor cell lines.

Taiwantaxin A (**2.1.25**) is a rearranged taxane with an opened oxetane moiety, and taiwantaxin B (**2.1.26**) and C (**2.1.27**) are both rearranged taxanes possessing a γ-lactone between C8 and C9. Taiwantaxin D (**2.1.28**) is an example of a taxane diterpenoid with a tetrahydrofuran ring comprising C2, C3, C4, and C20 (Figure 8.5) [29]. These compounds were isolated and identified in 2009, and they were tested against the human prostate cancer (PC-3) cell line. Only taiwantaxin B (**2.1.26**) exhibits cytotoxicity, with an IC_{50} of 8.9 μM, which is 20 times less active than paclitaxel in this cell line.

In 2009, four novel taxane diterpenoids (**2.1.29–2.1.32**) (Figure 8.6) possessing an unusual 6/12-membered ring system were discovered from an acetone extract of the leaves and twigs of *Taxus sumatrana* [30]. The preliminary *in vitro* biological evaluations revealed that taxoids **2.1.31** and **2.1.32** have modest cytotoxicity to human colon tumor cells with IC_{50} values of 3.95 and 2.80 μg/mL, respectively.

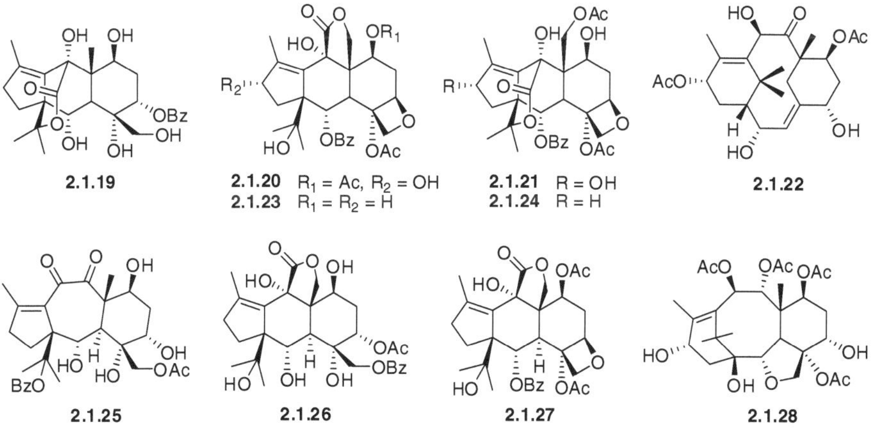

FIGURE 8.5 Tasumatrols U-Z and taiwantaxins A-D.

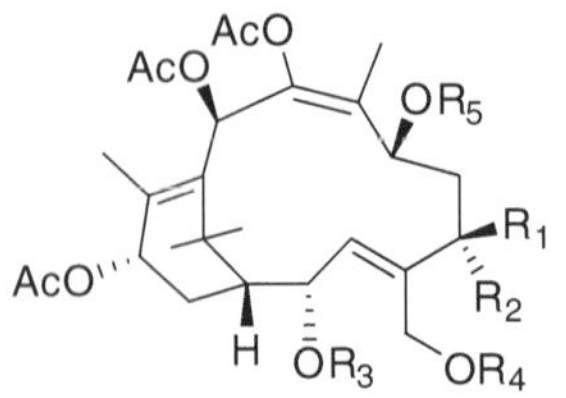

2.1.29 $R_1 + R_2 = O$, $R_3 = R_4 = Ac$, $R_5 = H$
2.1.30 $R_1 = H$, $R_2 = OH$, $R_3 = R_4 = Ac$, $R_5 = H$
2.1.31 $R_1 = H$, $R_2 = CinnO$, $R_3 = R_4 = H$, $R_5 = H$
2.1.32 $R_1 = H$, $R_2 = AcO$, $R_3 = Ac$, $R_4 = H$, $R_5 = Ac$

FIGURE 8.6 Tri- and bicyclic taxoids **2.1.29–2.1.32**.

Kosugi and coworkers discovered the two novel taxoids, taxinine NN-7 (**2.1.33**) and 3,11-cyclotaxinine NN-2 (**2.1.34**) (Figure 8.7), from the neutral fraction of the ethyl acetate extract of a mixture of needles and young stems of *Taxus cuspidata* in 2000 [31]. Both compounds showed activity as inhibitors of multidrug resistance (MDR) of tumor cells toward vincristine (VCR) accumulation in MDR tumor cells, indicating that the cinnamoyl-oxy group at C5 effectively increased the cellular accumulation of VCR in MDR tumor cells.

Zamir and coworkers isolated four novel taxane analogs (**2.1.35–2.1.38**) in 2001 from the needles of *Taxus canadensis* (Figure 8.7) [32]. Taxane **2.1.35**, or 9-dihydrotaxol, which had previously been prepared from 13-acetyl-9-dihydrobaccatin III isolated from *T. canadensis* [33], exhibited similar cytotoxicity to paclitaxel in the breast adenocarcinoma (MCF7) cell line with an IC_{50} value of 10 nM.

The taxine 5α-*O*-(3′-dimethylamino-3′-phenylpropionyl) taxinine M (**2.1.39**) (Figure 8.8) was isolated in 2001 by Konishi and his colleagues from the needles of the Himalayan yew, *Taxus wallichiana* Zucc [34]. This taxine was only weakly active against the A549 cell line, with an ED_{50} of 10.1 ± 2.4 μM.

In 2004, Shi and coworkers isolated three novel taxane-derived dipropellanes (**2.1.40–2.1.42**) from *T. canadensis* needles [35]. They were tested for *in vitro*

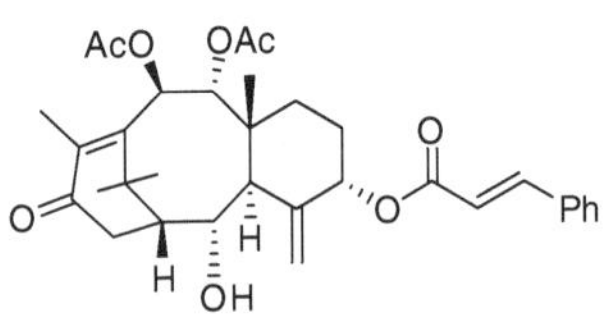

2.1.33

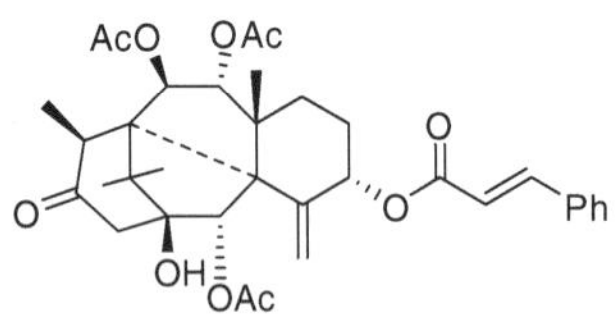

2.1.34

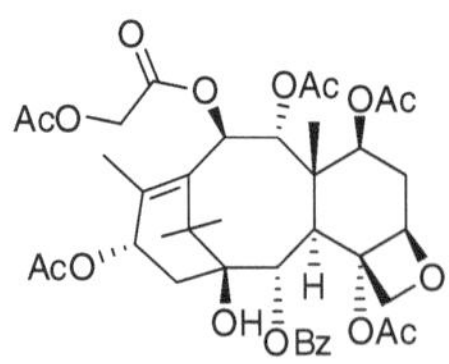

2.1.38

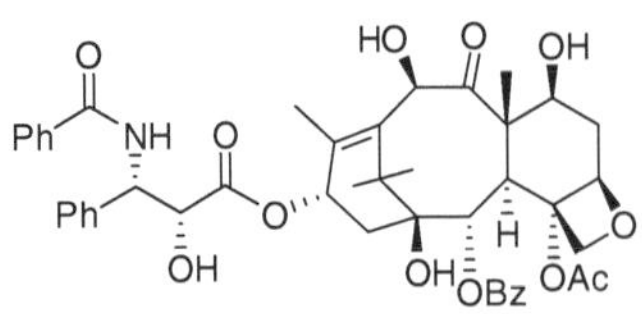

2.1.35

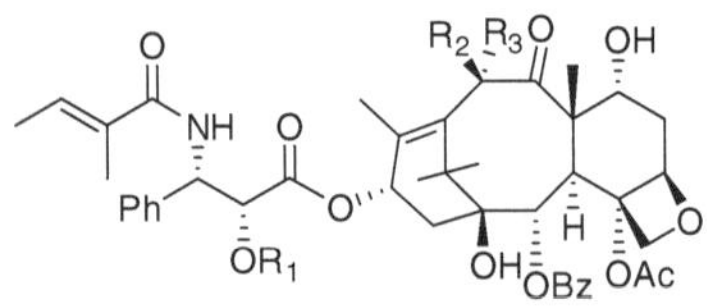

2.1.36 $R_1 = Ac$, $R_2 = OAc$, $R_3 = H$
2.1.37 $R_1 = H$, $R_2 + R_3 = O$

FIGURE 8.7 Taxanes **2.1.33–2.1.38**.

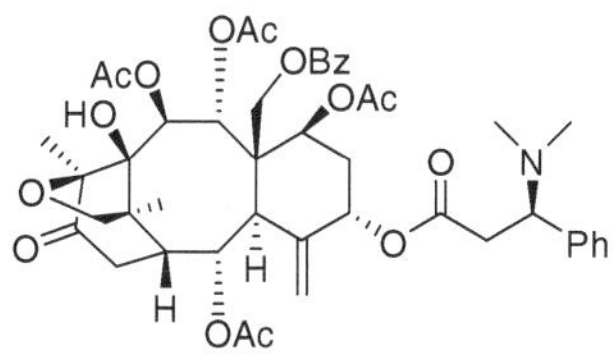

2.1.39

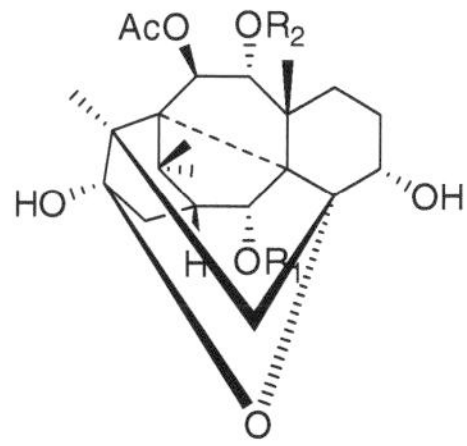

2.1.40 R_1 = Ac, R_2 = H
2.1.41 R_1 = H, R_2 = Ac
2.1.42 R_1 = Ac, R_2 = Ac

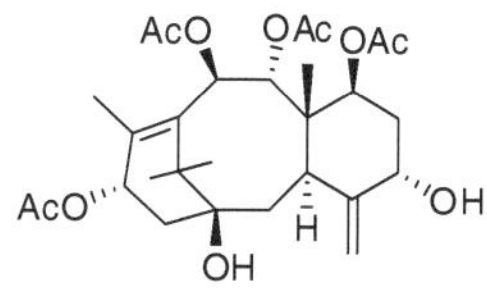

2.1.43

FIGURE 8.8 Taxanes **2.1.39–2.1.43**.

antileukemic activity against human HL-60 myeloid leukemic cells and in the L1210 leukemia mouse model, but showed no growth inhibition.

Chattopadhyay and coworkers re-isolated the taxane diterpenoid **2.1.43** (Figure 8.8) from the Himalayan yew, *T. wallichiana* [36] previously isolated by Barboni et al. [37]. It was found to possess dose-dependent cytotoxic activity against five human cancer cell lines (MCF-7, WRL-68, KB, PA-1, and Colo-320DM), and its cytotoxicity to some cell lines was equal to or better than those of the two clinically used drugs, vincristine and vinblastine. However, it was nontoxic to normal hepatocytes of mice, even at 50 μg/mL. The immunomodulating activity of this taxoid was evaluated using human lymphocytes and indicated that it had similar immunomodulatory properties to those shown by ConA.

Shi and coworkers identified a new taxane analog from rooted cuttings of the Canadian yew, *T. canadensis* [38]. Interestingly, the 1D and 2D-NMR spectra revealed that this taxane was composed of two rotamers (**2.1.44a** and **2.1.44b**) in a ratio of 2:1 in $CDCl_3$ (Figure 8.9). It is the first example of a taxane with an *N*-formyl group composed of *cis*- and *trans*-rotamers. This compound was investigated for the growth inhibitory activities against human tumor cells. The results indicated that it exhibited moderate activities against HMC-1, KT, T98, and MM1-CB cell lines, but it was inactive in the HeLa, HEC-1, SIN3, HOC-21, HAC-2, and U251-SP cells.

8.2.2 New Chemistry of Paclitaxel

The success of the taxoid drugs has led to a surge of interest in paclitaxel's chemistry and biology since the 1990s. A large number of studies have been carried out to understand paclitaxel's chemistry in order to prepare enhanced analogs for new generation taxoid drugs. Chemists have developed several total synthetic approaches and many semisynthetic approaches to obtain paclitaxel chemically, and numerous

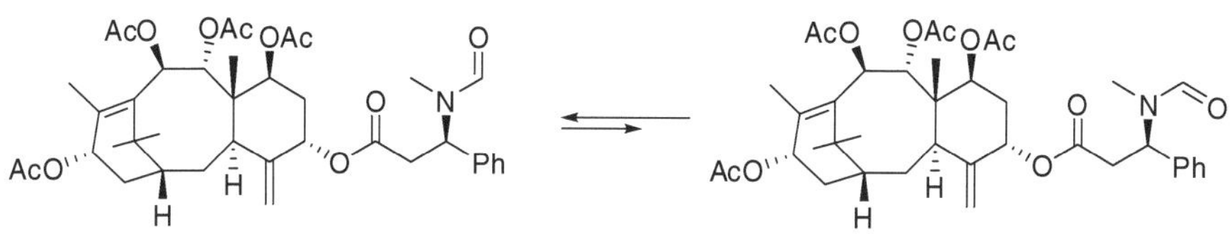

2.1.44a **2.1.44b**

FIGURE 8.9 Two rotamers **2.1.44a** and **2.1.44b**.

methods have been found to modify the complicated structure of paclitaxel at certain positions. This section summarizes the chemistry of paclitaxel published since 2000.

8.2.2.1 ***Total Synthesis.*** Numerous approaches to the total synthesis of paclitaxel have been made since 2000, although interest in this area as a whole has diminished somewhat following the successful syntheses of Holton [39,40], Nicolaou [41], Danishefsky [42], Wender [43,44], and Mukaiyama [45] in the 1990s and also since the "Taxol supply crisis" [46] was solved by the development of successful semi-synthetic and biosynthetic approaches to paclitaxel production. The only successful total synthesis of paclitaxel published in the past 10 years is that of Kuwajima and his collaborators (Scheme 8.1), who published their work in 2000 [47].

The desired ABC paclitaxel skeleton (**2.2.4**) was established via a coupling reaction of the A ring hydroxyaldehyde (**2.2.1**) with the aromatic C ring fragment (**2.2.2**) followed by Lewis acid-mediated B ring cyclization. The introduction of the requisite functionalities at the different positions of the ABC tricarbocyclic skeleton, the D ring cyclization by DBU, and the attachment of the C13 side chain completed the enantioselective total synthesis of paclitaxel in an overall yield of 0.06% over 46 steps.

Although no new total syntheses have appeared since the Mukaiyama work noted above, several authors have reported new approaches to the challenge. Thus, Paquette reported the construction of an oxetane D-ring in the framework of a highly functionalized taxane intermediate in 2002 [48]. This approach adopted a trimethylsilyl triflate promoted epoxide-opening step and a subsequent stereocontrolled, regioreversed oxirane formation and reductive transposition to establish the oxetane D-ring (**2.2.17**) (Scheme 8.2). They also found that OsO_4 oxidation opened up the C-ring, forming a rearranged taxane (**2.2.18**).

Malacria and coworkers developed a new efficient construction of the ABC taxoid skeleton via a combination of a cobalt(I)-[2+2+2] cyclization with a [4+2] reaction (Scheme 8.3) [49,50]. This route provided the pentacyclic structure **2.2.20** including the ABC taxoid core with an all-carbon D-ring in 65% overall yield, starting from an acyclic precursor **2.2.19**.

Another approach to assembling the ABC taxoid skeleton was achieved by Villalva-Servin and coworkers (Scheme 8.4) [51,52]. They employed sequential magnesium-mediated carbometallation of allyl-substituted propargyl alcohols followed by diastereoselective Lewis acid catalyzed intramolecular Diels–Alder reactions and ring-closing metathesis to establish the ABC taxane core structure **2.2.24**.

The assembly of the taxoid ABC skeleton oxa-bridged at the C2 and C9 position (**2.2.26**) from readily synthesized low-cost starting materials (**2.2.25**) was achieved by Hamon and coworkers (Scheme 8.5) [53]. This highly stereoselective strategy provided a taxoid diterpene framework in high-chemical yield and good stereocontrol on the basis of the simple diastereoselection offered by the enolate geometry and the feasibility of efficient cyclizations mediated by SmI_2. These ABC ring system taxoids could potentially serve as precursors in the synthesis of other taxanes.

A practical and convenient five-step protocol to synthesize the ABC taxoid ring system was established by Chavan and his collaborators (Scheme 8.6) [54]. They synthesized both the C-aromatic ABC skeleton **2.2.31** and the C-alicyclic ABC

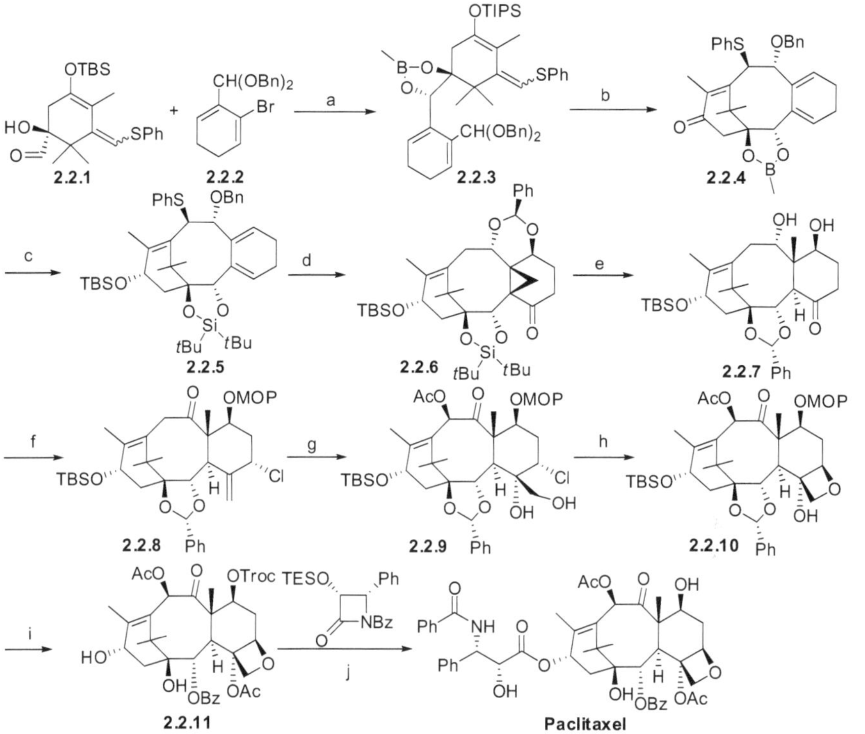

(a) i. *t*BuMgCl, **2.2.1**, THF, -78°C, then *t*BuLi, **2.2.2**, -78°C, 68%; ii. $(MeBO)_3$, pyridine, benzene, rt, 77%; (b) $TiCl_2(O\textit{i}Pr)_2$, DCM, -78°C to 0°C; (c) i. pinacol, DMAP, benzene, rt, 59% over 2 steps; ii. BuLi, *t*BuSiHCl, THF, -78°C to 0 °C; iii. DIBAL, DCM, -78°C; iv. TBSOTf, 2,6-lutidine, DCM, 62% over 3 steps; (d) i. hv, O_2, TPP, DCM, rt; ii. Bu_3SnH, AIBN, benzene, reflux, 80%; iii. Pd/C, H_2, EtOH, rt; iv. $PhCH(OMe)_2$, CSA, DCM, -23°C, 79%; v. Et_2Zn, $ClCH_2I$, toluene, 0°C, 66%; vi. Dess-Martin, DCM, rt, 77%; (e) i. H_2, $Pd(OH)_2$, EtOH, 84%, then triphosgene, pyridine, DCM, -45°C, then TBAF, AcOH,THF, 95%; ii. $PhCH(OMe)_2$, PPTS, benzene, reflux, 86%, then K_2CO_3, MeOH, THF, 100%; iii. SmI_2, THF, rt, 100%; iv. TBAF, THF-cat. BHT, then NaOMe, MeOH, degassed MeOH-cat. BHT, 45%; v. $PhB(OH)_2$, pyridine, DCM, then TBSOTf, lutidine, DCM, -45°C, then H_2O_2, $NaHCO_3$, H_2O-AcOEt,rt, 70%; (f) i. Dess-Martin, DCM, then PPTS, 2-methoxyprop-1-ene, DCM, 97%; ii. KHMDO, $PhNTf_2$, THF, -78°C, 89%; iii. $TMSCH_2MgCl$, $Pd(PPh_3)_4$, Et_2O, 91%; iv. NCS, MeOH, 88%, then PPTS, 2-methoxyprop-1-ene, DCM, 89%; (g) i. LDA, MoOPH, THF, -23°C, 80%, then Ac_2O, DMAP, DCM, 92%; ii. DBN, toluene, reflux, 68%; iii. OsO_4, pyridine, Et_2O, 86%; (h) DBU, toluene, reflux, 86%; (i) i. PPTS, MeOH, then TESCl, DMF, 97%; ii. $Pd(OH)_2$, H_2, EtOH, 97%, then triphosgene, pyridine, DCM, -78°C to 0°C, 94%, then Ac_2O, DMAP, DCM, 66%; iii. PhLi, THF, -78°C, 83%, then HF/pyridine, THF, 88%, then TrocCl, pyridine, DCM, 94%; iv. TASF, THF, 80%; (j) i. LiHMDS, β-lactam, THF, -78°C to 0°C, 77%; ii. Zn, AcOH, H_2O, 84%.

SCHEME 8.1 Kuwajima total synthesis of paclitaxel.

skeleton **2.2.34** by using the versatility of sulfur in its various oxidation states for C-C bond formation and cleavage. The synthetic schemes are shown in Scheme 8.6. This protocol has the potential to deliver a large number of advanced taxoids as well as fused ring systems leading to complex fused natural products.

The construction of the ABC taxoid skeleton **2.2.38** was achieved by Kakiuchi and coworkers (Scheme 8.7) [55]. The intermediate **2.2.37** that resulted from compound **2.2.36** underwent aldol reaction to establish the ABC skeleton of taxoids.

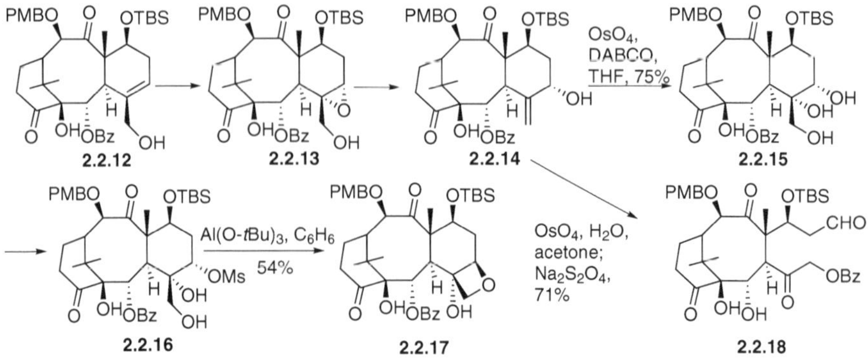

SCHEME 8.2 Paquette's approach to the D-ring of paclitaxel.

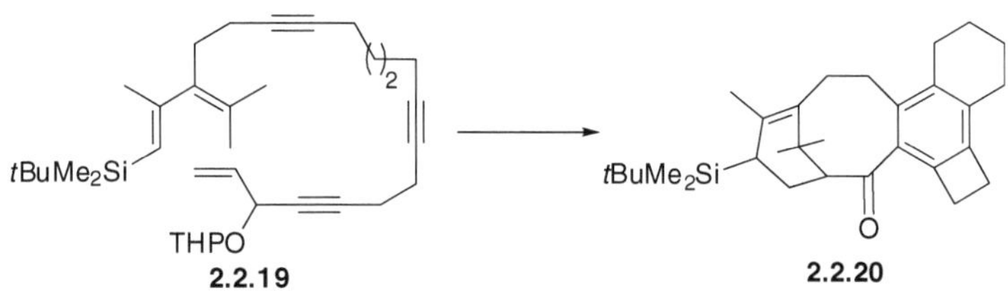

SCHEME 8.3 Formation of the ABC core of **2.2.20**.

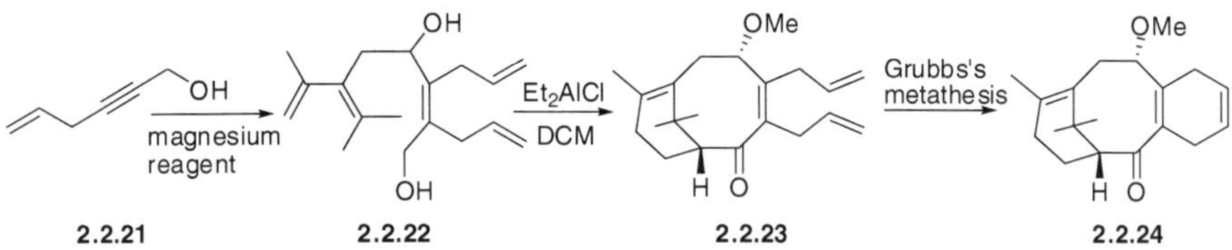

SCHEME 8.4 Formation of the ABC core of **2.2.24**.

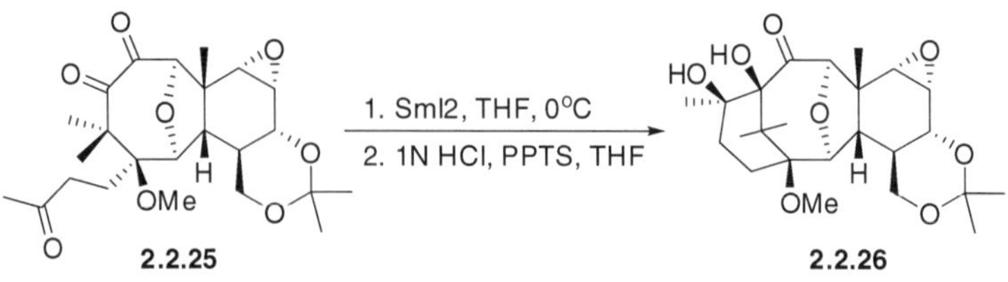

SCHEME 8.5 Formation of the ABC core of **2.2.26**

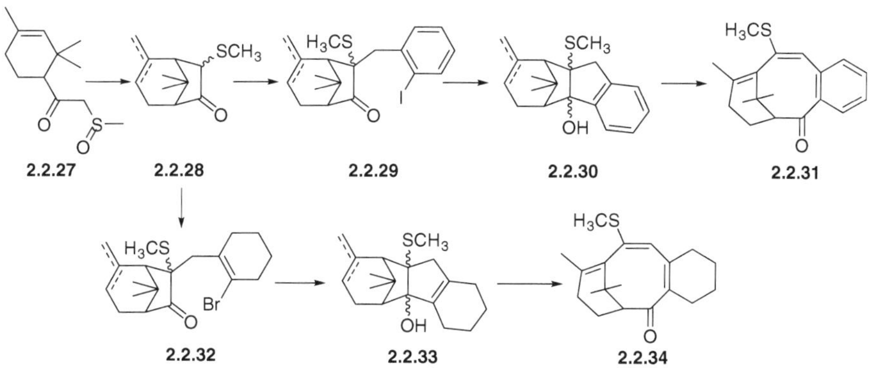

SCHEME 8.6 Formation of the ABC core of **2.2.31** and **2.2.34**.

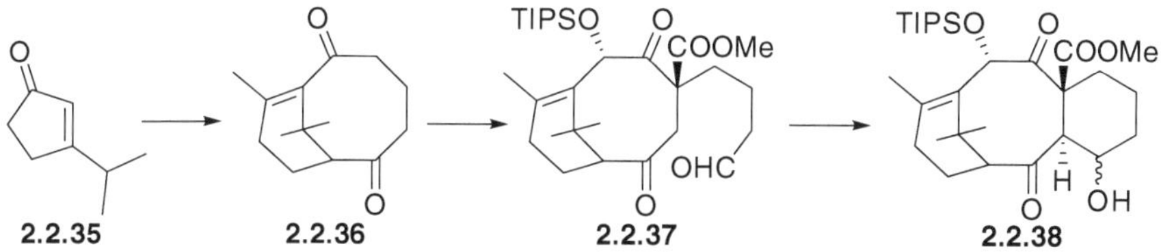

SCHEME 8.7 Formation of the ABC core of **2.2.38**.

A promising approach from the Pattenden group has also been reported [56]. A tricyclic product was prepared in six steps from a cyclohexene derivative and a stannane. The synthesis was then abandoned because of lack of material, but the authors point out that their advanced intermediate could be converted to an advanced intermediate of the successful Mukaiyama synthesis, so presumably this pathway could have resulted in another total synthesis.

8.2.2.2 ***Semisynthesis: Synthesis of the C13 Side Chain.*** In contrast to the relative paucity of new approaches to the total synthesis of paclitaxel, many new approaches to the synthesis of the paclitaxel side chain have been published and are described below.

Kim and collaborators reported a new synthesis of β-lactam **2.2.43**, a paclitaxel side chain precursor (Scheme 8.8). This approach started from L-tartaric acid and produced the desired β-lactam in a high yield and excellent diastereoselectivity [56a].

Wuts and coworkers synthesized the paclitaxel side chain **2.2.47** via the epoxide intermediate **2.2.46**, prepared from 2-chloro-3-phenyl-3-oxopropionate **2.2.44**

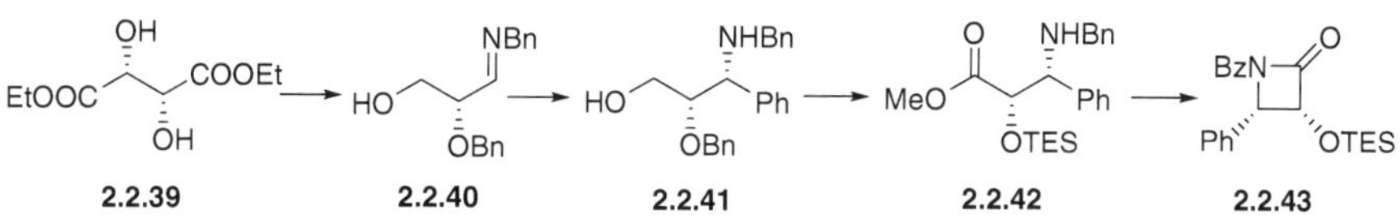

SCHEME 8.8 Synthesis of side chain precursor **2.2.43**.

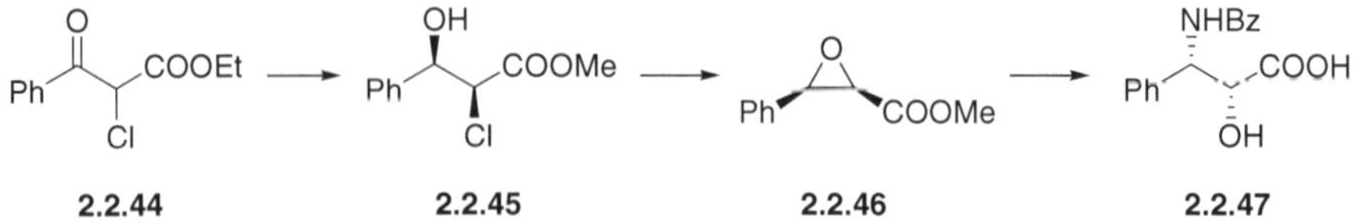

SCHEME 8.9 Synthesis of side chain **2.2.47**.

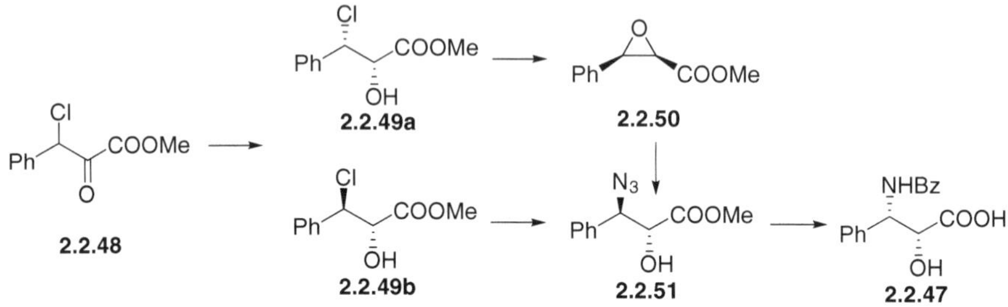

SCHEME 8.10 Synthesis of side chain **2.2.47**.

(Scheme 8.9) [57]. This strategy gave the paclitaxel side chain in high diastereoselectivity and enantioselectivity, although the overall yield was not high.

Tsuboi and collaborators adopted a chemoenzymatic approach to synthesize the two diastereomers **2.2.49a** and **2.2.49b**, which were subsequently converted to the paclitaxel C13 side chain in good yields and high stereoselectivities (Scheme 8.10) [58].

Aggarwal and coworkers developed a new approach to synthesize paclitaxel side chain ester **2.2.56** by utilizing their catalytic asymmetric sulfur ylide-mediated aziridination process of *N*-sulfonylimines (**2.2.52**) (Scheme 8.11) [59]. This approach is efficient and provides a high degree of enantioselectivity.

Choudary and collaborators described a convenient synthetic approach to the paclitaxel side chain (Scheme 8.12) [60]. They adopted a one-pot reaction mediated by the recyclable homogeneous bifunctional catalysts composed of OsO_4^{2-}–WO_4^{2-} and trifunctional catalysts comprising $PdCl_4^{2-}$–OsO_4^{2-}–WO_4^{2-} species to synthesize the chiral diol **2.2.59** in high yield and stereoselectivity; this was readily converted to the paclitaxel side chain ester **2.2.56**.

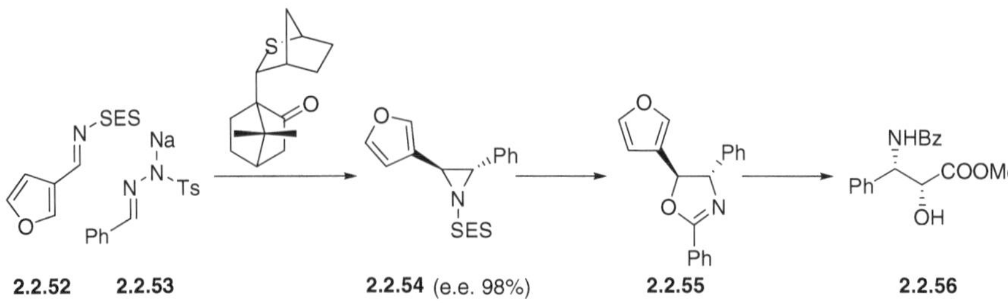

SCHEME 8.11 Synthesis of side chain **2.2.56**.

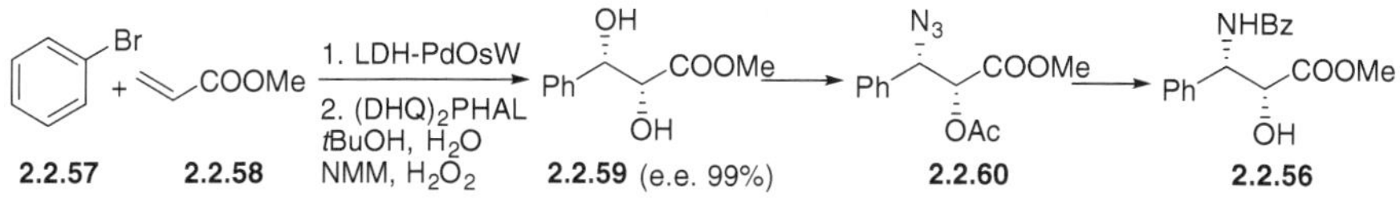

SCHEME 8.12 Synthesis of side chain **2.2.56**.

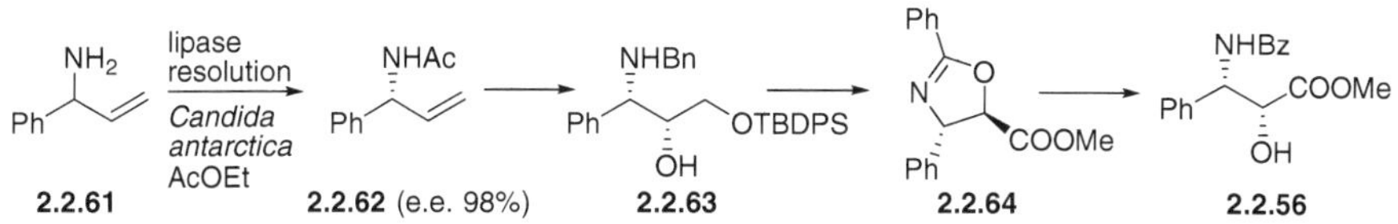

SCHEME 8.13 Synthesis of side chain **2.2.56**.

Corelli and coworkers developed a route to synthesize the paclitaxel side chain ester **2.2.56** via the oxazoline intermediate **2.2.64**, starting from the substituted 1-aryl-2-propenyl amine **2.2.62** prepared in high enantiomeric purity via lipase-catalyzed resolution of the corresponding racemates (Scheme 8.13) [61].

Barua and collaborators achieved a remarkably improved synthesis of the paclitaxel side chain **2.2.47** (Scheme 8.14) [62]. The approach is considerably simpler and shorter and goes in higher yield than other approaches. The key step of this approach involved Shibasaki's asymmetric Henry reaction by using the La-(*R*)-BINOL complex, which provided the desired stereochemistry (**2.2.67**) with high enantioselectivity.

An efficient protocol for the synthesis of the paclitaxel side chain **2.2.47** via a highly diastereoselective three-component coupling of imines, benzaldehyde, and EDA was developed by Somfai and coworkers (Scheme 8.15) [63]. This protocol provided a short enantioselective synthesis of the C13 paclitaxel side chain.

Qin and collaborators recently published their new approach to synthesize the paclitaxel side chain through a lithium enolate addition of the *O*-Boc-α-hydroxyacetate methyl ester **2.2.72** to benzylidene **2.2.73** in high yield with excellent

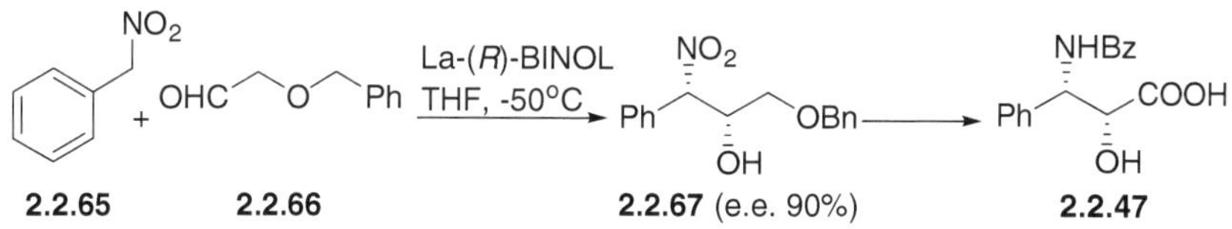

SCHEME 8.14 Synthesis of side chain **2.2.47**.

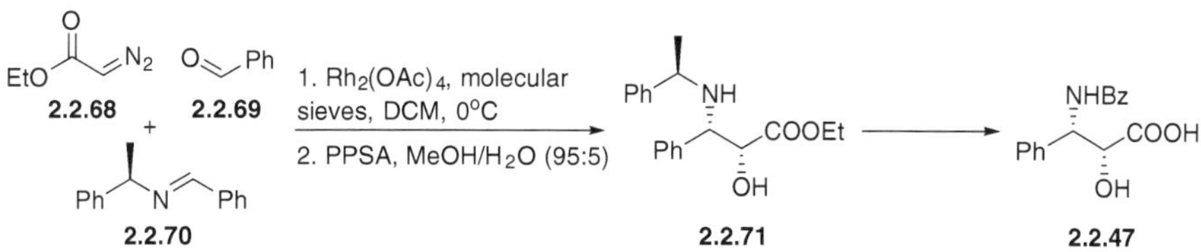

SCHEME 8.15 Synthesis of side chain **2.2.47**.

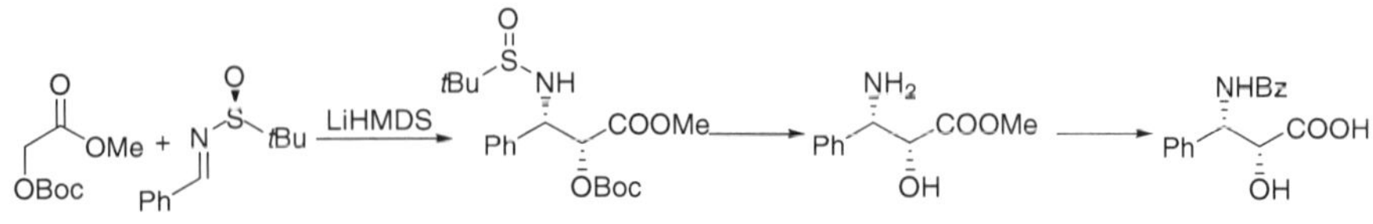

2.2.72 2.2.73 2.2.74 2.2.75 2.2.47

SCHEME 8.16 Synthesis of side chain **2.2.47**.

diastereoselectivity (Scheme 8.16) [64]. By using this strategy, a series of paclitaxel side chain derivatives were prepared for the semisynthesis of paclitaxel analogs.

A short, convenient and highly enantioselective catalytic route to the side chain of docetaxel **2.2.79** was developed by Armando Cordova and coworkers (Scheme 8.17) [65]. The key intermediate **2.2.78** was produced via the organocatalytic asymmetric reaction between protected α-oxyaldehyde (**2.2.77**) and *N*-Boc-protected aryl imine (**2.2.76**), and the oxidation of this intermediate by $NaClO_2$ gave the benzyl protected side chain of docetaxel **2.2.79**.

Soderquist and coworkers reported an asymmetric γ-methoxyallylation of an imine with the configurationally stable boron reagent **2.2.80** (Scheme 8.18) [66]. The key intermediate, nonracemic *threo*-β-methoxyhomoallylic amine **2.2.81** was produced in high yields with excellent stereoselectivity, and provided ready access to the paclitaxel side chain.

A convenient approach to the side chain **2.2.47** via chiral Brønsted acid-$Rh_2(OAc)_4$ co-catalyzed three-component reactions of diazoacetates **2.2.83** with alcohol **2.2.84** and imine **2.2.85** was achieved by Hu and coworkers (Scheme 8.19) [67]. This new synthetic strategy produced paclitaxel side chain in a good yield with moderate diastereoselectivity and high enantioselectivity.

8.2.2.3 ***The Synthesis of Structurally Modified Paclitaxels.*** The chemistry of paclitaxel is enormously rich and varied, and it is impossible to give a complete

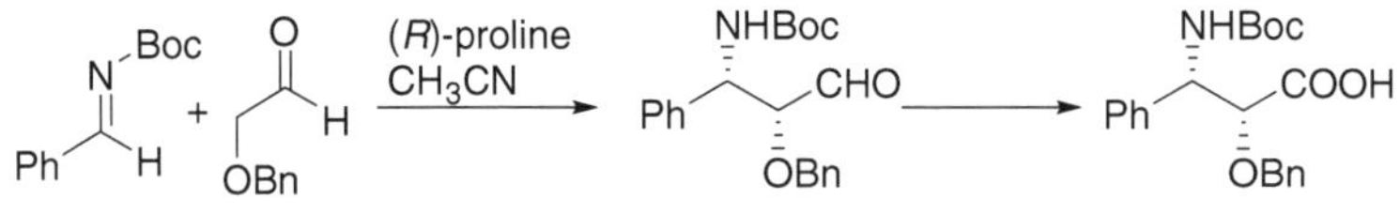

2.2.76 2.2.77 2.2.78 (e.e. 99%) 2.2.79

SCHEME 8.17 Synthesis of docetaxel side chain **2.2.79**.

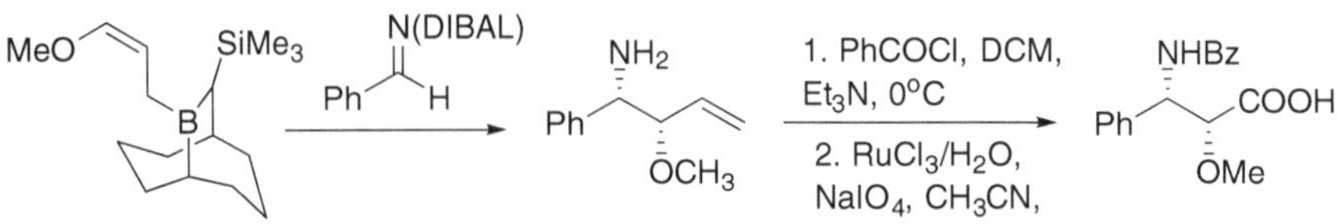

2.2.80 2.2.81 (e.e.86%) 2.2.82

SCHEME 8.18 Synthesis of paclitaxel side-chain analog **2.2.82**.

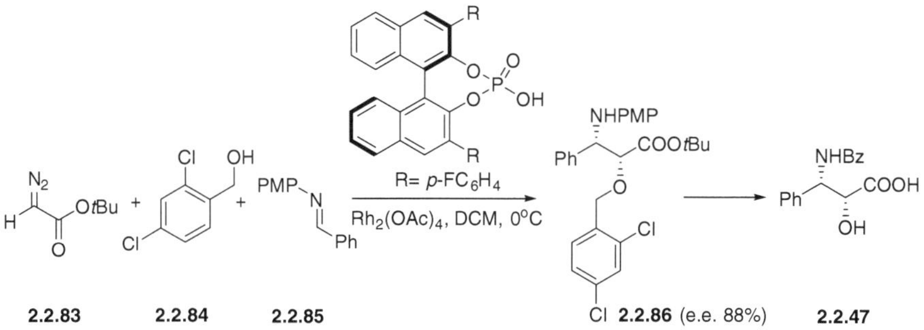

2.2.83 2.2.84 2.2.85 2.2.86 (e.e. 88%) 2.2.47

SCHEME 8.19 Synthesis of paclitaxel side chain **2.2.47**.

summary of it in this chapter. This section introduces the subject by giving some selected examples of methods used for the synthesis of modified paclitaxels over the last 10 years, while Section 8.2.3 will summarize the compounds made by these and other methods without any discussion of the synthetic pathways used.

Replacement of the oxygen atom at the C2 position with nitrogen to prepare the 2α-benzoylamido analog **2.2.92** of docetaxel was achieved by Fang and collaborators (Scheme 8.20) via azidation of the epoxide intermediate **2.2.88** to give the azide **2.2.89**. This was converted to side chain derivative **2.2.90**, which was subjected to hydrogenation to form the free amine **2.2.91** at the C2 position, followed by benzoylation and deprotection to give the final product **2.2.92** [68]. This product was evaluated for its cytotoxicity against three human cancer cell lines, A-549, KB

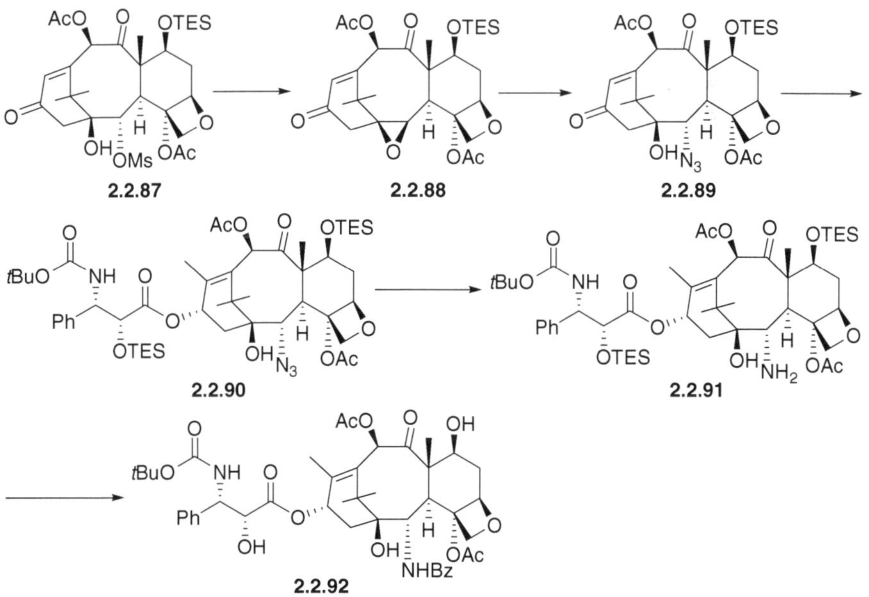

SCHEME 8.20 Synthesis of the C2 modified docetaxel analog **2.2.92**.

and A2780, and was found to exhibit less potency than paclitaxel, indicating that the ester linkage at the C2 position was crucial to maintain the cytotoxicity of paclitaxel.

C6- and C7-modified paclitaxel analogs (**2.2.95**, **2.2.96**, **2.2.98**, and **2.2.99**) were synthesized via the 6,7-olefin and the 6,7-diol derivatives of paclitaxel (Scheme 8.21) [69]. These analogs were evaluated for their cytotoxicities against human cancer cell lines, HCT116 and A2780. Interestingly, the C6 azido analog was found to be about three times more cytotoxic than paclitaxel, while the others exhibited less potency than paclitaxel.

The synthesis of a series of novel 7-deoxy-9-dihydro-9-(ω-nucleotidylsuccinyl)-paclitaxel analogs (**2.2.104**, **2.2.107**, **2.2.108**, and **2.2.109**) was achieved from a

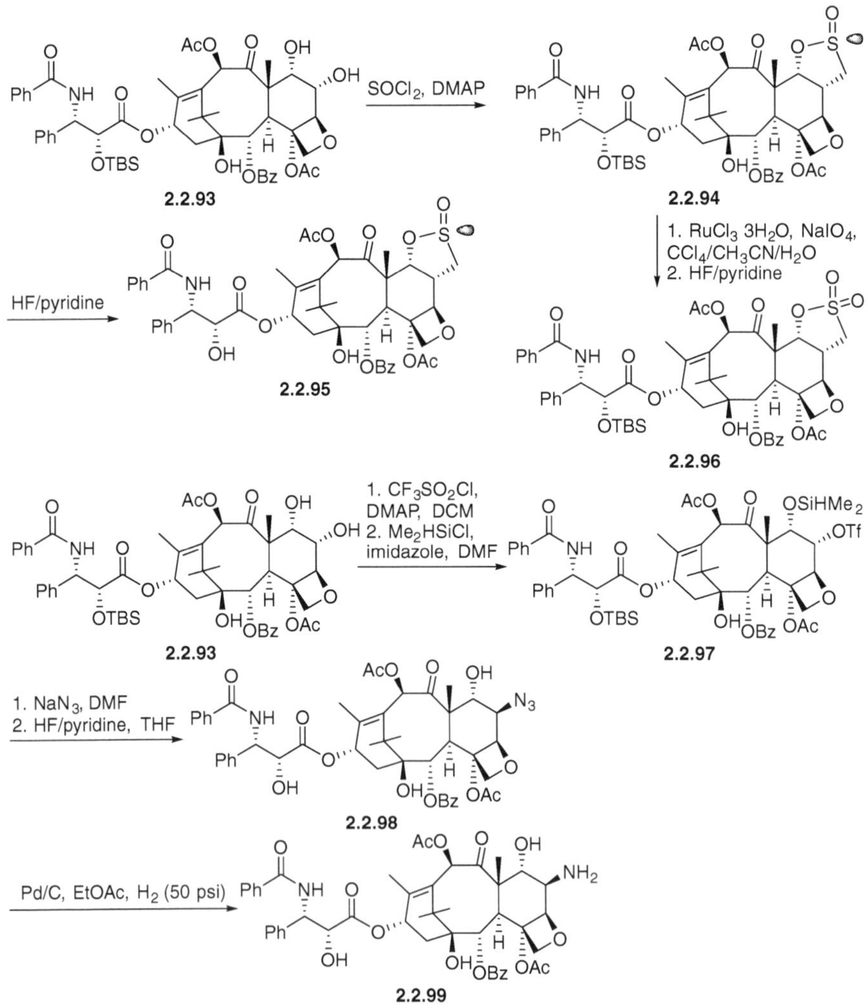

SCHEME 8.21 Synthesis of the C6–C7 modified paclitaxel analogs **2.2.95–2.2.99**.

natural taxoid by Cheng and coworkers (Scheme 8.22) [70]. The natural taxoid **2.2.100** was first converted to 7-deoxy-9-dihydrobaccatin, to which the functional heterocyclic fragments were connected at the C9 position through succinyl linkers.

Biological evaluation of the products indicated that three of the analogs had similar or superior activity to paclitaxel in BEL-7402 and ECa-109 cancer cell lines, although all of the analogs were less effective in promoting tubulin assembly than paclitaxel.

Danieli and coworkers synthesized bifunctional paclitaxel and docetaxel analogs (**2.2.110**, **2.2.111**, and **2.2.112**) with conjugated thiocolchicine at the C7, C10, and C2′ positions through succinyl linkers (Figure 8.10) [71]. These conjugates were tested in a tubulin assembly assay and evaluated for their cytotoxicities against human mammary cancer cell line MCF7. In the tubulin assembly assay, the C7 and C10 conjugates (**2.2.110** and **2.2.111**) exhibited a little higher potency than paclitaxel, and the C2′ conjugate **2.2.112** lost its activity, consistent with previous observations that

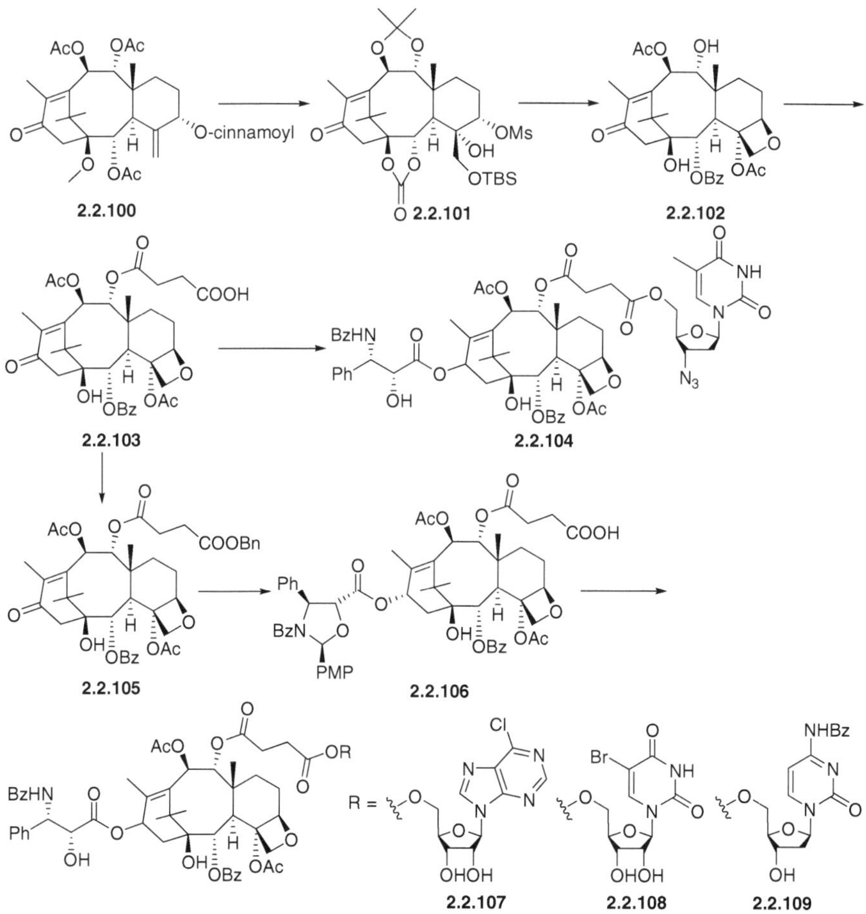

SCHEME 8.22 Synthesis of the C9-modified paclitaxel analogs.

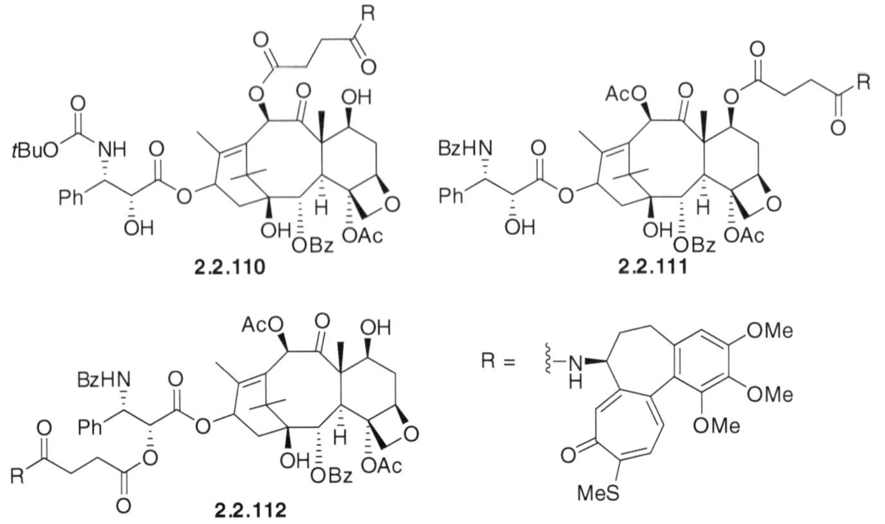

FIGURE 8.10 Taxane analogs **2.2.110**, **2.2.111**, and **2.2.112**.

the hydroxyl group at the C2′ position was crucial to maintain the tubulin assembly activity of paclitaxel [72]. In the cytotoxicity assay, all these compounds were less potent than paclitaxel.

Chen and coworkers synthesized a series of C3′-modified paclitaxel analogs (**2.2.116–2.2.126**, Scheme 8.23) [73]. Biological evaluation demonstrated that the 3′-*N*-thiocarbamate analogs **2.2.116–2.2.120** were more potent than paclitaxel in both tubulin assembly and *in vitro* cytotoxicity assays, and **2.2.118–2.2.120** exhibited *in vivo* activity. A smaller series of similar compounds was also prepared by Karliga et al., but these compounds did not show the same levels of activity [74].

8.2.3 Paclitaxel Analogs

This section seeks to cover the major groups of paclitaxel analogs prepared over the last decade. Since the ultimate goal of much of the work in this area is the development of new clinically active analogs, the section begins with a summary of the current status of those analogs, which are (or were) in clinical trials. The subject of drug delivery and prodrugs of paclitaxel will be discussed later in Section 8.4.

8.2.3.1 Simple Analogs in Clinical Development. Although paclitaxel and docetaxel are effective anticancer agents, they are not perfect, with both undesirable side effects as well as decreasing effectiveness due to drug resistance. Therefore, the structure–activity relationships of the taxane drugs have been extensively studied, and numerous new taxane analogs have been designed and synthesized to enhance activity against drug-resistant tumors, reduce side effects, and improve pharmacological properties. This section will summarize recent the preclinical or clinical development of new taxanes.

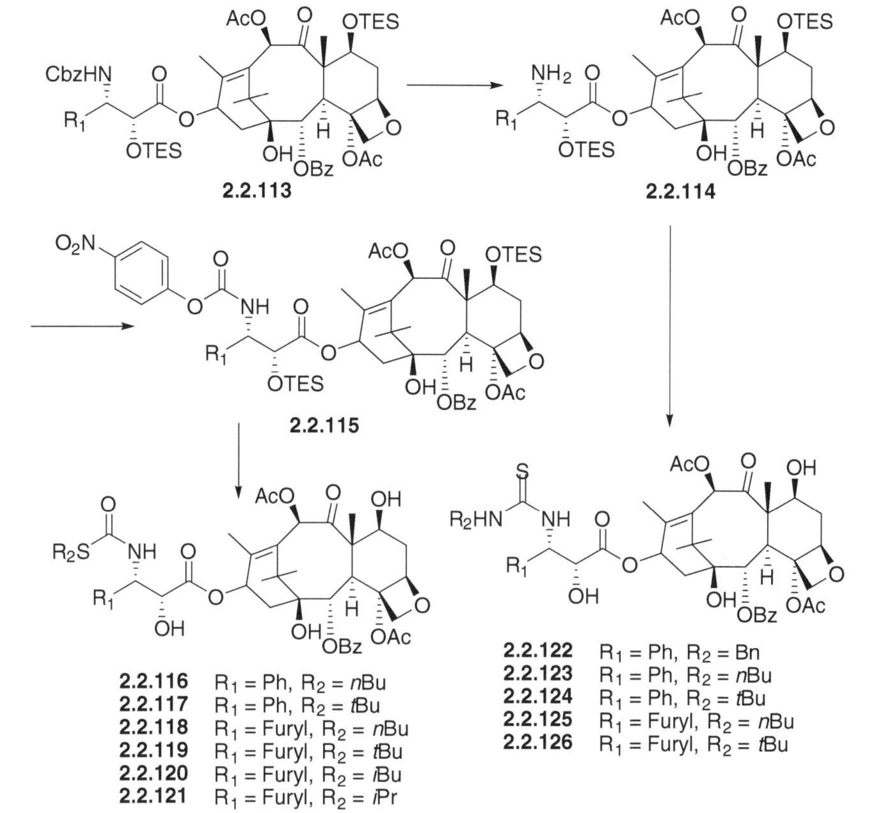

SCHEME 8.23 Taxane analogs **2.2.116–2.2.126**.

Cabazitaxel (**2.3.1**) and larotaxel (**2.3.2**) (Figure 8.11) are novel taxane analogs developed by Sanofi-Aventis [75–77]. They are efficient in promoting the assembly of tubulin and stabilizing microtubules against cold-induced depolymerization *in vitro* as potently as docetaxel [78–81]. They are semisynthesized from 10-deacetyl baccatin III, the major natural taxoid derived from the needles of *Taxus* species. Unlike paclitaxel and docetaxel, cabazitaxel and larotaxel are poor substrates to the ATP-dependent drug efflux pump, P-glycoprotein (encoded by the multidrug resistance gene ABCB1) [82]. Preclinical studies demonstrated that in docetaxel-sensitive

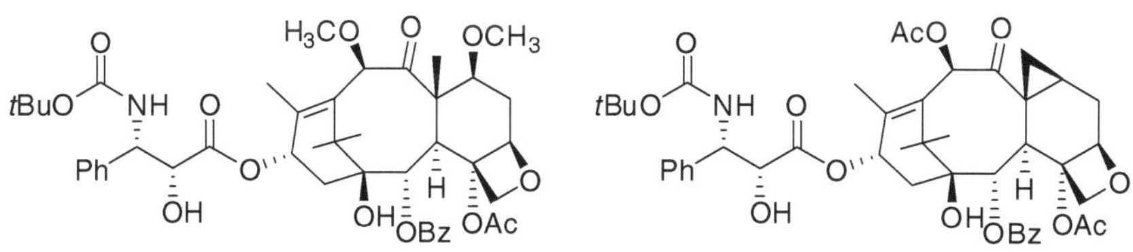

2.3.1 2.3.2

FIGURE 8.11 Cabazitaxeland larotaxel.

cancer cell lines, these two agents exhibited comparable cytotoxicity to docetaxel, with IC_{50} values ranging from 0.003 to 0.029 μM. More importantly, they were cytotoxic in cell lines resistant to doxorubicin, vinblastine, paclitaxel, and docetaxel [75,83]. Similarly, they have shown a broad *in vivo* spectrum of anticancer efficiency on docetaxel-sensitive and docetaxel-resistant models, such as B16/TXT melanoma [75,83]. Cabazitaxel and larotaxel also exhibited the ability to cross the blood–brain barrier, which might result from their poor affinity for P-glycoprotein [84]. Interestingly, larotaxel shows *in vivo* synergy in combination with doxorubicin, cisplatin, vinorelbine, and trastuzumab [85,86]. Cabazitaxel was approved by the U.S. Food and Drug Administration in 2009 for use in combination with prednisone for treatment of certain patients with hormone-refractory metastatic prostate cancer following docetaxel-based therapy [87]. Larotaxel is currently in Phase III trials in patients with advanced pancreatic cancer and in patients with advanced bladder cancer. It also has the potential to treat CNS cancers, since it can cross the blood–brain barrier. A recent review states "we believe that, in the near future, larotaxel may have an important role in taxane-based therapies. However, such a prospect will only be achieved after the evaluation of current and new trials with end points aimed at assessing the peculiarities of the drug, such as CNS penetration and different toxicity profile" [88].

BMS-275183 (**2.3.3**) (Figure 8.12) is a paclitaxel analog with modifications on the C-4 position and the C-13 side chain [89]. It was synthesized in an attempt to identify new taxanes with potential for oral use in the treatment of cancers [90]. BMS-275183 showed the ability to assemble tubulin with potency similar to that of paclitaxel [91]. The *in vitro* cytotoxicity studies demonstrated that BMS-275183 showed similar cytotoxicity to paclitaxel in HCT-116 and A2780 cancer cell lines [91]. In two paclitaxel-resistant cancer cell sublines, HCT-116/MDR and A2780/txl, the drug resistant was significantly reduced for BMS-275183 compared with that observed for paclitaxel [91]. More importantly, this C-4 methyl carbonate paclitaxel analog given p.o. was as effective as i.v. paclitaxel in five tumor models and showed good oral bioavailability in animals [89,92,93]. Its development has however been discontinued.

Milataxel (MAC-321) (**2.3.4**) (Figure 8.12) is a docetaxel analog, which is being developed by Taxolog, Inc. (Fairfield, NJ, USA) [94,95]. It exhibits similar activity and mechanism of action to paclitaxel and docetaxel and promotes the assembly of

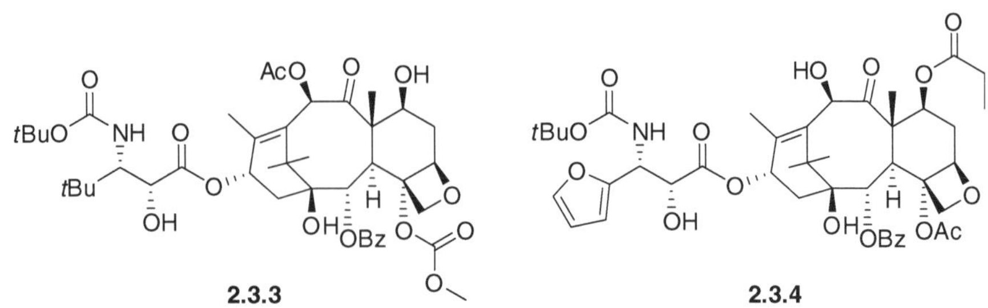

2.3.3 **2.3.4**

FIGURE 8.12 BMS-275183 and milataxel (MAC-321).

tubulin [96,97]. Preclinical studies demonstrated that milataxel inhibited proliferation of a panel of 14 tumor cell lines with IC_{50} values ranging from 0.6 to 5.3 nM, as potently as paclitaxel and docetaxel [97]. Unlike the commercial taxanes, milataxel demonstrated significant activity in paclitaxel-insensitive cancer cell lines such as colon cancer, pancreatic cancer, and melanoma [97]. More importantly, it showed remarkable antiproliferative effects against paclitaxel-resistant human cancer cell lines, due to its ability to overcome the P-glycoprotein mediated resistance [97]. Additionally, milataxel displayed superior *in vivo* efficacy and good oral bioavailability in early trials [97,98]. It is readily soluble and does not require the use of toxic vehicles, which allows milataxel to be administered orally with good bioavailability in animal models. However, a Phase II clinical trial found life-threatening side effects of neutropenic sepsis and diarrhea and a lack of clinical activity in patients with advanced previously treated colorectal cancer [99], and it appears that development has been discontinued.

BMS-188797 (**2.3.5**) (Figure 8.13) is the 4-desacetyl-4-methyl carbonate analog of paclitaxel. It was twice as potent as paclitaxel in tubulin assembly assay [100,101], and it displayed superior activity in four *in vivo* tumor models and against paclitaxel-resistant cell lines [101–105]. It also showed objective responses in a Phase I trial for 4 out of 16 patients [103]. BMS-184476 (**2.3.6**) (Figure 8.13) is a taxane analog characterized by the replacement of the 7-hydroxyl group of paclitaxel with a methylthiomethyl ether group. This analog showed efficacy equivalent or superior to paclitaxel in a number of paclitaxel-sensitive cancer cell lines [105], and more importantly, it was more active than paclitaxel in a P-glycoprotein-mediated multidrug resistant cell line and in the A2780/tax cell line [100]. Additionally, BMS-184476 was more soluble than paclitaxel in an aqueous cosolvent, which potentially eliminates the need for premedication [106,107]. No further reports on its clinical trials have appeared since 2005, however, and it appears that development has been discontinued.

The fluoropyridinyl derivative DJ-927 (Tesetaxel, **2.3.7**, Figure 8.14) was entered into clinical trials as an orally active compound for treatment of nonsmall cell lung cancer. It did not show any clear advantage over existing drugs in Phase II trials, and it was withdrawn from development in 2006, but Phase III trials have restarted under Genta [108].

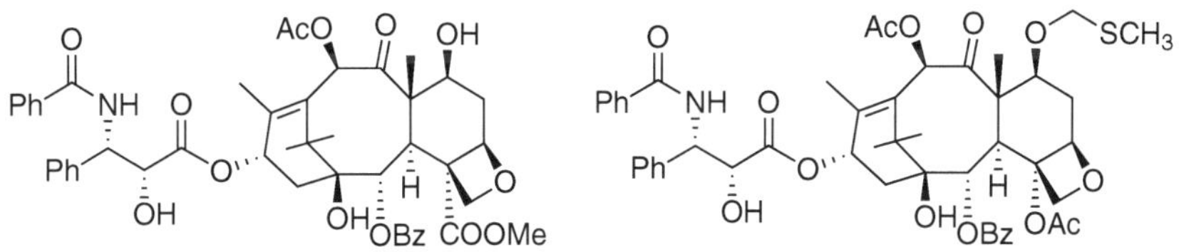

2.3.5 **2.3.6**

FIGURE 8.13 BMS-188797 and BMS-184476.

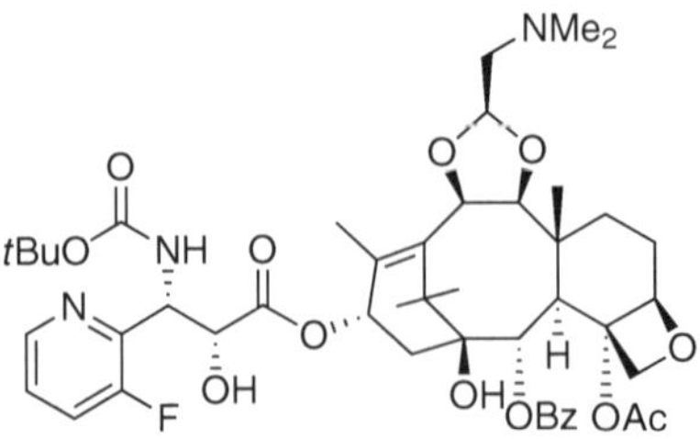

2.3.7 DJ-927

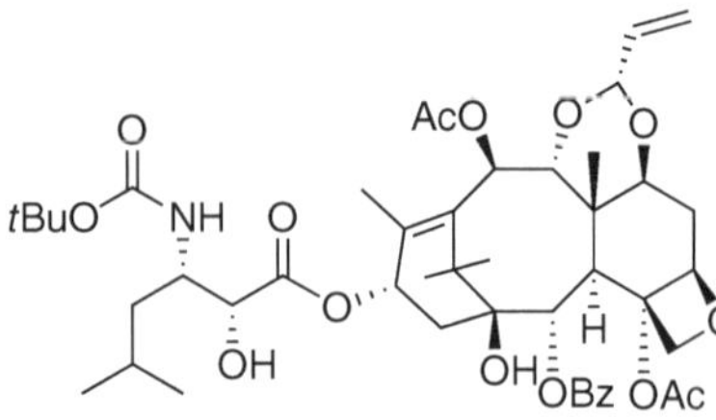

2.3.8 TPI 287

FIGURE 8.14 Clinical candidates **2.3.7–2.3.8**.

TPI-287 (**2.3.8**, Figure 8.15) is in Phase II trials for treatment of advanced pancreatic cancer and hormone-refractory prostate cancer, and initial results showed that it is well tolerated at a dose up to 56 mg/m^2 administered Q7D [109]. The thiophene derivative simotaxel (**2.3.9**) was in clinical development, but has been withdrawn, as has the furan derivative TL-310 (**2.3.11**) [7]. Ortataxel (**2.3.11**), however, is in Phase II trials for treatment of advanced pancreatic cancer and hormone-refractory prostate cancer (Figure 8.15) [4,110].

8.2.3.2 ***Other Paclitaxel Analogs.*** A series of 10-*C*-morpholinoethyl docetaxel analogs, exemplified by **2.3.12–2.3.13** (Figure 8.16) modified at the hydroxyl group and the C7 position were synthesized in Soga's group [111]. These analogs were evaluated for cytotoxicity *in vitro* against several cancer cell lines and antitumor

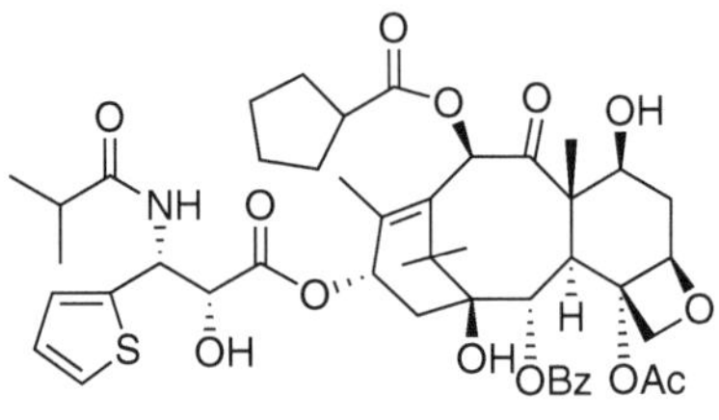

2.3.9 Simotaxel

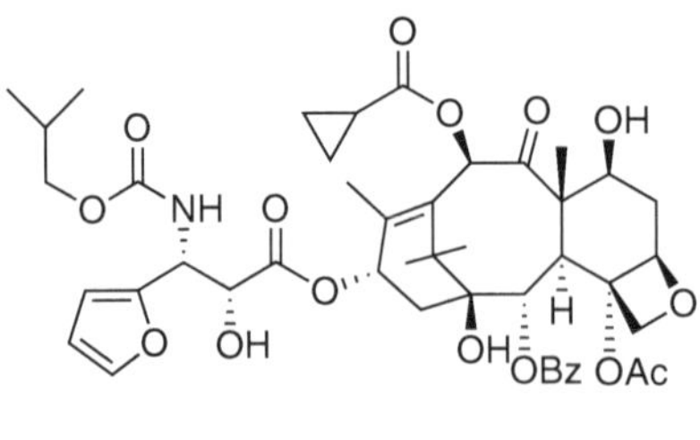

2.3.10 TL-310

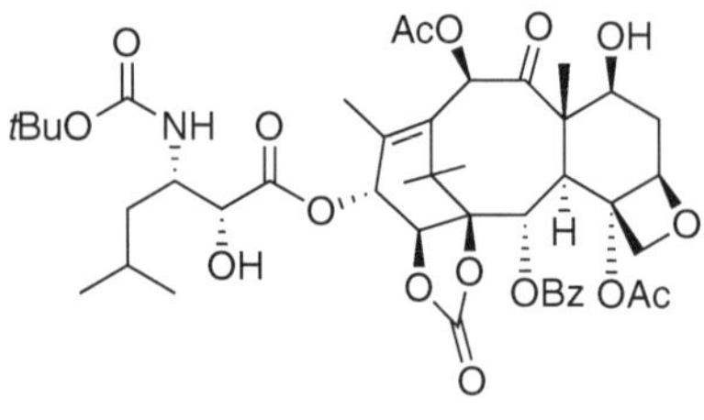

2.3.11 Ortataxel

FIGURE 8.15 Clinical candidates **2.3.9–2.3.12**.

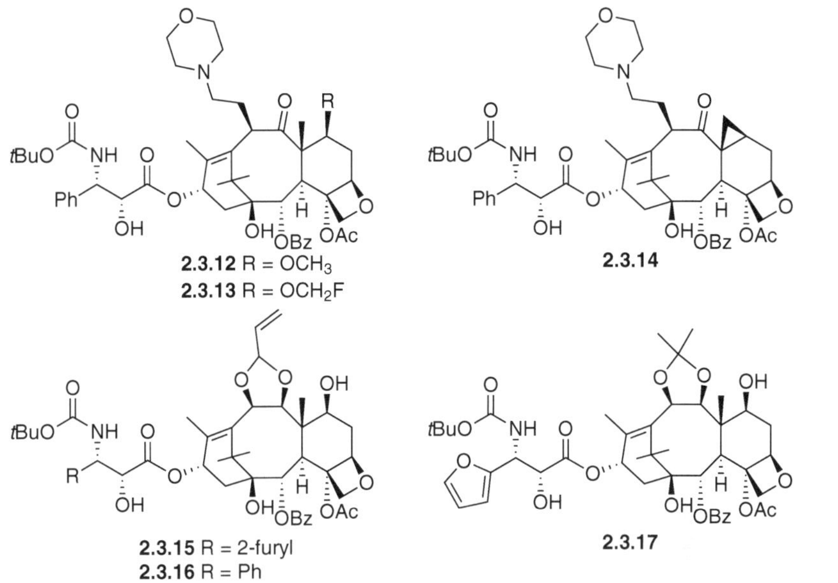

FIGURE 8.16 Docetaxel analogs **2.3.13–2.3.17**.

activity *in vivo* against B16 melanoma BL-6. The 7-methoxy analog **2.3.13** showed the highest potency *in vitro* and was highly active against BL6 intravenously and orally. In 2002, the same group synthesized a new series of 9β-dihydro-9,10-acetal docetaxel analogs including **2.3.15–2.3.17** (Figure 8.16) [112]. Their antitumor activities were tested *in vitro* against PC-6 cancer cell line as well as P-glycoprotein expressed PC-12 and PC-6/VCR cancer cell lines. The three analogs **2.3.15–2.3.17** were more potent than docetaxel, indicating the effectiveness of acetal groups in the C9 and C10 position and the importance of the β configuration of the hydroxyl group at the C9 position.

Encouraged by this result, they prepared the additional analogs **2.3.18–2.3.29** (Figure 8.17) [113,114]. Most of these analogs had more potent cytotoxicities than docetaxel, especially in resistant cancer cell lines expressing P-glycoprotein, and the 7-deoxy-3′-pyridyl analogs (**2.3.19** and **2.3.22–2.3.29**) showed the best activities, suggesting that the modifications of the region between the C7 and C10 position might disrupt the specific binding of the molecules to P-glycoprotein and reduce the P-glycoprotein-mediated efflux. Most of these 7-deoxy-3′-pyridyl analogs were more water soluble than docetaxel. They were also evaluated for their antitumor activity *in vivo*, and analog **2.3.27** showed the most potent antitumor effects over a wide dosage range intravenously and orally. In contrast, docetaxel administered orally gave no antitumor effect and no body weight loss at a dose of 600 mg/kg. However, this analog was found to be prone to be metabolized by human liver microsomes [115].

The major metabolite was identified as M-1 (**2.3.30**) (Figure 8.18), and so the new analogs **2.3.31–2.3.54** were designed (Figure 8.18) to improve the metabolic stability [115,116]. Some analogs (**2.3.36**, **2.3.39**, **2.3.46**, **2.3.48**, **2.3.51**, and **2.3.52–2.3.54**)

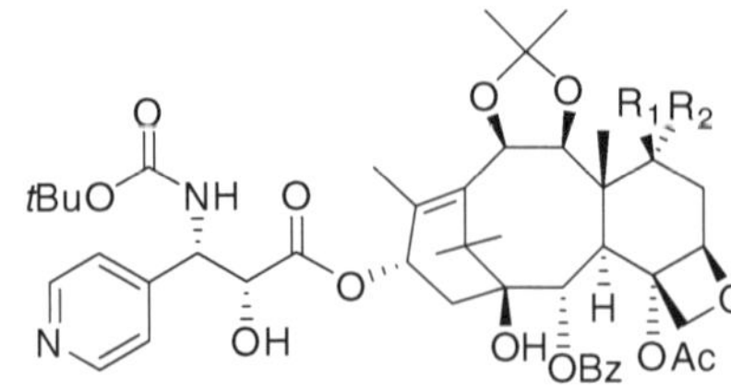

2.3.18 R_1 = OH, R_2 = H
2.3.19 R_1 = R_2 = H
2.3.20 R_1 = H, R_2 = F
2.3.21 R_1 = OCH_3, R_2 = H

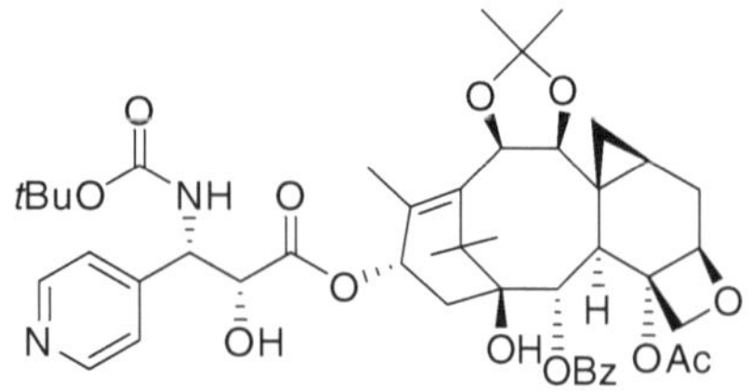

2.3.24

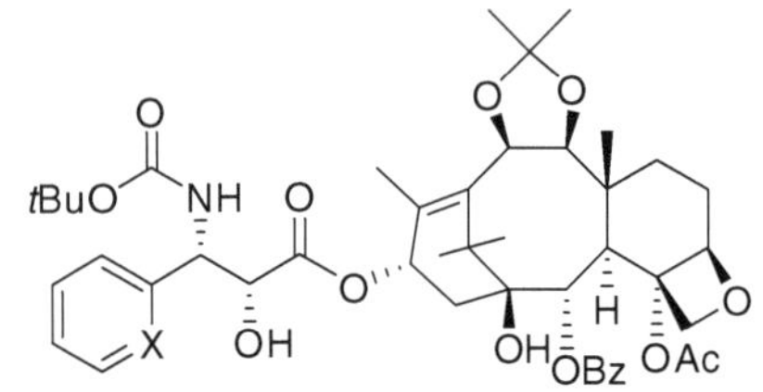

2.3.22 X = C
2.3.23 X = N

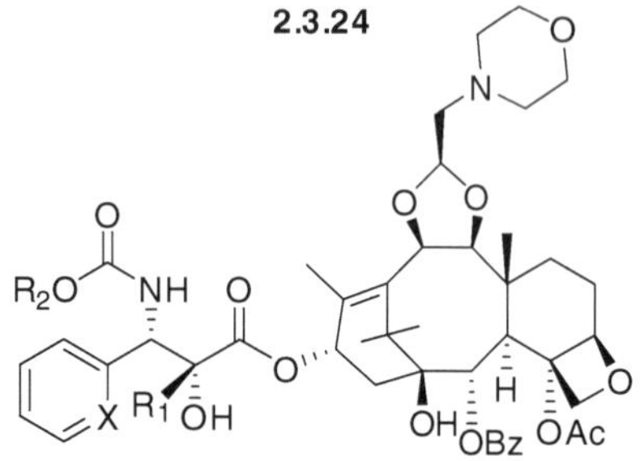

2.3.25 R_1 = H, R_2 = *t*Bu, X = C
2.3.26 R_1 = CH_3, R_2 = *t*Bu, X = C
2.3.27 R_1 = H, R_2 = *t*Bu, X = N
2.3.28 R_1 = CH_3, R_2 = *t*Bu, X = N
2.3.29 R_1 = CH_3, R_2 = Ph, X = N

FIGURE 8.17 Docetaxel analogs **2.3.18–2.3.29**.

had nearly the same biological properties as **2.3.27** and were metabolized much more slowly by human liver microsomes.

It was found that paclitaxel was able to inhibit angiogenesis by inhibiting endothelial cell motility [117]. Angiogenesis is important for the growth of a tumor mass and the formation of metastases, and great efforts have been made to develop therapeutic tools to block this process [118,119]. In 2002, Taraboletti and coworkers screened different taxanes and identified the C-secotaxane derivative, IDN 5390 (**2.3.55**, Figure 8.19) as a potent antiangiogenic agent with less cytotoxicity than paclitaxel [120]. In comparable experimental conditions, IDN 5390 inhibited endothelial cell migration without suppressing cell proliferation. *In vivo*, IDN 5390 reduced the size of metastases in mice bearing lung micrometastases from B16BL6 murine melanoma and slowed the s.c. growth of the paclitaxel-resistant human ovarian carcinoma, 1A9/PTX22, xenografted in nude mice. IDN 5390 thus represents the prototype of a novel class of taxanes with antiangiogenic activity. Further study of this analog demonstrated that it had oral efficacy and high tolerability in protracted daily treatment schedules [121,122]. Recently, Ferlini and coworkers observed that it had several fold better potency than paclitaxel against drug-resistant ovarian cancer cell lines overexpressing the Class III tubulin isotype [123]. They found that the open and flexible C ring and an acidic α-unsaturated enol-ketone moiety could fit at the binding site of Class III β-tubulin better than paclitaxel and overcome the drug resistance.

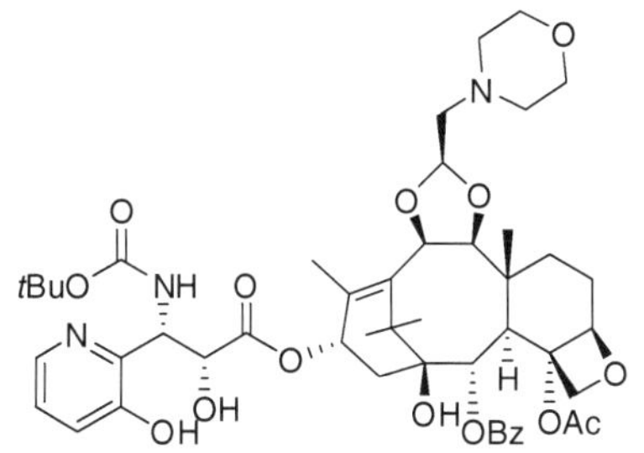

2.3.30 (M-1)

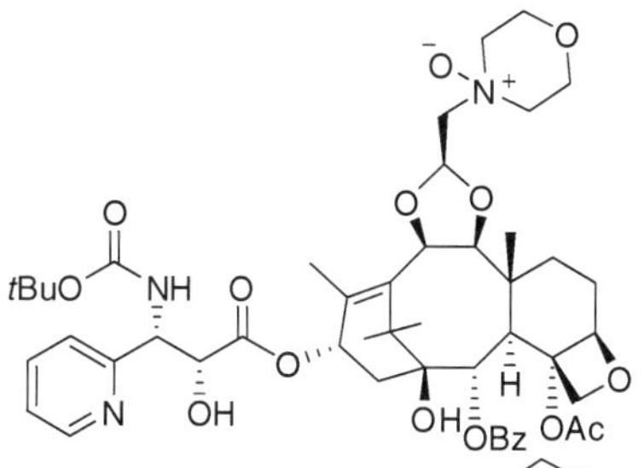

2.3.31

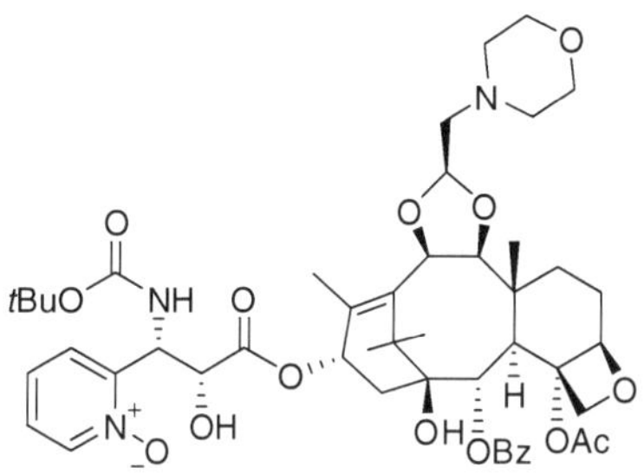

2.3.32

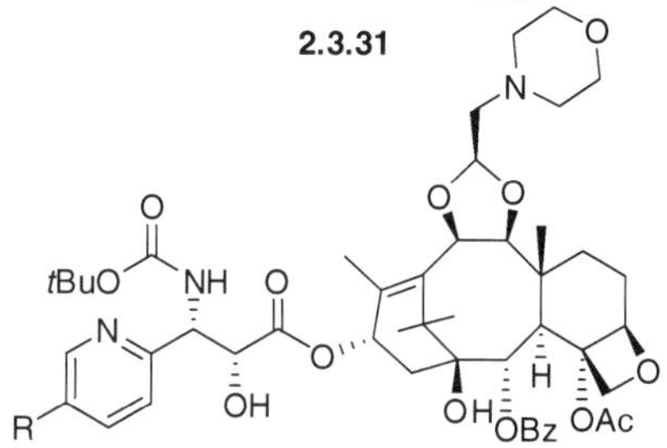

2.3.33 R = OH
2.3.34 R = CH_3
2.3.35 R = Et
2.3.36 R = F
2.3.37 R = Cl
2.3.38 R= CF_3
2.3.39 R = OCH_3

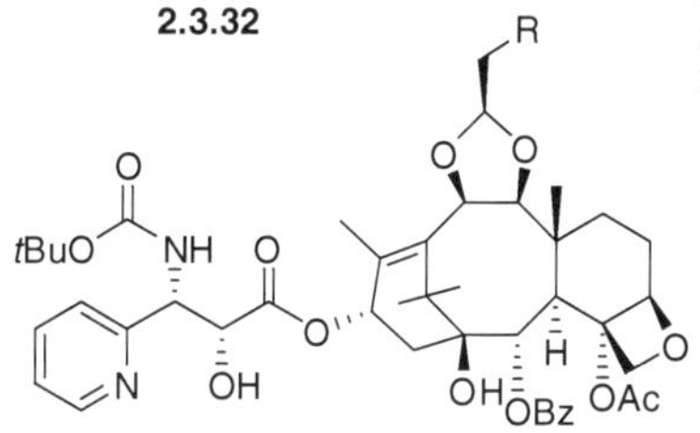

2.3.40 R = dimethylamino
2.3.41 R = methylamino
2.3.42 R = cyclopropylamino
2.3.43 R = cyclobutylamino

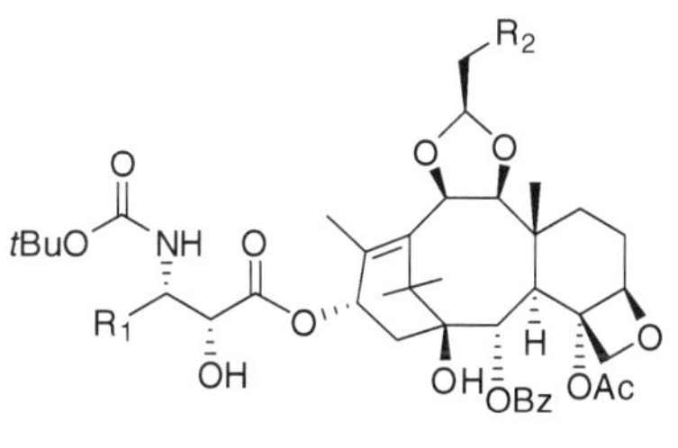

2.3.44 R_1 = *t*Bu, R_2 = morpholino
2.3.45 R_1 = *t*Bu, R_2 = dimethylamino
2.3.46 R_1 = 3-fluoro-2-pyridyl, R_2 = morpholino
2.3.47 R_1 = 3-fluoro-2-pyridyl, R_2 = dimethylamino
2.3.48 R_1 = 2-thiazolyl, R_2 = morpholino
2.3.49 R_1 = 2-thiazolyl, R_2 = dimethylamino
2.3.50 R_1 = 2-thiazolyl, R_2 = cyclopropylamino
2.3.51 R_1 = 4-thiazolyl, R_2 = morpholino
2.3.52 R_1 = 4-thiazolyl, R_2 = dimethylamino
2.3.53 R_1 = 5-thiazolyl, R_2 = morpholino
2.3.54 R_1 = 5-thiazolyl, R_2 = dimethylamino

FIGURE 8.18 Docetaxel analogs **2.3.30–2.3.54**.

Kingston and collaborators synthesized a series of C6- and C7-modified paclitaxel analogs (**2.3.56–2.3.59**, Figure 8.19) via the 6,7-diol derivative of paclitaxel [69]. Analog **2.3.58** was evaluated for its cytotoxicity against human cancer cell lines, HCT116 and A2780. Interestingly, it was found to be about three times more cytotoxic than paclitaxel.

Two docetaxel analogs (**2.3.60–2.3.61**, Figure 8.20) with modified D-rings were synthesized and tested by Dubois and coworkers [124]. They found that 5(20)-deoxydocetaxel (**2.3.61**) was as active as paclitaxel in promotion of microtubule

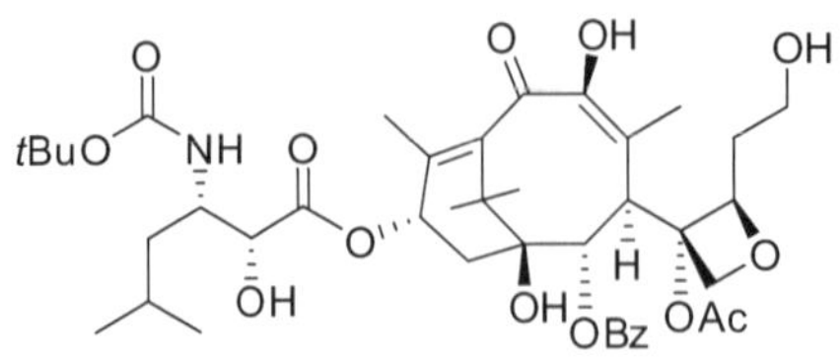

2.3.55 IDN 5390

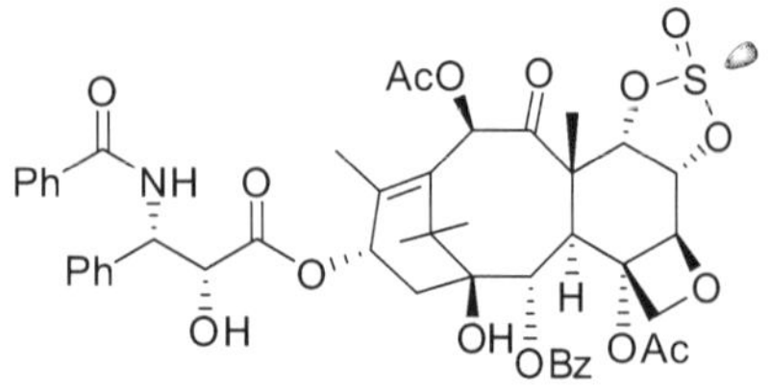

2.3.56

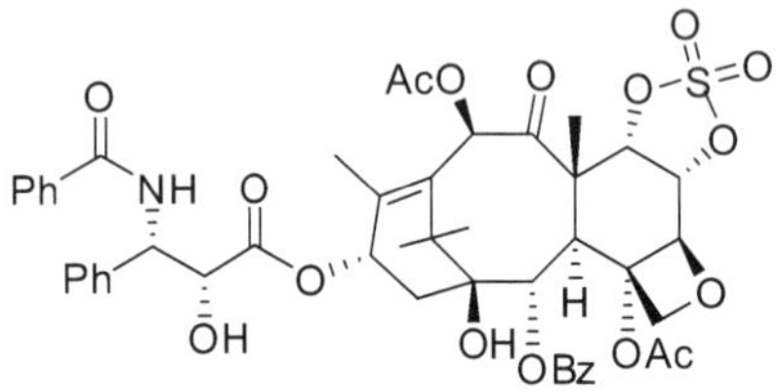

2.3.57

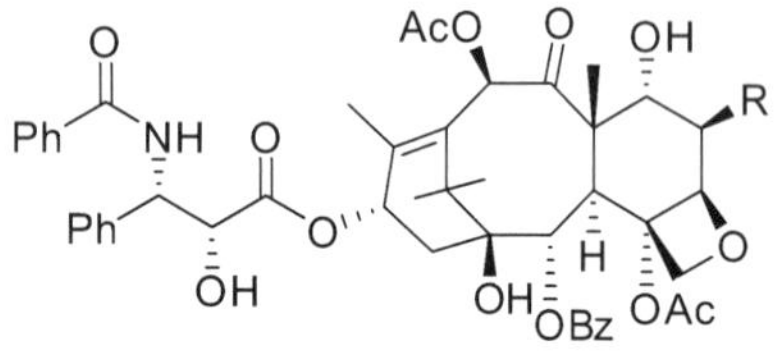

2.3.58 R = N_3
2.3.59 R = NH_2

FIGURE 8.19 Analogs **2.3.55–2.3.59**.

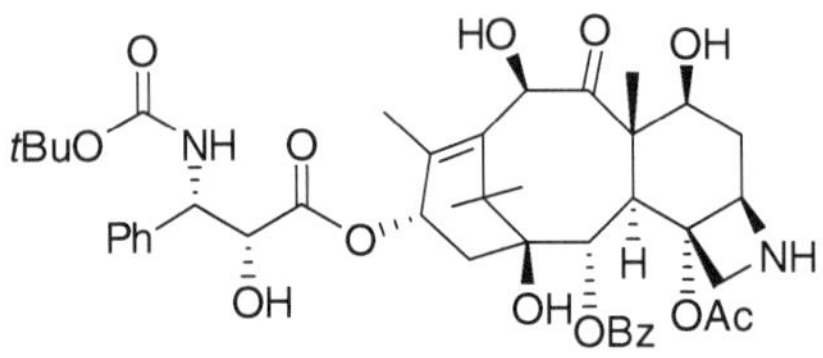

2.3.60

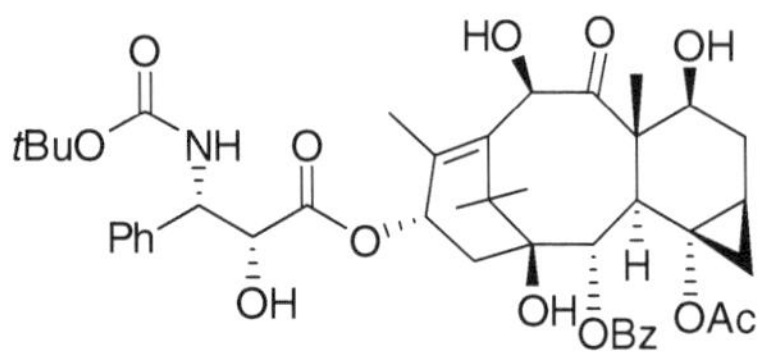

2.3.61

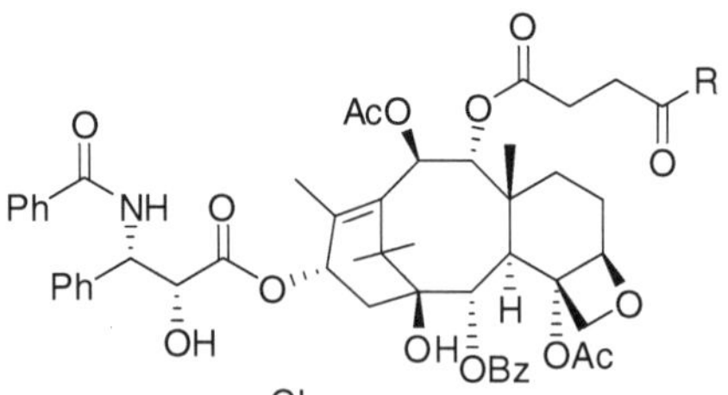

R =

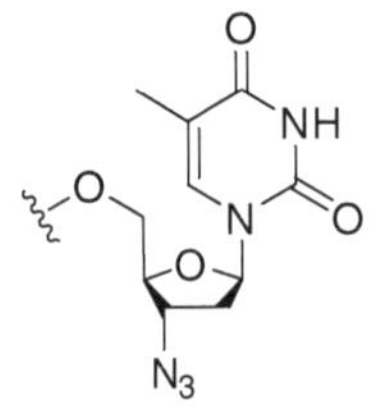

2.3.62

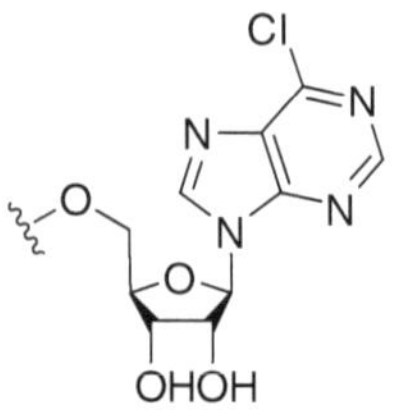

2.3.63

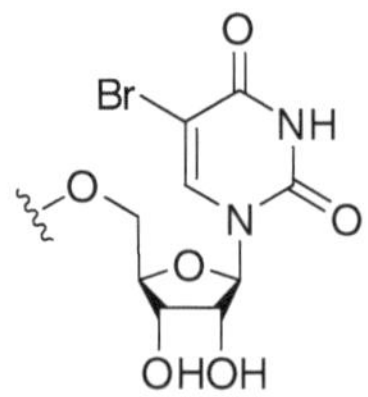

2.3.64

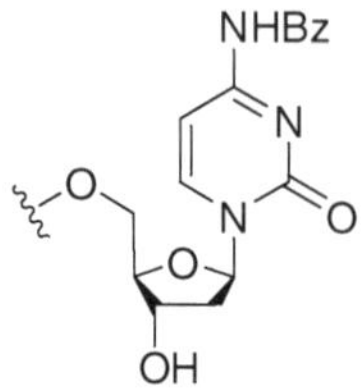

2.3.65

FIGURE 8.20 Analogs **2.3.60–2.3.65**.

assembly; this is the first example of a modified D-ring taxoid showing equivalent activity to paclitaxel.

Cheng and coworkers reported the synthesis of a series of novel 9-functional heterocyclic coupled 7-deoxy-9-dihydropaclitaxel analogs (**2.3.62–2.3.65**, Figure 8.20) [70]. These analogs were tested in a tubulin assembly assay and evaluated for their cytotoxicities against various human cancer cell lines. They found that three of the analogs exhibited similar to superior activity to paclitaxel in BEL-7402 and ECa-109 cancer cell lines, although all of the analogs were less effective in promoting tubulin assembly than paclitaxel.

In 2001, Dubois and coworkers reported the synthesis of two novel 5(20)-thia analogs of docetaxel (**2.3.66–2.3.67**, Figure 8.21) from 10-deacetylbaccatin III, similar to those prepared earlier by Kingston et al. [125,126]. The two analogs were less cytotoxic than paclitaxel against the KB cell line, indicating that the replacement of the oxygen at the D-ring with a sulfur atom leads to a decrease in cytotoxicity. The microtubule disassembly inhibition assay demonstrated that the C-7 dehydroxy analog showed much less inhibition than the one with the C-7 hydroxyl group, indicating the importance of the C-7 hydroxyl group in stabilizing microtubules in this series.

A series of 2′-methyl taxoids (**2.3.68–2.3.72**, Figure 8.21) was synthesized and evaluated for antitumor activity in various cancer cell lines including paclitaxel-resistant human ovarian cancer cell lines and human breast cancer lines [127]. These analogs were found to exhibit equivalent activity to paclitaxel in P-glycoprotein negative cancer cell lines and higher potency in drug-resistant cancer cell lines.

Kingston and coworkers synthesized a series of novel paclitaxel analogs with a cyclopropanated side chain (**2.3.73–2.3.76**, Figure 8.21) [128]. These analogs were

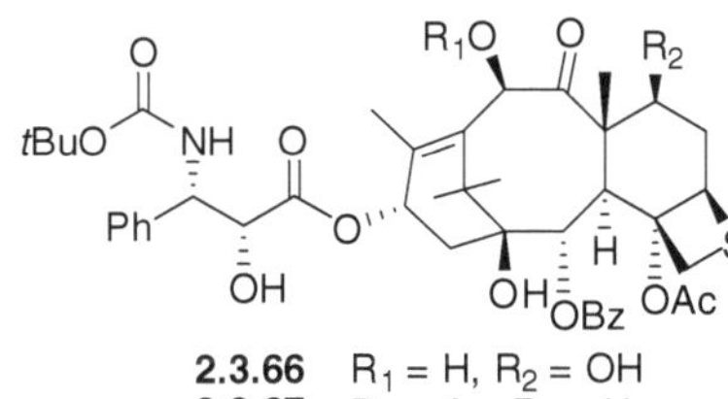

2.3.66 R_1 = H, R_2 = OH
2.3.67 R_1 = Ac, R_2 = H

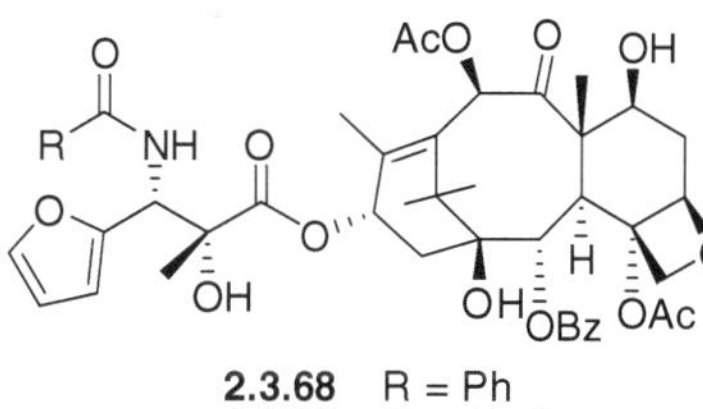

2.3.68 R = Ph
2.3.69 R = tBuO

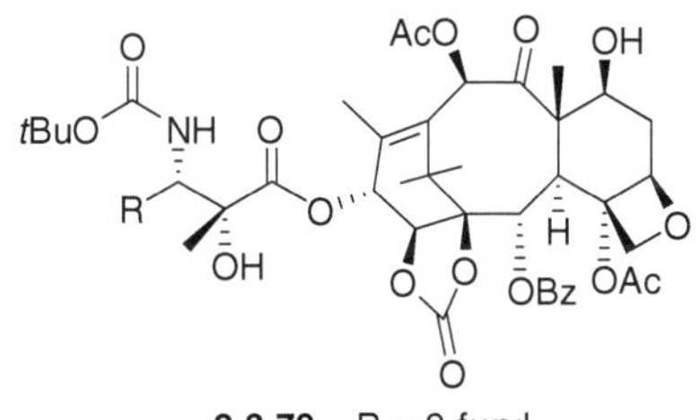

2.3.70 R = 2-furyl
2.3.71 R = Ph
2.3.72 R = CF_3

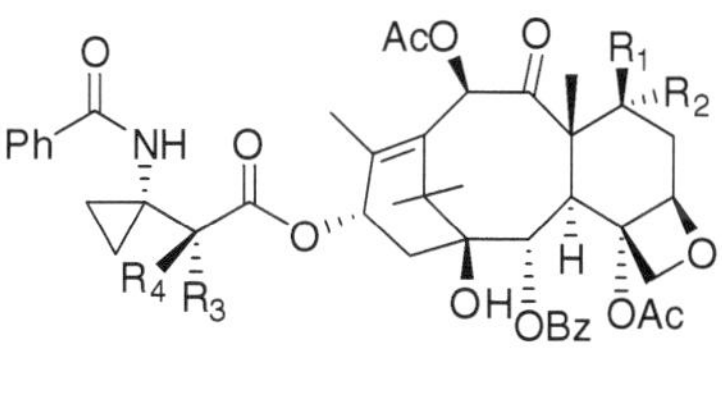

2.3.73 $R_1 = R_4$ = OH, $R_2 = R_3$ = H
2.3.74 $R_1 = R_3$ = H, $R_2 = R_4$ = OH
2.3.75 $R_1 = R_3$ = OH, $R_2 = R_4$ = H
2.3.76 $R_1 = R_4$ = H, $R_2 = R_3$ = OH

FIGURE 8.21 Analogs **2.3.66–2.3.76**.

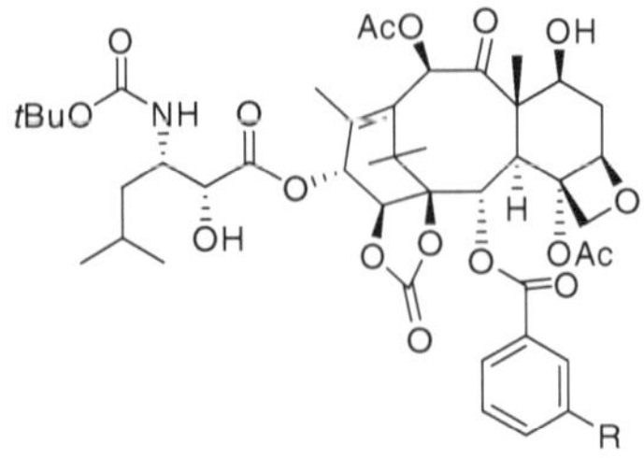

2.3.77 IDN 5109 R = H
2.3.78 R = OCH_3

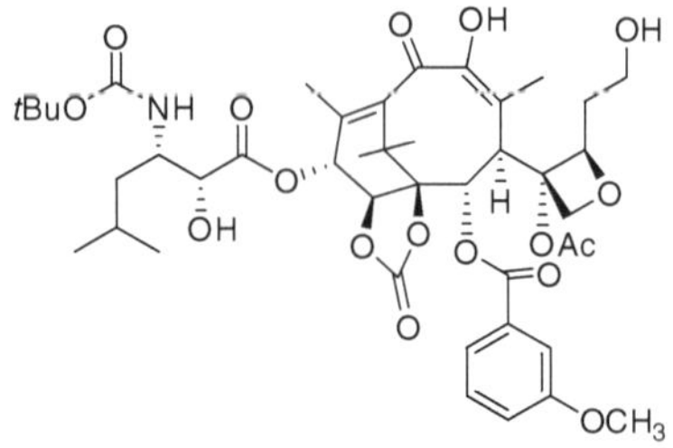

2.3.79

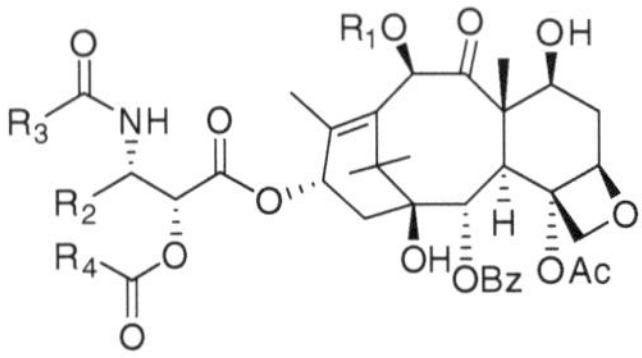

2.3.80 R_1 = Ac, R_2 = Ph, R_3 = Ph, R_4 = $C_{21}H_{31}$
2.3.81 R_1 = OH, R_2 = Ph, R_3 = *t*BuO, R_4 = $C_{21}H_{31}$
2.3.82 R_1 = EtCO, R_2 = isobutenyl, R_3 = *t*BuO, R_4 = $C_{21}H_{31}$
2.3.83 R_1 = EtCO, R_2 = isobutyl, R_3 = *t*BuO, R_4 = $C_{21}H_{31}$
2.3.84 R_1 = *c*-PrCO, R_2 = isobutenyl, R_3 = *t*BuO, R_4 = $C_{21}H_{31}$
2.3.85 R_1 = *c*-PrCO, R_2 = isobutyl, R_3 = *t*BuO, R_4 = $C_{21}H_{31}$
2.3.86 R_1 = Me_2NCO, R_2 = isobutenyl, R_3 = *t*BuO, R_4 = $C_{21}H_{31}$
2.3.87 R_1 = MeOCO, R_2 = isobutenyl, R_3 = *t*BuO, R_4 = $C_{21}H_{31}$
2.3.88 R_1 = EtCO, R_2 = isobutenyl, R_3 = *t*BuO, R_4 = $C_{17}H_{31}$
2.3.89 R_1 = EtCO, R_2 = isobutenyl, R_3 = *t*BuO, R_4 = $C_{17}H_{29}$

FIGURE 8.22 Analogs **2.3.77–2.3.89**.

evaluated for antiproliferative activity in the A2780 mammalian cell line. Analogs **2.3.75** and **2.3.76** with the same configuration at the C2′ position as paclitaxel exhibited better activity than the ones with inversed configuration, but inversion of configuration at the C7 position did not affect the activity significantly, as expected from previous work.

Barboni and coworkers synthesized two methoxylated analogs (**2.3.78–2.3.79**, Figure 8.22) of their norstatine esters IDN5190 (**2.3.77**, Figure 8.22) and IDN5390 (**2.3.55**, Figure 8.19), two interesting clinical candidates that overcome drug resistance and allow oral administration [129]. They demonstrated that the insertion of a *meta*-methoxy group on the C-2 benzoate resulted in an increase in the cytotoxicity of these newer generation taxanes, including those of the ring C-*seco* type, as had been previously observed for paclitaxel itself [130].

Taxoid conjugates with fatty acids linked at the C2′ position (**2.3.80–2.3.89**) (Figure 8.22) were synthesized by Ojima and coworkers to overcome MDR caused by overexpressing P-glycoprotein [131]. These conjugates were evaluated *in vivo* and exhibited strong activity against drug-resistant colon cancer and drug-sensitive ovarian cancer xenografts in mice. Analogs **2.3.82** and **2.3.84** brought about total regression of drug-resistant and drug-sensitive tumors, respectively, in animal models with reduced systemic toxicity.

Saicic and coworkers reported the synthesis and biology of a C,D-*seco*-taxoid (**2.3.90**, Figure 8.23) [132]. This analog had much lower tubulin stabilizing activity and cytotoxicity than paclitaxel, indicating once again the importance of the C,D-ring moiety for the bioactivity of taxoids.

Ojima and coworkers designed and synthesized a series of 3′-difluorovinyl-taxoids (**2.3.91–2.3.100**, Figure 8.23) to enhance metabolic stability and *in vivo* activity by

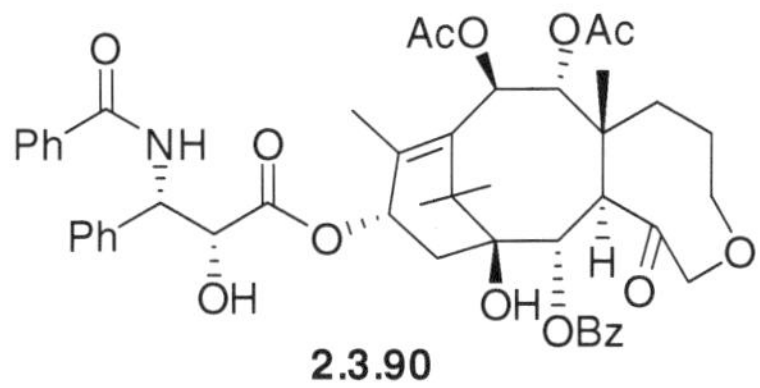

2.3.90

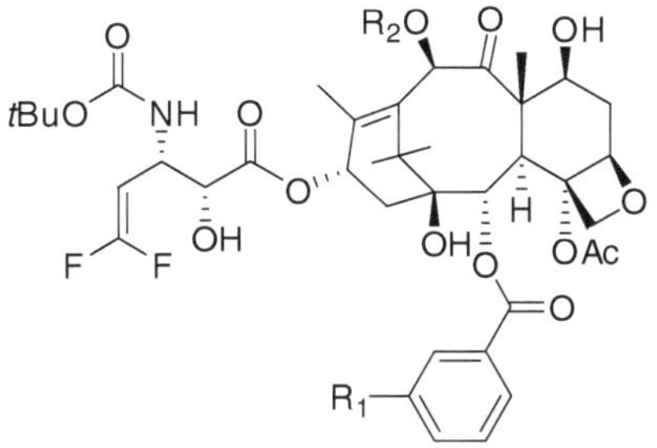

2.3.91 R_1 = H, R_2 = Ac
2.3.92 R_1 = H, R_2 = *c*-PrCO
2.3.93 R_1 = H, R_2 = EtCO
2.3.94 R_1 = H, R_2 = Me_2NCO
2.3.95 R_1 = MeO, R_2 = *c*-PrCO
2.3.96 R_1 = MeO, R_2 = EtCO
2.3.97 R_1 = MeO, R_2 = MeOCO
2.3.98 R_1 = N_3, R_2 = Ac
2.3.99 R_1 = N_3, R_2 = *c*-PrCO
2.3.100 R_1 = N_3, R_2 = MeOCO

FIGURE 8.23 Analogs **2.3.90–2.3.100**.

blocking allylic oxidation of the 3′-isobutenyl group of some previous analogs by CYP 3A4 [133]. These fluorinated analogs were highly potent compared to paclitaxel against various cancer cell lines, and the *meta*-substitution of the C2-benzoate affected the potency against drug-sensitive and drug-resistant MCF7 cell lines. Analogs with modifications at both the C2 and C10 positions also exhibited excellent potency.

Mikuni and coworkers reported the evaluation of the antitumor activity of taxoids **2.3.101–2.3.110** (Figure 8.24), possessing various sugar moieties to improve their water solubility [134]. They found that the galactose-bound taxoid **2.3.102** exhibited superior water solubility and antitumor activity, and its therapeutic efficiency in terms of *in vivo* antitumor activity was equivalent to that of docetaxel and better than that of paclitaxel. In serum, the sugar moiety was slowly hydrolyzed and a taxoid was released, indicating that the sugar taxoid conjugates acted as prodrugs.

A series of novel taxoids (**2.3.111–2.1.126** in Figure 8.24 and **2.3.127–2.3.165** in Figure 8.25) with systematic modifications at the C2, C10, and C3′*N* position were synthesized by Ojima and his colleagues [135], and were evaluated for their potency against drug-sensitive and drug-resistant cancer cell lines. Some of these analogs exhibited high potency against drug-resistant cancer cell lines, and others showed similar potency against both drug-sensitive and drug-resistant cancer cell lines. Analogs **2.3.126**, **2.3.133**, and **2.3.135** exhibited excellent activity against a paclitaxel-resistant ovarian cancer cell line, and they were found to possess exceptional activity in promoting tubulin assembly, forming numerous short microtubules similar to those formed by docetaxel. In addition, these three taxoids also exhibited high cytotoxicity against pancreatic cancer cell lines that expressed three to four multidrug-resistant genes. Taxoid **2.3.126** displayed remarkable antitumor activity *in vivo* against highly drug-resistant CFPAC-1 pancreatic and DLD-1 human colon tumor xenografts in mice.

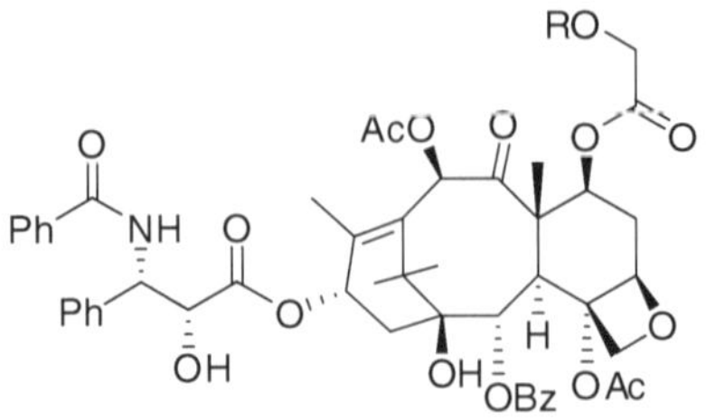

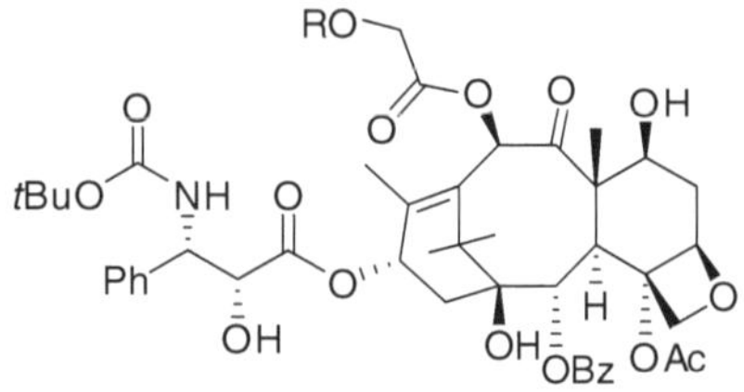

2.3.101 R = glucosyl
2.3.102 R = galactosyl
2.3.103 R = mannosyl
2.3.104 R = xylosyl

2.3.105 R = α-glucosyl
2.3.106 R = β-glucosyl
2.3.107 R = α-galactosyl
2.3.108 R = β-galactosyl
2.3.109 R = α-mannosyl
2.3.110 R = H

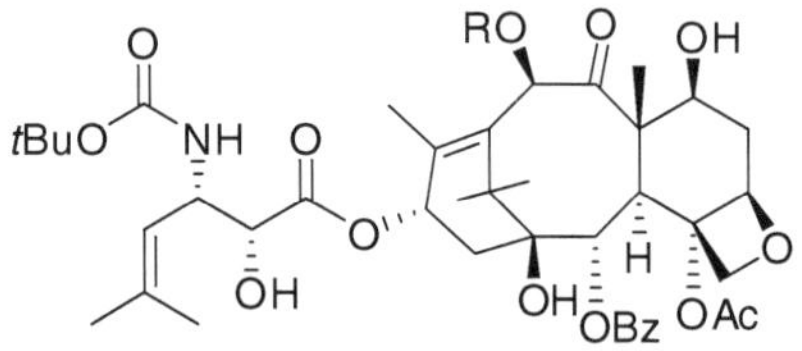

2.3.111 R = Bz
2.3.112 R = 2-MeO(C_6H_4)CO
2.3.113 R = 3-MeO(C_6H_4)CO
2.3.114 R = 4-MeO(C_6H_4)CO
2.3.115 R = 3,4-$(MeO)_2$(C_6H_3)CO
2.3.116 R = 1-naphthoyl
2.3.117 R = 2-naphthoyl
2.3.118 R = Cbz
2.3.119 R = 2-MeO(C_6H_4)CH_2CO
2.3.120 R = 3-MeO(C_6H_4)CH_2CO
2.3.121 R = 4-MeO(C_6H_4)CH_2CO
2.3.122 R = (C_6H_5)$(CH_2)_2$CO
2.3.123 R = 2-MeO(C_6H_4)$(CH_2)_2$CO
2.3.124 R = 3-MeO(C_6H_4)$(CH_2)_2$CO
2.3.125 R = 4-MeO(C_6H_4)$(CH_2)_2$CO
2.3.126 R =

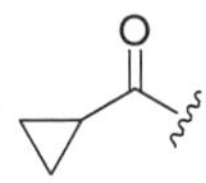

FIGURE 8.24 Analogs **2.3.101–2.3.126**.

Qu and coworkers designed and synthesized three novel 3′-*N*-*tert*-butylsulfonyl analogs of docetaxel (**2.3.166–2.3.168**, Figure 8.26) [136]. These compounds were tested *in vitro* against various human tumor cell lines. They found that the *N*-oxide pyridyl-substituted analogs **2.3.167** and **2.3.168** had more potent cytotoxicities than docetaxel and paclitaxel against several human cancer cell lines, including Eca-109, SKOV3, SMMC-7721, HCT-8, PC3, MCF-7, Hela, and KB.

A series of novel 3′-difluoromethyl and 3′-trifluoromethyl taxoids (**2.3.169–2.3.201**, Figure 8.26) were synthesized and evaluated for their *in vitro* cytotoxicities against various cancer cell lines by Ojima and coworkers [137]. They showed potencies against drug-sensitive cancer cell lines 7 to 20 times better than paclitaxel, and potencies over 100 times higher than paclitaxel in drug-resistant cancer cell lines. The modification at the C10 position was important for increasing activity again drug-resistant cancer cell lines.

In 2009, Ojima and coworkers reported the synthesis of novel C-*seco*-taxoids (**2.3.202–2.3.207**, Figure 8.27) and evaluated their potencies against drug-sensitive and drug-resistant cancer cell lines overexpressing Class III *β*-tubulin [138]. They

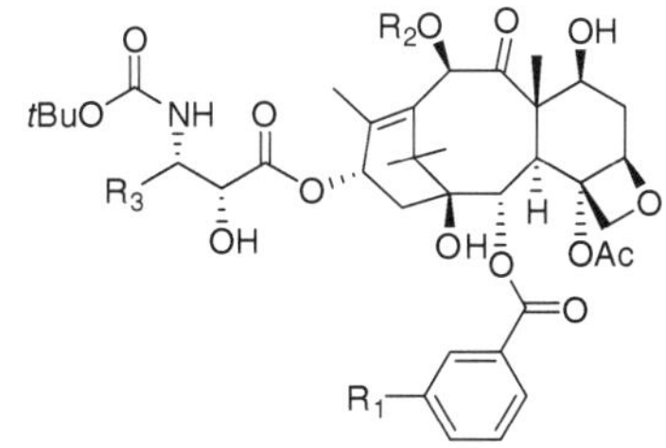

2.3.127 R_1 = Me, R_2 = MeCO, R_3 = Me_2C=CH-
2.3.128 R_1 = MeO, R_2 = MeCO, R_3 = Me_2C=CH-
2.3.129 R_1 = F, R_2 = EtCO, R_3 = Me_2C=CH-
2.3.130 R_1 = Cl, R_2 = EtCO, R_3 = Me_2C=CH-
2.3.131 R_1 = N_3, R_2 = EtCO, R_3 = Me_2C=CH-
2.3.132 R_1 = CH_2=CH_2-, R_2 = EtCO, R_3 = Me_2C=CH-
2.3.133 R_1 = MeO, R_2 = EtCO, R_3 = Me_2C=CH-
2.3.134 R_1 = MeO, R_2 = *c*-PrCO, R_3 = Me_2C=CH-
2.3.135 R_1 = MeO, R_2 = MeOCO, R_3 = Me_2C=CH-
2.3.136 R_1 = MeO, R_2 = $PhCH_2OCO$, R_3 = Me_2C=CH-
2.3.137 R_1 = MeO, R_2 = 2-MeO(C_6H_4)CO, R_3 = Me_2C=CH-
2.3.138 R_1 = MeO, R_2 = 4-MeO(C_6H_4)CO, R_3 = Me_2C=CH-
2.3.139 R_1 = MeO, R_2 = EtCO, R_3 = CH_2=$CHCH_2$-
2.3.140 R_1 = MeO, R_2 = EtCO, R_3 = (*E*)-CH_3CH=CH-
2.3.141 R_1 = MeO, R_2 = EtCO, R_3 = CH_2=$CH(CH_2)_2$-
2.3.142 R_1 = MeO, R_2 = EtCO, R_3 = (*S*)-2,2-Me_2-*c*-Pr-
2.3.143 R_1 = F, R_2 = EtCO, R_3 = Me_2CHCH_2-
2.3.144 R_1 = Cl, R_2 = EtCO, R_3 = Me_2CHCH_2-
2.3.145 R_1 = N_3, R_2 = EtCO, R_3 = Me_2CHCH_2-
2.3.146 R_1 = MeO, R_2 = EtCO, R_3 = Me_2CHCH_2-

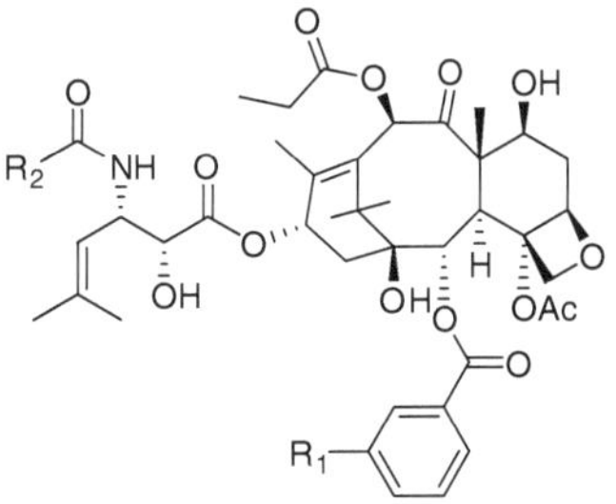

2.3.147 R_1 = H, R_2 = cyclobutyl
2.3.148 R_1 = H, R_2 = cyclopentyl
2.3.149 R_1 = H, R_2 = cyclohexyl
2.3.150 R_1 = H, R_2 = cyclopent-1-enyl
2.3.151 R_1 = H, R_2 = cyclohex-1-enyl
2.3.152 R_1 = H, R_2 = cyclopentyloxy
2.3.153 R_1 = H, R_2 = cyclohexyloxy
2.3.154 R_1 = OMe, R_2 = cyclopropyl
2.3.155 R_1 = OMe, R_2 = cyclobutyl
2.3.156 R_1 = OMe, R_2 = cyclopentyl
2.3.157 R_1 = OMe, R_2 = cyclohexyl
2.3.158 R_1 = OMe, R_2 = cyclohexyloxy

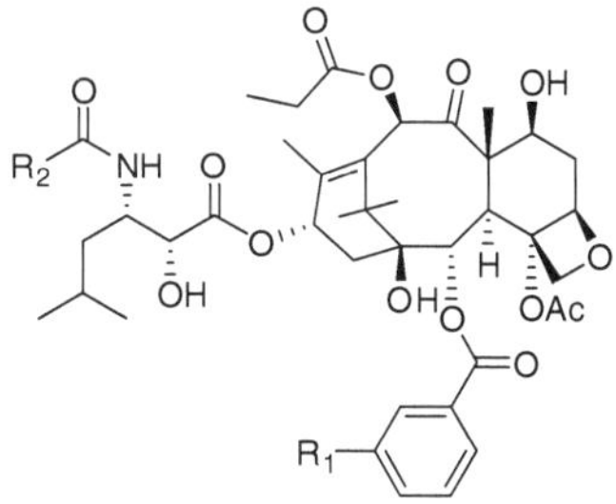

2.3.159 R_1 = H, R_2 = cyclopentyl
2.3.160 R_1 = H, R_2 = cyclohexyl
2.3.161 R_1 = H, R_2 = cyclopentyloxy
2.3.162 R_1 = H, R_2 = cyclohexyloxy
2.3.163 R_1 = MeO, R_2 = cyclopropyl
2.3.164 R_1 = MeO, R_2 = cyclohexyl
2.3.165 R_1 = MeO, R_2 = cyclopentyloxy

FIGURE 8.25 Analogs **2.3.127–2.3.165**.

found that the *meta*-substitution of the C2-benzoate moiety of C-*seco*-taxoids with appropriate groups significantly improved the interaction of C-*seco*-taxoids with Class III β-tubulin, which greatly reduced paclitaxel-resistance, since the over-expression of this subclass of β-tubulin leads to paclitaxel-resistance. Taxoid **2.3.206** was selected as the most promising drug candidate, as it always exhibited high potencies across different drug-resistant cancer cell lines.

Lu and coworkers reported the synthesis of a series of novel fluorinated docetaxel analogs in 2009 (**2.3.209–2.3.214**, Figure 8.27) [139]. These analogs were evaluated

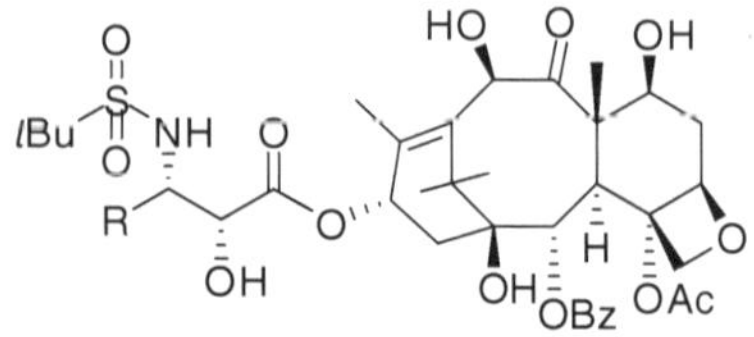

2.3.166 R = Ph
2.3.167 R = 2-Py N-oxide
2.3.168 R = 3-Py N-oxide

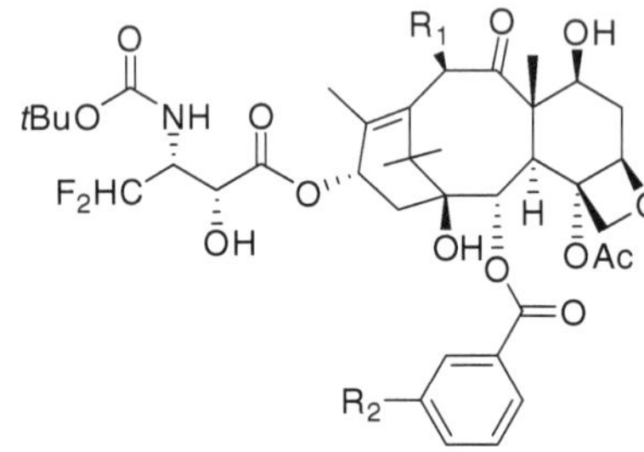

2.3.169 R_1 = Ac, R_2 = MeO
2.3.170 R_1 = EtCO, R_2 = MeO
2.3.171 R_1 = Me_2NCO, R_2 = MeO
2.3.172 R_1 = MeOCO, R_2 = MeO
2.3.173 R_1 = Ac, R_2 = F
2.3.174 R_1 = EtCO, R_2 = F
2.3.175 R_1 = Me_2NCO, R_2 = F
2.3.176 R_1 = MeOCO, R_2 = F
2.3.177 R_1 = Ac, R_2 = Cl
2.3.178 R_1 = EtCO, R_2 = Cl
2.3.179 R_1 = Me_2NCO, R_2 = Cl
2.3.180 R_1 = MeOCO, R_2 = Cl
2.3.181 R_1 = Ac, R_2 = N_3
2.3.182 R_1 = EtCO, R_2 = N_3
2.3.183 R_1 = Me_2NCO, R_2 = N_3
2.3.184 R_1 = MeOCO, R_2 = N_3

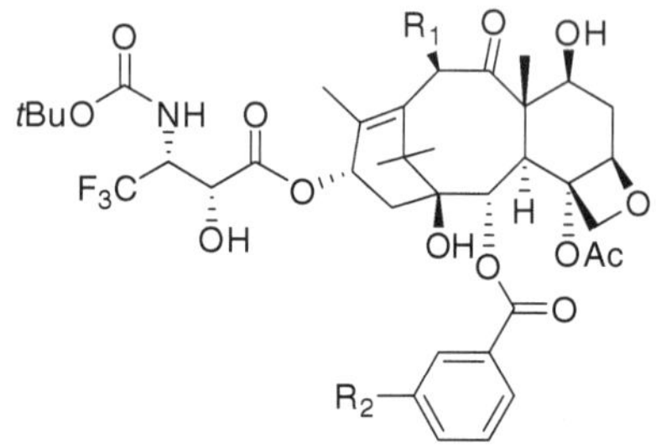

2.3.185 R_1 = Ac, R_2 = MeO
2.3.186 R_1 = EtCO, R_2 = MeO
2.3.187 R_1 = Me_2NCO, R_2 = MeO
2.3.188 R_1 = MeOCO, R_2 = MeO
2.3.189 R_1 = Ac, R_2 = F
2.3.190 R_1 = EtCO, R_2 = F
2.3.191 R_1 = Me_2NCO, R_2 = F
2.3.192 R_1 = MeOCO, R_2 = F
2.3.193 R_1 = Ac, R_2 = Cl
2.3.194 R_1 = EtCO, R_2 = Cl
2.3.195 R_1 = Me_2NCO, R_2 = Cl
2.3.196 R_1 = MeOCO, R_2 = Cl
2.3.197 R_1 = Ac, R_2 = N_3
2.3.198 R_1 = EtCO, R_2 = N_3
2.3.199 R_1 = Me_2NCO, R_2 = N_3
2.3.200 R_1 = MeOCO, R_2 = N_3
2.3.201 R_1 = H, R_2 = N_3

FIGURE 8.26 Analogs **2.3.166–2.3.201**.

for their activity *in vitro* and *in vivo*. All of them exhibited comparable or superior *in vitro* cytotoxicities to docetaxel against the human cancer cell line SK-OV-3, and two of them showed high *in vivo* activity in the human A549 nonsmall cell lung cancer xenograft in nude mice.

8.2.4 Biotechnological Production of Paclitaxel

The production of paclitaxel has evolved with the advancement of modern science and technology. Early supplies of paclitaxel relied solely on extraction from the bark of Pacific Yew trees, the harvest of which kills the plant in the process and gives only a low yield of drug. Ecological worries and the high demand for the supply of paclitaxel created a need for the development of more sustainable and efficient ways of

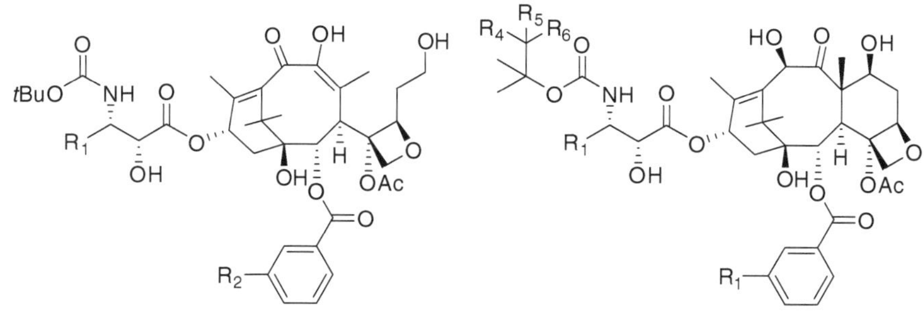

2.3.202	$R_1 = Me_2CHCH_2$, R_2 = MeO	**2.3.209**	$R_1 = R_3 = R_4$ = H, R_2 = F
2.3.203	$R_1 = Me_2CHCH_2$, R_2 = Cl	**2.3.210**	$R_1 = R_4$ = H, $R_2 = R_3$ = F
2.3.204	$R_1 = Me_2CHCH_2$, R_2 = F	**2.3.211**	R_1 = H, $R_2 = R_3 = R_4$ = F
2.3.205	$R_1 = Me_2C{=}CH$, R_2 = MeO	**2.3.212**	$R_1 = R_2$ = F, $R_3 = R_4$ = H
2.3.206	$R_1 = Me_2C{=}CH$, R_2 = Cl	**2.3.213**	$R_1 = R_2 = R_3$ = F, R_4 = H
2.3.207	$R_1 = Me_2C{=}CH$, R_2 = F	**2.3.214**	$R_1 = R_2 = R_3 = R_4$ = F

FIGURE 8.27 Analogs **2.3.202–2.3.214**.

paclitaxel production. It was not until the early 1990s that an efficient chemical semisynthesis of paclitaxel from the naturally occurring taxane 10-deacetylbaccatin III (10-DAB), which is produced by the European yew tree *Taxus baccata* in a much higher yield, was developed by Holton et al. [140]. This technique was utilized by Bristol-Myers Squibb for its commercial source of paclitaxel.

The biotechnological production of paclitaxel has become a promising alternative solution to the supply problem in recent years, and plant cell culture has been commercialized as an alternative source of paclitaxel in the pharmaceutical industry. Bristol-Myers Squibb now obtains its paclitaxel from Phyton Inc., which it produces by a plant cell culture process at its German subsidiary, Phyton Gesellschaft für Biotechnik mbH [141]. Paclitaxel is also produced (as Genexol) by plant tissue culture by Sam Yang Genex in Korea [142].

8.3 TUBULIN BINDING

8.3.1 Binding Conformation of Paclitaxel to Tubulin

The determination of the binding mode and binding conformation of paclitaxel on tubulin is important to fundamentally understand the molecular mechanism of this binding and to assist rational design of new antimitotic agents. However, it was not until recently that this goal was achieved. The difficulty was partly due to the structural complexity of the paclitaxel molecule, with four flexible side chains, which can generate numerous conformations, as well as the lack of a high-resolution structure of the binding site with enough resolution.

Paclitaxel was discovered to be a promoter of tubulin polymerization in 1979 [11]. Early photoaffinity labeling studies showed that paclitaxel bound to β-tubulin [143,144]. In one photoaffinity labeling study [143], 3′-*N*-(*p*-azidobenzoyl) paclitaxel photolabeled the N-terminal 31 amino acids of β-tubulin, while in another study [144] 2-debenzoyl-2-(*m*-azidobenzoyl)paclitaxel photolabeled amino acids 217–231 of β-tubulin.

An important discovery was made in 1998, when an atomic model of the α,β-tubulin dimer with a resolution of 3.7 Å was obtained by electron crystallography of zinc-induced tubulin sheets stabilized by paclitaxel [145], and this was later refined to 3.5 Å [146]. The X-ray crystallographic structure of docetaxel was docked into the electron density map of the tubulin. In this map, the electron density of the rigid taxane ring was found, but unfortunately that of the flexible side chains was not clear due to the low resolution. Nonetheless, the location of the drug was consistent with the photoaffinity labeling studies stated above. This discovery became a milestone in understanding the binding conformation of paclitaxel, in that it allowed computational studies integrating the experimentally obtained electron crystallographic data with proposed paclitaxel conformations.

Solution NMR was the most commonly applied spectroscopic method in early studies for the conformational assignment of paclitaxel in its uncomplexed state. Such studies in nonpolar [147–150] and polar solvents [149,151,152] yielded two proposed models, named the "nonpolar" and "polar" models, respectively. The former was characterized by hydrophobic interactions between the benzoyl group on C2 and the C-3′ benzamido phenyl. However, in polar solvents, hydrophobic collapse occurs between the C-3′ phenyl group and the C-2 benzoyl phenyl [153]. Both the nonpolar and polar conformations were proposed to be the bioactive conformations of paclitaxel.

In spite of these early NMR studies, it was pointed out by Snyder [154] that the common fundamental proposition underlying all these studies was that the molecule only adopts one single or largely dominant conformation in solution, and this structure leads to the experimental NMR data. This interpretation is incomplete, since with the existence of several easily rotatable single bonds the molecule is unlikely to adopt only one conformation in solution at ambient temperature. In a re-examination of the paclitaxel conformation in chloroform by the NMR analysis of molecular flexibility in solution (NAMFIS) methodology, Snyder and his coworkers identified several distinct conformers with populations from 1% to 35%. Among them was the so-called T-taxol conformation, which is a minor conformer at less than 3% in population [154].

The conformation of uncomplexed paclitaxel in the solid state was studied by X-ray crystallography [155,156]. The first crystallographic structure of the paclitaxel analog docetaxel was published by Guéritte-Voegelein [155] in 1990, before most of the NMR studies were carried out. This solid-state conformation of docetaxel was later found to resemble the nonpolar conformation of paclitaxel deduced by NMR studies [157,158]. In a similar study of paclitaxel in 1995, two different conformations were obtained in a single crystal unit cell [156]. One of them was consistent with the "polar" conformation in solution, while the other was a new conformer that had not been identified by previous studies. All these conformations, together with those derived from NMR studies, provided a large number of three-dimensional paclitaxel structures for molecular modeling studies [152,159] of the docking of paclitaxel into the β-tubulin binding site.

Solid-state rotational echo double resonance NMR (REDOR-NMR) and fluorescence spectroscopy are techniques capable of providing information on the orientation of paclitaxel and its conformation in the bound state on microtubules. In one study [160], a combination of these two approaches yielded data in support of a hydrophobically collapsed conformation. It is noticeable in this study that REDOR studies on an isotopically labeled tubulin-bound paclitaxel allowed the determination of the distances between ^{19}F at the *para*-position of the C-2 benzoyl group and ^{13}C atoms at the 3′-amide carbonyl and the C-3′methine carbons. The measured values were 9.8 ± 0.5 Å and 10.3 ± 0.5 Å, respectively, which are consistent with both the polar conformation and the T-taxol conformation (Figure 8.28).

The T-taxol model mentioned above is characterized by very similar distances between the C-2 benzoyl phenyl and both of the phenyl rings emanating from C-3′ (Figure 8.29) [152]. In a computational study, a large number of paclitaxel conformers were individually docked into the experimental density map of the tubulin–paclitaxel complex. Among these paclitaxel conformers, 26 were derived from the former X-ray crystallographic [161,162] and NMR studies [147–149,151,154,163,164] and the others were computer-generated. The T-taxol conformation gave the best fit to the experimental density. The intramolecular distances between both phenyls emanating from C-3′ and the C-2 benzoyl phenyl obtained in this study were similar to those obtained in the previous REDOR-NMR study on labeled tubulin-bound paclitaxel [160]. The T-taxol conformation was then found to be the best fit an extended series of REDOR NMR studies [165].

The Ojima group proposed an alternative REDOR-taxol model [159] also based on the results of the earlier incomplete REDOR NMR studies [160] mentioned above. They had previously shown that the C-7-benzodihydrocinnamoyl (C-7BzDC) derivative of paclitaxel formed a complex with tubulin, and that irradiation of this complex only labeled the Arg^{282} residue [166]. Using this information, 16 conformers derived from a Monte Carlo conformational search of 2-(*p*-fluorobenzoyl)-paclitaxel were docked into tubulin. The internuclear distances acquired from these data were then compared to the experimental distances given by the previous REDOR study [160]. The structure with the least deviation was considered to be the preferred

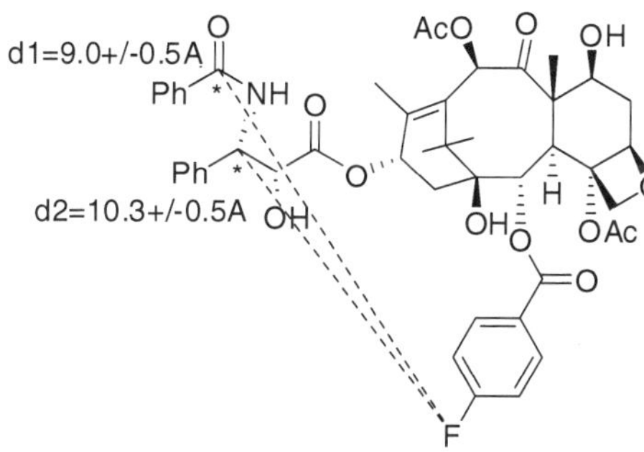

FIGURE 8.28 ^{13}C–^{19}FIntramolecular distances indicated by REDOR-NMR.

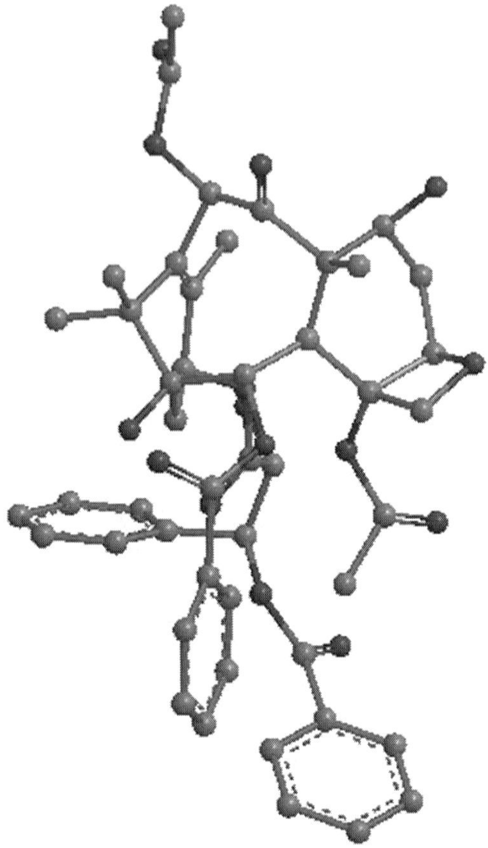

FIGURE 8.29 T-taxol conformation.

binding model and was named "REDOR-taxol." It was pointed out that this conformation differs from the T-taxol model in the orientation of the 2′-hydroxyl group. In REDOR-taxol, the 2′-hydroxyl group is H-bonded to His^{227}, while in T-taxol similar bonding occurs with Arg^{359} [159]. The relative merits of the T-taxol and the REDOR-taxol models have been argued [167,168].

8.3.2 Synthetic Efforts to Make Conformationally Restricted Analogs of Paclitaxel

An understanding of the bioactive tubulin-binding conformation of paclitaxel should, in principle, allow the design and synthesis of analogs designed to constrain paclitaxel to these conformations. Subsequent evaluation of the bioactivity of the analogs could, in turn, provide validation of the model being tested and information for better understanding of the tubulin-bound conformation and for designing better tubulin-binding agents.

8.3.2.1 Validation of the T-Taxol Model. After the proposal of the T-taxol model, its validity was debated for two reasons: the relatively low resolution of the tubulin structure used for docking, and the fact that the tubulin structure was of zinc-induced sheets rather than native microtubules. The obvious difference in the arrangement of protofilaments in the two types of tubulin aroused concerns that the binding conformation derived from the zinc-stabilized sheets may be different from the binding conformation in native microtubules [169]. To provide experimental evidence that T-taxol stabilizes genuine microtubules, the Kingston group has successfully constructed analogs constrained to the T-conformation [170,171]. In these studies, the T-conformation was achieved synthetically by bridging between the 4-OAc methyl and the C-3′-phenyl groups. This tethering strategy was developed by analysis of the 3D topographic diagram of the proposed T-taxol model, which displayed that the above two moieties are actually close to each other [171]. This indicates that tethering the two centers by a linker of the correct length should mimic the parent T-conformation.

Compound **3.2.1** (Figure 8.30) was the first analog based on this strategy. It used a 7-atom tether between the methyl and phenyl centers [170]. NMR/NAMFIS analysis in solution demonstrated that the compound exists in the T conformation with a population very similar to paclitaxel itself (5%). It was then concluded that the compound should be able to achieve the conformation as a low energy form. Nonetheless, the compound proved to be 8–10-fold less active than paclitaxel in both tubulin assembly and cytotoxicity assays. A substrate–protein steric interaction between His-227 of tubulin in the binding pocket and the lower part of the tether accounted for the absence of activity [170].

To overcome this problem, compounds **3.2.2** and **3.2.3** (Figure 8.31) were designed and synthesized [171]. In these compounds, the tethers contained fewer atoms and were attached to the *ortho* instead of the *meta* position of the 3′-phenyl, thus forcing the bridge inside the concave baccatin core. Because of these changes, NMR/NAMFIS analysis confirmed that the molecules retained the T-shape, with a population of over 80% T-taxol in solution. As predicted, compound **3.2.2** was twofold more active in the tubulin polymerization assay and equally potent against PC3 and A2780 cell lines, and compound **3.2.3** was 20 times more potent than paclitaxel in the A2780 cell line. These results prove that a compound constrained to the T-taxol

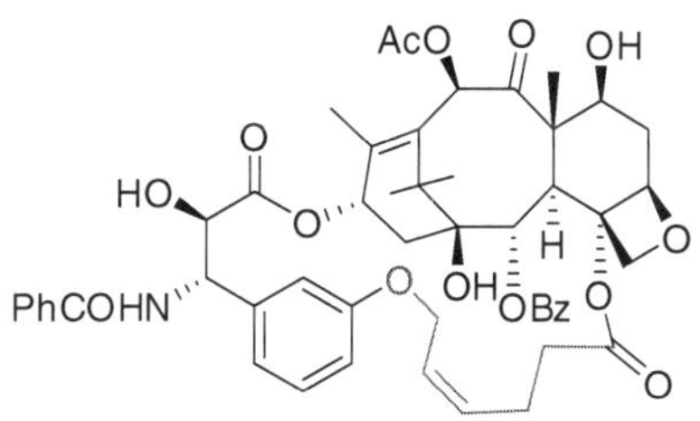

3.2.1

FIGURE 8.30 Compound **3.2.1**.

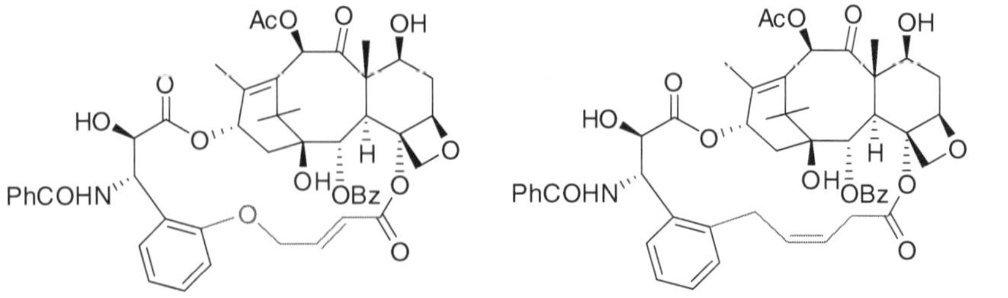

FIGURE 8.31 Compounds **3.2.2** and **3.2.3**.

conformation can bind to genuine microtubules, and thus that the T-taxol structure is an excellent model of the binding conformation of paclitaxel.

In a study by Appendino and Snyder [172], analogs **3.2.4–3.2.7** (Figure 8.32) with conformationally constrained C-13 side chains were synthesized. The NAMFIS analysis of these compounds resulted in the detection of seven conformations, among which "polar," "nonpolar," and T-conformations were all found. The latter was populated to the extent of 47% in compound **3.2.4**. While compounds **3.2.4** and **3.2.6** displayed tubulin binding and cytotoxicity values comparable to those of paclitaxel, the six-membered ring compounds **3.2.5** and **3.2.7** were 30–42 and 239–333 fold less active in tubulin binding and cytotoxicity, respectively. This sharp contrast could again be explained by the previously mentioned ligand–tubulin interaction.

8.3.2.2 Other Attempts to Make Constrained Analogs for Conformational Study. One of the earliest bridged paclitaxel analogs to mimic the "hydrophobically collapsed" conformation was prepared by the Georg group [173], who identified this conformation [151] in solution. Short tethers were bridged between the phenyl ring of the C-2 benzoyl group and the C-3′ phenyl to yield compounds **3.2.8–3.2.11** (Figure 8.33). Unfortunately, none of these compounds showed significant activity in a tubulin assembly assay. It was thus concluded that the bound conformation of paclitaxel is not the "hydrophobic collapse" conformation. This conclusion was supported by a study that showed that each of paclitaxel's hydrophobic centers engages in complementary interactions with tubulin, indicating that the polar "hydrophobically collapsed" conformation is unlikely to be correct [152].

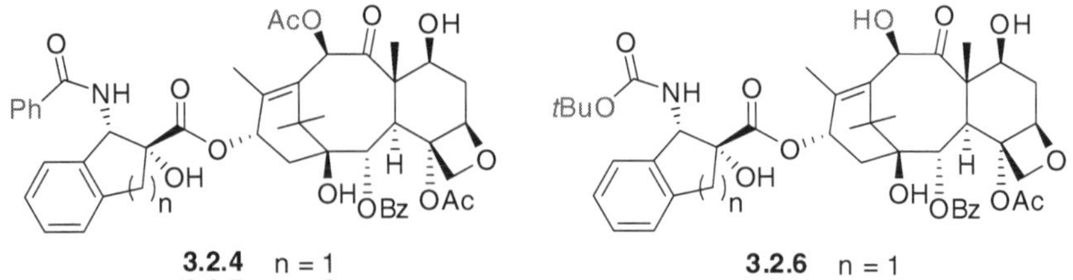

FIGURE 8.32 Compounds **3.2.4–3.2.7**.

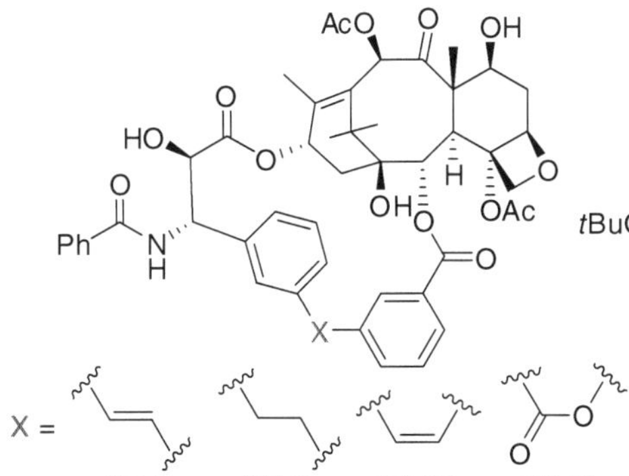

3.2.8 3.2.9 3.2.10 3.2.11

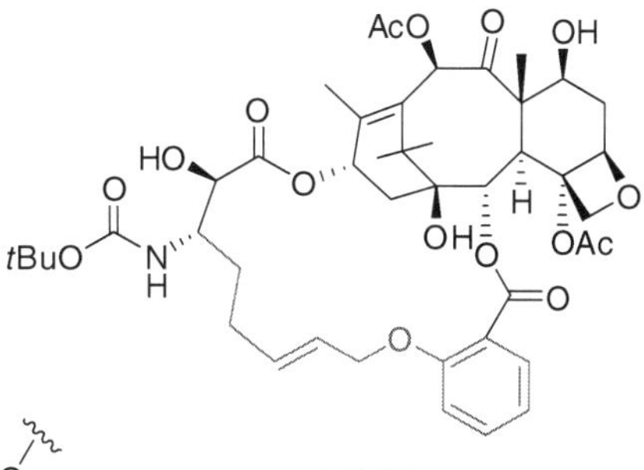

3.2.12

FIGURE 8.33 Compounds **3.2.8–3.2.12**.

In opposition to the study noted above, Ojima and coworkers obtained the tethered compound **3.2.12**, which was found to adopt predominantly the hydrophobically collapsed conformation in DMSO-d_6/D_2O, based on NMR analysis. This compound exhibited submicromolar level cytotoxicity against a human breast cancer cell line and 37% of the activity of paclitaxel in the tubulin polymerization assay [174]. In a subsequent paper [175], this group reported the preparation of a series of macrocyclictaxoids of general structure **3.2.13** tethered between C-3′ and C-2 with tethers of variable length and composition (Figure 8.34). Three of these compounds, including **3.2.14**, demonstrated submicromolar cytotoxicity against a human breast cancer cell line.

Constrained analogs were also achieved mimicking the "nonpolar" conformation, which was also formerly regarded as the tubulin bound structure. This included a first effort [176] to mimic the tubulin-bound docetaxel conformation based on Nogales's study [176,177] in which the analog was used as a surrogate for paclitaxel. Implanting linkers between the C-2 and C-3′ positions of the taxoid skeleton generated a group of *N*-linked macrocyclic taxoids illustrated by **3.2.15** (Figure 8.35). In the subsequent cytotoxicity evaluation, some analogs showed submicromolar-level activities. It is worth mentioning that the compounds containing linkers with *Z* double bonds were

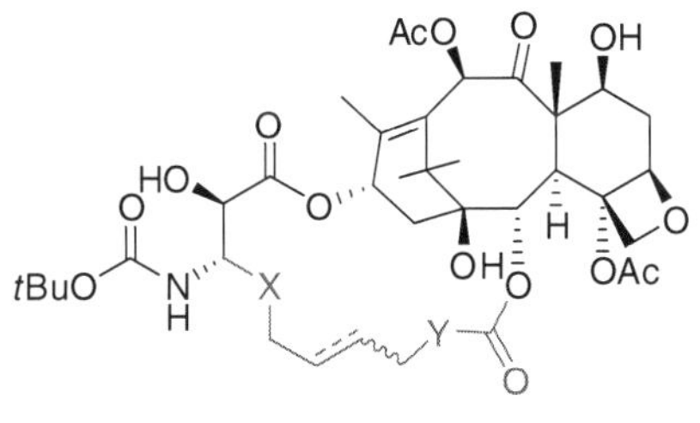

3.2.13

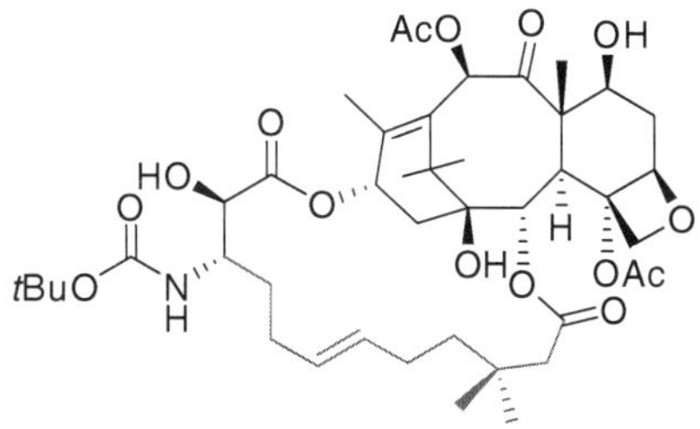

3.2.14

FIGURE 8.34 Compounds **3.2.13–3.2.14**.

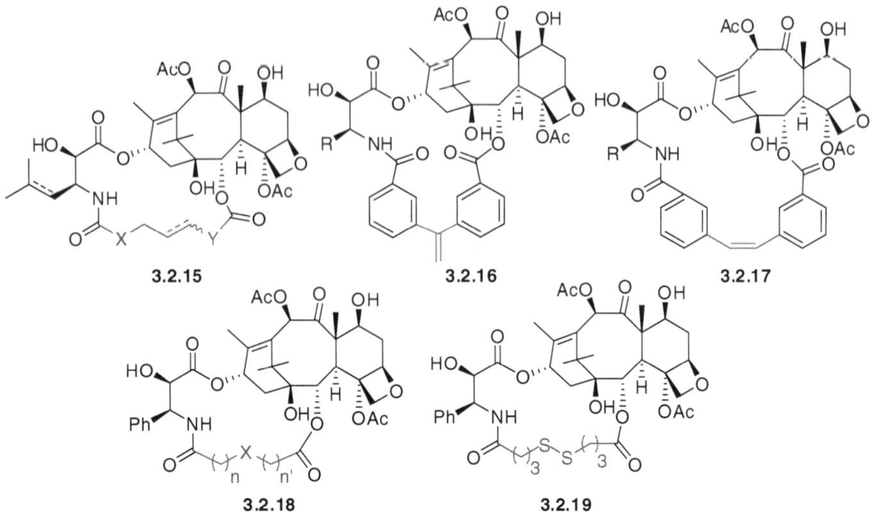

FIGURE 8.35 Compounds **3.2.15–3.2.19**.

more active than those containing linkers with *E* double bonds. In a continuing study [177] by the same group, a series of similar N-linked macrocyclic taxoids was prepared by intramolecular Heck coupling reactions. This series was reported to be more cytotoxic than the previous one. Also interestingly, exo compounds such as **3.2.16** tended to be more active than endo ones such as **3.2.17**. No conformational analysis was reported in these two studies.

Efforts to model the "nonpolar" conformation were also made by Dubois and coworkers. Three similar studies [178–180] were published describing several panels of N-linked macrocyclicpaclitaxel analogs that were designed and synthesized. The first series of compounds can be illustrated by structure **3.2.18**, in which *n* varies from 3 to 7 (Figure 8.35). Some of them displayed weak cytotoxicity against the KB cell line, and among these, the disulfide compounds **3.2.19** were most active. Also evaluated was the acyclic precursor of **3.2.19** which showed similar activity to **3.2.19**. This suggested that the presence of the macrocyclic disulfide did not account for the weak interaction with tubulin. However, a specific interaction between the disulfide and the protein was proposed to explain the difference in activity between analogs containing S–S bonds and C–S bonds [178]. A conformational study was not reported.

In a related study [179], macrocyclic analogs bearing only carbon tethers such as **3.2.20** were built because of a suspicion that sulfur may be deleterious to tubulin binding (Figure 8.36). Eleven cyclic compounds were evaluated for inhibition of microtubule disassembly, and among them compound **3.2.21** was only sevenfold less active than paclitaxel. All the other compounds were either 30- to 100-fold less active than paclitaxel or inactive. All compounds were at least 1000-fold less potent against the KB cell line. The conformations of these taxoids in solution were investigated by NMR and molecular modeling studies, which indicated that compound **3.2.21**

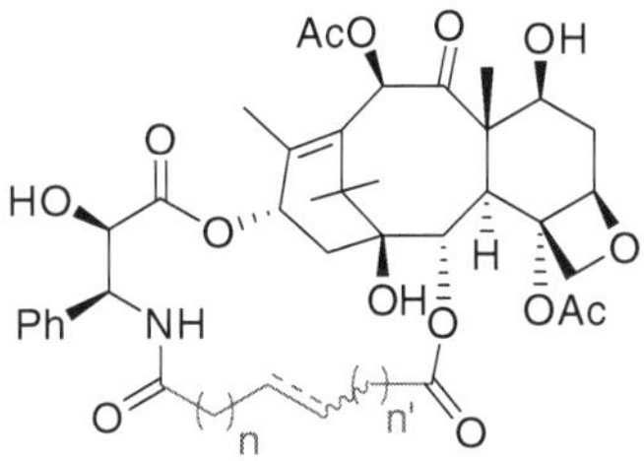

3.2.20

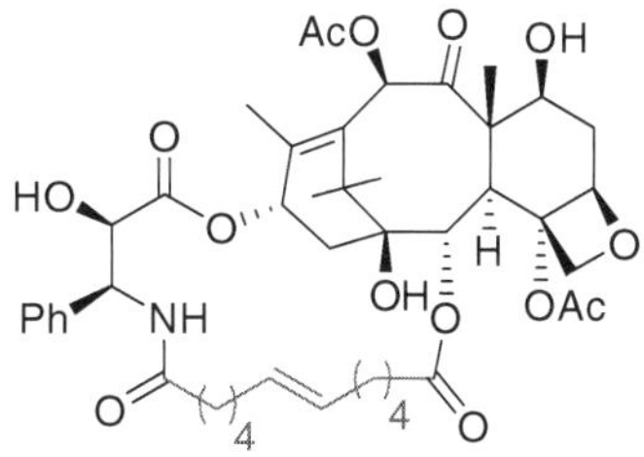

3.2.21

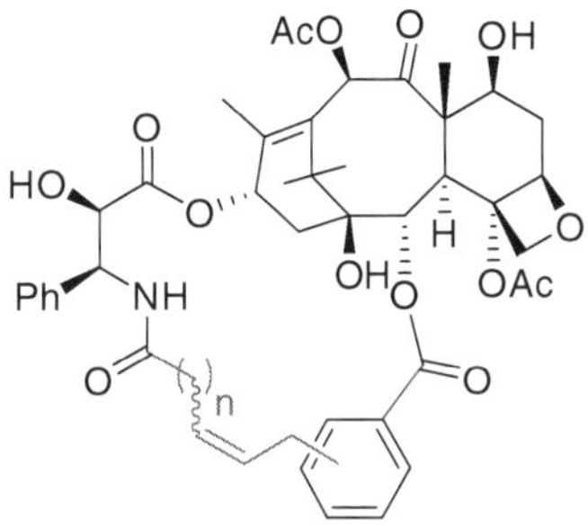

3.2.22

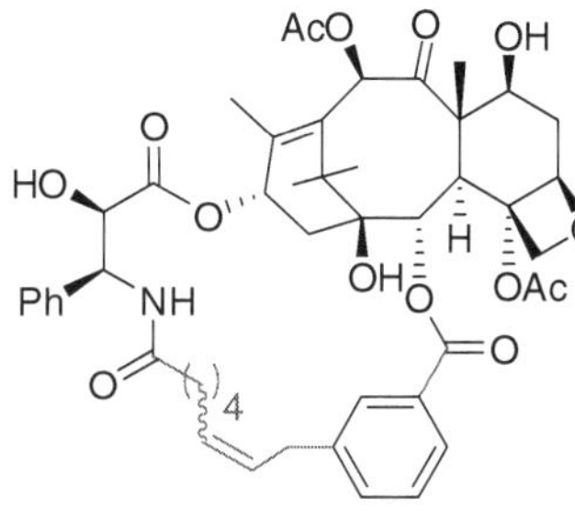

3.2.23

FIGURE 8.36 Compounds **3.2.20–3.2.22**.

adopted a conformation between "nonpolar" and T-shaped. Other analogs were shown to take conformations not recognized by tubulin.

A more recent endeavor to establish bridged analogs resembling tubulin-bound docetaxel was published by Dubois et al. in 2004 [180]. The macrocylic taxoids represented by **3.2.22** with varied ring sizes were constructed by inserting tethers between N-3′ and the *ortho*, *meta*, or *para* positions of the C-2 benzoate (Figure 8.36). Five compounds in this series were evaluated for inhibition of cold-induced microtubule disassembly and for cytotoxicity against the KB cell line. All compounds except one exhibited submicromolar IC_{50} values against the tumor cell line. While *Z*- and *E*-**3.2.23** proved to be equally as active as paclitaxel in inhibiting microtubule disassembly, analogs bearing linkers to the *para* positions were totally inactive. This difference of activity is consistent with the results obtained from modeling studies, which showed selective interaction between tubulin and *meta*-substituted compound. The possible presence of a π-stacking interaction observed between the imidazole of His-229 and the C-2 benzoyl group may account for the enhanced bioactivity of these compounds compared with those obtained in the two previous studies [178,179]. Computational studies suggested that the molecules mainly adopt the T-conformation in solution, and a modeling study verified the T-shape as the better docking conformation. Nevertheless, the corresponding acyclic compound proved to be as active as the cyclic compound **3.2.23**. It was pointed out later that the 22-membered ring in this compound is too flexible to be restricted to a specific conformation [153].

In the previously discussed study [159], identification of the REDOR-taxol structure by the Ojima group was accompanied by the synthesis of a corresponding

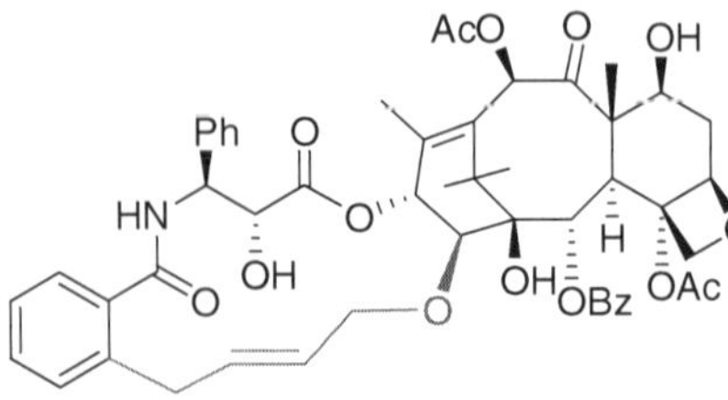

3.2.24

FIGURE 8.37 Compounds **3.2.24**.

constrained analog **3.2.24** designed to mimic the REDOR-taxol conformation (Figure 8.37). This compound proved to be comparable to paclitaxel in inducing microtubule polymerization and was only three times less cytotoxic than paclitaxel against the LCC6-WT cell line. Conformational studies gave C-13 side chain dihedral angle distributions, which were similar to those of REDOR-taxol than to those of T-taxol.

8.3.2.3 Synthetic Attempts to Make Simplified Paclitaxel Analogs Based on the Paclitaxel Pharmacophore. The complexity of the paclitaxel molecule has made its synthesis very difficult. Though a number of total syntheses of paclitaxel have been accomplished [39–45,47], none of these are currently applicable commercially. Therefore, it would be very desirable if structurally simplified analogs with retained or improved activity could be designed. Progress in the understanding of the SAR and tubulin-binding conformation of paclitaxel has spurred continuous attempts in this area. A major strategy is to replace the taxane polycyclic core with mimics with simple structures and subsequently to assemble the paclitaxel pharmacophore by putting side chains onto the core.

Two early studies in 1991 [181] and 1994 [182] produced two taxoids (**3.2.25–3.2.26**) with preservation of the taxane core but with fewer substituents (Figure 8.38). The C-13 side chain was retained based on the knowledge that this unit, and especially the C-2′ hydroxyl group, is essential to bioactivity [183,184]. These simplified analogs were reported to show reduced activity in inhibition of tubulin depolymerization, but no detailed data were presented.

A series of even more simplified paclitaxel analogs illustrated by **3.2.27** were designed by Klar and his coworkers in 1998 (Figure 8.39) [185]. Based on the SAR of

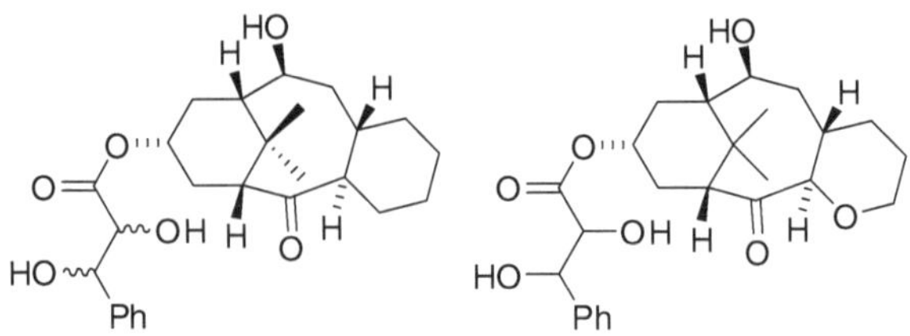

3.2.25 3.2.26

FIGURE 8.38 Compounds **3.2.25–3.2.26**.

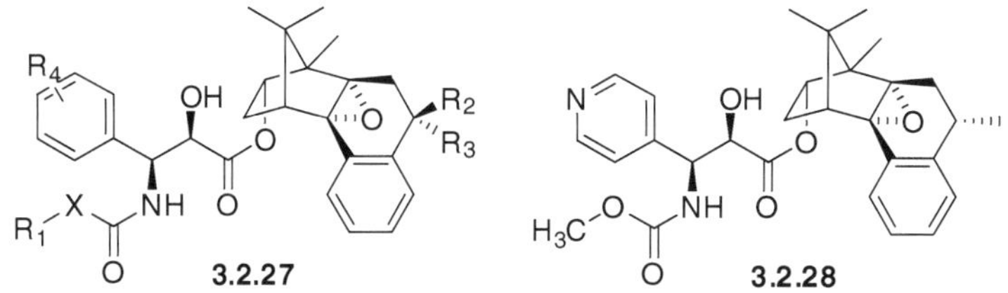

3.2.27 3.2.28

FIGURE 8.39 Compounds **3.2.27–3.2.28**.

paclitaxel showing that the C-13 side chain, an ester group at C-2 and the oxetane moiety at C-4/C-5 are essential to activity, and that the northern region has only a minor influence on activity, the polycyclic core of paclitaxel was replaced by the borneol skeleton in this strategy. SAR studies of this type of analog were addressed based on the outcome of the microtubule assay. The optimized structure **3.2.28** was reported to be superior to paclitaxel in terms of inhibition of tubulin depolymerization. Nevertheless, it was at least 100-fold less cytotoxic than paclitaxel against all the cell lines tested.

In a more recent effort [186], the baccatin core of paclitaxel was mimicked by a spirobicyclo-[2,2,2]-octane derivative designed by computer modeling. The target compound **3.2.29** was synthesized in six steps from a simple starting material (Figure 8.40). Compound **3.2.29** was subjected to a microtubule polymerization assay, but no obvious microtubule stabilization was shown. It was thus proposed that the absence of activity was due to the lack of counterparts of the paclitaxel C-2 benzoate and oxetane/4-OAc in this structure.

None of the above studies related either their structure design or analysis of activity to binding conformation, which unavoidably reduced the possibility of finding active analogs. The first conformationally restricted simplified analog was designed in a recent study [187]. An indolizidinone with hydroxyl groups at C-5 and C-8 was utilized to model the baccatin core, and moieties mimicking the paclitaxel C-2 and C-13 side chains were subsequently attached to the two hydroxyl groups to yield **3.2.30**. To constrain the side chain conformation, macrocyclic compound **3.2.31** with a 19-membered ring was prepared. Unfortunately, both of the compounds

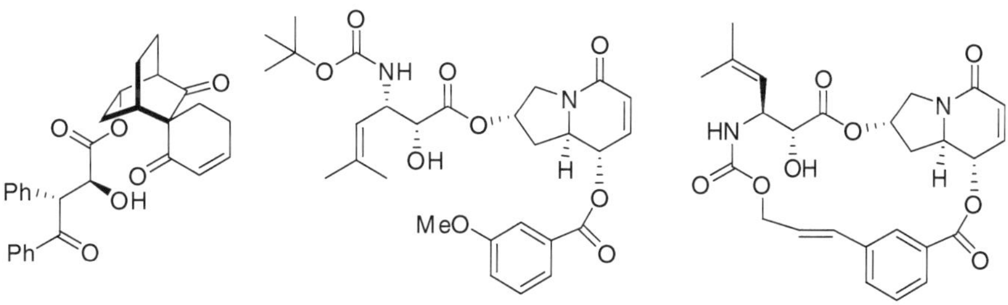

3.2.29 3.2.30 3.2.31

FIGURE 8.40 Compounds **3.2.29–3.2.30**.

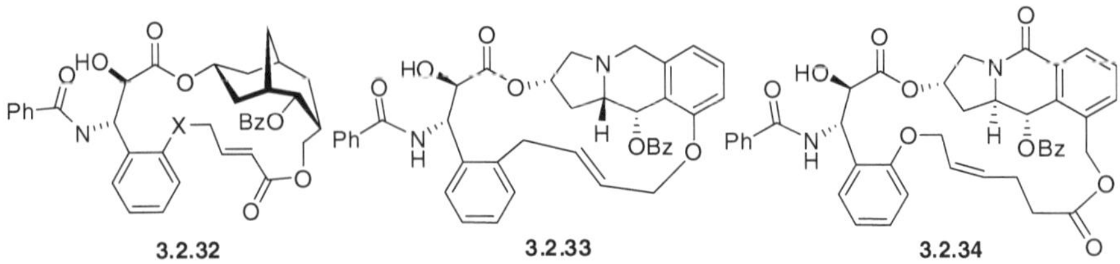

3.2.32 3.2.33 3.2.34

FIGURE 8.41 Compounds **3.2.32–3.2.34**.

proved to be inactive in promoting microtubule assembly. However, they displayed micromolar cytotoxicity against several human tumor cell lines.

As noted above, the Kingston group successfully built an active T-shaped paclitaxel analog, substantiating the T-taxol proposition [171]. Based on this understanding, a class of simplified analogs illustrated by **3.2.32** was designed to model the T-taxol conformation (Figure 8.41). In these compounds, the baccatin core of paclitaxel was replaced with a [3,3,1]-bicyclononane moiety. The T-conformation was maintained by a similar strategy used in designing the former bridged T-taxol analogs (**3.2.2** and **3.2.3** in Figure 8.31) [171]. A conformational study indicated that these molecules approximate the bioactive T-conformation with minor deviations due to the replacement of the baccatin core with a bicyclononane core [188]. Compounds **3.2.32** and its analogs exhibited moderate cytotoxicity, but they were also very insoluble in water, and this made it difficult to evaluate them. A compound that could be constrained to the T-taxol conformation but which showed significant water solubility remains a desirable goal.

In an approach to the objective of developing of a new-generation water-soluble simplified paclitaxel analogs, a series of analogs with structures represented by **3.2.33** have been designed (Figure 8.41). In these structures, an aza-tricyclic moiety mimics the baccatin core of paclitaxel, and the key side chains are maintained and constrained into the T-taxol conformation by a synthetically accessible bridge. Preliminary computer modeling has shown that the designed structures adopt a similar shape to the previous conformationally constrained compounds such as **3.2.3** when docked into tubulin. The compounds showed cytotoxicities at micromolar level against the A2780 cell line. The weaker activities relative to paclitaxel can, in part, be attributed to the complex conformational profile of the macrocyclic rings relative to the parent molecule [189]. Similar compounds, of which **3.2.34** is the closest example, have also been prepared by the Ojima group; these too were only weakly active [190].

8.4 PHARMACOLOGY OF PACLITAXEL

8.4.1 Prodrugs and Drug Delivery of Paclitaxel

Despite its great success in clinical use, paclitaxel suffers from several major drawbacks such as poor water solubility, multidrug resistance (MDR), and dose-limiting toxicities [191]. Due to the poor water solubility of paclitaxel, the detergent

Cremophor EL® is used in the drug formulation, which can cause severe adverse allergic reactions [192]. Moreover, systemic toxicity occurs due to the poor selectivity of the chemotherapeutic agent to tumor cells over normal cells, which often leads to severe adverse side effects such as mucositis, bone marrow suppression, febrile neutropenia, neurotoxicities, and ulceration [193]. One solution to these problems is the development of prodrugs with greater water solubility and/or lower systemic toxicity. Improved water solubility can be achieved by introducing a hydrophilic moiety to the prodrug, and toxicity can be lowered by tumor targeting prodrugs that only release paclitaxel at tumor sites.

8.4.1.1 ***Water-Soluble Prodrugs.*** Early studies mainly focused on the development of paclitaxel with improved water solubility, and many prodrugs bearing hydrophilic promoieties such as glutarate, succinate, sulfonate, phosphate, and amino acid functionalities were developed [194–200]. Several good reviews of this topic have been published [201–203]. Recent attempts toward this goal have yielded a broad range of new prodrugs. Damen et al. synthesized compound **4.1.1** with malic acid at the C-2′ position as a water-soluble moiety (Figure 8.42) [204]. This compound demonstrated a water solubility of 0.6 mg/mL, which is 2000 times greater than that of paclitaxel. It was stable during incubation in PBS buffer at pH 7.4 at 37 °C and released paclitaxel when incubated in human plasma with a half-life ($T_{1/2}$) of 4 h. This compound showed enhanced anticancer activity and lowered toxicity in a P388 murine leukemia *in vivo* model when compared to paclitaxel. In another report,

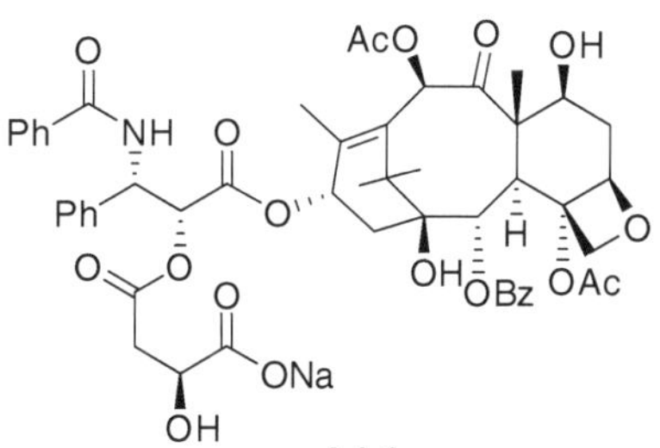

4.1.1

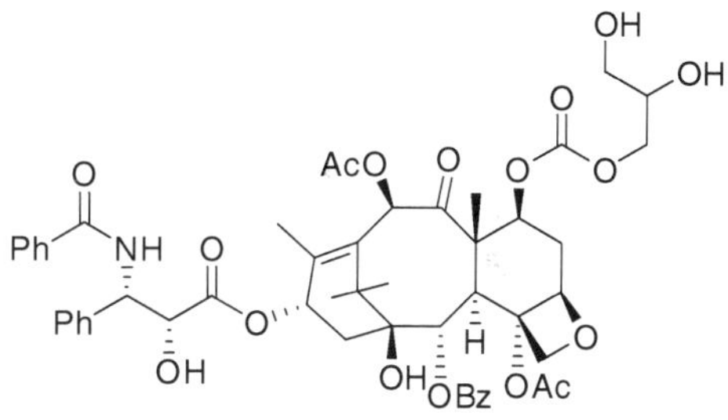

4.1.2

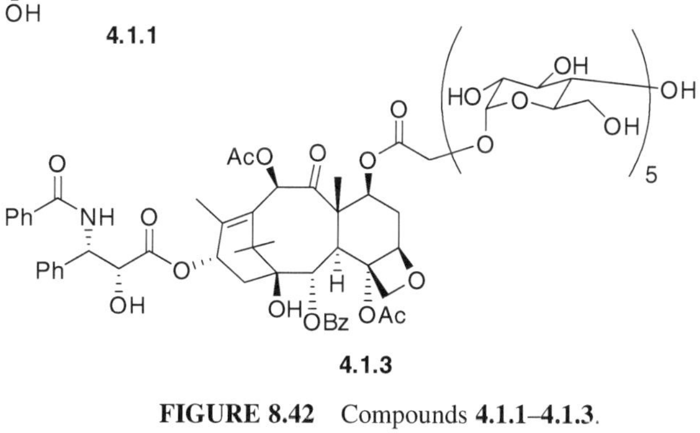

4.1.3

FIGURE 8.42 Compounds **4.1.1–4.1.3**.

Seligson et al. synthesized the C-7 carbonate derivative protaxel **4.1.2** with 200-fold greater water solubility (Figure 8.42) [205]. Compound **4.1.2** released paclitaxel by a pH-dependent intramolecular cyclization mechanism in human serum with a half-life of 10 min. This prodrug showed improved anticancer activity *in vivo*. Recently, an oligosaccharide-linked prodrug **4.1.3** of paclitaxel was synthesized by chemo-enzymatic procedures, including enzymatic transglycosylations with α-glucosidase and cyclodextrin glucanotransferase [206]. The compound exhibited 6800-fold greater solubility in water and was significantly less cytotoxic compared to paclitaxel.

Kirschberg et al. reported the synthesis of prodrug **4.1.4** that bears an oligomer of arginine with eight repeating unit at C-2′ (Figure 8.43) [207]. This amino acid oligomer was designed not only to enhance water solubility but also to increase cellular uptake of the prodrug, based on the rationale that short oligomers of arginine often show good membrane translocation activity [208,209]. Compound **4.1.4** demonstrated greatly enhanced water solubility (>1000-fold) and the ability to release paclitaxel with a half-life of 107 min at the physiological pH 7.4. Similarly, prodrug **4.1.5** with a C-7 spermine moiety was designed and synthesized as a substrate for polyamine-based transporter (Figure 8.43) [210]. This compound showed good water solubility and significantly lowered cytotoxicity compared to paclitaxel. No data showing enhanced cellular uptake were reported for either of the above prodrugs.

The "pro-dual-drug" **4.1.6** combined paclitaxel with the anticancer agent triphosphono-γ-(Z)-ethylidene-2,3-dimethoxybutenolide, which is an inhibitor of

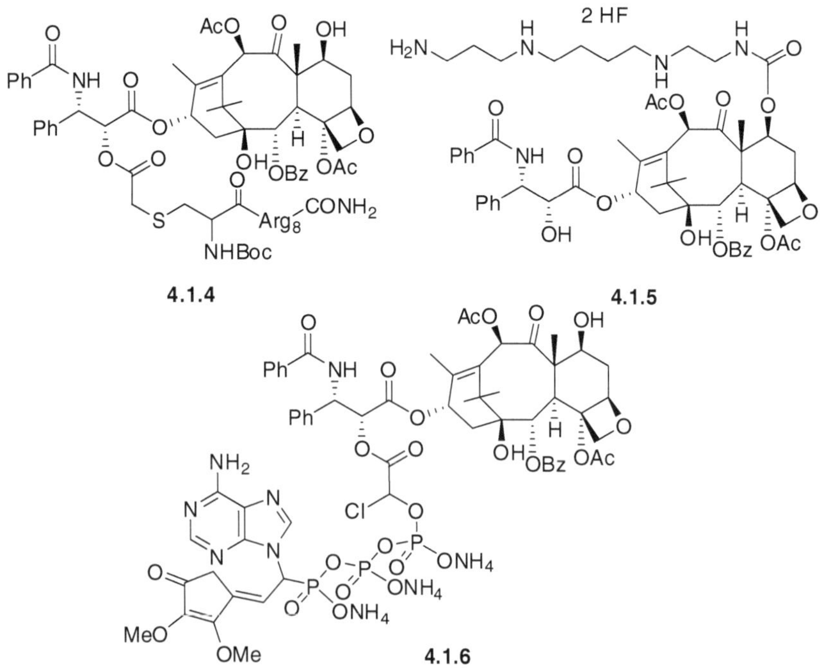

FIGURE 8.43 Compounds **4.1.4**–**4.1.6**.

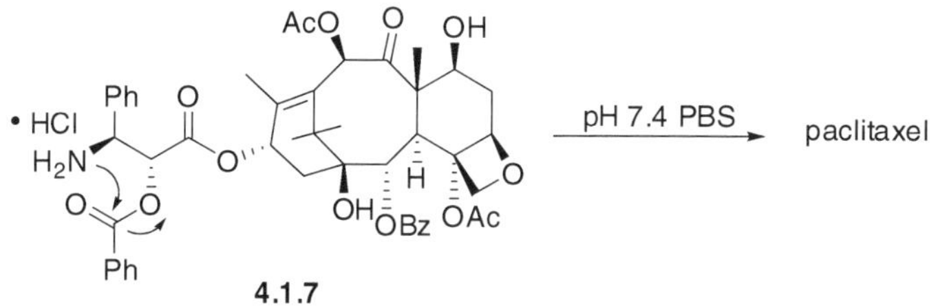

4.1.7

FIGURE 8.44 Compounds **4.1.7**.

ribonucleoside-diphosphate reductase (RDPR), through a glyoxylic acid linker at the C-2′ position (Figure 8.43) [211]. The resulting dual functional prodrug demonstrated about 1000-fold higher lipophilicity than the triphosphonate compound alone and 500-fold greater water solubility than paclitaxel. The optimized hydrophilic–lipophilic property led to higher bioavailability and better anticancer activity. The prodrug was completely converted to paclitaxel and triphosphono-γ-(Z)-ethylidene-2,3-dimethoxybutenolide in 0.4 h in human plasma.

Kiso et al. synthesized an interesting prodrug called isotaxel **4.1.7**, which showed 1800-fold greater water solubility than paclitaxel (Figure 8.44) [212]. This prodrug produces paclitaxel by an intramolecular acyl migration without generating any by-products and has a half-life of 15 min at the physiological pH 7.4.

8.4.1.2 ***Prodrugs Designed to Target Tumor Tissue.*** Tumor-targeting prodrugs are designed to achieve selective delivery of chemotherapeutics to tumor tissues and thus reduce the toxicity to normal tissues. Such prodrugs are usually conjugates of a therapeutic agent and a tumor-targeting agent. The latter could be hormones, peptides, polysaccharides, vitamins, nanoparticles, fatty acids, and antibodies, etc., that have a certain affinity to tumor cells. Ideally, the drug conjugate should be stable and nontoxic throughout the blood stream, bind selectively to cancer cells, and be cleaved readily to release the active drug upon reaching tumor cells.

Prodrug **4.1.8** was designed to target the enzyme plasmin that is overproduced by tumor cells (Figure 8.45) [213]. The prodrug exhibited plasmin-dependent paclitaxel release with a half-life of 42 min and was significantly less cytotoxic in various tumor cell lines without plasmin activation.

Bioreductive prodrugs **4.1.9** and **4.1.10** were designed to target the hypoxic environment of the tumor tissue due to fast proliferation of tumor cells (Figure 8.45). It was demonstrated that compound **4.1.9** can release the parent drug by chemical reduction with sodium borohydride [214], and it was significantly less cytotoxic against several cancer cell lines. Compound **4.1.10** can produce paclitaxel by reduction with dithiothreitol and showed 650-fold lower cytotoxicity than paclitaxel *in vitro* and superior tumor regression *in vivo* [215].

A series of estradiol–paclitaxel conjugates (represented by **4.1.11**) were designed to target the estrogen receptor (ER) (Figure 8.46) [216], which is overexpressed in breast, ovarian and gonad cancer cells [217]. Prodrug **4.1.11** was threefold more cytotoxic to the ER positive cell line than the ER negative cell line, although it was

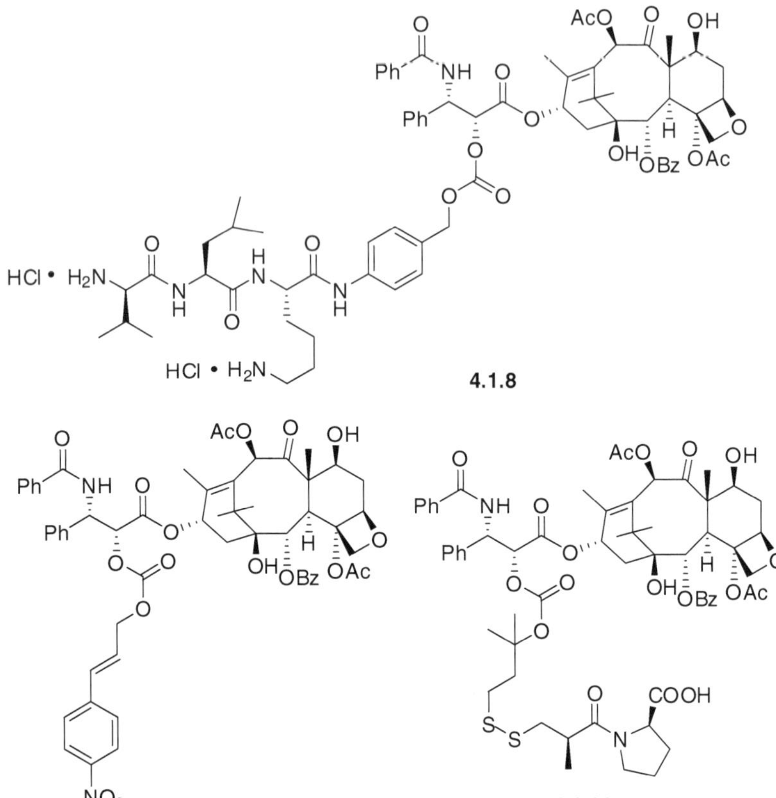

FIGURE 8.45 Compounds **4.1.8–4.1.10**.

much less active than paclitaxel against both of the cell lines. Several folic acid-paclitaxel conjugates such as **4.1.12** were designed to target the folate receptor (FR) [218], which was significantly overexpressed in a broad spectrum of human cancers [219]. The prodrugs retained the receptor-binding affinity of the folate ligand. However, they did not show selective cytotoxicity against cancer cell lines that express the folate receptor, and they were significantly less active *in vivo* compared to paclitaxel. Prodrug **4.1.13** (Figure 8.46) was designed to target the α_v integrin receptor, which is highly overexpressed in metastatic cancer cells [220,221]. It demonstrated improved tumor specificity and cytotoxicity *in vivo* with mice bearing the MDA-MB-435 breast carcinoma, which is known to overexpress integrin $\alpha_v\beta_3$.

Prodrug **4.1.14** (Figure 8.47) contained a docosahexaenoic acid (DHA) moiety as the targeting agent. It was designed based on the premise that tumor cells tend to uptake fatty acids as an energy source [222,223]. Compound **4.1.14** showed no microtubule assembly activity and no cytotoxicity before it was converted to paclitaxel, and it also exhibited better distribution and increased tolerance compared to paclitaxel. In several *in vivo* experiments, compound **4.1.14** demonstrated superior

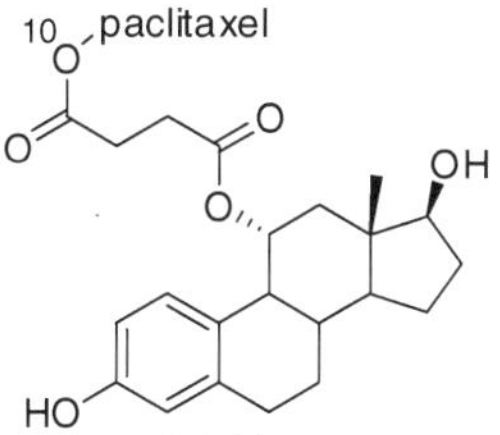

4.1.11

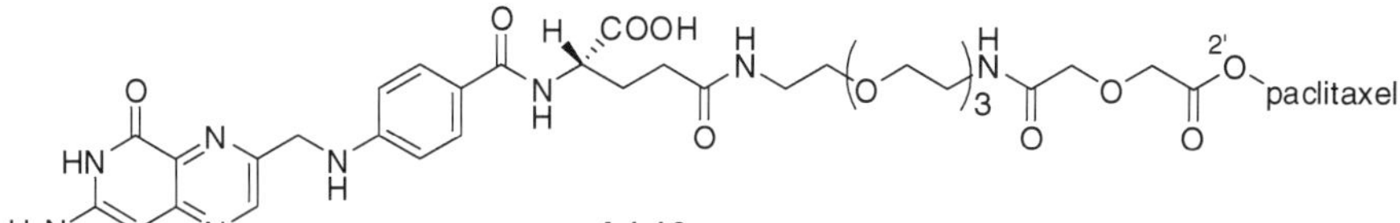

4.1.12

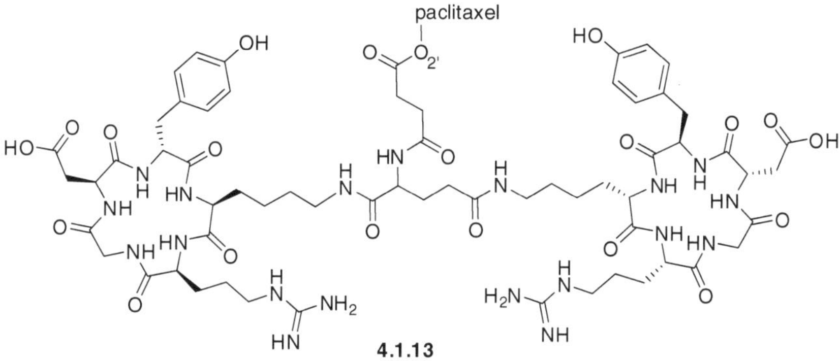

4.1.13

FIGURE 8.46 Compounds **4.1.11–4.1.13**.

antitumor activity, and it is currently in clinical trials; in a Phase I trial, it was found to be well tolerated by patients and to have a favorable toxicity profile [224].

The glycan-paclitaxel conjugate **4.1.15** (Figure 8.48) was designed to target tumor tissue via glucose transporters [225], a family of membrane proteins which are generally overexpressed in cancer cells compared with normal cells [226]. Selective cytotoxicity of this compound was observed against cancer cells over normal cells.

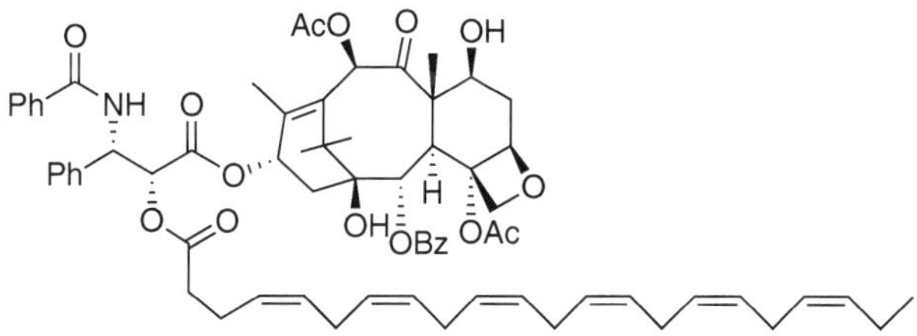

4.1.14

FIGURE 8.47 Compounds **4.1.14**.

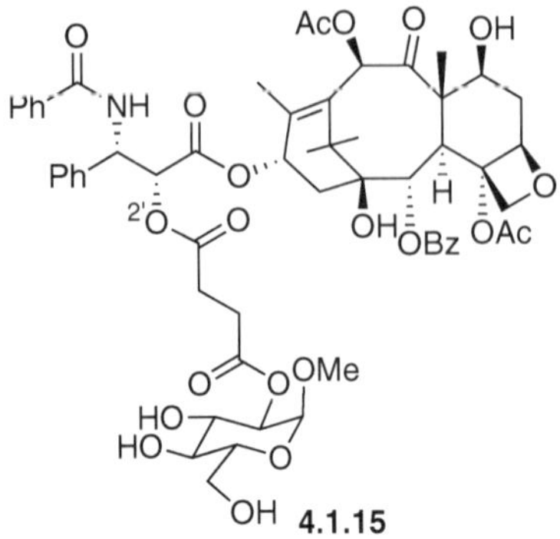

4.1.15

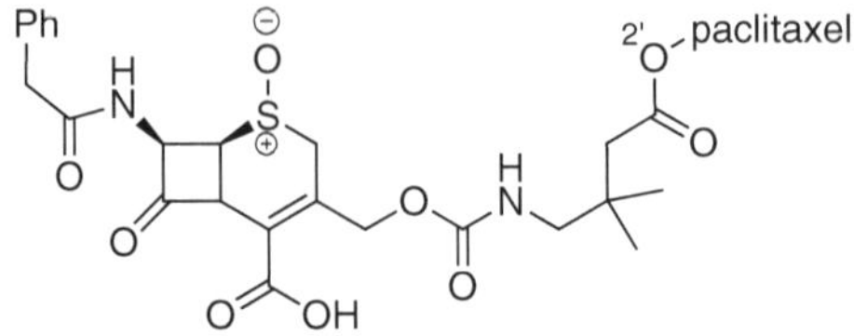

4.1.16

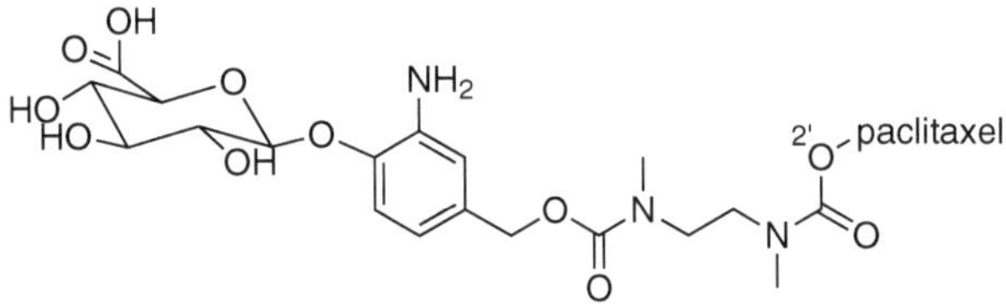

4.1.17

FIGURE 8.48 Compounds **4.1.15–4.1.17**.

It was also less cytotoxic against cancer cell lines that express low levels of glucose transporters.

Prodrugs **4.1.16–4.1.17** (Figure 8.48) were designed on the basis of antibody-directed enzyme prodrug therapy (ADEPT). In this technique, an antibody conjugated with a prodrug-activating enzyme is directed against a tumor-specific antigen located on the surface of tumor cells, followed by the administration of a nontoxic prodrug, which can only be activated by the aforementioned enzyme. Compound **4.1.16**, which can be activated by β-lactamase [227], was developed by Bristol-Myers Squibb. This compound released paclitaxel sevenfold faster in the presence of β-lactamase than that in human serum. The prodrug was 12-fold less active *in vitro* against 3667 melanoma cells compared to paclitaxel. However, it was similarly active against cancer cells presaturated with monoclonal antibody-β-lactamase fusion protein L-49-sFv-bL. Prodrug **4.1.17** was designed to be activated by human β-glucuronidase [228]. This compound showed enhanced water solubility and released paclitaxel with a half-life of 10 min at an enzyme concentration of 2.5 μg/mL.

Prodrug **4.1.18** (Figure 8.49) contained a photolabile unit, 7-*N*,*N*-diethylamino-4-hydroxymethyl coumarin (DECM), which released paclitaxel by minimal tissue-damaging 365 nm UV light irradiation at low power [228a]. Tumor targeting could thus be achieved by selective delivery of light to the tumor after administration.

8.4.1.3 Macromolecular Prodrugs. Macromolecules can serve as excellent pro-moieties for prodrug development. Macromolecules accumulate preferentially in tumor tissue through a mechanism called the enhanced permeability and retention (EPR) effect. Besides this tumor-targeting feature, some hydrophilic macromolecules

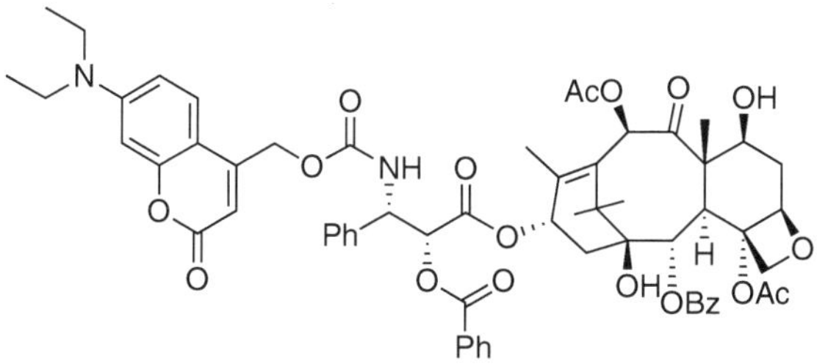

4.1.18

FIGURE 8.49 Compound **4.1.18**.

can increase the water solubility and enhance the bioavailability of prodrugs. For example, in prodrug **4.1.19**, paclitaxel is conjugated with a highly hydrophilic polyethylene glycol (PEG) polymer [229], and in **4.1.20** (Figure 8.50) it is conjugated with poly(vinyl alcohol) (PVA) units [230]. In both cases, the water solubility was greatly enhanced and both prodrugs allowed fast release of the parent.

Another advantage of macromolecular prodrugs is that macromolecules can have multiple reaction sites. This feature allows a high load of active compound and makes possible the development of multifunctional prodrugs. Compounds **4.1.21** (Figure 8.50) and **4.1.22** (Figure 8.51) are both water-soluble conjugates [231,232] of paclitaxel with *N*-(2-hydroxylpropyl)methacrylamide (HMPA) copolymer, which is biocompatible and nonimmunogenic. Paclitaxel was connected to the polymer through a peptide linker, which can be cleaved by lysosomal proteases [232,233]. In the case of **4.1.21**, a Phase I clinical trial revealed that this water-soluble conjugate had a high maximum tolerated dose (MTD) and superior activity in patients with advanced breast cancer. However, a further study was discontinued because of serious neurotoxicity. Compound **4.1.22** was designed as a triple-payload version through a self-immolative dendritic linker that acts as a molecular amplifier. It was reported to exhibit enhanced cytotoxicity compared to **4.1.21** [232].

Compound **4.1.23** is another successful macromolecular prodrug (Figure 8.52) that also entered clinical trials [234,235]. Paclitaxel was attached to a water-soluble and nonimmunogenic poly(L-glutamic acid) scaffold that can be degraded by cathepsin B to produce the parent drug and the nontoxic amino acid. The prodrug has shown comparable efficacy to paclitaxel with decreased side effects. The prodrug is currently in Phase III clinical trials.

Compound **4.1.24** (Figure 8.53) was synthesized as a conjugate of paclitaxel with hyaluronic acid (HA) [236], a naturally occurring linear polysaccharide, which is often found in elevated levels in tumors [237]. In addition, most malignant cell types overexpress cell-membrane-localized HA receptors CD44 and RHAMM [238,239]. The prodrug was internalized into cancer cells by receptor-mediated endocytosis, and paclitaxel was then released inside. The compound showed selective cytotoxicity against cancer cells that overexpressed HA receptors.

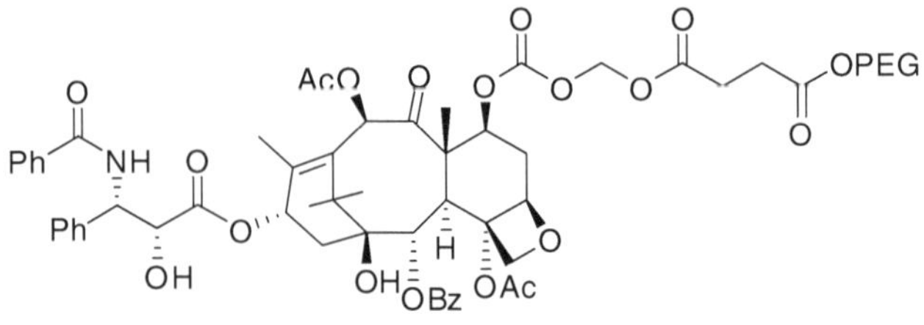

4.1.19

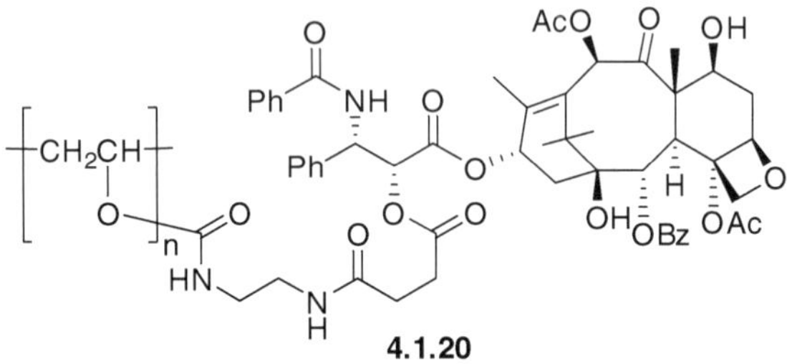

4.1.20

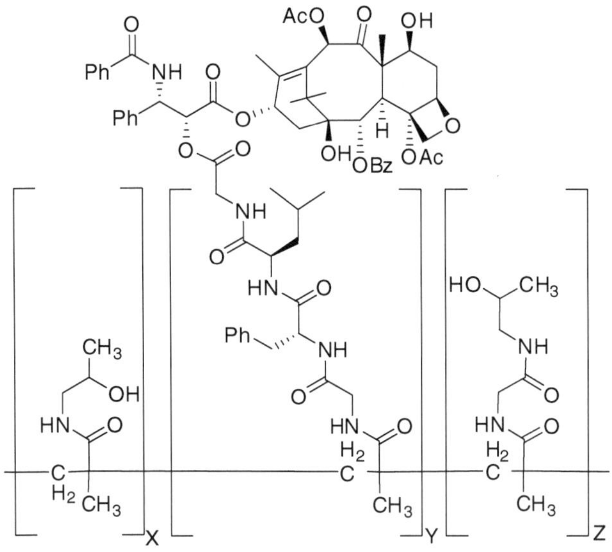

4.1.21

FIGURE 8.50 Conjugates **4.1.19–4.1.21**.

A potential oral delivery platform of paclitaxel **4.1.25** (Figure 8.54) was developed recently through chemical conjugation of PTX to a low-molecular-weight chitosan (LMWC) [240]. The water-soluble prodrug contained ~12 wt% paclitaxel and showed comparable *in vitro* cytotoxicity and *in vivo* inhibition of tumor growth with the parent paclitaxel. Pharmacokinetic studies demonstrated high oral bioavailability of the prodrug, which was likely absorbed mainly from the ileum and reached the blood as the intact conjugate.

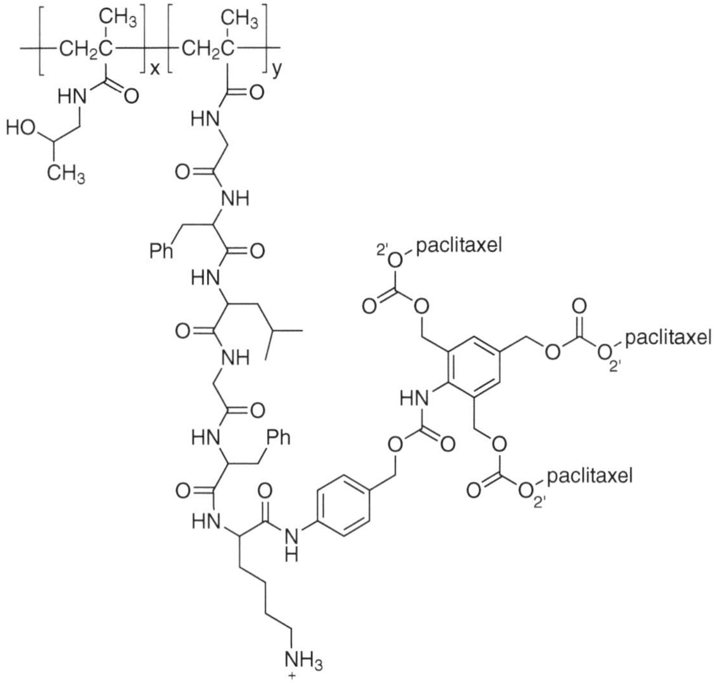

4.1.22

FIGURE 8.51 Conjugate **4.1.22**.

A poly(amidoamine) (PAMAM) dendrimer-based multifunctional conjugate **4.1.26** (Figure 8.55) was designed and synthesized by Majoros et al. [241]. The PAMAM used contained 110 surface amine groups, which allowed the conjugation of multiple functional molecules including folic acid (FA, to target overexpressed folate

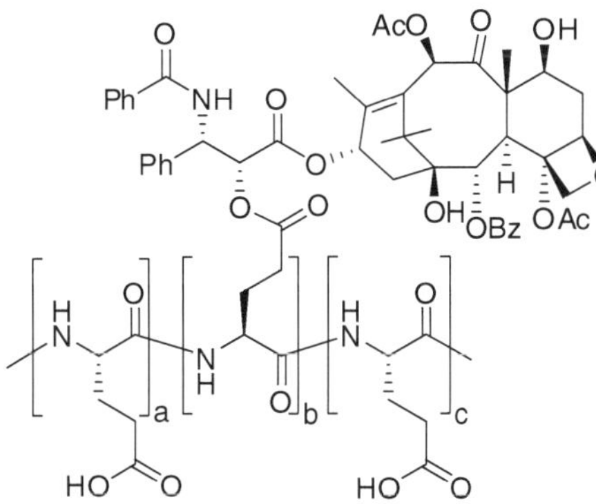

4.1.23

FIGURE 8.52 Conjugate **4.1.23**.

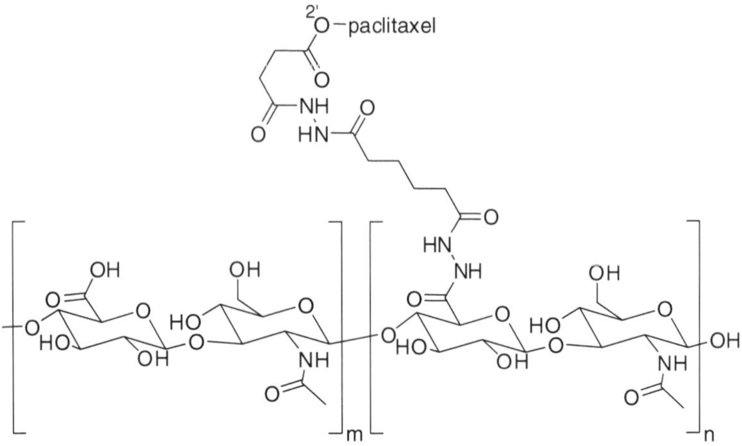

4.1.24

FIGURE 8.53 Conjugate **4.1.24**.

receptors (FR) on specific cancer cells [219]), fluorescein isothiocyanate (FITC, an imaging agent), and paclitaxel (the chemotherapeutic drug) to the dendrimer. The average number of paclitaxel molecules loaded on one conjugate was determined to be 3 by gel permeation chromatography (GPC). Fluorescently traced assay showed that the conjugate was internalized only by the FR positive cancer cells. The conjugate was selectively cytotoxic to FR positive cancer cells at low concentrations (50 nM and 100 nM) *in vitro*. However, at high concentrations (200 nM), it was equally active to both the FR-positive and FR-negative cancer cells. No *in vivo* data were reported.

8.4.1.4 Paclitaxel–Nanoparticle Conjugates. Paclitaxel conjugated with iron oxide nanoparticles (Fe-NPs) and gold nanoparticles (Au-NPs) have been reported as

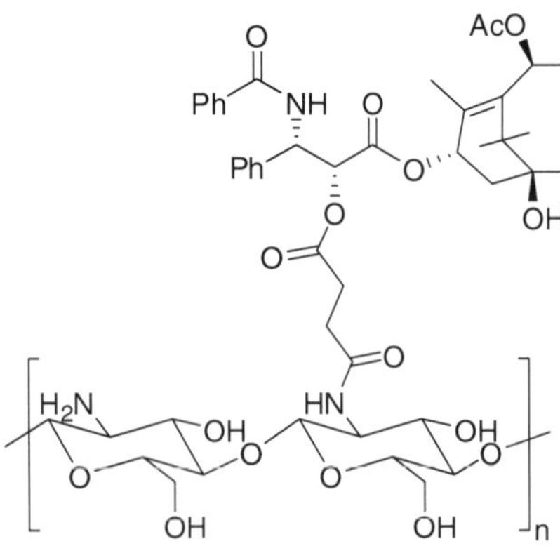

4.1.25

FIGURE 8.54 Conjugate **4.1.25**.

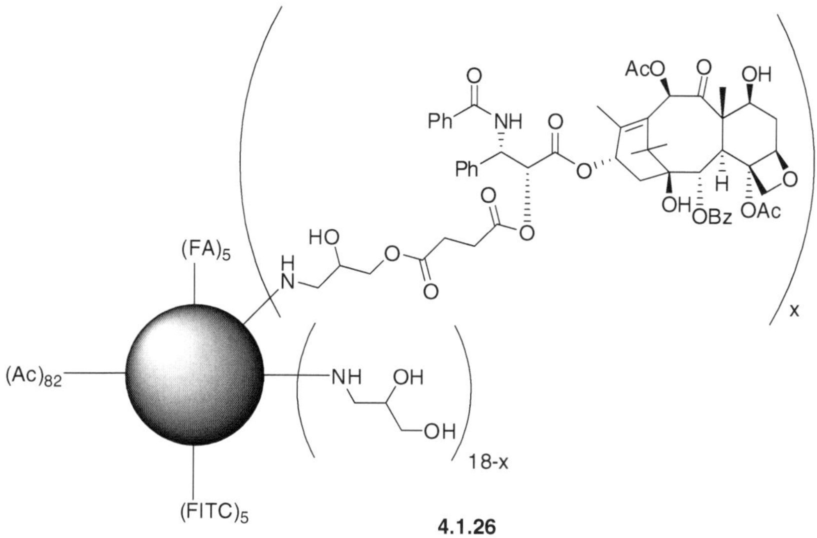

4.1.26

FIGURE 8.55 Conjugate **4.1.26**.

alternative drug delivery systems. Fe-NPs exhibit strong magnetization and little to no toxicity *in vivo*. Besides, the superparamagnetic Fe-NPs can be delivered to the desired target area by an external magnetic field. Au-NPs can serve as good drug delivery scaffolds because of their biocompatibility, tunable size, and strong reactivity with sulfur-containing compounds, and controllable drug release. Paciotti et al. developed a gold nanoparticle-based drug delivery system **4.1.27** (Figure 8.56) with multiple components attached on the particle surface, which includes tumor necrosis factor alpha (TNFα) as a tumor targeting agent, polyethylene glycol (PEG) to shield the system from being taken up by the immune system, and paclitaxel as the anticancer therapeutic agent [242]. It was demonstrated that the parent drug can be released in the presence of reducing agents such as dithiothreitol and cysteamine.

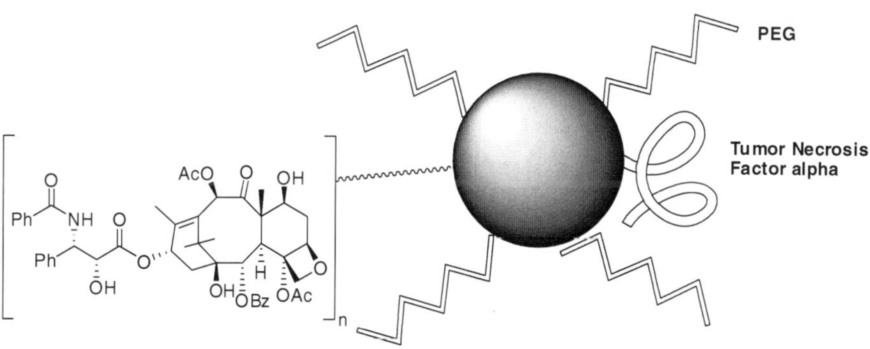

4.1.27

FIGURE 8.56 Conjugate **4.1.27**.

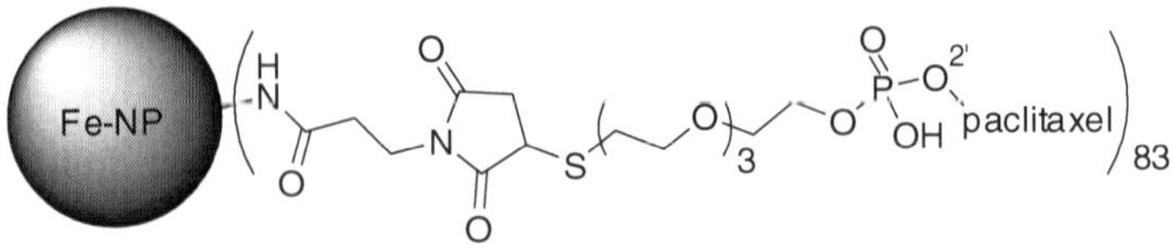

4.1.28

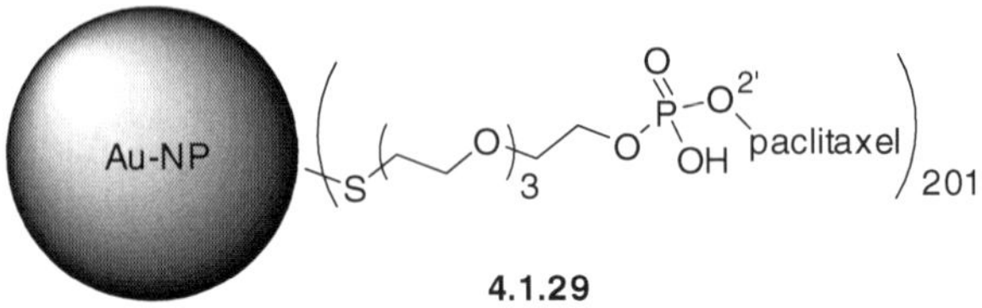

4.1.29

FIGURE 8.57 Conjugates **4.1.28–4.1.29**.

Recently, conjugates of paclitaxel with Fe-NPs **4.1.28** and Au-NPs **4.1.29** (Figure 8.57) were reported [243]. It was expected that hydrolysis of the phosphodiester moieties with phosphodiesterase could release paclitaxel from both nanoparticles. Conjugate **4.1.28** indeed produced paclitaxel at the presence of phosphodiesterase in 10 days. However, in the case of conjugate **4.1.29**, less than 5% of paclitaxel was liberated under the same conditions. It was demonstrated that conjugate **4.1.28** was 10^4-fold more cytotoxic against a human cancer cell line (OECM1) than a human normal cell line (HUVEC).

Although not strictly speaking chemical conjugates, two nanoparticle formulations of paclitaxel were introduced into clinical practice over the last decade. The albumin-bound nanoparticle formulation of taxol known as Abraxane® (ABI-007, American Bioscience Inc.) has a shorter infusion time than paclitaxel and a lower systemic toxicity, so it offers significant clinical benefits. Its development and use have been reviewed [244,245]. The similar formulation Nanoxel has been introduced in India [246]. The micellar taxol formulation Genexol-PM® uses polyethylene glycol (PEG) and poly(D,L)-lactic acid, and gave high response rates in patients with metastatic breast cancer; it has been approved by the FDA for this indication [247,248].

8.5 CONCLUSIONS

As can be seen from the results described above, work on the chemistry of paclitaxel continued at a brisk pace during the period 2000–2010. From the taxane single agent clinical therapy point of view, the most important advance was the introduction of cabazitaxel into clinical practice, but new paclitaxel delivery methods such as Abraxane have made and will continue to make an impact on cancer treatment. In the future, one or more targeted derivatives of paclitaxel will almost certainly come into clinical use, and the combination of potent new analogs of paclitaxel with drug targeting is also likely to be important. The utility of paclitaxel and its analogs as anticancer drugs is thus assured for many years to come.

REFERENCES

[1] Fu, Y., Li, S., Zu, Y., Yang, G., Yang, Z., Luo, M., Jiang, S., Wink, M., Efferth, T. (2009). Medicinal chemistry of paclitaxel and its analogues. *Current Medicinal Chemistry*, *16*, 3966–3985.

[2] Ojima, I., Das, M. (2009). Recent advances in the chemistry and biology of new generation taxoids. *Journal of Natural Products*, *72*, 554–565.

[3] Saloustros, E., Mavroudis, D., Georgoulias, V. (2008). Paclitaxel and docetaxel in the treatment of breast cancer. *Expert Opinion on Pharmacotherapy*, *9*, 2603–2616.

[4] Ferlini, C., Gallo, D., Scambia, G. (2008). New taxanes in development. *Expert Opinion on Investigational Drugs*, *17*, 335–347.

[5] Hudis, C., Dang, C. (2008). The taxane limbo: How long can we go? *Journal of the National Cancer Institute*, *100*, 761–763.

[6] Kingston, D.G.I. (2001). Taxol, a molecule for all seasons. *Chemical Communications*, 867–880.

[7] Kingston, D.G.I. (2011). Taxol and Tbs analogs In: *Anticancer Agents from Natural Products* Ed. by Gragg, G.C., Kingston, D.G.I., Newman, D.J. CRC Press, Boca Raton, FL pp. 123–175.

[8] Wall, M.E., Wani, M.C., Recent progress in plant anti-tumor agents. In: 153rd National Meeting of the American Chemical Society, Miami Beach, FL, Abstr. M006, 1967.

[9] Wani, M.C., Taylor, H.L., Wall, M.E., Coggon, P., McPhail, A.T. (1971). Plant antitumor agents: VI. The isolation and structure of taxol, a novel antileukemic and antitumor agent from *Taxus brevifolia*. *Journal of the American Chemical Socisty*, *93*, 2325–2327.

[10] Suffness, M., Wall, M.E., Discovery and development of taxol. In: Suffness, M., editor. *Taxol: Science and Applications*, CRC Press, Boca Raton, FL, 1995, pp. 3–25.

[11] Schiff, P.B., Fant, J., Horwitz, S.B. (1979). Promotion of microtubule assembly *in vitro* by taxol. *Nature*, *277*, 665–667.

[12] McGuire, W.P., Rowinsky, E.K., Rosenshein, N.B., Grumbine, F.C., Ettinger, D.S., Armstrong, D.K., Donehower, R.C. (1989). Taxol: A unique antineoplastic agent with significant activity in advanced ovarian epithelial neoplasms. *Annals of Internal Medicine*, *111*, 273–279.

[13] Holmes, F.A., Walters, R.S., Theriault, R.L., Forman, A.D., Newton, L.K., Raber, M.N., Buzdar, A.U., Frye, D.K., Hortobagyi, G.N. (1991). Phase II trial of taxol, an active drug in the treatment of metastatic breast cancer. *Journal of the National Cancer Institute*, *83*, 1797–1805.

[14] Gueritte-Voegelein, F., Senilh, V., David, B., Guenard, D., Potier, P. (1986). Chemical studies of 10-deacetyl baccatin: III. Hemisynthesis of taxol derivatives. *Tetrahedron*, *42*, 4451–4460.

[15] Denis, J.-N., Greene, A.E., Guenard, D., Gueritte-Voegelein, F., Mangatal, L., Potier, P. (1988). A highly efficient, practical approach to natural taxol. *Journal of the American Chemical Society*, *110*, 5917–5919.

[16] Clarke, S.J., Rivory, L.P. (1999). Clinical pharmacokinetics of docetaxel. *Clinical Pharmacokinetics*, *36*, 99–114.

[17] Kant, J., O'Keeffe, W.S., Chen, S.-H., Farina, V., Fairchild, C., Johnston, K., Kadow, J.F., Long, B.H., Vyas, D. (1994). A chemoselective approach to functionalize the C-10 position of 10-deacetylbaccatin: III. Synthesis and biological properties of novel C-10 taxol analogues. *Tetrahedron Letters*, *35*, 5543–5546.

[18] Denduluri, N., Swain, S.M. (2008). Ixabepilone for the treatment of solid tumors: A review of clinical data. *Expert Opinion on Investigational Drugs*, *17*, 423–435.

[19] Baloglu, E., Kingston, D.G.I. (1999). The taxane diterpenoids. *Journal of Natural Products*, *62*, 1448–1472.

[20] Shi, Q.-W., Kiyota, H. (2005). New natural taxane diterpenoids from *Taxus* species since 1999. *Chemistry and Biodiversity*, *2*, 1597–1623.

[21] Shen, Y.-C., Chen, C.-Y., Hung, M.-C. (2000). Taxane diterpenoids from seeds of *Taxus mairei. Chemical and Pharmaceutical Bulletin*, *48*, 1344–1346.

[22] Shen, Y.-C., Prakash, C.V.S., Chen, Y.-J., Hwang, J.-F., Kuo, Y.-H., Chen, C.-Y. (2001). Taxane diterpenoids from the stem bark of *Taxus mairei. Journal of Natural Products*, *64*, 950–952.

[23] Shen, Y., Wang, S., Pan, Y., Lo, K., Chakraborty, R., Chien, C., Kuo, Y., Lin, Y. (2002). New taxane diterpenoids from the leaves and twigs of *Taxus sumatrana. Journal of Natural Products*, *65*, 1848–1852.

[24] Galm, U., Hager, M.H., Van Lanen, S.G., Ju, J., Thorson, J.S., Shen, B. (2005). Antitumor antibiotics: Bleomycin, enediynes, and mitomycin. *Chemical Reviews*, *105*, 739–758.

[25] Torregiani, E., Rafaiani, G., Barboni, L., Appendino, G. (1995). A serendipitous clue to the biogenesis of taxchins. *Tetrahedron Letters*, *36*, 7127–7128.

[26] Shen, Y.-C., Lin, Y.-S., Cheng, Y.-B., Cheng, K.-C., Khalil, A.T., Kuo, Y.-H., Chien, C.-T., Lin, Y.-C. (2005). Novel taxane diterpenes from *Taxus sumatrana* with the first C-21 taxane ester. *Tetrahedron*, *61*, 1345–1352.

[27] Shen, Y.-C., Lin, Y.-S., Hsu, S.-M., Khalil, A.T., Wang, S.-S., Chien, C.-T., Kuo, Y.-H., Chou, C.-H. (2007). Tasumatrols P-T, five new taxoids from *Taxus sumatrana. Helvetica Chimica Acta*, *90*, 1319–1329.

[28] Shen, Y.-C., Wang, S.-S., Chien, C.-T., Kuo, Y.-H., Khalil, A.T. (2008). Tasumatrols U-Z, taxane diterpene esters from *Taxus sumatrana. Journal of Natural Products*, *71*, 576–580.

[29] Wang, S.-S., El-Razek, M.H.A., Chen, Y.-C., Chien, C.-T., Guhb, J.-H., Kuod, Y.-H., Shen, Y.-C. (2009). *Abeo*-taxane diterpenoids from the Taiwanese Yew *Taxus sumatrana. Chemistry and Biodiversity*, *6*, 2255–2262.

[30] Luh, L.-J., El-Razek, M.H.A., Liaw, C.-C., Chen, C.-T.A., Lin, Y.-S., Kuo, Y.-H., Chien, C.-T., Shen, Y.-C. (2009). Tri- and bicyclic taxoids from the Taiwanese Yew *Taxus sumatrana. Helvetica Chimica Acta*, *92*, 1349–1358.

[31] Kosugi, K., Sakai, J.-I., Zhang, S., Watanabe, Y., Sasaki, H., Suzuki, T., Hagiwara, H., Hirata, N., Hirose, K., Ando, M., Tomida, A., Tsuruo, T. (2000). Neutral taxoids from *Taxus cuspidata* as modulators of multidrug-resistant tumor cells. *Phytochemistry*, *54*, 839–845.

[32] Zhang, J., Sauriol, F., Mamer, O., You, X.-L., Alaoui-Jamali, M.A., Batist, G., Zamir, L.O. (2001). New taxane analogues from the needles of *Taxus canadensis. Journal of Natural Products*, *64*, 450–455.

[33] Klein, L.L. (1993). Synthesis of 9-dihydrotaxol: A novel bioactive taxane. *Tetrahedron Letters*, *34*, 2047–2050.

[34] Prasain, J.K., Stefanowicz, P., Kiyota, T., Habeichi, F., Konishi, Y. (2001). Taxines from the needles of *Taxus wallichiana*. *Phytochemistry*, *58*, 1167–1170.

[35] Shi, Q.W., Sauriol, F., Lesimple, A., Zamir, L.O. (2004). First three examples of taxane-derived di-propellanes in *Taxus canadensis* needles. *Chem. Comm.*, 544–545.

[36] Chattopadhyay, S.K., Pal, A., Maulik, P.R., Kaur, T., Garg, A., Khanuja, S.P.S. (2006). Taxoid from the needles of the Himalayan yew *Taxus wallichiana* with cytotoxic and immunomodulatory activities. *Bioorganic and Medicinal Chemistry Letters*, *16*, 2446–2449.

[37] Barboni, L., Pierluigi, G., Torregiani, E.G.A., Gabetta, B., Zini, G., Bombardelli, E. (1993). Taxanes from the needles of *Taxus Wallichiana*. *Phytochemistry*, *33*, 145–150.

[38] Zhang, M.-L., Dong, M., Li, Z.-N., Li, L.-G., Sauriol, F., Huo, C.-H., Shi, Q.-W., Gu, Y.-C., Kiyota, H., Cong, B. (2008). A new taxane composed of two *N*-formyl rotamers from *Taxus canadensis*. *Tetrahedron Letters*, *49*, 3405–3408.

[39] Holton, R.A., Somoza, C., Kim, H.-B., Liang, F., Biediger, R.J., Boatman, P.D., Shindo, M., Smith, C.C., Kim, S., Nadizadeh, H., Suzuki, Y., Tao, C., Vu, P., Tang, S., Zhang, P., Murthi, K.K., Gentile, L.N., Liu, J.H. (1994). First total synthesis of taxol: 1. Functionalization of the B ring. *Journal of the American Chemical Society*, *116*, 1597–1598.

[40] Holton, R.A., Kim, H.B., Somoza, C., Liang, F., Biediger, R.J., Boatman, P.D., Shindo, M., Smith, C.C., Kim, S., Nadizadeh, H., Suzuki, Y., Tao, C., Vu, P., Tang, S., Zhang, P., Murthi, K.K., Gentile, L.N., Liu, J.H. (1994). First total synthesis of taxol: 2. Completion of the C and D rings. *Journal of the American Chemical Society*, *116*, 1599–1600.

[41] Nicolaou, K.C., Yang, Z., Liu, J.J., Ueno, H., Nantermet, P.G., Guy, R.K., Claiborne, C.F., Renaud, J., Couladouros, E.A., Paulvannan, K., Sorensen, E.J. (1994). Total synthesis of taxol. *Nature*, *367*, 630–634.

[42] Danishefsky, S.J., Masters, J.J., Young, W.B., Link, J.T., Snyder, L.B., Magee, T.V., Jung, D.K., Isaacs, R.C.A., Bornmann, W.G., Alaimo, C.A., Coburn, C.A., Di Grandi, M.J. (1996). Total synthesis of baccatin III and taxol. *Journal of the American Chemical Society*, *118*, 2843–2859.

[43] Wender, P.A., Badham, N.F., Conway, S.P., Floreancig, P.E., Glass, T.E., Granicher, C., Houze, J.B., Janichen, J., Lee, D., Marquess, D.G., McGrane, P.L., Meng, W., Mucciaro, T.P., Muhlebach, M., Natchus, M.G., Paulsen, H., Rawlins, D.B., Satkofsky, J., Shuker, A.J., Sutton, J.C., Taylor, R.E., Tommoka, K. (1997). The pinene path to taxanes: 5. Stereocontrolled synthesis of a versatile taxane precursor. *Journal of the American Chemical Society*, *119*, 2755–2756.

[44] Wender, P.A., Badham, N.F., Conway, S.P., Floreancig, P.E., Glass, T.E., Houze, J.B., Krauss, N.E., Lee, D., Marquess, D.G., McGrane, P.L., Meng, W., Natchus, M.G., Shuker, A.J., Sutton, J.C., Taylor, R.E. (1997). The pinene path to taxanes: 6. A concise stereocontrolled synthesis of taxol. *Journal of the American Chemical Society*. *119*, 2757–2758.

[45] Mukaiyama, T., Shiina, I., Iwadare, H., Saitoh, M., Nishimura, T., Ohkawa, N., Sakoh, H., Nishimura, K., Tani, Y.-I., Hasegawa, M., Yamada, K., Saitoh, K. (1999). Asymmetric total synthesis of taxol. *Chemistry – A European Journal*, *5*, 121–161.

[46] Cragg, G.M., Schepartz, S.A., Suffness, M., Grever, M.R. (1993). The taxol supply crisis. New NCI policies for handling the large scale production of novel natural product anticancer and anti-HIV agents. *Journal of Natural Products*, *56*, 1657–1668.

[47] Kusama, H., Hara, R., Kawahara, S., Nishimori, T., Kashima, H., Nakamura, N., Morihira, K., Kuwajima, I. (2000). Enantioselective total synthesis of (−)-taxol. *Journal of the American Chemical Society*, *122*, 3811–3820.

[48] Paquette, L.A., Lo, H.Y. (2003). Chemical modification of a highly functionalized taxane. The consequences of an absent bridgehead double bond on oxetane D-ring construction. *Journal of Organic Chemistry*, *68*, 2282–2289.

[49] Petit, M., Chouraqui, G., Phansavath, P., Aubert, C., Malacria, M. (2002). New efficient construction of the ABC core of the taxoids via a sequence of consecutive cobalt(I)-mediated [2+2+2] and [4+2] cyclizations. *Organic Letters*, *4*, 1027–1029.

[50] Chouraqui, G., Petit, M., Phansavath, P., Aubert, C., Malacria, M. (2006). From an acyclic, polyunsaturated precursor to the polycyclic taxane ring system: The [4+2]/[2+2+2] and [2+2+2]/[4+2] cyclization strategies. *European Journal of Organic Chemistry*, *6*, 1413–1421.

[51] Laurent, A., Villalva-Servin, N.P., Forgione, P., Wilson, P.D., Smil, D.V., Fallis, A.G. (2004). Part 1: Efficient strategies for the construction of variably substituted bicyclo[5.3.1]undecenones (AB taxane ring systems). *Canadian Journal of Chemistry*, *82*, 215–226.

[52] Villalva-Servin, N.P., Laurent, A., Fallis, A.G. (2004). Part 2: Efficient strategies for the construction of variably substituted bicyclo[5.3.1]undecenones (AB-taxane ring systems) and their conversion to tricyclo[9.3.1.0]pentadecenones (ABC taxane ring systems) and bicyclo[2.2.2]octanones. *Canadian Journal of Chemistry*, *82*, 227–239.

[53] Hamon, S., Birlirakis, N., Toupet, L., Arseniyadis, S. (2005). Stereoselective construction of a highly functionalized taxoid ABC-ring system: The C2-C9 oxa-bridge approach. *European Journal of Organic Chemistry*, *2005*, 4082–4092.

[54] Chavan, S.P., Chavan, S.P., Sonawane, H.R., Kalkote, U.R., Sudrik, S.G., Gonnade, R.G., Bhadbhade, M.M. (2007). Utilization of the versatility of sulfur in C–C bond formation and cleavage: Synthesis of ABC taxoid skeletons. *European Journal of Organic Chemistry*, *2007*, 3277–3280.

[55] Enomoto, T., Morimoto, T., Ueno, M., Matsukubo, T., Shimada, Y., Tsutsumi, K., Shirai, R., Kakiuchi, K. (2008). A novel route for the construction of taxol ABC-ring framework: Skeletal rearrangement approach to AB-ring and intramolecular aldol approach to C-ring. *Tetrahedron*, *64*, 4051–4059.

[56] Goldring, W.P.D., Pattenden, G., Rimmington, S.L. (2009). Synthesis of a tetraoxy-bis-nortaxadiene, en route to taxol, using a cascade radical cyclisation sequence. *Tetrahedron*, *65*, 6670–6681.

[56a] Kim, S.-K., Kim, H.-K. (2000). Synthesis of new Taxol side chain precursor from L-tartaric ester. *Bulletin of the Korean Chemical Society*, *21*, 1047–1048.

[57] Wuts, P.G.M., Gu, R.L., Northuis, J.M. (2000). Synthesis of (2*R*, 3*S*)-isobutyl phenylisoserinate, the taxol side chain, from ethyl benzoylacetate. *Tetrahedron: Asymmetry*, *11*, 2117–2123.

[58] Hamamoto, H., Mamedov, V.A., Kitamoto, M., Hayashi, N., Tsuboi, S. (2000). Chemoenzymatic synthesis of the C-13 side chain of paclitaxel (Taxol) and docetaxel (Taxotere). *Tetrahedron: Asymmetry*, *11*, 4485–4497.

[59] Aggarwal, V.K., Vasse, J.-L. (2003). Asymmetric sulfur ylide mediated aziridination: Application in the synthesis of the side chain of taxol. *Organic Letters*, *5*, 3987–3990.

[60] Choudary, B.M., Chowdari, N.S., Madhi, S., Kantam, M.L. (2003). A trifunctional catalyst for one-pot synthesis of chiral diols via heck coupling-*N*-oxidation-asymmetric dihydroxylation: Application for the synthesis of diltiazem and taxol side chain. *Journal of Organic Chemistry*, *68*, 1736–1746.

[61] Castagnolo, D., Armaroli, S., Corelli, F., Botta, M. (2004). Enantioselectiv synthesis of 1-aryl-2-propenylamines: A new approach to a stereoselective synthesis of the taxol side chain. *Tetrahedron*, *15*, 941–949.

[62] Borah, J.C., Gogoi, S., Boruwa, J., Kalita, B., Barua, N.C. (2004). A highly efficient synthesis of the C-13 side-chain of taxol using Shibasaki's asymmetric Henry reaction. *Tetrahedron Letters*, *45*, 3689–3691.

[63] Torssell, S., Kienle, M., Somfai, P. (2005). 1,3-Dipolar cycloadditions of carbonyl ylides to aldimines: A three-component approach to *syn*-a-hydroxy-b-amino esters. *Angewandte Chemie International Edition*, *44*, 3096–3099.

[64] Wang, Y., He, Q., Wang, H., Zhou, X., Huang, Z., Qin, Y. (2006). Highly diastereoselective enolate anddition of *O*-protected a-hydroxyacetate to (S_R)-*tert*-butanesulfinylimines: Synthesis of taxol side chain. *Journal of Organic Chemistry*, *71*, 1588–1591.

[65] Dziedzic, P., Vesely, J., Cordova, A. (2008). Catalytic asymmetric synthesis of the docetaxel (Taxotere) side chain: Organocatalytic highy enantioselective synthesis of esterification-ready a-hydroxy-b-amino acids. *Tetrahedron Letters*, *49*, 6631–6634.

[66] Munoz-Hernandez, L., Soderquist, J.A. (2009). Asymmetric g-methoxyallylation with the robust 10-TMS-9-borabicyclo[3.3.2]decanes. *Organic Letters*, *11*, 2571–2574.

[67] Qian, Y., Xu, X., Jiang, L., Prajapati, D., Hu, W. (2010). A strategy to synthesize taxol side chain and (−)-*epi* cytoxazone via chiral bronsted acid-$Rh_2(OAc)_4$ co-catalyzed enantioselective three-component reactions. *Journal of Organic Chemistry*, *75*, 7483–7486.

[68] Fang, W.-S., Fang, Q.-C., Liang, X.-T. (2001). Synthesis of the 2a-benzoylamido analogue of docetaxel. *Tetrahedron Letters*, *42*, 1331–1333.

[69] Yuan, H., Fairchild, C.R., Liang, X., Kingston, D.G.I. (2000). Synthesis and biological activity of C-6 and C-7 modified paclitaxels. *Tetrahedron*, *56*, 6407–6414.

[70] Cheng, Q., Oritani, T., Horiguchi, T., Yamada, T., Mong, Y. (2000). Synthesis and biological evaluation of novel 9-functional heterocyclic coupled 7-deoxy-9-dihydropaclitaxel analogue. *Bioorganic and Medicinal Chemistry Letters*, *10*, 517–521.

[71] Danieli, B., Giardini, A., Lesma, G., Passarella, D., Silvani, A., Appendino, G., Noncovich, A., Fontana, G., Bombardelli, E., Sterner, O. (2004). Synthesis and biological evaluation of paclitaxel-thiocolchicine hybrids. *Chemistry and Biodiversity*, *1*, 327–345.

[72] Mellado, W., Magri, N.F., Kingston, D.G.I., Garcia-Arenas, R., Orr, G.A., Horwitz, S.B. (1984). Preparation and biological activity of taxol acetates. *Biochemical and Biophysical Research Communication*, *124*, 329–336.

[73] Xue, M., Long, B.H., Fairchild, C., Johnston, K., Rose, W.C., Kadow, J.F., Vyas, D.M., Chen, S.-H. (2000). Structure–activity relationships study at the 3′-*N* position of paclitaxel. Part 2: Synthesis and biological evaluation of 3′-N-thiourea- and 3′-N-thiocarbamate-bearing paclitaxel analogues. *Bioorganic and Medicinal Chemistry Letters*, *10*, 1327–1331.

[74] Karliga, B., Schilling, J.K., Kingston, D.G.I., Bane, S., Ravindra, R., Talinli, N. (2006). Synthesis and biological evaluation of *N*-debenzoyl-*N*-(arylthio)carbonyl paclitaxel derivatives. *Chemistry and Biodiversity*, *3*, 396–404.

[75] Gelmon, K.A., Latreille, J., Tolcher, A., Genier, L., Fisher, B., Forand, D., D'Aloisio, S., Vernillet, L., Daigneault, L., Lebecq, A., Besenval, M., Eisenhauer, E. (2000). Phase I dose-finding study of a new taxane, RPR 109881A, administered as a one-hour intravenous infusion days 1 and 8 to patients with advanced solid tumors. *Journal of Clinical Oncology*, *18*, 4098–4108.

[76] Mita, A.C., Denis, L.J., Rowinsky, E.K., Debono, J.S., Goetz, A.D., Ochoa, L., Forouzesh, B., Beeram, M., Patnaik, A., Molpus, K., Semiond, D., Besenval, M., Tolcher, A.W. (2009). Phase I and pharmacokinetic study of XRP6258 (RPR116258A), a novel taxane, administered as a 1-hour infusion every 3 weeks in patients with advanced solid tumors. *Clinical Cancer Research*, *15*, 723–730.

[77] Pivot, X., Koralewski, P., Hidalgo, J.L., Chan, A., Goncalves, A., Schwartsmann, G., Assadourian, S., Lotz, J.P. (2008). A multicenter phase II study of XRP6258 administered as a 1-h i.v. infusion every 3 weeks in taxane-resistant metastatic breast cancer patients. *Annals of Oncology*, *19*, 1547–1552.

[78] Pivot, X., Asmar, L., Buzdar, A.U., Valero, V., Hortobagyi, G. (2000). A unified definition of clinical anthracycline resistance breast cancer. *British Journal of Cancer*, *83*, 529–534.

[79] Engels, F.K., Sparreboom, A., Mathot, R.A., Verweij, J. (2005). Potential for improvement of docetaxel-based chemotherapy: A pharmacological review. *British Journal of Cancer*, *93*, 173–177.

[80] Cisternino, S., Bourasset, F., Archimbaud, Y., Semiond, D., Sanderink, G., Sherrmann, J.-M. (2003). Nonlinear accumulation in the brain of the new taxoid TXD258 following saturation of P-glycoprotein at the blood–brain barrier in mice and rats. *British Journal of Pharmacology*, *138*, 1367–1375.

[81] Dykes, D.J., Sarsat, J.P., Bissery, M.C., Efficacy evaluation of TXD258, a taxoid compound against orthotopic and subcutaneous glioblastomas. In: *Proceeding of the 91th American Association for Cancer Research*, San Francisco, CA, 2000.

[82] Brooks, T., Minderman, H., O'Loughlin, K.L., Pera, P., Ojima, I., Baer, M.R., Bernacki, R.J. (2003). Taxane-based reversal agents modulate drug resistance mediated by P-glycoprotein, multidrug resistance protein, and breast cancer resistance protein. *Molecular Cancer Therapeutics*, *2*, 1195–1205.

[83] Bissery, M.-C., Bouchard, H., Riou, J.F., Vrignaud, P., Combeau, C., Bourzat, J.D., Commercon, A., Lavelle, F. (2000). Preclinical evaluation of TXD258, a new taxoid. *Proceedings of the American Association for Cancer Research, 41*, 214

[84] Sessa, C., Cuvier, C., Caldiera, S., Bauer, J., Van den Bosch, S., Monnerat, C., Semiond, D., Perard, D., Lebecq, A., Besenval, M., Marty, M. (2002). Phase I clinical and pharmacokinctic studics of the taxoid derivative RPR109881A administered as a 1-hour or 3-hour infusion in patients with advanced solid tumors. *Annals of Oncology*, *13*, 1140–1150.

[85] Vrignaud, P., Lejeune, P., Bissery, M.C., In vivo synergy between doxorubicin, cisplatin or vinorelbine and RPR109881A, a new taxoid. In: *Procedings of the 96th Annual AACR Meeting*, Anaheim, CA, 2005.

[86] Vrignaud, P., Chiron, M., Stackhouse, M.A., Zuany-Amorim, C., Goulaouic, H., Bissery, M.C. *In vivo* synergy between trastuzumab (Herceptin) and larotaxel (RPR109881A), a new taxoid. In: *Proceedings of the 98th Annual AACR Meeting*, Los Angeles, CA, 2007.

[87] Galsky, M.D., Dritselis, A., Kirkpatrick, P., Oh, W.K. (2010). Cabazitaxel. *Nature Reviews Drug Discovery*, *9*, 677.

[88] Metzger-Filho, O., Moulin, C., Azambuja, E.D., Ahamad, A. (2009). Larotaxel: Broadening the road with new taxanes. *Expert Opinion on Investigational Drugs*, *18*, 1183–1189.

[89] Mastalerz, H., Cook, D., Fairchild, C.R., Hansel, S., Johnson, W., Kadow, J.F., Long, B.-H., Rose, W.C., Tarrant, J., Wu, M.-J., Xue, M.Q., Zhang, G., Zoeckler, M., Vyas, D.M. (2003). The discovery of BMS-275183: An orally efficacious novel taxane. *Bioorganic and Medicinal Chemistry*, *11*, 4315–4323.

[90] Bröker, L.E., Veltkamp, S.A., Heath, E.I., Kuenen, B.C., Gall, H., Astier, L., Parker, S., Kayitalire, L., Lorusso, P.M., Schellens, J.H., Giaccone, G. (2007). A phase I safety and pharmacologic study of a twice weekly dosing regimen of the oral taxane BMS-275183. *Clinical Cancer Research*, *13*, 3906–3912.

[91] Rose, W.C., Long, B.H., Fairchild, C.R., Lee, F.Y.F., Kadow, J.F. (2001). Preclinical pharmacology of BMS-275183, an orally active taxane. *Clinical Cancer Research*, *7*, 2016–2021.

[92] Zhang, D., Ly, V.T., Lago, M., Tian, Y., Gan, J., Humphreys, W.G., Comezoglu, S.N. (2009). CYP3A4-mediated ester cleavage as the major metabolic pathway of the oral taxane 3′-*tert*-butyl-3′-*N*-*tert*-butyloxycarbonyl-4-deacetyl-3′-dephenyl-3′-*N*-debenzoyl-4-*O*-methoxycarbonyl-paclitaxel (BMS-275183). *Drug Metabolism and Disposition*, *37*, 710–718.

[93] Ly, V.T., Caceres-Cortes, J., Zhang, D., Humphreys, W.G., Ekhato, I.V., Everett, D., Comezoglu, S.N. (2009). Metabolism and excretion of an oral taxane analog, [^{14}C] 3′-*tert*-butyl-3′-*N*-*tert*-butyloxycarbonyl-4-deacetyl-3′-dephenyl-3′-*N*-debenzoyl-4-*O*-methoxycarbonyl-paclitaxel (BMS-275183), in rats and dogs. *Drug Metabolism and Disposition*, *37*, 1115–1128.

[94] Lockhart, A.C., Bukowski, R., Rothenberg, M.L., Wang, K.K., Cooper, W., Grover, J., Appleman, L., Mayer, P.R., Shapiro, M., Zhu, A.X. (2007). Phase I trial of oral MAC-321 in subjects with advanced malignant solid tumors. *Cancer Chemotherapy and Pharmacology*, *60*, 203–209.

[95] Ramanathan Ramesh, K., Picus, J., Raftopoulos, H., Bernard, S., Lockhart, A.C., Frenette, G., Macdonald, J., Melin, S., Berg, D., Brescia, F., Hochster, H., Cohn, A. (2008). A phase II study of milataxel: A novel taxane analogue in previously treated patients with advanced colorectal cancer. *Cancer Chemotherapy and Pharmacology*, *61*, 453–458.

[96] Dumontet, C., Sikic, B.I. (1999). Mechanisms of action of and resistance to antitubulin agents: Microtubule dynamics, drug transport, and cell death. *Journal of Clinical Oncology*, *17*, 1061–1070.

[97] Sampath, D., Discafani, C.M., Loganzo, F., Beyer, C., Liu, H., Tan, X., Musto, S., Annable, T., Gallagher, P., Rios, C., Greenberger, L.M. (2003). MAC-321, a novel taxane with greater efficacy than paclitaxel and docetaxel *in vitro* and *in vivo*. *Molecular Cancer Therapeutics*, *2*, 873–884.

[98] Safarpour, H., Connolly, P., Tong, X., Bielawski, M., Wilcox, E. (2009). Overcoming extractability hurdles of a ^{14}C labeled taxane analogue milataxel and its metabolite from xenograft mouse tumor and brain tissues. *Journal of Pharmaceutical and Biomedical Analysis*, *49*, 774–779.

[99] Ramanathan, R.K., Picus, J., Raftopoulos, H., Bernard, S., Lockhart, A.C., Frenette, G., Macdonald, J., Melin, S., Berg, D., Brescia, F., Hochster, H., Cohn, A. (2008). A phase II study of milataxel: A novel taxane analogue in previously treated patients with advanced colorectal cancer. *Cancer Chemotherapy and Pharmacology*, *61*, 453–458.

[100] Rose, W.C., Fairchild, C., Lee, F.Y.F. (2001). Preclinical antitumor activity of two novel taxanes. *Cancer Chemotherapy and Pharmacology*, *47*, 97–105.

[101] Fishman, M.N., Garrett, C.R., Simon, G.R., Chiappori, A.A., Lush, R.M., Dinwoodie, W.R., Manhany, J.J., Dellaportas, A.M., Cantor, A., Gollerki, A., Cohen, M.B., Sullivan, D.M. (2006). Phase I study of the taxane BMS-188797 in combination with carboplatin administered every 3 weeks in patients with solid malignancies. *Clinical Cancer Research*, *12*, 523–528.

[102] Advani, R., Fisher, G.A., Lum, B.L., Jambalos, C., Cho, C.D., Cohen, M., Gollerkeri, A., Sikic, B.I. (2003). Phase I and pharmacokinetic study of BMS-188797, a new taxane analog, administered on a weekly schedule in patients with advanced malignancies. *Clinical Cancer Research*, *9*, 5187–5194.

[103] du Bois, A., Jung, B., Loehr, A., Schaller-Kranz, T., Cohen, M., Frickhofen, N. (2006). A phase I and pharmacokinetic study of novel taxane BMS-188797 and cisplatin in patients with advanced solid tumours. *British Journal of Cancer*, *94*, 79–84.

[104] Garrett, C.R., Fishman, M.N., Rago, R.R., Williams, C.C., Dellaportas, A.M., Mahany, J.J., Lush, R.M., Dalton, W.S., Gollerkeri, A., Cohen, M.B., Sullivan, D.M. (2005). Phase I study of a novel taxane BMS-188797 in adult patients with solid malignancies. *Clinical Cancer Research*, *11*, 3335–3341.

[105] Kim, J., Amorino, G.P., Pyo, H., Cao, Q., Price, J.O., Choy, H. (2001). The novel taxane analogs, BMS-184476 and BMS-188797, potentiate the effects of radiation therapy *in vitro* and *in vivo* against human lung cancer cells. *International Journal of Radiation Oncology, Biology, Physics*, *51*, 525–534.

[106] Plummer, R., Ghielmini, M., Calvert, P., Voi, M., Renard, J., Gallant, G., Gupta, E., Calvert, H., Sessa, C. (2002). Phase I and pharmacokinetic study of the new taxane analog BMS-184476 given weekly in patients with advanced malignancies. *Clinical Cancer Research*, *8*, 2788–2797.

[107] Camps, C., Felip, E., Sanchez, J.M., Massuti, B., Artal, A., Paz-Ares, L., Carrato, A., Alberola, V., Blasco, A., Baselga, J., Astier, L., Voi, M., Rosell, R. (2005). Phase II trial of the novel taxane BMS-184476 as second-line in non-small-cell lung cancer. *Annals of Oncology*, *16*, 597–601.

[108] Roche, M., Kyriakou, H., Serden, M. (2006). Drug evaluation: tese taxel-an oral Semisynthetic taxane derivative. *Current Opinion Investigational Drugs*, *7*, 1092–1099.

[109] Hwang, J.J., Marshall, J.L., Malik, S., Chun, H., Ahmed, T., Pilia, P.A. (2006). A Q7D phase I study of TPI 287, a novel taxane, in advanced malignancies: Imaging studies,

pharmacokinetics (PK) and circulating tumor cell (CTC) quantitation. *Journal of Clinical Oncology*, *24*, 12008.

[110] Beer, M., Lenaz, L., Amadori, D. (2008). Phase II study of ortataxel in taxane-resistant breast cancer. *Journal of Clinical Oncology*, *26*, Abstract 1066.

[111] Iimura, S., Uoto, K., Ohsuki, S., Chiba, J., Yoshino, T., Iwahana, M., Jimbo, T., Terasawa, H., Soga, T. (2001). Orally active docetaxel analogue: Synthesis of 10-deoxy-10-*C*-morpholinoethyl docetaxel analogues. *Bioorganic and Medicinal Chemistry Letters*, *11*, 407–410.

[112] Ishiyama, T., Iimura, S., Ohsuki, S., Uoto, K., Terasawa, H., Soga, T. (2002). New highly active taxoids from 9b-dihydrobaccatin-9,10-acetals. *Bioorganic and Medicinal Chemistry Letters*, *12*, 1083–1086.

[113] Ishiyama, T., Iimura, S., Yoshino, T., Chiba, J., Uoto, K., Terasawa, H., Soga, T. (2002). New highly active taxoids from 9b-dihydrobaccatin-9,10-acetals: Part 2. *Bioorganic and Medicinal Chemistry Letters*, *12*, 2815–2819.

[114] Takeda, Y., Yoshino, T., Uoto, K., Chiba, J., Ishiyama, T., Iwahana, M., Jimbo, T., Tanaka, N., Terasawa, H., Soga, T. (2003). New highly active taxoids from 9b-dihyrobaccatin-9,10-acetals: Part 3. *Bioorganic and Medicinal Chemistry Letters*, *13*, 185–190.

[115] Takeda, Y., Uoto, K., Chiba, J., horiuchi, T., Iwahana, M., Atsumi, R., Ono, C., Terasawa, H., Soga, T. (2003). New highly active taxoids from 9b-dihydrobaccatin-9,10-acetals: Part 4. *Bioorganic and Medicinal Chemistry Letters*, *11*, 4431–4447.

[116] Takeda, Y., Uoto, K., Iwahana, M., Jimbo, T., Nagata, M., Atsumi, R., Ono, C., Tanaka, N., Terasawa, H., Soga, T. (2004). New highly active taxoids from 9b-dihydrobaccatin-9,10-acetals: Part 5. *Bioorganic and Medicinal Chemistry Letters*, *14*, 3209–3215.

[117] Carmeliet, P., Jain, R.K. (2000). Angiogenesis in cancer and other diseases. *Nature*, *407*, 249–257.

[118] Brower, V. (1999). Tumor angiogenesis – New drugs on the block. *Nature Biotechnology*, *17*, 963–968.

[119] Giavazzi, R., Taraboletti, G. (1999). Angiogenesis and angiogenesis inhibitors in cancer. *Forum-(Genova)*, *9*, 261–272.

[120] Taraboletti, G., Micheletti, G., Rieppi, M., Poli, M., Turatto, M., Rossi, C., Borsotti, P., Roccabianca, P., Scanziani, E., Nicoletti, M. I., Bombardelli, E., Morazzoni, P., Riva, A., Giavazzi, R. (2002). Antiangiogenic and antitumor activity of IDN 5390, a new taxane derivative. *Clinical Cancer Research*, *8*, 1182–1188.

[121] Pratesi, G., Laccabue, D., Lanzi, C., Cassinelli, G., Supino, R., Zucchetti, M., Frapolli, R., D'Incalci, M., Bombardelli, E., Morazzoni, P., Riva, A., Zunino, F. (2003). IDN 5390: An oral taxane candidate for protracted treatment schedules. *British Journal of Cancer*, *88*, 965–972.

[122] Frapolli, R., Marangon, E., Zaffaroni, M., Colombo, T., Falcioni, C., Bagnati, R., Simone, M., D'Incalci, M., Manzotti, C., Fontana, G., Morazzoni, P., Zucchetti, M. (2006). Pharmacokinetics and metabolism in mice of IDN 5390 (13-(*N*-Boc-3-*i*-butylisoserinoyl)-*C*-7,8-*seco*-10-deacetylbaccatin III), a new oral C-seco-taxane derivative with antiangiogenic property effective on paclitaxel-resistant tumors. *Drug Metabolism and Disposition*, *34*, 2028–2035.

[123] Ferlini, C., Raspaglio, G., Mozzetti, S., Cicchillitti, L., Filippetti, F., Gallo, D., Gattorusso, C., Campiani, G., Scambia, G. (2005). The *seco*-taxane IDN5390 is able

to target class III b-tubulin and to overcome paclitaxel resistance. *Cancer Research*, *65*, 2397–2405.

[124] Dubois, J., Thoret, S., Gueritte, F., Guenard, D. (2000). Synthesis of 5(20)deoxydocetaxel, a new active docetaxel analogue. *Tetrahedron Letters*, *41*, 3331–3334.

[125] Merckle, L., Dubois, J., Place, E., Thoret, S., Gueritte, F., Guenard, D., Poupat, C., Ahond, A., Potier, P. (2001). Semisynthesis of D-ring modified taxoids: Novel thia derivatives of docetaxel. *Journal of Organic Chemistry*, *66*, 5058–5065.

[126] Gunatilaka, A.A.L., Ramdayal, F.D., Sarragiotto, M.H., Kingston, D.G.I., Sackett, D.L., Hamel, E. (1999). Synthesis and biological evaluation of novel paclitaxel (taxol) D-ring modified analogues. *Journal of Organic Chemistry*, *64*, 2694–2703.

[127] Battaglia, A., Bernacki, R.J., Bertucci, C., Bombardelli, E., Cimitan, S., Ferlini, C., Fontana, G., Guerrini, A., Riva, A. (2003). Synthesis and biological evaluation of 2′-methyl taxoids derived from baccatin III and 14b-OH-baccatin III 1,14-carbonate. *Journal of Medicinal Chemistry*, *46*, 4822–4825.

[128] Liu, C., Tamm, M., Notzel, M.W., de Meijere, A., Schilling, J.K., Kingston, D.G.I. (2003). Synthesis and bioactivities of paclitaxel analogs with a cyclopropanated side-chain. *Tetrahedron Letters*, *44*, 2049–2052.

[129] Barboni, L., Ballini, R., Giarlo, G., Appendino, G., Fontana, G., Bombardelli, E. (2005). Synthesis and biological evaluation of methoxylated analogs of the newer generation taxoids IDN5109 and IDN5390. *Bioorganic and Medicinal Chemistry Letters*, *15*, 5182–5186.

[130] Chaudhary, A.G., Gharpure, M.M., Rimoldi, J.M., Chordia, M.D., Gunatilaka, A.A.L., Kingston, D.G.I., Grover, S., Lin, C.M., Hamel, E. (1994). Unexpectedly facile hydrolysis of the 2-benzoate group of taxol and synthesis of analogs with increased activities. *Journal of the American Chemical Society*, *116*, 4097–4098.

[131] Kuznetsova, L., Chen, J., Sun, L., Wu, X., Pepe, A., Veith, J. M., Pera, P., Bernacki, R.J., Ojima, I. (2006). Syntheses and evaluation of novel fatty acid-second-generation taxoid conjugates as promising anticancer agents. *Bioorganic and Medicinal Chemistry Letters*, *16*, 974–977.

[132] Ferjancic, Z., Matovic, R., Cekovic, Z., Jiang, Y., Snyder, J.P., Trajkovic, V., Saicic, R.N. (2006). Synthesis, biology, and modeling of a C-4 carbonyl C,D-*seco*-taxoid. *Tetrahedron*, *62*, 8503–8514.

[133] Ojima, I., Kuznetsova, L.V., Sun, L., Organofluorine chemistry at the biomedical interface: A case study on fluoro-taxoid anticancer agents. In: Soloshonok, V.A., Mikami, K., Yamazaki, T., Welch, J.T., Honek, J., editors. *Current Fluoroorganic Chemistry*, Vol. 949, American Chemical Society, Stony Brook, NY, USA, 2007, pp. 288–304.

[134] Mikuni, K., Nakanishi, K., Hara, K., Hara, K., Iwatani, W., Amano, T., Nakamura, K., Tsuchiya, Y., Okumoto, H., Mandai, T. (2008). *In vivo* antitumor activity of novel water-soluble taxoids. *Biological and Pharmaceutical Bulletin*, *31*, 1155–1158.

[135] Ojima, I., Chen, J., Sun, L., Borella, C.P., Wang, T., Miller, M.L., Lin, S., Geng, X., Kuznetsova, L., Qu, C., Gallager, D., Zhao, X., Zanardi, I., Xia, S., Horwitz, S.B., Mallen-St. Clair, J., Guerriero, J.L., Bar-Sagi, D., Veith, J.M., Pera, P., Bernacki, R.J. (2008). Design, synthesis, and biological evaluation of new-generation taxoids. *Journal of Medicinal Chemistry*, *51*, 3203–3221.

[136] Ke, B., Qin, Y., Zhao, F., Qu, Y. (2008). Synthesis and biological evaluation of novel 3′-*N-tert*-butylsulfonyl analogues of docetaxel. *Bioorganic and Medicinal Chemistry Letters*, *18*, 4783–4785.

[137] Kuznetsova, L.V., Pepe, A., Ungureanu, I.M., Pera, P., Bernacki, R.J., Ojima, I. (2008). Syntheses and structure–activity relationships of novel 3′-difluoromethyl and 3′-trifluoromethyl-taxoids. *Journal of Fluorine Chemistry*, *129*, 817–828.

[138] Pepe, A., Sun, L., Zanardi, I., Wu, X., Ferlini, C., Fontana, G., Bombardelli, E., Ojima, I. (2009). Novel *C-seco*-taxoids possessing high potency against paclitaxel-resistant cancer cell lines overexpressing class III b-tubulin. *Bioorganic and Medicinal Chemistry Letters*, *19*, 3300–3304.

[139] Lu, H., Sun, X., Xu, L., Lou, L., Lin, G. (2009). Design, synthesis and biological evaluation of novel fluorinated docetaxel analogues. *European Journal of Medicinal Chemistry*, *44*, 482–491.

[140] Holton, R.A. (1992: 29 Dec). Method for Preparation of Taxol Using b-lactam. Florida State University, US. US Patent 5175315.

[141] Exposito, O., Bonfill, M., Moyano, E., Onrubia, M., Mirjalili, M.H., Cusido, R.M., Palazon, J. (2009). Biotechnological production of taxol and related taxoids: Current state and prospects. *Anti-Cancer Agents in Medicinal Chemistry*, *9*, 109–121.

[142] Available at http://www.genexol.com/eng/app//manufacture/default.asp.

[143] Rao, S., Krauss, N.E., Heerding, J.M., Swindell, C.S., Ringel, I., Orr, G.A., Horwitz, S.B. (1994). 3′-(*p*-Azidobenzamido)taxol photolabels the *N*-terminal 31 amino acids of b-tubulin. *Journal of Biological Chemistry*, *269*, 3132–3134.

[144] Rao, S., Orr, G.A., Chaudhary, A.G., Kingston, D.G.I., Horwitz, S.B. (1995). Characterization of the taxol binding site on the microtubule. *Journal of Biological Chemistry*, *270*, 20235–20238.

[145] Nogales, E., Wolf, S.G., Downing, K.H. (1998). Structure of the ab tubulin dimer by electron crystallography. *Nature 39*, 199–203.

[146] Lowe, J., Li, H., Downing, K.H., Nogales, E. (2001). Refined structure of b-tubulin at 3.5 A resolution. *Journal of Molecular Biology*, *313*, 1045–1057.

[147] Dubois, J., Guenard, D., Gueritte-Voeglein, F., Guedira, N., Potier, P., Gillet, B., Betoeil, J.-C. (1993). Conformation of taxotere and analogues determined by NMR spectroscopy and molecular modeling studies. *Tetrahedron*, *49*, 6533–6544.

[148] Ojima, I., Kuduk, S.D., Chakravarty, S., Ourevitch, M., Begue, J.-P. (1997). A novel approach to the study of solution structures and dynamic behavior of paclitaxel and docetaxel using fluorine-containing analogs as probes. *Journal of the American Chemical Society*, *119*, 5519–5527.

[149] Williams, H.J., Scott, A.I., Dieden, R.A., Swindell, C.S., Chirlian, L.E., Francl, M.M., Heerding, J.M., Krauss, N.E. (1993). NMR and molecular modeling study of the conformations of taxol and of its side chain methylester in aqueous and non-aqueous solution. *Tetrahedron*, *49*, 6545–6560.

[150] Milanesio, M., Ugliengo, P., Viterbo, D. (1999). *Ab initio* conformational study of the phenylisoserine side chain of paclitaxel. *Journal of Medicinal Chemistry*, *42*, 291–299.

[151] Vander Velde, D.G., Georg, G.I., Grunewald, G.L., Gunn, C.W., Mitscher, L.A. (1993). "Hydrophobic collapse" of Taxol and Taxotere solution conformations in mixtures of water and organic solvent. *Journal of the American Chemical Society*, *115*, 11650–11651.

[152] Snyder, J.P., Nettles, J.H., Cornett, B., Downing, K.H., Nogales, E. (2001). The binding conformation of taxol in b-tubulin: A model based on electron crystallographic density. *Proceedings of the National Academy of Sciences of the U.S.A.*, 98, 5312–5316.

[153] Kingston, D.G.I., Bane, S., Snyder, J.P. (2005). The taxol pharmacophore and the T-taxol bridging principle. *Cell Cycle*, *4*, 279–289.

[154] Snyder, J.P., Nevins, N., Cicero, D.O., Jansen, J. (2000). The conformations of taxol in chloroform. *Journal of the American Chemical Society*, *122*, 724–725.

[155] Gueritte-Voegelein, F., Guenard, D., Mangatal, L., Potier, P., Guilhem, J., Cesario, M., Pascard, C. (1990). Structure of a synthetic taxol precursor: *N-tert*-butoxycarbonyl-10-deacetyl-*N*-debenzoyltaxol. *Acta Crystallographica*, *C46*, 781–784.

[156] Mastropaolo, D., Camerman, A., Luo, Y., Brayer, G.D., Camerman, N. (1995). Crystal and molecular structure of paclitaxel (Taxol). *Proceedings of the National Academy of Sciences of the U.S.A.*, *92*, 6920–6924.

[157] Williams, H.J., Moyna, G., Scott, A.I., Swindell, C.S., Chirlian, L.E., Heerding, J.M., Williams, D.K. (1996). NMR and molecular modeling study of the conformations of taxol 2′-acetate in chloroform and aqueous dimethyl sulfoxide solutions. *Journal of Medicinal Chemistry*, *39*, 1555–1559.

[158] Chmurny, G.N., Hilton, B.D., Brobst, S., Look, S.A., Witherup, K.M., Beutler, J.A. (1992). ^{1}H- and ^{13}C-NMR assignments for Taxol, 7-epi-Taxol, and Cephalonannine. *Journal of Natural Products*, *55*, 414–423.

[159] Geney, R., Sun, L., Pera, P., Bernacki, R.J., Xia, S., Horwitz, S.B., Simmerling, C.L., Ojima, I. (2005). Use of the tubulin bound paclitaxel conformation for structure-based rational drug design. *Chemical Biology*, *12*, 339–348.

[160] Li, Y., Poliks, B., Cegelski, L., Poliks, M., Cryczynski, Z., Piszczek, G., Jagtap, P.,G., Studelska, D.R., Kingston, D.G.I., Schaefer, J., Bane, S. (2000). Conformation of microtubule-bound paclitaxel determined by fluorescence spectroscopy and REDOR NMR. *Biochemistry*, *39*, 281–291.

[161] Gao, Q., Wei, J.-M., Chen, S.-H. (1995). Crystal structure of 2-debenzoyl, 2-acetoxy paclitaxel (Taxol): Conformation of the paclitaxel side-chain. *Pharmaceutical Research*, *12*, 337–341.

[162] Gao, Q., Chen, S.-H. (1996). An unprecedented side chain conformation of paclitaxel (Taxol): Crystal structure of 7-mesylpaclitaxel. *Tetrahedron Letters*, *37*, 3425–3428.

[163] Cachau, R.E., Gussio, R., Beutler, J.A., Chmurny, G.N., Hilton, B.D., Muschik, G.M., Erickson, J.W. (1994). Solution structure of taxol determined using a novel feedback-scaling procedure for NOE-restrained molecular dynamics. *International Journal of High Performance Computing Applications*, *8*, 24–34.

[164] Paloma, L.G., Guy, R.K., Wrasidlo, W., Nicolaou, K.C. (1994). Conformation of a water-soluble derivative of taxol in water by 2D-NMR spectroscopy. *Chemical Biology*, *1*, 107–112.

[165] Park, Y., Yang, C., Metaferia, B., Tang, S., Bane, S., Ravindra, R., Shanker, N., Alcaraz, A.A., Johnson, S.A., Schaefer, J., O'Connor, R.D., Cegeleski, L., Snyder, J.P., Kingston, D.G.I. (2007). Rotational-echo double-resonance NMR distance measurements for the tubulin-bound paclitaxel conformation. *Journal of the American Chemical Society*, *129*, 361–370.

[166] Rao, S., He, L., Chakravarty, S., Ojima, I., Orr, G.A., Horwitz, S.B. (1999). Characterization of the taxol binding site on the microtubule. *Journal of Biological Chemistry*, *274*, 37990–37994.

[167] Sun, L., Simmerling, C., Ojima, I. (2009). Recent advances in the study of the bioactive conformation of taxol. *ChemMedChem*, *4*, 719–731.

[168] Yang, Y., Alcaraz, A.A., Snyder, J.P. (2009). The tubulin-bound conformation of paclitaxel: T-taxol vs "PTX-NY". *Journal of Natural Products*, *72*, 422–429.

[169] Ivery, M.T., Le, T. (2003). Modeling the interaction of paclitaxel with b-tubulin. *Oncology Research*, *14*, 1–19.

[170] Metaferia, B.B., Hoch, J., Glass, T.E., Bane, S.L., Chatterjee, S.K., Snyder, J.P., Lakdawala, A., Cornett, B., Kingston, D.G.I. (2001). Synthesis and biological evaluation of novel macrocyclic paclitaxel analogues. *Organic Letters*, *3*, 2461–2464.

[171] Ganesh, T., Guza, R.C., Bane, S., Ravindra, R., Shanker, N., Lakdawala, A.S., Snyder, J.P., Kingston, D.G.I. (2004). The bioactive taxol conformation on b-tubulin: Experimental evidence from highly active constrained analogs. *Proceedings of the National Academy of Sciences of the U.S.A.*, 101, 10006–10011.

[172] Barboni, L., Lambertucci, C., Appendino, G., Vander Velde, D.G., Himes, R.H., Bombardelli, E., Wang, M., Snyder, J.P. (2001). Synthesis and NMR-driven conformational analysis of taxol analogues conformationally constrained on the C13 side chain. *Journal of Medicinal Chemistry*, *44*, 1576–1587.

[173] Boge, T.C., Wu, Z.-J., Himes, R.H., Vander Velde, D.G., Georg, G.I. (1999). Conformationally restricted paclitaxel analogues: Macrocyclic mimics of the "hydrophobic collapse" conformation. *Bioorganic and Medicinal Chemistry Letters*, *9*, 3047–3052.

[174] Chakravarty, S., Lin, S., Inoue, T., Horwitz, S.B., Kuduk, S.D., Danishefsky, S.J., Ojima, I., A common pharmacophore for cytotoxic microtubule-stabilizing agents: Interpretation of SAR studies. In: *217th ACS National Meeting*, Anaheim, CA, 1999.

[175] Ojima, I., Lin, S., Inoue, T., Miller, M.L., Borella, C.P., Geng, X., Walsh, J.J. (2000). Macrocycle formation by ring-closing metathesis. Application to the syntheses of novel macrocyclic taxoids. *Journal of the American Chemical Society*, *122*, 5343–5353.

[176] Ojima, I., Geng, X., Lin, S., Pera, P., Bernacki, R.J. (2002). Design, synthesis and biological activity of novel C2-C3′N-linked macrocyclic taxoids. *Bioorganic and Medicinal Chemistry Letters*, *12*, 349–352.

[177] Geng, X., Miller, M.L., Lin, S., Ojima, I. (2003). Synthesis of novel C2-C3′N-linked macrocyclic taxoids by means of highly regioselective Heck macrocyclization. *Organic Letters*, *5*, 3733–3736.

[178] Querolle, O., Dubois, J., Thoret, S., Dupont, C., Gueritte, F., Guenard, D. (2003). Synthesis of novel 2-O,3′-N-linked macrocyclic taxoids with variable ring size. *European Journal of Organic Chemistry*, *2003*, 542–550.

[179] Querolle, O., Dubois, J., Thoret, S., Roussi, F., Montiel-Smith, S., Gueritte, F., Guenard, D. (2003). Synthesis of novel macrocyclic docetaxel analogues. Influence of their macrocyclic ring size on tubulin activity. *Journal of Medicinal Chemistry*, *46*, 3623–3630.

[180] Querolle, O., Dubois, J., Thoret, S., Roussi, F., Guéritte, F., Guénard, D. (2004). Synthesis of C2–C3′N-linked macrocyclic taxoids. Novel docetaxel analogues with high tubulin activity. *Journal of Medicinal Chemistry*, *47*, 5937–5944.

[181] Blechert, S., Kleine-Klausing, A. (1991). Synthesis of a biologically active taxol analogue. *Angewndte Chemie International Edition*, *30*, 412–414.

[182] Blechert, S., Jansen, R., Velder, J. (1994). Synthesis of new taxoids. *Tetrahedron*, *50*, 9649–9656.

[183] Senilh, V., Gueritte, F., Guenard, D., Colin, M., Potier, P. (1984). Hemisynthese de nouveaux analogues du taxol. Etude de leur interaction avec la tubuline. *Comptes-rendus des Séances de l'Acadėmie de Sciences*, *299*, 1039–1043.

[184] Kingston, D.G.I., Hawkins, D.R., Ovington, L. (1982). New taxanes from *Taxus brevifolia*. *Journal of Natural Products*, *45*, 466–470.

[185] Klar, U., Graf, H., Schenk, O., Rohr, B., Schulz, H. (1998). New synthetic inhibitors of microtubule depolymerization. *Bioorganic and Medicinal Chemistry Letters*, *8*, 1397–1402.

[186] Almqvist, F., Manner, S., Thornqvist, V., Berg, U., Wallin, M., Frejd, T. (2004). Spirobicyclo[2.2.2]octane derivatives: Mimetics of baccatin III and paclitaxel (Taxol). *Organic and Biomolecular Chemistry*, *2*, 3085–3090.

[187] Geng, X., Geney, R., Pera, P., Bernacki, J., Ojima, I. (2004). Design and synthesis of de novo cytotoxic alkaloids through mimicking taxoid skeleton. *Bioorganic and Medicinal Chemistry Letters*, *14*, 3491–3494.

[188] Ganesh, T., Norris, A., Sharma, S., Bane, S., Alcaraz, A.A., Snyder, J.P., Kingston, D.G.I. (2006). Design, synthesis, and bioactivity of simplified paclitaxel analogs based on the T-taxol bioactive conformation. *Bioorganic and Medicinal Chemistry*, *14*, 3447–3454.

[189] Zhao, J., Bane, S., Snyder, J.P., Hu, H., Kingston, D.G.I. (2011). Design and synthesis of simplified paclitaxel analogs based on the T-taxol bioactive conformation. *Bioorganic and Medicinal Chemistry*, *19*, 7664–7678.

[190] Sun, L., Veith, J.M., Pera, P., Bernacki, R.J., Ojima, I. (2010). Design and synthesis of de novo cytotoxic alkaloids by mimicking the bioactive conformation of paclitaxel. *Bioorganic and Medicinal Chemistry*, *18*, 7101–7112.

[191] Giannakakou, P., Sackett, D.L., Kang, Y.-K., Zhang, Z., Buters, J.T.M., Fojo, T., Poruchynsky, M.S. (1997). Paclitaxel-resistant human ovarian cancer cells have mutant b-tubulins that exhibit impaired paclitaxel-driven polymerization. *Journal of Biological Chemistry*, *272*, 17118–17125.

[192] Sharma, A., Mayhew, E., Bolcsak, L., Cavanaugh, C., Harmon, P., Janoff, A., Bernacki, R.J. (1997). Activity of paclitaxel liposome formulations against human ovarian tumor xenografts. *International Journal of Cancer*, *71*, 103–107.

[193] Michaud, L.B., Valero, V., Hortobagyi, G. (2000). Risks and benefits of taxanes in breast and ovarian cancer. *Drug Safety*, *23*, 401–428.

[194] Deutsch, H.M., Glinski, J.A., Hernandez, M., Haugwitz, R.D., Narayanan, V.L., Suffness, M., Zalkow, L.H. (1989). Synthesis of congeners and prodrugs: 3. Water-soluble prodrugs of taxol with potent antitumor activity. *Journal of Medicinal Chemistry*, *32*, 788–792.

[195] Zhao, Z., Kingston, D.G.I., Crosswell, A.R. (1991). Modified taxols, 6. Preparation of water-soluble prodrugs of taxol. *Journal of Natural Products*, *54*, 1607–1611.

[196] Mathew, A.E., Mejillano, M.R., Nath, J.P., Himes, R.H., Stella, V.J. (1992). Synthesis and evaluation of some water-soluble prodrugs and derivatives of taxol with antitumor activity. *Journal of Medicinal Chemistry*, *35*, 145–151.

[197] Vyas, D.M., Wong, H., Crosswell, A.R., Casazza, A.M., Knipe, J.O., Mamber, S.W., Doyle, T.W. (1993). Synthesis and antitumor evaluation of water soluble taxol phosphates. *Bioorganic and Medicinal Chemistry Letters*, *3*, 1357–1360.

[198] Ueda, Y., Mikkilineni, A.B., Knipe, J.O., Rose, W.C., Casazza, A.M., Vyas, D.M. (1993). Novel water soluble phosphate prodrugs of taxol possessing *in vivo* antitumor activity. *Bioorganic and Medicinal Chemistry Letters*, *3*, 1761–1766.

[199] Ueda, Y., Matiskella, J.D., Mikkilineni, A.B., Farina, V., Knipe, J.O., Rose, W.C., Casazza, A.M., Vyas, D.M. (1995). Novel, water-soluble phosphate derivatives of 2′-ethoxy carbonylpaclitaxel as potential prodrugs of paclitaxel: Synthesis and antitumor evaluation. *Bioorganic and Medicinal Chemistry Letters*, *5*, 247–252.

[200] Golik, J., Wong, H.S.L., Chen, S.-H., Doyle, T.W., Wright, J.J.K., Knipe, J., Rose, W.C., Casazza, A.M., Vyas, D.M. (1996). Synthesis and antitumor evaluation of paclitaxel phosphonooxymethyl ethers: A novel class of water soluble paclitaxel pro-drugs. *Bioorganic and Medicinal Chemistry Letters*, *6*, 1837–1842.

[201] Skwarczynski, M., Hayashi, Y., Kiso, Y. (2006). Paclitaxel prodrugs: Toward smarter delivery of anticancer agents. *Journal of Medicinal Chemistry*, *49*, 7253–7269.

[202] Ganesh, T. (2007). Improved biochemical strategies for targeted delivery of taxoids. *Bioorganic and Medicinal Chemistry*, *15*, 3597–3623.

[203] Safavy, A. (2008). Recent developments in taxane drug delivery. *Current Drug Delivery*, *5*, 42–54.

[204] Damen, E.W.P., Wiegerinck, P.H.G., Braamer, L., Sperling, D., de Vos, D., Scheeren, H.W. (2000). Paclitaxel esters of malic acid as prodrugs with improved water solubility. *Bioorganic and Medicinal Chemistry*, *8*, 427–432.

[205] Seligson, A.L., Terry, R.C., Bressi, J.C., Douglass, J.G., Sovak, M. (2001). A new prodrug of paclitaxel: Synthesis of protaxel. *Anti-Cancer Drugs*, *12*, 305–313.

[206] Shimoda, K., Hamada, H., Hamada, H. (2008). Chemo-enzymatic synthesis of ester-linked taxol-oligosaccharide conjugates as potential prodrugs. *Tetrahedron Letters*, *49*, 601–604.

[207] Kirschberg, T.A., VanDeusen, C.L., Rothbard, J.B., Yang, M., Wender, P.A. (2003). Arginin-based molecular transporters: The synthesis and chemical evaluation of releasable taxol-transporter conjugates. *Organic Letters*, *5*, 3459–3462.

[208] Futaki, S., Goto, S., Sugiura, Y. (2003). Membrane permeability commonly shared among arginine-rich peptides. *Journal of Molecular Recognition*, *16*, 260–264.

[209] Rothbard, J.B., Kreider, E., VanDeusen, C.L., Wright, L., Wylie, B.L., Wender, P.A. (2002). Arginine-rich molecular transporters for drug delivery: Role of backbone spacing in cellular uptake. *Journal of Medicinal Chemistry*, *45*, 3612–3618.

[210] Battaglia, A., Guerrini, A., Baldelli, E., Fontana, G., Varchi, G., Samori, C., Bombardelli, E. (2006). Synthesis of 7- and 10-spermine conjugates of paclitaxel and 10-deacetyl-paclitaxel as potential prodrugs. *Tetrahedron Letters*, *47*, 2667–2670.

[211] Moosavi-Movahedi, A.A., Hakimelahi, S., Chamani, J., Khodarahmi, G.A., Hassanzadeh, F., Luo, F.-T., Ly, T.W., Shia, K.-S., Yen, C.-F., Jain, M.L., Kulatheeswaran, R., Xue, C., Pasdar, M., Hakimelahi, G.H. (2003). Design, synthesis, and anticancer activity of phosphonic acid diphosphate derivative of adenine-containing

butenolide and its water-soluble derivatives of paclitaxel with high antitumor activity. *Bioorganic and Medicinal Chemistry*, *11*, 4303–4313.

[212] Skwarczynski, M., Sohma, Y., Kimura, M., Hayashi, Y., Kimura, T., Kiso, Y. (2003). O–N Intramolecular acyl migration strategy in water-soluble prodrugs of taxoids. *Bioorganic and Medicinal Chemistry*, *13*, 4441–4444.

[213] de Groot, F.M.H., van Berkom, L.W.A., Scheeren, H.W. (2000). Synthesis and biological evaluation of 2′-carbamate-linked and 2′-carbonate-linked prodrugs of paclitaxel: Selective activation by the tumor-associated protease plasmin. *Journal of Medicinal Chemistry*, *43*, 3093–3102.

[214] Damen, E.W.P., Nevalainen, T.J., van den Bergh, T.J.M., de Groot, F.M.H., Scheeren, H.W. (2002). Synthesis of novel paclitaxel prodrugs designed for bioreductive activation in hypoxic tumour tissue. *Bioorganic and Medicinal Chemistry*, *10*, 71–77.

[215] Vrudhula, V.M., MacMaster, J.F., Li, Z., Kerr, D.E., Senter, P.D. (2002). Reductively activated disulfide prodrugs of paclitaxel. *Bioorganic and Medicinal Chemistry Letters*, *12*, 3591–3594.

[216] Liu, C., Strobl, J.S., Bane, S., Schilling, J.K., McCracken, M., Chatterjee, S.K., Rahim-Bata, R., Kingston, D.G.I. (2003). Design, synthesis, and bioactivities of steroid-linked taxol analogues as potential targeted drugs for prostate and breast cancer. *Journal of Natural Products*, *67*, 152–159.

[217] Park, W.-C., Jordan, V.C. (2002). Selective estrogen receptor modulators (SERMS) and their roles in breast cancer prevention. *Trends in Molecular Medicine*, *8*, 82–88.

[218] Lee, J.W., Lu, J.Y., Low, P.S., Fuchs, P.L. (2002). Synthesis and evaluation of taxol-folic acid conjugates as targeted antineoplastics. *Bioorganic and Medicinal Chemistry*, *10*, 2397–2414.

[219] Wang, S., Low, P.S. (1998). Folate-mediated targeting of antineoplastic drugs, imaging agents, and nucleic acids to cancer cells. *Journal Controlled Release*, *53*, 39–48.

[220] Chen, X., Plasencia, C., Hou, Y., Neamati, N. (2005). Synthesis and biological evaluation of dimeric RGD peptide-paclitaxel conjugate as a model for integrin-targeted drug delivery. *Journal of Medicinal Chemistry*, *48*, 1098–1106.

[221] Ryppa, C., Mann-Steinberg, H., Biniossek, M.L., Satchi-Fainaro, R., Kratz, F. (2009). *In vitro* and *in vivo* evaluation of a paclitaxel conjugate with the divalent peptide E-[c(RGDfK)2] that targets integrin avb3. *International Journal of Pharmacy*, *368*, 89–97.

[222] Bradley, M.O., Swindell, C.S., Anthony, F.H., Witman, P.A., Devanesan, P., Webb, N.L., Baker, S.D., Wolff, A.C., Donehower, R.C. (2001). Tumor targeting by conjugation of DHA to paclitaxel. *Journal of Controlled Release*, *74*, 233–236.

[223] Bradley, M.O., Webb, N.L., Anthony, F.H., Devanesan, P., Witman, P.A., Hemamalini, S., Chander, M.C., Baker, S.D., He, L., Horwitz, S.B., Swindell, C.S. (2001). Tumor targeting by covalent conjugation of a natural fatty acid to paclitaxel. *Clinical Cancer Research*, *7*, 3229–3238.

[224] Wolff, A.C., Donehower, R.C., Carducci, M.K., Carducci, M.A., Brahmer, J.R., Zabelina, Y., Bradley, M.O., Anthony, F.H., Swindell, C.S., Witman, P.A., Webb, N.L., Baker, S.D. (2003). Phase I study of docosahexaenoic acid-paclitaxel: A taxane-fatty acid conjugate with a unique pharmacology and toxicity profile. *Clinical Cancer Research*, *9*, 3589–3597.

[225] Lin, Y.-S., Tungpradit, R., Sinchaikul, S., An, F.-M., Liu, D.-Z., Phutrakul, S., Chen, S.-T. (2008). Targeting the delivery of glycan-based paclitaxel prodrugs to cancer cells via glucose transporters. *Journal of Medicinal Chemistry*, *51*, 7428–7441.

[226] Nguyen, X.C., Lee, W.W., Chung, J.H., Park, S.Y., Sung, S.W., Kim, Y.K., So, Y., Lee, D.S., Chung, J.K., Lee, M.C., Kim, S.E. (2007). FDG uptake, glucose transporter type 1, and Ki-67 expressions in non-small-cell lung cancer: Correlations and prognostic values. *European Journal of Radiology*, *62*, 214–219.

[227] Vrudhula, V.M., Kerr, D., Siemers, O., Dubowchik, G.M., Senter, P.D. (2003). Cephalosporin prodrugs of paclitaxel for immunologically specific activation by L-49-sFv-b-lactamase fusion protein. *Bioorganic and Medicinal Chemistry Letters*, *13*, 539–542.

[228] Bouvier, E., Thirot, S., Schmidt, F., Monneret, C. (2003). A new paclitaxel prodrug for use in ADEPT strategy. *Organic and Biomolecular Chemistry*, *1*, 3343–3352.

[228a] Skwarczynski, M., Noguchi, M., Hirota, S., Sohma, Y., Kimura, T., Hayashi, Y., Kiso, Y. (2006). Development of first photoresponsive prodneg of paclitaxel. *Bioorganic and Medicinal Chemistry Letters*, *16*, 4492–4496.

[229] Ryu, B.-Y., Sohn, J.-S., Hess, M., Choi, S.-K., Choi, J.-K., Jo, B.-W. (2008). Synthesis and anti-cancer efficacy of rapid hydrolysed water-soluble paclitaxel pro-drugs. *Journal of Biomaterials Science, Polymer Edition*, *19*, 311–324.

[230] Kakinoki, A., Kaneo, Y., Tanaka, T., Hosokawa, Y. (2008). Synthesis and evaluation of water-soluble poly(vinyl alcohol)-paclitaxel conjugate as a macromolecular prodrug. *Biological and Pharmaceutical Bulletin*, *31*, 963–969.

[231] Meerum Terwogt, J.M., ten Bokkel Huinink, W.W., Schellens, J.H.M., Schot, M., Mandjes, I.A.M., Zurlo, M.G., Rocchetti, M., Rosing, H., Koopman, F.J., Beijnen, J.H. (2001). Phase I clinical and pharmacokinetic study of PNU166945, a novel water-soluble polymer-conjugated prodrug of paclitaxel. *Anti-Cancer Drugs*, *12*, 315–323.

[232] Erez, R., Segal, E., Miller, K., Satchi-Fainaro, R., Shabat, D. (2009). Enhanced cytotoxicity of a polymer–drug conjugate with triple payload of paclitaxel. *Bioorganic and Medicinal Chemistry*, *17*, 4327–4335.

[233] Rihova, B., Kubackova, K. (2003). Clinical implications of *N*-(2-hydroxy-propyl) methacrylamide copolymers. *Current Pharmaceutical Biotechnology*, *4*, 311–322.

[234] Singer, J.W. (2005). Paclitaxel poliglumex (XYOTAX(TM), CT-2103): A macromolecular taxane. *Journal of Controlled Release 109*, 120–126.

[235] Veronese, M.L., Flaherty, K., Kramer, A., Konkle, B.A., Morgan, M., Stevenson, J.P., O'Dwyer, P.J. (2005). Phase I study of the novel taxane CT-2103 in patients with advanced solid tumors. *Cancer Chemotherapy and Pharmacology*, *55*, 497–501.

[236] Luo, Y., Ziebell, M.R., Prestwich, G.D. (2000). A hyaluronic acid-taxol antitumor bioconjugate targeted to cancer cells. *Biomacromolecules*, *1*, 208–218.

[237] Yeo, T., Nagy, J., Yeo, K., Dvorak, H., Toole, B. (1996). Increased hyaluronan at sites of attachment to mesentery by CD44-positive mouse ovarian and breast tumor cells. *American Journal of Pathology*, *148*, 1733–1740.

[238] Toole, B.P. (1997). Hyaluronan in morphogenesis. *Journal of Internal Medicine*, *242*, 35–40.

[239] Turley, E.A., Belch, A.J., Poppema, S., Pilarsk, L.M. (1993). Expression and function of a receptor for hyaluronan-mediated motility on normal and malignant B lymphocytes. *Blood*, *81*, 446–453.

[240] Lee, E., Lee, J., Lee, I.-H., Yu, M., Kim, H., Chae, S.Y., Jon, S. (2008). Conjugated chitosan as a novel platform for oral delivery of paclitaxel. *Journal of Medicinal Chemistry*, *51*, 6442–6449.

[241] Majoros, I.J., Myc, A., Thomas, T., Mehta, C.B., Baker, J.R. (2006). PAMAM dendrimer-based multifunctional conjugate for cancer therapy: Synthesis, characterization, and functionality. *Biomacromolecules*, *7*, 572–579.

[242] Paciotti, G.F., Kingston, D.G.I., Tamarkin, L. (2006). Colloidal gold nanoparticles: A novel nanoparticle platform for developing multifunctional tumor-targeted drug delivery vectors. *Drug Development Research*, *67*, 47–54.

[243] Hwu, J.R., Lin, Y.S., Josephrajan, T., Hsu, M.-H., Cheng, F.-Y., Yeh, C.-S., Su, W.-C., Shieh, D.-B. (2008). Targeted paclitaxel by conjugation to iron oxide and gold nanoparticles. *Journal of the American Chemical Society*, *131*, 66–68.

[244] Hawkins, M.J., Soon-Shiong, P., Desai, N. (2008). Protein nanoparticles as drug carriers in clinical medicine. *Advanced Drug Delivery Reviews*, *60*, 876–85.

[245] Henderson, I.C., Bhatia, V. (2007). Nab-paclitaxel for breast cancer: A new formulation with an improved safety profile and greater efficacy. *Expert Review of Anticancer Therapy*, *7*, 919–943.

[246] Ron, N., Cordia, J., Yang, A., Ci, S., Nguyen, P., Hughs, M., Desai, N. (2008). Comparison of physicochemical characteristics and stability of three novel formulations of paclitaxel: Abraxane, Nanoxel, and Genexol PM. *Proceedings of the American Association for Cancer Research*, 49, Abst. 5622.

[247] Oerlemans, C., Bult, W., Bos, M., Storm, G., Nijsen, J.F.W., Hennick, W.E. (2010). Polymeric micelles in anticancer therapy: Targeting, imaging and triggered release. *Pharmaceutical Research*, *27*, 2569–2589.

[248] Aapro, M. (1996). The scientific rationale for developing taxoids. *Anti-Cancer Drugs*, *7*, 33–36.

9

CANCER CHEMOPREVENTIVE ACTIVITY OF HIGHER PLANTS

A. Douglas Kinghorn, Yulin Ren, Jie Li, and Chung Ki Sung

9.1 INTRODUCTION

Cancer is a complex disease state and major public health problem affecting the inhabitants of both developed and developing countries. In the United States for the year 2010, it was estimated that over 1.5 million new cases of cancer would be diagnosed, and that there would be more than half a million fatalies as a result of cancer [1]. Moreover, cancer now accounts for more deaths than heart disease in the United States in persons under 85 years of age [1]. However, cancer incidence rates have declined somewhat recently for several major types in both males and females, and it is recommended that further progress in ameliorating this group of diseases can be made through newly available discoveries in detection, prevention, and treatment [1].

"Cancer chemoprevention" refers to the use of specific nontoxic synthetic or naturally occurring chemicals to reverse, suppress, or prevent the phases of initiation, promotion, or progression of cancer [2,3]. The field of chemoprevention is multidisciplinary, involving both epidemiological observations and laboratory studies, with the latter enhanced by advances in phytochemistry and the refinement of relevant *in vitro* and *in vivo* bioassay procedures. Determinations of the putative mechanisms of action of potential cancer chemopreventive agents and the search for possible biomarker substances related to this type of activity have been aided by developments in molecular biology [4–7]. Although only a relatively small number of cancer chemopreventive agents are used as prescription drugs for this purpose, a wide range of synthetic and naturally occurring compounds has chemopreventive potential,

Plant Bioactives and Drug Discovery: Principles, Practice, and Perspectives, Fourth Edition.
Edited by Valdir Cechinel-Filho.

representing varied structural types and mechanisms of action, and currently several of these are undergoing preclinical or clinical (human intervention) trials under the auspices of the U.S. National Cancer Institute, Bethesda, Maryland [8].

There has been a long-standing interest in the biological properties of phenolic [9,10] and other constituents of fruits, vegetables, teas, spices, and medicinal herbs [11,12]. Numerous contemporary reviews are now available in the literature in relation to the association of the consumption of certain fruits and vegetables (dietary and other edible plants) and/or their purified constituents with a decreased risk of acquiring specific cancers, for example [13–24]. Compounds from plants have been found to interact with all of the three overall steps of the carcinogenesis phenomenon (intiation, promotion, and progression) [14]. Moreover, they have been found to act as both "blocking agents" (i.e., by preventing carcinogens from reaching target sites, such as through inhibiting metabolic activation) and "suppressing agents" (i.e., inhibiting the malignant transformation of initiated cells in either the promotion or progression stage) [14]. The consumption of certain dietary plants or their bioactive "secondary metabolite" principles has been correlated with the chemoprevention of selected types of cancer, such as of colon [25], oropharyngeal [26], and prostate [27]. Strategies for the discovery of novel naturally occurring small organic molecule cancer chemopreventive agents, using the technique of "activity-guided" chromatographic fractionation, have been developed [19,28].

In this chapter, summaries of the recent literature on selected dietary plants and their potential cancer chemopreventive constituents will be provided, with pertinent new information on their mechanisms of action briefly mentioned. Owing to the now very large literature on potential cancer chemopreventive agents from plants, no attempt will be made to provide a comprehensive treatment of this very active research area, with only a few relevant citations cited per plant/compound to be discussed. However, it is hoped that the higher plants and their constituents that have been selected will provide the reader with a useful overview of the structural range of the secondary metabolites of current interest in chemoprevention, especially those compounds under clinical evaluation.

9.2 POTENTAL CANCER CHEMOPREVENTIVE AGENTS FROM SELECTED DIETARY HIGHER PLANTS

As mentioned above, dietary agents may play an important role in investigations on cancer chemoprevention. These agents often possess antioxidant effects and may be safe and economical to use. A tabulation of selected higher plants and their potential chemopreventive components, based on extensive *in vitro* and *in vivo* biological testing, is summarized in Table 9.1. The relevant structures are shown in Figure 9.1. In the following paragraphs, these will be discussed in turn, with emphasis on the more recent relevant literature.

The bulbs of *Allium sativum* L. (Alliaceae) (garlic) are used as a culinary and medicinal plant and contain a complex profile of organosulfur compounds, of which some are labile to enzymatic action during crushing of the cloves [29]. The potential

TABLE 9.1 Higher Plants with Potential Cancer Chemopreventive Activity and Their Constituent(s) of Interest

Species (family)	Compound (code)	References
Allium sativum L. (Alliaceae)	Diallyl sulfide (**1**)	[30–36]
	Diallyl disulfide (**2**)	
Brassica oleracea L. (Brassicaceae)	Brassinin (**3**)	[37–57]
	Indole-3-carbinol (**4**)	
	3,3′-diindolylmethane (**5**)	
	Allyl isothiocyanate (**6**)	
	Phenethyl isothiocyanate (**7**)	
	Sulforaphane (**8**)	
Camellia sinensis (L.) Kuntze (Theaceae)	Epigallocatechin gallate (**9**)	[58–63]
Capsicum annuum L. (Solanaceae)	Capsaicin (**10**)	[64–67]
Citrus species (Rutaceae)	Nobiletin (**11**)	[68–77]
	Apigenin (**12**)	
Curcuma longa L. (Zingiberaceae)	Curcumin (**13**)	[78–81]
	Demethoxycurcumin (**14**)	
	Bisdemethoxycurcumin (**15**)	
Glycine max (L.) Merr. (Fabaceae)	Genistein (**16**)	[82–87]
Glycyrrhiza species (Fabaceae)	Isoliquiritigenin (**17**)	[88–94]
Rosmarinus officinalis L. (Lamiaceae)	Carnosol (**18**)	[59,95–97]
Silybum marianum (L.) Gaertn. (Asteraceae)	Silibinin (**19**)	[98–103]
	Silychristin (**20**)	
	Silydianin (**21**)	
Solanum lycopersicum L. (Solanaceae)	Lycopene (**22**)	[104–109]
Vaccinium myrtillus L. (Ericaceae)	Cyanidin 3-*O*-glucoside (**23**)	[121–124]
Vitis vinifera L. (Vitaceae)	Resveratrol (**24**)	[125–131]
	trans-Astringin (**25**)	
	trans-Piceid (**26**)	
Zingiber officinale Roscoe (Zingiberaceae)	[6]-gingerol (**27**)	[132–137]

of certain constituents of garlic in cancer chemoprevention has been demonstrated some time ago [30], and relevant reviews have appeared recently [31,32]. Of the naturally occurring or artifactual constituents associated with *A. sativum*, two bioactive substances associated with chemopreventive effects are diallyl sulfide (**1**) and diallyl disulfide (**2**). For example, diallyl sulfide (**1**) was found in earlier work to inhibit benz[*a*]pyrene-induced neoplasia of the forestomach and lungs in female A/J mice [30]. Mechanistically, this compound down-regulates levels of dimethyl[*a*] benzanthracene (DMBA)-induced p21/ras protein in mouse skin tumors [33]. Compound **1** was also studied for its mode of action against Colo 320 DM colon cancer cells and was found to upregulate nuclear factor kappa B (NF-κB) and to promote the expression of caspase-3 and the suppression of extracellular regulatory kinase-2 (ERK-2) [34]. In turn, administration of diallyl disulfide (**2**) in a rat prostate cancer model induced with testosterone, and *N*-methyl-*N*-nitrosourea resulted in a reduced incidence of prostatic intraepithelial neoplasia, when compared with controls [35]. When evaluated against breast, lung, and prostate human cancer cells, compound **2** caused apoptosis through a Bax-triggered mitochrondrial pathway, independent of caspase-3 [36].

Consumption of cruciferous vegetables, principally *Brassica oleracea* L. (Brassicaceae) and its cultivars (Brussels sprouts, cabbage, cauliflower, collard greens, kale, kohlrabi), has been associated with the reduction of the risk of several cancers (e.g., bladder, breast, colon, lung, and prostate) [37–39]. Plants in the family Brassicaceae accumulate glucosinolates, responsible in part for their pungent-tasting properties, which are glucosides subject to enzymatic hydrolysis [29]. The biological effects of cruciferous vegetables have been attributed to both indole derivatives, such as brassinin [*N*-(1*H*-indol-3-yl-methyl)-carbamodithioic acid methyl ester, **3**], indole-3-carbinol (1*H*-indole-3-methanol, **4**) and 3,3′-diindolylmethane (**5**), and also isothiocyanates, inclusive of allyl isothiocyanate (3-isothiocyanato-1-propene, **6**), phenethyl isothiocyanate [2-(isothiocyanatoethyl)-benzene, **7**], and sulforaphane [1-isothiocyanato-4-(methylsufinyl)-butane, **8**]. These compounds will be mentioned briefly in turn.

Brassinin (**3**) is a phytoalexin from cabbage and is thus produced in the plant as a stress metabolite. A chemically synthesized form of this substance showed dose-dependent inhibition of preneoplastic alveolar lesions induced by DMBA in a mouse mammary carcinoma *ex vivo* assay, followed up by inhibition of skin tumors in a full-term tumorigenesis study in mouse skin, as initiated by DMA and promoted by 12-*O*-tetradecanoylphorbol 13-acetate [40]. In a more recent study, a mechanism for the cancer chemopreventive activity of compound **3** was found to be inhibition of indoleamine 2,3-dioxygenase (IDO) [41]. Indole-3-carbinol (**4**, I3C), a product of the hydrolysis of glucobrassicin found in cabbage, has been associated with the chemoprevention of various types of cancer for some time [14,37,42,43]. This compound was found to influence the NF-κB pathway in human breast cancer cells in a cell-specific manner, not related to apoptosis [44]. 3,3′-Diindolylmethane (**5**, DIM) is a further product of glucobrassicin, with its immediate precursor being compound **4** [44]. In an early experiment, compound **5** was shown to reduce the incidence of DMBA-induced mammary tumors in rats [43]. This compound inhibits DNA synthesis and cell proliferation in MCF-7 and MDA-MB-231 human breast cancer cells and induces apoptosis independent of estrogen receptor status, in a manner mediated by the expression of the Bax/Bcl-2 apoptotic regulatory factors [45]. Recently, compound **5** was shown to selectively induce proteasome-mediated degradation of class I histone deacetylases (HDACs) without affecting the class II HDAC proteins, both *in vitro* and *in vivo* in tumor xenografts [46].

Allyl isothiocyanate (**6**) is a further volatile hydrolysis product found in cruciferous vegetables, inclusive of black mustard (*Brassica nigra* W.J.D. Koch) [29]. This compound showed an antineoplastic effect by retarding the growth of PC-3 human prostate xenografts *in vivo*, when administered (intraperitoneal injection, 10 μM) at the time of tumor implantation and three times a week subsequently until the completion of the experiment [47]. Compound **6** was found to be active in an othotopic model of rat bladder cancer at an oral dose of 1 mg/kg, but not active in subcutaneous xenograft of the same cells, also in rats [48]. Phenethyl isothiocyanate (**7**) was found to be effective in a tumorigenesis inhibition study in a mouse colon cancer model initiated with azoxymethane and promoted with dexrtan sodium sulfate [49]. Several reviews are available on the potential cancer chemopreventive effects of compounds **6** and **7** and their putative cellular targets [50–53].

Sulforaphane (**8**) is a further isothiocyanate with potential cancer chemopreventive activity and has been obtained from broccoli, with its glucosinolate precursor being glucoraphanin [29]. This compound was shown initially to induce phase 2 detoxification enzymes in murine hepatoma cells, and also to block the formation of

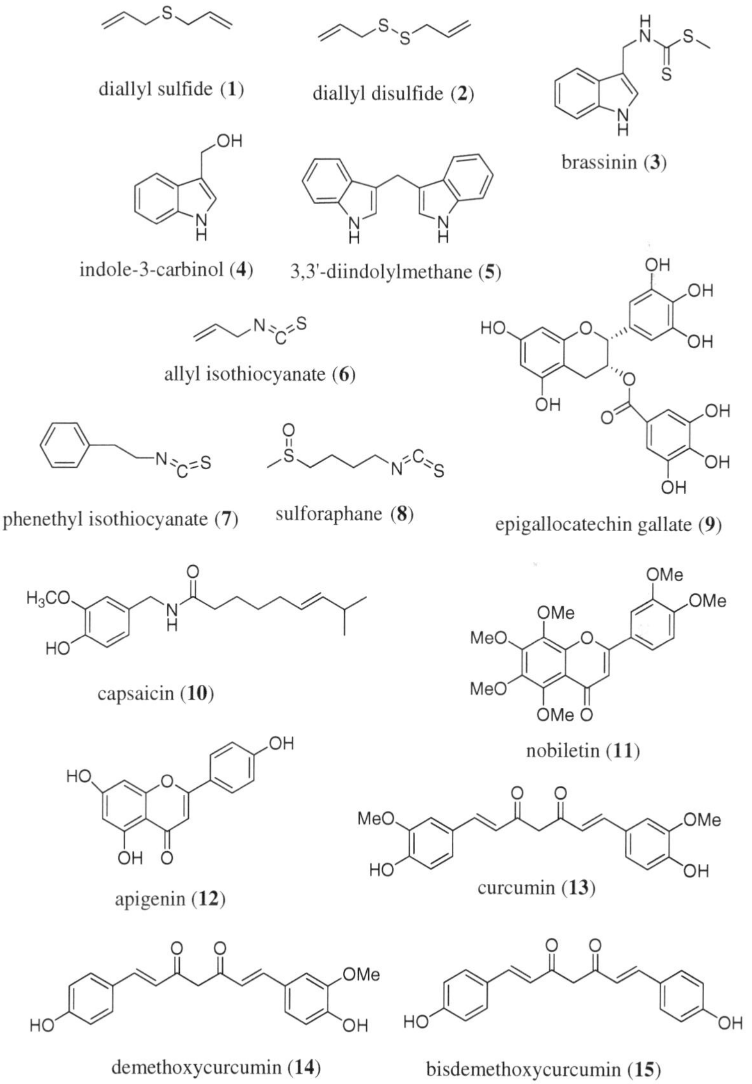

FIGURE 9.1 Structures of plant-derived compounds with potential cancer chemopreventive activity.

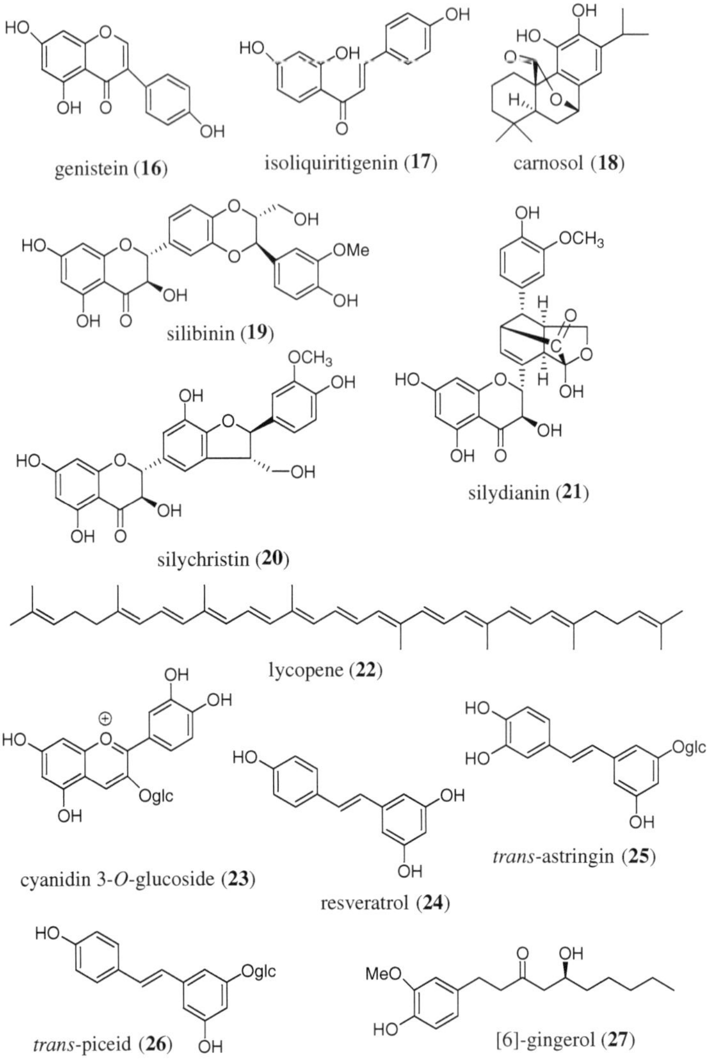

FIGURE 9.1 *(Continued)*

mammary tumors in rats treated with DMBA [54]. There is now a very extensive literature on this compound, with a number of reviews being available on its anticarcinogenic activity and mechanisms, for example [52,55]. The compound shows considerable promise for being an effective chemopreventive agent for several different types of cancer, for example [56,57].

Green or unfermented tea [*Camellia sinensis* L. Kuntze (Theaceae)] is one of the most well documented of potential natural product cancer chemopreventive agents, with a voluminous literature in support of such effects, inclusive of epidemiological studies, for example [14,27,58]. Green tea is produced by steaming and drying the leaves to prevent oxidation and therefore has a different profile of phenolic constituents to black tea, which is fermented, thereby permitting enzymatic oxidation of these compounds [29]. Green tea phenols (also known as polyphenols) include (−)-epigallocatechin 3-gallate (EGCG, **9**) as the major such compound in terms of percentage yield [59]. Compound **9** has been reviewed as a purified entity relative to its potential as a chemopreventive and/or health-promoting chemical, for example [60,61]. *In vitro* and *in vivo* evidence for the chemopreventive action of EGCG in the laboratory setting has been summarized [62,63]. Numerous mechanisms have been proposed for the activity of this polyphenol, including the induction of apoptosis and cell cycle arrest, and the inhibition of the metabolic activation of carcinogens and scavenging of reactive oxygen species. Compound **9** also inhibits NF-κB and activator protein 1 (AP-1) and modulates the activity of receptor tyrosine kinases (RTKs) [14,61–63].

Capsaicin (8-methyl-*N*-vanillyl-6(*E*)-nonenamide, **10**), a major pungent-tasting amide constituent of *Capsicum annuum* L. (Solanaceae) (chilli pepper), was suspected in early work as being carcinogenic or co-carcinogenic because of its irritant properties [14,64]. However, more recently, this amide has become of some interest as a potential cancer chemopreventive agent [14,64]. This substance induces apoptotic cell death in HT-29 human colon cancer cells by activation of the peroxisome proliferator-activated receptor (PPAR) γ pathway [65]. Capsaicin (**10**) treatment showed a protective role in terms of lowering tumor incidence in a mouse model of lung cancer induced by benz(*a*)pyrene and reduced the levels of several marker enzymes [66] and has been found to suppress chemical carcinogen-induced oxidative damage in rats [67].

There is an increasing interest on the possible chemopreventive activity of plants of the genus *Citrus* (Rutaceae), with considerable attention focusing on the methoxylated flavone, nobiletin (**11**) in this regard [68,69]. Nobiletin (5,6,7,8,3′,4′-hexamethoxyflavone), isolated initially from *Citrus nobilis* Lour. (king orange), is also a constituent of *C. aurantium* L. (bitter orange). This compound suppressed the expression of cyclooxygenase-2 (COX-2) and inducible NO synthase proteins and prostaglandin E2 release and inhibited DMBA-induced and 12-*O*-tetradecanoylphorbol-13-acetate (TPA)-promoted skin tumor formation *in vivo* in mice [70]. Another *in vivo* study showed that feeding nobiletin (**11**) to male ICR mice suppressed the incidence and numbers of colonic malignancies induced by a combination of azoxymethane and dextran sulfate sodium. These effects of compound **11** have been related to the regulation of leptin levels [71]. Walle has pointed out that methoxylated flavonoids like nobiletin (**11**) show a greater oral bioavailability than the more common hydroxylated flavonoids found throughout the plant kingdom, which enhances their potential value as cancer chemopreventives [69].

Apigenin (5,7,4′-trihydroxyflavone, **12**) is a component of not only *Citrus* species such as *Citrus aurantium* L. (bitter orange) and *C. limon* (L.) Burm.f. (lemon) (Rutaceae), but has been found in other dietary plants. This compound has been of interest as a potential cancer chemopreventive for some time and still remains of interest in this regard [72,73]. In earlier work, Birt and associates showed that apigenin (**12**) could inhibit UV-induced mouse skin carcinogenesis when applied topically, with this compound inducing a reversible G2/M arrest in cultured keratinocytes, partially due to the perturbation of cyclin B1 levels [74,75]. More recently, apigenin has been demonstrated to inhibit migration and invasion of human ovarian cancer A2780 cells through focal adhesion kinase (FAK) [76] and to retard oxidative stress-induced DNA damage in hepatocellular carcinoma in Wistar albino rats induced with *N*-nitrosodiethylamine (NDEA) and phenobarbital [77].

There is an intense interest in the cancer chemopreventive potential of the spice *Curcuma longa* L. (Zingiberaceae) (turmeric) and its major diarylheptanoid (curcuminoid) yellow pigment constituent, curcumin [(1*E*,6*E*)-1,7-bis(4-hydroxy-3-methoxyphenyl)-1,6-heptadiene-3,5-dione, **13**], with the latter being one of the most highly investigated natural products in the biomedical literature [14,78–80]. However, much of the preclinical and clinical evaluation work that has been performed on commercially available curcumin has contained smaller quantities of the related curcuminoids, demethoxycurcumin [(1*E*,6*E*)-1-(4-hydroxyphenyl)-7-(4-hydroxy-3-methoxyphenyl)-1,6-heptadiene-3,5-dione, **14**] and bisdemethoxycurcumin [(1*E*,6*E*)-1,7-bis(4-hydroxyphenyl)-1,6-heptadiene-3,5-dione, **15**]. As an example of an earlier study on the chemopreventive properties of curcumin, this compound was found to be highly effective in a rat colon cancer model with tumor induction by azoxymethane (AOM), but only weakly active in a DMBA-induced rat mammary gland tumorigenesis protocol [81]. There are now several thousand preclinical investigations on curcumin reported in the literature, and this substance has been found to interact with a large number of molecular targets germane to its potential cancer chemopreventive activity. Some well-known examples of the cellular targets of curcumin (**13**) are inhibition of phorbol myristate acetate-induced activation of NF-κB and AP-1, and inhibition of epidermal extracellular-signal-regulated kinase (Erk1/2) [14]. While concern has been expressed about the incomplete absorption and low bioavailability of curcumin (**13**), it does appear that sufficient concentration levels of this compound reach the target organs in various human clinical trials [80].

Glycine max L. (Fabaceae) (soybean) is more commonly used in Asian cuisine than in the Western diet, and its ingestion has been linked to a reduced risk of certain types of cancer, based on epidemiological studies [14,82,83]. The possible benefits to health of soybeans are due to the presence of isoflavonoids, with the aglycone genistein (5,7,4′-trihydroxyisoflavone, **16**) being prominent in this regard [84,85]. In an early study, genistein (**16**) was found to be effective in reducing tumor multiplicity, with a lesser effect on tumor incidence, in a rat mammary cancer model induced with *N*-methyl-*N*-nitrosourea (NMU) [86]. Genistein (**16**) acts on mammalian cancer cells by modulating genes related to the cell cycle and apoptosis, and inhibits the activation of the NF-κB and Akt signaling pathways [14,87]. However, genistein (**16**) is also a known phytoestrogen that interacts with androgen- and estrogen-mediated signaling pathways [85,87].

Licorice species (*Glycyrrhiza glabra* L. and *G. uralensis* Fisch. ex DC.; Fabaceae) are very well-investigated plants used in sweetening and flavoring, and for medicinal purposes [88,89]. Glycyrrhizin is a major component of the roots of both species and is an oleanane-type triterpenoid glycoside with sweet-tasting and flavor-modifying properties [88,89]. A second major group of licorice root constituents are flavonoids, of which many are prenylated molecules [90]. A licorice flavonoid constituent that has emerged as a potential cancer chemopreventive agent is the chalcone, isoliquiritigenin (4,2′,4′-trihydroxychalcone, **17**). Thus, in an initial study, this compound caused apoptosis and suppressed colon carcinoma cell growth and azoxymethane (AOM)-induced colon carcinogenesis in ddY mice and was found to inhibit the induction of preneoplastic aberrant crypt foci (ACF) in the male F344 rat colon [91]. In work performed at The Ohio State University, compound **17** was synthesized and evaluated in a 1,2-dimethylhydrazine (DMH)-induced mouse colon and lung model, and tested at various dose levels in the diet. It was found that at a 300 mg/kg dose, when administered under a standard protocol, the multiplicity of both colon and lung cancers was reduced on a statistical basis relative to controls [92]. Very recently, in a full-term tumorigenesis study using DMBA in female Sprague-Dawley rats, isoliquiritigenin (**17**) was reported to slightly increase tumor latency but to have a negative effect on mammary tumor incidence after about 65 days of carcinogen administration [93]. This compound induced quinone reductase-1 activity in the colon as well as glutathione and glutathione *S*-transferase, in the liver of the test animals [93]. Isoliquiritigenin (**17**) has been found to restrain cell-cycle progression at the G2/M phase and enhance the expression of p21CIP1/WAF1 in human lung cancer cells [94].

The culinary herb, rosemary (*Rosmarinus officinalis* L.; Lamiaceae) is a widely used spice in food processing, and it contains several phenolic diterpenoid constituents with antioxidant properties, inclusive of carnosol (7β,11,12-trihydroxy-13-isopropyl-podocarpa-8,11,13-trien-17-oic acid-17,7-lactone, **18**), along with other chemical classes of secondary metabolites such as triterpenoids [59]. Both rosemary extract and carnosol (**18**) have been found to inhibit DMBA-induced mammary carcinogenesis in female rats and to retard DMBA-DNA adduct formation *in vivo* [95]. Mechanistically, carnosol (**18**) suppresses inducible nitric oxide synthase by down-regulating NF-κB in mouse macrophages [96]. It has also been found that compound **18** activates human peroxisome proliferator-activated receptor gamma (PPARγ) [97].

The medicinal plant, *Silybum marianum* Gaertn. (Asteraceae) (milk thistle), has long been of interest as a potential hepatoprotective agent [98]. However, more recently, the flavonolignans in this plant have attracted attention as potential cancer chemopreventive agents, especially in terms of potential effects on prostate cancer [99,100]. "Silymarin" is a standardized extract of the seeds of *S. marianum* and contains silibinin {also known as silybin; 2-[(2*R*,3*R*)-2,3-dihydro-3-(4-hydroxy-3-methoxyphenyl)-2-(hydroxymethyl)-1,4-benzodioxin-6-yl]-2,3-dihydro-3,5,7-trihydroxy-(2*R*,3*R*)-4*H*-1-benzopyran-4-one, **19**}, silychristin {(2-[(2*R*,3*S*)-2,3-dihydro-7-hydroxy-2-(4-hydroxy-3-methoxyphenyl)-3-(hydroxymethyl)-5-benzofuranyl]-2,3-dihydro-3,5,7-trihydroxy-(2*R*,3*R*)-4*H*-1-benzopyran-4-one, **20**}, and silydianin {4-[(2*R*,3*R*)-3,4-dihydro-3,5,7-trihydroxy-4-oxo-2*H*-1-benzopyran-2-yl]-2,3,3a,7a-tetrahydro-7a-hydroxy-8-(4-hydroxy-3-methoxyphenyl)-(3*R*,3a*R*,6*R*,7a*R*,8*R*)-3,6-methanobenzofuran-7(6*H*)-one, **21**}, as the main flavonolignan constituents, in a ratio of 3:1:1 [98]. Silybinin itself is a mixture of the diastereomeric pair of compounds, silybin A and silybin B, and separation methodology has been developed to resolve the latter substances and other isomeric components [101,102]. In a carcinogenesis inhibition study, it was found that preapplication of silymarin in a DMBA-initiated and 12-*O*-tetradecanoylphorbol 13-acetate-promoted mouse skin tumorigenesis model resulted in reductions of tumor incidence and tumor multiplicity [103]. Silymarin induces cyclin-dependent kinase inhibitors (e.g., p15, p21, and p27), downregulates antiaptotic gene products such as Bcl2), and inhibits cell-survival kinases (akt, POKC, and MAPK) and the inflammatory transcription factor, NF-κB [100].

Solanum lycopersicum L. (Solanaceae) (tomato) is a very important dietary plant worldwide, with a number of perceived health benefits [14,104]. Of the tomato constituents with potential chemopreventive activity, the red-colored carotenoid, *trans*-lycopene [(all-*E*)-2,6,10,14,19,23,27,31-octamethyl-2,6,8,10,12,14,16,18,20, 22,24,26,30-dotriacontatridecaene, or ω,ω-carotene, **22**] has attracted considerable attention, particularly in terms of preventing prostate cancer [105–107]. Lycopene (**22**), an all-*trans* double bond carotene derivative, is known to protect the plants in which it occurs against photo-oxidative damage, and it is able to quench singlet oxygen and peroxy radicals, to minimize cell damage [29]. The lipid antioxidant activity of lycopene (**22**) is thus thought to provide the basis for the efficacy of this substance as a chemopreventive agent, as exemplified by the reduction of DMBA-induced buccal pouch squamous cell carcinomas in a hamster model [108]. There is some debate as to whether tomato paste and other foods containing lycopene are more efficacious than purified lycopene itself, when used in cancer chemoprevention supplementation studies [109].

There is now a very large literature awareness of the potential cancer chemopreventive properties of various types of berries used in the diet, for example [22,110–112]. In addition to commonly consumed fruits with potent antioxidant activities such as bilberry, blackberry, blueberry, cranberry, raspberry (both black and red), and strawberry, a number of exotic fruits such as açai, mangosteen, noni, and wolfberry (goji berry) are also of great current interest [112–116]. Attention has been focused on the potential cancer chemopreventive properties of anthocyanins, the main antioxidant and colored flavonoid glycoside components of berries, which show potential cancer chemopreventive activity both *in vitro* and *in vivo* [117–120]. Bilberry (*Vaccinium myrtillus* L.; Ericaceae) may be taken as an example of a berry for the purposes of this relatively short chapter. Thus, an anthocyanin-rich extract of bilberry was found to inhibit the growth of HT-29 human colon cancer cells when tested in a sulforhodamine B assay, while the growth of the nontumorigenic NCM460 colon cell culture was not affected by this extract [121]. It was found in an ApcMin mouse model that the development of intestinal adenoma formation was reduced either by the ingestion of either a standardized anthocyanin mixture from bilberries or its principal anthocyanidin constituent, cyanidin 3-*O*-glucoside (**23**) [122]. Mechanistically, an anthocyandin-enriched extract of bilberry suppressed the kinase activity of receptor tyrosine kinases inclusive of epidermal growth factor receptor (EGFR) and vascular endothelial growth factor 2 (VEGFR-2) [123]. It has been concluded that the glycosidic anthocyanins from bilberries and related fruits offer better prospects for development as cancer chemopreventive agents than their more chemically unstable anthocyanidin aglycone derivatives [124].

There is a very great interest in the potential health benefits of consuming grapes (*Vitis vinifera* L. (Vinaceae) and certain red wines, due to the presence of the stilbenoid constituent, *trans*-resveratrol (3,5,4′-trihydroxy-,*trans*-stilbene, **24**) [14]. A number of recent reviews have appeared on the potential cancer chemopreventive activity of resveratrol [125–128]. In collaborative work carried out at the University of Illinois at Chicago over a decade ago, resveratrol was found to act as an antioxidant, and to inhibit cyclooxygenase-2 (COX-2), and also to inhibit preneoplastic alveolar lesion formation in carcinogen-treated mouse mammary glands in organ culture [129]. Furthermore, in a full-term tumorigenesis study using a mouse skin model, resveratrol (**24**) was found to show significant inhibitory activity [129]. In a follow-up study, 11 other phenolic constituents of grapes, of the flavan and stilbenoid structural types, were compared with compound **24** for their effects in inhibiting cyclooxygenase-1 (COX-1), COX-2, and DMBA-induced alveolar lesions of the mouse mammary glands. It was found that while resveratrol (**24**) was active in all three assays, the stilbenoids *trans*-astringin (3,4,3′,5′-tetrahydroxy-*trans*-stilbene 3′-glucoside, **25**) and *trans*-piceid (3,5,4′-trihydroxy-,*trans*-stilbene 3-glucoside, **26**) inhibited alveolar lesions in the mouse mammary gland organ culture assay by 68.8% and 76.9% at a dose of 10 μg/ml [130]. Mechanistically, resveratrol interacts with numerous other cellular targets, including AP1 transcriptional activity and the activation of protein tyrosine kinase C (PKC), and NF-κB [14]. In addition to the stilbenoid constituents of grapes, the polyphenols in grape seed extract are also of interest as a potential chemopreventive agent [131].

The rhizomes of *Zingiber officinale* Roscoe (Zingiberacaeae) (ginger) and the major constituent, [6]-gingerol [5(*S*)-hydroxy-1-(4-hydroxy-3-methoxyphenyl)-3-decanone, **27**], are well known for their potential cancer chemopreventive activity, with such activity linked to various possible mechanisms [14,18]. In some recent work performed, ginger was administered orally at various stages of the induction of colon carcinogenesis by 1,2-dimethylhydrazine (DMH) in male Wistar rats. This supplementation by ginger led to reduced plasma lipid peroxidation and a decreased cancer incidence [132]. In a follow-up study using the same experimental model, it was concluded that colon cancer occurrence in DMH-induced rats may be reduced by the hypolipidemic and antioxidative effects of ginger [133]. Furthermore, it was demonstrated that ginger dietary supplementation reduced carcinogenesis in DMH-induced rats by lowering the activities of the microbial enzymes β-glucuronidase and mucinase [134]. Ginger supplementation also suppressed liver carcinogenesis by scavenging the free radical formation and reducing lipid peroxidation in ethionine induced male Wistar rats [135]. In turn, [6]-gingerol (**27**) was found to suppress tumor growth in HCT116 colorectal cancer cells and in an *in vivo* model in nude mice, as a result of the inhibition of leukotriene A4 hydrolase (LTA4H) activity [136]. A review has appeared summarizing the presently known molecular targets of [6]-gingerol (**27**) in relation to its anticarcinogenic and other biological activities, inclusive of effects on the expression of inducible nitric oxide synthase (iNOS), tumor necrosis factor alpha (TNF-α), 1-kappaB alpha (IκBα) phosphorylation, and on the translocation of NF-κB [137].

9.3 CONCLUSIONS

It is documented in this chapter that selected fruits, vegetables, spices, and herbal remedies, as well as their constituents that are representative of considerable structural diversity, exhibit promising cancer chemopreventive activity in *in vivo* tumorigenesis models, and have varied mechanisms of action. Some are currently in clinical trial as potential cancer chemopreventive agents under the auspices of the U.S. National Cancer Institute, including the carotenoid, lycopene (**22**, prostate cancer), the curcuminoid mixture, curcumin (**13**), demethoxycurcumin (**14**), and bis-demethoxycurcumin (**15**, colon cancer), the green tea catechins, inclusive of epigallocatechin gallate (**9**, colon, lung, prostate, and skin cancers), the indoles, indole-3-carbinol (**3**) and 3,3′-diindolylmethane (**5**, colon cancer), the isothiocyanate, phenethyl isothiocyanate (**7**, colon cancer), the stilbenoid, resveratrol (**24**, colon cancer), and the soy isoflavones, inclusive of genistein (**16**, breast and colon cancers) [8]. Previous work directed toward the discovery of novel cancer chemopreventive agents from edible plants has suggested that secondary metabolites of several other structural classes may ultimately afford compounds suitable for further clinical trials, such as certain substances of the alkaloid, amide, porphyrin, rotenoid, triterpenoid, and withanolide classes [19,29]. Moreover, organisms other than plants, such as terrestrial fungi [138] and marine microbes [139], may also afford future molecules with potential as cancer chemopreventives. Interest in the field of cancer chemoprevention is well established in the general biomedical area as well as in the more specialized field of natural products chemistry. Therefore, it is to be hoped that additional promising examples of natural product-derived constituents will be evaluated preclinically and clinically for this purpose.

ACKNOWLEDGMENTS

Certain phytochemical and biological work described in this chapter that was carried out in the laboratory of the senior author (A.D.K.) was aided by faculty start-up funding from the Molecular Carcinogenesis and Chemoprevention Program of The Ohio State University Comprehensive Cancer Center. C.K.S. was supported while on a sabbatical year at The Ohio State University by a stipend from Chonnam National University.

REFERENCES

[1] Jemal, A., Siegel, R., Xu, J., Ward, E. (2010). Cancer statistics, 2010. *CA – A Cancer Journal for Clinicians*, *60*, 277–300.

[2] Sporn, M.B., Dunlop, N.M., Newton, D.L., Smith, J.M. (1976). Prevention of chemical carcinogenesis by vitamin A and its synthetic analogs (retinoids). *Federation Proceedings*, 35, 1332–1338.

[3] Wattenburg, L.W. (1985). Chemoprevention of cancer. *Cancer Research*, *45*, 1–8.

[4] Sporn, M.B., Suh, N. (2000). Chemoprevention of cancer. *Carcinogenesis*, *21*, 525–530.

[5] Levi, M.S., Borne, R.F., Williamson, J.S. (2001). A review of cancer chemopreventive agents. *Current Medicinal Chemistry*, *8*, 1349–1362.

[6] Kelloff, G.J., Hawk, E.T., Sigman, C.C., Cancer chemoprevention. *Promising Cancer Chemopreventive Agents*, Vol. *1*, Humana Press, Inc., Totowa, NJ, 2004.

[7] Kelloff, G.J., Hawk, E.T., Sigman, C.C., Cancer chemoprevention. *Strategies for Cancer Chemoprevention*, Vol. *2*, Humana Press, Inc., Totowa, NJ, 2005.

[8] National Cancer Institute, Division of Cancer Prevention. Chemopreventive Agents. (Accessed November 30, 2010). Available at http://prevention.cancer.gov/programs-resources/resources/agents.

[9] Ho, C.-T., Lee, C.Y., Huang, M.-T., Phenolic Compounds in Food and Their Effects on Health: I. Analysis, Occurrence, and Chemistry. *Symposium Series 506*, American Chemical Society Books, Washington, DC, 1992.

[10] Huang, M.-T., Ho, C.-T., Lee, C.Y., Phenolic Compounds in Food and Their Effects on Health: II. *Antioxidants and Cancer Prevention. Symposium Series 507*, American Chemical Society Books, Washington, DC, 1992.

[11] Huang, M.-T., Osawa, T., Ho, C.-T., Rosen, R.T., Food Phytochemicals for Cancer Prevention: I. Fruits and Vegetables. Symposium Series 546, American Chemical Society Books, Washington, DC, 1994.

[12] Ho, C.-T., Osawa, T., Huang, M.-T., Rosen, R.T., Food Phytochemicals for Cancer Prevention: II. Teas, Spices, and Herbs. Symposium Series 547, American Chemical Society Books, Washington, DC, 1994.

[13] Reddy, L., Odhav, B., Bhoola, K.D. (2003). Natural products for cancer prevention: A global perspective. *Pharmacology and Therapeutics*, *99*, 1–13.

[14] Surh, Y.-J. (2003). Cancer chemoprevention with dietary phytochemicals. *Nature Reviews Cancer*, *3*, 768–780.

[15] Dorai, T., Aggarwal, B.B. (2004). Role of chemopreventive agents in cancer therapy. *Cancer Letters*, *215*, 129–140.

[16] Howells, L.M., Manson, M.M. (2005). Prospects for plant-derived chemopreventive agents exhibiting multiple mechanisms of action. *Current Medicinal Chemistry: Anti-Cancer Agents*, *5*, 201–213.

[17] Khan, N., Adhami, V.M., Mukhtar, H. (2008). Apoptosis by dietary agents for prevention and treatment of cancer. *Biochemical Pharmacology*, *76*, 1333–1339.

[18] Aggarwal, B.B., Kunnumakkara, A.B., Harikumar, K.B., Tharakan, S.T., Sung, B., Anand, P. (2008). Potential of spice-derived phytochemicals for cancer prevention. *Planta Medica*, *74*, 1560–1569.

[19] Kinghorn, A.D., Su, B.-N., Jang, D.S., Chang, L.C., Lee, D., Gu, J.-Q., Carcache-Blanco, E.J., Pawlus, A.D., Lee, S.K., Park, E.J., Cuendet, M., Gills, J.J., Bhat, K., Park, H.-S., Mata-Greenwood, E., Song, L.L., Jang, M., Pezzuto, J.M. (2004). Natural inhibitors of carcinogenesis. *Planta Medica*, *70*, 691–705.

[20] Stan, S.D., Kar, S., Stoner, G.D., Singh, S.V. (2008). Bioactive food components and cancer risk reduction. *Journal of Cellular Biochemistry*, *104*, 339–356.

[21] Ruhul Amin, A.R.M., Kucuk, O., Khuri, F.R., Shin, D.M. (2009). Perspectives for cancer prevention with natural compounds. *Journal of Clinical Oncology*, *27*, 2712–2725.

[22] Stoner, G.D. (2009). Foodstuffs for preventing cancer: The preclinical and clinical development of berries. *Cancer Prevention Research*, *2*, 187–194.

[23] Gullett, N.P., Ruhul Amin, A.R.M., Bayraktar, S., Pezzuto, J.M., Shin, D.M., Khuri, F.R., Aggarwall, B.B., Surh, Y.-J., Kucuk, O. (2010). Cancer prevention with natural compounds. *Seminars in Oncology*, *37*, 258–281.

[24] Mehta, R.G., Murillo, G., Naithani, R., Peng, X. (2010). Cancer chemoprevention by natural products: How far have we come? *Pharmaceutical Research*, *27*, 950–961.

[25] Rajamanickam, S., Agarwal, R. (2008). Natural products and colon cancer: Current status and future prospects. *Drug Development Research*, *69*, 460–471.

[26] Davidson, P.G., Touger-Decker, R. (2009). Chemopreventive role of fruits and vegetables in oropharyngeal cancer. *Nutrition in Clinical Practice*, *24*, 250–260.

[27] Syed, D.N., Suh, Y., Afaq, F., Mukhtar, H. (2008). Dietary agents for chemoprevention of prostate cancer. *Cancer Letters*, *265*, 167–176.

[28] Kinghorn, A.D., Su, B.-N., Lee, D., Gu, J.-Q., Pezzuto, J. M. (2003). Cancer chemopreventive agents discovered by activity-guided fractionation: An update. *Current Organic Chemistry*, *7*, 213–226.

[29] Dewick, P.M., *Medicinal Natural Products – a Biosynthetic Approach*, John Wiley & Sons, Inc., New York, 2002.

[30] Sparnins, V.L., Barany, G., Wattenberg, L.W. (1988). Effects of organosulfur compounds from garlic and onions on benzo[*a*]pyrene-induced neoplasia and glutathione *S*-transferase activity in the mouse. *Carcinogenesis*, *9*, 131–134.

[31] Shukla, Y., Kalra, N. (2007). Cancer chemoprevention with garlic and its constituents. *Cancer Letters*, *247*, 167–181.

[32] Nagini, S. (2008). Cancer chemoprevention by garlic and its organosulfur compounds – panacea or promise? *Anti-Cancer Agents in Medicinal Chemistry*, *8*, 313–321.

[33] Arora, A., Kalra, N., Shukla, Y. (2006). Regulation of p21/ras protein expression by diallyl sulfide in DMBA induced neoplastic changes in mouse skin. *Cancer Letters*, *242*, 28–36.

[34] Sriram, N., Kalayarasan, S., Ashokkumar, P., Sureshkumar, A., Sudhandiran, G. (2008). Diallyl sulfide induces apoptosis in Colo 320 DM human colon cancer cells: Involvement of caspase-3, NK-κB, and ERK-2. *Molecular and Cellular Biochemistry*, *311*, 157–165.

[35] Arunkumar, A., Vijayababu, M.R., Venkataraman, P., Senthilkumar, K., Arunakaran, J. (2006). Chemoprevention of rat prostate carcinogenesis by diallyl disulfide, an organosulfur compound of garlic. *Biological and Pharmaceutical Bulletin*, *29*, 375–379.

[36] Nagaraj, N.S., Anilakumar, K.R., Singh, O.V. (2010). Diallyl disulfide causes caspase-dependent apoptosis in human cancer cells through a Bax-triggered mitochondrial pathway. *Journal of Nutritional Biochemistry*, *21*, 405–412.

[37] Ashok, B.T., Tiwari, R.K. (2004). Cruciferous vegetables and cancer chemoprevention. *Recent Research Developments in Nutrition*, *6*, 83–94.

[38] Tang, L., Zirpoli, G.R., Guru, K., Moysich, K.B., Zhang, Y., Ambrosone, C.B., McCann, S.E. (2008). Consumption of raw cruciferous vegetables is inversely associated with bladder cancer risk. *Cancer Epidemiology, Biomarkers and Prevention*, *17*, 938–944.

[39] Vasanthi, H.R., Mukherjee, S., Das, D.K. (2009). Potential health benefits of broccoli – a chemico-biological overview. *Mini-Reviews in Medicinal Chemistry*, *9*, 749–759.

[40] Mehta, R.G., Liu, J., Constantinou, A., Thomas, C.F., Hawthorne, M., You, M., Gerhauser, C., Pezzuto, J.M., Moon, R.C., Moriarty, R.M. (1995). Cancer chemopreventive activity of brassinin, a phytoalexin from cabbage. *Carcinogenesis*, *16*, 399–404.

[41] Banerjee, T., DuHadaway, J.B., Gaspari, P., Sutanto-Ward, E., Munn, D.H., Mellor, A.L., Malachowski, W.P., Prendergast, G.C., Muller, A.J. (2008). A key *in vivo* antitumor mechanism of action of natural product-based brassinins is inhibition of indoleamine 2, 3-dioxygenase. *Oncogene*, *27*, 2851–2857.

[42] Rogan, E.G. (2006). The natural chemopreventive compound indole-3-carbinol: State of the science. *In Vivo, 20*, 221–228.

[43] Wattenberg. L.W., Loub, W.D. (1978). Inhibition of polycyclic aromatic hydrocarbon-induced neoplasia by naturally occurring indoles. *Cancer Research*, *38*, 1410–1413.

[44] Moiseeva, E.P., Heukers, R. (2008). Indole-3-carbinol-induced modulation of NF-κB signalling is breast cancer cell-specific and does not correlate with cell death. *Breast Cancer Research and Treatment*, *109*, 451–462.

[45] Hong, C., Firestone, G.L., Bjeldanes, L.F. (2002). Bcl-2 family-mediated apoptotic effects of 3,3′-diindolymethane (DIM) in human breast cancer cells. *Biochemical Pharmacology*, *63*, 1085–1097.

[46] Li, Y., Li, X., Guo, B. (2010). Chemopreventive agent 3, 3′-diindolylmethane selectively induces proteasomal degradation of class I histone deacetylases. *Cancer Research*, *70*, 646–654.

[47] Srivastava, S.K., Xiao, D., Lew, K.L., Hershberger, P., Kokkinakis, D.M., Johnson, C.S., Trump, D.L., Singh, S.V. (2003). Allyl isothiocyanate, a constituent of cruciferous vegetables, inhibits growth of PC-3 human prostate cancer xenografts *in vivo*. *Carcinogenesis*, *24*, 1665–1670.

[48] Bhattacharya, A., Tang, L., Li, Y., Geng, F., Paonessa, J.D., Chen, S.C., Wong, M.K.K., Zhang, Y. (2010). Inhibition of bladder cancer development by allyl isothiocyanate. *Carcinogenesis*, *31*, 281–286.

[49] Cheung, K.L., Khor, T.O., Huang, M.-T., Kong, A.-N. (2010). Differential *in vivo* mechanism of chemoprevention of tumor formation in azoxymethane/dextran sodium sulfate mice by PEITC and DBM. *Carcinogenesis*, *31*, 880–885.

[50] Conaway, C.C., Yang, Y.-M., Chung, F.-L. (2002). Isothiocyanates as cancer chemopreventive agents: Their biological activities and metabolism in rodents and humans. *Current Drug Metabolism*, *3*, 233–255.

[51] Zhang, Y. (2009). Allyl isothiocyanate as a cancer chemopreventive phytochemical. *Molecular Nutrition and Food Research*, *54*, 127–135.

[52] Cheung, K.L., Kong. A.-N. (2010). Molecular targets of dietary phenethyl isothiocyanate and sulforaphane for cancer chemoprevention. *AAPS Journal*, *12*, 87–97.

[53] Wang, L.G., Chaio, J.W. (2010). Prostate cancer chemopreventive activity of phenethyl isothiocyanate through epigenetic regulation. *International Journal of Oncology*, *37*, 533–539.

[54] Zhang, Y., Kensler, T.W., Cho, C.-G., Posner, G.H., Talalay, P. (1994). Anticarcinogenic activities of sulforaphane and structurally related synthetic norbornyl isothiocyanates. *Proceedings of the National Academy of Sciences of the United States of America, 91* 3147–3150.

[55] Juge, N., Mithen, R.F., Traka, M. (2007). Molecular basis for chemoprevention by sulforaphane: A comprehensive review. *Cellular and Molecular Life Sciences*, *64*, 1105–1127.

[56] Cornblatt, B.S., Ye, L., Dinkova-Kostova, A.T., Erb, M., Fahey, J.W., Singh, N.K., Chen, M.-S.A., Stierer, T., Garrett-Mayer, E., Argani, P., Davidson, N.E., Talalay, P.,

Kensler, T.W., Visvanathan, K. (2007). Preclinical and clinical evaluation of sulforaphane for chemoprevention in the breast. *Carcinogenesis*, *28*, 1485–1490.

[57] Dickinson, S.E., Melton, T.F., Olson, E.R., Zhang, J., Saboda, K., Bowden, G.T. (2009). Inhibition of activator protein-1 by sulforaphane involves interaction with cysteine in the cFos DNA-binding domain: Implications for chemoprevention of UVB-induced skin cancer. *Cancer Research*, *69*, 7103–7110.

[58] Butt, M.S., Sultan, M.T. (2009). Green tea: Nature's defense against malignancies. *Critical Reviews in Food Science and Nutrition*, *49*, 463–473.

[59] Ho, C.-T., Ferraro, T., Chen, Q., Rosen, R.T., Huang, M.-T., Phytochemicals in teas and rosemary and their cancer-preventive properties. In: Ho, C.-T., Osawa, T., Huang, M.-T., Rosen, R. T., editors. *Food Phytochemicals and Cancer Prevention. II. Teas, Spices, and Herbs.* Symposium Series 547, American Chemical Society Books, Washington, DC, 1994, pp. 2–19.

[60] Nagle, D.G., Ferreira, D., Zhou, Y.-D. (2006). Epigallocatechin-3-gallate (EGCG): Chemical and biomedical perspectives. *Phytochemistry*, *67*, 1849–1855.

[61] Tachibana, H. (2009). Molecular basis for cancer chemoprevention by green tea polyphenol EGCG. *Forum of Nutrition*, *61*, 156–169.

[62] Surh, Y.-J. (2006). NF-κB and AP-1 as molecular targets for chemoprevention with EGCG, a review. *Environmental Chemistry Letters*, *4*, 137–141.

[63] Shimizu, M., Shirakami, Y., Moriwaki, H. (2008). Targeting receptor tyrosine kinases for chemoprevention by green tea catechin, EGCG. *International Journal of Molecular Sciences*, *9*, 1034–1049.

[64] Surh, Y.-J., Lee, S.S. (1995). Capsaicin, a double-edged sword: Toxicity, metabolism, and chemopreventive potential. *Life Sciences*, *56*, 1845–1855.

[65] Kim, C.-S., Park, W.-H., Park, J.-Y., Kang, J.-H., Kim, M.-O., Kawada, T., Yoo, H., Han, I.-S., Yu, R. (2004). Capsaicin, a spicy component of hot pepper, induces apoptosis by activation of the peroxisome proliferator-activated receptor γ in HT-29 human colon cancer cells. *Journal of Medicinal Food*, *7*, 267–273.

[66] Anandakumar, P., Kamaraj, S., Ramakrishnan, G., Jagan, S., Devaki, T. (2009). Chemopreventive task of capsaicin against benzo(*a*)pyrene-induced lung cancer in Swiss albino mice. *Basic and Clinical Pharmacology and Toxicology*, *104*, 360–365.

[67] Yu, R., Choi, M.-A., Kawada, T., Kim, B.-S., Han, I.S., Yoo, H. (2002). Inhibitory effect of capsaicin against carcinogen-induced oxidative damage in rats. *Journal of Food Science and Nutrition*, *7*, 67–71.

[68] Tanaka, T., Kohno, H. (2002). Cancer chemoprevention by citrus components and juice. *Recent Research Developments in Cancer*, *4*, 439–451.

[69] Walle, T. (2007). Methoxylated flavones, a superior cancer chemopreventive flavonoid subclass? *Seminars in Cancer Biology*, *17*, 354–362.

[70] Murakami, A., Nakamura, Y., Torikai, K., Tanaka, T., Koshiba, T., Koshimizu, K., Kuwahara, S., Takahashi, Y., Ogawa, K., Yano, M., Tokuda, H., Nishino, H., Mimaki, Y., Sashida, Y., Kitanaka, S., Ohigashi, H. (2000). Inhibitory effect of citrus nobiletin on phorbol ester-induced skin inflammation, oxidative stress, and tumor promotion in mice. *Cancer Research*, *60*, 5059–5066.

[71] Miyamoto, S., Yasui, Y., Tanaka, T., Ohigashi, H., Murakami, A. (2008). Suppressive effects of nobiletin on hyperleptinemia and colitis-related colon carcinogenesis in male ICR mice. *Carcinogenesis*, *29*, 1057–1063.

[72] Patel, D., Shukla, S., Gupta, S. (2007). Apigenin and cancer chemoprevention: Progress, potential and promise. *International Journal of Oncology*, *30*, 233–245.

[73] Shukla, S., Gupta, S. (2010). Apigenin: A promising molecule for cancer prevention. *Pharmaceutical Research*, *27*, 962–978.

[74] Li, B., Birt, D.F. (1996). *In vivo* and *in vitro* percutaneous absoprtion of cancer preventive flavonoid apigenin in different vehicles in mouse skin. *Pharmaceutical Research*, *13*, 1710–1715.

[75] Lepley, D.M., Li, B., Birt, D.F., Pelling, J.C. (1996). The chemopreventive flavonoid apigenin induces G2/M arrest in keratinocytes. *Carcinogenesis*, *17*, 2367–2375.

[76] Hu, X.-W., Meng, D., Fang, J. (2008). Apigenin inhibited migration and invasion of human ovarian cancer A2780 cells through focal adhesion kinase. *Carcinogenesis*, *29*, 2369–2376.

[77] Jeyabal, P.V.S., Syed, M.B., Venkataraman, M., Sambandham, J.K., Sakthisekaran, D. (2005). Apigenin inhibits oxidative stress-induced macromolecular damage in *N*-nitrosodiethylamine (NDEA)-induced hepatocellular carcinogenesis in Wistar albino rats. *Molecular Carcinogenesis*, *44*, 11–20.

[78] Stoner, G.D., Mukhtar, H. (1995). Polyphenols as cancer chemopreventive agents. *Journal of Cellular Biochemistry*, *22*, 169–180.

[79] Aggarwal, B.B., Shishodia, S. (2006). Molecular targets of dietary agents for prevention and therapy of cancer. *Biochemical Pharmacology*, *71*, 1397–1421.

[80] Shehzad, A., Wahid, F., Lee, Y.S. (2010). Curcumin in cancer chemoprevention: Molecular targets, pharmacokinetics, bioavailability, and clinical trials. *Archiv der Pharmazie*, *343*, 489–499.

[81] Pereira, M.A., Grubbs, C.J., Barnes, L.H., Li, H., Olson, G.R., Eto, I., Juliana, M., Whitaker, L.M., Kelloff, G.J., Steele, V.E., Lubet, R.A. (1996). Effects of the phytochemicals, curcumin and quercetin, upon azoxymethane-induced colon cancer and 7, 12-dimethylbenz[*a*]anthracene-induced mammary cancer in rats. *Carcinogenesis*, *17*, 1305–1311.

[82] McCue, P., Shetty, K. (2004). Health benefits of soy isoflavonoids and strategies for enhancement: A review. *Critical Reviews in Food Science and Nutrition*, *44*, 361–367.

[83] Messina, M. (2010). Insights gained from 20 years of soy research. *Journal of Nutrition*, *140*, 2289S–2295S.

[84] Barnes, S. (1997). The chemopreventive properities of soy isoflavonoids in animal models of breast cancer. *Breast Cancer Research and Treatment*, *46*, 169–179.

[85] Dixon, R.A., Ferreira, D. (2002). Genistein. *Phytochemistry*, *60*, 205–211.

[86] Constantinou, A.I., Mehta, R.G., Vaughan, A. (1996). Inhibition of *N*-methyl-*N*-nitrosourea-induced mammary tumors in rats by the soybean isoflavones. *Anticancer Research*, *16*, 3293–3298.

[87] Banerjee, S., Li, Y., Wang, Z., Sarkar, F.H. (2008). Multi-targeted therapy of cancer by genistein. *Cancer Letters*, *269*, 226–242.

[88] Fenwick, G.R., Lutomski, J., Nicman, C. (1990). Liquorice, *Glycyrrhiza glabra* L. Composition, uses and analysis. *Food Chemistry*, *38*, 119–143.

[89] Kitagawa, I. (2002). Licorice root. A natural sweetener and an important ingredient in Chinese medicines. *Pure and Applied Chemistry*, *74*, 1189–1198.

[90] Nomura, T., Fukai, T., Akiyama, T. (2002). Chemistry of phenolic compounds of licorice (*Glycyrrhiza* species) and their estrogenic and cytotoxic activities. *Pure and Applied Chemistry*, *74*, 1199–1206.

[91] Takahashi, T., Takasuka, N., Iigo, M., Baba, M., Nishino, H., Tsuda, H., Okuyama, T. (2004). Isoliquiritigenin, a flavonoid from licorice, reduces prostaglandin E2 and nitric oxide, causes apoptosis, and suppresses aberrant crypt foci development. *Cancer Science*, *95*, 448–453.

[92] Chin, Y.-W., Jung, H.-A., Liu, Y., Su, B.-N., Castoro, J. A., Keller, W. J., Pereira, M. A., Kinghorn, A. D. (2007). Anti-oxidant constituents of the roots and stolons of licorice (*Glycyrrhiza glabra*). *Journal of Agricultural and Food Chemistry*, *55*, 4691–4697.

[93] Cuendet, M., Guo, J., Luo, Y., Chen, S., Oteham, C.P., Moon, R.C., van Breemen, R.B., Marler, L.E., Pezzuto, J.M. (2010). Cancer chemopreventive activity and metabolism of isoliquiritigenin, a compound found in licorice. *Cancer Prevention Research*, *3*, 221–232.

[94] Ii, T., Satomi, Y., Katoh, D., Shimada, J., Baba, M., Okuyama, T., Nishino, H., Kitamura, N. (2004). Induction of cell cycle arrest and p21CIP1/WAF1 expression in human lung cancer cells by isoliquiritigenin. *Cancer Letters*, *207*, 27–35.

[95] Singletary, K., MacDonald, C., Wallig, M. (1996). Inhibition by rosemary and carnosol of 7,12-dimethylbenz[*a*]anthracene (DMBA)-induced rat mammary tumorigenesis and *in vivo* DMBA-DNA adduct formation. *Cancer Letters*, *104*, 43–48.

[96] Lo, A.-H., Liang, Y.-C., Lin-Shiau, S.-Y., Ho, C.-T., Lin, J.-K. (2002). Carnosol, an antioxidant in rosemary, suppresses indicible nitric oxide synthase through down-regulating nuclear factor-κB in mouse macrophages. *Carcinogenesis*, *23*, 983–991.

[97] Rau, O., Wurglics, M., Paulke, A., Zitzkowski, J., Meindl, N., Bock, A., Dingermann, T., Abdel-Tawab, M., Schubert-Zsilavecz, M. (2006). Carnosic acid and carnosol, phenolic diterpene compounds of the labiate herbs rosemary and sage, are activators of human peroxisome proliferator-activated receptor gamma. *Planta Medica*, *72*, 881–887.

[98] Begum, S.A., Sahai, M., Ray, A.B., Non-conventional lignans: Coumarinolignans, flavonolignans, and stilbenolignans. In: Kinghorn, A.D., Falk, H., Kobayashi, J., series editors. *Progress in the Chemistry of Organic Natural Products*, Springer-Verlag, Vienna, 2010, Vol. *93*, pp. 1–70.

[99] Katiyar, S.K. (2005). Silymarin and skin cancer prevention: Anti-inflammatory, antioxidant, and immunomodulatory effects. *International Journal of Oncology*, *26*, 169–176.

[100] Agarwal, R., Agawal, C., Ichikawa, H., Singh, R.P., Aggarwal, B.B. (2006). Anticancer potential of silymarin: From bench to bedside. *Anticancer Research*, *26*, 4457–4498.

[101] Kim, N.-C., Graf, T.N., Sparacino, C.M., Wani, M. C, Wall, M.E. (2003). Complete isolation and characterization of silybins and isosilybins from milk thistle (*Silybum marianum*). *Organic and Biomolecular Chemistry*, *1*, 1684–1689.

[102] Lee, D.Y.-W., Liu, Y. (2003). Molecular structure and stereochemistry of silybin A, silybin B, isosilybin A, and isosilybin B, isolated from *Silybum marianum* (milk thistle). *Journal of Natural Products*, *66*, 1171–1174.

[103] Zi, X., Muktar, H., Agarwal, R. (1997). Novel cancer chemopreventie effects of a flavonoid antioxidant silymarin: Inhibition of mRNA expression of an endogenous tumor promoter TNF alpha. *Biochemical and Biophysical Research Communications*, *239*, 334–339.

[104] Nohara, T., Ono, M., Ikeda, T., Fujiwara, Y., El-Aasr, M. (2010). The tomato saponin, esculeoside A. *Journal of Natural Products*, *73*, 1734–1741.

[105] Cohen, L.A. (2002). A review of animal model studies of tomato carotenoids, lycopene, and cancer chemoprevention. *Experimental Biology and Medicine*, *227*, 864–868.

[106] Dahan, K., Fennal, M., Kumar, N.B. (2008). Lycopene in the prevention of prostate cancer. *Journal of the Society for Integrative Oncology*, *6*, 29–36.

[107] van Breemen, R.B., Pajkovic, N. (2008). Multitargeted therapy of cancer by lycopene. *Cancer Letters*, *269*, 339–351.

[108] Bhuvaneswari, V., Velmurugan, B., Balasenthil, S., Ramachandran, C.R., Nagini, S. (2001). Chemopreventive efficacy of lycopene on 7,12-dimethylbenz[*a*]anthracene-induced hamster buccal pouch carcinogenesis. *Fitoterapia*, *72*, 865–874.

[109] Konijeti, R., Henning, S., Moro, A., Sheikh, A., Elashoff, D., Shapiro, A., Ku, M., Said, J.W., Heber, D., Cohen, P., Aronson, W.J. (2010). Chemoprevention of prostate cancer with lycopene in the TRAMP model. *Prostate*, *70*, 1547–1554.

[110] Kresty, L.A., Morse, M.A., Morgan, C., Charlton, P.S., Lu, J., Gupta, A., Blackwood, M., Stoner, G.D. (2001). Chemoprevention of esophageal tumorigenesis by dietary administration of lyophilized black raspberries. *Cancer Research*, *61*, 6112–6119.

[111] Stoner, G.D., Wang, L.-S., Casto, B.C. (2010) Laboratory and clinical studies of cancer chemoprevention by antioxidants in berries. *Carcinogenesis*, *29*, 1665–1674.

[112] Stoner, G.D., Wang, L.-S., Seguin, C., Rocha, C., Stoner, K., Chiu, S., Kinghorn, A.D. (2010). Multiple berry types prevent *N*-nitrosomethylbenzylamine-induced esophageal cancer in rats. *Pharmaceutical Research*, *27*, 1138–1145.

[113] Pawlus, A.D., Kinghorn, A.D. (2007). Review of the ethnobotany, chemistry, biological activity and safety of the botanical dietary supplement *Morinda citrifolia* (noni). *Journal of Pharmacy and Pharmacology*, *59*, 1587–1609.

[114] Chin, Y.-W., Chai, H.-B., Keller, W.J., Kinghorn, A.D. (2008). Lignans and other constituents of the fruits of *Euterpe oleracea* (açai) with antioxidant and cytoprotective activities. *Journal of Agricultural and Food Chemistry*, *56*, 7759–7764.

[115] Chin, Y.-W., Kinghorn, A.D. (2008). Structural characterization, biological effects, and synthetic studies on xanthones from mangosteen (*Garcinia mangostana*), a popular botanical dietary supplement. *Mini-Reviews in Organic Chemistry*, *5*, 355–364.

[116] Potterat, O. (2010). Goji (*Lycium barbatum* and *L. chinense*): Phytochemistry, pharmacology, and safety in the perspective of traditional uses and recent popularity. *Planta Medica*, *76*, 7–19.

[117] Roy, S., Khanna, S., Alessio, H.M., Vider, J., Bagchi, D., Bagchi, M., Sen, C.K. (2002). Anti-angiogenic property of edible berries. *Free Radical Research*, *36*, 1023–1031.

[118] Lala, G., Malik, M., Zhao, C., He, J., Kwon, Y., Giusti, M.M., Magnuson, B.A. (2006). Anthocyanin-rich extracts inhibit multiple biomarkers of colon cancer in rats. *Nutrition and Cancer*, *54*, 84–93.

[119] Wang, L.-S., Stoner, G.D. (2008). Anthocyanins and their role in cancer prevention. *Cancer Letters*, *269*, 281–290.

[120] Thomasset, S., Berry, D.P., Cai, H., West, K., Marczylo, T.H., Marsden, D., Brown, K., Dennison, A., Garcea, G., Miller, A., Hemingway, D., Steward, W.P., Gescher, A.J. (2009). Pilot study of oral anthocyanins for colorectal cancer chemoprevention. *Cancer Prevention Research*, *2*, 625–633.

[121] Zhao, C., Giusti, M.M., Malik, M., Moyer, M.P., Magnuson, B.A. (2004). Effects of commercial anthocyanin-rich extracts on colonic cancer and nontumorigenic colonic cell growth. *Journal of Agricultural and Food Chemistry*, *52*, 6122–6128.

[122] Cooke, D., Schwarz, M., Boocock, D., Winterhalter, P., Steward, W.P., Gescher, A.J., Marczylo, T.H. (2006). Effect of cyanidin-3-glycoside and an anthocyanidin mixture

from bilberry on adenoma development in the Apc mouse model of intestinal carcinogenesis – relationship with tissue anthocyanin levels. *International Journal of Cancer*, *119*, 2213–2220.

[123] Teller, N., Thiele, W., Marczylo, T.H., Gescher, A.J., Boettler, U., Sleeman, J., Marko, D. (2009). Suppression of the kinase activity of receptor tyrosine kinases by anthocyanin-rich mixtures extracted from bilberries and grapes. *Journal of Agricultural and Food Chemistry*, *57*, 3094–3101.

[124] Thomasset, S., Teller, N., Cai, H., Marko, D., Berry, D.P., Steward, W.P., Gescher, A.J. (2009). Do anthocyanins and anthocyanidins, cancer chemopreventive pigments in the diet, merit development as potential drugs? *Cancer Chemotherapy and Pharmacology*, *64*, 201–211.

[125] Athar, M., Back, J.H., Tang, X., Kim, K.H., Kopelovich, L., Bickers, D.R., Kim, A.L. (2007). Resveratrol: A review of preclinical studies for human cancer prevention. *Toxicology and Applied Pharmacology*, *224*, 274–283.

[126] Goswami, S.K., Das, D.K. (2009). Resveratrol and chemoprevention. *Cancer Letters*, *284*, 1–6.

[127] Bishayee, A., Politis, T., Darvesh, A.S. (2010). Resveratrol in the chemoprevention and treatment of hepatocellular carcinoma. *Cancer Treatment Reviews*, *36*, 43–53.

[128] Li, N.-G., Shi, Z.-H., Tang, Y.-P., Yang, J.-P., Wang, Z.-J., Song, S.-L., Lu, T.-L., Duan, J.-A. (2010). Targeting the development of resveratrol as a chemopreventive agent. *Drug Development Research*, *71*, 335–350.

[129] Jang, M., Cai, L., Udeani, G.O., Slowing, K.V., Thomas, C.F., Beecher, C.W.W., Fong, H.H.S., Farnsworth, N.R., Kinghorn, A.D., Mehta, R.G., Moon, R.C., Pezzuto, J.M. (1997). Cancer chemopreventive activity of resveratrol, a natural product derived from grapes. *Science*, *275*, 218–220.

[130] Waffo-Teguo, P., Hawthorne, M.E., Cuendet, M., Merillon, J.-M., Kinghorn, A.D., Pezzuto, J.M., Mehta, R.G. (2001). Potential cancer chemopreventive activities of wine stilbenoids and flavans extracted from grape (*Vitis vinifera*) cell cultures. *Nutrition and Cancer*, *40*, 173–179.

[131] Kaur, M., Agarwal, C., Agarwal, R. (2009). Anticancer and cancer chemopreventive potential of grape seed extract and other grape-based products. *Journal of Nutrition*, *139*, 1806S–1812S.

[132] Manju, V., Nalini, N. (2005). Chemopreventive efficacy of ginger, a naturally occurring anticarcinogen during the initiation, post-initiation stages of 1, 2-dimethylhydrazine-induced colon cancer. *Clinica Chimica Acta*, *358*, 60–67.

[133] Manju, V., Viswanathan, P., Nalini, N. (2006). Hypolipidemic effect of ginger in 1, 2-dimethyl hydrazine-induced experimental colon carcinogenesis. *Toxicology Mechanisms and Methods*, *16*, 461–472.

[134] Manju, V., Nalini, N. (2006). Effect of ginger on bacterial enzymes in 1, 2-dimethylhydrazine induced experimental colon carcinogenesis. *European Journal of Cancer Prevention*, *15*, 377–383.

[135] Yusof, Y.A.M., Ahmad, N., Das, S., Sulaiman, S., Murad, N.A. (2009). Chemopreventive efficacy of ginger (*Zingiber officinale*) in ethionine induced rat hepatocarcinogenesis. *African Journal of Traditional, Complementary and Alternative Medicines*, *6*, 87–93.

[136] Jeong, C.-H., Bode, A.M., Pugliese, A., Cho, Y.-Y., Kim, H.-G., Shim, J.-H., Jeon, Y.-J., Li, H., Jiang, H., Dong, Z. (2009). [6]-Gingerol suppresses colon cancer growth by targeting leukotriene A4 hydrolase. *Cancer Research*, *69*, 5584–5591.

[137] Oyagbemi, A.A., Saba, A.B., Azeez, O.I. (2010). Molecular targets of [6]-gingerol: Its potential roles in cancer chemoprevention. *BioFactors*, *36*, 169–178.

[138] Akihisa, T., Nakamura, Y., Tokuda, H., Uchiyama, E., Suzuki, T., Kimura, Y., Uchikura, K., Nishino, H. (2007). Triterpene acids from *Poria cocos* and their anti-tumor-promoting effects. *Journal of Natural Products*, *70*, 948–953.

[139] Asolkar, R.N., Freel, K.C., Jensen, P.R., Fenical, W., Kandratyuk, T.P., Park, E.-J., Pezzuto, J.M. (2009). Arenamides A-C, cytotoxic NF-κB inhibitors from the marine actinomycete *Salinispora arenicola*. *Journal of Natural Products*, *72*, 396–402.

10

MEDICINAL PLANTS AND PHARMACEUTICAL TECHNOLOGY

RUTH M. LUCINDA DA SILVA, ANGÉLICA G. COUTO, AND
TANIA M.B. BRESOLIN

10.1 INTRODUCTION

There are intrinsic (genetic) and extrinsic (environment, cultivation, harvesting, and postharvest processes, transport, and storage practices) factors that affect the safety and quality of raw medicinal plant materials and finished products. The contamination by microbial or chemical agents during the plant material production stages can decrease their safety and quality. Another aspect is the origin of the medicinal plants, which may be contaminated by other species or plant parts through misidentifications, accidental contamination, or intentional adulteration, leading to unsafe consequences [1].

Adherence to good cultivation practices (GCP) and good manufacturing practices (GMP) is very important for the sustainability of the phytopharmaceuticals industry and definition of formulations, together with an appropriate pharmaceutical technology and analytical practices. The whole phytopharmaceutical production chain must be aligned with the legal regulations of the governmental health agencies.

Pharmaceutical technology includes knowledge of production techniques, formulations, pharmaceutical dosage forms, and quality control. Consistency in the production of the herbal medicinal products is guided by the same principles as those used for synthetic medicines, and the reproducibility of pharmaceutical dosage forms is necessary for quality assurance in relation to their safety and efficacy [2].

Plant Bioactives and Drug Discovery: Principles, Practice, and Perspectives, Fourth Edition.
Edited by Valdir Cechinel-Filho.

Many different denominations are used in this field; therefore, the terms adopted in this chapter are listed below:

Active principle: a substance or chemical class (e.g., alkaloid, flavonoid, fatty acid, etc.) with the known pharmacological activity that is wholly or partially responsible for the therapeutics effects.

Adjuvant: a substance of natural or synthetic source that can be added to the pharmaceutical dosage form to prevention its alteration, correct and/or improve its organoleptic, biopharmaceutical, or technological characteristics. Adjuvants are also termed aids or inert material by some authors.

Alcoholate: a liquid preparation usually obtained from fresh plants, occasionally from dried plants or drugs, by maceration with ethanol at room temperature. Historically, alcoholates are known as mother tinctures, a homeopathic input, as used by Hahnemann.

Aqueous extract: a liquid preparation obtained by maceration, infusion or decoction with water, which should be prepared just before use, due to its high susceptibility to degradation. Suitable antimicrobial preservatives should be added.

Dry extract: a solid preparation with a dry residue of not less than 95% by mass, obtained by complete evaporation of the solvent used for extraction. Suitable inert materials may be added for drying or standardization to a defined content of constituents.

Essential oil/spirit: a volatile compound dissolved in alcohol or hydroalcoholic solution. It is prepared by mixing aromatic plants with alcohol, and then the volatile compounds are recovered by vapor drag.

Extract: a concentrated preparation of liquid, solid, or intermediate consistency, usually obtained from dried vegetable or animal matter.

Fluid extract: a liquid extract in which one part by mass or volume of a liquid extract is equivalent to one part by mass of the original dried drug.

Herbal drug: a medicinal plant, whether cultivated or wild, or its parts, mainly whole, fragmented or cut, in an unprocessed state, usually in dried form but sometimes fresh.

Herbal medicinal product: a pharmaceutical dosage form, also termed phytomedicine and phytopharmaceutical, exclusively containing one or more herbal drugs or herbal preparations as active substances.

Herbal preparation: the extraction product of a herbal drug, its extracts, tinctures, oil, wax, exudate, juice, or others.

Herbal tea: one or more herbal drugs, intended for oral aqueous preparation by means of decoction, infusion, or maceration.

Juice: the extraction product obtained by squeezing fresh citrus fruits, or by careful concentration of aqueous extracts.

Liquid extract: a preparation obtained by one of the extraction methods, such as percolation or maceration. In general, liquid extracts are first prepared using water and ethanol mixtures. They may be prepared by dissolving a soft or dry extract in an appropriate solvent. These preparations are adjusted, if necessary, to meet the requirements for solvent, constituent or dry residue contents. Suitable antimicrobial preservatives should be added.

Marker: a component or chemical class (e.g., alkaloid, flavonoid, fatty acid, etc.), present in the vegetable raw material, ideally the active principle, and preferably correlated with therapeutic effect, used as reference in the quality control of vegetal raw material and herbal medicinal products.

Soft extract: an intermediate consistency preparation with a dry residue of not less than 70% by mass, obtained by partial evaporation of the solvent used for extraction.

Tincture: a liquid extracts usually obtained from one part drug and 15 or 10 parts extraction solvent, since one part of drug is extracted with more than two parts, but at most 10 parts of extraction solvent. Therefore, they are less concentrated than fluid extracts. They are usually clear, but a slight sediment is acceptable as long as the composition is not affected.

Vegetal (or herbal) raw material: the fresh medicinal plant, herbal drug, or herbal preparation.

10.2 SUPPLY OF HERBAL MATERIALS

10.2.1 Cultivation of Medicinal Plants

In recent years, investments and partnerships for the development of new drugs from medicinal plants have increased. One of the key items to ensure the production of the industrial herbal medicine is the source of raw material, which needs to be produced in sufficient quantity, quality, and consistency.

In the preparation of a finished herbal product, the cultivation stage is considered the most influential, both quantitatively and qualitatively. The cultivation of medicinal plants is recommended, especially when domestication of native species or acclimatization of exotic plants are necessary, in order to meet market demand, guarantee sustainability, and counteract the great variability in herbal drug content. In such cases, agronomic experimentation must be accompanied by phytochemical monitoring [3].

General technical guidance on obtaining herbal materials of good quality for the sustainable production of finished herbal products is given in the *WHO guidelines on good agricultural and collection practices (GACP) for medicinal plants*. These guidelines may need to be adjusted according to the situation of each country. The European Union, China, Japan and several other countries have developed regional and national guidelines for good agricultural and collection practices for medicinal plants [1].

The production of secondary metabolites, and the resulting safety and quality of plant material, are strongly influenced by extrinsic factors, such as environment, light, temperature, wind, soil conditions, pest attack, collection methods, cultivation, postharvest processing, transport and storage practices, as well as ontogenetic factors, such as plant age and development [3].

Thus, herbal drug cultivation involves the stages of seedling production, fertilization, monitoring cultivation, harvesting, and processing. Also, particular needs of each species must be considered, in order to select appropriate procedures.

The selection of plants or seeds with desirable characteristics, and obtaining of homogeneous genotypes, contributes to reducing the genetic and active principles content variability of naturally occurring plant and the impact on seed dormancy, morphology, plant development, and resistance to pests and diseases [4].

The seedlings can be produced by cuttings, rhizomes, bulbs, corms or seedlings, clump division and tissue culture (asexual reproduction), or by use of seed (sexual reproduction). The propagation form for starting cultivation should be chosen by criteria of practicality and economy, aiming at sustainable production. But regardless of the form selected, the matrix must be free of contamination and disease in order to protect the plant development. Propagation by cutting has less variability between individuals, since the seedlings are genetically identical to the mother-plants [4].

For optimum selection, the matrices should not be in flower, and their removal in the winter must be avoided. Thereafter, small portions of 10 to 15 cm of stem, leaves, or roots can be placed in a substrate, such as a mixture of equal parts of sand, soil and humus; kept in shade at temperatures around 20°C to 27°C for rooting.

In the cultivation of medicinal plants the addition of pesticides must be avoided, due to the presence of heavy metals and permanent toxic waste, and the lack of products that are free of this toxicity. Although medicinal plants usually have high resistance to diseases and pests, some precautions can be helpful, such as careful selection of the cultivation area, soil management, culture (species) rotation, selection of cultivation season, plant spacing, and the use of healthy seeds, seedlings and cuttings [4].

Indeed, intercropping of two or more species instead of monocultures reduces the risk of emergence and rapid development of specific pests and diseases. For example, the sole crop of *Phyllanthus niruri* and *Phyllanthus fraternus* was severely attacked by pest; therefore, intercropping with other herbs was recommended [5]. Biological control of infestations can be accomplished using plants that are repellant to ants and insects, such as *Tanacetum vulgare* and *Mentha* sp. Herbal extracts, broths and solutions, such as homogenized tobacco and *Artemisia absinthium* infusion can also be used in emergency cases [6].

The use of insecticidal substances extracted from wild plants has many advantages over synthetic ones, as they are less expensive, renewable, easily biodegradable, safe for operators, do not leave residues in foods, and cause slow insect resistance to these substances. Several species of Meliaceae are highlighted among the plant families with insecticidal activity. Among them, *Azadirachta indica* A. Juss., neem tree, is the most widely studied, and is distinguished by its high efficiency in controlling insects, mites, phytophagous nematodes, fungi, and bacteria [7]. Neem derivatives do not kill directly, but by physiological pressure and starvation of the insects, invariably altering their reproduction cycle [7]. Authors have found that derivatives of *A. indica*, oil (1% and 2%) or aqueous extract, were more efficient than other plant extracts in controlling fall armyworm *Spodoptera frugiperda* in corn [8].

10.3 HARVEST AND POSTHARVEST PROCESSING

10.3.1 Harvest

Appropriate agricultural practices include technical factors such as season, mode of harvesting, transport, drying and storage, which should be suitable for each species, their active ingredients and their purpose of use [9].

For the development of the herbal medicinal product, the quality concept of harvesting involves compliance with agronomic, botanical, and phytochemical specifications related to the purpose of use. Ideally, harvest procedure protocols must adhere to an agronomic calendar, in accordance with the optimum harvest time, age, and stage of plant development, in order to maximize the yield of active principles and biomass production, while preserving the physical and chemical properties [9].

The amount of serpentine in *Catharanthus roseus* roots, for example, is 1% to 1.2% in plants at 5 months, but decreases to 0.5% after this age. Other plants show fluctuations in active content over their life cycle, such as the cardiotonic glycoside content of the leaves of *Digitalis lanata*, which reaches its peak at the start of flowering [3].

Harvesting should be conducted carefully, ensuring that the integrity of the plant parts is protected against the loss of active principles until the next stage of production. The sensitivity of the species to harvesting is well characterized in plants that produce essential oils. The location of secretory structures and essential oil storage facilities usually determine the harvesting method. Species that store essential oil in glandular hairs on the surface of leaves or flowers (e.g., peppermint and chamomile) require more care than crop species in which oil accumulates in internal structures (e.g., cinnamon-sassafras and fennel) [9].

The herbal drug should preferably be harvested in dry weather, after the dew has evaporated. Harvesting after a prolonged period of rain is not recommended, as the active ingredient content can decrease as the plant moisture content to increase. Furthermore, the increased humidity hinders drying, increases the possibility of fungus growth, and raises the costs of transport and drying [9].

During harvesting, procedures should be carried out to minimize contamination from microbes or residues from other plants and any foreign material. Direct contact with the ground should be avoided, to prevent soil particles in the plants. The use of tools is then advised (pruning shears, scythe, etc), as well as clean, appropriate machinery, or equipment. The collection of foreign plant material should be avoided, as well as any damaged parts of the plant, and insects [1,9,10].

10.3.2 Postharvest Treatment

After collection, the herbal materials may be subjected to a preliminary processing stage that includes the elimination of undesirable materials and contaminants, washing (to remove soil excess), classification, and cutting. The herbal materials should be protected from insects, rodents, birds, and other pests, and also from livestock and domestic animals. And if the collection site is located far from the processing facilities, it is advisable to air- or sun-dry the herbal materials prior to transport. Separate harvesting and transport in different containers may be done if there is more than one medicinal plant part is to be collected. Cross-contamination should be avoided at all times [1].

Health criteria should be established during the post-harvesting process, seeking to minimize the microbial contamination, since the preparations obtained from herbal drugs often have high levels of microbial contamination. Such procedures include cleaning, choice of drying mechanisms, extraction of essential oils by steam distillation, storage and appropriate packing conditions and disinfection of the herbal material [1]. Besides cleaning and removal of foreign material, some operations, such as fragmentation of the material by erasing, cutting, and separating leaves from stems, increase the drying efficiency by reducing time and energy in the process and obtaining a homogeneous product.

The moisture content of the harvested plant parts usually is high, around 60% to 80%. Drying should be done until the plant reaches 5% to 12% of moisture, according to the species and plant part. The acceptable moisture content levels can be obtained from pharmacopoeias or other authoritative monographs. The drying process prevents degradation through metabolic processes, and prevents microbial development [10].

The drying methods are divided into natural or artificial. Natural drying methods are: exposure to fresh air (shaded from direct sunlight), placed in thin layers on drying frames, or by direct sunlight, where appropriate. Artificial methods are: drying with cold or hot air; in rooms or buildings with wire screens; in solar dryers, for indirect fire; roasting in metal or clay pots, as for the seeds of *Paullinia cupana* [11], in a forced air oven, in vacuum, by freeze drying, by microwave and by infrared.

The drying temperature may vary from 35°C to 70°C, depending on the part of the plant and sensitivity of the active principles [10]. For the leaves, a temperature range of 20°C to 40°C is recommend, without direct heat, preferably under butane, propane, or natural gas. The material must be placed in thin layers of 3 cm, to ensure it is completely reached by the circulating air [1].

The packaging material must be of a type that prevents moisture absorption from the air after drying. Packages with one or two outer layers of Kraft paper are recommended, to protect from light, plus an inner layer of nontoxic polyethylene, to prevent rehydration of the product. The packaging depends on the type of drug, quantity, transport, distance, and specific requirements of the buyer. The most commonly used packaging is paper or plastic bags, paper plus polyethylene bags, cardboard boxes, or polyethylene bags or cardboard drums. In general, large volumes of compressible plant materials, such as leaves, from 60 to 100 kg, are baled by machine and wrapped in a cloth or polyethylene. Uncompressible materials, like roots, are placed in large bags, also called bales. Herbal materials with high density, such as seeds and fruits, are packaged in smaller bags, and those sensitive to rough handling during transport, such as chamomile flowers and mint leaves, can be packed in cardboard boxes [9].

10.3.3 Pulverization

Powdered herbal drugs are obtained through fragmentation of raw material into small particles by the application of mechanical force. This action may be preceded by cutting, shredding, softening, cleaning, and drying in order to improve milling efficiency.

The choice of milling method and equipment depends on the properties of the herbal material, for example easily friable, hard, brittle or soft, and also on the fibrous stem and fat resin content [12].

Fibrous materials cannot be crushed by pressure or impact, and must be cut. Brittle materials are fractionated into defined layers and can be crushed by friction processes (drag), impact, or pressure. In industrial mills, except for knife mills, only brittle materials can be processed. Materials that exhibit plasticity when subjected to deformation processes can be ground in some cases, if resources such as drying or cryogenic techniques are used [12].

Thermal sensitivity and chemical reactivity of the sample, such as the characteristics of gums, waxes, and resins, should also be considered [13]. Cooling of sample and mill, with liquid nitrogen or carbon dioxide can be used to prevent loss of volatility, especially for the production of oleoresins [3].

When the herbal material is intended for industrial production of herbal extracts, *coarsely powdered* ($<710\,\mu m$) or *moderately fine* ($<355\,\mu m$) powders are used as starting material. *Fine powders* ($<105\,\mu m$) are used for the production of sachets, while coarsely powdered materials are used to prepare medicinal teas packed in bulk [3,14].

10.4 EXTRACTION OF HERBAL DRUGS

10.4.1 Extraction

One of the first steps of the process for obtaining medicines from herbs is extraction [12]. This is a dynamic process that involves physical and chemical phenomena: wetting, penetration and swelling in the initial stage by solvent contact with the plant substrate; lixiviation of substances from the surface of broken cells, and dissolution of inner cell substances, with simultaneous diffusion to the external environment, depending on the chemical affinity of the substances for the extractor solvent.

Extraction operation can be influenced by three groups of factors related to:

(a) *Herbal material*: physical properties, such as size and particle surface area, porosity, and chemical composition, such as viscosity compounds.
(b) *Process*: equipment parameters, such as agitation, temperature, and their variables such as stirring speed and time.
(c) *Product specifications*: the nature of the solvent, drug:solvent ratio to obtain standardized extracts.

The extraction operation has been optimized, in order to study the influence of variables on technological characteristics and/or physical chemical properties of the liquid extracts, and their pharmacological properties. The choice of method, solvent, and drug:solvent ratio should be based not only on the efficiency of the process to maximize these properties, but also on the ease of operation, safety, and physiological and environmental expenditures of time and energy [12].

Input variables for the optimization study can be related to the extraction process, herbal material, and solvent. The herbal drug is normally dry and ground, but in specific cases it is fresh, such as for the extraction of volatile compounds, or when the drying affects the stability of the active principles [14]. The solvents commonly used as liquid extractors are water, alcohol, glycerin, propylene glycol, or mixtures thereof, and their choice is based on selectivity, safety, and ease of removal.

The correct choice of solvent and method is crucial for achieving the desired therapeutic goal. Once the substances or classes related to the pharmacological activity are known, the goal is to obtain them in higher yields, through extraction. A combination of this knowledge allows the quality control to be planned in each stage of the production cycle, in order to ensure to check the stability of the bioactive compounds [13].

10.4.2 Extraction Methods

In general, extraction methods can be classified as static, dynamic (where agitation of the medium and/or renewal of the solvent occur), hot (where heat is applied), and cold (where no heat is applied). The main methods can be also grouped according the ability to achieve total phytochemical extraction from the herbal drug (Figure 10.1).

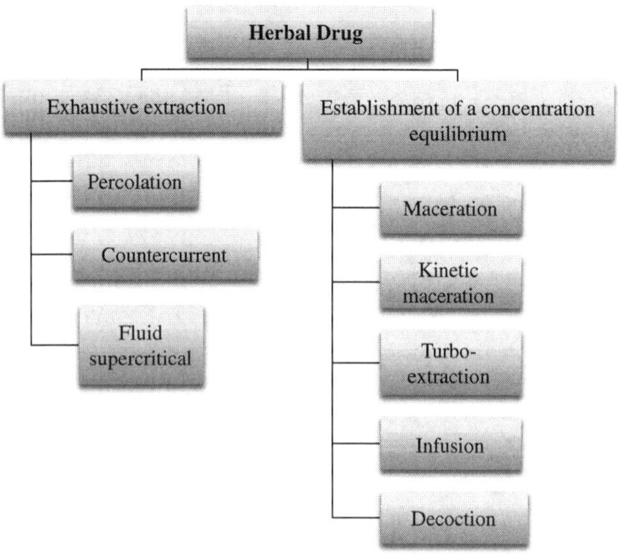

FIGURE 10.1 Extraction methods for herbal drug.

10.4.2.1 Maceration. Maceration is the operation whereby the herbal drug is held in a closed container at different temperatures and for different periods. Simple maceration, kinetic or dynamic maceration, infusion, digestion, decoction, and turbo-extraction are some of the derivatives of the maceration method.

In *simple* or *static* maceration, the herbal drug is in contact with the solvent for a period of 5 to 14 days, at room temperature with sporadic stirring. This method can achieve drug and solvent equilibrium, but not exhaustive extraction. Re-maceration is a variable method, in which the solvent is renewed by a pure one, allowing total extraction of soluble herbal drug constituents by the solvent used.

In the *dynamic* or *kinetic* maceration, herbal drug, and solvent are maintained under constant stirring. The extraction time may be much shorter (1 to 12 h) than the extraction time by static maceration (5 at 14 days), depending mainly on the equipment and stirring rate.

Digestion is an aqueous extraction process that occurs at temperature of 40°C to 60°C for a period to 2–6 h. *Infusion* and *decoction* also are derivatives of maceration, using water as extractor solvent and a higher temperature. In *infusion*, boiling water is poured over the herbal drug, usually leaves, flowers, bark, roots, or rhizomes, left to stand 5 to 10 min, and then filtered. This is a traditional method of preparing herbal tea or infusion in which the proportion of herbal drug is usually 5% for species containing volatile or thermolabile substances. In *decoction*, the fresh herb or herbal drug is placed in cold water and then heated to boiling for 10 to 15 min, under reflux to prevent loss on evaporation. This method can be useful for herbal drugs with hard extraction, or extraction of thermal resistant chemical classes such as tannins, flavonoids, and triterpene derivatives.

Another derivative method of maceration is *turbo extraction*. In this process, the extraction is usually conducted with the cold solvent under high shear forces (50,000 to 200,000 rpm). This process leads to a simultaneous reduction in the particle size, cell disintegration, and temperature increase, which must be controlled to avoid filtration problems and thermolabile decomposition of compounds, respectively [15].

The maceration method does not, by nature, lead to exhaustion of herbal material, either due to saturation of the liquid extractor or to the establishment of diffusion and solvent equilibrium within the cell. Generally, herbal drugs recommended for extraction by maceration are those rich in active substances that have no cellular structures, such as gums, resins, and alginates [12].

For industrial extractive process, the extractors are fixed or mobile tanks, with or without a steam jacket. They may differ in size, geometric shape, and the presence of inner agitation elements, which in turn, differ in terms of geometry, provision, and number. An example that combines the highest number of features that allow greater versatility in operation is the Nauta extractor. The tank is fixed and tapered and fitted with a steam jacket, and the stirring element is like a worm that moves the sound from the bottom to the top, enabling waste to be discharged easily [3].

10.4.2.2 Percolation and Derivate Methods. *Percolation* is a dynamic operation that is mainly indicated for the extraction of highly active substances present in small quantities, poorly soluble, or when the price of the herbal drug is relevant. In percolation, the ground herbal drug is placed in a conical or cylindrical container (percolator) made of glass or metal, provided with a tap at the bottom to regulate the solvent flow. A height and column diameter ratio of 5:1 is recommended for the percolator design [3]. The product is called lixiviate [3,16]. Pharmacopoeias describe *simple percolation* and *percolation fractionation* as usual percolation methods.

In the latter method, a series of percolator and more diluted extracts are used to extract portions of the new drug, making a battery of percolators in which the solvent is recirculated by pump.

In the *countercurrent method* of extraction, a fresh plant is extracted by a solvent that moves in the opposite direction to the raw material. The extraction process may vary according to the movement of herbal drug and solvent. In the continuous process, the solvent and herbal drug move in opposite directions. In the discontinuous process, the solvent and herbal drug motion is discontinuous [15].

The extraction method by *fluid supercritical* is used for extraction of essential oil constituents. The plant parts are placed in a tank and the supercritical fluid is injected. The main fluid used is supercritical carbon dioxide that reaches the fourth physical state, under extreme pressure of 200 atmospheres and temperatures above 31°C, under which conditions CO_2 has a viscosity similar to a gas, but its solubility capacity is very high. After extraction, the pressure is reduced and the gas returns to a gaseous state, leaving no solvent residue [15].

10.4.2.3 Alcoholates and Tinctures. Some officially known liquid extracts are distinguished depending on the type of solvent, method, and solvent:drug ratio. Actual a*lcoholates* are obtained by maceration, over 10 days, from cut fresh drugs, in alcohol 90°GL, 80°GL, or 75°GL, in a closed vessel. High alcohol graduation is used due to the aqueous content of fresh drugs. The ratio of drug to alcohol is from 1:1 to 1:2, and is the same as that established for tinctures, since 1000 parts of fresh drug give 200 parts of dried drug.

Tinctures are classified as simple or compound, depending on the number of herbal drugs used as starting materials. As a general rule, if the drug is rich in hydrosoluble substances, such as alkaloids, salts, heterosides, and tannins, lower alcohol graduation is recommended. Conversely, if the drug has poorly soluble active substances, such as alkaloid base, essences, resins, and balsams, a higher alcohol graduation is preferred. Some exceptions occur when the active principles are alcohol soluble and water hydrolysable, as is the case with aconite when extracted with alcohol 90°GL, and which is easily decomposed in acetic acid, benzoic acid, and aconin.

If minimal extraction of chlorophyll is of interest, alcohol 50°GL or less than 42°GL is recommended, as this solvent is more specific for extraction of tannins, alkaloid salts, and heterosides. There are a few examples involving the use of lower graduations of alcohol, such as 30°GL for opium and saffron tinctures and 8–10°GL for apple rust. Tinctures obtained from highly active drugs, such as aconite, belladonna, coca, strophanthus, hemp, foxglove, jimson weed, ipecac, lobelia, henbane, nux vomica, and opium are defined as heroic tinctures. These are prepared by percolation of one part of the drug with 10 parts of alcohol 70% (v/v) [12,15].

10.4.3 Filtration

After extraction, the residue can be removed by pressing or centrifugation. The simplest model consists of a wire basket with perforated walls and a plunger, operated manually by a screw. Other improved models have a hydraulic drive system that enables increased pressure on the residue [3].

10.4.4 Clarification

After extraction, filtration or cooling, some waste may still remain in the extracts in the form of fine particles of the drug, or as flocculent precipitate, generally consisting of inert material. This residue can be separated after decanting or siphoning, industrially, through filtration in cellulose or synthetic filter presses, or by centrifugation [3].

10.4.5 Concentration: Partial Removal of the Solvent

By partially removing the solvent, the solid content increases for a given amount of dry matter. This is important for the production of soft extract, or as a preliminary step for the production of dry extract. Examples of equipment used in industry include: (a) Roberts concentrator consisting of an array of heated thin tubes, through which the solution to be concentrated is passed and vaporized;(b) film evaporator, ideal for thermolabile substances, since evaporation occurs from a very thin layer of liquid, reducing the heat exposure time [3].

10.4.6 Decontamination by Ionizing Radiation

The irradiation process is a method of preservation and decontamination, which uses ionizing radiation source (gamma rays, X-rays, or accelerated electrons), reducing the microbial load without affecting the principles of the herbal medicines. The method is legally accepted by the European Union, with specific regulations to prevent toxicological risks [17]. This alternative, which minimizes the risk of microbiological contamination and avoids the use of chemicals, has been used in several countries, including Brazil, where this technology is laid down in legislation [18].

Some works on gamma irradiation have been carried out, to detect the reduction of microbial flora, as well as evaluating the therapeutic quality [19]. In a study by Koseki et al. [20], plants were irradiated with cobalt 60, at doses ranging from 10, 20, and 30 kGy and no changes were observed. Although considered an alternative method due to its applicability in the absence of water and high temperatures, possible incompatibilities, and undesirable chemical changes need to be evaluated [17,19].

10.5 DRY EXTRACTS

Taking into account their low stability, handling difficulty, and growth of microorganisms, soft extracts have been completely replaced by dry extracts. Useful techniques for completely removing the liquid from the extracts include those which use heat, such as oven, drum, belt, fluid bed, spray and spout bed drying, and others that work by cooling, freezing, or congealing, such as spray, drum and freeze drying (Table 10.1).

All the drying techniques can be compared from technological, industrial, and economic development points of view, but there is no doubt that the drying temperature, time, operating cost, and characteristics of the final product are decisive when selecting the most appropriate method. The ideal method is the one that combines low temperature and shorter time.

Compared to freeze drying and other techniques, spray drying is less time-consuming, and despite using high temperatures, still proves to be feasible for drying thermolabile products, taking into account the rapid heat exposure. The management of drying conditions, such as the mechanism for heating the air, can be used as a resource to reduce operating costs [12,31].

Simple methods of drying by adsorption of liquid plant extract on inert excipients have been reported, such as fumed silica, lactose, cellulose, starch, and dicalcium phosphate [23,30]; along with proper selection of adsorbent, to obtain a dry extract with good flow properties. The adsorption of the concentrated extract of *G. lotoides* on Aeroperl®300 Pharma (30% w/w) resulted in a free-flowing nonadherent powder, using a vacuum oven at 40°C and 125 mbar for 24 h.

Freeze or belt dried products are usually soft, friable, porous, and bulky, while oven dried or fluidized bed dried products are hard and brittle, requiring postdrying operations, for example, mechanical grinding. On the other hand, spray drying is 30–50 times less expensive than freeze drying [33], and the dried product is directly obtained as a powder. Three distinct categories of particle morphology have been related to spray dried particles, namely crystalline, skin-forming, and agglomerate [34]. Although material-specific, the morphology also depends on feed specifications, such as concentration, temperature, and degree of aeration, as well as the dryer design and drying conditions [34].

Therefore, spray drying has been accepted as more favorable, especially when uniformity of size and form of particles are desired [12], making this technology the most widely used industrial process for particle formation and drying.

The spray-drying process mainly involves atomization of a feed liquid preparation, optionally concentrated prior to its introduction into the spray dryer chamber; droplet–air contact with hot gas in the chamber, resulting in droplet drying by evaporation of about 95% of the water contained in the droplets in a matter of a few seconds, followed by separation of the final product with the aid of cyclones, bag filters, and electrostatic precipitators, which may also be used [35].

The dryer design involves the type of atomizer, air flow, and spray-drying chamber. There are atomizers that act by air atomization or two fluid nozzles, airless atomization nozzles, pressure nozzles, rotary or disk nozzles, and ultrasonic nozzles. The air flow may be co-current flow, where the spray is directed into the hot air entering the dryer and both pass through the chamber in the same direction; counter-current flow, where the atomizer is positioned at the top and the air enters at the bottom; and mixed flow, combining both concurrent and counter current flow. In

TABLE 10.1 Drying Methods of Herbal Drug Extracts

Drying Methods	Operating Features	Dry Extract Examples
Spray drying	Pumpable suspensions and feed fluid are broken into small droplets by a rotating or pneumatic atomizing device; meet a stream of hot air (120–220°C), the liquid is rapidly evaporated and powder is separated from the moist air in cyclones by centrifugal action.	*Panax notoginseng* [21] *Opuntia ficus-indica* [22] *Bauhinia forficata* [23]
Spouted bed drying	Inert material is spouted by injected heated air at the base of the bed, while a feed fluid liquid or paste of a concentrated extract is broken by an atomizing air and heat is transferred from gas stream to solid particles.	*Passion flower* [24] *Maytenus ilicifolia* [25] *Bauhinia forficata* [26]
Vacuum belt drying	Heat transfer by contact at low temperatures (40–70°C) into liquid bubbles for a long-term exposure under reduced pressure (70–300 mbar), suitable for dry sticky or high viscous products.	*Panax notoginseng* [21]
Oven drying	Evaporation for a long exposure time to a temperature of around 40°C and optional reduced pressure	*Bauhinia forficata* [23]
Freeze drying	Congealing, primary drying and successive stages of drying by sublimation of water	*Panax notoginseng* [21]
Fluid bed drying	Heat transfer by convection at a temperature around 40°C in a granulated or powdery wet product, which is fluidized by an air that flows through a porous or perforated distributor to an upward direction in the core region	*Opuntia ficus-indica* [22,27]
Drum drying	Heat or cooling transfer by conduction through a rotating metal cylinder where a liquid, molten or paste film is fed in, dried and scraped off.	*Dioscorea* spp flours [28]
Supercritical fluid spray-particles from gas saturated solutions (*PGSS drying process*)	A high-pressure pump doses the sample solution into a static mixer, where compressed and preheated carbon dioxide is added. Fine droplets are formed when the mixture is rapidly depressurized via a nozzle into a spray tower at ambient pressure. The solvent can be withdrawn with the expanded carbon dioxide and a dry powder of the extract is obtained. Low temperature (30–60°C) and an inert, oxygen-free atmosphere.	green tea – *Camellia sinensis* [29]
Vacuum oven drying	Adsorption of liquid plant extract on inert excipients.	*Glinus lotoides* [30]

TABLE 10.2 Influence of Spray Drying Conditions on Process Performance and Finished Product Features

Drying Parameters Settings	Prospects
Higher speed rate of rotary atomizer; compressed air flow and feed stock ratio for pneumatic nozzle	Reduction of size particle [31]; insufficient aspersion and poor droplets shape [44]
Higher inlet temperature	Little effect on crystalline materials; size particle reduction of film forming materials; lower dissolution velocity [44]; improved compressibility [45]; higher process yield and powder hygroscopicity and lower moisture content [38,46].
Higher outlet temperature	Lower powder hygroscopicity [12]
Higher difference between inlet and outlet temperature	Greater evaporation ability [12]
Air pressure flutuations	Size particle variation [12]
Higher feed flow rate and viscosity	Lower process yield and higher moisture content [38]
Low glass transition temperature particles on feed stock, such as low molecular weight sugars and acids	Stickiness on the dryer chamber wall during drying, leading to low product yield, hygroscopicity and operational problems [47]

addition, the spray-drying system can be modified or adapted by preliminary air dehumidification [36].

Spray-drying technology requires adequate adjustment of the operating conditions and composition of the feed solution (Table 10.2). In this field, many studies combine input variables, such as air inlet temperature (120–202°C), atomization air flow rate, feed flow rate, preliminary air dehumidification [36], aspiration capacity, atomization speed [37], solid concentration in liquid preparation, and drying aid/extract content ratio [38–42], to optimize the spray-drying process and products in terms of moisture content, flowability, hygroscopicity, water solubility, particle size distribution, morphology, crystalline and porous structure, compatibility, process yield, caking tendency, bulk density, and color characteristics, as well as the recovery of substances of interest in the dry extract.

The way in which changes in these processing parameters affect some of the desirable powder specifications for further phytomedicine and nutraceutical industrial production has been investigated, in order to optimize the functionality and cost of the process. Since the number of variables is relatively high, fractional factorial statistical designs have been applied to determine up the best operating conditions for obtaining high quality powders and process yields [42].

Part of the problems related to low yield of dry extracts can be resolved by adding carrier agents, such as polymers and gums, to the product before it is atomized [47]. Drying aids for spray-drying have been reported in many fields of application. For products for food consumption, gelatin [48], modified starches [49,50], whey protein [49], gum arabic [50], starches, corn syrup solids, and maltodextrins [41,51]; cashew tree gum [39], Glucose syrup [40], and lactose-maltodextrin [52] have all been used.

For the development of phytomedicines, extracts have been dried with the aid of starch and modified starch derivatives [53]; arabic gum (ratio of 0.2–2.0% (w/w) arabic gum: dried residue [54], cyclodextrin – ratio of 30% (w/w) Kleptose: dried residue [46]; cellulose and derivates – Hydroxyethylcellulose at 1.5% [55]; gelatin derivative or hydroxylapatit [55]; colloidal silicon dioxide – ratio of 0.5:1; 1:1 [56]; 20:80; 25:75; and 30:70 (w/w) Aerosil 200: dried residue [46].

Colloidal silicon dioxide has widely been used as a carrier for herbal extracts, particularly due to its adsorbent properties and high specific surface area (200 m^2/g) overcoming stickiness on the drying chamber wall, and powder hygroscopicity, and also because it improves the flow properties of the end product. Added to its high purity, chemical inertness, and safety [57], these characteristics make this adjuvant superior to others. High concentrations of colloidal silicon dioxide, together with low atomization air flow rates, improved the flowability of extract of spray dried *Rhamnus purshiana*, increased the glass transition temperature of the product, and decreased its moisture content and hygroscopicity [43].

An appropriate selection of drying methods requires weighing all the relevant factors, such as production rate, capital and operating cost and drying time, as well as the physical and chemical properties of the end products.

10.6 PHYTOPHARMACEUTICAL DOSAGE FORMS

Phytopharmaceutical dosage forms can be processed from herbal drugs into liquid, solid, and semisolid preparations, with proper adjuvants, technology, and quality control.

10.6.1 Semisolid Dosage Forms

Semisolid herbal preparations include ointments, pastes, poultices, gels, lotion, and cream emulsions for application to the skin or mucous membranes, due to their therapeutic or protective action or the cosmetic function. They are commonly obtained by incorporating a liquid, soft or dry extract, previously mixed with a co-solvent, into a hydrophilic or lipophilic base. Propylene glycol, polyethylene glycol, ethanol, and glycerin are typical co-solvents used to improve the miscibility of the extracts in gel or O/A emulsions.

The cutaneous absorption of active principles from the bases depends on the physical and chemical properties of each, and the possible interactions between them, as well as on the physiological conditions of the topical route. Thus, the type of extract and base formula chosen are crucial for phytopharmaceutical development, not only to promote the penetration of active principles in the skin, but also to guarantee the safety and stability of the finished product. In other words, an appropriate type of extract and base for each kind of plant should be chosen.

TABLE 10.3 Examples of Medicinal Plants for Semisolid Preparations

Plant	Therapeutic Use	Dosage Form	Reference
Aloe barbadensis	Acne	Cream and gel	[58]
Azardirachta indica	Acne	Cream and gel	[58]
Calendula officinalis	Inflammation and wound healing	Unguent and cream	[59,60]
Cordia verbenacea	Inflammation	Cream	[61]
Curcuma longa	Acne	Cream and gel	[58]
Glaucium grandiflorum	Inflammation and analgesia	Cream	[59]
Glycyrrhiza glabra	Dermatits	Unguent and cream	[62,63]
Glycyrrhiza uralensis	Inflammation		
Hamamelis virginiana	Pruritus, dermatitis, inflammation and acne	Cream	[64–66]
Mandevilla illustris	Inflammation	Cream and gel	[67]
Ocimum gratissimum	Acne	Cream	[68]
Verbena officinalis	Inflammation and analgesia	Cream	[69]
Anogeissus latifolia	Wound healing	Ointment	[70]

Many semisolid herbal preparations have been developed for the treatment of topical and systemic diseases, particularly dermatitis, pain, and inflammation (Table 10.3). Studies have been carried out to evaluate the physical and chemical stability and therapeutic activity of herbal extracts incorporated into semisolid bases.

Arús-Pampin et al. [71] evaluated the stability of the cream base of cetyl alcohol, petrolatum, glycerin monostearate, and polysorbate containing 1.2% dry extract of *Mangifera indica* L. by quantification of total flavonoids. They observed a reduction of more than 15% in flavonoids after 3 months of conditioning at 45°C, as well as changes in pH. The authors recommended 3 months at room temperature, and 18 months under refrigeration (2–8°C) as storage conditions for stability.

Di Mambro and Fonseca [72] prepared creams containing extracts of *Glycyrrhiza glabra*, *Ginkgo biloba*, *Arctium majus*, *Glycine soja*, and *Nelumbium speciosum*, using a base of self-emulsifying wax, macadamia oil, propylene glycol and carbomer, and evaluated the antioxidant activity using DL-α-tocopherol as positive control. The formulations showed pseudoplastic behavior with a reduction in thixotropic viscosity after accelerated stability study at 40°C/70% RH, the pH remaining constant during the study. The formulations containing *G. glabra* and *G. biloba* showed greater antioxidant effect.

Bilia and collaborators [73] developed and evaluated the stability of different semisolid formulations (cetomacrogol, polysorbate, Polawax, Natrosol, and Sepigel) containing 10% *Arnica montana* extracts. Cetomacrogol formulations had a better release profile *in vitro*, while polysorbate and Polawax formulations showed the best release *in vivo*. Of the tested gels, Sepigel was more stable than Natrosol. In cream formulations, a reduction in sesquiterpenes of up to 10% was observed after two months at 40°C during the accelerated stability study.

Zanella [66] evaluated the activity of anti-inflammatory creams and hydrophilic gel bases with permeation adjuvants containing extract of *Mandevilla illustris*. Both herbal preparations produced dose-dependent inhibition of edema, but the permeation adjuvants did not increase the pharmacological effect.

Orafidiya et al. [68] evaluated the antimicrobial activity of semisolid preparations containing oil of *Ocimum gratissimum* (0.5%, 1%, 2%, and 5%) in different bases (polysorbate 80, cetomacrogol, petrolatum, and alcohol) using 10% benzoyl peroxide as positive control, in the treatment of acne lesions over a 4-week period. The authors observed that the effect was dependent on the base and the dose. Alcohol and cetomacrogol formulations with 5% oil were more effective than the positive control. However, cetomacrogol preparations containing 2% oil were the formulation of choice, since they were similar to the positive control and caused less irritation.

10.6.2 Solid Dosage Forms

Solid herbal preparations include powders, granules, capsules, and tablets, mostly for oral use. Powders are the simplest preparation, and consist of the milled and dried plant, packaged as sachets, for the preparation of teas. Powders for this purpose may also take the form of a soluble dry extract.

However, for industrial manufacture of phytomedicines, dried and milled plant have been mostly used as raw material for further operations such as extraction, concentration and drying, resulting in an intermediate product, so that granules, tablets or capsules can be produced.

In practice, dry extracts are the main raw material used to produce solid dosage forms. One of most important problems is their hygroscopicity, along with the high chemical diversity and complexity, favoring degradation reactions such as hydrolyses and oxidations. Thus, the research for the development and optimization of herbal drugs in solid dosage form is focused not only on the formulation, but also on the process conditions.

To overcome these problems, dry extracts have been produced by co-drying the liquid extracts with one or more adjuvants [43]. Thus, co-processing can provide raw materials containing a high content of dry extract of the plant that is suitable for other operations, such as direct compression, since the selection of excipients and process conditions have been properly combined.

Although the direct compression method is the simplest way to obtain tablets, it is necessary for the formulation to meet the minimum requirements for the process, such as the flow and compressibility, leading to an increase in the cost of the raw material itself, in comparison to the whole process. Thus, it is assumed that investments are needed for the development of promising technologies that result in these herbal raw materials for direct compression.

Numerous reports have addressed techniques to solve these problems. Investment in granulation is justified by the improved flow properties and density, favoring compression. Dry granulation has been the most commonly used method, through techniques such as slugging and roller compaction, for which formulation and compaction force variables are input [30,74]. Wet granulation was reported to obtain *Rauwolfia vomitoria* granules [75].

Compared to powder dry extract, granules show greater stability to ambient humidity, but do not completely fill this need. A more appropriate degree of protection is conferred by coating. The coating of granules containing dry extract of plantago reduced moisture and instability and improved the rheological properties, such as free flow, plasticity, and low friability, but influenced the dissolution rate [76].

However, tablets obtained from dry extracts may have some problems, such as their high hardness and long disintegration time. The compression method should also be consistent with the limit of the proportion at which the extracts can be incorporated into the final mixture for tablet compression, since active components of the extracts are diluted by co-extracted substances, and high dosages are required [77].

In addition, tableting machines require specific rheological properties of the material to be compressed, such as free flow, compressibility, appropriated cohesiveness, and lubricity. Most dry extracts lack sufficient flow properties for processing by direct compression, requiring the addition of a series of adjuvants, which can be classified according to their function in the tablet.

The literature contains few studies on direct compression of dry plant extracts. In these studies, dry extracts have been used of between 20% and 58% (w/w) of the tablet weight, with varying compression force or formulation [56,77,78]. In fact, the poor rheological characteristics and the high hygroscopicity shown by the majority of dry extracts make it difficult to obtain tablets by the direct route. Thus, the development of a formulation for direct compression depends on the properties of the adjuvants, ensuring that the proportion of the active substance does not exceed around 10% of the total weight of the formulation. Otherwise, characteristics of the formulation will be determined by its active component properties.

In the technology for compression of herbal drug extraction products, studies related to adjuvants effects on formulation performance have been focused on, in order to overcome or minimize problems such as flow, cohesiveness, compression sensibility, disintegration, dissolution, hygroscopicity, hardness, and friability.

Adjuvants used in the development of the tablet core containing herbal dry extracts include binders, lubricants, and disintegrants. Microgranulated lactose, microcrystalline cellulose, modified starch, dicalcium phosphate, Lactose and co-processed lactose–cellulose have been reported as binder adjuvants, and colloidal silicon dioxide, magnesium stearate and talc are the most commonly used lubricants. As disintegrants, croscarmellose sodium, crospovidone, sodium starch glycolate, and other starch derivatives such as pregelatinized starch and physically modified starch, are preferable, due to their superdisintegrant property, which enables their use at minimal concentrations in dosage forms with high herbal preparation content.

Studies have already shown that replacing Avicel PH 101 with PH 102, and adding magnesium stearate after Aerosil, improved the flow properties [79]; tablets with low friability were obtained from the mixture of dicalcium phosphate and Avicel PH 101 [80]; microcrystalline cellulose was superior to lactose in terms of friability and disintegration time [81]; faster disintegration of the tablets was obtained after dry granulation [82]; and Avicel PH101 and Lactose CD formulation showed the best pharmaceutical performance when compared to Emcompress and Avicel or Lactose mixture [32].

In general, tablets containing dry herbal extracts are highly hygroscopic, requiring a protective coating against moisture. Therefore, other studies have focused on the use of polymeric coatings to increase the stability and biopharmaceutical characteristics of dry extract derivatives [30,76,83,84].

These studies have shown some potential problems in terms of compatibilities and operations. For example, ethylcellulose (Ethocel Premium Standard-7) and derived basic polymethacrylic acid (Eudragit E-100), and hypromellose (Pharmacoat -603), and copolymers polyvinyl acetate – povidone (Kollidon-VA-64) are able to improve the stability of herbal chemicals, such as matricine, while carboxylic groups of Polymethacrylate (Eudragit-L-100) caused rapid decomposition of that same chemical [83]. For larger scale production, mixing Aquacoat E30D has been preferred, for example, because of the potential risk of clotting in the preparation of the suspension of Eudragit L30D and hypromellose [84].

10.7 QUALITY ASSURANCE AND QUALITY CONTROL OF HERBAL DRUGS AND PHYTOPHARMACEUTICALS

Quality control refers to the monitoring compliance with the material specifications. It is, therefore, an essential part of the good manufacturing practices (GMP) in the area of quality assurance. A specification is defined as a list of tests, references, analytical and biological procedures, and appropriate acceptance criteria [85]. Specifications are part of a total control strategy for the herbal preparation and herbal medicinal product, designed to ensure product quality and consistency. Other parts of this strategy are adherence to good agricultural and collection practices (GAP) [1,85,86], and a validated manufacturing process [85].

In the case of the phytomedicines, the overall quality control must be applied in each manufacturing stage, including the vegetable raw materials or *herbal drugs* (and excipients), in-process testing in the development of the intermediate products (or *herbal preparations*), and the dosage form (or *herbal medicinal product*), as well as the process validation, stability testing, and testing for batch consistency [85].

Monographs published in a pharmacopeia are the most practical statement for quality control of herbal drugs, and there are many available. The Brazilian Pharmacopeia [87] has completed monographs including botanical, macro- and microscopic drawings, among other specifications, for more than 40 medicinal plants with traditional use in Brazil.

When pharmacopeial monographs are unavailable, development and analytical validation of procedures must be carried out by the manufacturer. Since specifications are chosen to confirm the quality rather than to characterize the product, the manufacturer should provide the rationale and justification for including and/or excluding testing for specific quality attributes. Also, specifications should be based on data obtained from the batches used to demonstrate manufacturing consistency [85].

According to the World Health Organization (WHO) [88] reasonable limits may be established using simple statistical methods, for example, control chart techniques.

Analytical results from about 20 successive batches are pooled together, and the overall average and "three sigma limits" (±3 standard deviations from the overall average) are calculated. These calculations are applicable when more than one individual or independent sample per batch is analyzed [1,89].

Prior to performing the quality control test, there is an essential step of sampling the plant material, since the reliability of any conclusions drawn from the analysis of a sample will depend upon how closely the sample represents the whole batch [90].

10.7.1 Recommended Procedures of Sampling of Material in Bulk

Initially, inspection of each container or packaging unit is recommended, to ensure conformity with the pharmacopoeia monographs, or other requirements for packaging and labeling, to check the condition of the package and the presence of any defects that could influence the quality or stability of the contents (physical damage, moisture, etc.). If the initial inspection indicates that the batch is uniform, the sampling may be set based on the container number, or the division degree and the quantity [87,90,91]. The European Pharmacopoeia [91] recently introduced a new text (2.8.20 Herbal drugs: sampling and sample preparation).

10.7.2 Parameters for Quality Control of Herbal Drugs

A definition is needed of the qualitative statement of the botanical source, the plant part used, its state (whole, reduced, powdered, fresh, dry, etc), the geographical source(s) and the conditions under which the herbal drug was obtained [85].

10.7.2.1 Identification. The identification tests not only give subjective or objective information about the authenticity of the material, but also relate to the purity of the samples, or the presence of contaminants, which can affect the expected result in comparison with authentic material.

The analysis of visual aspect, color, consistency, odor, or taste of herbal materials is an important initial subjective method for determining their identity and purity, in comparison with authentic material. This simple analysis enables damage to the samples to be detected, such as mold contamination, odor, and color changes to the herbal drug as a result of storage conditions or shelf life [92].

Botanical identification requires analysis of different parts of the plant, including the flowers, fruits, leaves, and bark comparing the results with the specific literature [92]. These analyses include features that distinguish the herbal drug from potential adulterants [85].

A primary visual evaluation (macroscopic) can be used to ensure that the plant is of the required species, and that the right part of the plant is being used. Macroscopic identity of medicinal plant materials is based on shape, size, color, surface characteristics, texture, fracture characteristics, and cut surface appearance. A detailed procedure for analysis of size, color, surface, texture, and fracture characteristics is described [90]. Microscopic analysis is also required to determine the correct species and/or correct part of the plant in the sample [93]. Also, microscopy of powder herbal drug is very useful considering that is the most commonly available form of commercialized herbal drugs [87].

A simple chromatographic technique such as thin layer chromatography (TLC) may provide valuable additional information for establishing the identity of the plant material. Also, there are reaction tests to identify classes of components, such as color reactions, or fluorescence emission, among others, which may play an important role as auxiliary tests to determine the authenticity of the herbal drug. Based on the concept of phytoequivalence, chromatographic fingerprinting of herbal medicines can be used as part of the quality control. Methods based on information theory, similarity estimation, chemical pattern recognition, spectral correlative chromatograms, multivariate resolution, and a combination of chromatographic fingerprinting and chemometric evaluation of fingerprints are all powerful tools for the quality control of herbal products [93].

10.7.2.2 Purity Tests

10.7.2.2.1 Determination of Foreign Matter. The herbal drug should be free of stones, sand, poisonous, molds, insects, and other animal contaminations (foreign elements). Additionally, the presence of other parts of the plant beyond what is stated as part of herbal drug is also considered foreign matter (foreign organs). In general no more than 2% of foreign matter (w/w) is acceptable [87,94]. Macroscopical examination can be easily employed, although microscopy is indispensable in certain special cases (for example, starch deliberately added to "dilute" the plant material) or the use of TLC to detect chemical residues [93].

10.7.2.2.2 Ash Content. Ash represents the nonvolatile inorganic components of the herbal drug when the plant material is burned at high temperatures (>450°C). The residual ash is measured as total ash and includes the ash derived from plant tissue itself, and "nonphysiological" ash, which is the residue of the foreign matter (e.g., sand and soil) adhering to the surface of the plant. For herbal drugs, it is also necessary to determine the acid-insoluble ash content, that is, the residue obtained after boiling the total ash with dilute hydrochloric acid and igniting the remaining insoluble matter to measure the amount of silica, present as sand and siliceous earth [87].

The differences among pharmacopeias and the recommended procedures for herbal drugs are discussed by Zhi-Cen [95].

10.7.2.2.3 Loss on Drying and Water Content. The quality issues associated with the excess of water in herbal materials are microbial growth, the presence of fungi or insects, and subsequent deterioration by hydrolysis. Limits for water content should therefore be established for each type of plant material [90]. The usual range established in the pharmacopeias is 8–14% [92].

Various methodologies have been established in the pharmacopoeias [87,91] and WHO guidelines [90] such as: (i) The azeotropic method; (ii) Karl Fischer's volumetric water assay, which specifically determines the water content in the sample; and (iii) the test for loss on drying, which determines both water and volatile matter, at 100–105°C until constant weight, or for 3 h; or in a desiccator over phosphorus pentoxide under atmospheric pressure or reduced pressure, at room temperature, for a specified period of time The room temperature desiccation method is especially useful for materials that melt to a sticky mass at high temperatures, or those with a high content of volatile oil [90,92].

10.7.2.2.4 Inorganic Impurities, Toxic Metals. Contamination of medicinal plant materials by heavy metals can be attributed to many causes, including environmental pollution and traces of pesticides [88]. According to EMEA [85], the need for inclusion of these tests and acceptance criteria should be studied during the herbal drug development, and should be based on knowledge of the plant species, the cultivation, and manufacturing processes, and safety.

DeStefano et al. [96] adopted the term "elemental impurities" as an alternative to the poorly defined term "heavy metals," since the first term includes various transition metals and metalloids. In addition to catalyst or reagents normally used in chemical synthesis, sources of elemental impurities in pharmaceuticals may include minerals used in the manufacture of excipients, container-closure systems, and other contact surface products.

The pharmacopeial forum [97] classifies the elemental impurities in Class 1 and Class 2. The former are elements that should be essentially absent; these are known or strongly suspected to be human toxicants, and environmental hazards (Table 10.4). Class 2 impurities are those that should be limited, and are less toxic than Class 1 or are deliberately (chromium, copper, manganese, molybdenum, nickel, palladium, platinum, vanadium, osmium, rhodium, ruthenium, and iridium). The European Pharmacopoeia [98] has issued a draft monograph entitled *Herbal drugs,* in which the limits shown in Table 10.4 are proposed.

10.7.2.2.5 Microbial Limits. Herbs and herbal materials normally carry a large number of bacteria and molds, often originating in soil or derived from manure. While a large range of bacteria and fungi form the naturally occurring microflora of medicinal plants, aerobic spore-forming bacteria are often predominant [88].

Proliferation of microorganisms may result from failure to control the moisture levels of herbal drugs during transport and storage. The presence of *Escherichia*

TABLE 10.4 Elemental Limits in Dietary Supplements [97] or Herbal Drugs [90,98]

	Pharmacopeial Forum [97]	European Pharmacopoeia [98]	WHO [90]
Element	Limit (μg/g)	Limit (μg/g)	Limit (μg/g)
Arsenic (inorganic)	1.5	nd	nd
Cadmium	0.5	0.5	0.3
Lead	1.0	5.0	10
Mercury (total)	1.5	0.1	nd
Methylmercury (as Hg)	0.2	nd	nd

TABLE 10.5 Current Limits for Microbiological Quality of Medicinal Herbs and Herbal Products According to the European Pharmacopoeia [98]

	TAMC	TYMC	*E. coli*	*Salmonella spp.*	BTGNB
		CFU/g or CFU/mL			
Category A	10^7 (max. 5×10^7)	10^5 (max. 5×10^5)	10^3	Absence in 25 g mL^{-1}	NS
Category B	10^4 (max. 5×10^4)	10^2 (max. 5×10^2)	Absence in 1 g/mL	Absence in 25 g/mL	10^2
Category C	10^5 (max. 5×10^5)	10^4 (max. 5×10^4)	Absence in 1 g/mL	Absence in 25 g/mL	10^4

Abbreviations: TAMC = total aerobic microbial count; TYMC = total yeast and mold count; BTGNB = bile-tolerant Gram-negative bacteria; CFU = colony-forming units; NS = not set.

coli, *Salmonella* spp. and molds may indicate poor quality in the production and harvesting practices. This should be controlled by implementing best practice guidelines such as GACP and GMP [88,99]. Kosalec et al. [99] mention the lack of data for comparing biological contamination between wild and cultivated medicinal herbs, and recommend the use of GACP and GMP, to reduce the overall risk of contamination.

The microbial limits or absence of specified microorganisms in herbal pharmaceuticals have recently been harmonized [98,99], as shown in Table 10.5. Category A includes herbal medicinal products that contain herbal drugs with or without excipients and are intended for preparation in infusions or decoctions with boiling water (e.g., brewed tea). Category B includes extracts and/or herbal drugs that have been pretreated to reduce microbial contamination. If pretreatment (processing or extraction with low strength alcohol or nonboiling water) does not meet the Category B criteria of decontamination, the product falls into Category C (Table 10.5).

Brazilian Pharmacopeia [87] recently established the microbial limits (in CFU/g or CFU/mL) to the herbal drugs that should be submitted to hot (A) or cold (B) extractive process, as:

A: TAMC: 10^7; TYMC: 10^5; 2 *Escherichia coli*: 10^2 in 1 g; BTGNB: 10^4 in 1 g; absence of Salmonella in 10 g

B: TAMC: 10^5; TYMC: 10^3; 2 *Escherichia coli*: 10^1 in 1 g; BTGNB: 10^3 in 1 g; absence of Salmonella in 10 g

10.7.2.2.6 Mycotoxins. Mycotoxins are secondary metabolites produced by molds, and can be toxic to humans. Aflatoxins are natural mycotoxins produced by *Aspergillus flavus* and *Aspergillus parasiticus* that are common and widespread in nature and are most often found when certain grains are grown under conditions of stress, such as high moisture content and high temperatures. Thus, herbal drug that are subject to contamination by aflatoxins are tested by a validated method such as liquid chromatography, as proposed by the European Pharmacopoeia, [91]. Recently, the European Pharmacopoeia [100] has set limits for aflatoxin 1 (2 μg/kg) and the sum of

B_1, B_2, G_1, and G_2 aflatoxins (4 μg/kg) for some herbal drugs. A recent review show the results of herbal drug samples analyzed for molds that produce mycotoxin [99].

10.7.2.2.7 Pesticides, Fumigation Agents. Pesticides are chemical compounds used to control or eradicate pests. Based on intended use, they are grouped as insecticides, fungicides, nematicides, herbicides, rodenticides, and others [101]. Chapter 2.8.13 of the European Pharmacopoeia [91] indicates the limits that apply to various pesticides, with an equation to calculate the limits for other pesticides and additionally the procedures test for analysis of organochlorine, organophosphorus, and pyrethroid insecticides. Kosalec et al. [99] compared the pesticide limits between the European Pharmacopoeia, the United States Pharmacopeia, and the Codex Alimentarius Commission for medicinal plant material, and discuss the regional limits and indicated methodologies for determining pesticide residues. Chromatography is recommended as the principal method for determining pesticide residues. However, these techniques are not universally applicable. The WHO [90] discussed main problems related to analytical methodologies, including stability of analytes and solvent effect, among others.

Fumigation agents, such as ethylene oxide, methyl bromide, and phosphine, are considered potential contaminants of herbal drugs [102]. The use of ethylene oxide for the decontamination of herbal drugs is prohibited in Europe [103]. It is recommended that manufacturers consider alternative strategies for pest control of herbal substances used in herbal medicinal products, and that this treatment should be carried out at the earliest possible stage, by adequately trained and qualified personnel, according to the specific recommendations for use [104].

10.7.2.2.8 Radioactive Contamination. Many source of ionization radiation, including radionuclides, occur naturally in the environment, and a certain degree of exposure is inevitable. But dangerous levels of contamination can be the consequence of a nuclear accident [93]. The amount of exposure to radiation depends on the intake of radionuclides, and its significance depends, in turn, on other variables such as age, metabolic kinetics and the weight of the individual who ingests them (also known as the dose conversion factor). Therefore, at present, no limits are proposed for radioactive contaminations [88,105].

10.7.2.3 Assay. The marker assay should consider the active markers, defined by the constituents or groups of constituents that are generally accepted as contributing to the therapeutic activity, and the analytical markers. The latter is represented by classes that serve analytical purposes. Both are chemically defined constituents of interest for quality control purposes [85,106]. In the case of herbal drug with constituents of the known therapeutic activity or with active markers, assays of their content are required, with details of the analytical method and including, where possible, a specific, stability-indicating assay method (SIAM) [85].

The importance has been pointed out of the use of SIAMs for evaluating plant materials, especially herbal preparations and herbal medicinal products. Among them, the stability-indicating assay methods required by the International Conference on Harmonization (ICH) guidelines are more easily applied to well established or existing synthetic drug substances and products. The development and validation of SIAMs for combination preparations of two or more drugs, and thus, to the inherent complex herbal medicinal product, are very difficult and there may be no single SIAM that profiles their stability characteristics, unlike those of chemical drugs [85,107]. Thus, the manufacturer should propose a series of SIAMs, to ensure that changes in product quality during its shelf-life will be detected, according to the respective guidelines [108–110].

Bakshi and Singh [107] discuss the difference between specific stability, indicating the assay method (specific SIAM) and a selective indicating assay method (selective SIAM). The authors suggest that the former can be defined as "a method that is able to unequivocally measure the drug(s) in the presence of all degradation product, excipients and additives expected to be present in the formulation." The latter, on the other hand, can be defined as "a method that is able to unequivocally measure the drug(s) *and* all degradation products, in the presence of excipients and additives expected to be present in the formulation." Thus, the selective SIAM is the method that is selective to the drug (markers and other constituents) as well its degradation products, and is also specific to all the cited components.

For herbal drug in which the active constituents are unknown, the assay of analytical markers or other justified determinations are required [85]. Scientists and regulatory agencies have paid attention to the selection of chemical markers in quality control [111]. Srinivasan [112] proposed the following four markers categories: active principles, active markers, analytical markers, "phantom" markers, and negative markers. Li et al. [111] proposed eight marker categories, and discussed their applications and the common problems involved. The authors also point out the lack of chemical and pharmacological data of chemical markers and the challenges involved in their production.

10.7.2.4 Other Tests

10.7.2.4.1 Extractable Matter. According to the WHO [90], this method determines the amount of active constituents extracted with solvents from a given amount of medicinal plant material. It is used for materials for which no suitable chemical or biological assay yet exists. There are two methods: by hot extraction, using water as solvent; and by cold maceration with the appropriate solvent. The Brazilian Pharmacopeia [87] describes three methods, by soxhlet (with ethanol as solvent), by hot extraction (with water as solvent) and by cold extraction (using an appropriate solvent).

10.7.2.4.2 Estimate of Particle Size and Distribution. Medicinal plant materials are used either whole, or in cut or powdered form. Cut medicinal plant materials are prepared by cutting or crushing the plant into small pieces [90]. The degree of fineness of a powder may be expressed by reference to sieves and is defined in relation to the sieve number(s) used either by means of the following terms, or, where such terms cannot be used, by expressing the fineness of the powder as a percentage m/m passing the sieve used [91] in Chapter 2.9.12 (volume 1) and also in Chapters 2.9.35 and 2.9.36 (supplement 6.2 of 6th edition). The classification of powders by degree of fineness is given in the supplement. When the powder is <75 μm, light diffraction can be used for measuring the particle size [91].

10.7.2.4.3 Other Appropriate Tests. Swelling index, hemolytic index, bitterness value, foaming index, volatile and fixes oils, stomata and stomatal index determination are other appropriate tests described in the pharmacopeias for herbal drugs, which may be useful for establishing supplementary specifications.

10.7.3 Parameters for Quality Control of Herbal Preparations

For the herbal preparations, according to their physical state (powders, tinctures, oils, resins, etc) an appropriate specification must be developed based on recent scientific data [85], in view of the following requirements.

(a) *Definition*: the botanical source, the type of preparation (e.g., dry or liquid extract) and the ratio of the herbal drug to the genuine herbal preparation.
(b) *Characters*: a qualitative statement about the organoleptic characters. In addition to the organoleptic analysis, physicochemical characterization is necessary, according the physical state of the herbal preparation [87,91] such as liquid extract: dry residue, pH, alcohol content, density; dry extract: loss on drying, solubility, apparent density; to essential oils: density, solubility in alcohol, refraction index, residue on evaporation, optical rotation; fixes oils: acid index, ester index, and iodine index; among others.
(c) *Identification*: specific tests for the herbal preparation in view of discriminate the possible adulterants that are like to occur (e.g., the combination of HPLC/UV-diode array, HPLC-TLC-densitometry, HPLC/MS or GC/MS, may be acceptable).
(d) *Purity tests*: water content, residual solvents, inorganic impurities, toxic metal, microbial limits, mycotoxins, pesticides, and fumigation agents, among others.

The water content is important in the case of hygroscopic preparation, and its acceptance criteria may be justified with data on the effects of hydration or moisture absorption [85].

The test and acceptance criteria for inorganic impurities and toxic metals should be studied during the development, and based on knowledge of the plant species. If the latter reduces the burden of toxic residues, the tests with the herbal drug may be sufficient. The procedures and acceptance criteria for sulfated ash/residue on ignition should follow pharmacopeial precedents. Other inorganic impurities may be determined by other appropriate procedures, for example, atomic absorption spectroscopy [85].

The procedure and microbial limits should comply with those in the statement in the European Pharmacopoeia (Table 10.5) [98]. The Brazilian Pharmacopeia [87] establishes the microbial limits (in CFU/g or CFU/mL) for the following herbal preparations: dried extract (A) or tinctures/fluid extract (B), as: (A) = TAMC: 10^4; TYMC: 10^3; absence of *Salmonella* and *Escherichia coli* in 10 g; (B) = TAMC: 10^4; TYMC: 10^3.

The potential for contamination by mycotoxins, pesticides, and fumigant agent, among other potential residues, should be fully considered, and appropriate validated methods should be used for their control, as stated in the European Pharmacopoeia [91,98].

Depending on the herbal preparation, there are specific purity tests (as for essential oils for measuring fatty and resinified oils, water, or foreign esters in essential oils, etc. [91].

(e) *Assay*: in the case of herbal preparations with constituents of known therapeutic activity or with active markers, assays of their content are required, with details of the analytical procedure. Where possible, a specific, stability-indicating procedure should be included to determine the content of the herbal drug in the herbal preparation. In cases where a nonspecific assay is used (such as a UV/VIS spectrophotometric assay) a combination of the assay and a suitable identification test (such as fingerprint chromatography) can be used, as well as other supporting analytical procedures, to determine the overall specificity.

10.7.4 Parameters for Quality Control of Herbal Medicinal Product

The tests applied to the herbal drug and herbal preparation may be adapted, based on the excipients of pharmaceutical dosage form, in order to analyze the qualitative aspects of the herbal medicinal product, such as the identification of the markers, chemical reactions and chromatographic analysis, among others. Also, to analyze the purity aspects, the previous tests of herbal preparation may be adapted. However the pharmaceutical dosage form should be taken into account.

The following tests and acceptance criteria are considered generally applicable to all herbal medicinal products [85].

(a) *Description*: a qualitative description of the dosage form should be provided (e.g., size, shape, and color). The acceptance criteria should include the final acceptable appearance at the end of the shelf-life. If the color changes occur during storage, a quantitative procedure may be appropriate.

(b) *Impurities*:

- If impurities arising from the herbal drug (s) and/or herbal preparations, for example, contaminants such as pesticide/fumigant residues, and heavy metals are controlled during the testing of the herbal substance/preparation, it is not necessary to test for these in the herbal medicinal product.
- Similarly, residual solvent arising from the manufacture of the herbal preparation (e.g., an extract) need not be controlled in the herbal medicinal

product provided it is appropriately controlled in the extract specification. However, solvents used in tablet coating, for example, need to be controlled in the dosage form.

- In cases where degradation products of the herbal substance/preparation are evident (e.g., aglycones from hydroxyanthracene glycosides), these should be monitored in the herbal medicinal product. The acceptance limits should be stated for such degradation products, but when it has been demonstrated conclusively that the herbal drug and/or herbal preparation do not degrade in the specific formulation and under the specific storage conditions proposed in the marketing authorization, the testing may be reduced or eliminated, with the approval of the regulatory authorities. *Microbial limits*: these limits should comply with the European Pharmacopoeia (Table 10.5) and the frequency of testing should be justified. Follow Brazilian Pharmacopeia [87], the limits (in CFU/g or CFU/mL) to oral herbal medicinal products are TAMC: 10^4; TYMC: 10^2; absence of *Escherichia coli* and *Staphylococous aureus* in 1 g, or 1 mL; absence of *Salmonella* in 10 g or 10 mL and BTGNB: 10^2 in 1 g or 1 mL.

(c) *Assay*: the studies of analytical development applied to the herbal preparations may be used for herbal medicinal products.

(d) *Other tests*: in addition the physical tests and other tests related to the specific performance of the pharmaceutical dosage form must be carried out (Table 10.6).

TABLE 10.6 Analytical Tests to Herbal Medicinal Product [85]

	Pharmaceutical Dosage Forms			
Tests	Tablet	Capsules	Oral Liquid	Semisolid
Dissolution	x	x	x	x
Disintegration	x	x		
Hardness	x			
Friability	x			
Uniformity of unitary doses	x	x	x	
Microbial limits	x	x	x	x
pH			x	x
Antimicrobial and antioxidante preservative content			x	x
Extractables			x	x
Alcohol content			x	
Particle size distribution			x	
Redispersibility			x	
Rheological properties			x	x
Specific gravity			x	x
Reconstitution time			x	
Water content	x	x	x	

Phytopharmaceutical manufacturing is a relevant tool of economic growth, especially for developing countries, and represents an alternative for the management of health problems. Among the challenges for the improvement of the phytopharmaceutical global market is the need of quality acceptance criteria and harmonization of regulatory aspects. Last but by no means least each country must support the improvement of a productive chain, applying appropriate sustainable practices for the use of their biodiversity.

REFERENCES

[1] World Health Organization (WHO), Guidelines on agricultural and collection practices (GACP) for medicinal plants, WHO, Geneva, 2003, 80 p. WHO guidelines on good agricultural and collection practices (GACP) for medicinal plants.

[2] Siani, A.C., *Desenvolvimento Tecnológico de Fitoterápicos: Plataforma Metodologica – Relatorio PPA/MCT 2000–2003*. 1 st. ed., Scriptorio, Rio de Janeiro, Brazil, 2003.

[3] Sharapin, N., Matérias-primas vegetais para a indústria de fitofármacos. In: Sharapin, N., editor. *Fundamentos de Tecnologia de Produtos Fitoterápicos*. Quebecor-Impreandes, Santafé de Bogotá, 2000, pp. 17–26.

[4] Martins, E.R., Figueiredo, L.S., Cultivo de plantas medicinais. In: Leite, J.P.V., editor. *Fitoterapia*. Atheneu, São Paulo, 2009, pp. 143–168.

[5] Magalhães, P.M., *O Caminho Medicinal das Plantas*. UNICAMP, Campinas, 1997.

[6] Martins, E.R., Castro, D.M., Castellani, D.C., Dias, J.E., *Plantas Medicinais*. 5th ed., UFV, Viçosa, 2003.

[7] Saxena, R.C., Insecticides from neem. In: Arnason, J.T., Philogene, B.J.R., Morand, P., editors. *Insecticides of Plant Origin*. ACS, Washington, 1989, pp. 110–129.

[8] Oliveira, M.S.S., Roel, A.R., Arruda, E.J., Marques, A.S. (2007). Efficiency of extracts of plantas in control of fall armyworm in corn *Spodoptera frugiperda* (J.E. Smith, 1797) (Lepidoptera: Noctuidae). *Ciência Agrotécnica*, *31*, 326–331.

[9] Scheffer, M.C., Corrêa Júnior, C., *Boas Práticas Agrícolas (BPA) de Plantas Medicinais, Aromáticas e Condimentares (Plantas Medicinais & Orientações Gerais para o Cultivo)*. MAPA/SDC, Brasília, 2006.

[10] Marchese, J.A., Figueira, G.M. (2005). O uso de tecnologias pré e pós-colheita e boas práticas agrícolas na produção de plantas medicinais e aromáticas. *Revista Brasileira de Plantas Medicinais*, *7*, 86–96.

[11] Ushirobira, T.M.A., Yamaguti, E., Uemura, L.M., Audi, E. A., de Mello, J.P.C. (2004). Avaliação físico-química de sementes de guaraná secas por diferentes métodos. *Revista Brasileira de Farmacognosia*, *14*, 15–20.

[12] List, P.H., Schmidt, P.C., *Phytopharmaceutical Technology*. CRC, Boca Raton, FL, 1989.

[13] Sonaglio, D., Ortega, G.G., Petrovick, P.R., Bassani, V.L. In: Simões, C.M.O., Schenkel, E.P., Gosmann, G., Mello, J.C.P., Mentz, L.A., Petrovick, P.R., editors. *Farmacognosia: Da Planta ao Medicamento*. UFRGS, UFSC, Porto Alegre, Florianópolis, 2003, pp. 289–326.

[14] Santos, E.V.M., Extração de matérias-primas vegetais. In: Sharapin, N., editor. *Fundamentos de Tecnologia de Produtos Fitoterápicos*. Quebecor-Impreandes, Santafé de Bogotá, 2000, pp. 27–61.

[15] Marques, L.C., Vigo, C.L.S., Preparação e padronização de extratos vegetais. In: Leite, J.P.V., editor. *Fitoterapia*. Atheneu, São Paulo, 2009, pp. 169–204.

[16] Voigt, R., *Pharmazeutische Technologie*. 9th ed. Deutscher Apotheker, Ausgart, 2000.

[17] Fang, X., Wu, J. (1998). Feasibility of sterilizing traditional medicines by gamma-irradiation. *Radiation Physics and Chemistry*, *52*, 53–58.

[18] Brasil, (2001). RDC n. 21, de 26 de janeiro de 2001 (Aprova o regulamento técnico para irradiação de alimentos). *Diário Oficial da União*, 29.01.2001.

[19] Owczarczyk, H.B., Migdal, W., Kedzia, B. (2000). The pharmacological activity of medical herbs after microbiological decontamination by irradiation. *Radiation Physics and Chemistry*, *57*, 331–335.

[20] Lacaz, C.S., Porto, E., Martins, J.E.C., Vaccari, E.M.H., Melo, N.T., *Tratado de Micologia Médica*. 9th ed. Sarvier, São Paulo, 2002.

[21] Liu, X., Qiu, Z., Wang, L., Chen, Y. (2011). Quality evaluation of *Panax notoginseng* extract dried by different drying methods. *Food and Bioproducts Processing*, *89*, 10–14.

[22] León-Martínez, F.M., Méndez-Lagunas, L.L. Rodríguez-Ramírez, J. (2010). Spray drying of nopal mucilage (*Opuntia ficus-indica*): Effects on powder properties and characterization. *Carbohydrate Polymers*, *81*, 864–870.

[23] Cunha, A.M., Menon, S., Menon, R., Couto, A.G., Bürger, C., Biavatti, M.W. (2010). Hypoglycemic activity of dried extracts of *Bauhinia forficata* Link. *Phytomedicine*, *17*, 37–41.

[24] Runha, F.P., Cordeiro, D.S., Pereira, C.A.M., Vilegas, J., Oliveira, W.P. (2001). Production of dry extracts of medicinal Brazilian plants by spouted bed process: Development of the process and evaluation of thermal degradation during the drying operation. *Trans IChemE, 79*, Part C.

[25] Cordeiro, D.S., Oliveira, W.P. (2005). Technical aspects of the production of dried extract of *Maytenus ilicifolia* leaves by jet spouted bed drying. *International Journal of Pharmaceutics*, *299*, 115–126.

[26] Souza, C.R.F., Oliveira, W.P. (2005). Spouted bed drying of *Bauhinia forficate* link extract: The effects of feed atomizer position and operating conditions on equipment performance and product properties. *Brazilian Journal of Chemical Engineering*, *22*, 239–247.

[27] Temple, S.J., Tambala, S.T., van Boxtel, A.J.B. (2000). Monitoring and control of fluid-bed drying of tea. *Control Engineering Practice*, *8*, 165–173.

[28] Hsu, C.L., Chen, W., Weng, Y.M., Tseng, C.Y. (2003). Chemical composition, physical properties, and antioxidant activities of yam flours as affected by different drying methods. *Food Chemistry*, *83*, 85–92.

[29] Meterc, D., Petermann, M., Weidner, E. (2008). Drying of aqueous green tea extracts using a supercritical fluid spray process. *Journal of Supercritical Fluids*, *45*, 253–259.

[30] Endale, A., Gebre-Mariam, T., Schmidt, P.C. (2008). Granulation by roller compaction and enteric coated tablet formulation of the extract of the seeds of *Glinus lotoides* loaded on Aeroperl® 300 Pharma. *AAPS PharmSciTech*, *9*, 31–38.

[31] Masters, K., *Spray Drying Handbook*. Longman Scientific and Technical, UK, 1991.

[32] Palma, S.D., Manzo, R.H., Allemandi, D.A. (1999). Dry plant extract loaded on fumed silica for direct compression: Preparation and preformulation. *Pharmaceutical Development Technology*, *44*, 523–530.

[33] Oomah, B.D., Mazza, G. (2001). Optimization of a spray drying process for flaxseed gum. *International Journal of Food Science and Technology*, *36*, 135–143.

[34] Walton, D.E., Mumford, C.J. (1999). Spray dried products characterization of particle morphology. *Trans IChemE*, *77*, Part A.

[35] Patel, R.P., Patel, M. P., Suthar, A.M. (2009). Spray drying technology: An overview. *Indian Journal of Science and Technology*, *2*, 44–47.

[36] Goula, A.M., Adamopoulos, K.G. (2005). Spray drying of tomato pulp in dehumidified air: II. The effect on powder properties. *Journal of Food Engineering*, *66*, 35–42.

[37] Leon-Martinez, F.M., Rodriguez-Ramirez, J., Medina-Torres, L.L., Lagunasa, L.L.M., Bernad-Bernad, M.J. (2011). Effects of drying conditions on the rheological properties of reconstituted mucilage solutions (*Opuntia ficus-indica*). *Carbohydrate Polymers*, *84*, 439–445.

[38] Tonon, R.V., Brabet, C., Hubinger, M.D. (2008). Influence of process conditions on the physicochemical properties of açai (*Euterpe oleracea* Mart.) powder produced by spray drying. *Journal of Food Engineering*, *88*, 411–418.

[39] Moreira, G.E.G., Costa, M.G.M., de Souza, A.C.R., de Brito, E.S., de Medeiros, M.F.D., de Azeredo, H.M.C. (2009). Physical properties of spray dried acerola pomace extract as affected by temperature and drying aids. *LWT – Food Science and Technology*, *42*, 641–645.

[40] Obón, J.M., Castellar, M.R., Alacid, M., Fernández-López, J.A. (2009). Production of a red-purple food colorant from *Opuntia stricta* fruits by spray drying and its application in food model systems. *Journal of Food Engineering*, *90*, 471–479.

[41] Kha, T.C., Nguyen, M.H., Roach, P.D. (2010). Effects of spray drying conditions on the physicochemical and antioxidant properties of the Gac (*Momordica cochinchinensis*) fruit aril powder. *Journal of Food Engineering*, *98*, 385–392.

[42] Beirão-da-Costa, S., Duarte, C., Moldão-Martins, M., Beirão-da-Costa, M. L. (2011). Physical characterization of rice starch spherical aggregates produced by spray-drying. *Journal of Food Engineering*, *104*, 36–42.

[43] Gallo L., Llabot, J. M., Allemandi, D., Bucalá, V., Piña, J. (2011). Influence of spray-drying operating conditions on *Rhamnus purshiana* (Cáscara sagrada) extract powder physical properties. *Powder Technology*, *208*, 205–214.

[44] Wan, L.S., Heng, P.W., Chia, C.G. (1991). Preparation of coated particles using a spray-drying process with an aqueous system. *International Journal of Pharmacy*, *77*, 183–191.

[45] Chan, T.M. (1987). The effect of the specific surface area of heavy magnesium carbonate on its tableting properties. *Drug Development and Industrial Pharmacy*, *13*, 1989–2015.

[46] Vasconcelos, E.A.F., Medeiros, M.G.F., Raffin, F.N., Moura, T.F.A.L. (2005). Influência da temperatura de secagem e da concentração de Aerosil®200 nas características dos extratos secos por aspersão da *Schinus terebinthifolius* Raddi (Anacardiaceae). *Brazilian Journal of Pharmacognosy*, *15*, 243–249.

[47] Bhandari, B.R., Senoussi, A., Dumoulin, E.D., Lebert, A. (1993). Spray drying of concentrated fruit juices. *Drying Technology*, *11*, 1081–1092.

[48] Chiu, Y.T., Chiu, C.P., Chien, J.T., Ho, G.H., Yang, J., Chen, B.H. (2007). Encapsulation of lycopene extract from tomato pulp waste with gelatin and poly(γ-glutamic acid) as carrier. *Journal of Agricultural and Food Chemistry*, *55*, 5123–5130.

[49] Brückner, M., Bade, M., Kunz, B. (2007). Investigations into the stabilization of avolatile aroma compound using a combined emulsification and spray drying process. *European Food Research and Technology*, *226*, 137–146.

[50] Kanakdande, D., Bhosale, R., Singhal, R.S. (2007). Stability of cumin oleoresin microencapsulated in different combination of gum arabic, maltodextrin and modified starch. *Carbohydrate Polymers*, *67*, 536–541.

[51] Gharsallaoui, A., Roudaut, G., Chambin, O., Voilley, A., Saurel, R. (2007). Applications of spray-drying in microencapsulation of food ingredients: An overview. *Food Research International*, *40*, 1107–1121.

[52] Ruiz, C.M.A., Espinosa-Muñoz, L.C., Aviles-Aviles, C., González-García, R., Moscosa-Santillán, M., Grajales-Lagunes, A., Abud-Archila, M. (2009). Spray-drying of passion fruit juice using lactose–maltodextrin blends as the support material. *Brazilian Archives of Biology and Technology*, *52*, 1011–1018.

[53] Daiuto, E.R., Cereda, M.P. (2006). Influência da granulometria de grânulos de amido sobre a densidade aparente de extratos atomizados. *Revista de Ciências Farmacêuticas Básica e Aplicada*, *27*, 51–56.

[54] Valduga, A.T., Battestin, V., Finzer, J.R.D. (2003). Secagem de extratos de erva-mate em secador por atomização. *Ciência e Tecnologia de Alimentos*, *23*, 184–189.

[55] Silva Júnior, J.O.C., Vieira, J.L.F., Barbosa, W.L.R., Pereira, N.L. (2006). Caracterização físico-química do extrato fluido e seco por nebulização de *Symphytum officinale* L. *Revista Brasileira de Farmacognosia*, *16*, 671–677.

[56] González Ortega, G., Schmidt, P.C. (1995). Obtención de Comprimidos conteniendo Extractos Atomizados de Flor de la Pasión (*Passiflora incarnata* L.). *Acta Pharmaceutica Bonaerense*, *14*, 173–180.

[57] Rowe, R. C., Sheskey, P. J., Owen, S. C., *Handbook of Pharmaceutical Excipients*. 5th ed. American Pharmaceutical Association, PhP Pharmaceutical Press, Washington, 2006.

[58] Lalla, J.K., Nandedkar, S.Y., Paranjape, M.H., Talreja, N.B. (2001). Clinical trials of Ayurvedic formulations in the treatment of acne vulgaris. *Journal of Ethnopharmacology*, *78*, 99–102.

[59] Blumenthal, M. *The Complete German Commission E Monographs – Therapeutic Guide to Herbal Medicines*. American Botanical Council, Boston, MA, EUA, 1998, p. 685.

[60] Brown, D.J., Dattner, A.M. (1998). Phytotherapeutic approaches to common dermatologic conditions. *Archives of Dermatology*, *134*, 1401–1404.

[61] Calixto, J.B. (2003). Biodiversidade como fonte de medicamentos. *Ciência e Cultura* [online], *55*, 37–39.

[62] Morteza-Semnani, K., Saeedi, M., Hamidian, M. (2004). Anti-inflammatory and analgesic activity of the topical preparation of *Glaucium grandiflorum*. *Fitoterapia*, *75*, 123–129.

[63] Akamatsu, H., Komura, J., Asada, Y., Niwa Y. (1991). Mechanism of anti-inflammatory action of glycyrrhizin: Effect on neutrophil functions including reactive oxygen species generation. *Planta Medica*, *57*, 119–121.

[64] Azimov, M.M., Zakirov, U.B., Radzhapova, ShD. (1988). Pharmacological study of the anti-inflammatory agent glyderinine. *Farmakol Toksikol*, *51*, 90–93.

[65] Korting, H.C., Schäfer-Korting, M., Hart, H., Laux, P., Schmid, M. (1993). Anti-inflammatory activity of hamamelis distillate applied topically to the skin. *European Journal of Clinical Pharmacology*, *44*, 315–318.

[66] Hughes-Formella, B.J., Bohnsack, K., Rippke, F., Benner, G., Rudolph, M., Tausch, I., Gassmueller, J. (1998). Anti-inflammatory effect of hamamelis lotion in a UVB erythema test. *Dermatology*, *196*, 316–322.

[67] Zanella, A.H.; *Monografia (Graduação)*, Universidade do Vale do Itajaí, Brazil, 2005.

[68] Orafidiya, L.O., Agbani, E.O., Oyedele, A.O., Babalola, O.O., Onayemi, O. (2002). Preliminary clinical tests on topical preparations of *Ocimum gratissimum* linn leaf essential oil for the treatment of *Acne Vulgaris*. *Clinical and Drug Investigation*, *22*, 313–319.

[69] Calvo, M.I. (2006). Anti-inflammatory and analgesic activity of the topical preparation of *Verbena officinalis* L. *Journal of Ethnopharmacology*, *107*, 380–382.

[70] Govindarajan, R., Vijayakumar, M., Rao, C.V., Shirwaikar, A., Mehrotra, S. Pushpangadan, P. (2004). Healing potential of *Anogeissus latifolia* for dermal wounds in rats. *Acta Pharmaceutica*, *54*, 331–338.

[71] Arus-Pampin, L., Quevedo, A.P., Pulpeiro, O. G., Rivera, A.B., Martinez, L.S., Selles, A.N. (2003). Estabilidad de la crema elaborada a partir del extracto seco de la corteza de *Mangifera indica* L.(Vimang). *Acta Farmaceutica Bonaerense*, *22*, 335–338.

[72] Di Mambro, V.M., Fonseca, M.J.V. (2005). Assays of physical stability and antioxidant activity of a topical formulation added with different plant extracts. *Journal of Pharmacological and Biomedical Analysis*, *37*, 287–295.

[73] Bilia, A.R., Bergonzi, M.C., Mazzi, G., Vincieri, F.F. (2006). Development and stability of semisolid preparations based on a supercritical CO_2 Arnica extract. *Journal of Pharmacological and Biomedical Analysis*, *41*, 449–454.

[74] Soares, L.A.L., Ortega, G.G., Petrovick, P.R., Schmidt, P.C. (2005). Optimization of tablets containing a high dose of spray-dried plant extract: A technical note. *AAPS PharmSciTech*, *6*, E367–E371.

[75] Onunkwo, G.C., Udeala, O.K. (1995). Studies on *Rauwolfia vomitoria* root: III. Flow properties of *R. vomitoria* granulations. *STP Pharma Sciences*, *5*, 296–301.

[76] Diaz, L.M., Cal, S., Souto, S., Conchero, A., Gomez-Amoza, J.L., Martinez-Pacheco, R. (1997). Eudragit RS as major excipient in controlled release tablets for plant dry extracts. *Pharmazie*, *52*, 466–470.

[77] Vennat, B., Gross, D., Pourrat, A., Pourrat, H. (1993). Tablets of hamamelis dry extract by direct compression: comparative study of natural starches and starch derivatives. *Drug Development and Industrial Pharmacy*, *19*, 1357–1368.

[78] Linden, R., González Ortega, G., Petrovick, P.R., Bassani, V.L. (2000). Response surface analysis applied to the preparation of tablets containing high concentrations of vegetable spray-dried extract. *Drug Development and Industrial Pharmacy*, *26*, 441–446.

[79] Plaizier-Vercammen, J.A., Bruwier, C. (1986). Evaluation of excipients for direct compression of the spray-dried extract of *Harpagophyton procumbens*. *STP Pharma Sciences*, *2*, 525–530.

[80] Plaizier-Vercammen, J.A., Bourgeois, A., De Boeke, L. (1991). Evaluation of Emcocel®50 and Emcocel®90, a new excipient in direct compression. *Drug Development and Industrial Pharmacy*, *17*, 763–776.

[81] De Souza, T.P., Gonzáles Ortega, G., Bassani. V.L., Petrovick, P.R. (2000). Avaliação da viabilidade de compressão direta de formulações contendo alto teor de produto seco nebulizado de *Maytenus ilicifolia*. *Acta Farmaceutica Bonaerense*, *19*, 91–98.

[82] Rocksloh, K., Rapp, F.R., Abu Abed, S., Müller, W., Reher, M., Gauglitz, G., Schmidt, P.C. (1999). Optimization of crushing strength and disintegration time of a high-dose plant extract tablet by neural networks. *Drug Development and Industrial Pharmacy*, *25*, 1015–1025.

[83] Schmidt, P.C., Soyke, B. (1992). Development of a chamomile preparation containing matricine. *Pharmazie*, *47*, 516–518.

[84] Jekö, Zs.B., Máté, M., Krausz, E., Bene, M. (1999). Development and scale up of a new film-coated tablet containing dry herba extract. *Pharmazie*, *54*, 148–150.

[85] European Medicines Agency Inspections (EMEA). Guidelines on specifications: Test procedures and acceptance criteria for herbal substances, herbal preparations and herbal medicinal products/traditional herbal medicinal products, Rev. 1, March 2006, p. 21.

[86] AHPA-AHP. American Herbal Products Association. Good Agricultural and Collection Practice for Herbal Raw Materials, December 2006, p. 32.

[87] *Farmacopeia Brasileira*. Anvisa, Brasília, 2010, Vol. 1, 5th ed., p. 515.

[88] World Health Organization (WHO). Guidelines for Assessing Quality of Herbal Medicines with References to Contaminants and Residue. WHO, Geneva, 2007, p. 105.

[89] *The International Pharmacopeia*, Vols. 1–2. WHO, Geneva, 2006.

[90] World Health Organization (WHO). *Quality Control Methods for Medicinal Plant Materials*. WHO, Geneva, 1998, p. 117.

[91] *European Pharmacopoeia*. 6th ed., Council of Europe, Strasbourg, 2009, p. 4.392.

[92] Farias, M.R., Avaliação da qualidade de matérias-primas vegetais. In: Simões, C.M.O., Schenkel, E.P., Gosmann, G., Mello, J.C.P., Mentz, L.A., Petrovick, P.R., editors. *Farmacognosia: da Planta ao Medicamento*. Porto Alegre, Florianópolis, UFRGS, UFSC, 1999, pp. 197–220.

[93] Bandaranayake, W.M., Quality Control, Screening, Toxicity, and Regulation of Herbal Drugs. In: Ahmad, I., Aqil, F., Owais, M., editors. Modern Phytomedicine. Wiley, New York, 2006, pp. 25–57.

[94] *British Pharmacopeia*. HMSO, London, 2010, p. 3828.

[95] Zhi-Cen, L., *General Methods for Vegetable Drugs*. World Health Organization, Geneva, 1980.

[96] Destefano, A.J., Zaidj, K., Cecil, T.L., Giancaopro, G.I. (2010). Elemental impurities – information. *Pharmacopeial Forum*, *36*(1), 2–9.

[97] Pharmacopeial Forum. (2010). (232) Elemental impurities-limits. *The United States Pharmacopeial Convention*, *36*(1), 2010, pp. 1–4.

[98] Herbal Drugs Monograph 1433. *Pharmaeuropa*. 2008. 20 2), pp. 302–303.

[99] Kosalec, I., Cvek, J., Tomic, S. (2009). Contaminants of medicinal herbs and herbal products. *Archives of Industrial Hygiene and Toxicology*, *60*, 485–501.

[100] European Pharmacopoeia. Directorate for the Quality of Medicines and HealthCare of the Council of Europe (EDQM). Determination of ochratoxin A in herbal drugs. In: *European Pharmacopoeia*, 6th ed., EDQM, Strasbourg, 2007, pp. 256–257.

[101] Britt, J.K., Properties and effects of pesitcides. In: Williams, P.L., James, R.C., Roberts, S.M., editors. *Principles of Toxicology: Environmental and Industrial Applications*. John Wiley and Sons Inc., New York, 2000, pp. 345–366.

[102] Mosihuzzaman, M., Choudhary, M.I. (2008). Protocols on safety, efficacy, standardization, and documentation of herbal medicine (IUPAC technical report). *Pure and Applied Chemistry*, *80*, 2195–2230.

[103] The Council of the European Communities. Directive 86/355/EEC, 21 July, 1986.

[104] European Medicines Agency Inspections (EMEA); Reflection Paper on the Use of Fumigants. EMEA/HMPC/125562/06, 2006, p. 3.

[105] WHO Traditional Medicine Strategy: 2002–2005. Geneva, World Health Organization, 2002 (WHO/EDM/TRM/2002.1), p. 63.

[106] European Medicines Agency Inspections (EMEA) Guidelines on declaration of herbal substances and herbal preparations in herbal medicinal products/traditional herbal medicinal products. EMA/HMPC/CHMP/CVWP/287539/2005.Rev. 1, March 2010, p. 18.

[107] Bakshi, M., Singh, S. (2002). Development of validated stability-indicating assay methods – critical review. *Journal of Pharmaceutical and Biomedical Analysis*, *28*, 1011–1040.

[108] European Medicines Agency Inspections (EMEA) Note for Guidance on Stability Testing: Stability Testing of New Drug Substances and Products. Rev. 2, CPMP/ICH/2736/99. August, 2003, p. 20.

[109] European Medicines Agency Inspections (EMEA) Guidelines on Stability Testing of Existing Active Substances and Related Finished products. CPMP/QWP/122/02. December, 2003, p. 18.

[110] European Medicines Agency Inspections (EMEA) Concept Paper on the Revision of the Note for guidance on Stability Testing of Existing Active Substances and Related Finished products. April, 2007, p. 3.

[111] Li, S., Han, Q., Qiao, C., Song, J., Cheng, C.L., Xu, H. (2008). Chemical markers for the quality control of herbal medicines: an overview. *Chinese Medicine*, *3*, 1–16.

[112] Srinivasan, V.S. (2006). Challenges and scientific issues in the standardization of botanicals and their preparations. United States Pharmacopeia's Dietary Supplement Verification Program – a Public Health Program. *Life Sciences*, *78*, 2039–2043.

11

NATURAL PRODUCTS IN CLINICAL TRIALS

SIGRUN CHRUBASIK

11.1 THE QUALITY OF CLINICAL TRIALS

The International Conference on Harmonization of Technical Requirements for Registration of Pharmaceuticals for Human Use (ICH) has brought together the regulatory authorities and pharmaceutical industry of Europe, Japan, and the United States to discuss scientific and technical aspects of drug registration (www.ich.org/home.html). Since its inception in 1990, the guidelines are implemented into the law in various countries, among them in Germany. The "Efficacy Guidelines" cover the design, conduct, safety, and reporting of clinical trials. Natural products are handled like synthetic drugs. However, in order to be able to repeat studies with phytomedicines, their description must include the plant part, kind of preparation (powder, extract, tincture etc.), in case of extract also the solvent and the drug extract ratio (how many parts of crude plant material are necessary to produce one part of extract) and, the content of marker substance in the daily dosage (if possible the mg of active principle per day or mg of co-active ingredients per day) and the recommended dose of extract.

Prior to carry out a study, a protocol needs to be prepared that considers (i) the predefined quality items of the guidelines, for example, the inclusion and exclusion criteria of the population to be studied (the population should be as homogenous as possible), (ii) that the study groups are comprehensively described to confirm their comparability, (iii) statistics employed is adequate and comprehensive including a multivariable analysis to detect any confounders and intention-to-treat analysis (those not finishing the study being considered as treatment failures), (iv) drop-outs and

Plant Bioactives and Drug Discovery: Principles, Practice, and Perspectives, Fourth Edition.
Edited by Valdir Cechinel-Filho.

adverse events are explicitly described, that validated outcome measures are used, and (v) that the expectations on primary outcome measures are defined in advance.

This allows us to calculate the numbers of patients to be included (with a power of at least 80%). The more of the 18 quality items of the efficacy guideline are considered, the better the quality of the clinical trial. Only studies with predefined expectations and a worthwhile result are confirmatory, that means proving the effectiveness of the natural product, all other studies are explanatory, just showing a trend of effectiveness. For a good internal validity, a study needs not necessarily to be randomized and double-blind with potential confounders equally distributed among the groups. If cohort studies consider potential confounders and adjust them in analysis, the internal validity of such studies may also be good.

Depending on the quality items considered, the internal validity is graded good, fair, or poor. Likewise, the quality of evidence is graded "good," "fair," and "poor." Only confirmatory studies with a good internal validity are included in the quality of evidence evaluation.

If the quality of evidence is good, the benefit harm ratio needs to be assessed. Benefit and harm are graded into "substantial," "moderate," "small," and "zero/negative." Even a small relative impact on a frequent condition with a substantial population burden (e.g., low back pain) is great benefit. So even if small effect sizes are achieved in general with natural products, they may contribute in improving the population burden. Unfortunately, the quality of most studies investigating natural products so far is insufficient to draw definitive conclusions. When the net benefit magnitude is assessed, the net harm must be outweighed. In general, the net harm of natural products is less than that of synthetic drugs due to a lower incidence of adverse events.

Ethics committees serve as a control for the quality of clinical trials. Patients have to give their informed written consent if they want to participate in a clinical trial.

11.2 EXAMPLES OF CLINICAL STUDIES WITH NATURAL PRODUCTS

11.2.1 An Observational Study with Potato Juice

The Swiss physician Dr. Maximilian Bircher-Benner (1867–1939) recommended freshly squeezed juice from raw potatoes when patients suffered from heartburn or other dyspeptic complaints. He had observed that immediate consumption of freshly squeezed juice was helpful. If the juice was left for a while the juice started to get brown, the darker brown as the more time passed. This is due to oxidation process which have a negative impact on the taste of the juice selected for the study.

Once suffering from heartburn, I bought a bottle of freshly squeezed potato juice (Biotta®). After 3 days my complaints had gone. Was this a true effect or did the juice help because I wanted so? The next step was a plan for a pilot study, an open, observational uncontrolled study.

Information leaflets needed to be prepared to inform the patients about the study goal, the treatment schedule, and about possible risks when participating in the study. Potato peel contains glycol-alkaloids, for example, solanin, the toxic dose being around 2–5 mg/kg of body weight (140–350 mg/70 kg), the fatal dose around 3–6 mg/kg of body weight (210–420 mg/70 kg) [1]. In a variable dose study, none of the volunteers who were given total glycol-alkaloid doses of 0.3–0.7 mg/kg developed symptoms. One of those receiving 1.25 mg/kg of glycol-alkaloids became nauseated and started vomiting about 4 h following the meal, possibly due to glycoalkaloid toxicity [2]. The juice contained less than 10 mg glycol-alkaloids per 100 ml.

The juice (Figure 11.1) was produced by Biotta AG, Tägerwilen/Switzerl and from Solani tuberosi tuber recens harvested in the area of Tägerwilen and the south of Germany close to the Swiss border. The potatoes were cleaned of soil, washed, selected, mashed, and squeezed, and then treated with amylases (to initiate metabolism into polysaccharides and improve the taste) and lactic acid (acidifier, to decrease the pH and allowing pasteurization instead of sterilization). After further treatment with amylases, the juice was pasteurized (89°C) and immediately bottled.

There is no general recommendation for the juice as to how often and how much to drink. Empirically it was decided to drink 100 ml of juice (one cup) half an hour before breakfast and at night before going to sleep. In case of complaints, three to four additional cups were allowed.

The observational study was in conformity with the regulatory requirements in Switzerland. Informed written consent was obtained from 45 patients suffering from dyspeptic symptoms that had lasted more than 12 weeks in the preceding year. Exclusion criteria comprised current or recent participation in a clinical study, serious concurrent illness, a history of drug or alcohol abuse, pregnancy or lactation, known allergy to potatoes, and anticipated difficulties with language or cooperation.

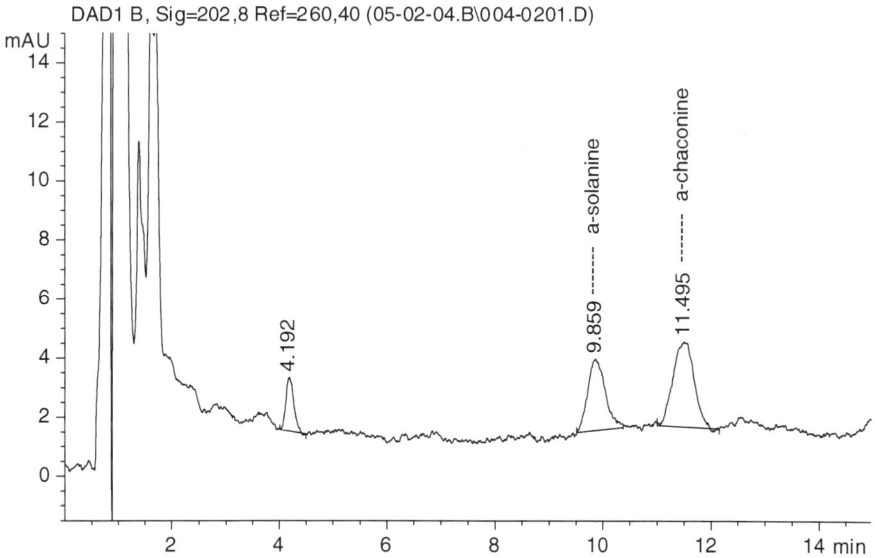

FIGURE 11.1 HPLC-fingerprint of the proprietary potato juice (Column LiChrospher RP 8, endcapped, 5 μm 125 × 4 mm; 0.02 M phosphate buffer pH 7.0/acetonitrile 70 : 30; 40°C; 1.0 ml/min, 202 nm).

Once their baseline characteristics had been recorded, patients were given 4 × 500 ml of fresh potato juice. They were advised to continue their current dyspepsia medication if they took any.

Three outcome measures were recorded, the gastrointestinal symptom profile (GIS [3], a 10 symptom profile, each symptom scored 0 to 4 (max score 40); and a recently validated disease-specific quality of life measure (QOLRAD [4], a 25 item questionnaire, each item scored 0 to 7 (max. 175) and, self-rated treatment success (Efficacy) on a 0 to 4 Likert scale (0 best, 4 worst).

Statistical analyses were carried out using SPSS (SPSS Inc. Chicago, IL, 60611). One patient withdrew consent and was not included in the analysis. All variables were tested for normal distribution by the Kolmogorov–Smirnof test. If the distribution differed significantly from normal, only nonparametric tests were applied. Where scores on a 5-point scale were used, again nonparametric tests were employed. Correlations of parametric variables were tested using Pearson's product moment correlation and those of nonparametrics by Kendall's tau-b. The before and after QOLRAD and GIS were converted to a percent change scale, (value before – value after/value before)*100. Drop-outs were included in the analyses with their initial values used for after treatment outcomes (for "intention-to-treat," ITT analysis). Per protocol (PP) data for the 42 patients are also included. Changes in GIS or QOLRAD were tested against individual symptoms, diurnal distribution of the symptoms, the presence or absence of reflux, reflux medications, other GI medications, other synthetic medications as a collective group, family history of GI disease, *Helicobacter* infection, concurrent disease or irritants (alcohol, tobacco or spices). Treatment success was defined as "very good," "good," "moderate," "slight," and "poor" by using >80%, 60–79.99%, 40–59.99%, 20–39.99, and <19.99% for the percentage change in the outcomes. In the categorical classification of outcomes the drop-outs were included in the "poor" group. The confidence level for rejecting null hypotheses was taken as 95% ($p < 0.05$).

One patient withdrew her permission to use the data after successful treatment on her family doctor advice. This patient was excluded from all results and treated as if she had not been enrolled. The study was completed by 42 patients. Two patients dropped out because of insufficient relief of complaints (one suffered from intolerable heartburn after consumption of the juice). The baseline characteristics of the 44 patients (10 men, 34 women) who entered the study are summarized in Table 11.1 together with the duration of their symptoms. Forty-two patients suffered from concomitant reflux symptoms such as retrosternal discomfort and regurgitation but not dominating the clinical picture. Thirteen patients received acid suppressive therapy (proton pump inhibitors $n = 10$ (one took in addition an antacid, one a prokinetic), antacids $n = 2$, H_2-blocker $n = 1$).

Contrary to instructions, seven patients discontinued their current dyspepsia medication. In five of these patients symptoms had improved on treatment with potato juice. One of the two patients who consumed only one cup of juice per day instead of two or more cups was satisfied with the treatment. During the potato juice treatment, the GIS and QOLRAD scores improved significantly by 41.9 ± 40.8/44.0 ± 40.7% (mean ± SD; CI_{95} 31.2 to 56.5) and 50.7 ± 36.1/53.1 ± 35.1%

TABLE 11.1 Baseline Characteristics of the 44 Patients

Age[a] (years)	64 ± 11
Weight[a] (kg)	70 ± 12
Size[a] (cm)	168 ± 9
Duration (months) of	
Chronic disease[a]	156 ± 142
Acute exacerbation[a]	22 ± 29
Diurnal distribution of symptoms	
Morning[b]	10
Noon[b]	28
Afternoon[b]	4
Evening[b]	19
At night[b]	12
Continuous complaints[b]	5
Intermittent complaints[b]	40
Risk factors	
Regular alcohol consumption[b]	11
Regular nicotine consumption[b]	7
Regular consumption of spices[b]	2
Use of synthetic medications[b]	18
Intercurrent diseases[b]	34
1° relative GI complaints[b]	15
Family history of stomach Ca[b]	9
Family history of colon/rectum Ca[b]	4

[a] Mean $\pm$ SD.
[b] Number of patients.

(CI_{95} 41.9 to 63.9), respectively (ITT/PP, both $p < 0.001$). The absolute changes are shown in Figure 11.2. Concomitant reflux symptoms had improved in 19 of the 41 patients. The frequency distributions of the outcome classes are listed in Table 11.2: 17, 18, and 25 patients rated the effectiveness of the treatment as very good or good on the GIS, QOLRAD (improvements > 60%), and on the self-rated efficacy, respectively. Poor treatment success was achieved in 12 (GIS), 13 (QOLRAD), and 11 (Efficacy) patients, not necessarily in the same subjects. In the poor outcome class, four subjects had greater GIS and four greater QOLRAD scores after treatment than before and, three (GIS) and one (QOLRAD) subjects had no change in the scores.

None of the individual GIS symptoms showed significantly greater improvement than others, of the irritants, alcohol, and tobacco did not affect the outcome variables significantly. Only the use of spices (regularly used by two patients) had a significant effect, affecting the percentage change in QOLRAD significantly ($p < 0.02$; the percentage change in GIS was not significant ($p = 0.168$). Because only two subjects used spices regularly, this "significant" result must be treated cautiously. By testing for 40 possible differences, the probability of rejecting once a true null hypothesis was about 0.81. This problem of multiple tests is well recognized when performing ANOVAs, but it also applies here.

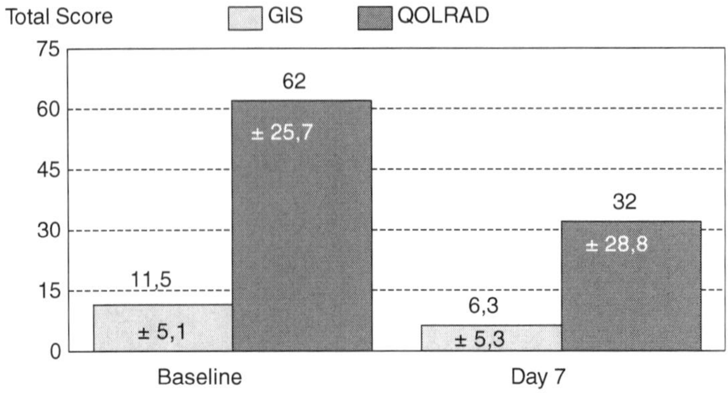

FIGURE 11.2 The total GIS and QOLRAD scores before and after 7 days potato juice therapy, mean ± SD. The absolute values decreased significantly (both $p < 0.001$). (*See the color version of this figure in Color Plates section*).

Efficacy self-rating scores correlated poorly with objective symptom scores or QOLRAD or with tolerability or taste. Not surprisingly, the taste and tolerability of the juice correlated a little better (Table 11.3). All these correlations were significant $p < 0.01$. The various cross tabulations of self-rated efficacy, tolerability, and taste and further details are placed on the webpage www.uniklinik-freiburg.de/rechtsmedizin/live/forschung/phytomedicine.html.

The reports of adverse reactions were scrutinized by an independent investigator. Eight patients suffered from adverse events, which were possibly related to the potato juice. One complained of intolerable heartburn, three of meteorizm one of whom also developed diarrhea and one complained of increased mucus production. Three patients suffered from upper respiratory tract infections during the study.

Table 11.4 summarizes the quality items considered in this observational study. The internal validity of the study was good. The study is not proving that potato juice

TABLE 11.2 Frequency Distribution of Outcome Classes for All Three Outcome Variables (GIS and QOLRAD Percentage Change and Efficacy)[a]

	GIS group		QOLRAD group		Efficacy	
	Number	%	Number	%	Number	%
Very good	8	18 (19)	12	27 (29)	9	20 (21)
Good	8	18 (19)	6	14 (14)	17	39 (40)
Moderate	8	18 (19)	6	14 (14)	6	14 (14)
Slight	7	16 (38)	9	20 (21)	2	5 (5)
Poor	13 (11)	30 (31)	11 (9)	25 (21)	10 (8)	23 (24)
Total	44 (42)		44 (42)		44 (42)	

[a] Percent changes in GIS and QOLRAD are classed as poor: 0 to 19.99, slight: 20 to 39.99, moderate: 40 to 59.99, good: 60 to 79.99, and very good: 80 to 100%. Self-rated efficacy was rated on a very good = 0, poor = 4, scale. The two subjects who dropped out are classed as "poor" responders on all scales (ITT). Forty four subjects are included. The numbers in brackets refer to the same data, but on a "per protocol" treatment of 42 subjects.

TABLE 11.3 Correlations of the Outcome Variables (GIS and QOLRAD% Change and Efficacy) and the Variables, Efficacy, Taste, and Tolerability[a]

(a)	GIS Group	QOLRAD Group	Efficacy
GIS Group		0.626	0.623
QOLRAD Group	<0.001		0.702
efficacy	<0.001	<0.001	

(b)	Efficacy	Tolerability	Taste
Efficacy		0.493	0.498
Tolerability	0.001		0.630
Taste	0.001	<0.001	

[a] The data shown in the "north-east" triangle are the Kendall's correlation coefficients while those in the south-west are their significance. The fraction of the variance explained by the correlation is equal to the square of the correlation coefficient.

TABLE 11.4 Quality Items of the Observational Study on the Effectiveness of Potato Juice for Heartburn

Reference	*Phytomedicine*. 2006 *13* (1–2), 11–15.
Centers	1
Medication	Characterized
(standardized on)	Solanin
Number of patients	45
Study period	1 week
Baseline data	Sufficient
Randomized	No
Double-blind	No
Control	No
Patient	Inclusion criteria: defined
Population	Exclusion criteria: defined
Outcome	2 validated instruments
measures	1 nonvalidated instrument
Hypothesis	No
Power	No
Analysis (ITT/PP)	Yes
Statistical tcst	Appropriate
Multivariable analysis	Yes
Drop-outs described	Yes
Adverse events described	Yes
Effect size	Worthwhile

is effective in treating heartburn. But a trend of effectiveness that potato juice may help to alleviate dyspeptic complaints and improve the quality of life was demonstrated. Two-thirds of the patients experienced at least some improvement and about half-benefited considerably. In its present formulation, the juice may have a place in self-treatment of dyspeptic symptoms, before starting synthetic medication; this is especially important to patients who prefer a natural product to a synthetic one. Since complete relief of complaints was achieved in only 20% of the patients, greater concentration of the active principle of the potato juice for a proprietary preparation may be desirable and create a product that might compete with synthetic medication (see below). Although two-thirds of the patients rated taste as excellent ($n = 12$) or good ($n = 20$) and the taste of the juice had no statistical impact on its effectiveness, improvement in taste of the juice may increase its acceptability to patients. The fact that 7 of the 15 patients on reflux medications discontinued the synthetics on their own may indicate that there is still a need for effective dyspepsia phytomedicines. The observational study provided, thus, useful data, to carry out a definitive study (a randomized, placebo-controlled double-blind study with predefined expectations and calculated numbers of patients to be included with a power of at least 80%). Dyspepsia has shown to impair health-related quality of life compared in the general healthy population [5]. Since the different dyspepsia outcome measures (GIS, QOLRAD, and self-rated efficacy) reflect different dimensions of the complaints, all three measures and in particular a quality of life measure should be included when evaluating the effectiveness of phytomedicines for upper gastrointestinal complaints.

Biotta AG Tägerwilen produced a fivefold concentrate of the potato juice for a pilot study carried out in Dniepropetrovsk, Ukraine. After the State Committee on Food Supplements approved the study, 6 men and 12 women suffering from dyspeptic complaints (Table 11.5) volunteered to take the potato juice concentrate, starting with two teaspoons per day (morning and evening or more, if necessary). Two patients dropped out, a patient who had no gastroscopic diagnosis withdrew his consent and another patient who suffered from dry cough prior to the trial, and the first dose of the concentrate was associated with a coughing attack, such that she decided not to continue with the treatment.

Three of the patients took two teaspoons/day of the concentrate, nine took three teaspoons/day and five consumed three tablespoons/day. Three outcome indicators were used. The GIS score, improvement in the worst-rated GIS symptom (where more than one component had the same worst score before treatment, the one with the least improvement was chosen as the outcome variable) and self-rated efficacy. After 6 days, the GIS score had improved by $73 \pm 32\%$ (ITT, mean $\pm$ SD $p < 0.001$). The absolute GIS values improved from 12.8 ± 4.3 to 3.1 ± 3.7, the subjects' worst GIS symptom improved in 15 by a median score of 2. Of the three nonimprovers, two were the patients who withdrew, and in the remaining one, of six GIS components that were rated as two at commencement of treatment, five had improved. The doses of the medication used did not correlate with either outcome variable, neither did age, nor sex, weight, BMI, or the duration of acute or chronic symptoms. Fourteen patients improved their GIS by 60% or more and 10 patients rated the efficacy of the treatment as very good or good. Failure of the treatment was not observed in any of the patients.

TABLE 11.5 Baseline Characteristics of the 18 Patients (Mean ± SD, or Number of Patients)

Age (years)	48 ± 15
Weight (kg)	70 ± 18
Height (cm)	167 ± 9
Duration of chronic symptoms (months)	105 ± 120
Duration of acute exacerbation (weeks)	33 ± 51
Diurnal distribution of symptoms:	
Morning	5
Noon	10
Afternoon	1
Evening	10
At night	2
Continuous complaints	3
Intermittent complaints	15
Risk factors	
Regular alcohol consumption	2
Regular nicotine consumption	3
Regular consumption of spices	3
Use of synthetic medications	3
Endoscopic diagnoses (multiple):	
Gastroduodenitis	4
Reflux esophagitis	11
Duodenal ulcer	2
Deformation after duodenal ulcer	1
Gastritis	2
No gastroscopy	1

In this pilot trial, patients endoscopically verified gastrointestinal inflammations were treated with a potato concentrate. The results indicate that a fivefold concentrate of potato juice may be more effective in alleviating dyspeptic complaints than ordinary potato juice. A confirmatory clinical study should now be carried out to compare their effectiveness.

11.2.2 A Randomized Double-Blind Study and a Cohort Study Investigating Two Doses of a Proprietary Willow Bark Extract in Low Back Pain Exacerbations

11.2.2.1 The Randomized Double-Blind Study. The study was approved by the Human Ethics Committee of the Rappaport Faculty of Medicine, Technion, Haifa, Israel. Patients suffering from exacerbations of chronic low back pain with or without radiation to one or both legs were invited to participate. Patients received randomly and double-blind a standardized willow bark extract (drug extract ratio 8–14 : 1, solvent 70% ethanol, and 15% salicin) with either 120 or 240 mg salicin per day or a placebo. The medication to be studied contained approximately 0.153 mg of salicin per milligram of extract and was manufactured by Plantina GmbH Munich, Germany.

Patients were allowed tramadol as the only rescue analgesic medication, and the prospectively chosen principal outcome measure was the proportion of patients free of pain without tramadol for at least 5 days in the fourth and final week of treatment. Subsidiary outcome measures were to be the number of patients requiring tramadol and the change in the Arhus low back pain index [6] from baseline.

The estimated minimum number of patients required in each group was 70, based on conventional calculations for 90% power for detecting differences in proportions of pain-free patients between 5% in patients receiving placebo, and 15% and 25% in those receiving daily doses of willow bark extract with 120 and 240 mg salicin, when 95% confidence was required to reject the null hypothesis that there would be no trend detectable by a one-sided Cochrane–Armitage test.

Public advertisements were used to recruit suitable patients from the Haifa area. The selection criteria were (i) age 18 to 75 years; (ii) a background of at least 6 months susceptibility to low back pain that was not attributable to identifiable causes such as disc prolapse, hip disease, spondylolisthesis, osteomalacia, or inflammatory arthritis; (iii) a current exacerbation of their complaint that was affecting both rest and movement, was giving rise to subjective pain of at least 5 out of 10 on a visual analog scale, and was expected by the physician as well as the patient to require at least 4 weeks of symptomatic treatment. Exclusion criteria were (i) current or recent (within 30 days) participation in any other clinical study; (ii) serious organic illness affecting any organ system; (iii) a history of drug or alcohol abuse or requirement for psychotherapeutic agents; (iv) pregnancy, actual or possible (unreliable contraceptive practice), or lactation; (v) known allergy to any of the proposed trial medications; (vi) difficulties with language or anticipated co-operation. An account of the plan and an invitation to participate (conditional on the results of an assessment in clinic) was given to patients who met these criteria. The first 210 who gave written consent to participate were entered into the study. The study medication was allocated to the treatment groups according to a computerized randomization list (ratio 1 : 1 : 1) in a random-permuted block design of length three. No access was possible to the list by persons involved in the study conduct. Consecutive numbers of the randomization list corresponded to patient entry in chronological order.

To record their baseline state, the patients completed a standardized questionnaire about their general health, daily activity, and the particular characteristics of their pain. They then underwent a clinical examination with recording of vital signs and were asked to give a venous blood sample for screening baseline variables of the organ system function. They also underwent the questioning and examination for completing the Arhus score. This was modified for this study by the exclusion of the items relating to analgesic medications, because the patients were to be allowed only tramadol as rescue medication. The modifications gave the score a maximum of 120–60 for pain, 30 for disability, and 30 for physical impairment.

Once their baseline characteristics had been recorded, the patients received their coded trial medication that consisted of pairs of identical sugar-coated pills of red color to be taken twice a day for 4 weeks. In the placebo group (P), both tablets were lactose; in the group to receive a daily dose of extract equivalent to 120 mg salicin (low dose) one tablet contained 393.24 mg dry willow bark extract and the other lactose, and in the group to receive a daily extract dose with 240 mg salicin (high dose) both, tablets contained 393.24 mg dry extract. The coding of the tablets concealed their identity from both, the patients and the investigators. Patients were allowed to supplement their trial medication with tramadol liquid (2.5 mg/ml) in doses from 50 to 400 mg per day.

Once a week the patients were contacted by telephone. This enabled the clinical investigator to record the number of patients who were with or without pain and the doses of rescue medication on each day of the preceding week; the investigator also recorded the occurrence of adverse events and encouraged the patients to keep taking the study medication. At the end of the 4-week study period, the Arhus scoring and the vital signs were repeated in the same format as at baseline.

The principal outcome measure was the proportion of patients who "responded" to treatment by being pain free without tramadol for at least 5 days of the last week. The principal analysis of effectiveness was carried out on the principle of "intention to treat," with patients who did not complete the study after randomization ("drop-outs") being included as "nonresponders." A one-sided test Cochrane–Armitage was used to test the null hypothesis that there was no monotonic increase in the proportion of responders with increasing daily dose of willow bark. The principal analysis was backed up by a sensitivity analysis in which the "drop-outs were simply excluded." These and other analyses were undertaken using the procedures available in the Statistical Analysis System Software package (SAS Institute Inc., Cary, NC). Categorical data were examined in contingency tables, with inferential testing by Fisher's exact test. Ordinal or continuously distributed data were summarized as median and quartiles, with inferential testing by Kruskal–Wallis/Mann–Whitney–Wilcoxon (paired as appropriate for differences between baseline and end of study). To investigate differences in the Arhus index and tramadol requirement, the one-sided Jonkheere–Terpstra test for monotonic trend of effect with dose was employed. The confidence level for rejecting null hypotheses was taken as 95% ($p < 0.05$).

Table 11.6 lists, by the treatment group, some of the characteristics of the 210 patients who entered the trial, 70 in group P, 70 in the low dose and 70 in the high-dose willow bark extract groups. The randomization had clearly produced matching of the groups for age, height, weight, and gender but not for their presenting pain. An additional set of more than 100 features were recorded including indicators of general health, professional and employment status, sporting activity, further details of the low back pain and results of attempted treatments and circulatory measurements. There were only six features in around 110 in which the observed intergroup differences would have reached "significance" in single isolated comparisons. The groups receiving willow bark extract had a "significantly" higher Arhus low back pain score than the patients receiving placebo. Furthermore, in "significantly" more patients receiving extract with 240 mg salicin per day, the propensity to low back pain had persisted for more than 6 years and osteoporosis contributed to the global low back pain; more patients (5) in the high dose willow bark extract group admitted to

TABLE 11.6 Physical Characteristics and Initial Features and Associations of the Pain in the Patients Who Entered the Study[a]

		Willow Bark Extract		
	Placebo	Low Dose	High Dose	p
	$n = 70$	$n = 70$	$n = 70$	
Characteristics of the patients	Median (25;75)	Median (25;75)	Median (25;75)	
Age (years)	56 (43;68)	55 (46;63)	58 (45;72)	0.28
Size (cm)	168 (160;172)	170 (160;176)	165 (160;172)	0.14
Weight (kg)	73 (67;85)	79 (68;90)	75 (63;85)	0.32
Men	29	35	32	0.61
Characteristics of the pain				
Duration of low back pain > 6 years	$n = 39$ (56%)	$n = 46$ (66%)	$n = 53$ (76%)	0.050
Duration of present complaints				0.65
<7 days	$n = 11$ (16%)	$n = 9$ (13%)	$n = 6$ (9%)	
>7 days	$n = 10$ (14%)	$n = 7$ (10%)	$n = 10$ (14%)	
>90 days	$n = 49$ (70%)	$n = 54$ (77%)	$n = 54$ (77%)	
Radiation into leg(s)	$n = 54$ (77%)	$n = 53$ (76%)	$n = 51$ (73%)	0.88
Neurological accompaniments				
Deficits in straight leg raising	15 (22%)	12 (17%)	14 (20%)	0.78
Motor deficit	1 (1%)	2 (3%)	1 (1%)	1.0
Sensory deficit	5 (7%)	10 (14%)	7 (10%)	0.42
Normal reflexes	68 (99%)	64 (91%)	65 (93%)	0.16
Beck depression inventory	6 (2;10)	7 (5;10)	8 (5;14)	0.024
Arhuse low back pain index				
(1) Pain index	34 (24;44)	40 (25;56)	44 (28;57)	0.010
(2) Invalidity index	18 (14;23)	23 (17;26)	24 (17;27)	<0.001
(3) Physical impairment index	18 (12;24)	22 (16;30)	24 (18;30)	<0.001
Arhuse low back pain index (1 + 2 + 3)	67 (54;84)	80 (65;103)	88 (69;108)	<0.001

[a] The p values are indicative only.

depression than in the other two groups ($N = 0$ in both), and the Beck scores for depression also tended to be higher in the high dose willow bark extract group; more had previously tried opioids for their pain and a third of them had changed their occupation due to intolerable pain.

The other items were well matched. The current exacerbation had lasted for more than 3 months in 75% of the patients. The pain radiated into the buttock in 68% and into one or both legs in 75% and it was exacerbated by physical activity in 87%. About 75% of patients had suffered physical impairment for more than 14 days in the previous 6 months, and their condition had affected their work in 58% of cases.

Specific causes were generally difficult to identify, though a variety of possible causes was advanced by the attending physicians, the most common of which were "arthrosis," "spondylosis," "scoliosis," and "disc protrusion." Pain tended to be maximum in the morning in 28% of patients, in the evening in 28% and showed no pattern of intensity in 35%.

About 20% of all patients had a positive straight-leg raising test and 94% had normal reflexes. About 2% had some motor impairment and 11% had some sensory deficit. Vital signs were well matched. Almost all patients had previously resorted to some sort of treatment for the chronic condition either regularly or irregularly. These had achieved variable success for an unspecified period but had been discontinued for a variety of reasons, some of which may well have been financial.

The trial was completed by 191 patients: there were 19 drop-outs, eight because of insufficient pain relief (all in group P), four because of noncompliance for reasons unspecified (one from group P, one from group low dose, and two from group high dose willow bark extract), two (from group high dose willow bark extract) because they left the area unexpectedly, two (from group P) because of gastrointestinal complaints, one (from group low dose willow bark extract) because of an allergic reaction, one (from group P) because of laboratory-confirmed anaemia and another (from group P) who complained of unbearable adverse effects related to tramadol intake.

The number of patients who were pain free without tramadol for at least 5 days in the fourth week of treatment ("responders") was four in group P, 15 in group low dose willow bark extract and 27 in group high dose willow bark extract ($p < 0.001$, Cochrane–Armitage test for trend on basis of intention to treat. Sixteen of the n=61 patients with any sort of neurological deficit were included in the responders (one in group P, seven in group low dose, and eight in group high dose willow bark extract). The proportion of responders with pain radiating into one or both legs (all four responders in group P, 13 of the 15 in the low dose, and 21 of the 27 in the high-dose willow bark extract group) was not significantly different from the proportions in the sample as a whole ($p = 0.25$). The same was true of the proportions whose chronic condition had lasted more than 6 years or whose exacerbations had lasted more than 3 months. A significant increase in the number of responders in group high-dose willow bark extract was already apparent after only 1 week of treatment (Table 11.7a) and, in all 4 weeks of the study period, significantly more patients in the placebo group required tramadol (Table 11.7a). The median relative change in the overall Arhus index was significantly larger in the high dose than in the low-dose willow bark extract group in a subsidiary intention to treat analysis in which drop-outs were assigned no change from initial value. The change in the overall Arhus index stemmed largely from a significant effect on the pain component (Table 11.7b).

The reports of adverse reactions were scrutinized by an independent investigator. There was only one case of allergy (exanthema, swollen eyes, and pruritus) that could be attributed to willow bark extract (in group low-dose willow bark extract). Symptoms disappeared 2 days after discontinuation of treatment. The two patients with short-lasting adverse events in group high-dose willow bark extract (dizziness attributed to tramadol, dizziness, and fatigue on day 1) later dropped out (one because

TABLE 11.7a Taking Tramadol and Relative Change in the Arhus Index ([before-end]/ before) and its Components (median and range)[a]

	Placebo	W 120	W 240	*p*1
Patients pain-free	$n=70$	$n=70$	$n=70$	
Week 1	2 (3%)	1 (1%)	6 (9%)	0.048
Week 2	5 (7%)	9 (13%)	17 (24%)	0.002
Week 3	6 (9%)	10 (14%)	26 (37%)	<0.001
Week 4	4 (6%)	15 (21%)	27 (39%)	<0.001
Requiring T				
Week 1	$n=45$	$n=21$	$n=16$	<0.001
Week 2	$n=35$	$n=14$	$n=6$	<0.001
Week 3	$n=35$	$n=12$	$n=7$	<0.001
Week 4	$n=33$	$n=10$	$n=3$	<0.001
AS change				
1+2+3	0 (−13;5)	44 (18;60)	54 (19;90)	<0.001
(1) Pain	0 (−9;13)	40 (0;63)	67 (24;100)	<0.001
(2) Invalidity	0 (−14;23)	46 (12;78)	57 (0;89)	<0.001
(3) PHI[b]	0 (−15;0)	27 (0;75)	41 (0;86)	<0.001
AS change				*p*2
1+2+3	0 (−13;5)	44 (18;60)	54 (19;90)	0.022
(1) Pain	0 (−9;13)	40 (0;63)	67 (24;100)	<0.001
(2) Invalidity	0 (−14;23)	46 (12;78)	57 (0;89)	0.40
(3) PHI	0 (−15;0)	27 (0;75)	41 (0;86)	0.77

[a] *p*1 is the *p* value of the one-sided Jonckheere test for monotonic increase with dose; *p*2 is the *p* value of the Kruskal–Wallis test for any intergroup difference.
[b] Physical impairment.

of insufficient pain relief and the other because of failure to comply for unspecified reasons). The adverse events reported by six patients in group P were mild: in three cases, the patients attributed them to the tramadol rescue medication (dizziness/headache, dizziness/vomiting/diarrhea, and dry mouth); the remaining three suffered from mild stomach pain with or without diarrhea, but two of them discontinued the study on the first day of treatment.

This study reflects the gold standard. All quality items had been considered, the study proves the dose-dependent effectiveness of the proprietary willow bark extract in the treatment of acute low back pain exacerbations.

11.2.2.2 ***The Cohort Study.*** At the same time while carrying out the study in Haifa, an open controlled study was running in Germany, in outpatients of randomly chosen general practitioners (GP, $n=3$) and orthopedists (ORTH, $n=3$). Patients suffering from acute exacerbations of chronic nonspecific low back pain were included in the study and examined by three of the authors before and during treatment. Selection criteria included, age 18 to 80 years; a low back pain disease lasting at least 6 months

TABLE 11.7b Calculation with Drop-Outs Excluded (Sensitivity Analysis)[a]

	Placebo	W 120	W 240	*p*1
Patients pain-free	$n = 59$	$n = 67$	$n = 65$	
Week 1	2 (3%)	1 (1%)	6 (9%)	0.12
Week 2	5 (8%)	9 (13%)	17 (26%)	0.007
Week 3	6 (10%)	10 (15%)	26 (40%)	<0.001
Week 4	4 (7%)	15 (22%)	27 (42%)	<0.001
Requiring T				
Week 1	$n = 45$	$n = 21$	$n = 16$	<0.001
Week 2	$n = 35$	$n = 14$	$n = 6$	<0.001
Week 3	$n = 35$	$n = 12$	$n = 7$	<0.001
Week 4	$n = 33$	$n = 10$	$n = 3$	<0.001
AS change				
1+2+3	−3 (−16;11)	46 (29;62)	57 (29;92)	<0.001
(1) Pain	0 (−11;29)	44 (0;63)	73 (41;100)	<0.001
(2) Invalidity	0 (−20;31)	46 (17;79)	64 (8;90)	<0.001
(3) PHI[b]	0 (−33;0)	31 (7;75)	48 (0;90)	<0.001
AS change				*p*2
1+2+3	−3 (−16;11)	46 (29;62)	57 (29;92)	0.010
(1) Pain	0 (−11;29)	44 (0;63)	73 (41;100)	<0.001
(2) Invalidity	0 (−20;31)	46 (17;79)	64 (8;90)	0.24
(3) PHI	0 (−33;0)	31 (7;75)	48 (0;90)	0.55

[a] *p*1 is the *p* value of the two-sided Jonckheere test for monotonic trend with dose; *p*2 is the *p* value of the two-sided Mann–Whitney–Wilcoxon test (willow bark intergroup test).
[b] Physical impairment index.

that was not attributable to identifiable causes such as disc prolapse, spondylolisthesis, osteomalacia, or inflammatory arthritis; acute complaints that affected both rest and movement, which was giving rise to subjective pain of at least 5 out of 10 on a visual analog scale and was expected to require at least 4 weeks of symptomatic treatment. Exclusion criteria were current or recent participation in any other clinical study; serious organic illness affecting any organ system; a history of drug or alcohol abuse or requirement for psychotherapeutic agents; pregnancy or lactation; known allergy to salicylates; difficulties with language or cooperation. A total of 451 patients were enrolled in the study. According to the German monograph, 112 GP patients received two capsules willow bark extract per day (W_{120}, extract with 120 mg salicin/day) as mono- or co-treatment; according to the ESCOP monograph, subsequent 115 GP patients received four capsules willow bark extract per day (W_{240}, extract with 240 mg salicin/day); 224 ORTH patients received conventional treatment (C). Before, after 2 weeks (if possible) and after 4 weeks the patients underwent questioning and examination for completing the Arhus index and the total pain index. The Arhus index was modified (see above). The mean of five components (subjective pain while sitting, lying, in upright position, during movement, and at night) was defined as the total pain index. After 1 and 3 weeks, the patients were contacted by telephone to record their global pain on a visual rating scale (VRS 0–4) in order to assess the number of pain-free patients and to document their well being, the occurrence of adverse events and any change in the treatment of the low back pain. The patients could reach the physicians at any time by telephone.

Statistical analysis was done with the procedures available in the package of statistical analysis system software (SAS Institute Inc., Cary, NC). Categorical data were examined in contingency tables, with inferential testing by Chi square or Fisher's exact test. Ordinal or interval data were summarized as median and quartiles, with intergroup testing by the Kruskal Wallis and Mann–Whitney–Wilcoxon tests. Spearman Rank correlation was used to relate physicians' and patients' judgment of treatment success. To examine for possible confounding on the observed treatment effects, the explanators severity, duration of the acute exacerbation and radiation of pain into leg(s) were offered along with dummy variables for W_{120} and W_{240} to multiple regressions on the number of pain-free patients (logistic regression), Arhus component pain index and total pain index (linear regression). The confidence level for rejecting null hypothesis was taken as 95% ($p < 0.05$).

Table 11.8 lists, by treatment group, the characteristics of the 451 patients who entered the study. Patient groups were well matched with respect to the biometrical data. However, in group W_{240} the sex ratio was in favor of females. Whereas the duration of the chronic low back pain disease was similar in all groups, the acute

TABLE 11.8 Biometrical and Anamnestic Data and Low Back Pain Scores (Arhus Index) of the Patients Receiving Willow Bark Extract with 120 or 240 mg Salicin as Co-treatment (W240, W120) (Median, 25% Q; 75%Q), *p* Kruskal–Wallis Test for Intergroup Differences

	Group W240	Group W120	Group C	*p*
Number of patients	$n = 112$	$n = 115$	$n = 224$	
Age (years)	57 (50;63)	59 (52;64)	57 (50;63)	0.20
Size (cm)	168 (164;177)	170 (165;177)	170 (165;178)	0.42
Weight (kg)	74 (64;85)	77 (68;85)	78 (67;87)	0.51
BMI	26 (23;29)	26 (24;28)	26 (24;29)	0.73
Sex (M:F)	35:77	56:59	100:124	0.018
Duration of low back pain				
Chronic LBP (months)	120 (60;216)	120 (60;204)	90 (48;180)	0.27
Acute LBP (weeks)	8 (2;24)	8 (2;24)	4 (2;16)	<0.001
Acute LBP (>6 weeks)	$n = 71$ (63%)	$n = 58$ (51%)	$n = 85$ (38%)	<0.001
Acute LBP (>3 months)	$n = 53$ (47%)	$n = 54$ (47%)	$n = 70$ (31%)	0.002
Pain radiating into leg(s)	$n = 19$ (17%)	$n = 38$ (33%)	$n = 28$ (13%)	
(1) Pain index	20 (16;24)	18 (15;22)	22 (19;26)	<0.001
(2) Invalidity index	16 (13;19)	14 (10;16)	18 (14;21)	<0.001
(3) Physical impairment index	19 (16;21)	18 (16;20)	20 (18;22)	<0.001
Arhus index 1+2+3	54 (46;63)	50 (44;56)	60 (52;67)	<0.001
Total pain score	5.3 (4.0;5.9)	5.1 (4.0;5.9)	5.8 (5.0;6.9)	$p < 0.001$

exacerbation of pain was shorter in the group C patients that also had a 10% higher baseline Arhus index and total pain index.

The study was completed by 439 patients: there were 12 drop-outs, $n = 7$ because of social reasons ($n = 6$ in W_{240}, $n = 1$ in C), $n = 4$ because of adverse events (W_{240}: $n = 2$, W_{120}: $n = 2$, all suffering from stomach complaints, causality: $n = 3$ possible, $n = 1$ unlikely), $n = 1$ because of allergic skin reaction in W_{120} (causality likely). The conventional treatments consisted mainly of nonsteroidal anti-inflammatory drugs (oral or parenteral NSAIDs), nerve blocks, manual, and electrical therapy (Table 11.9). The patients receiving willow bark extract had significantly less conventional co-treatments; median consumption of diclofenac equivalents was zero in both willow bark extract groups (Table 11.9). Only one patient of each group had taken paracetamol. Treatment outcome as assessed on the Arhus component pain index, total pain index, and the patients' subjective well-being improved and the number of pain-free patients increased in the course of treatment in all groups. At the end of the study period, the Arhus pain index, the total pain index relative change and the number of pain-free patients were highest in group W_{240} and similar in groups W_{120} and C except for the Arhus component pain score (Table 11.10). Multivariate analysis including possible explanators (severity, duration of the acute exacerbation, radiation of pain into leg(s), dummy variables for treatment groups) showed that the Arhus pain component scores and the total pain score at the beginning of treatment were prognostic factors ($p < 0.01$) for treatment response; in contrast duration of the acute exacerbation and pain radiation into leg(s) had no prognostic impact ($p > 0.20$). Despite the initial imbalance between the groups, the treatment response in terms of pain-free ($p < 0.001$) or change in total pain index ($p = 0.008$) but not in terms of the

TABLE 11.9 Applied Conventional Treatments (Median (25Q;75Q) M (25;75))

Prescribed treatments (n)	Group W240 +	Group W120 +	Group C	p^a
Exercise	11 (10%)	17 (15%)	31 (14%)	<0.001
Massage	1 (1%)	2 (2%)	9 (4%)	<0.001
Electrical therapy	2 (2%)	3 (3%)	127 (57%)	<0.001
Heat application	1 (1%)	4 (4%)	32 (14%)	<0.001
Paravertebral nerve blocks	2 (2%)	2 (2%)	148 (66%)	<0.001
Cryotherapy	0	0	21 (9%)	<0.001
Manual therapy	0	1 (1%)	139 (62%)	<0.001
Extension	0	1 (1%)	12 (5%)	0.007
TENS	2 (2%)	1 (1%)	24 (11%)	<0.001
NSAIDs oral	13 (12%)	16 (14%)	88 (39%)	<0.001
NSAIDs parenteral	1 (1%)	3 (3%)	173 (77%)	<0.001
Diclofenac Equ. (mg)	0 (0;0)	0 (0;0)	225 (75;450)	<0.001[b]
Muscle Relaxants	0	3 (3%)	7 (3%)	<0.001
Cortison	0	0	10 (4%)	0.006
Other	6	20	14	

[a] p: Chi square test.
[b] Wilcoxon rank sum test.

TABLE 11.10 Relative Change of Arhus index ((AIBegin–AIEnd)/AIBegin) × 100 and its Component s, of the Total Pain Index Two and Four Weeks After Start of Treatment, Number of Patients Without Pain

	Group W240	Group W120	Group C	*P*
Relative index change (%)	M (25;75)	M (25;75)	M (25;75)	
after 2 weeks[a]	$n=87$	$n=55$	$n=204$	
(1) Pain	24 (5;50)	16 (5;35)	14 (5;28)	
(2) Invalidity	18 (6;37)	6 (−7;28)	14 (0;28)	
(3) Physical impairment	0 (0;13)	0 (0;6)	6 (0;11)	
Arhus index 1+2+3	16 (5;32)	9 (2;17)	11 (4;23)	
Total pain index	38 (7;63)	20 (5;56)	18 (6;33)	
after 4 weeks	$n=104$	$n=112$	$n=223$	
(1) Pain	35 (9;65)	18 (0;42)	25 (10;47)	0.0062
(2) Invalidity	26 (11;49)	7 (−7;35)	20 (0;40)	<0.001
(3) Physical impairment	9 (0;17)	0 (−7;9)	9 (0;18)	<0.001
Arhus index 1+2+3	22 (10;39)	11 (0;23)	17 (6;33)	<0.001
Total pain index	51 (11;79)	22 (3;43)	27 (5;58)	<0.001
Subgroups[a]	M (25;75)	M (25;75)	M (25;75)	
Radiating into leg(s)	$n=16$	$n=37$	$n=28$	
(1) Pain index	38 (26;69)	41 (13;56)	26 (13;39)	
Total pain index	48 (44;71)	35 (19;60)	24 (2;39)	
Local pain	$n=88$	$n=75$	$n=195$	
(1) Pain index	33 (6;64)	13 (0;33)	25 (10;48)	
Total pain index	52 (9;80)	14 (0;37)	27 (6;37)	
Pain-free patients	$n=$	$n=$	$n=$	
Week 1	4 (107) 4%	1 (114) 1%	2 (223) 1%	0.13
Week 2[b]	14 (57)	29 (87)	24 (204)	
Week 3	20 (101) 20%	6 (112) 5%	14 (224) 6%	<0.001
Week 4	42 (104) 40%	21 (112) 19%	40 (223) 18%	<0.001
Well being	%	%	%	
Week 1 (vw/w/mo/mi)	4/54/35/7	0/42/55/3	0.5/32/52/16	
Week 3 (vw/w/mo/mi)	16/63/19/2	5/60/32/4	5/42/49/4	

[a] Median (25Q;75Q) M (25;75)), *p* intergroup difference between all groups W240, W120, and C.
[b] No *p* value due to possible bias (small sample size).

Arhus component pain index ($p=0.20$) was significantly better in the treatment group W_{240}, whereas between group W_{120} and group C there was no significant difference in the three outcome variables. The relative changes of the invalidity and physical impairment indices were similar in groups W_{240} and C, but significantly less in group W_{120} (Table 11.10). Treatment success as judged by physicians and patients agreed by 91% ($r=0.90$).

The reports of adverse reactions were scrutinized by an independent investigator. There were three cases of allergy (exanthema, pruritus) that could be attributed to willow bark extract (Table 11.3), 16 cases of gastrointestinal complaints (patients receiving willow bark extract as co-treatment $n=4$, as mono-treatment $n=2$ and in

TABLE 11.11 Quality Items of the Two Studies with Willow Bark (One Confirmatory, One Exploratory Study)

Reference	*American Journal of Medicines* 2000, *109*(1), 9–14.	*Phytomedicine*. 2001, *8*(4):241–2451.
Centers	1	6
Medication (standardized on)	Characterized Salicin	Characterized Salicin
No of Patients	210	451
Study period	4 weeks	4 weeks
Baseline data	Sufficient	Sufficient
Randomized	Yes	No
Double-blind	Yes	No
Control	Yes	Yes
Patient population	Inclusion criteria: defined Exclusion criteria: defined	Inclusion criteria: defined Exclusion criteria: defined
Outcome measures	Validated and Nonvalidated instruments	Validated and Nonvalidated instruments
Hypothesis	Yes	No
Power	Yes	No
Analysis (ITT/PP)	Yes	Yes
Statistical test	Appropriate	Appropriate
multivariable analysis	Yes	Yes
Drop-outs described	Yes	Yes
Adverse events described	Yes	Yes
Effect size	Clinically relevant	Clinically relevant

group C during NSAID or NSAID + cortison treatment $n = 10$). All symptoms disappeared uneventfully. Three of these cases dropped out. Four patients of group C suffered from treatment-related adverse effects.

Table 11.11 summarizes the quality items of the two studies with willow bark. Similar results were achieved with the definitive study and the cohort study. About 40% of the patients taking extract with 240 mg salicin per day were pain-free compared to about 20% of the patients taking half dose. This confirms that cohort studies considering all quality items possible may reflect true effectiveness.

11.3 EVIDENCE OF EFFECTIVENESS

Systematic reviews may consider all published material or – as the Cochrane group – only randomized controlled (placebo or active control) parallel and crossover trials (7). In case of phyromedicines for treating osteoarthritis (OA) – as an example – only studies that considered the criteria of the American college of Rheumatologists were included. Studies with samples defined according to vague descriptions (e.g., "joint pain") were not considered. Studies with participant samples defined according to incomplete or partial ACR criteria were included, and notes provided to identify possible weaknesses in sample selection in these studies.

Any form of herbal intervention compared with an inert (placebo) or active control, via any route of administration, was included in the OA review (8). Herbal therapy used in conjunction with other treatments or combined with a nonherbal substance were also included if the effect of the nonherbal intervention was consistent among all groups and quantifiable. Herbal intervention included any plant preparation (whole, powder, extract, and standardized mixture) but excluded homeopathy, aromatherapy, or any preparation of synthetic origin.

Primary outcomes included: changes in assessed clinical measures of effectiveness (e.g., WOMAC, Lequesne index); changes in self-reported measures of effectiveness (e.g., pain, use of medication); and any adverse reaction. Secondary outcomes included: general well-being or satisfaction indicators. Unfortunately, the consensus criterion recommended by the orthopedic associations OMERACT and OARSI [9,10] were not considered in any of the studies. This composite index permits presentation of results of symptom modifying clinical trials in osteoarthritis based on individual patient responses (responder yes/no, Figure 11.3). If this instrument would be additionally used in all studies, the effectiveness of different herbal preparations could be compared. Figure 11.4 shows the OMERACT–OARSI response of three open, uncontrolled surveys over 1 year in patients with acute exacerbations of nonspecific low back pain, two investigated an aqueous extract of *Harpagophytum procumbens* with 60 mg of harpagoside in the daily dosage [11,12] and one investigating a powder of rosehip and seed with 303 mg galactolipid per 1 kg of powder [13]. The figure shows that the effect sizes of both phytomedicines are very close.

For the systematic review on osteoarthritis and phytomedicines we did a search in the following electronic databases: cochrane musculoskeletal group register, cochrane complementary medicine field register, cochrane controlled trials register (CCTR), MEDLINE, EMBASE, CISCOM, AMED, CINAHL, Dissertation

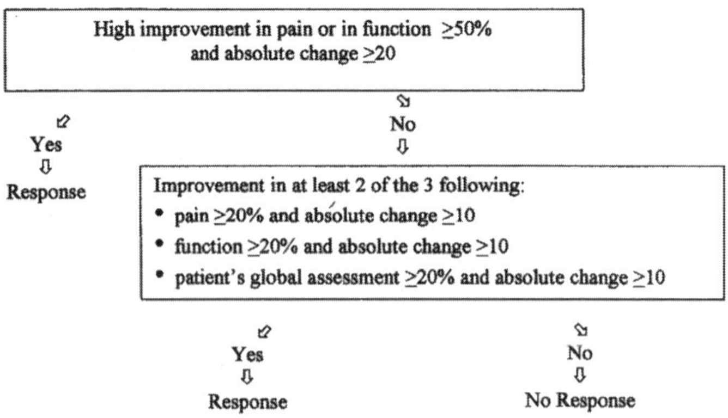

FIGURE 11.3 OMERACT–OARSI set of responder criteria (modified after [9]).

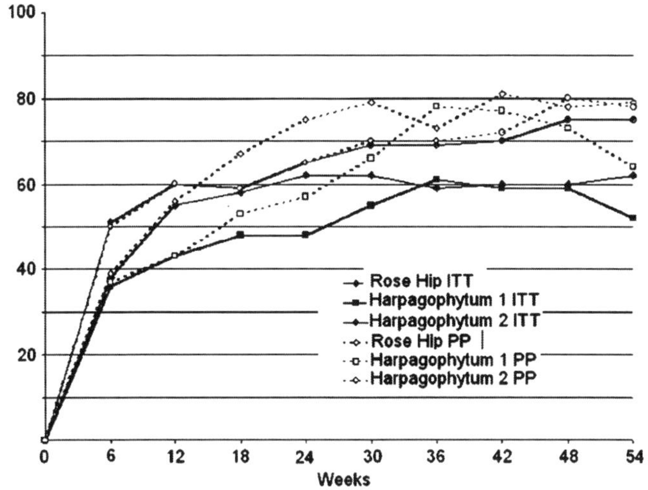

FIGURE 11.4 ITT and PP analyses of the OMERACT–OARSI responder rates in the survey investigating rose hip powder [study 13] and the aqueous *Harpagophytum* extract (study 1 [11], study 2 [12]) (modified after [13]).

Abstracts, and BIDS ISI. Thesaurus and free text searches was also performed performed across each database to combine the terms arthritis and herbal medicine. The general structure of the search strategy was arthritis (or synonyms) and herb (or synonyms). No methodological filter was applied and the search was not limited by language.

All titles and abstracts identified from electronic databases and other searches were independently examined by two investigators. A full manuscript was retrieved for each record that had the possibility of meeting the review criteria. Three reviewers independently assessed eligibility of retrieved studies for review according to the inclusion criteria. Five reviewers contributed to data extraction. Data were extracted from each eligible study by two reviewers acting independently. Where a study was defined as a crossover trial, data were extracted only up to the point of crossover in order that these data could be compared with those derived from parallel trials.

Two review authors independently assessed the risk of bias of each included trial, against key criteria: random sequence generation; allocation concealment; blinding of participants, personnel and outcomes; incomplete outcome data; selective outcome reporting; and other sources of bias, in accordance with methods recommended by the cochrane collaboration [14]. Each of these criteria were explicitly judged using: A = yes (low risk of bias); B = no (high risk of bias); C = unclear (either lack of information or uncertainty over the potential for bias). Potential disagreements were discussed and resolved by referring to the original protocol and, if necessary, arbitration by member(s) of the Cochrane steering group.

Thirty-five studies evaluating the effectiveness of 22 herbal medicinal products were included. However, due to differing preparations, interventions, comparators, and outcome measures, meta-analysis was restricted to data from studies of three products: topical capsaicin, avocado-soybean unsaponifiables, and the Chinese herbal mixture SKI306X showed benefit in the alleviation of osteoarthritic pain.

Several studies investigating products from devil's claw, and a powder from rose hip and seed, reported favorable effects on osteoarthritic pain, whereas two studies of a willow bark extract returned disparate results. Three studies of Phytodolor N® were of limited use because doses and measures were inconsistent among trials. The remaining single studies for each herbal medicinal product provided moderate evidence of effectiveness. No serious side effects were reported with any herbal intervention [7].

Despite some evidence, the effectiveness of none of the phytomedicines in the treatment of OA is proven beyond doubt. The obvious potential benefits of these preparations in the treatment of osteoarthritis may reduce reliance on synthetic medications and the associated risks of harmful adverse events, but further clinical trials are necessary before herbal medicinal products can be adopted in osteoarthritis treatment guidelines.

Systematic reviews are also useful to collect data on adverse events [15]. For *Harpagophytum* preparations, for example, 28 clinical trials were identified of which 20 stated adverse events. In none of the double-blind studies was the incidence of adverse events during treatment with *Harpagophytum procumbens* higher than during placebo treatment. Minor adverse events occurred in around 3% of the patients, mainly gastrointestinal adverse events. Since the dosage used in most of the studies is at the lower limit and since long-term treatment with *Harpagophytum* products is advisable, more safety data are urgently needed.

REFERENCES

[1] Friedman, M. (2006). Potato glycoalkaloids and metabolites: Roles in the plant and in the diet. *Journal of Agricultural and Food Chemistry*, *54*, 8655–8681.

[2] Mensinga, T.T., Sips, A.J., Rompelberg, C.J., van Twillert, K., Meulenbelt, J., van den Top, H.J., van Egmond, H.P. (2005). Potato glycoalkaloids and adverse effects in humans: An ascending dose study. *Regulatory Toxicology and Pharmacology*, *41*, 66–72.

[3] Holtmann, G., Adam, B., Frote, E., Saadat-Gilani, K., Vinson, B. (2004). Validation of the gastrointestinal symptom score (GIS) in patients with functional dyspepsia. *Zeitschrift für Gastroenterologie*, *42*, T1179.

[4] Madisch, A., Kulich, K.R., Malfertheimer, P., Ziegle, K., Bayerdörffer, E., Miehlke, S., Labenz, J., Carlsson, J., Wiklund, I.K. (2003). Impact of reflux disease on general and disease-related quality of life – evidence from a recent comparative methodological study in Germany. *Zeitschrift für Gastroenterologie*, *41*, 1137–1143.

[5] Chang, L. (2004). Review article: Epidemiology and quality of life in functional gastrointestinal disorders. *Alimentary Pharmacology and Therapeutics*, *7*, 31–39.

[6] Manniche, C., Asmussen, K., Lauritsen, B., Vinterberg, H., Kreiner, S., Jordan, A. (1994). Low Back Rating Scale: Validation of a tool for assessment of low back pain. *Pain*, *57*, 317–326.

[7] Cameron, M., Gagnier, J.J., Chrubasik, S. (2011). Herbal therapy for treating rheumatoid arthritis. *Cochrane Database Syst Rev*, *16*(2):CD002948.

[8] Cameron, M., Gagnier, J.J., Little, C.V., Parsons, T.J., Blümle, A., Chrubasik, S. (2009). Evidence of effectiveness of herbal medicinal products in the treatment of arthritis. Part I: Osteoarthritis. *Phytother Res*, *23*, 1497–515.

[9] Pham, T., Van Der Heijde, D., Lassere, M., Altman, R.D., Anderson, J.J., Bellamy, N., Hochberg, M., Simon, L., Strand, V., Woodworth, T., Dougados, M. (2003). Outcome variables for osteoarthritis clinical trials: The OMERACT–OARSI set of responder criteria. *Journal of Rheumatology*, *7*, 1648–1654.

[10] Pham, T., van der Heijde, D., Altman, R.D., Anderson, J.J., Bellamy, N., Hochberg, M., Simon, L., Strand, V., Woodworth, T., Dougados, M. (2004). OMERACT–OARSI initiative: Osteoarthritis Research Society International set of responder criteria for osteoarthritis clinical trials revisited. *Osteoarthritis Cartilage*, *12*, 389–399.

[11] Chrubasik, S., Künzel, O., Thanner, J., Conradt, C., Black, A. (2005). A 1-year follow-up after a pilot study with Doloteffin for low back pain. *Phytomedicine*, *12*, 1–9.

[12] Chrubasik, S., Chrubasik, C., Künzel, O., Black, A. (2007). Patient-perceived benefit during one year of treatment with Doloteffin. *Phytomedicine*, *14*, 371–376.

[13] Chrubasik, C., Wiesner, L., Black, A., Müller-Ladner, U., Chrubasik, S. (2008). A one-year survey on the use of a powder from Rosa canina lito in acute exacerbations of chronic pain. *Phytotherapy Research*, *22*, 1141–1148.

[14] Higgins, J.P.T., Green, S., The Cochrane Handbook for Systematic Reviews of Interventions Version 5.0.0. The Cochrane Collaboration. Available at www.cochrane-handbook.org, 2008.

[15] Vlachojannis, J., Roufogalis, B.D., Chrubasik, S. (2008). Systematic review on the safety of Harpagophytum preparations for osteoarthritic and low back pain. *Phytotherapy Research*, *22*, 149–152.

12

THE INFLUENCE OF BIOTIC AND ABIOTIC FACTORS ON THE PRODUCTION OF SECONDARY METABOLITES IN MEDICINAL PLANTS

Dayana R. Gouvea, Leonardo Gobbo-Neto, and Norberto P. Lopes

12.1 INTRODUCTION

On a field work conducted in the beginning of the 1990s in the western region of Amapá State (Brazil) at approximately 100 km from the city Serra do Navio, the use of volatile constituents from the leaves of the adult plant *Virola surinamensis* by the Waiãpi tribe was verified (Figure 12.1). Shaman Camupi described the utilization of the plant's leaves in immersion and steam inhalation obtained from ebullition of the plant's leaves, for the intense fever crisis of malaria. According to the shaman, the steam did not cure, but reduced the crisis and the clinical picture of this illness. This process was openly discussed, as the shaman stated the existence of a procedure, a cultural/spiritual field, regarding the collection of the leaves, and that without this information the former would have no meaning to anybody. It is obvious that this information was important to the shaman and his community, and besides that, shaman Camupi reinforced that the collection of leaves in wrong days and hours could lead the ill to hallucination outbreaks. The chemical analysis of these leaves' essential oil showed that during the blooming season (June), the highest concentration of monoterpenes happens with the near absence of circadian variations; however, in February (when the plant has no flowers or fruit) and the concentration of

Plant Bioactives and Drug Discovery: Principles, Practice, and Perspectives, Fourth Edition.
Edited by Valdir Cechinel-Filho.

FIGURE 12.1 Adult plants of *Virola surinamensis* (a), fruits details (b), and Waiãpi tribe (c) (*See the color version of this figure in Color Plates section*).

monoterpenes is lower, a clear circadian variation, was observed between the monoterpenes and phenylpropanoids, with eminence to the substance elemi – a resin obtained from such plants. At the end of the Amazonian winter, with the arrival of rain, the plant starts to form its first green fruit and a new substantial alteration in the oil composition happens; when the monoterpene concentration drops to half its value, the phenylpropane' concentration reaches nearly zero while the concentration of sesquiterpenes becomes fivefold [1]. Discussions apart, a point that drew our attention was to imagine, even in a romantic way, how shaman Camupi could know about the presence of this metabolic rhythm and yet how could it affect the treatments of his people. Posterior research confirmed what was already expected. When the essential oil presented a higher concentration of nerolidol, it was more active because the parasite used for the synthesis of dolicol the nerolidol instead of using the substance farnesol, and by doing this, the parasite died from the damaged structure of dolicol. Conferred accentuation is given to the elemi substance with its hallucinogenic potential, which confirms shaman Camupi observations.

Based on this report, what can we expect regarding a vegetable secondary metabolism? What are its rhythms? When qualitative and quantitative variations take place? Professor Otto Richard Gottlieb, the greatest Natural Products chemist who worked in Brazil, often says "it is important to understand nature so we can use it." In order for us to embark in this field, we should first consider that vegetables are organisms exposed to many and different environmental stimuli. Such stimuli are classified into biotic and abiotic factors; according to their origin and action they can also be understood in different ways by the organism, being it ignored or causing heavy stress. Stress can lead these vegetables to adapt in many ways, be it morphologically, physiologically, or biochemically, as they possess a great phenotypic plasticity [2]. This process can be compared to the classic phenotypic definition, where $F = G + \text{MA}$; in other words, the sum of the genetic potential of an organism and the environment actions suffered by it. Let us consider our individual, in this case, a plant; one of the factors that need to be considered is that many times the chemical adaptation is easier than the morphological one, mainly when we talk about the proportion of actives or secondary metabolites. The secondary metabolites are special components of the vegetable metabolism often related to ecological and evolutional aspects of species, as they perform an important role regarding the plant–environment interaction, being it able to protect the vegetables against a wide variety of microorganisms (viruses, bacteria, and fungi) and herbivores (vertebral arthropods) and act upon the attraction of pollinators among other functions that guarantee the survival of such species in the ecosystem [3].

The presence and the quantity of these compounds are affected by the genetic load and expression, the biochemical and physiological processes of the plant besides the influence of many environmental variations as discussed previously. The fluctuations in the concentration and relative proportions of such metabolites can happen temporally and spatially at different levels: seasonally and daily; within plants; within species and interspecies [4–8].

Many factors are essential for the maintenance of vegetable homeostasis, such as temperature, hydric availability, ultraviolet radiation intensity, presence of atmospheric gas, and type of soil and nutrients, besides the interaction of plant–microorganism and herbivores as well as other mechanical stimuli. Change in such factors can often lead to stress and affect the vegetable chemical composition. Stress can be responsible, for example, for the increase of volatile compounds' emission in plants. It is known that exposition to higher temperatures often increases the emission of volatile compounds, which is in part a physical operation, due to increase in steam of such compounds; however, this can be explained better than the changes in steam pressure, as a higher biosynthetic rate of the compounds can occur in this condition. Among other abiotic factors, a high light intensity also stimulates the emission of volatile compounds in a variety of different plant species and its development growth. Hydric, saline, and oxidative stresses are also referred to as factors that can cause the increase in volatile organic compounds' emission [9] as well as affect production of other types of secondary metabolites. It is important to point out that variables that affect the secondary metabolites many times do not act alone, but as a group, or a variable can be the product of another such as the pluviometric rate and seasoning or temperature and altitude, which often makes difficult the study of wild plants individually.

The Waiãpi's example and the discussions presented so far served as stimuli to some Brazilian researches to venture into this investigation field, analyzing mainly the secondary metabolic variations in species applied as medicinal. In the following section, our intention is to try to organize a selection of scientific articles that can affect the secondary metabolism of plants, followed by discussions regarding variations that can be observed due to these stimuli.

12.2 BIOTIC AND ABIOTIC FACTORS THAT CAN AFFECT BIOSYNTHESIS AND/OR METABOLITES ACCUMULATION

As previously discussed, many factors can affect biosynthesis and accumulation of natural products, and in many cases the combined action of factors is yet more important. Despite that, we will show many results separately, trying to turn this chapter presentation more instructive and easier to read.

12.2.1 The Atmosphere's Chemical Composition

The actual changes in the world climate are being mainly assigned to human actions such as fossil fuel burning, deforestation, and forest fire resulting in significant quantities of carbon in the atmosphere, due to the increase in carbon dioxide (CO_2). High level of CO_2, followed by methane and ozone, is considered to be one of the main causes of climate change regarded today as global warming, amount of rainfall, rise of the sea level, and other "extreme climate events." There is much information available regarding global climate change; however, little is known about how such changes can affect the ecosystems, especially the interaction between live organisms [10].

It is known that the levels of CO_2 and O_3 can cause change in the vegetable chemical composition once they alter the availability of substrates and affect biochemical/physiological processes such as photosynthesis besides means of defense signalization. Plants that are grown in CO_2-rich environment usually present a higher C/N relation, due to the decline in the foliar nitrogen and rise of the carbohydrates and phenol compounds. Ozone can increase the antioxidant plants' defense leading to the production of ordinary phenolic compounds as well as flavonoids [11].

An example of a secondary metabolite that has its content affected due to atmospheric composition is digoxin, a cardio-active heteroside used in congestive cardiac insufficiency therapeutics present in the species *Digitalis lanata*. A production 3.5 times higher per hectare of this substance was verified in plants produced in greenhouses with CO_2-enriched atmosphere (1000 ppm) than those grown in field atmosphere [12]. Meanwhile as far as the iridoid levels are concerned, it has been reported that such substances do not suffer alterations or are present in smaller amounts in the species *Plantago lanceolata* grown in atmosphere with 700 $\mu L\ L^{-1}$ of CO_2, when compared to plants grown in normal environmental conditions where the level of CO_2 is half [13].

Ozone is considered as an abiotic factor that induces the plants' defense [14]. It has been verified in vegetation houses that the species *Fagus sylvatica* L., when submitted to ozone exposition, presented a gene transcription reprogramming in the chiquimato paths, which have consequences upon secondary metabolites; therefore, there is a significant rise in the levels of the conjugates of salicylic (SA) and gentisic acids (GA) [15].

Qualitative and quantitative changes can also happen in resins besides the increase in volatile terpenoids after treatment with SO_2 or O_3 [16]. O_3, for example, increased the concentration of terpenes and decreased the concentration of phenolics on the leaves of *Ginkgo biloba* [17]. When exposed to ozone treatment, plants of the genre *Populus* presented higher emission of some isoprenes and monoterpenes with antioxidant functions, reducing the oxidative damage on the plants. Although the mechanism of such protection is not clear yet, some studies propose that the isoprene can stabilize the cell, particularly the chloroplasts [18].

It has also been reported that the fumigation of the soya bean leaves (*Glycine max*) with O_3 besides causing foliar damage has led to a higher concentration of isoflavonoids such as coumestrol, daidzein, and sojagol, given that for such isoflavonoids, a small accumulation caused by NO_2 and SO_2 has been reported [19].

Studing the O_3 exposition in *Pinus sylvestris*, Rosemann et al. [20] found an dosage and time dependant increase of thousand folds for the stilbenes pinosilvine and pinosilvine-3-methyl-ethyl, acconpanied by the increase on the enzime stilbene synthase concentrations. No significant alterations were found on pinocembrin and catechin concentrations. [20]. The content of polyphenols, such as *p*-hydroxy-acetophenone, picein, and kaempferol 3-*O*-glucoside on the abies' leaves (*Picea abies*), is also affected by ozone exposition [20,21].

In oilseed rape (*Brassica napus*), it has been noted that with elevated O_2, there is increase in the foliaceous biomass, while with the increased levels of O_3 a decrease in the plants' growth was observed, and still high concentrations of CO_2 and O_3 were responsible for the decrease in the concentration of indolic glucosinolates, compounds involved in the direct chemical defense of Brassicaceae against herbivores [22].

This set of data therefore shows that there is no unique answer regarding exposition of gas in the atmosphere, but undoubtedly, the increase in the levels of some of these can cause alterations on secondary metabolism.

12.2.2 Pathogen Attacks, Mechanic Stimulus, and Herbivory

Plants react to stimuli caused by trauma, for example, rain, hail, wind, sand, and also herbivory. Such mechanical phenomenon and interaction with microorganisms can directly affect secondary metabolism, as biochemical reactions are induced [23–25]. The induction level varies according to causing agent and factors such as the presence of estimulating substances in the insect saliva [26,27]. This can occur only on the damaged vegetable organ or create a more wide and general response [28–30].

Substances can be produced, for example, to obstruct the plant's palatability, inhibiting herbivory, as happens with tomatoes and potatoes, in answer to the alimentation by insects which produce peptides that are inhibitors of protease [30–33] be the damaged caused on the vegetable tissue a higher production or an accumulation of secondary metabolites, which lead to animals' flight [30,33].

Regarding illnesses caused by fungi, bacteria, and viruses, a signalization cascade is generated and it can lead to long-time resistance, with changes in the composition of the cellular wall, protein structures such as chitin and glucanase, and the synthesis of phytoalexins associated with this resistance. Thus, when a second stimuli is caused, the reaction will occur faster than the primary infection, allowing a more efficient reaction to the new infection and less acceptability of the organ or of the whole organism to future attacks [30,34]. The phytoalexins are compounds normally derived from phenylpropanoids and also terpenoids and polyethylene, generated in response to pathogen attack [30,35–37].

Opitz et al,. (2008) verified on cotton leaves (*Gossypium hirsutum*) a powerful increase in the production of terpenoids and the glands that store them on plants treated with jasmonic acid, a vegetable hormone known for defense reaction mediation to biotic and abiotic stresses. Therefore, it has been shown that the increased levels of terpenoids on the cotton leaves after real and simulated herbivory represent a general reaction that is mediated by jasmonic acid, which indicates that such substances can act synergically on the defense against herbivores or pathogens [38].

It is known that jasmonic acid is a regulator of biosynthetic routes, it being partly responsible for the signaling mechanism, which induces *again* the increase of nicotine in the roots of *Nicotiana sylvestris* where it is then distributed to the entire plant; a positive correlation is observed between the number of trauma on the leaves, endogenous jasmonic acid, and nicotine accumulation in the entire plant [38]. As for the leaves of the shrub *Eriodictyon californicum*, a study done on artificial trauma observed a decrease of up to 63% in the content of phenolic resins [39].

The toxic cardenolides are accumulated rapidly after trauma suffered by the leaves of *Asclepias syriaca* reaching the maximum point 24 h after the trauma and the decrease of cardenolides can be verified practically in elemental levels after 148 h [40]. In pumpkins curcubits, it was observed that an hour after the trauma, there is an induction of such substances which return to the elemental levels less than 10 h [41]. There are also reports of increase in the content of tannins [42] and iridoids [7] in reaction to the trauma caused by caterpillars besides accumulation of condensed tannin in quaking aspen (*Populus tremuloides*) after trauma caused by herbivores [43]. For plantules of *Catharanthus roseus,* an artificially caused trauma led to an increase of nearly 100% after 48 h in the contents of the alkaloids ajmalicine and vindoline; no significant change in the catharanthine content has been observed [23].

In many plants, the volatile compounds can act as direct defense due to their toxicity and/or indirect defense, by attracting natural enemies of the insects that attack them [44]. Thus, biotic stress can also induce the emission of volatile compounds, for example damaged caused by herbivory that can cause the increase upon the liberation of such compounds that could be the result of a rupture of pre-existent structures where the volatile compounds are stored as in the trichomes [9].

12.2.3 Temperature

Temperature and its variation along time and space form a factor that has a significant effect upon vegetable development, hence affecting the production of secondary metabolites [45]. For example, Albert and collaborators [46] showed that temperature is the key factor in the alterations of the phenolic compounds in Arnica (*Arnica montana*) at higher altitudes, as temperature drops with altitude [46]; see Figure 12.2. According to Bilger et al. [47] temperature is an important factor in acclimation of epidermal UV transmittance under environmental conditions in many species of plants; it is observed that in the lack of UV-B radiation, the low temperatures induced an increase in epidermal screening. As this increase is often associated with the synthesis of compounds that absorb UV-B and UV-A radiation, the authors argue that flavonoids could be responsible for this increase. Therefore, low temperature would induce chalcone synthase (CHS) as probably phenylalanine ammonia lyase (PAL). These results indicate that temperature could control the flavonoids biosynthesis [47,48].

A great abundance of mRNA for PAL and CHS has been verified over corn (*Zea mays*) scion, besides anthocyanins according to the intensity and duration of the cold weather that the corn scions have been exposed to [48]. On tobacco leaves (*Nicotiana tabacum*) after submission to low temperatures, an increase of four to five times in the levels of scopolin, genic chlorine acid, and its isomers has been observed [49,50].

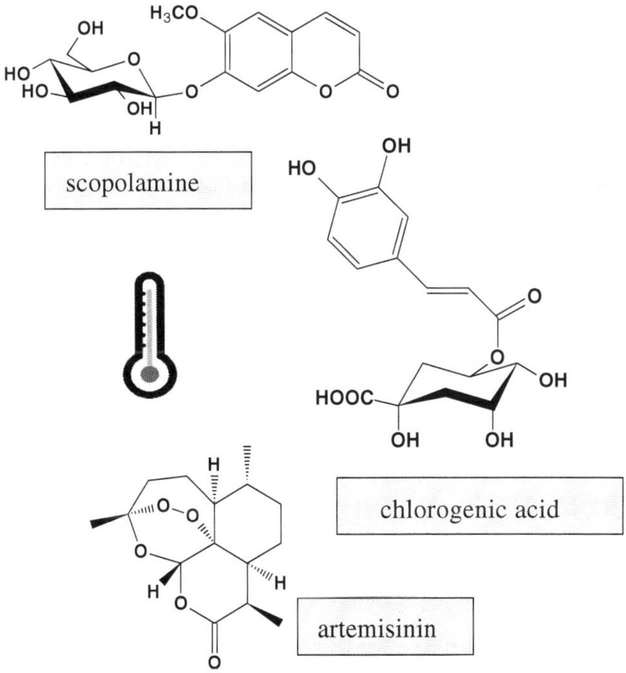

FIGURE 12.2 Examples of secondary metabolites that have their content altered due to temperature effects.

In respect to genic chlorine acid and anthocyanins, increased concentration of such substances have already been reported due to low temperature on the leaves of *Mahonia repens*; however, in this case, they could be linked to seasonality [50,54].

Artemisia annua, after suffering a metabolic stress due to frost, showed an increase of nearly 60% in the levels of artemisinin, which possesses activity against susceptible and resistant cepas from *Plasmodium falciparum*, the causing agent of malaria; concomitantly a decrease in its biosynthetic precursor, dihydroartemisinic acid, has been observed, revealing a fast mobilization of this acid into artemisinin [51].

Regarding the formation of volatile oils, it was verified that hot days can result in an excessive loss of such metabolites; however, in general, their formation takes place in higher temperatures [45]. Nevertheless, the opposite has also been reported, as in the case of the essential oil of *Santolina rosmarinifolia,* where a decrease in the concentration of some substances has been observed as temperature increased [52].

12.2.4 Ultraviolet Radiation

Although different plant species are adaptable to an enormous variation in intensity and quantity of luminous incidence [30,35,45], there is a rising concern with the increased effects of ultraviolet radiation (UV-B, 280–320 nm), resulting from the depletion of the ozone layer, as the concentration and/or composition of many secondary metabolites can be affected by light intensity [30], for example: terpenoids [53,54], alkaloids [55,57], cyanogenic glycosides [58], flavonoids [59,60], tannins [56,61], and anthocyanins [50,62,63]. In Figure 12.3 secondary metabolites whose content is affected due to UV radiation have been exemplified.

The jasmonates can mediate the effects of UV-B radiation upon the plant's defense mechanism; in the recent study it has been verified that UV-B radiation doses affect the important defense mechanisms in *Nicotiana attenuata,* verifying that the solar UV-B radiation does not cause damage to the accumulation of jasmonate, but increases the susceptibility to it, presumably through a regulation of the expression or the stability of some of the compounds involved in this signalization. Therefore, the solar UV-B energy would use dependent and independent jasmonate paths on the elicited phenolic compounds, and it would increase the susceptibility to the jasmonates [64].

A number of studies have already shown that the production of phenolic compounds is directly affected by the intensity of solar radiation [65], as it in the case of flavonoids [66–68] besides tannins [56,61] and anthocyanins [56,61,65]. Flavonoids and correlative phenylpropanoids are recognized by the protection they offer against photo-destruction as they can absorb and/or dissipate the solar energy, lowering the damage that UV-B radiation can cause to vegetable tissues [69–71].

Flavonoids absorb UV-B radiation without altering the active photosynthetic radiation and for that reason they are mainly accumulated in superficial tissues exposed to direct luminous incidence, such as epidermis, subepidermis, fur, cuticle, and epicuticle material [72,73]. It is known that besides acting as filters for such UV substances, they could also have oxidant and radical scavenger functions [69,74].

Phenylalanine ammonia lyase (PAL) and chalcone synthase (CHS) are enzymes of the biosynthetic route of the phenylpropanoids known for being induced

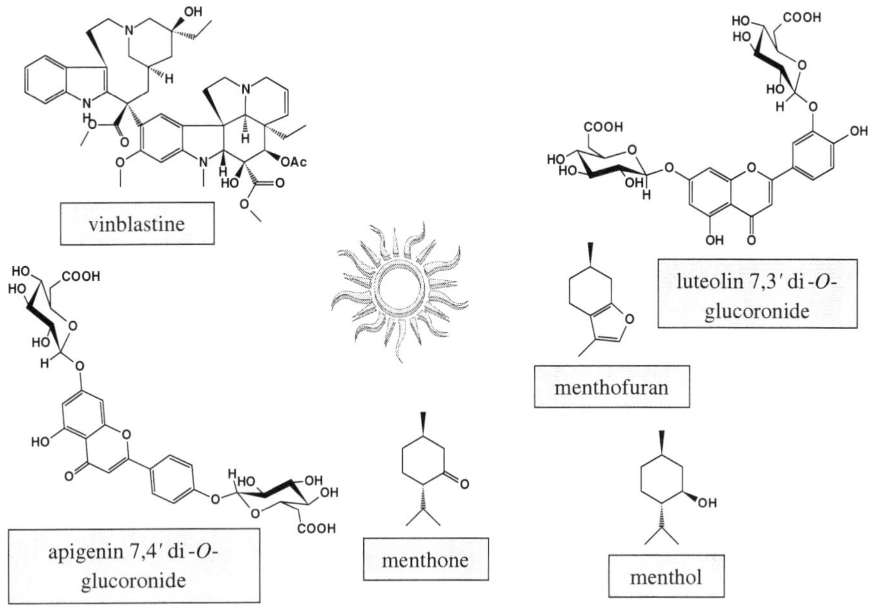

FIGURE 12.3 Examples of secondary metabolites that have their content affected by UV radiation.

by UV-B radiation, as well as other enzymes of this route. In parsley (*Petroselinum crispum*), for example, it has been observed that genes of various important enzymes involved in the first steps in flavonoid biosynthesis, such as glucose 6 phosphate desidrogenase, 3-deoxy-D-arabino heptulosonate 7-phosphate synthase, and *S*-adenosylhomocysteine, also have their control affected by UV radiation [75–82].

The increase in incident UV-B radiation in *Marchantia polymorpha* leads to a rise in the proportion of luteolin glycosides/apigenin glycosides, which, can increase the antioxidant defense levels, as luteolin derivatives are more effective than those of apigenin to consume the absorbed energy [68]; this also happens to the flavonoids/hydroxy-cinnamate relation, being verified an increase of this proportion due to luminous induction [66].

Different doses of UV-A and UV-B solar energy can regulate the genic expression and the accumulation of flavonoids on the leaves of *Betula pendula* where the *in vivo* epidermic flavonoids have decreased when UV-B radiation was excluded; beyond this PAL and HYH transcriptions were expressed in lower levels. A strong positive correlation between the PAL expression and the accumulation of four flavonols caused by the treatments with UV was also observed [59].

Studies made with basil (*Ocimum basilicum*) and thyme (*Thymus vulgaris*), two plants used in folk medicine and as condiments, verified that light directly affects the production and storage of essential oils because the development of glandular trichomes, where the oils are stored, is light dependent [54,83].

It has been verified on mint leaves (*Mentha piperita*) that when they are kept under long-day conditions, the menthol and menthone are observed; as for menthofuran only traces are presented; under short-day conditions, the menthofuran is observed as a major constituent of the essential oil [84]. Light also affects the formation of sesquiterpenes lactones in sunflowers' trichomes (*Helianthus annuus*), as a new intensive exposition increases the biosynthesis of such substances [53].

Regarding alkaloids, it has been observed on the species *Tabernaemontana pachysiphon* that light exposure lowers the concentration of these compounds [85] whereas in *Atropa belladonna*, *Datura stramonium*, and *Cinchona ledgeriana* light exposion results in higher concentration of these compounds. [45]. Catharanthine and vindoline alkaloids when dimerized form vimblastine that is used in cancer treatments. In *C. roseus*, it has been observed that such dimerization increased UV irradiation [57]. On the leaves of *Hypericum perforatum*, the high UV-B radiation increases the concentration of flavonoids and tannins, yet the levels of hypericin are lower on the leaves exposed to a greater UV-B light intensity than those exposed to lower intensity [86].

Kuhlmann and Müller's [2] study on a broccoli (*Brassica oleracea*) observed that high levels of UV radiation led to the increase in the concentration of flavonoids, but a decrease in the accumulation of biomass; however, the changes induced by UV did not alter the capacity to attract herbivore insects. Thus, the authors conclude that plants react to different abiotic and biotic factors with high specificity, showing that perception as well as stress factors is processed separately.

12.2.5 Hydric Stress

Hydric stress can affect secondary metabolism of plants, which already has influence upon important physiologic processes [87]; for example, during a long dry spell, the abscisic acid (ABA) causes closure of the stomas decreasing light exposition and hence reducing photosynthesis and leading to an oxidative stress, smaller protein synthesis, inactivity of enzymes present in the chloroplasts, damage to electrons' transport, increase in membrane permeability, besides an increase in the enzyme activities of H_2O_2 scavenger system [88].

Thus, hydric stress can have a great effect upon the concentration of secondary metabolites in plants, and there are many reports that such conditions usually lead to an increase in the production of various secondary metabolites [89] such as glycosides cyanogenics, glucosinolates [90], some terpenoids [91], anthocyanins [85], and alkaloids [55,85].

Palá-Paúl and coworkers [52] found a positive correlation in relation to some of the components of the essential oil of *S. rosmarinifolia* and the rain [52]. In *Phytolacca dodecandra*, there is a lower production of saponins, such as lemmatoxin when there is a higher level of rain [93].

About St. John's wort, it was observed on its leaves, grown under hydric stress, that the concentrations- of hypericin such as pseudohypericin reduces with time; however, it was noted that the concentration of hyperforin increased significantly, doubled in comparison with the controlled one [95].

In a study done on the evergreen Mediterranean oak *Quercus ilex*, it was verified that monoterpenes' emission was lower due to a more pronounced dry spell in summer. Besides that, it was verified that irrigated trees emitted 82% more monoterpenes during the dry spell period than the trees that received other treatment, which led the authors to conclude that dry spells reduce significantly the emission of monoterpenes in the atmosphere due to the lack of primary substrates derived from photosynthetic processes [96].

As for the phenolic compounds, there is no well-defined correlation between the concentration and the osmotic stress [95]. There are cases, such as in the relation of alkaloids in *C. roseus,* where there is no alteration in the accumulation of metabolites independent of hydric variations [103]; however, the hydric stress usually has consequences upon the production of many secondary metabolites [94,96].

Some parameters must be taken into consideration when studying the effect of hydric stress on the content of secondary metabolites (Figure 12.4), such as the period and degree to which this stress occurs, as a short-term dry spell can lead to an increased production of metabolites; nevertheless, in the long term a contrary effect can be observed [65,99,102,104].

Thus, rain effects must be taken into consideration related to annual rate, its distribution on a year, and effect on the capacity of water absorption from the soil [45]. Another important factor is the fact that rain can carry the hydrosoluble substances of the leaves and roots by lixiviation, as it is the case of alkaloids, glycosides, and volatile oils inclusive [65,99, 102,104].

FIGURE 12.4 Example of secondary metabolites that have their content altered due to changes on hydric availability.

12.2.6 The Soil Influence and Its Nutrients

The soil and its nutrients are extremely important in the growth and development of a vegetable, and therefore they can significantly affect in an expressive way the secondary metabolism of a vegetable. For example, in *Cistus monspeliensis*, the total and individual emissions of terpenes were affected by the soil type; the siliceous substrate was seven times greater than the calcareous substrate. Considering that siliceous substrate has less nutritional content in relation to calcareous soil, the authors argued that when all factors are favorable to growth, the growth processes predominate upon the production of secondary compounds, but when there is scarcity of nutrients such as nitrogen, the plant will allocate proportional resources into secondary compounds, such as the terpenes [105].

Regarding the relation of micronutrients to the production of secondary metabolites in plants, there are few studies, many of them having controversial results, for example, about the secondary metabolite production with cardio-active activity and the effect of manganese [106–108] and other nutrients present in the leaves [106–108]. A study carried out with the foxglove (*Digitalis grandiflora*), where the leaves were sprinkled with manganese and molybdenum solution, showed that the cardioactive heterosides had their concentration increased more than twice [110].

In a study about the essential oils from the leaves of *Eugenia uniflora,* the spathulenol and the caryophyllene oxide presented a strong relation with the nutrients' balance (S, Ca, Fe) and the foliar phenolics, whilst the selenin epoxide-1,3,7 (11)trien-8-one was correlated to the contents of K, Cu, and Mn [97]. In this case, the authors emphasize the important role that the micronutrients play as enzymatic factors in the secondary metabolite biosynthesis, for example, in the case of the sesquiterpene synthase that has divalent metallic ions as cofactors [97].

The nutritional stress causes an increase in the concentration of secondary metabolites with exception in the case of sulfur and nitrogen deficiency [111]. The increase in nutrients, such as nitrogen, is used with the intention of producing biomass in agriculture; however, the nutrients affect, besides the primary metabolism, the secondary one [111]. Many studies correlate light incident intensity and the 'nutrients present in the soil to discuss the relation between light and the availability of nitrogen and carbon [13,112]. These studies do not present clear results; however, there are attempts of hypothesis' refinement upon the carbon/nutrients (C/N) relation [13,30,97,112–115]. Secondary metabolism is usually increased with the increase in the proportion (C/N), with exception in the case of nitrogenic compounds. Therefore, nutrient-deficient soils can result in vegetables with a lower growth rate and a higher secondary metabolite production, especially phenolic derivatives [30,56,97,113,116].

As for nitrogen metabolites such as alkaloids, glycosides cyanogenics, and glucosinolates, an increase in production is usually observed when more nitrogen is present in the soil [45,91], leading to the dilution of such substances; a greater vegetable biomass is also observed. Still related to the nitrogen compounds, there is a secondary metabolism effect when the levels of nitrogen and potassium are altered [45].

It is important to point out that the production of secondary metabolites is affected by the quantity of nitrogen, which is incorporated in the vegetable tissues; thus, the determinant is not the available environment quantity only [56]. There is less conversion of ammonia to nitrate, in acid soils for example, which inhibits nitrogen incorporation, thus explaining why studies with plants that grow in this kind of soil have greater quantities of secondary metabolites, especially phenolic compounds.

As for derivatives of ordinary chiquimic acid, hydrolyzable and condensed tannins, it is known that low levels of nitrogen, phosphorus, sulfur, and potassium can cause an increase in the production of these substances. Still, the metabolites derived from mevalonate were not affected regarding the availability of these nutrients [56].

The amount of the essential oil of *O. basilicum* increased significantly when increased the soil salinity. On the contrary, even increasing the linalool levels, a decrease was observed in the eugenol contents [117].

Ajmalicine is an alkaloid of commercial interest present in the species *C. roseus*; a study carried out its cellular suspension treatment with cadmium revealing that not only the production of ajmalicine can be increased up to three times but also its excretion to culture medium. Another study carried out with cellular culture medium was done with transformed roots of *A. belladonna* bearing that in this case, the cadmium or copper treatment affected the excretion of alkaloids into the cellular culture medium [118]. Regarding the phenolic compounds, a study carried out in palms showed that boron insufficiency reduces the production of these compounds; tannin production is increased when tin alloy and bismuth are offered in *Acacia catechu* [118,119].

Thus, the six sets of different environmental effects presented earlier (2.1 to 2.6) plants are subject to will lead to secondary metabolism variations spatially and temporally. In order to better understand this process, the following observed variation types, such as eco-physiologic variations upon plants' secondary metabolite content, variations of secondary metabolite contents according to the altitude of collection points, and finally rhythmical variations and ontogenesis, are exemplified.

12.3 TYPES OF OBSERVED VARIATIONS ON SECONDARY METABOLITES CONTENT

12.3.1 Eco-Physiologic Variations on Secondary Metabolites Content

Taking into account that the previously described factors are totally interconnected and vary according to the collection point, it is expected that many plants' population species can show variation in one or more chemical constituent classes [118,119].

This happens, for example, with the polyphenols of *Hypericum perforatum*: many studies done with this plant from different provenances showed its enormous chemical composition variation. Despite the use of plants in different blooming

Another type of variation that can take place is the one that regards the plant's ontogenesis; the age and the development level of the vegetable are essential factors in the determination of the content of secondary metabolites. It is obvious that such content can vary according to the different vegetable organs, not only in relation to the total quantity, but also in relation to the components' proportion [7,45,181–191].

A decrease in the production of secondary metabolites often exists in periods of fast tissue growth [192–194]. Being that, younger tissues usually present a higher biosynthetic rate of secondary metabolites [195]; some examples can be cited: essential oils [89,196], sesquiterpenes lactones [53], phenolic acids [49], alkaloids [191], and flavonoids and stilbenes [189].

A work done by Kuhlmann and Müller [2] showed that the impact of UV radiation upon the growth parameters of broccoli (*B. oleracea*) is highly dependent upon the development stage of the plant when it is exposed to stress. The glucosinolates, characteristic compounds of the family Brassicaceae, remained unaltered by UV exposition in all the plants, showing that the regulation paths are independent of such metabolites. However, the infestation of insects that feed themselves on phloem, Aleyrodidae and Aphididae, were highly accentuated in plants exposed to +UV, while the cellular content feeders, such as Thripidae, were more abundant in plants exposed to –UV [2].

A study on the damage to the young leaves of vinca reports the increase in indolic alkaloids, suggesting that younger tissues present higher protection in relation to more developed ones, as in adult leaves it was not possible to verify alterations of phytochemical content [103].

In species of the genus *Hypericum*, the highest concentration of hypericines has been reported during blooming than fruiting [197]; it was also observed that the monoterpenes and the alcoois alifaticos levels varied according to the phenolic cycle and the number of detected compounds increased during ontogenesis in *H. perforatum* [198].

As for *Pituranthos chloranthus*, it was possible to distinguish the different phases of plant development by the presence of specific compounds; for example, the *p*-cymenene was only detected during the budding period, while the high quantities of exo-2-hydroxycineole and exo-2-hydroxycineole acetate were specific to the blooming period; the carvacrol was mainly characteristic to the fruiting period, while the vegetative stage was distinguished from the other by the presence of alpha- and beta-pinene-*p*-cymene; high quantities of exo-2-hydroxycineole and exo-2-hydroxycineole were specific to the flowering period, while the vegetative stage was distinguished from the others by the presence of alpha- and beta-pinenes [199].

In a study on the Mexican arnica, *Heterotheca inuloides*, the concentrations of anti-inflammatory sesquiterpenes were 60% higher during blooming of 15-month-old plants than for 4- to 8-month-olds, independently of the number of collections made and the fertilization regime [200]. In European beech (*F. sylvatica* L.) grown under greenhouse conditions, it was observed that the total solvable phenolic present in the leaves strongly decreased during the leaves' development, independently of ozone treatment [15]. In Table 12.2, other cases are exemplified in which variation upon the secondary metabolites were observed according to the vegetable's development proportion.

TABLE 12.2 Examples of Content Alteration of Secondary Metabolites in Relation to Ontogenesis

Species	Secondary Metabolite	Description	References
Gentiana lutea	*C*-glycosides	The leaves are rich in *C*-glycosides such as the mangiferin, during its blooming stage.	[145]
	O-glycosides	*O*-glycosides, such as isoorientin, are mainly accumulated before the flowers' development.	
Papaver somniferum (poppy)	Morphine	Increase of less than 20 μg/g on the 50th day after germination to more than 120 µg/g on the 75th day.	[190]
	Codeine	Practically constant.	[190]
Tanacetum parthenium	Parthenolide	During the plant's development, higher concentration (before the emergence of the stem on the top where the leaves will come from) During the plant's growth: the total obtained quantity per plant rises circa of 10 to 20 mg, due to the leaves appearing (which, on the contrary to the stems, have more than 1% of parthenolide)	[181]
Digitalis obscura	Lanatoside A digitoxin	Considerably increase with the plant's development	[106,140]
Arnica montana	Sesquiterpene lactones	Young plants: accumulate in majority derivates of helenalin. After approximately six counted weeks from the formation of the leaves, the concentration of these compounds is reduced to practically zero and the compounds of the dihydrohelenalin type increase significantly and then remained constant for a long period.	[140]

12.4 CONCLUSIONS

In this chapter, we presented in a descriptive way the biotic and abiotic factors that can affect the biosynthesis and/or accumulation of secondary metabolites, and at the end, also in a descriptive manner, the main types of observed variations of secondary metabolites content. The majority of the examples are from cultivated plants, but many wild species were also presented. On both models, we observed the variations that can happen on a big scale. In some of the cases, we can even affirm that the constant of secondary metabolites' concentrations is practically an exception in some while it is a rule in others.

The exposed factors in this revision, as well as other factors that can affect the final content of secondary metabolites in medicinal plants, such as collection conditions, stabilization, and storage, can have a great effect on the quality and consequently on the therapeutic value of phytotherapeutic preparation. Taking into account the quality control and the standardization of phytotherapeutic, the constant refinement and investment in domestic studies, biotechnical production, and genetic improvement of medicinal plants are important, instead of using wild plants collected directly from fields. Such precautions can give more guarantee to the obtainment of uniform and high-quality raw material, and it could in part decrease the intrapopulational chemical variations. Knowing about the various factors that can lead to variations of the secondary metabolites' content, it is clear the need of studies that try to detect the conditions and season for the plants' cultivation and/or collection that would lead to a vegetable raw-material with desirable concentration of active principles. A rigorous quality control is also necessary by means of analytic modern techniques, to guarantee a constant in the composition of secondary metabolites in phytotherapeutic preparations on an industrial scale. In addition, the acknowledgment and the comprehension of such variations can, in the future, help in understanding the ecological interactions of a vegetable with its environment.

REFERENCES

[1] Lopes, N.P., Kato, M.J., Andrade, E.H.D.A., Maia, J.G.S., Yoshida, M. (1997). Circadian and seasonal variation in the essential oil from *Virola surinamensis* leaves. *Phytochemistry*, *46*, 689–693.

[2] Kuhlmann, F., Muller, C. (2009). Development-dependent effects of UV radiation exposure on broccoli plants and interactions with herbivorous insects. *Environmental and Experimental Botany*, *66*, 61–68.

[3] Bennett, R.N., Wallsgrove, R.M. (1994). Secondary metabolites in plant defence mechanisms. *New Phytologist*, *127*, 617–633.

[4] Jamieson, M.A., Bowers, M.D. (2010). Iridoid glycoside variation in the invasive plant dalmatian toadflax, *Linaria dalmatica* (Plantaginaceae), and sequestration by the biological control agent. Calophasia lunula. *Journal of Chemical Ecology*, *36*, 70–79.

[5] Gobbo-Neto, L., Lopes, N.P. (2007). Medicinal plants: Factors of influence on the content of secondary metabolites. *Quimica Nova*, *30*, 374–381.

[6] Darrow, K., Bowers, M.D. (1997). Phenological and population variation in iridoid glycosides of *Plantago lanceolata* (Plantaginaceae). *Biochemical Systematics and Ecology*, *25*, 1–11.

[7] Bowers, M.D., Stamp, N.E. (1993). Effects of plant-age, genotype, and herbivory on *Plantago* performance and chemistry. *Ecology*, *74*, 1778–1791.

[8] Lindroth, R.L., Hsia, M.T.S., Scriber, J.M. (1987). Seasonal patterns in the phytochemistry of 3 *Populus* species. *Biochemical Systematics and Ecology*, *15*, 681–686.

[9] Holopainen, J.K., Gershenzon, J. (2010). Multiple stress factors and the emission of plant VOCs. *Trends in Plant Science, 15*, 176–184.

[10] Bidart-Bouzat, M.G., Imeh-Nathaniel, A. (2008). Global change effects on plant chemical defenses against insect herbivores. *Journal of Integrative Plant Biology, 50*, 1339–1354.

[11] Lindroth, R.L. (2010). Impacts of elevated atmospheric CO_2 and O_3 on forests: Phytochemistry, trophic interactions, and ecosystem dynamics. *Journal of Chemical Ecology, 36*, 2–21.

[12] Evans, W.C., *Trease and Evans' Pharmacognosy*, W B Saunders Company Ltd., London, 1996.

[13] Fajer, E.D., Bowers, M.D., Bazzaz, F.A. (1992). The effect of nutrients and enriched CO_2 environments on production of carbon-based allelochemicals in *Plantago* – a test of the carbon nutrient balance hypothesis. *American Naturalist, 140*, 707–723.

[14] Sandermann, H., Ernst, D., Heller, W., Langebartels, C. (1998). Ozone: An abiotic elicitor of plant defence reactions. *Trends in Plant Science, 3*, 47–50.

[15] Betz, G.A., Gerstner, E., Stich, S., Winkler, B., Welzl, G., Kremmer, E., Langebartels, C., Heller, W., Sandermann, H., Ernst, D. (2009). Ozone affects shikimate pathway genes and secondary metabolites in saplings of European beech (*Fagus sylvatica* L.) grown under greenhouse conditions. *Trees-Structure and Function, 23*, 539–553.

[16] Riemer, J., Whittaker, J.B., Air pollution and insect herbivores: Observed interactions and possible mechanisms. In: Bernays, E.A. editor. *Insect–Plant Interactions*. CRC Press, Boca Raton, FL, 1989, pp. 73–105.

[17] He, X.Y., Huang, W., Chen, W., Dong, T., Liu, C.B., Chen, Z.J., Xu, S., Ruan, Y.N. (2009). Changes of main secondary metabolites in leaves of *Ginkgo biloba* in response to ozone fumigation. *Journal of Environmental Sciences – China, 21*, 199–203.

[18] Chen, F., Liu, C.J., Tschaplinski, T.J., Zhao, N. (2009). Genomics of secondary metabolism in *Populus*: Interactions with biotic and abiotic environments. *Critical Reviews in Plant Sciences, 28*, 375–392.

[19] Keen, N.T., Taylor, O.C. (1975). Ozone injury in soybeans – isoflavonoid accumulation is related to necrosis. *Plant Physiology, 55*, 731–733.

[20] Rosemann, D., Heller, W., Sandermann, H. (1991). Biochemical-plant responses to ozone: induction of stilbene biosynthesis in scots pine (*Pinus sylvestris* L) seedlings. *Plant Physiology, 97*, 1280–1286.

[21] Heller, W., Rosemann, D., Osswald, W.F., Benz, B., Schonwitz, R., Lohwasser, K., Kloos, M., Sandermann, H. (1990). Biochemical response of norway spruce (*Picea abies* (L) Karst) towards 14-month exposure to ozone and acid mist: Effects on polyphenol and monoterpene metabolism. *Environmental Pollution, 64*, 353–366.

[22] Himanen, S.J., Nissinen, A., Auriola, S., Poppy, G.M., Stewart, C.N., Holopainen, J.K., Nerg, A.M. (2008). Constitutive and herbivore-inducible glucosinolate concentrations in oilseed rape (*Brassica napus*) leaves are not affected by Bt Cry1Ac insertion but change under elevated atmospheric CO_2 and O_3. *Planta, 227*, 427–437.

[23] Vazquez-Flota, F., Carrillo-Pech, M., Minero-Garcia, Y., Miranda-Ham, M.D. (2004) Alkaloid metabolism in wounded *Catharanthus roseus* seedlings. *Plant Physiology and Biochemistry, 42*, 623–628.

[24] Cipollini, D.F. (1997). Wind-induced mechanical stimulation increases pest resistance in common bean. *Oecologia, 111*, 84–90.

[59] Morales, L.O., Tegelberg, R., Brosche, M., Keinanen, M., Lindfors, A., Aphalo, P.J. (2010). Effects of solar UV-A and UV-B radiation on gene expression and phenolic accumulation in *Betula pendula* leaves. *Tree Physiology*, *30*, 923–934.

[60] Markham, K.R., Ryan, K.G., Bloor, S.J., Mitchell, K.A. (1998). An increase in the luteolin: apigenin ratio in *Marchantia polymorpha* on UV-B enhancement. *Phytochemistry*, *48*, 791–794.

[61] Dudt, J.F., Shure, D.J. (1994). The influence of light and nutrients on foliar phenolics and insect herbivory. *Ecology*, *75*, 86–98.

[62] Jeong, S.T., Goto-Yamamoto, N., Kobayashi, S., Esaka, A. (2004). Effects of plant hormones and shading on the accumulation of anthocyanins and the expression of anthocyanin biosynthetic genes in grape berry skins. *Plant Science*, *167*, 247–252.

[63] Pare, P.W., Tumlinson, J.H. (1997). *De novo* biosynthesis of volatiles induced by insect herbivory in cotton plants. *Plant Physiology*, *114*, 1161–1167.

[64] Demkura, P.V., Abdala, G., Baldwin, I.T., Ballare, C.L. (2010). Jasmonate-dependent and -independent pathways mediate specific effects of solar ultraviolet B radiation on leaf phenolics and antiherbivore defense. *Plant Physiology*, *152*, 1084–1095.

[65] Waterman, P.G., Mole, S., Why are phenolic compounds so important? In: Waterman, P.G., Mole, S. editors. *Analysis of Phenolic Plant Metabolites*, Blackwell Scientific Publications, Oxford, UK, 1994, pp. 44–55.

[66] Tattini, M., Galardi, C., Pinelli, P., Massai, R., Remorini, D., Agati, G. (2004). Differential accumulation of flavonoids and hydroxycinnamates in leaves of *Ligustrum vulgare* under excess light and drought stress. *New Phytologist*, *163*, 547–561.

[67] Cuadra, P., Harborne, J.B., Waterman, P.G. (1997). Increases in surface flavonols and photosynthetic pigments in Gnaphalium luteo-album in response to UV-B radiation. *Phytochemistry*, *45*, 1377–1383.

[68] Markham, K.R., Tanner, G.J., Caasi-Lit, M., Whitecross, M.I., Nayudu, M., Mitchell, K.-A. (1998). Possible protective role for 3′, 4′-dihydroxyflavones induced by enhanced UV-B in a UV-tolenant rice cultivar. *Phytochemistry*, *49*, 1913–1919.

[69] Grace, S.C., Logan, B.A. (2000). Energy dissipation and radical scavenging by the plant phenylpropanoid pathway. *Philosophical Transactions of the Royal Society of London Series B – Biological Sciences*, *355*, 1499–1510.

[70] Delucia, E.H., Day, T.A., Vogelman, T.C. (1992). Ultraviolet-B and visible-light penetration into needles of 2 species of sub-alpine conifers during foliar development. *Plant Cell and Environment*, *15*, 921–929.

[71] Alenius, C.M., Vogelmann, T.C., Bornman, J.F. (1995). A 3-dimensional representation of the relationship between penetration of UV-B radiation and UV-screening pigments in leaves of *Brassica napus*. *New Phytologist*, *131*, 297–302.

[72] Bieza, K., Lois, R. (2001). An *Arabidopsis* mutant tolerant to lethal ultraviolet-B levels shows constitutively elevated accumulation of flavonoids and other phenolics. *Plant Physiology*, *126*, 1105–1115.

[73] Martyn M. Caldwell, R.R., Stephan D. Flint. (1983). Internal filters: Prospects for UV-acclimation in higher plants. *Physiologia Plantarum*, *58*, 445–450.

[74] Larson, R.A. (1995). Plant defenses against oxidative stress. *Archives of Insect Biochemistry and Physiology*, *29*, 175–186.

[75] Zhang, W.J., Bjorn, L.O. (2009). The effect of ultraviolet radiation on the accumulation of medicinal compounds in plants. *Fitoterapia*, *80*, 207–218.

[76] Chappell, J., Hahlbrock, K. (1984). Transcription of plant defense genes in response to UV-light or fungal elicitor. *Nature*, *311*, 76–78.

[77] Chappell, J., Hahlbrock, K., Boller, T. (1984). Rapid induction of ethylene biosynthesis in cultured parsley cells by fungal elicitor and its relationship to the induction of phenylalanine ammonia-lyase. *Planta*, *161*, 475–480.

[78] Logemann, E., Tavernaro, A., Schulz, W.G., Somssich, I.E., Hahlbrock, K. (2000). UV light selectively coinduces supply pathways from primary metabolism and flavonoid secondary product formation in parsley. *Proceedings of the National Academy of Sciences of the United States of America*, 97, 1903–1907.

[79] Jones, D.H. (1984). Phenylalanine ammonia-lyase – regulation of its induction and its role in plant development. *Phytochemistry*, *23*, 1349–1359.

[80] Zucker, M. (1972). Light and Enzymes. *Annual Review of Plant Physiology*, *23*, 133–156.

[81] Schmelzer, E., Jahnen, W., Hahlbrock, K. (1988). *In situ* localization of light-induced chalcone synthase messenger-RNA, chalcone synthase, and flavonoid end products in epidermal-cells of Parsley leaves. *Proceedings of the National Academy of Sciences of the United States of America*, 85, 2989–2993.

[82] Bharti, A.K., Khurana, J.P. (1997). Mutants of *Arabidopsis* as tools to understand the regulation of phenylpropanoid pathway and UVB protection mechanisms. *Photochemistry and Photobiology*, *65*, 765–776.

[83] Ioannidis, D., Bonner, L., Johnson, C.B. (2002). UV-B is required for normal development of oil glands in *Ocimum basilicum* L. (sweet basil). *Annals of Botany*, *90*, 453–460.

[84] Voirin, B., Brun, N., Bayet, C. (1990). Effects of daylength on the monoterpene composition of leaves of *Mentha x piperita*. *Phytochemistry*, *29*, 749–755.

[85] Hoft, M., Verpoorte, R., Beck, E. (1996). Growth and alkaloid contents in leaves of *Tabernaemontana pachysiphon* Stapf (Apocynaceae) as influenced by light intensity, water and nutrient supply. *Oecologia*, *107*, 160–169.

[86] Germ, M., Stibilj, V., Kreft, S., Gaberscik, A., Kreft, I. (2010). Flavonoid, tannin and hypericin concentrations in the leaves of St. John's wort (Hypericum perforatum L.) are affected by UV-B radiation levels. *Food Chemistry*, *122*, 471–474.

[87] Bazzaz, F.A., Chiariello, N.R., Coley, P.D., Pitelka, L.F. (1987). Allocating resources to reproduction and defense. *Bioscience*, *37*, 58–67.

[88] Langebartels C., Schraudner M., Heller W., Ernst D.H.S., Oxidative stress and defense reactions in plants exposed to air pollutants and UV-B radiation. *Oxidative Stress in Plants*. Taylor & Francis, London, 2002.

[89] Gershenzon, J., Maffei, M., Croteau, R. (1989). Biochemical and histochemical-localization of monoterpene biosynthesis in the glandular trichomes of spearmint (*Mentha spicata*). *Plant Physiology*, *89*, 1351–1357.

[90] Blua, M.J., Hanscom, Z., Collier, B.D. (1988). Glucocapparin variability among 4 populations of *Isomeris arborea* Nutt. *Journal of Chemical Ecology*, *14*, 623–633.

[91] Lokar, L.C., Maurich, V., Mellerio, G., Moneghini, M., Poldini, L. (1987). Variation in terpene composition of *Artemisia alba* in relation to environmental conditions. *Biochemical Systematics and Ecology*, *15*, 327–333.

[92] Jung, V., Olsson, E., Caspersen, S., Asp, H., Jensen, P., Alsanius, B.W. (2004). Response of young hydroponically grown tomato plants to phenolic acids. *Scientia Horticulturae*, *100*, 23–37.

[93] Ndamba, J., Lemmich, E., Molgaard, P. (1994). Investigation of the diurnal, ontogenic and seasonal variation in the molluscicidal saponin content of *Phytolacca dodecandra* aqueous berry extracts. *Phytochemistry*, *35*, 95–99.

[94] Gray, D.E., Pallardy, S.G., Garrett, H.E., Rottinghaus, G.E. (2003). Acute drought stress and plant age effects on alkamide and phenolic acid content in purple coneflower roots. *Planta Medica*, *69*, 50–55.

[95] Zobayed, S.M.A., Afreen, F., Kozai, T. (2007). Phytochemical and physiological changes in the leaves of St. John's wort plants under a water stress condition. *Environmental and Experimental Botany*, *59*, 109–116.

[96] Lavoir, A.V., Staudt, M., Schnitzler, J.P., Landais, D., Massol, F., Rocheteau, A., Rodriguez, R., Zimmer, I., Rambal, S. (2009). Drought reduced monoterpene emissions from the evergreen Mediterranean oak *Quercus ilex*: Results from a throughfall displacement experiment. *Biogeosciences*, *6*, 1167–1180.

[97] Gershenzon, J. (1984). Changes in the levels of plant secondary metabolites under water and nutrient stress. *Recent Advances in Phytochemistry*, *18*, 273–320.

[98] Cooperdriver, G., Finch, S., Swain, T., Bernays, E. (1977). Seasonal-variation in secondary plant compounds in relation to palatability of *Pteridium aquilinum*. *Biochemical Systematics and Ecology*, *5*, 177–183.

[99] Horner, J.D. (1990). Nonlinear effects of water deficits on foliar tannin concentration. *Biochemical Systematics and Ecology*, *18*, 211–213.

[100] Dement, W.A., Mooney, H.A. (1974). Seasonal-variation in production of tannins and cyanogenic glucosides in chaparral shrub, *Heteromeles arbutifolia*. *Oecologia*, *15*, 65–76.

[101] Guinn, G., Eidenbock, M.P. (1982). Catechin and condensed tannin contents of leaves and bolls of cotton in relation to irrigation and boll load. *Crop Science*, *22*, 614–616.

[102] Mattson, W.J., Haack, R.A. (1987). The role of drought in outbreaks of plant-eating insects. *Bioscience*, *37*, 110–118.

[103] Frischknecht, P.M., Battig, M., Baumann, T.W. (1987). Effect of drought and wounding stress on indole alkaloid formation in *Catharanthus roseus*. *Phytochemistry*, *26*, 707–710.

[104] Medina, E., Olivares, E., Diaz, M. (1986). Water-stress and light-intensity effects on growth and nocturnal acid accumulation in a terrestrial cam bromeliad (*Bromelia humilis* Jacq) under natural conditions. *Oecologia*, *70*, 441–446.

[105] Rivoal, A., Fernandez, C., Lavoir, A.V., Olivier, R., Lecareux, C., Greff, S., Roche, P., Vila, B. (2010). Environmental control of terpene emissions from *Cistus monspeliensis* L. in natural Mediterranean shrublands. *Chemosphere*, *78*, 942–949.

[106] Gavidia, I., PerezBermudez, P. (1997). Cardenolides of *Digitalis obscura*: The effect of phosphate and manganese on growth and productivity of shoot-tip cultures. *Phytochemistry*, *45*, 81–85.

[107] Hagimori, M., Matsumoto, T., Obi, Y. (1983). Studies on the production of *Digitalis* cardenolides by plant-tissue culture: Effects of mineral salts, initial pH and precursors on digitoxin formation by shoot-forming cultures of *Digitalis purpurea* L grown in liquid-media. *Agricultural and Biological Chemistry*, *47*, 565–571.

[108] Ohlsson, A.B., Berglund, T. (1989). Effects of high $MnSO_4$ levels on cardenolide accumulation by *Digitalis lanata* tissue-cultures in light and darkness. *Journal of Plant Physiology*, *135*, 505–507.

[109] Roca-Perez, L., Boluda, R., Perez-Bermudez, P. (2004). Soil-plant relationships, micronutrient contents, and cardenolide production in natural populations of *Digitalis obscura*. *Journal of Plant Nutrition and Soil Science-Zeitschrift Fur Pflanzenernahrung Und Bodenkunde*, *167*, 79–84.

[110] Letchamo, W. (1986). Effect of trace-elements on cardiac-glycosides accumulation in *Digitalis grandiflora*. *Planta Medica*, *5*, 421–421.

[111] Costa, D.P., Santos, S.C., Seraphin, J.C., Ferri, P.H. (2009). Seasonal variability of essential oils of Eugenia uniflora leaves. *Journal of the Brazilian Chemical Society*, *20*, 1287–1293.

[112] Koricheva, J., Larsson, S., Haukioja, E., Keinanen, M. (1998). Regulation of woody plant secondary metabolism by resource availability: Hypothesis testing by means of meta-analysis. *Oikos*, *83*, 212–226.

[113] Bryant, J.P., Chapin, F.S., Klein, D.R. (1983). Carbon Nutrient Balance of boreal plants in relation to vertebrate herbivory. *Oikos*, *40*, 357–368.

[114] Jones, C.G., Hartley, S.E. (1999). A protein competition model of phenolic allocation. *Oikos*, *86*, 27–44.

[115] Koricheva, J. (2002). The Carbon-Nutrient Balance Hypothesis is dead; long live the carbon-nutrient balance hypothesis? *Oikos*, *98*, 537–539.

[116] Coley, P.D., Bryant, J.P., Chapin, F.S. (1985). Resource availability and plant anti-herbivore defense. *Science*, *230*, 895–899.

[117] Said-Al Ahl, H.A.H., Meawad, A.A., Abou-Zeid, E.N., Ali, M. (2010). Response of different basil varieties to soil salinity. *International Agrophysics*, *24*, 183–188.

[118] Lee, K.T., Yamakawa, T., Kodama, T., Igarashi, Y., Shimomura, K. (1998). Effects of aeration on tropane alkaloid production by transformed root of *Atropa belladonna* in flask cultures. *Journal of Fermentation and Bioengineering*, *86*, 614–616.

[119] Rajaratnam, J.A., Hock, L.I. (1975). Effect of boron nutrition on intensity of red spider-mite attack on oil palm seedlings. *Experimental Agriculture*, *11*, 59–63.

[120] Laitinen, M.L., Julkunen-Tiitto, R., Rousi, M. (2000). Variation in phenolic compounds within a birch (*Betula pendula*) population. *Journal of Chemical Ecology*, *26*, 1609–1622.

[121] Mabry, T.J. (1973). The chemistry of geographical races. *Pure and Applied Chemistry*, *34*, 377–400.

[122] Wilt, F.M., Geddes, J.D., Tamma, R.V., Miller, G.C., Everett, R.L. (1992). Interspecific variation of phenolic concentrations in persistent leaves among 6 taxa from subgenus *Tridentatae* of *Artemisia* (Asteraceae). *Biochemical Systematics and Ecology*, *20*, 41–52.

[123] Bruni, R., Sacchetti, G. (2009). Factors affecting polyphenol biosynthesis in wild and field grown St. John's wort (Hypericum perforatum L. Hypericaceae/Guttiferae). *Molecules*, *14*, 682–725.

[124] Gobbo-Neto, L., Guaratini, T., Pessoa, C., de Moraes, M.O., Costa-Lotufo, L.V., Vieira, R.F., Colepicolo, P., Lopes, N.P. (2010). Differential metabolic and biological profiles of *Lychnophora ericoides* Mart. (Asteraceae) from different localities in the Brazilian "campos rupestres". *Journal of the Brazilian Chemical Society*, *21*, 750–759.

[125] Sobrinho, T.J.S.P., Cardoso, K.C.M., Gomes, T.L.B., Albuquerque, U.P., Amorim, E.L.C. (2009). Analysis of rainfall and edge-effects on flavonoid levels in *Bauhinia cheilantha* (Bong.) Steud., Fabaceae. *Revista Brasileira de Farmacognosia*, *19*, 740–745.

[126] Seemann, A., Wallner, T., Poschlod, P., Heilmann, J. (2010). Variation of sesquiterpene lactone contents in different *Arnica montana* populations: Influence of ecological parameters. *Planta Medica*, *76*, 837–842.

[127] Carey, D.B., Wink, M. (1994). Elevational variation of quinolizidine alkaloid contents in a lupine (*Lupinus argenteus*) of the rocky-mountains. *Journal of Chemical Ecology*, *20*, 849–857.

[128] Spitaler, R., Schlorhaufer, P.D., Ellmerer, E.P., Merfort, I., Bortenschlager, S., Stuppner, H., Zidorn, C. (2006). Altitudinal variation of secondary metabolite profiles in flowering heads of *Arnica montana* cv. ARBO. *Phytochemistry*, *67*, 409–417.

[129] Spitaler, R., Winkler, A., Lins, I., Yanar, S., Stuppner, H., Zidorn, C. (2008). Altitudinal variation of phenolic contents in flowering heads of *Arnica montana* cv. ARBO: A 3-year comparison. *Journal of Chemical Ecology*, *34*, 369–375.

[130] Ganzera, M., Guggenberger, M., Stuppner, H., Zidorn, C. (2008). Altitudinal variation of secondary metabolite profiles in flowering heads of *Matricaria chamomilla* cv. BONA. *Planta Medica*, *74*, 453–457.

[131] Zidorn, C. (2010). Altitudinal variation of secondary metabolites in flowering heads of the Asteraceae: Trends and causes. *Phytochemistry Reviews*, *9*, 197–203.

[132] Alonso-Amelot, M.E., Oliveros-Bastidas, A., Calcagno-Pisarelli, M.P. (2007). Phenolics and condensed tannins of high altitude *Pteridium arachnoideum* in relation to sunlight exposure, elevation, and rain regime. *Biochemical Systematics and Ecology*, *35*, 1–10.

[133] Zidorn, C., Stuppner, H. (2001). Chemosystematics of taxa from the *Leontodon* section Oporinia. *Biochemical Systematics and Ecology*, *29*, 827–837.

[134] Bachereau, F., Marigo, G., Asta, J. (1998). Effect of solar radiation (UV and visible) at high altitude on CAM-cycling and phenolic compound biosynthesis in *Sedum album*. *Physiologia Plantarum*, *104*, 203–210.

[135] Yang, H.L., Duan, Y.W., Hu, F.Z., Liu, J.Q. (2004). Lack of altitudinal trends in phytochemical constituents of *Swertia franchetiana* (Gentianaceae). *Biochemical Systematics and Ecology*, *32*, 861–866.

[136] Robinson, T. (1974). Metabolism and function of alkaloids in plants. *Science*, *184*, 430–435.

[137] Fairbairn, J.W., Suwal, P.N. (1961). The alkaloids of hemlock (*Conium maculatum* L): Evidence for a rapid turnover of the major alkaloids. *Phytochemistry*, *1*, 38–46.

[138] Angelopoulou, D., Demetzos, C., Perdetzoglou, D. (2002). Diurnal and seasonal variation of the essential oil labdanes and clerodanes from *Cistus monspeliensis* L. leaves. *Biochemical Systematics and Ecology*, *30*, 189–203.

[139] Pitarevic, I., Kuftinec, J., Blazevic, N., Kustrak, D. (1984). Seasonal-variation of essential oil yield and composition of Dalmatian sage, Salvia officinalis. *Journal of Natural Products*, *47*, 409–412.

[140] Schmidt, T.J., Bomme, U., Alfermann, A.W. (1998). Sesquiterpene lactone content in leaves of *in vitro* and field cultivated *Arnica montana*. *Planta Medica*, *64*, 268–270.

[141] Yao, L.H., Caffin, N., D'Arcy, B., Jiang, Y.M., Shi, J., Singanusong, R., Liu, X., Datta, N., Kakuda, Y., Xu, Y. (2005). Seasonal variations of phenolic compounds in Australia-grown tea (*Camellia sinensis*). *Journal of Agricultural and Food Chemistry*, *53*, 6477–6483.

[142] Brooks, J.S., Feeny, P. (2004). Seasonal variation in *Daucus carota* leaf-surface and leaf-tissue chemical profiles. *Biochemical Systematics and Ecology, 32*, 769–782.

[143] Atkinson, P., Blakeman, J.P. (1982). Seasonal occurrence of an antimicrobial flavanone, sakuranetin, associated with glands on leaves of *Ribes nigrum. New Phytologist, 92*, 63–74.

[144] Wilt, F.M., Miller, G.C. (1992). Seasonal-variation of coumarin and flavonoid concentrations in persistent leaves of Wyoming Big Sagebrush (*Artemisia tridentata* ssp *wyomingensis,* Asteraceae). *Biochemical Systematics and Ecology, 20*, 53–67.

[145] Menkovic, N., Savikin-Fodulovic, K., Savin, K. (2000). Chemical composition and seasonal variations in the amount of secondary compounds in *Gentiana lutea* leaves and flowers. *Planta Medica, 66*, 178–180.

[146] Lobstein, A., Rietschjako, L., Haagberrurier, M., Anton, R. (1991). Seasonal-variations of the flavonoid content from *Ginkgo biloba* leaves. *Planta Medica, 57*, 430–433.

[147] Clark, L.E., Clark, W.D. (1990). Flavonoid aglycones of the tribe Astereae: Seasonal-variation in leaf exudate flavonoids of *Isocoma acradenia* (Asteraceae). *Biochemical Systematics and Ecology, 18*, 145–148.

[148] Jalal, M.A.F., Read, D.J., Haslam, E. (1982). Phenolic composition and its seasonal-variation in *Calluna vulgaris. Phytochemistry, 21*, 1397–1401.

[149] Kim, S.K., Sakamoto, I., Morimoto, K., Sakata, M., Yamasaki, K., Tanaka, O. (1981). Seasonal-variation of saponins, sucrose and monosaccharides in cultivated *Ginseng* roots. *Planta Medica, 42*, 181–186.

[150] Elgorashi, E.E., Drewes, S.E., Van Staden, J. (2002). Organ-to-organ and seasonal variation in alkaloids from *Crinum macowanii. Fitoterapia, 73*, 490–495.

[151] Wu, T.S., Wang, M.L., Wu, P.L. (1996). Seasonal variations of carbazole alkaloids in *Murraya euchrestifolia. Phytochemistry, 43*, 785–789.

[152] Bandopadhyay, J., De, B. (1997). Seasonal variation of strychnine and brucine in vegetative parts of *Strychnos nux-vomica.* International *Journal of Pharmacognosy, 35*, 349–353.

[153] Roca-Perez, L., Boluda, R., Gavidia, I., Perez-Bermudez, P. (2004). Seasonal cardenolide production and Dop5Br gene expression in natural populations of *Digitalis obscura. Phytochemistry, 65*, 1869–1878.

[154] Salminen, J.P., Ossipov, V., Haukioja, E., Pihlaja, K. (2001). Seasonal variation in the content of hydrolysable tannins in leaves of *Betula pubescens. Phytochemistry, 57*, 15–22.

[155] Osier, T.L., Hwang, S.Y., Lindroth, R.L. (2000). Within- and between-year variation in early season phytochemistry of quaking aspen (*Populus tremuloides* Michx.) clones. *Biochemical Systematics and Ecology, 28*, 197–208.

[156] Jonasson, S., Bryant, J.P., Chapin, F.S., Andersson, M. (1986). Plant phenols and nutrients in relation to variations in climate and rodent grazing. *American Naturalist, 128*, 394–408.

[157] Feeny, P.P., Bostock, H. (1968). Seasonal changes in tannin content of Oak leaves. *Phytochemistry, 7*, 871–880.

[158] Faini, F., Labbe, C., Coll, J. (1999). Seasonal changes in chemical composition of epicuticular waxes from the leaves of *Baccharis linearis. Biochemical Systematics and Ecology, 27*, 673–679.

[159] Bowers, M.D., Collinge, S.K., Gamble, S.E., Schmitt, J. (1992). Effects of genotype, habitat, and seasonal-variation on iridoid glycoside content of *Plantago lanceolata* (Plantaginaceae) and the implications for insect herbivores. *Oecologia*, *91*, 201–207.

[160] Hogedal, B.D., Molgaard, P. (2000). HPLC analysis of the seasonal and diurnal variation of iridoids in cultivars of *Antirrhinum majus*. *Biochemical Systematics and Ecology*, *28*, 949–962.

[161] Kaplan, M.A.C., Figueiredo, M.R., Gottlieb, O.R. (1983). Variation on cyanogenesis in plants with season and insect pressure. *Biochemical Systematics and Ecology*, *11*, 367.

[162] Agerbirk, N., Olsen, C.E., Nielsen, J.K. (2001). Seasonal variation in leaf glucosinolates and insect resistance in two types of *Barbarea vulgaris* ssp *arcuata*. *Phytochemistry*, *58*, 91–100.

[163] Rodman, J.E., Louda, S.M. (1985). Seasonal flux of isothiocyanate-yielding glucosinolates in roots, stems and leaves of *Cardamine cordifolia*. *Biochemical Systematics and Ecology*, *13*, 405–412.

[164] Pio, C.A., Silva, P.A., Cerqueira, M.A., Nunes, T.V. (2005). Diurnal and seasonal emissions of volatile organic compounds from cork oak (*Quercus suber*) trees. *Atmospheric Environment*, *39*, 1817–1827.

[165] Staudt, M., Mandl, N., Joffre, R., Rambal, S. (2001). Intraspecific variability of monoterpene composition emitted by *Quercus ilex* leaves. *Canadian Journal of Forest Research-Revue Canadienne De Recherche Forestiere*, *31*, 174–180.

[166] Curado, M.A., Oliveira, C.B.A., Jesus, J.G., Santos, S.C., Seraphin, J.C., Ferri, P.H. (2006). Environmental factors influence on chemical polymorphism of the essential oils of *Lychnophora ericoides*. *Phytochemistry*, *67*, 2363–2369.

[167] Southwell, I.A., Bourke, C.A. (2001). Seasonal variation in hypericin content of *Hypericum perforatum* L. (St. John's wort). *Phytochemistry*, *56*, 437–441.

[168] Beek, T.A., Bombardelli, E., Morazzoni, P., Peterlongo, F. (1998). *Ginkgo biloba* L. *Fitoterapia*, *59*, 195.

[169] Beppu, H., Kawai, K., Shimpo, K., Chihara, T., Tamai, I., Ida, C., Ueda, M., Kuzuya, H. (2004). Studies on the components of *Aloe arborescens* from Japan – monthly variation and differences due to part and position of the leaf. *Biochemical Systematics and Ecology*, *32*, 783–795.

[170] Park, M.K., Park, J.H., Kim, N.Y., Shin, Y.G., Choi, Y.S., Lee, J.G., Kim, K.H., Lee, S.K. (1998). Analysis of 13 phenolic compounds in aloe species by high performance liquid chromatography. *Phytochemical Analysis*, *9*, 186–191.

[171] Bos, R., Woerdenbag, H.J., van Putten, F.M.S., Hendriks, H., Scheffer, J.C. (1998). Seasonal variation of the essential oil, valerenic acid and derivatives, and valepotriates in *Valeriana officinalis* roots and rhizomes, and the selection of plants suitable for phytomedicines. *Planta Medica*, *64*, 143–147.

[172] Hook, I., Poupat, C., Ahond, A., Guenard, D., Gueritte, F., Adeline, M.T., Wang, X.P., Dempsey, D., Breuillet, S., Potier, P. (1999). Seasonal variation of neutral and basic taxoid contents in shoots of European Yew (*Taxus baccata*). *Phytochemistry*, *52*, 1041–1045.

[173] Vesela, D., Saman, D., Valterova, I., Vanek, T. (1999). Seasonal variations in the content of taxanes in the bark of *Taxus baccata* L. *Phytochemical Analysis*, *10*, 319–321.

[174] Vance, N.C., Kelsey, R.G., Sabin, T.E. (1994). Seasonal and tissue variation in taxane concentrations of *Taxus brevifolia*. *Phytochemistry*, *36*, 1241–1244.

[175] ElSohly, H.N., Croom, E.M., Kopycki, W.J., Joshi, A.S., McChesney, J.D. (1997). Diurnal and seasonal effects on the taxane content of the clippings of certain *Taxus* cultivars. *Phytochemical Analysis*, *8*, 124–129.

[176] Glowniak, K., Mroczek, T., Zobel, A.M. (1999). Seasonal changes in the concentrations of four taxoids in *Taxus baccata* L. during the autumn-spring period. *Phytomedicine*, *6*, 135–140.

[177] Wheeler, N.C., Jech, K., Masters, S., Brobst, S.W., Alvarado, A.B., Hoover, A.J., Snader, K.M. (1992). Effects of genetic, epigenetic, and environmental factors on taxol content in *Taxus brevifolia* and related species. *Journal of Natural Products*, *55*, 432–440.

[178] Silva, M.G.D., Craveiro, A.A., Matos, F.J.A., Machado, M.I.L., Alencar, J.W. (1999). Chemical variation during daytime of constituents of the essential oil of *Ocimum gratissimum* leaves. *Fitoterapia*, *70*, 32–34.

[179] Itenov, K., Molgaard, P., Nyman, U. (1999). Diurnal fluctuations of the alkaloid concentration in latex of poppy *Papaver somniferum* is due to day-night fluctuations of the latex water content. *Phytochemistry*, *52*, 1229–1234.

[180] Fairbairn, J.W., Wassel, G. (1964). The alkaloids of *Papaver somniferum* L.: Evidence for a rapid turnover of the major alkaloids. *Phytochemistry*, *3*, 253–258.

[181] Hendriks, H., AndersonWildeboer, Y., Engels, G., Bos, R., Woerdenbag, H.J. (1997). The content of parthenolide and its yield per plant during the growth of *Tanacetum parthenium*. *Planta Medica*, *63*, 356–359.

[182] Jenks, M.A., Tuttle, H.A., Feldmann, K.A. (1996). Changes in epicuticular waxes on wildtype and *Eceriferum* mutants in *Arabidopsis* during development. *Phytochemistry*, *42*, 29–34.

[183] Buta, J.G., Spaulding, D.W. (1997). Endogenous levels of phenolics in tomato fruit during growth and maturation. *Journal of Plant Growth Regulation*, *16*, 43–46.

[184] Doan, A.T., Ervin, G., Felton, G. (2004). Temporal effects on jasmonate induction of anti-herbivore defense in *Physalis angulata*: Seasonal and ontogenetic gradients. *Biochemical Systematics and Ecology*, *32*, 117–126.

[185] Vogt, T., Gulz, P.G. (1994). Accumulation of flavonoids during leaf development in *Cistus laurifolius*. *Phytochemistry*, *36*, 591–597.

[186] Scora, R.W., Kumamoto, J., Horner, P.F., Hollenberg, J.L. (1984). Ontogenetic variation and diurnal study in the composition of essential oil in *Artemisia douglasiana*. *Journal of Natural Products*, *47*, 279–284.

[187] Wisdom, C.S., Rodriguez, E. (1983). Seasonal age-specific measurements of the sesquiterpene lactones and chromenes of *Encelia farinosa*. *Biochemical Systematics and Ecology*, *11*, 345–352.

[188] Goralka, R.J.L., Schumaker, M.A., Langenheim, J.H. (1996). Variation in chemical and physical properties during leaf development in California Bay Tree (*Umbellularia californica*): Predictions regarding palatability for deer. *Biochemical Systematics and Ecology*, *24*, 93–103.

[189] Slimestad, R. (1998). Amount of flavonols and stilbenes during needle development of *Picea abies*; variations between provenances. *Biochemical Systematics and Ecology*, *26*, 225–238.

[190] Williams, R.D., Ellis, B.E. (1989). Age and tissue distribution of alkaloids in *Papaver somniferum*. *Phytochemistry*, *28*, 2085–2088.

[191] Hoft, M., Verpoorte, R., Beck, E. (1998). Leaf alkaloid contents of *Tabernaemontana pachysiphon* as influenced by endogenous and environmental factors in the natural habitat. *Planta Medica*, *64*, 148–152.

[192] Crankshaw, D.R., Langenheim, J.H. (1981). Variation in terpenes and phenolics through leaf development in *Hymenaea* and its possible significance to herbivory. *Biochemical Systematics and Ecology*, *9*, 115–124.

[193] Palo, R.T., Sunnerheim, K., Theander, O. (1985). Seasonal-variation of phenols, crude protein and cell-wall content of Birch (*Betula pendula* Roth) in relation to ruminant *in vitro* digestibility. *Oecologia*, *65*, 314–318.

[194] Lilov, D., Angelova, Y. (1987). Changes in the content of some phenolic compounds in connection with flower-formation and fruit-formation of vine. *Biologia Plantarum*, *29*, 34–37.

[195] Hartmann, T. (1996). Diversity and variability of plant secondary metabolism: A mechanistic view. *Entomologia Experimentalis Et Applicata*, *80*, 177–188.

[196] Hall, G.D., Langenheim, J.H. (1986). Temporal changes in the leaf monoterpenes of *Sequoia sempervirens*. *Biochemical Systematics and Ecology*, *14*, 61–69.

[197] Xenophontos, M., Stavropoulos, E., Avramakis, E., Navakoudis, E., Dornemann, D., Kotzabasis, K. (2007). Influence of the developmental stage on the (proto)hypericin and (proto)pseudohypericin levels of *Hypericum* plants from Crete. *Planta Medica*, *73*, 1309–1315.

[198] Schwob, I., Bessiere, J.M., Masotti, W., Viano, J. (2004). Changes in essential oil composition in Saint John's wort (*Hypericum perforatum* L.) aerial parts during its phenological cycle. *Biochemical Systematics and Ecology*, *32*, 735–745.

[199] Neffati, A., Hennequin, D., Basset, B., Chekir-Ghedira, L., Ghedira, K., Barillier, D., Ledauphin, J. (2009). Influence of growth phase and geographic origin on the essential oil composition of *Pituranthos chloranthus* from Tunisia. *Natural Product Communications*, *4*, 1585–1594.

[200] Maldonado-Lopez, Y., Linares-Mazari, E., Bye, R., Espinosa-Garcia, F.J., Delgado, G. (2008). Mexican arnica anti-inflammatory action: Plant age is correlated with the concentration of anti-inflammatory sesquiterpenes in the medicinal plant *Heterotheca inuloides* Cass. (Asteraceae). *Economic Botany*, *62*, 161–170.

13

PRODUCTION OF BIOACTIVES COMPOUNDS: THE IMPORTANCE OF PICTET–SPENGLER REACTION IN THE XXI CENTURY

PILAR MENENDEZ, ILARIA D'ACQUARICA, GIULIANO DELLE MONACHE, FRANCESCA GHIRGA, ANDREA CALCATERRA, MARCO BARBA, ALBERTO MACONE, ALBERTO BOFFI, ALESSANDRA BONAMORE, AND BRUNO BOTTA

13.1 INTRODUCTION

Since its appearance in the literature, the Pictet–Spengler (P-S) reaction has become one of the most powerful reactions, not only for the preparation of tetrahydro-β-carbolines (THBCs) and tetrahydroisoquinolines (THIQs), but also for the construction of different scaffolds.

The reaction is named after Amè Pictet and Theodor Spengler, who in 1911 synthesized 1,2,3,4-tetrahydroisoquinoline (**3a**, Scheme 13.1) by cyclocondensation of β-phenylethylamine (**1a**) with formaldehyde dimethyl acetal (**2**, $R^1 = H$), in the presence of hydrochloric acid [1].

The Pictet–Spengler two-component reaction (PS-2CR) can be considered an intramolecular variant of the Mannich three-component reaction [2]. After being used for two decades exclusively for the synthesis of tetrahydroisoquinolines, the process was extended to indole bases in 1928 by Tatsui [3], who prepared 1-methyl-1,2,3,4-tetrahydro-β-carboline (**5**, Scheme 13.1) from tryptamine (**4**) and an aldehyde.

Plant Bioactives and Drug Discovery: Principles, Practice, and Perspectives, Fourth Edition.
Edited by Valdir Cechinel-Filho.

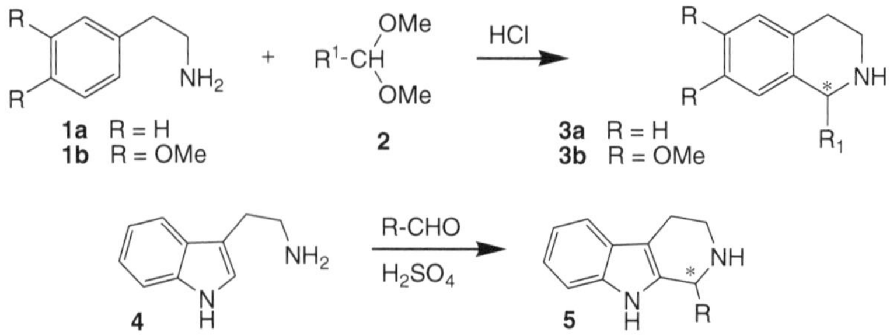

SCHEME 13.1 Pictet–Spengler (P-S) reaction.

The mechanism of the Pictet–Spengler condensation is essentially based on two steps (Scheme 13.2):

1. the acid catalyzed condensation of the amine and the aldehyde to form the iminium intermediate **4a**;
2. the intramolecular attack of the new double bond to the pyrrole ring or the activated aromatic ring [4]. The final deprotonation of the intermediate **4b** gives the THBC.

Both carbons 2 and 3 of tryptamine are nucleophilic. Therefore, after iminium formation, P-S reactions that utilize indole amine substrates can proceed either by attack of carbon 2 to yield directly the six-membered ring intermediate **4b** or by attack of carbon 3 to generate a spiroindolenine intermediate **4c**, which then undergoes a 1,2-alkyl shift to form the product (Scheme 13.2). The pathway of the direct attack from C-2 position is considered by many authors as the most plausible key step in the condensation mechanism.

Essentially, the reaction is an electrophilic substitution on aromatic rings such as indole or pyrrole. Less nucleophilic aromatic rings as phenyl or 3,4-dimethoxyphenyl (**1b**, R = OMe) require higher temperature and strong acids like hydrochloric acid, trifluoroacetic acid, or superacids [5].

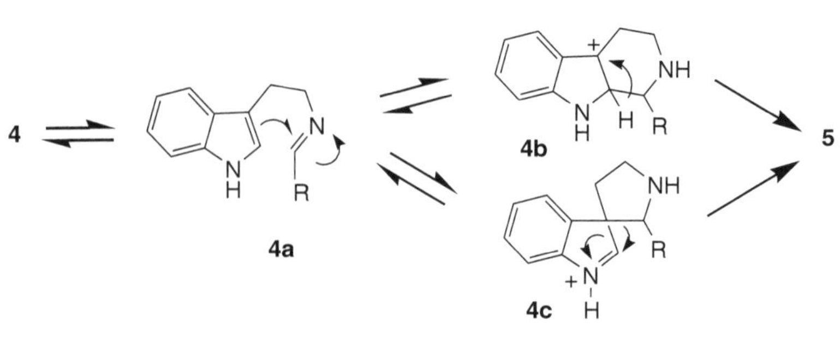

SCHEME 13.2 P-S condensation mechanism.

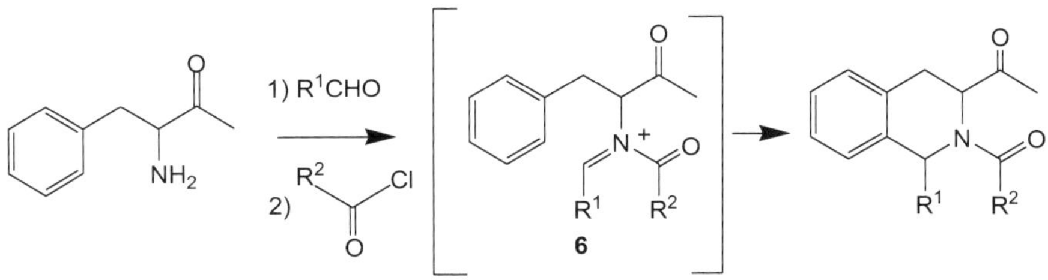

SCHEME 13.3 *N*-acyl P-S reaction.

13.2 VARIANTS AND APPLICATIONS

In a century of life the Pictet–Spengler reaction underwent different modifications. The *N*-acyl Pictet–Spengler and the oxa-Pictet–Spengler reaction are two of the most important variations.

The enhancement of the electrophilicity of the iminium ion by the formation of an *N*-acyliminium intermediate (**6**) allows cyclization of phenyl and aryl ring systems under mild conditions with good yields (Scheme 13.3). The cyclization of *N*-acyliminium ions has been reviewed by Marianoff et al. [6], and the synthesis of Taladafil (Cialis®) (**5**) is one of its prominent results [7].

More recently, in a study related to a biomimetic approach to manzamine alkaloids, Marazano et al. discovered that the electrophilicity of 5-aminopenta-2,4-dienal substrates was dramatically increased by *O,N*-bistrifluoro-acetylation [8]. Aminopentadienal **7a**, derived from tryptamine and 2-methylglutaconaldehyde potassium salt in the presence of TFA, was treated with trifluoroacetic anhydride (TFAA) at room temperature, to give the *N*-acylated tetrahydro-β-carboline **8a** (Scheme 13.4) with

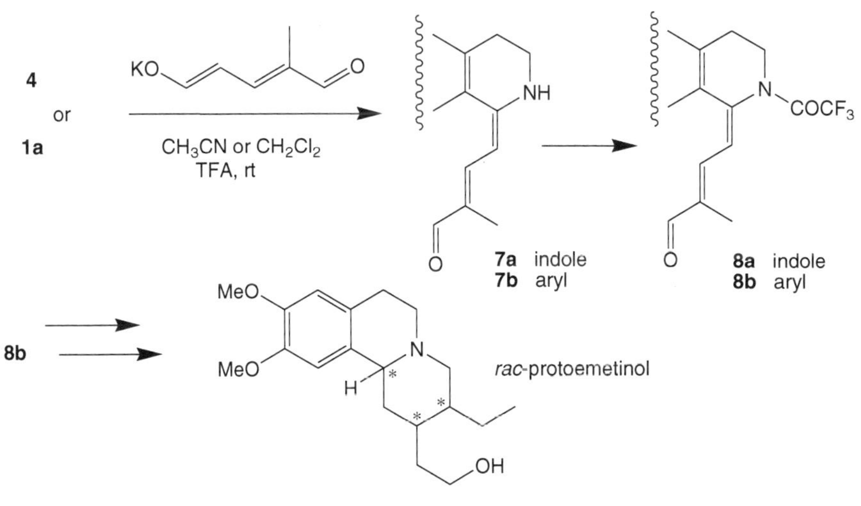

SCHEME 13.4 Synthesis of protoemetinol by P-S reaction.

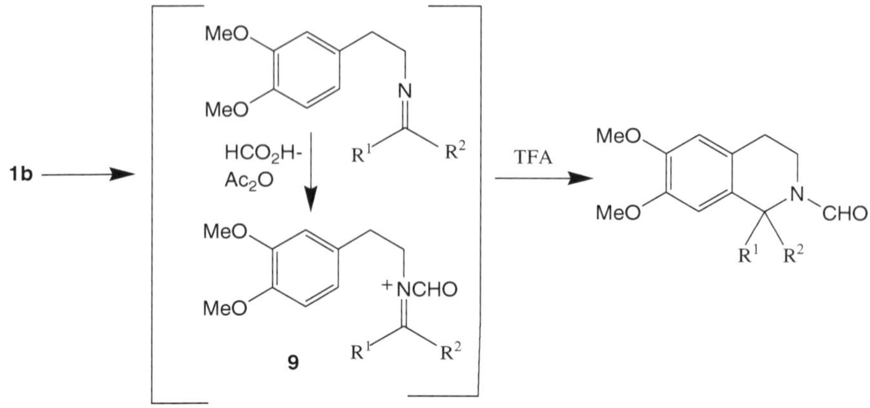

SCHEME 13.5 P-S reaction via *N*-formyliminium ion formation.

an enal function as the exclusive (*E*)-isomer. Notably, the yield was 47%, when 1 equivalent anhydride was used, but increased to 82% with 2 equivalent suggesting that a diacylated species with an *N*-acyliminium group is involved in the P-S reaction. Aminopentadienal **7b**, readily prepared from homoveratrylamine (**1b**), led in the same conditions to tetrahydroisoquinoline **8b** with 78% yield. The methodology was applied to the total synthesis of the THIQ alkaloid (±)-protoemetinol.

Another variant of PS-2CR cyclization exploits the reactivity of the *N*-formyliminium ion **9** formed *in situ* from amine or arylethylamine using acetic-formic anhydride [9a] (Scheme 13.5). Several 1-alkyl-THIQs, as well as their *N*-propargylamino analogous derivatives, were synthesized by this protocol to be biochemically evaluated for their neuroprotective effect [9b].

An example of catalyzed Acyl-Pictet–Spengler reaction has been reported by Youn, who used the couple $AuCl_3$/AgOTf [10]. Examples of Acyl-PS-3CR diastereoselective cyclization will be reported later.

The oxa-Pictet–Spengler condensation, using alcohols instead of amines, provides a convenient and widely used entry for pyran ring closure in the synthesis of natural products, such as pyrano-naphthoquinones [11,12] and isochromans [13,14], as well in the preparation of lipophilic catechin derivatives [15]. Notably, an oxa-Pictet–Spengler cyclization directly provided with high diastereoselectivity dihydronaphthopyran **10a** (Scheme 13.6) with the natural (3a,5)-*trans*-substitution, in the key step of the synthesis of (+)-7-deoxyfrenolicin B and (+)-7-deoxykalafungin [11].

The reaction was also used to improve the lipophilicity of catechin derivatives [15] and plant tannins [16]. The catechins were modified as shown in Scheme 13.7, while the tannins were endowed of alkyl chains by esterification with fatty acids and oxa-PS-2CR coupling with methyl ketones of variable length, not loosing however their anti-oxidative activity.

The most recent report on the oxa-Pictet–Spengler reaction concerns the syntheses of isochromans and/or isothiochromans (thia-PS-2CR) in water (Scheme 13.8) with excellent yields [14].

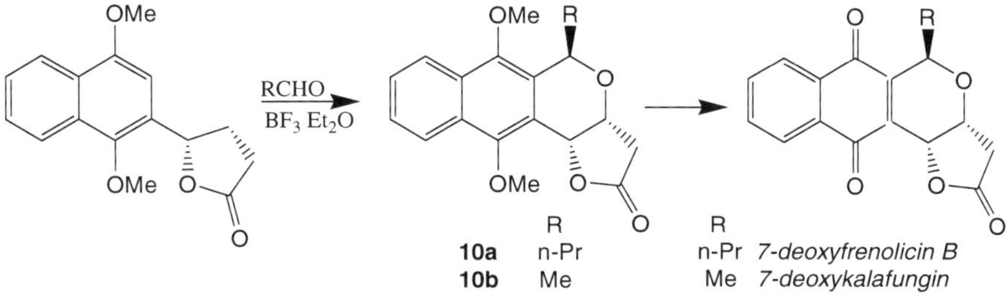

SCHEME 13.6 Oxa-P-S cyclization.

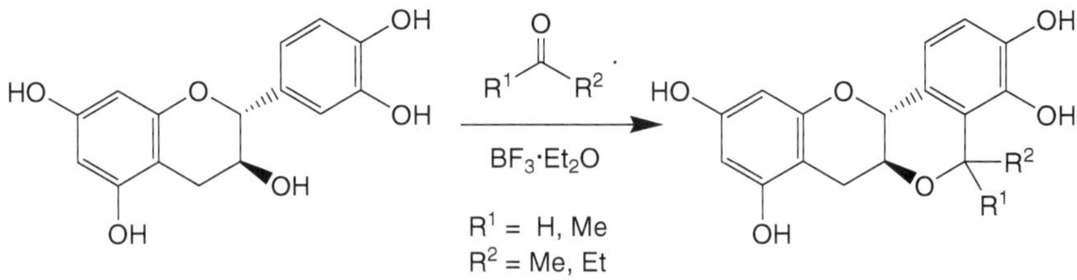

SCHEME 13.7 P-S condensation of cathechin derivatives.

13.3 ASYMMETRIC SYNTHESIS

The largest mess of the literature on the Pictet–Spengler reaction has been dedicated to the asymmetric synthesis of THIQs and THBCs. When the P-S reaction is run with an aldehyde other than formaldehyde, a new stereogenic center is created and *cis*/*trans* isomers are expected for 1,3-disubstituted-tetrahydro-β-carboline, as an example. Stereoselectivity and the ratio of isomers in the synthesis of isoquinoline alkaloids and THBCs have been a long-standing challenge for organic chemists.

Different conditions, such as temperature, solvents, and achiral carbonyl substrates, have been studied to improve the stereoselectivity of the P-S reaction [17]. For instance, it was found that, when the reactions were carried out at low temperature, the major product was the *cis*-isomer, whereas N_b-benzyl or N_b-diphenyl methyl substituted tryptophan esters favored the formation of the *trans*-isomer [4,18]. On the other hand, it has been shown that the P-S reaction of tryptophan allyl ester is *cis*-specific [19].

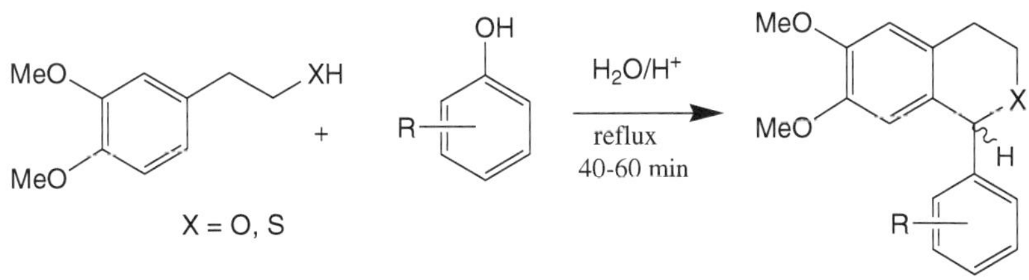

SCHEME 13.8 Synthesis of isochromane derivatives via P-S condensation.

Stereoselectivity is important also in the synthesis of analogs of protein aromatic amino acids (Phe, Tyr, Trp, and His): the incorporation of constrained amino acids with limited rotational freedom into a peptide sequence reduces the flexibility, avoids unselective interaction with multiple receptors, and is thus used to probe the bioactive conformation [20]. 1,2,3,4-Tetrahydro-β-carboline-3-carboxylic acid (Tcc) is one of these successfully applied analogs of constrained tryptophan and was prepared by the P-S cyclization of a chiral amino ester and formaldehyde. Analogously, the *trans* stereoisomer of THBC derivatives can serve as β-turn dipeptide mimetic, as it was shown for similar tetrahydroisoquinoline compounds by Grieco et al. [21a] and Lesma et al. [21b].

Recently, Misicka reported a diastereoselective P-S condensation between tryptophan and α-amino aldehydes derived from L- or D-amino acids, leading to *cis* or *trans* disubstituted THBC derivatives with general structure **11** (Scheme 13.9) [22]. The α-amino aldehydes were prepared via the Fehrentz–Castro procedure [23] and protected immediately with a benzyloxycarbonyl group (Cbz). The diastereomeric (*cis/trans*) tetrahydro-β-carboline analogs were easily separable, and the ratio before purification and after silica gel column was comparable.

The configuration of products, determined by NMR analysis, showed that the P-S reaction of L-tryptophan methyl ester (L-Trp-OMe) with L-amino aldehydes preferentially provide the *trans* isomers, whereas in the case of D-amino aldehydes reactions gave selectively the *cis* isomers and no *trans* products were observed. The P-S cyclization was also performed with the D-Trp-OMe and both Cbz-L-Leu-H and Cbz-D-Leu-H. From the reaction with L-amino aldehyde, only one isomer was obtained and was designated as the *cis* isomer. In the case of D-amino aldehydes, the *trans* isomer was instead the major product (the *cis* compound was the minor product). In general, when both substrates for P-S cyclization have the same configuration (L,L or D,D) a

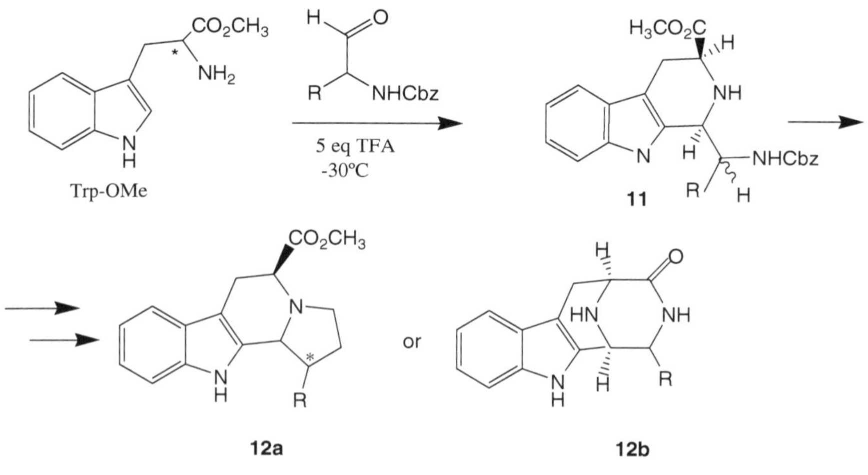

SCHEME 13.9 Diastereoselective P-S condensation between amino acids and α-amino aldehydes.

diastereoisomeric mixture of *cis* and *trans* 1,3-disubstituted-1,2,3,4-tetrahydro-β-carbolines is produced ("mismatched" situation) and the major compound is the *trans* diastereoisomer (65 – 100%). In contrast, when the configuration of the stereogenic centers in the substrates is opposite (L,D or D,L), the P-S reaction led only to the *cis* isomer ("matched" situation). The authors hypothesized that in the reaction with L-amino aldehydes the nucleophilic attack occurs preferentially from the face opposite to the ester group. A hydrogen bond is formed between the iminium and benzyloxycarbonyl groups and stabilizes such pseudocyclic structure, explaining the predomination of the *trans* product. On the other hand, for the *cis* isomer the nucleophilic attack occurs from the same face as the ester group, and there is a slight unfavorable steric interaction between R and the iminium nitrogen. The smaller the R, the higher is the amount of *cis* isomer that is observed: this explains the highest amount of *cis* compound obtained using Cbz-L-Ala-H. For β-branched amino aldehydes, the steric hindrance results in unfavorable interactions, and consequently the *cis* isomer is not formed. For D-amino aldehydes, attack from the same face as the ester group is preferred. Because the hydrogen bond stabilizes the intermediate structure, only the *cis* compound is formed. This is because, in the intermediate leading to the *trans* diastereoisomer, there would be unfavorable interactions, and the transition state would be less stable and in effect *trans* isomer is not formed.

Derivative **11** can be utilized for building more rigid tetracyclic structures such as **12a** and **12b** (Scheme 13.9), which have been proposed as β-turn mimetics [24]. This approach allows creating new structural diversity from a common scaffold in "a library from library" concept.

13.4 CHIRAL AUXILIARY AND ENANTIOSELECTIVE CATALYSIS

Other synthetic strategies used to influence the stereoselectivity of the P-S reaction include the use of optically active carbonyl compound [25], chiral auxiliaries, and enantioselective catalysts. Two remarkable examples of PS-2CR activation by chiral auxiliaries are reported in the papers of Schmidt [26] and Koomen [27]. In the first one, benzyloxycarbonyl (Z)- or 9-fluorenyl-methoxycarbonyl (Fmoc)-protected prolines were initially investigated as chiral auxiliary group. The iminium intermediate **13** was generated from tryptamine and aldehyde in the presence of a dehydrating agent and treated with N-protected acid chlorides **14** in the presence of a base or a Lewis acid, or even without any further additive (Scheme 13.10). In general, the isomeric ratio of THBC **15** was lower when the reaction was performed without addition of Lewis acid, and the highest selectivities and yields were recorded in the presence of titanium alkoxides.

However, the proline-derived chiral auxiliary (**14a**) led to only moderate levels of stereoselectivity; moreover, the Pictet–Spengler adducts were obtained as racemic mixtures not separable by crystallization neither by chromatographic techniques. The research addressed thus to bulky amino acids (**14b**) with sterically more demanding side chains (e.g., *iPr* in valine or *tBu* in leucine), which were protected with *N*-ethoxycarbonylphthalimide and treated with titanium alkoxide in dichloromethane,

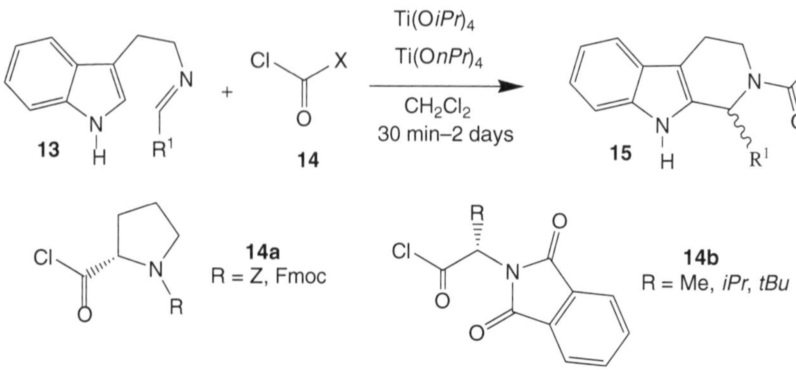

SCHEME 13.10 Use of chiral auxiliaries in the P-S reaction.

prior to the addition of the Schiff base. The desired THBCs were obtained in satisfactory yields (44%) and with good stereoselectivity (89%) [26].

The stereoselectivity in the P-S cyclizations with *N,N*-phthaloyl-protected amino acid chlorides was dependent mainly on the steric demand of the amino acid side chain, the nature of the imine substituent, and the alcohol incorporated into the titanium Lewis acid. With aromatic imines, the diastereomeric ratio increased by the size of the amino acid side chain, while the opposite was observed for aliphatic electrophiles with larger alkyl substituents. The presence of titanium alkoxides is needed for reaching not only good reaction yields, but also high levels of diastereoselectivity. Actually, in the reaction of a valine derivative the isomeric ratio dropped from 99% to 84%, when titanium tetra-*n*-propylate was not included. Moreover, titanium tetra-*n*-propylate gave better results than the isopropylate with aromatic imines, except for *p*-chlorobenzaldimine. The main advantage in the use of α-*N*,*N*-phthaloyl amino acids is the very high diastereoselectivity, but the chiral auxiliary used can be removed only under relatively harsh reductive conditions.

Koomen and coworkers investigated instead the influence of *N*-sulfinyl chiral auxiliaries on the reactivity and diastereoselectivity of the P-S condensation of tryptamine-derived sulfinamines [27]. Enantiopure (*R*)-**16a** was prepared using the commercially available Andersen reagent (1*R*,2*S*,5*R*)-(*S*)-menthyl *p*-toluensulfonate, whereas the (*S*)-**16a** enantiomer was obtained in comparable yield (68 vs. 73%) starting from (1*S*,2*R*,5*S*)-(*R*)-menthyl *p*-toluenesulfonate. To improve the diastereoselectivity and overcome the problem of hydrolysis of the nitrogen–sulfur bond, the P-S condensation was performed at low temperature. The best results for the preparation of tolyl-sulfinamine **17** were obtained with 10-camphorsulfonic acid (CSA) as the acid catalyst in a 1:1 mixture of dry dichloromethane and chloroform at −78°C. Finally, removal of the chiral auxiliary using HCl in EtOH gave in high yield the enantiopure tetrahydro-β-carboline **18** (Scheme 13.11) [27]. As compared with the previous method, lower selectivity is observed, but the diastereoisomers were readily separable and the sulfinyl auxiliary can be cleaved easily by mild hydrolysis.

Following the useful substrate- or auxiliary-controlled diastereoselective versions [26,27], the first example of an enantioselective reagent-controlled

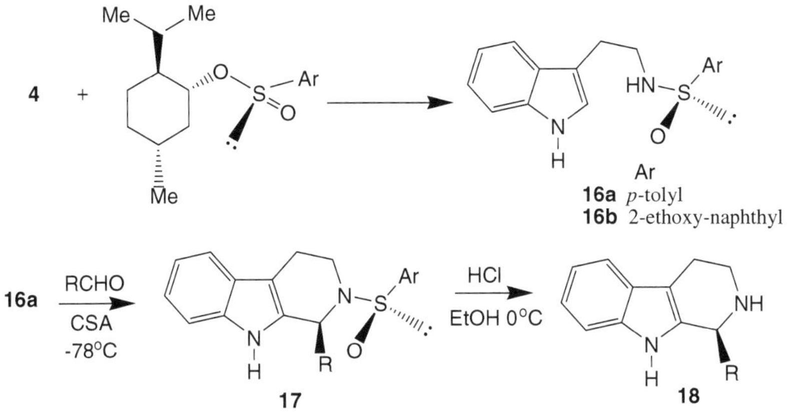

SCHEME 13.11 Use of *N*-sulfinyl chiral auxiliaries in P-S condensation.

Pictet–Spengler-type cyclization was reported by Nakagawa et al. and concerns a chiral Lewis acid-promoted enantioselective cyclization of nitrones **19** (Scheme 13.12), leading to optically active N_b-hydroxytetrahydro-β-carbolines **20** [28].

Nitrones, prepared from N_b-hydroxytryptamine and different chiral aldehydes, resulted stable and could be purified by either recrystallization or column chromatography. The reaction of nitrones in dichloromethane or acetonitrile at −78°C with (+)-diisopinocamphenychloroboran (Ipc_2BCl) as a chiral Lewis acid gave the corresponding tetrahydro-β-carboline in 97% yield with excellent stereoselectivity (up to 90% enantiomeric excess, e.e.). The stereochemical outcome was explained by assuming a transition state involving the nucleophilic attack of indole to the C=N double bond from the less hindered side.

The authors used also chiral binaphthol-based Brønsted acid-assisted Lewis acids (BLAs) for enantioselective Pictet–Spengler reactions [28]. The reaction, performed in dichloromethane at room temperature under argon atmosphere, led to the (*S*)-isomer in 81% yield, with 73% e.e. The (*R*)-isomer was obtained with comparable yield and selectivity. Notably, in all the reactions, 70–80% binaphthol was recovered in yield. The stereochemical results were explained by assuming a model in which the oxygen of nitrone is coordinated to the boron of BLA. The availability of the enantiomeric pair of both Ipc_2BCl and the binaphthol catalyst is a typical advantage

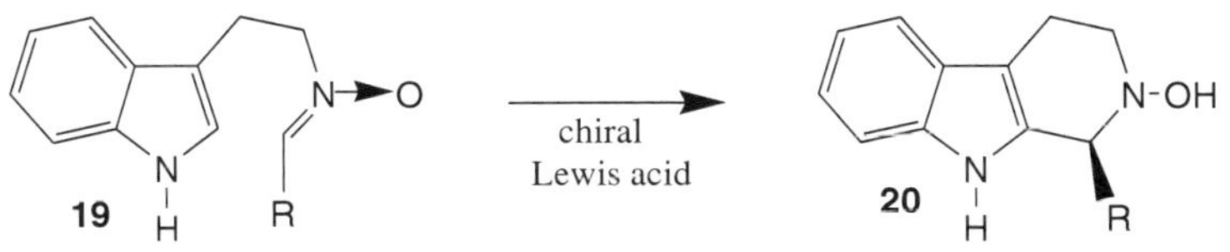

R = Ph, p-MeO-C_6H_4, p-O_2N-C_6H_4, 1-naphthyl, Me, *t*-Bu

SCHEME 13.12 Enantioselective reagent-controlled P-S-type cyclization.

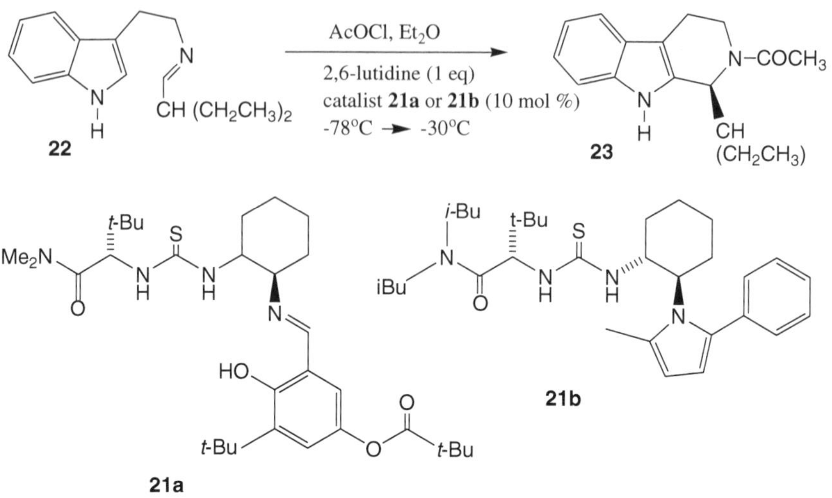

SCHEME 13.13 Use of Lewis or Brønsted acid catalysts in the P-S reaction.

of this approach, which however requires superstoichiometric amounts of chiral Lewis acid reagent [28].

Taylor and Jacobsen [29] reported the first truly catalytic approach by developing an elegant organocatalytic acyl-PS reaction. The authors investigated the possibility that the reaction might be catalyzed by relatively mild Lewis or Brønsted acid catalysts, such as the chiral thiourea **21a**, which had been shown to promote highly enantioselective additions of nucleophiles to *N*-alkyl [30a] and *N*-*tert*-butoxycarbonyl (Boc) imines [30b]. From preliminary screening experiments emerged the strong dependence of the enantioselectivity upon the structure of the acylating agent, as well as the reaction solvent and temperature. As a result, tryptamine-derived imine **22** (Scheme 13.13) underwent cyclization in the presence of acetyl chloride, 2,6-lutidine, and catalyst **21a** in diethyl ether at $-30°C$ to give the N_{β}-acetyl-tetrahydro-β-carboline **23** with 59% e.e. and 60% yield. Taking advantage of the modular structure of **21a**, the catalyst was optimized by several variations to the *N,N*-diisobutylamide **21b**, which allowed to obtain the reaction product with 93% e.e. in 70% yield. The new methodology provided access to a range of substituted THBCs in high enantiomeric excess. Limitations include substrate derived from aromatic aldehydes or trimethylacetaldehyde, which provided cyclized products with lower enantioselectivity.

List et al. reported an enantioselective Brønsted acid-catalyzed Pictet–Spengler cyclization of substituted tryptamines to the corresponding tetrahydro-β-carbolines using a chiral phosphoric acid catalyst [31], which has been shown very recently by Terada et al. [32] and Akiyama et al. [33] to accelerate Mannich-type reactions in highly enantiomeric fashion. These compounds of general structure **24** (Scheme 13.14) were chosen because of their unique characteristics: (i) the tetradentate structure around the phosphorus (V) atom should prevent the free rotation at α of the phosphorus center by formation of a ring structure. This feature cannot be found

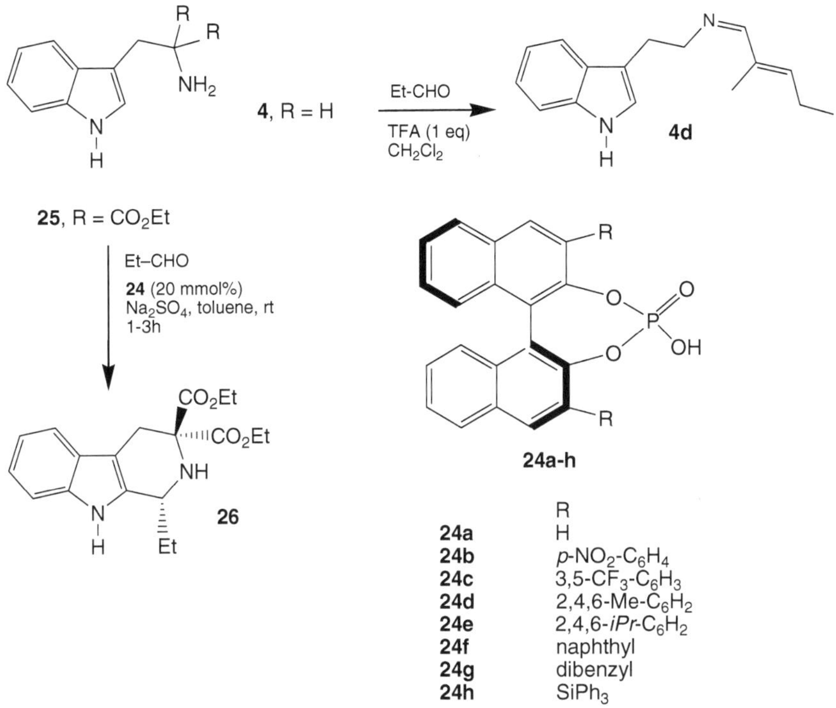

SCHEME 13.14 Chiral phosphoric acids catalyzed P-S reaction.

in other candidates Brønsted acids, such as carboxylic, sulfonic acids, etc.; (ii) the appropriate acidity [34] should catch up the imine through the formation of hydrogen bonds without loose ion-pair formation; (iii) the phosphoryl oxygen should act as a Lewis basic site, and thus a phosphoric acid could function as a bifunctional catalyst. List et al. hypothesized that easily accessible geminally disubstituted tryptamines, such as **25** (R = CO_2Et, Scheme 13.14), could be promising substrates for electronic reasons and might favor cyclization by virtue of a Thorpe-Ingold effect. Indeed, treatment of **25** (R = CO_2Et) with propionaldehyde (EtCHO) and TFA in dichloromethane provided the desired product **26** in high yield. It should be noted that the treatment of tryptamine itself (**4**) with EtCHO in the same conditions gave only compound **4d**, resulting from homo-aldol condensation and imine formation [31]. After screening the series of substituted chiral phosphoric acids **24a**–**f** as asymmetric catalysts, the highest enantioselectivity (66% e.e.) was found for catalyst **24e** in the presence of Na_2SO_4. The enantiomeric excess was further improved (88% e.e.) reducing the reaction temperature to -30°C. Other tryptamines, substituted in the aromatic ring, gave generally high yields (94 – 98%) and good enantiomeric excesses (86 – 90%), under the same reaction conditions. Remarkably, the reaction tolerates a variety of different aldehydes with good results: both unbranched and branched aliphatic aldehydes gave the products in reasonable to excellent yields (50 – 98%) and in high e.e. (72 – 94%) [31]. Although both methods ([29] and [31]) are

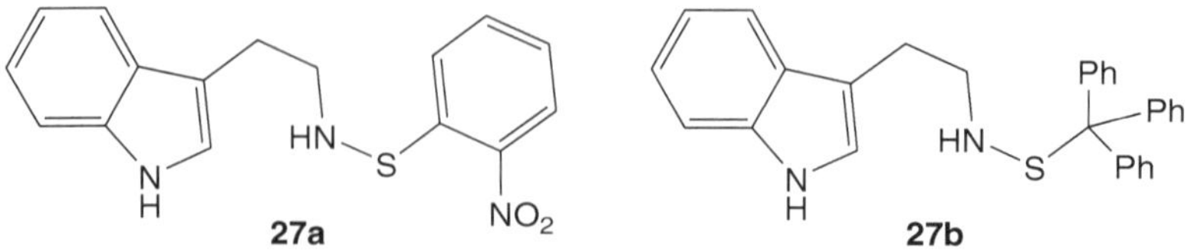

27a 27b

Figure 13.1 *N*-Sulfenyliminium ions as intermediates in the P-S condensation.

impressive examples of the power of the asymmetric organocatalysis, the former has the disadvantage that the *N*-acetyl group is difficult to remove and the latter is clearly limited by the requirement of two ester functions.

More recently, Hiemstra et al. [35] considered the use of *N*-sulfenyliminium ions as intermediates in the P-S condensation. The sulfenyl substituent was expected to stabilize the iminium ion and thus favor Pictet–Spengler cyclization over undesired enamine formation [31]. Another advantage of the sulfenyl group is that it can be easily removed after the cyclization. On the other hand, they focused on phosphoric acids derived from binaphthol as chiral catalysts, which had been highly successful for the addition of carbon-centered nucleophiles to prochiral iminium intermediates [31–33].

The Pictet–Spengler precursors **27a** and **27b** (Figure 13.1) were readily prepared by treating tryptamine with the appropriate sulfenyl chlorides in the presence of a base. When *N*-(*o*-nitrophenyl)sulfenyl tryptamine **27a** was stirred with *n*-hexanal in chloroform in the presence of the (*R*)-binaphthylphosphoric acid **24a** (Scheme 13.14), the corresponding tetrahydro-β-carboline was obtained. The reaction was repeated with a range of chiral phosphoric acid catalysts, such as **24b** – **e**, **24g**, and **24h**: in all cases, excellent yields were observed (92–98%), revealing that the formation of *N*-sulfenyliminium ions intermediates facilitates the reaction. However, the enantioselectivities for the cyclization step were still unsatisfactory, reaching in the best case (**24c**) a modest 31% e.e. The Pictet–Spengler reaction of the bulkier *N*-tritylsulfenyl substituted tryptamine **27b** (Figure 13.1) with *n*-hexanal required optimized conditions, such as the addition of a radical scavenger to prevent the decomposition of the THBC product, the choice of the solvent (chloroform instead of toluene) and the lowering of the reaction temperature to 0°C. Importantly, the addition of powdered molecular sieves to remove the water released during the formation of the iminium salt led to an improvement in both yields and enantioselectivity. The best result of the screening of chiral Brønsted acids was again obtained with catalyst **24c** in the presence of which the THBC was formed with a satisfactory 88% e.e. To avoid the isolation of the labile *N*-tritylsulfenyl product, a one-pot procedure for the synthesis of the corresponding tetrahydro-β-carboline was developed. Thus, the molecular sieves were removed by filtration after the reaction completion. Then, thiophenol and HCl in dioxane were added to the filtrate, and the resulting mixture was stirred overnight to give the hydrochloride salt of the pure tetrahydro-β-carboline as a precipitate. This salt was collected by filtration and then neutralized to produce the pure tetrahydro-β-carboline in 87% yield and with an 84% e.e. of the (*S*)-isomer. The investigation of the effect of aldehyde substrates on the reaction revealed that fast reactions with fair enantioselectivities occurred when unbranched aldehydes were used, except in the case of acetaldehyde. The best result was obtained with phenylacetaldehyde, the reaction of which led to the corresponding product in 90% yield and with 87% e.e. Aromatic aldehydes used reacted more slowly; nevertheless, the corresponding products were obtained in satisfactory yields and with good e.e. values. The ease of the introduction of the sulfenyl substituent and its acid-mediated removal make this method very attractive for the catalytic asymmetric synthesis of tetrahydro-β-carbolines [35].

Catalysts such as Brønsted and Lewis acids, chiral thiourea, and molecular iodine are highly corrosive in nature and cannot be regenerated for further use. Moreover, tedious work setup is required for treating the large quantities of metal salt wastes. For this reason, mesoporous molecular sieves have attracted significant attention in the fields of catalysis and separation. A novel, convenient and simple one-pot procedure has been developed by Vinu et al. [36] for the synthesis of THBCs and THIQs: tryptamine, aryl aldehydes, and benzyl chloride undergo smooth coupling in the presence of mesoporous A1SBA-15 catalyst in acetonitrile at 80°C to give THBCs in good yields and with a high selectivity. AlSBA-15 is a highly acidic catalyst with well-ordered porous structure, which can be recovered and recycled. The presence of both catalyst and benzyl chloride is essential to accomplish the reaction. When compared to other methods for the synthesis of THBCs, the solid acid-induced reaction tolerates the sensitive indole unit well. Another interesting feature of this method is that aryl aldehydes bearing either electron-withdrawing or electron-donating groups successfully undergo the P-S reaction with equal ease. The reaction also proceeds with homoveratrylamine (**1b**) to give the corresponding THIQs. Recently, a similar procedure using H^+-montmorillonite heterogeneous acid catalyst has been reported [37]. The reaction of β-phenylethylamine (PEA) with *para*-substituted benzaldehyde derivatives afforded the corresponding condensation products in the presence of H^+-montmorillonite with tolerance of functional groups. The highest yields (94 – 97%) were obtained with electron-rich aryl ethylamines as substrate rather than electron neutral phenylethylamine. Aliphatic aldehydes also gave the corresponding THIQs with very high yields (94 – 96%). In these environmentally friendly processes, the use of organic solvents was avoided and the catalyst was recycled with maintenance of high catalytic activity. Few reports have been published on Pictet–Spengler cyclization in water or aqueous media: both THBCs and THIQs as well as isochromans were synthesized in the presence of Brønsted acid-surfactant combined catalyst [38].

13.5 ENZYMATIC CATALYSIS

Pictet–Spengler condensation is the key step both in many biosynthetic pathways [39] and consequently in the synthesis of natural products. For instance, dopargine (**28**, Figure 13.2), isolated from black cohosh (*Cimicifuga racemosa*), is the presumed Pictet–Spengler reaction THIQ product of dopamine and γ-guanidinobutyraldehyde (**29a**, Figure 13.2), co-occurring in the plant [40]. A similar pathway can be postulated for the THBC 6-hydroxytrypargine (**30**), a toxin from the venom of the colonial spider

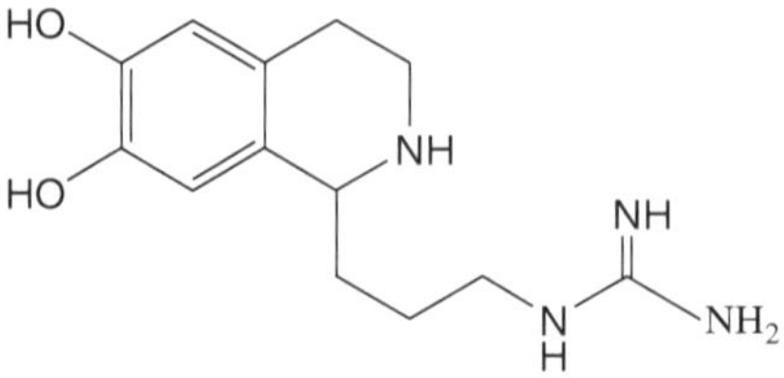

28

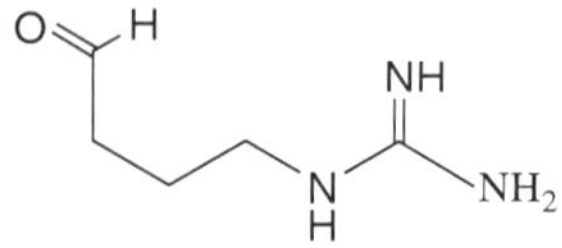

29a

Figure 13.2 P-S condensation in the synthesis of natural products.

Parawixia bistriata, which was synthesized by P-S cyclization of 5-hydroxytryptamine **4e** with **29b** (Scheme 13.15) and resolved by enantioselective chromatography. Notably, the (−)-enantiomer was twofold more potent in assays of monoaminoxidase A and B inhibition than the (+)-enantiomer [41].

Enzymatic catalysis has been recognized as a practical alternative to traditional organic synthesis not only for its chemo-, regio-, and enantioselectivity, but also for the advantage of *green* synthetic processes in the absence of contaminants, such as organic solvents and metal catalysts [42]. Moreover, many enzymes act as bifunctional catalysts, being able to simultaneously bind and activate two reacting substrates, using amino acid residues at the active site both as Lewis acids and Lewis bases [43]. Bifunctional biocatalysts, such as hydrolases, racemases, lipases, isomerases, etc., are currently used for a broad range of selective organic transformations [44–46]. Enzymes catalyzing the P-S reaction, namely "Pictet–Spenglerases" have been isolated from several plants [47–49]. Among them, strictosidine synthase is the key enzyme in the biosynthesis of the monoterpene indole alkaloids, a large and structurally varied family of natural products, including strychnine, ajmalicine (**31a**), catharanthine (**31b**), and vindoline (**31c**) (Figure 13.3) [50]. Strictosidine synthase catalyzes diastereoselectively the condensation between tryptamine **4** and secologanin (**32**) to the THBC strictosidine (**33**), the biosynthetic precursor to terpene indole alkaloids (Scheme 13.16). A sample of strictosidine synthase from *Rauvolfia serpentina* has been co-crystallized in the presence of both secologanin and tryptamine and X-ray analysis of the crystals allowed the first insights into the substrate binding orientation and the enzymatic mechanism [51]. The structural information provided by the crystal structure, the data derived from kinetic isotope effects (KIE) and rate dependence on pH, and the comparison to a nonenzymatic Pictet–Spengler

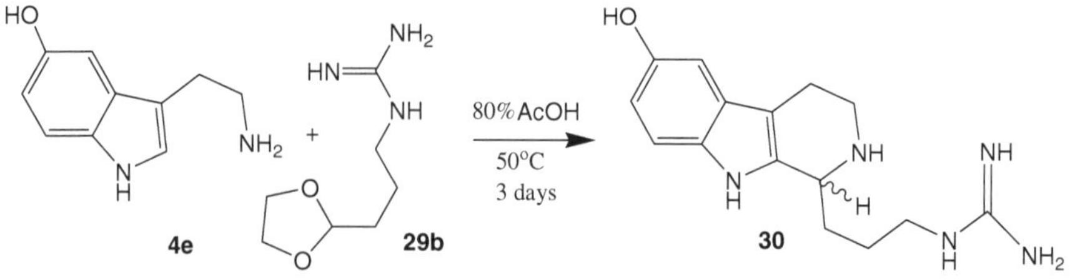

4e 29b 30

SCHEME 13.15 P-S reaction in the synthesis of 6-hydroxytrypargine.

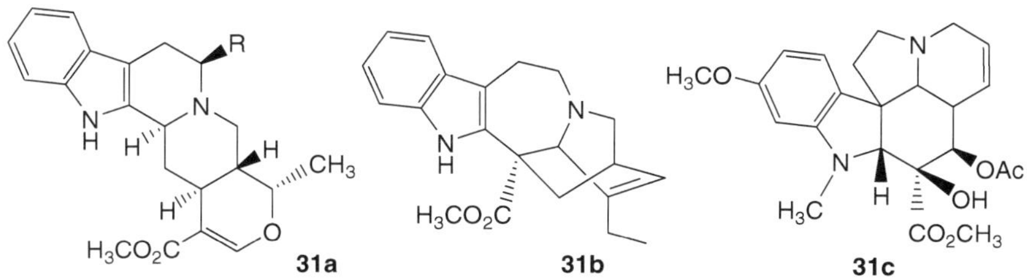

Figure 13.3 Monoterpene indole alkaloids synthesis by "Pictet-Spenglerase."

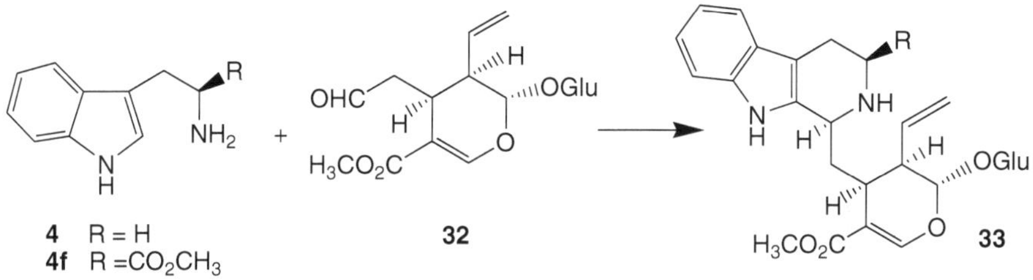

SCHEME 13.16 P-S condensation catalyzed by strictosidine synthase.

reaction suggested the involvement of a first acid-catalyzed step regarding the iminium formation, and a second base-catalyzed step concerning the final deprotonation [51]. Another important member of "Pictet–Spenglerase" family is Norcoclaurine Synthase (NCS), the enzyme that catalyzes the Pictet–Spengler condensation of dopamine with 4-hydroxy-phenylacetaldehyde (4-HPAA) to yield (*S*)-norcoclaurine (**34**, Scheme 13.17) [52,53], the central precursor of the biosynthetic pathway leading to the diverse benzylisoquinoline alkaloids. The interest in (*S*)-norcoclaurine is paramount, due to its well-established sympathomimetic activity and a safe pharmacological profile [54]. For its activities, the drug can advantageously compete with racemic dobutamine, the current "best in class" among sympathomimetic drugs [55]. To obtain the biologically active (*S*)-isomer, diverse synthetic strategies have been employed so far. Such strategies are based on asymmetric catalytic approaches involving enantioselective hydrogenation of the corresponding dihydroquinoline intermediates by appropriate metal catalysts [56]. The crystallographic structure of

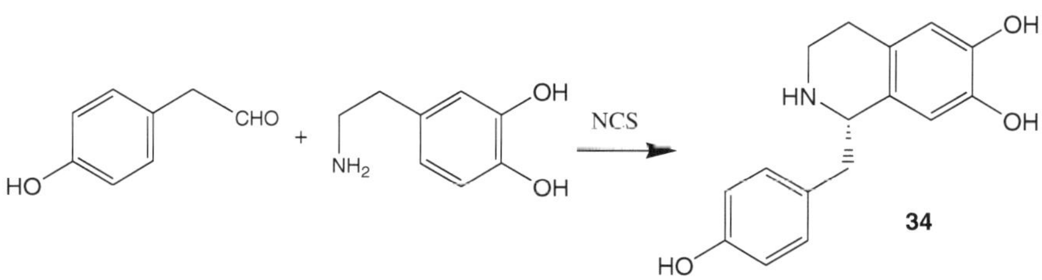

SCHEME 13.17 NCS catalyzed P-S condensation of 4-HPAA and dopamine.

NCS from *Thalictrum flavum* has recently been solved [57]. The crystallographic data obtained in complex with dopamine, the natural substrate, and *p*-hydroxybenzaldehyde, a nonreactive substrate analog, provided a snapshot of the initial step of the reaction mechanism [58]. Notably, the reaction mechanism was in agreement with the reaction scheme proposed by Luk et al. [59], based on KIE studies.

13.6 THE PICTET–SPENGLER REACTION AT PRESENT

After almost a century, the investigation on the Pictet–Spengler reaction is not arrived at the end and a large number of citations are present in the literature of the last two years (2009–2010). THIQs and THBCs are structural motifs in a diverse array of biologically active compounds: both of these ring systems are popular choices for combinatorial libraries targeted at drug discovery. Moreover, new generations of substrates have arisen to testify the popularity of this reaction.

13.6.1 Tetrahydroisoquinoline (THIQ) Family

The tetrahydroisoquinoline nucleus is often part of a more complex molecule: for instance, the condensation of the aminophenol **35** to the Pictet–Spengler adduct **36** (Scheme 13.18) was a key step for the construction of the lemonomycin framework. The reaction occurred in dichloromethane in the presence of acetic acid and 4Å molecular sieves, and **36** was obtained as a single diastereoisomer [60]. Phtalascidin (**37a**, Figure 13.4) and renieramycin G (**37b**) are two representative compounds of a class of antitumor alkaloids belonging to the THIQ family and containing five condensed six-membered rings, four of which are two THIQ moieties (either as quinones or hydroquinones) and one a pyrazine subunit.

In a project addressed to the synthesis of analogs of these compounds, a Spanish group prepared the target compound **39a** with very good yields by *N*-acyliminium cyclization of the intermediate (3*R**, 6*R**, 11a*S**)-3-arylmethyl-pyrazino[1,2-b]isoquinoline-4-dione (**38a**). The key step was a Pictet–Spengler-type reaction and led to the formation of a second THIQ ring (Scheme 13.19) [61]. Since the reaction failed with all-*cis*-isomers, the same group proposed for the synthesis of **39b** a new protocol based on a different substrate **38b** and the use of trifluoromethanesulfonic acid as superacid catalyst (Scheme 13.19) [62].

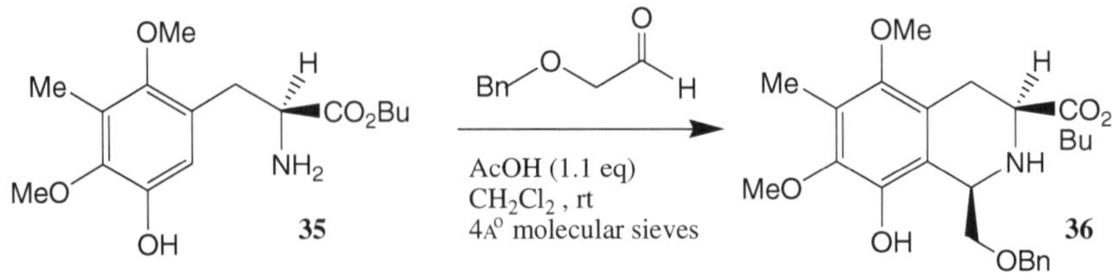

SCHEME 13.18 P-S adduct for the synthesis of the lemonomycin scaffold.

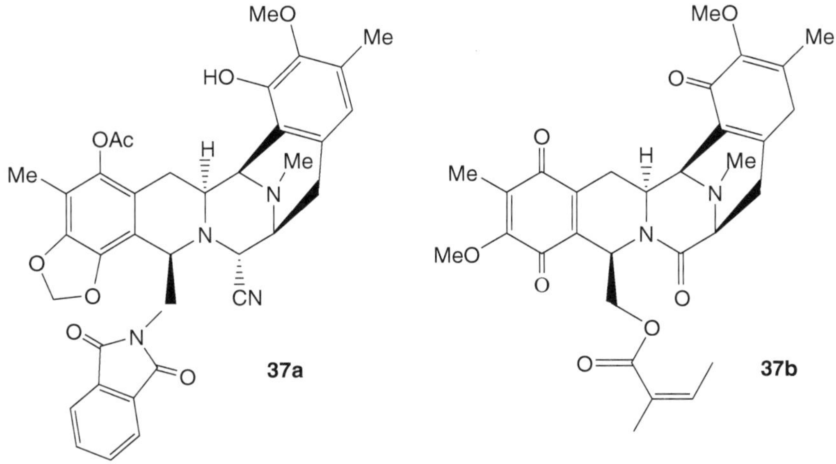

Figure 13.4 Phtalascidin (**37a**) and renieramycin G (**37b**) structures.

In the recently reported total synthesis of renieramycin G (**37b**, Figure 13.4) starting from L-tyrosine [60], the Pictet–Spengler reaction is again first carried out for the intermolecular cyclization leading to the THIQ unit and second for the intramolecular ring closure leading to the pentacyclic framework. Benzyloxyacetaldehyde and acetic acid were the protagonist of the first reaction (Scheme 13.20), whereas the second cyclization was achieved again with CF_3SO_3H [63].

Schulzeines A, B, and C, a new class of marine alkaloids from the sponge *Penares schulzi,* have a THIQ core with a C_{28} amide side chain. The construction of *trans-* and

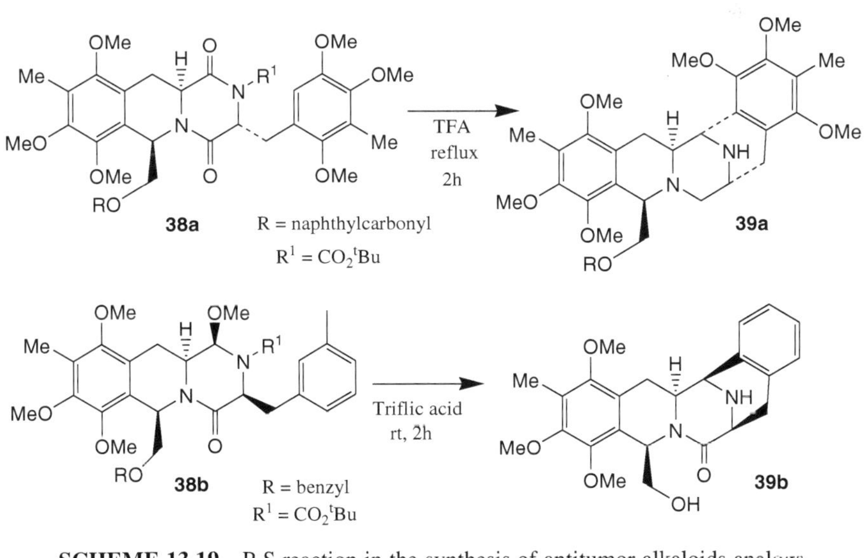

SCHEME 13.19 P-S reaction in the synthesis of antitumor alkaloids analogs.

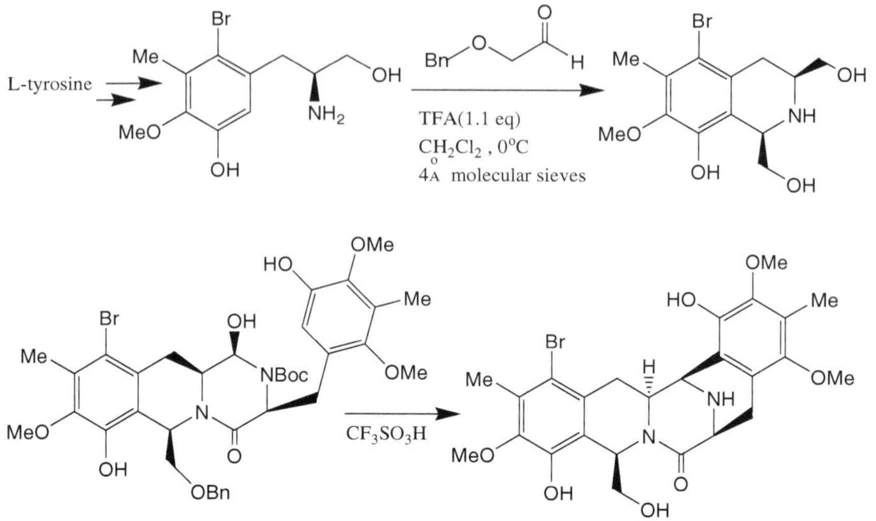

SCHEME 13.20 Total synthesis of renieramycin G.

cis-3-aminobenzo[a]quinolizidine subunits from L-glutamate derivatives **40** and **41** (both R = Phth) required the presence of TFA (in dichloromethane at −78°C) and AcOH (in chloroform at reflux), respectively, as P-S promoters (Scheme 13.21). The *cis* and *trans* diastereomers **42a** (R = Phth) were separated after hydrazinolysis of the phthalimide group and formation of *cis*- and *trans*-**42b** (R = Boc). Treatment of each stereoisomer with TFA afforded the desired *tran*s- and *cis*-subunits **42c** (R = H) [64]. In the enantioselective total synthesis of schulzeines B and C by another group [65], the tricyclic core of the target molecules was prepared by the intermediate *trans*-**44a** and *cis*-THIQ **44b** (Scheme 13.22). The mixture of the two stereoisomers was obtained by P-S condensation of the amine **1d** with the masked bis-homoserine aldehyde **43** in glacial AcOH with heating. **44a** and **44b** were separated by flash chromatography. A series of (hetero)arenopyrrolopyridines **46a–e** have been

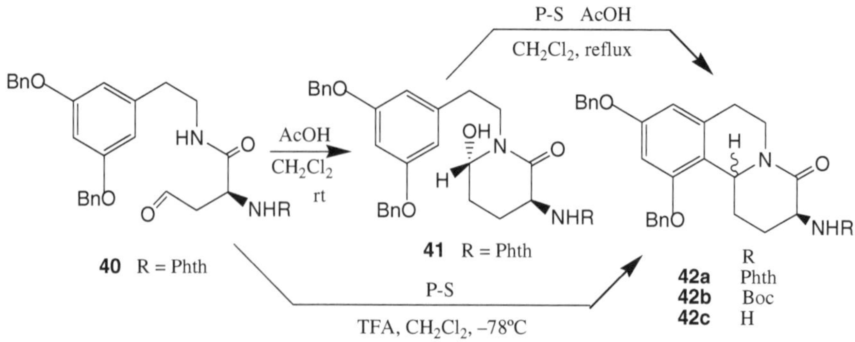

SCHEME 13.21 Synthesis of schulzeines A, B, and C.

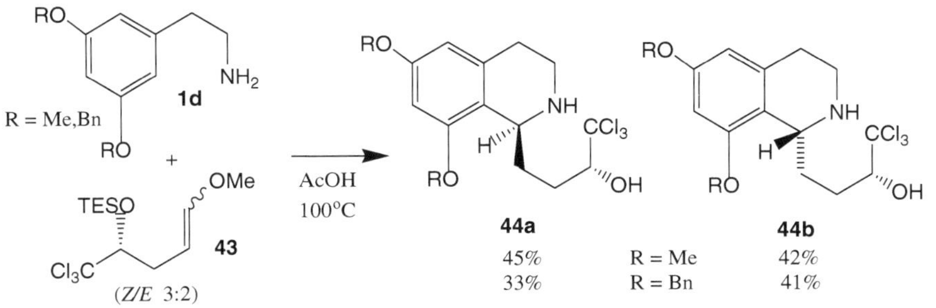

SCHEME 13.22 Enantioselective total synthesis of schulzeines B and C.

synthesized by a classical P-S cyclization of β-arylpyrrolidones **45** (Scheme 13.23). Compounds **46a–c** feature the triciyclic skeleton of a THIQ-fused pyrrolidine, whereas **46d** and **46e** contain THBC and furan moieties, respectively. Since **46c** and **46d** had biological activity, the five scaffolds were further derivatized to make a 32-membered amide library of potentially pharmacological important compounds [66].

In the synthesis of the highly functionalized β-lactam substituted pyrroloisoquinoline **48a,** the final step is the P-S cyclization of the aromatic ring A in **47a** to a THIQ moiety (Scheme 13.24). From the intermediate (not shown) with an indole nucleus instead of the aromatic ring, a classical P-S condensation afforded the analogous indolizinindole with a THBC nucleus [67].

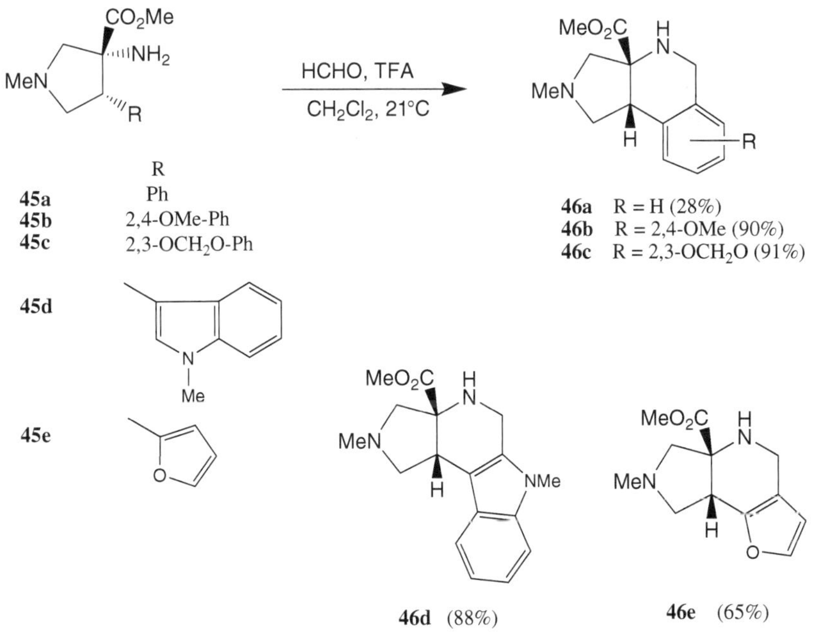

SCHEME 13.23 Synthesis of (hetero)arenopyrrolopyridines by P-S reaction.

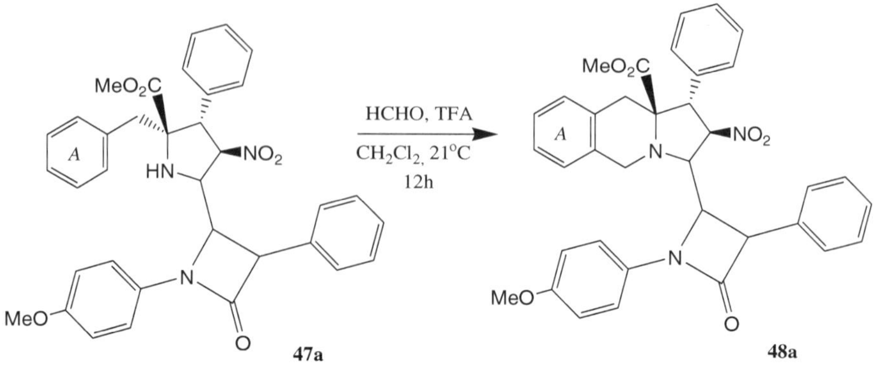

SCHEME 13.24 Synthesis of β-lactam substituted pyrroloisoquinoline.

13.6.2 Tetrahydro-β-Carbolines and Indole-Related Alkaloids

Even in the last years the search to improve the stereoselectivity of tetrahydro-β-carbolines continues actively. A Chinese group demonstrated that the P-S reaction of D-tryptophan methyl ester hydrochloride with piperonal could give high selectivities and yields via a crystallization-induced asymmetric transformation (CIAT) in the appropriate solvent. The CIAT process exploited the large difference of solubility of the hydrochloride salts of *cis*- and *trans*-1,3-substituted THBCs [68]. When the study was extended to various aldehydes, high stereoselectivities and good yields were obtained for aromatic aldehydes by tuning the ratios of a toluene/nitromethane mixture. Since aliphatic aldehydes were instead giving moderate yields, a two-step procedure was preferred: first the P-S condensation of a D-tryptophan ester ($R^1 = Me$, Et) and an aliphatic aldehyde R^2-CHO in 2-propanol afforded a mixture of the epimeric THBC **49a**-HCl salts, which in turn was processed by CIAT in a mixed solvent with different ratios of toluene and nitromethane, to give either the pure *cis*-**49a**-HCl or *trans*-**49a**-HCl (Scheme 13.25) [69]. To obtain the *trans*-1,3-disubstituted tetrahydro-β-carboline **49b**-HCl, a common chiral synthon of mitotic kinesin Eg5 inhibitor HR22C-16 and its analogs, Shi and Xiao treated L-tryptophan-HCl with protected *m*-hydroxybenzaldehyde. The CIAT process of the diastereomeric mixtures gave the *cis*-THBC **49b** for R = H and Ac, respectively, and the desired *trans*-THBC **49b** for R = Bz (Scheme 13.25) [70].

To prepare a series of THBCs (**51**) for SAR studies by P-S reaction (Scheme 13.26), He et al. used the intermediate α-siloxy α,β-unsaturated esters **50**, as surrogates of the α-ketoesters, bypassing the isolation of the latter [71].

The preparation of the α,β-unsaturated ester **54a** via the PS-2CR reaction was the pivotal step of the synthesis of biologically active *Corinanthe* indole alkaloids, such as 9-methoxygeissoschizol (**55**, Scheme 13.27) and its *N*-methyl derivative. The condensation of the 4-methoxytryptophan derivative **52** and aldehyde **53** in acetic acid gave a mixture of *cis* and *trans* isomers in a 1:3 ratio. Epimerization of *cis*-isomer with TFA provided the desired intermediate **54a**, which in turn yielded the precursor **54b** [72].

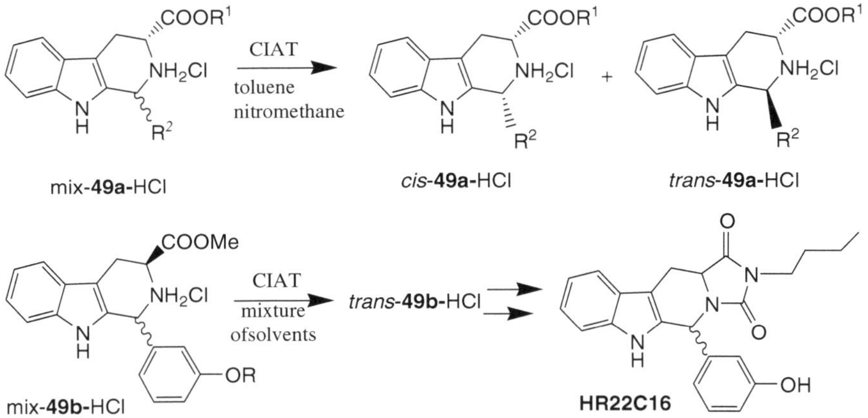

SCHEME 13.25 P-S reaction in the synthesis of tetrahydro-β-carbolines.

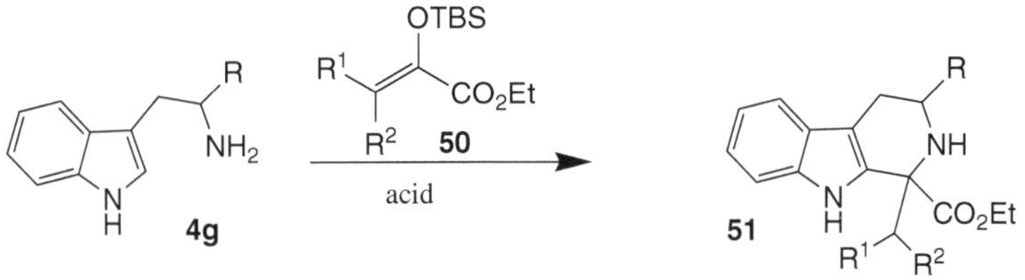

SCHEME 13.26 α,β-unsaturated esters as intermediates in the P-S reaction.

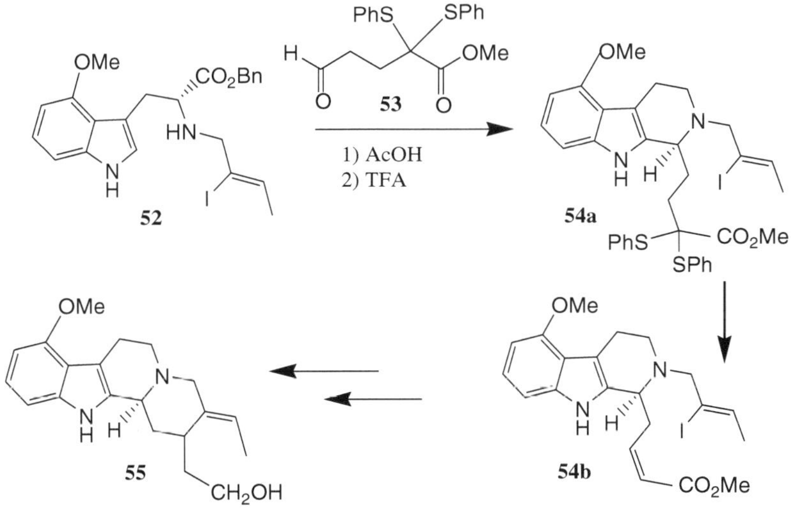

SCHEME 13.27 Synthesis of *Corinanthe* indole alkaloids.

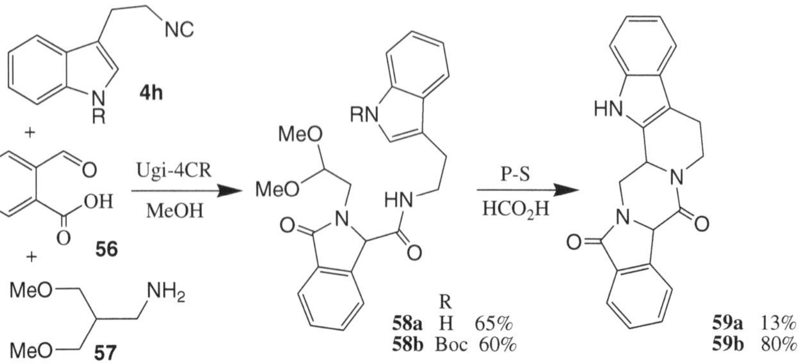

SCHEME 13.28 MCR and P-S condensation in the synthesis of indole alkaloids.

In the search for an efficient synthesis of the basic skeleton of indole alkaloids, the chemists took into consideration the technique of multicomponent reaction (MCR), initiated by Dömling and Ugi [73] and combined with the P-S condensation to give different types of scaffolds [74].

More recently, Liu and Dömling developed a new two-step sequence, which comprises an Ugi-4CR followed by a P-S ring closure. Using tryptophan-derived isocyanide **4h**, the bifunctional 2-formyl-benzoic acid **56**, and the amine **57** under HCO_2H-mediated conditions, the authors obtained the polycyclic indole alkaloids **59a–b** via the Ugi intermediates **58 a–b** (Scheme 13.28) [75a].

A study on the scope and limitations of the reaction afforded *inter alia* the complex hexacyclic molecule **60** (Figure 13.5), containing four heterocycles (pyrrole, piperazinone, hydropyridine, and γ-lactam) and two carbocycles (benzene and cyclopentane) [75b]. Also the orexin I antagonist, almorexant® (**61**, Figure 13.5), with a THIQ skeleton and currently undergoing Phase III clinical trials for sleeping disorders, has been discovered by a combination of the Ugi-3CR and a subsequent PS-2CR pathway [76].

The formation of indoloquinolizidine via an iminium ion intermediate by P-S cyclization is well documented in the literature [77]. Recently, a Chinese group

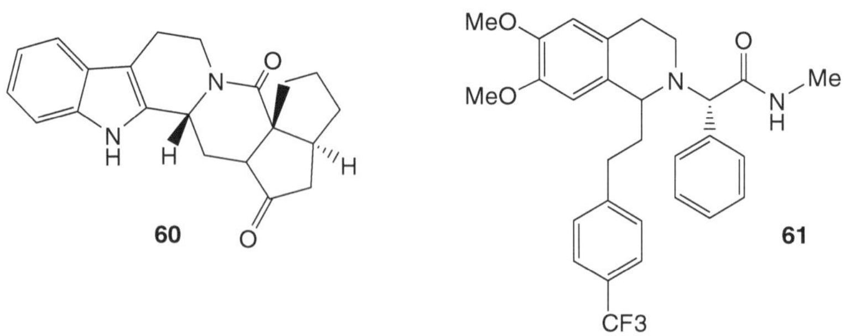

Figure 13.5 Hexacyclic molecule (**60**) and orexin I antagonist (**61**) synthesized via the P-S reaction.

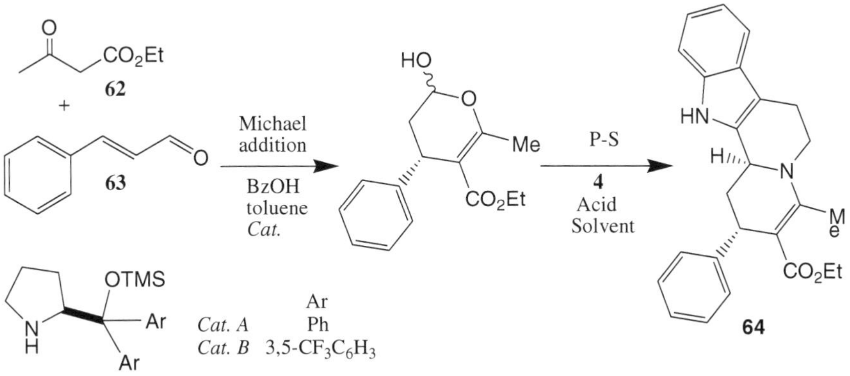

SCHEME 13.29 P-S enantioselective synthesis of substituted indoloquinolizidine.

reported a facile and rapid one-pot access to highly substituted indoloquinolizidines using an enantioselective three-component Michael addition-Pictet–Spengler-sequence of β-ketoester **62**, α,β-unsaturated aldehyde **63** and tryptamine **4** (Scheme 13.29). The first step of the reaction was conducted at 10°C using an organo-catalyst (A or B), benzoic acid as an additive and toluene as solvent. The second step consisted in the addition of **4** and a stoichiometric amount of benzoic acid to the reaction mixture diluted with toluene. The mixture was heated at 50°C for 24 h. The desired indoloquinolizidine **64** was isolated as a single diastereoisomer in moderate yield and good selectivity (Scheme 13.29) [78].

13.6.3 Novel Scaffolds and New Generation Substrates

Traditionally, the substrates (first-generation substrates) used for the P-S reaction such as tryptamine/tryptofan, histidine/histamine, and dimethoxy-phenylethylamines are based on aliphatic amines that act as a source for iminium/electrophilic partner. In 2005, Kundu and coworker [79a,b] presented a modified and successful strategy demonstrating *endo*-cyclization in substrates with aryl amines attached to the activated heterocyclic ring (**65a**, second-generation substrates, Figure 13.6).

The modified strategy has also been used successfully by others with the substrates **65b** [80a] and **65c** [80b] (Figure 13.6), in which the imidazole, the phenyl tethered to NH and the indole serve as π-nucleophiles. Further investigation revealed that the PS-2CR reaction could be affected even when the aryl amine is linked to a deactivated (poorly nucleophilic) heterocyclic ring, such as quinoxalines, triazoles, and tetrazoles, and categorized as third-generation substrates [81]. In contrast with the second-generation substrates [79c], these ones had limitation for aldehydes with strong electron-donating groups. To expand the repertoire of the third-generation substrates with abilities to facilitate *endo*-cyclization with all classes of aromatic aldehydes, the pyrimidine nucleus was identified as another type of π-nucleofile after being linked to an aryl amine. The three 2-(2,6-disubstituted-pyrimidin-4-yl)phenylamines **66a–c** (Scheme 13.30) were synthesized as the probable substrates for the application of the P-S reaction: in the three compounds, an aryl amine is located in C-4 as a source for

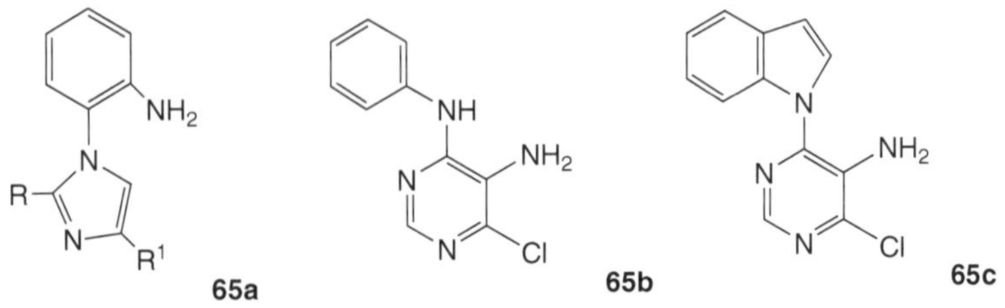

Figure 13.6 Arylamines substrates for the P-S reaction.

electrophilic partner, while the C-2 and C-6 positions have been used for introducing diversity, and C-5, the nucleophilic carbon, has been kept free for π-cyclizations.

The reaction was optimized and substrates **66a,b** underwent 6-*endo*-cyclization with a variety of aldehydes in DMF in the presence of 1% triflic acid to furnish pyrimido[5,4-c]quinolines **67a,b** (Scheme 13.30) in good to excellent yields. Substrate **66c** gave no reaction with 2-substituted aldehydes, while reacted with 4-substituted aldehydes with only about 50% yield. Finally, attempts to further extend this strategy on new structurally analogous substrate, involving the pyridine ring as nucleophilic partner, failed, and thus limiting the scope of the reaction [82].

The same strategy was applied to the synthesis of new imidazopyrazine-based polycyclic structures using tethered heteroarylamines **68a**,**b** (instead of **65a**) as starting material. By treatment with various aldehydes (best results for $R^3 = EtOC_6H_5$) and 1% TFA in toluene at reflux, **68a** gave a mixture of compounds **69a** and **70**, but the latter were obtained as the sole products by treatment with DDQ of the P-S reaction mixture (Scheme 13.31).

Substitution (R^1,R^2) on substrate **68a** had no effect on *endo*-cyclization and provided the cyclized products **70** in excellent yields within 1–5 h, with aldehydes containing both withdrawing and electro-donating groups, except propionaldehyde (8 h). The *endo*-cyclization of **68a** with both aliphatic and aromatic ketones in the same conditions gave products (not shown) similar to **69a** in excellent yield.

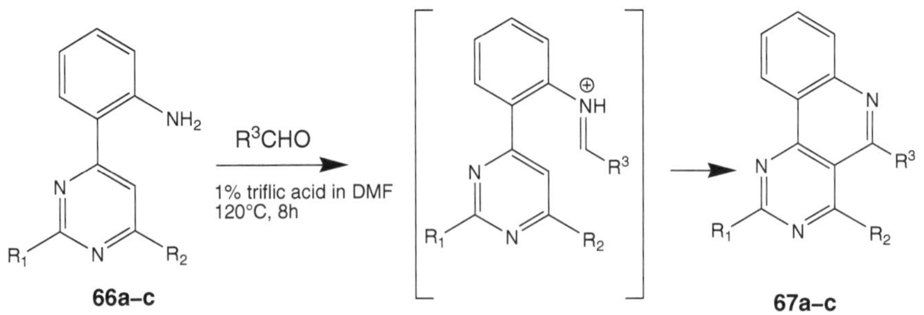

a = $R_1 = NH_2$; $R_2 = H$
b = $R_1 = NHCH_2C_6H_5$; $R_2 = H$
c = $R_1 = NH_2$; $R_2 = C_6H_5$

SCHEME 13.30 General strategy for the P-S reaction involving pyrimidine derivatives.

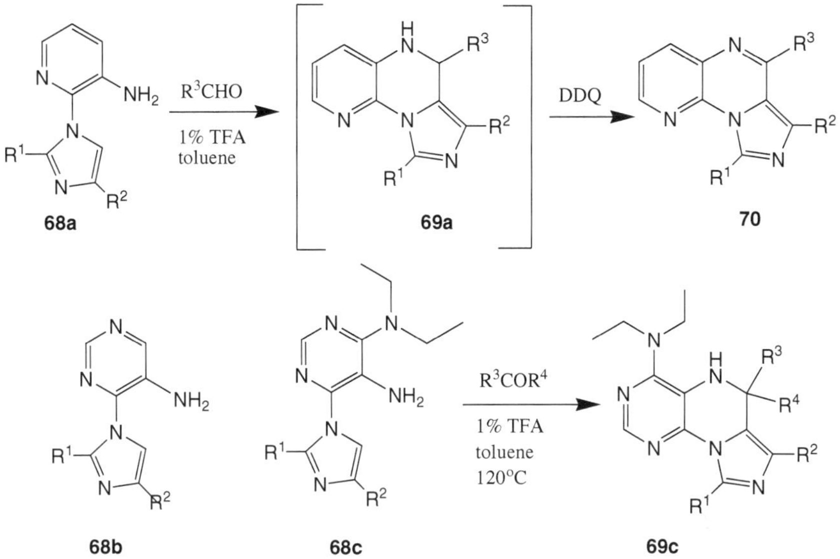

SCHEME 13.31 Synthesis of imidazopyrazine-based polycyclic structures using heteroarylamines.

Since the synthesis of the pyrimidine derivative **68b** gave a mixture of products, a different substrate **68c** was used instead: by the same condensation protocol, **68c** gave the pyrimido-imidazopyrazine analog to **70** with aldehydes and the expected *endo*-cyclization product **69c** with ketones (Scheme 13.31) [83].

Ageladine A (**73a**, Scheme 13.32) is a unique imidazol-pyrrole-based marine natural product, which was isolated from the sponge *Agelas nakamurai* and was shown to be an inhibitor of matrix metallo-proteinases. In an effort to discover new anticancer lead structures, a series of ageladine A analogs was synthesized via P-S cyclization of 2-aminoimidazoles (**71**) and subsequent oxidation of cyclo-compounds

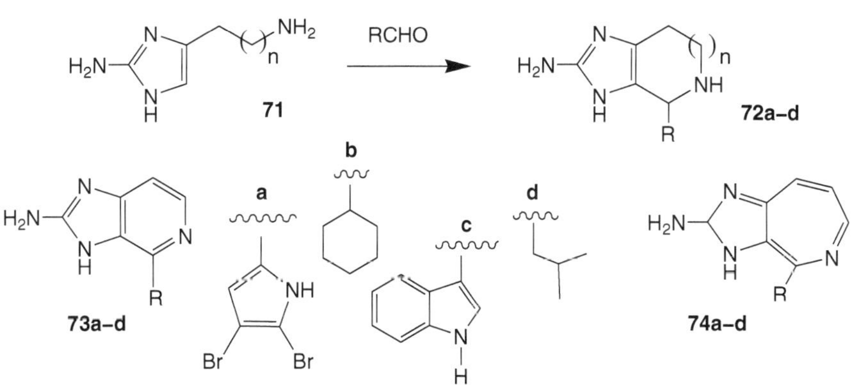

SCHEME 13.32 Synthesis of agelanin A analogs via P-S cyclization of 2-aminoimidazoles.

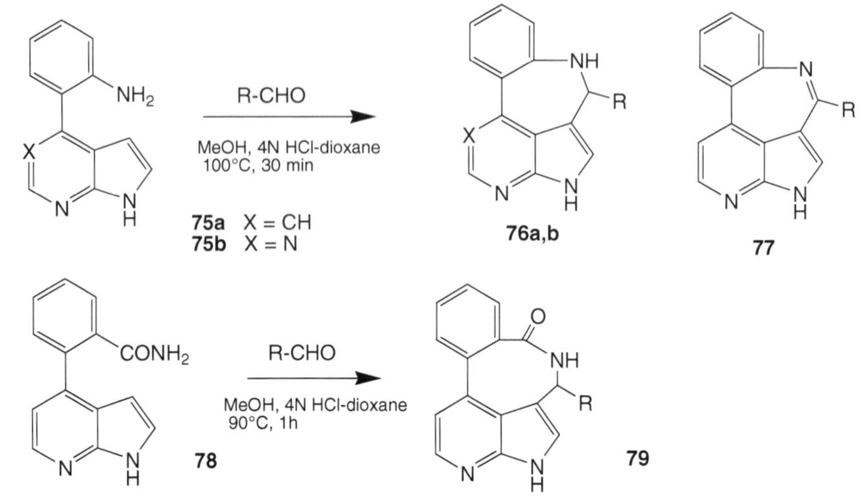

SCHEME 13.33 Synthesis of seven- and eight-membered fused cyclic compounds.

72 to the requisite aromatic systems (**73**). Together with 2-amino-3H-imidazo [4,5-c] pyridine (six-membered rings) derivatives **73a–d**, the collection included 2-aminoimidazo-[4,5-c]azepines (seven-membered rings) **74a–d** (Scheme 13.32). Only the compounds possessing the aromatic azepine core showed anticancer activity with IC_{50} values in the micromolar range [84].

P-S condensation of aldehydes or α-ketoesters with 4-(2-anilinophenyl)-7-azaindole **75a** or deazapurine **75b** gave high yields of seven-membered-3,4-fused cyclic compounds **76a,b** (Scheme 13.33). When the reaction was performed in methanol and held at reflux in an open flask for two days, the imino derivative **77** was isolated in good yield. The methodology was further extended to include synthesis of an eight-membered-3,4-fused ring (**79** from **78**). SAR studies on the synthesized compounds led to the discovery of a series of potent JAK2 kinase inhibitors [85]. The novel dihydrobenzo[f]naphtha[2,3-b][1,4]thiazepine-6,11-diones, the seven-membered analogs (**80a,b**) of the biologically active thiazine derivative, were obtained by the P-S cyclization of **81a,b** with paraformaldehyde using $BF_3{\cdot}Et_2O$ (Scheme 13.34) as a catalyst [86].

13.7 CONCLUSIONS

As conclusive remarks two papers published in 2010 can be highlighted. In the first one [87], the total synthesis of the enantiopure indole alkaloid phalarine seems an opportunity to discuss the mechanistic uncertainty concerning whether the P-S cyclization occurs at C-2 or first takes place at C-3 position. The authors demonstrated that at least whit an aryl group at indole C-2 and no tosyl group at nitrogen N_a, cyclization of **82** occurs more rapidly at C-3 (Scheme 13.35) and the “indolenine

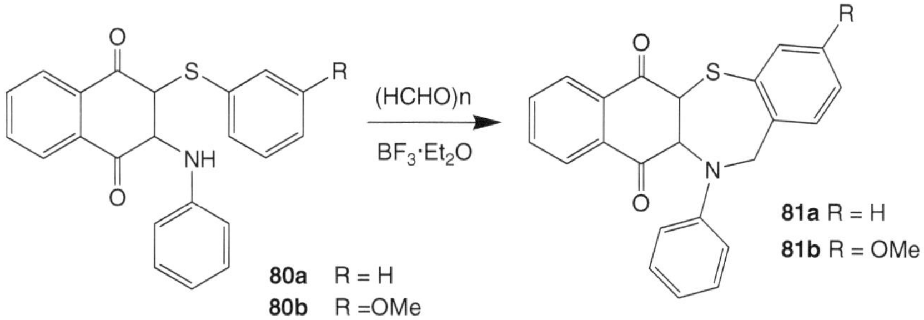

80a R = H
80b R =OMe

81a R = H
81b R = OMe

SCHEME 13.34 Synthesis of an active thiazine derivative analog via P-S cyclization.

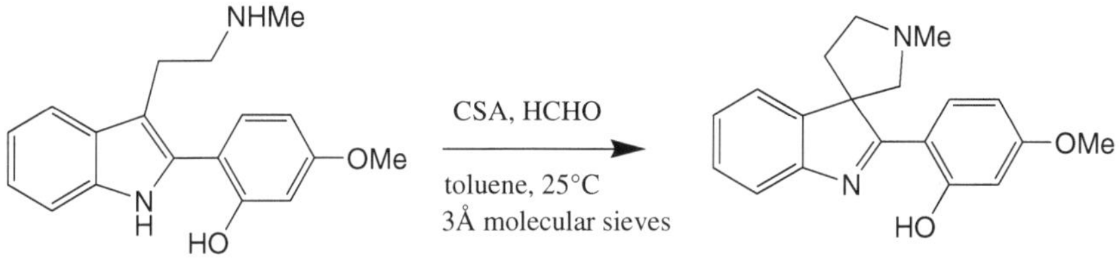

82

83 (86%)

SCHEME 13.35 Synthesis of a prephalarine intermediate by P-S rearrangement.

moiety" **83**, a prephalarine intermediate, is actually quite stable, giving only the starting material by treatment with CSA and methanol. In the second one [88], the very classic P-S reaction of formation of THBC is optimized by the conversion of the THBC in β-carbolines **84** with tetracyanuric acid in the one-pot reaction (Scheme 13.36).

We may conclude that the P-S reaction of Scheme 13.1 is still brilliant on stage in Scheme 13.36, no better demonstration of its actuality. Moreover, an interesting paper on the application of P-S reactions for the synthesis of pharmaceutically relevant heterocycles has just been published [89], clearly demonstrating that the P-S reaction, known for approximately one century, remains a particularly popular cyclization method.

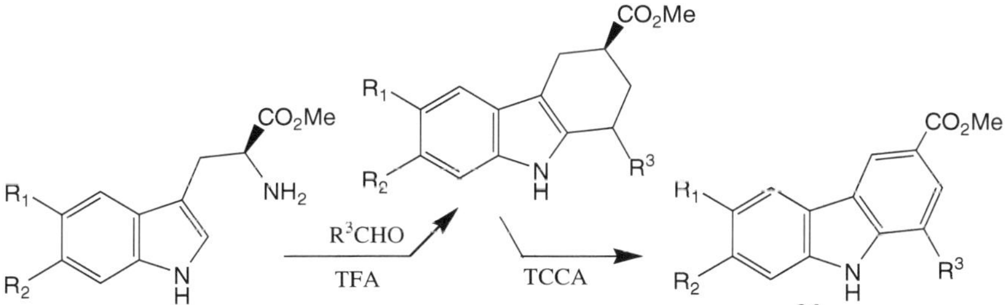

84

SCHEME 13.36 Use of tetracyanuric acid in the one-pot P-S reaction.

ACKNOWLEDGMENT

We thank financial support from Fondazione Roma (Roma, Italy) and Bilateral Projects Italy-Uruguay, sponsored by Sapienza Università di Roma (Italy), and Universidad de la República, Montevideo (Uruguay).

REFERENCES

[1] (a) Pictet, A., Spengler T. (1911). Formation of isoquinoline derivatives by the action of methylal on phenylethylamine, phenylalanine and tyrosine. *Berichte der Deutschen Chemischen Gesellschaft*, *44*, 2030–2036; (b) Whaley, W.M., Govindachari, T.R. (1951). The Pictet–Spengler synthesis of tetrahydroisoquinolines and related compounds. *Organic Reaction, 6*, 74–150.

[2] (a) Mannich, C., Krösche, W. (1912). Über ein Kondensations product aus Formaldehyd, Ammoniak und Antipyrin, *Archiv der Pharmazie*, *250*, 647–667; (b) Blicke, F.F., (1942). Organic reactions: I. *Mannich Reaction*, *1*, 303–341.

[3] Tatsui, G. (1928). Preparation of 1-methyl-1,2,3, 4-tetrahydro-β-carbolines. *Journal of the Pharmaceutical Society of Japan*, *48*, 92.

[4] Cox, E.D., Cook, J.M. (1995). The Pictet–Spengler condensation: A new direction for an old reaction. *Chemical Reviws*, *95*, 1797–1842.

[5] Yokoyama, A. (1999). Prototype Pictet–Spengler reaction catalyzed by superacids. Involvement of dicationic superelectrophiles. *Journal of Organic Chemistry*, *64*, 611–617.

[6] Maryanoff, B.E., Zhang, H.-C., Cohen, J.H., Turchi, I.J., Maryanoff, C.A. (2004). Cyclizations of *N*-acyliminium ions. *Chemical Reviews*, *104*, 1431–1628.

[7] Bonnet, D., Ganesan, A. (2002). Solid-phase synthesis of tetrahydro-β-carbolinehydantoins via the *N*-acyliminium Pictet–Spengler reaction and cyclative cleavage. *Journal of Combinatorial Chemistry*, *4*, 546–548.

[8] Nuhant, P., Raikar, S.B., Wypych, J.-C., Delpech, B., Marazano, C. (2009). Enhancement of 5-aminopenta-2,4-dienals electrophilicity via activation by *O,* N-bistrifluoroacetylation. Application to N-acyl Pictet–Spengler reaction. *Journal of Organic Chemistry*, *74*, 9413–9421.

[9] (a) Horiguchi, Y. Kodama, H., Nakamura, M., Yoshimura, T., Hanezi, K., Hamada, H., Saitoh, T., Sano, T. A. (2002). Convenient synthesis of 1,1-disubstituted-1,2,3, 4-tetrahydroisoquinolines via Pictet–Spengler reaction using titanium(IV) isopropoxide and acetic-formic anhydride. *Chemical and Pharmaceutical Bulletin*, *50*, 253–257; (b) Kitabatake, M., Nagai, J., Abe, K., Tsuchiya, Y., Ogawa, K., Yokohama, T., Mohri, K., Taguchi, K., Horiguchi, Y. (2009). Facile synthesis and *in vitro* properties of 1-alkyl- and 1-alkyl-*N*-propargyl-1,2,3, 4-tetrahydroisoquinoline derivatives on PC12 cells. *European Journal of Medicinal Chemistry*, *44*, 4034–4043.

[10] Youn, S.W. (2006). Development of Pictet–Spengler reaction catalyzed by $AuCl_3$/AgOTf. *Journal of Organic Chemistry*, *71*, 2521–2523.

[11] Eid, C.N., Shim, J., Bikker, J., Lin, M. (2009). Direct oxa-Pictet–Spengler cyclization to the natural (3a, 5)-trans-stereochemistry in the syntheses of (+)-7-deoxyfrenolicin B and (+)-7-deoxykalafungin. *Journal of Organic Chemistry*, *74*, 423–426.

[12] Sawant, R.T., Waghmode, S.B. (2009). Organocatalytic enantioselective formal synthesis of HRVC 3C-protease inhibitor (1*R*, 3S)-thysanone. *Tetrahedron*, *65*, 1599–1602.

[13] (a) Bianchi, D.A., Rúa, F., Kaufman, T.S. (2004) Studies on the intramolecular oxa-Pictet–Spengler rearrangement of 5-aryl-1, 3-dioxolanes to 4-hydroxy-isochromans. *Tetrahedron Letters*, *45*, 411–415; (b) Bianchi, D.A., Cipulli, M.A., Kaufman, T.S. (2003). Model studies towards stephaoxocanes: Construction of the 2-oxa-4-aza-phenalene core of stephaoxocanidine and eletefine. *European Journal of Organic Chemistry*, 2003, 4731–4736; (c) Guiso, M., Marra, C., Cavarischia, C. (2001). Isochromans from 2-(3', 4'-dihydroxy)phenylethanol. *Tetrahedron Letters*, *42*, 6531–6538.

[14] Saeed, A. (2010). Oxa-Pictet–Spengler reaction in water. Synthesis of some (±)-1-aryl-6,7-dimethoxyisochromans. *Chinese Chemical Letters*, *21*, 261–264.

[15] Poaty, B., Dumarcay, S., Perrin, D. (2009). New lipophilic catechin derivatives by oxa-Pictet–Spengler reaction. *European Food Research and Technology*, *230*, 111–117.

[16] Poaty, B., Dumarçay, S., Gérardin, P., Perrin, D. (2010). Modification of grape seed and wood tannins to lipophlic antioxidant derivative. *Industrial Crops and Products*, *31*, 509–515.

[17] (a) Bailey, P.D., Hollinshead, S.P., McLay, N.R., Morgan, K.M., Palmer, S.I., Prince, S.N., Reynolds, C.D., Wood, S.D. (1993). Diastereo- and enantio-selectivity in the Pictet–Spengler reaction. *Journal of the Chemical Society, Perkin Transactions I*, 431–439; (b) Bailey, P.D., Moore, M.H., Morgan, K.M., Smith, D.I., Vernon, J.M. (1994). Enhancing the yield and the diastereoselectivity of the Pictet–Spengler reaction: A highly efficient route to *cis*-1, 3-disubstituted tetrahydro-β-carbolines. *Tetrahedron Letters*, *35*, 3587–3588, and references cited therein.

[18] Cox, E.D., Hamaker, L.K., Li, J., Yu, P., Czerwinski, K.M., Deng, L., Bennet, D.W., Cook, J.M., Watson, W.H., Kraviec, M. (1997). Enantiospecific formation of *trans*-1, 3-disubstituted tetrahydro-β-carbolines by the Pictet–Spengler reaction and conversion of cis-diastereomers into their trans counterpart by scission of the C1-N2 bond. *Journal of Organic Chemistry*, *62*, 44–61, and references cited therein.

[19] Alberch, L., Bailey, P.D., Clingan, P.D., Mills, T.J., Price, R.A., Pritchard, R.G. (2004). The *cis*-specific Pictet–Spengler reaction. *European Journal of Organic Chemistry*, *2004*, 1887–1890.

[20] (a) Tourwe, D., Verschueren, K., Frycia, A., Davis, P., Porreca, F., Hruby, V.J., Toth, G., Jaspers, H., Verheyden, P., Van Binst, G. (1996). Conformational restriction of Tyr and Phe side chains in opioid peptides: Information about preferred and bioactive side-chain topology. *Biopolymers*, *38*, 1–12; (b) Gibson, S.E., Guillo, N., Tozer, M.J. (1999). Towards control of χ-space: Conformationally constrained analogues of Phe, Tyr, *Trp and His. Tetrahedron*, *55*, 585–615.

[21] (a) Grieco, P., Campiglia, P., Gomez-Monterrey, I., Novellino, E. (2002). Synthesis of new β-turn dipeptide mimetic based on tetrahydroisoquinoline moiety. *Tetrahedron Letters*, *4336*, 6297–6299; (b) Lesma, G., Meschini, E., Recca, T., Sacchetti, A., Silvani, A. (2007). Synthesis of tetrahydroisoquinoline-based pseudopeptides and their characterization as suitable reverse-turn mimetics. *Tetrahedron*, *63*, 5567–5578.

[22] Pulka, K., Kulis, P., Tymecka, D., Frankiewicz, L., Wilczek, M., Kozminski, W., Misicka, A. (2008). Diastereoselective Pictet–Spengler condensation of tryptophan with α-amino aldehydes as chiral carbonyl components. *Tetrahedron*, *64*, 1506–1514.

[23] Fehrentz, J.A., Castro, B. (1983). An efficient synthesis of optically active α-(*t*-butoxycarbonylamino)-aldehydes from α-amino acids. *Synthesis*, *8*, 676–679.

[24] Pulka, K., Feytens, D., Misicka, A., Tourwé, D. (2010). New tetracyclic tetrahydro-β-carbolines as tryptophan-derived peptidomimetics. *Molecular Diversity*, *14*, 97–108.

[25] Larghi, E.L., Amongero, M., Bracca, A.B.J., Kaufman, T.S. (2005). The intermolecular Pictet–Spengler condensation with chiral carbonyl derivatives in the stereoselective synthesis of optically-active isoquinoline and indole alkaloids. *ARKIVOC*, *12*, 98–153.

[26] Schmidt, G., Waldmann, H., Henke, H., Burkard, M. (1996). Asymmetric control in the Pictet–Spengler reaction by means of N-protected amino acids as chiral auxiliary groups. *Chemistry – A European Journal*, *2*, 1566–1571.

[27] Gremmen, C.B., Willemse, B., Wanner, M.J., Koomen, G. (2000). Enantiopure tetra-hydro-β-carbolines via Pictet–Spengler reactions with *N*-sulfinyl tryptamines. *Journal of Organic Letters*, *2*, 1955–1958.

[28] Yamada, H., Kawate, T., Matsumizu, M., Nishida, A., Yamaguchi, K., Nakagawa, M. (1998). Chiral Lewis acid-mediated enantioselective Pictet–Spengler reaction of N_b-hydroxytryptamine with aldehydes. *Journal of Organic Chemistry*, *63*, 6348–6354.

[29] Taylor, M.S., Jacobsen, E.N. (2004). Highly enantioselective acyl-Pictet–Spengler reactions. *Journal of the American Chemical Society*, *126*, 10558–10559.

[30] (a) Sigman, M.S., Jacobsen, E.N. (1998). Schiff base catalysts for the asymmetric strecker reaction identified and optimized from parallel synthetic libraries. *Journal of the American Chemical Society*, *120*, 4901–4902; (b) Wenzel A.G., Jacobsen, E.N. (2002). Asymmetric catalytic Mannich reactions catalyzed by urea derivatives: Enantioselective synthesis of β-aryl-β-amino acids. *Journal of the American Chemical Society*, *124*, 12964–12965.

[31] Seayad, J., Seayad, A.M., List, B. (2006). Catalytic asymmetric Pictet–Spengler reaction. *Journal of the American Chemical Society*, *128*, 1086–1087.

[32] Uraguchi, D., Terada, M. (2004). Chiral Brønsted acid-catalyzed direct Mannich reactions via electrophilic activation. *Journal of the American Chemical Society*, *126*, 5356–5357.

[33] Akiyama, T., Itoh, J., Yokota, K., Fuchibe, K. (2004). Enantioselective Mannich-type reaction catalyzed by a chiral Brønsted acid. *Angewndte Chemie International Edition*, *43*, 1566–1568.

[34] Quin, L.D., *A Guide to Organophosphorus Chemistry*, John Wiley & Sons, New York, 2000, pp. 133–165.

[35] Wanner, M.J., van der Haas, R.N., de Cuba, K.R., van Maarseveen, J.H., Hiemstra, H. (2007). Catalytic asymmetric Pictet–Spengler reactions via sulfenyliminium ions. *Angewndte Chemie International Edition*, *46*, 7485–7487.

[36] Vinu, A., Kalita, P., Samie, L., Chari, A.M., Pal, R., Subba Reddy B.V. (2010). Novel synthesis of tetrahydro-β-carbolines and tetrahydroisoquinolines via three-component reaction using hexagonally ordered mesoporous A1SBA-15 catalysts. *Tetrahedron Letters*, *51*, 702–706.

[37] Wang, Y.F., Song Z.B., Chen, C.X., Peng J.S. (2010). Pictet–Spengler condensation reactions catalyzed by a recyclable H^+-montmorillonite as a heterogeneous Brønsted acid. *Science in China: Chemistry*, *53*, 562–568.

[38] (a) Saito, A., Numaguchi, J., Hanzawa, Y. (2007). Pictet–Spengler reactions catalyzed by Brønsted acid-surfactant-combined catalyst in water or aqueous media. *Tetrahedron Letters*, *48*, 835–839; (b) Saha, B., Sharma, S., Sawant, D., Kundu, B. (2007). Water as an

efficient medium for the synthesis of tetrahydro-β-carbolines via Pictet–Spengler reactions. *Tetrahedron Letters, 48,* 1370–1383; (c) Saito, A., Takayama, M., Yamazaki, A., Numaguchi, J., Hanzawa, Y. (2007). Synthesis of tetrahydroisoquinolines and isochromans via Pictet–Spengler reactions catalyzed by Brønsted acid-surfactant combined catalyst in aqueous media. *Tetrahedron, 63,* 4039–4047.

[39] (a) Han, D., Fösterling, F.H., Deschamps, J.R., Parrish, D., Lin, X., Yin, W., Huang, S., Cook, J.M. (2007). Conformational analysis of the *cis-* and *trans*-adducts of the Pictet–Spengler reaction. Evidence for the structural basis for the C(1)–N(2) scission process in the cis- to trans-isomerization. *Journal of Natural Products, 70,* 75–82; (b) Ma, J., Yin, W., Zhou, H., Cook, J.M. (2007). Total synthesis of the opioid agonistic indolealkaloid mitragynine and the first total synthesis of 9-methoxygeissoschizol and 9-methoxy-N_b-methylgeissoschizol. *Organic Letters, 9,* 3491–3494; (c) Sewgobind, N.V., Wanner, M.J., Ingemann, S., de Gelder, R., van Maarsaven, J.H., Hiemstra, H. (2008). Enantioselective DINOL-phosphoric acid catalyzed Pictet–Spengler reactions of *N*-benzyltryptamine. *Journal of Organic Chemistry, 73,* 6405–6408, and references cited therein.

[40] Gödecke, T., Lankin, D.C., Nikolic, D., Chen, S.-N., van Breemen, R.B., Farnsworth, N., Pauli, G.F. (2009). Guanidine alkaloids and Pictet–Spengler adducts from Black Cohosh (*Cimicifuga racemosa*). *Journal of Natural Products, 72,* 433–437.

[41] Saidemberg, D.M., Ferreira, A.B., Takahashi, T.N., Gomes, P.C., Cesar-Tognoli, L.M.M., da Silva Filho, L.C., Tormena, C.F., da Silva, J.V.J. (2009). Monoamine oxidase inhibitory activities of indolalkaloid toxins from the Venom of the Colonial Spider (*Parawixia bistriata*): Functional characterization of PwTX-I. *Toxicon, 54,* 717–724.

[42] Koeller, K.M., Wong, C.H. (2001). Enzymes for chemical synthesis. *Nature, 409,* 232–240.

[43] Cucinotta, C.S., Kosa, M., Melchiorre, P., Cavalli, A., Gervasio, F. (2009). Bifunctional catalysis by natural cinchona alkaloids: A mechanism explained. *Chemistry – A European Journal, 15,* 7913–7921.

[44] Schafer, S.L., Barrett, W.C., Kallarakal, A.T., Mitra, B., Kozarich, J.W., Gerlt, J.A. (1996). Mechanism of the reaction catalyzed by mandelate racemase: Structure and mechanistic properties of the D270N mutant. *Biochemistry, 35,* 5662–5669.

[45] Rodríguez, S.M., Las Heras-Vázquez, F.J., Mingorance-Cazorla, L., Clemente-Jiménez, J.M., Rodríguez-Vico, F. (2004). Molecular cloning, purification, and biochemical characterization of hydantoin racemase from the legume symbiont *Sinorhizobium meliloti* CECT 4114. *Applied and Environmental Microbiology, 70,* 625–630.

[46] Strauss, U.T., Faber, K. (1999). Deracemization of (±)-mandelic acid using a lipase-mandelate racemase two-enzyme system. *Tetrahedron: Asymmetry, 10,* 4079–4081.

[47] Samanani, N., Liscombe, D.K., Facchini, P.J. (2004). Molecular cloning and characterization of norcoclaurine synthase, an enzyme catalyzing the first committed step in benzylisoquinoline alkaloid biosynthesis. *Plant Journal, 40,* 302–313.

[48] McKnight, T.D., Roessner, C.A., Devagupta, R., Scott A.I., Nessler, C.L. (1990). Nucleotide sequence of a cDNA encoding the vacuolar protein strictosidine synthase from *Catharanthus Roseus. Nucleic Acids Research, 18,* 4939.

[49] De Eknamkul, W., Suttipanta, N., Kutchan, T.M. (2000). Purification and characterization of deacetylipecoside synthase from *Alangium Lamarckii* Thw. *Phytochemistry, 55,* 177–181.

[50] Stadler, R., Zenk, M.H. (1990). A revision of the generally accepted pathway for the biosynthesis of the benzyltetrahydroisoquinoline alkaloid reticuline. *Liebigs Annalen der Chemie*, *6*, 555–562.

[51] Ma, X., Panjikar, S., Koepke, J., Loris, E., Stöckigt, J. (2006). The structure of *Rauvolfia serpentina* strictosidine synthase is a novel six-bladed β-propeller fold in plant proteins. *Plant Cell*, *18*, 907–920.

[52] Stadler, R., Kutchan, T.M., Zenk, M.H. (1989). (*S*)-Norcoclaurine is the central intermediate in benzylisoquinoline alkaloid biosynthesis. *Phytochemistry*, *28*, 1083–1086.

[53] Facchini, P.J., St-Pierre, B. (2005). Synthesis and trafficking of alkaloid biosynthetic enzymes. *Current Opinion in Plant Biology*, *8*, 657–666.

[54] (a) Park, C.W., Chang, K.C., Lim, J.K. (1984). Effects of higenamine on isolated heart adrenoceptor of rabbit. *Archives Internationales de Pharmacodynamie*, *167*, 279–288; (b) Chang, K.C., Chong, W.S., Lee, I.J. (1994). Different pharmacological characteristics of structurally similar benzylisoquinoline analogs, papaverine, higenamine, and GS 839, on isolated rat aorta and heart. *Canadian Journal of Physiology and Pharmacology*, *72*, 327–334; (c) Chong, W.S., Lee, Y.S., Kang, Y.J., Lee, D.H., Ryu, J.C., Yun-Choi, H.S., Chang, K.C. (1998). Comparison of inodilator effect of higenamine, YS-49, YS-51, tetrahydroisoquinoline analogs, and dobutamine in the rat. *Korean Journal of Physiology and Pharmacology*, *2*, 323–340; (d) Yun-Choi, H.S., Pyo, M.K., Park, K.M., Chang, K.C., Lee, D.H. (2001). Antithrombotic effects of higenamine. *Planta Medica*, *67*, 619–622.

[55] (a) Bao, Y.X., Yu, G.R., Xu, J.M., Xu, Y.Q., Bian, Y.T., Zheng D.S. (1982). Effect of acute higenamine administration on bradyarrhythmias and his bundle: A clinical study of 14 cases and animal experiment on dogs. *Chinese Medical Journal*, *95*, 781–784; (b) Zhang, Z., Liu, X., Tao, Z., Shi, R., Zhang, X, Liu, Y., Zhu, K., Chen, B. (2002). Effects of higenamine on hemodinamics and its tolerability and safety, an experimental study. *Zhonghua Yi Xue Za Zhi*, *82*, 352–355.

[56] Pyo, M.K., Lee, D.H., Kim, D.H., Lee, J.H., Moon, C.H., Chang, K.C., Yun-Choi, H.S. (2008). Enantioselctive synthesis of (*R*)-(+)- and (*S*)-(−)-higenamine and their analogues with effects on platelet aggregation and experimental animal model of disseminated intravascular coagulation. *Bioorganic and Medicinal Chemistry Letters*, *18*, 4110–4114.

[57] (a) Pasquo, A., Bonamore, A., Franceschini, S., Macone, A., Boffi, A., Ilari, A. (2008). Cloning, expression, crystallization and preliminary X-ray data analysis of norcoclaurine synthase from *Thalictrum Flavum*. *Acta Crystallographica Section F: Structural Biology and Crystallization Communications*, *64*, 281–283; (b) Ilari, A., Franceschini, S., Bonamore, A., Arenghi, F., Botta, B., Macone, A., Pasquo, A., Bellucci, L., Boffi, A. (2009). Structural basis of enzymatic (*S*)-norcoclaurine biosynthesis. *Journal of Biological Chemistry*, *284*, 897–904.

[58] Bonamore, A., Barba, M., Botta, B., Boffi, A., Macone, A. (2010). Norcoclaurine synthase: Mechanism of an enantioselective Pictet–Spengler catalyzing enzyme. *Molecules*, *15*, 2070–2078.

[59] Luk, L.Y., Bunn, S., Liscombe, D.K., Facchini, P.J., Tanner, M.E. (2007). Mechanistic studies on norcoclaurine synthase of benzylisoquinoline alkaloid biosynthesis: An enzymatic Pictet–Spengler reaction. *Biochemistry*, *46*, 10153–10161.

[60] Liao, X.W., Liu, W., Dong, W.F., Guan, B.H., Chen, S.Z., Liu, Z.Z. (2009). Total synthesis of (−)-renieramycin G from L-tyrosine. *Tetrahedron*, *65*, 5709–5715.

[61] Ortín, I., González, J.F., de la Cuesta, E., Avendaño, C. (2009). Synthesis of a novel tetrahydroisoquinoline pentacyclic framework. *Tetrahedron*, *65*, 2201–2211.

[62] Ortín, I., González, J.F., de la Cuesta, E., Avendaño, C. (2009). C(3)-Alkylation and cyclization of pirazino[1, 2-*b*]isoquinolin-4-ones. *Tetrahedron*, *65*, 9944–9951.

[63] Wu, Y.C., Bernadat, G., Masson, G., Couturier, C., Schlama, T., Zhu, J. (2009). Synthetic studies on (−)-lemonomycin: An efficient asymmetric synthesis of lemonomycinone amide. *Journal of Organic Chemistry*, *74*, 2046–2052.

[64] Bowen, E.G., Wardrop, D.J. (2009). Total synthesis of the α-glucosidase inhibitors schulzeine A, B, and C and a structural revision of schulzeine A. *Journal of the American Chemical Society*, *131*, 6062–6063.

[65] Liu, G., Romo, D. (2009). Enantioselective synthesis of schulzeines B and C via a β-lactone-derived surrogate for bishomoserine aldehyde. *Organic Letters*, *11*, 1143–1146.

[66] Blanco-Ania, D., Hermkens, P.H.H., Sliedregt, L.A.J.M., Scheeren, H.W., Rutjes, F.P.J.T. (2009). Synthesis of tetrahydro-β-carboline and tetrahydroisoquinolines fused to pyrrolidines and solution-phase parallel acylation. *Journal of Combinatorial Chemistry*, *11*, 539–546.

[67] Arumugam, N., Raghunathan, R. (2010). Synthesis of highly functionalized β-lactam substituted pyrroloisoquinoline and indolizinoindole system by sequential intermolecular 1, 3-dipolar cyclo-addition reaction and Pictet–Spengler cyclization. *Tetrahedron*, *66*, 969–975.

[68] Shi, X.-X., Liu, S.-L., Xu, W., Xu, Y.-L. (2008). Highly stereoselective Pictet–Spengler reaction of D-tryptophan methyl ester with piperonal: Convenient synthesis of cialis (Tadalafil), 12a-epi-cialis and their deuterated analogs. *Tetrahedron: Asymmetry*, *19*, 435–442.

[69] Xiao, S., Lu, X., Shi, X.-X., Sun, Y., liang, L.L., Yu, X.-H., Dong, J. (2009). Syntheses of chiral 1, 3-disubstituted tetrahydro-β-carbolines via CIAT process: Highly stereoselective Pictet–Spengler reaction of D-tryptophan ester hydrochlorides with various aldehydes. *Tetrahedron: Asymmetry*, *20*, 430–439.

[70] Xiao, S., Shi, X.-X. (2010). The first highly stereoselective approach to the mitotic kinesin Eg5 inhibitor HR22C16 and its analogues. *Tetrahedron: Asymmetry*, *21*, 226–231.

[71] He, S., Lai, Z., Yang, D.X., Hong, Q., Reibarkh, M., Nargund, R.P., Hagmann, W.K. (2010). Synthesis of β-carbolines from aldehydes and ketones via the α-siloxy α, β-unsaturated esters. *Tetrahedron Letters*, *51*, 4361–4364.

[72] Ma, J., Yin, W., Zhou, H., Liao, X., Cook, J.M. (2009). General approach to the total synthesis of 9-methoxy-substituted indole alkaloids: Synthesis of mitragynine as well as 9-methoxygeossischizol and 9-methoxy-N_b-methyl-geossischizol. *Journal of Organic Chemistry*, *74*, 264–273.

[73] Dömling, A., Ugi, I. (2000). Multicomponent reactions with isocyanides. *Angewandte Chemie International Edition*, *39*, 3168–3120.

[74] Sapi, J., Laronze, J.-Y. (2004). Indole based multicomponent reactions towards functionalized heterocycles. *ARKIVOC*, *7*, 208–222.

[75] (a) Liu, H., Dömling, A. (2009). Efficient and diverse synthesis of indole derivatives. *Journal of Organic Chemistry*, *74*, 6895–6898; (b) Wang, W., Herdtweck, E., Dömling, A. (2010). Polycyclic indole alkaloid-type compounds by MCR. *Chemical Communications*, *46*, 770–772.

[76] Boss, C., Brisbare-Roch, C., Jenck, F. (2009). Biomedical application of orexin/hypocretin receptor ligands in neuroscience. *Journal of Medicinal Chemistry*, *52*, 891–903.

[77] Royer, J., Bonin, M., Micouin, L. (2004). Chiral heterocycles by iminium ion cyclization. *Chemical Reviews*, *104*, 2311–2352.

[78] Wu, X., Dai, X., Nie, L., Fang, H., Chen, J., Ren, Z., Cao, W., Zhao, G. (2010). Organocatalized enantioselective one-pot three-component access to indoloquinolizidines by a Michael addition-Pictet–Spengler sequence. *Chemical Communications*, *46*, 2733–2735.

[79] (a) Kundu, B., Sawant, D., Chhabra, R. (2005). A modified strategy for Pictet–Spengler reaction leading to the synthesis of imidazoquinoxalines on solid phase. *Journal of Combinatorial Chemistry*, *7*, 317–321; (b) Kundu, B., Sawant, D., Partani, P., Kesarwani, A.P. (2005). New application of Pictet–Spengler reaction leading to the synthesis of 7-membered heterocyclic ring system. *Journal of Organic Chemistry*, *70*, 4889–4892; (c) Duggineni, S., Sawant, D., Kundu, B. (2006). Application of modified Pictet–Spengler reaction for the synthesis of thiazolo- and pyrazolo-quinones. *Tetrahedron*, *62*, 3228–3241; (d) Sharma, S., Saha, B., Sawant, D., Kundu, B. (2007). Synthesis of novel N-rich polycyclic skeletons based on azoles and pyridines. *Journal of Combinatorial Chemistry*, *9*, 783–792.

[80] (a) Duncton, M.A.J., Smith, I.M., Burdzovic-Wizeman, S., Burns, A., Liu, H., Mao, Y., Wong, W.C., Kiselyov, A.S. (2005). Preparation of substituted pyrimido-[4,4-*b*]-1,4-benzoxazepines, thiazepines and diazepines via Pictet–Spengler cyclization. *Journal of Organic Chemistry*, *70*, 9629–9631; (b) Zheng, L., Xiang, J., Dang, Q.D., Bai, X. (2005). Novel heterocyclic scaffold consisting of indole fused pteridine. *Journal of Combinatorial Chemistry*, *7*, 813–815.

[81] Saha, B., Sharma, S., Sawant, D., Kundu, B. (2008). Application of the Pictet–Spengler reaction to aryl amine substituents linked to deactivated aromatic heterosystems. *Tetrahedron*, *64*, 8676–8684.

[82] Agarwal, P.K., Sharma, S.K., Savant, D., Kundu, B. (2009). Application of the Pictet–Spengler reaction to aryl amine-based substrates having pyrimidine as a π-nucleophile: Synthesis of pyrimidoquinolines with structural analogy to benzonaphthyridines present in alkaloids. *Tetrahedron*, *65*, 1153–1161.

[83] Sharma, S., Kundu, B. (2009). Application of the modified Pictet–Spengler cyclization reaction for the preparation of an imidazopyrazine ring: Synthesis of new pyrido- and pyrimido-imidazo-pyrazines. *Journal of Combinatorial Chemistry*, *11*, 720–731.

[84] Ma, Y., Nam, S., Jove, R., Yakushijin, K., Horne, D.A. (2010). Synthesis and anticancer activities of ageladine A and structural analogs. *Bioorganic and Medicinal Chemistry Letters*, *20*, 83–86.

[85] Wang, T., Ledeboer, M.W., Duffy, J.P., Pierce, A. J, Zuccola, H.J., Block, E., Shlyakter, D., Hogan, J.K., Bennani, Y.L. (2010). A novel chemotype of chinase inhibitors: Discovery of 3, 4-ring fused-7-azaindoles and deazapurine as potent JAK2 inhibitors. *Bioorganic and Medicinal Chemistry Letters*, *20*, 153–156.

[86] Tandon, V.K., Maurya, H.K., Mishra, N.N., Shukla, P.K. (2009). Design, synthesis and biological evaluation of novel nitrogen and sulfur containing hetero-1, 4-naphthoquinones as potent antifungal and antibacterial agents. *European Journal of Medicinal Chemistry*, *44*, 3130–3137.

[87] Tzupek, J.D., Lee, D., Crowley, B.M., Marathias, V.M., Danishefsky, S.J. (2010). Total synthesis of enantiopure phalarine via a stereospecific Pictet–Spengler reaction: Traceless transfer of chirality from L-tryptophan. *Journal of the American Chemical Society*, *132*, 8506–8512.

[88] Tang, J. G, Liu, H., Zhou, Z.U., Liu J.K. (2010). Facile and efficient one-pot synthesis of β-carbolines. *Synthetic Communications*, *40*, 1411–1417.

[89] Pulka, K. (2010). Pictet–Spengler reactions for the synthesis of pharmaceutically relevant heterocycles. *Current Opinion in Drug Discovery and Development*, *13*, 669–684.

14

SCREENING METHODS FOR DRUG DISCOVERY FROM PLANTS

ALAN L. HARVEY

14.1 FROM TRADITIONAL TO PHENOTYPIC SCREENING

Many early pharmacological studies with plant extracts involved injecting samples into rats or mice. The so-called hippocratic screening [1] was widely used. It typically involves administration of several different doses of test substance to small groups of animals and then observing behavior and looking for obvious symptoms at regular intervals for a period of up to 24 h. A systematic system for scoring observations relating to toxicity and to effects on the CNS has been developed as the Irwin test [2]. Such assays are expensive in terms of time and consumables, and they also require relatively large amounts of plant extract. They are more likely to be useful in assessing potential toxic effects than in revealing novel drug effects.

Model organisms can be used instead of rodents. Nematode worms (*Caenorhabditis elegans*), fruit flies (*Drosophila melanogaster*), and zebrafish (*Danio rerio*) have all been used in drug discovery (see [3] for a review). Extracts of some traditionally used medicinal plants have been tested on nematodes for antioxidant activities and effects on aging [4–6]. Zebrafish embryos are attracting attention because they are vertebrates and many stages of their development can be readily visualized (e.g., [7–9]). They are suitable for screening plant extracts in 96-well microtiter plates [10]. Embryos can be exposed to test substances and observed for the development of unusual phenotypes, or disease models can be created in the embryos and samples tested for their ability to modify the expression of the disease (see [3] for examples of disease models in such organisms). However, assays

Plant Bioactives and Drug Discovery: Principles, Practice, and Perspectives, Fourth Edition.
Edited by Valdir Cechinel-Filho.

involving the assessment of phenotypic changes in model animals are still hampered by the absence of systems that can automatically quantify effects. Furthermore, the concentration of substances reaching their sites of action will also be uncertain: negative results might not be due to inherent lack of activity but to poor bioavailability.

14.2 MOLECULAR AND CELLULAR ASSAYS

To move beyond observations of effects in whole animals, researchers previously used classical pharmacological preparations for functional or mechanistic studies of plant components (for an earlier review of methods used, see [11]). Such methods are time- and sample consuming, and are not, therefore, suitable for large-scale screening of collections of plant extracts or compounds. For such screening, approaches used by pharmaceutical companies for the initial stages of drug discovery can be adapted for use with natural products. Although generically referred to as high throughput screening (HTS), the approach refers not only to the automated and rapid testing of very large numbers of test samples, but also to the application of robust conditions in assay development to provide methods that can be run in multiwell formats where interactions between drug and target are detected by sensitive readouts. The readouts can be radiometric but are more generally light based, whether by absorbance, fluorescence, or chemiluminescence.

While drug discovery laboratories in companies will typically use assays formatted for 384- or 1536-well plates in order to screen hundreds of thousands of compounds, the numbers of plant samples available for testing are generally much smaller than those in synthetic chemical libraries. Hence, assays can be run in 96-well plates, obviating the need for very expensive high-precision liquid dispensing equipment. Otherwise, the approach to establishing an assay for use in routine screening campaigns is the same for natural products as it is for synthetic compounds. Care must be taken to optimize the concentrations of reagents and the time course of the assay in order to maximize the signal to background ratio. Assays should include, wherever possible, positive and negative controls in order to demonstrate the reliability of the assay in each plate. Any assay can be susceptible to false positives, that is, "hits" that are not due to a specific interaction between the test substance and the biological target. Some substances can interfere with the readout (for example, by fluorescing at the wavelength being monitored) and some can disrupt a biological component in the assay (for example, by precipitating a protein or acting like a detergent to disrupt membranes). It is very useful if back-up assays using a different readout are available. Also, assays can be run in the presence of proteins such as BSA or with low concentrations of detergent to lessen the risk of nonspecific binding causing spurious hits. Much work has been done with collections of pure compounds to define the problem of "promiscuous inhibitors" (e.g., see [12]), but less attention has been given to compounds isolated from plants or to plant extracts. It is generally assumed that polyphenolic tannins can precipitate proteins and, therefore, are likely to be the source of many false positives when screening crude extracts. Routine detannification is often carried out to avoid this [13], knowing that polyphenolic compounds are not regarded as promising starting points for drug development. However, detannification is not appropriate if the sample being tested is a traditional herbal medicine that is used as a total extract, and some experimental evidence indicates that polyphenolics are not generally reactive in screening assays [14].

HTS approaches work well when there is a defined biological target such as a receptor protein or an enzyme (for a recent review, see [15]). However, there are often concerns about the importance of interfering with a single protein in the context of a disease that most probably involves several biochemical pathways and interacting networks of cells. Moreover, natural product extracts are suggested to be more likely to cause false positives in assays based on isolated proteins than in cell-based assays [16].

Nevertheless, a large number of human membrane-bound receptors have been cloned and expressed and there are corresponding sets of labeled ligands that can be used for binding assays. Given that 50% of currently used drugs are thought to act through G-protein-coupled receptors and that there are still many orphan GPCRs whose function remains to be uncovered, there should be plenty of scope for continued drug discovery based on receptor binding assays. In addition, there are nuclear receptors that can also be assayed with classical binding methods. As well as radiolabeled ligands, there are fluorescently tagged compounds that can be used to study interactions at receptors. Fluorescence-based methods can be more sensitive than radiometric assays and, when labeled compounds with long emission times are used in time-resolved fluorescence, there is no need to separate bound from unbound compounds in the assay mixture (see [17] for examples of different varieties of receptor assays used with plant extracts).

Many enzymes, like receptors, are available as cloned and expressed proteins that can be used in HTS assays. With the current interest in protein kinases as targets for new anticancer agents, many technologies have been developed to enable routine screening of compound collections for kinase inhibitors. The pan-kinase inhibitor staurosporine is a natural product (a bis-indole alkaloid isolated from a *Streptomyces*), and other compounds that are more selective in their activities have been discovered from screening: for example, the homoisoflavonoid hematoxylin was found to be a potent inhibitor of the c-Src kinase activity [18].

Cell-based assays are growing in popularity for drug discovery (e.g., [19–21]). These have advantages over molecular assays in that they can detect the activity against a functioning pathway within a cell without assuming a specific target or mechanism of action. Conversely, they are open to confounding results from effects unrelated to the pathway being studied. If the site of action is within the cell, then positive hits are presumably caused by compounds that can penetrate the cell membrane; however, impermeable compounds that might be active on the molecular target will be missed; such compounds might be capable of being chemically modified to improve their bioavailability.

The simplest cell-based assays are those that measure cell viability or cell proliferation. There are many commercially available kits to detect cell multiplication and cell death. Other cell-based assays involve engineering cells so that they express a reporter gene when a particular pathway is activated, for example, where stimulation of NFκB signaling results in the expression of luciferase, and that can be detected by a standard light reaction. Care must be taken to make sure that any putative hits are not interfering with the reporter gene or its readout.

More recent developments with cell-based assays are in the domain of high content screening (HCS) [20]. In such assays, complex changes to cells (such as neurite formation) are followed, or several biochemical pathways are monitored simultaneously by means of parallel detection systems. As can be imagined, the information content from the assays is high, and there is generally a need for sophisticated imaging microscopes and large amounts of computing power. HCS is probably not at the stage where it is feasible for routine screening of massive collections of compounds, but the approach can be used with smaller numbers of samples with several readouts combined in a multiplexed format.

14.3 DISEASE-SPECIFIC ASSAYS

Although medicines in current use act through only a few hundred different molecular targets, there are potentially thousands of additional drug targets under exploration or still to be found [22]. Consequently, it is not possible in one chapter to cover all assay targets that could be used for plant-based drug discovery. Instead, some examples will be given in some key therapeutic areas: cancer, diabetes, infectious disease, and inflammation. Other chapters give additional examples, and for recent reviews relating to screening for cardiovascular diseases, see [23] and [24] for cardiac arrhythmias and other conditions in zebrafish, and [25] for an overview of phenotypic assays for atherosclerosis and heart failure.

14.3.1 Anticancer Drug Discovery

There are many cell lines that have been established from human cancers. An obvious approach to screening for the potential anticancer activity is to test samples for the ability to slow the proliferation of such cells in culture or to induce cell death [26]. However, cancer cell lines may not faithfully mimic the behavior of cancer cells *in situ* and, in particular, they may be more susceptible to drugs than cells in real life. Various approaches have been used to try to make *in vitro* screening more predictive, including growing cells in nutrient-poor conditions or in three-dimensional assemblies and by using primary cells from biopsies [26,27]. Despite the limitations of the conventional assays, screening of natural products continues to provide leads for new anticancer agents, including several that are in clinical development [28].

More complex phenotypic screens can be performed using zebrafish embryos [8,9,29], either by looking for phenotypes that result from disrupting a developmental pathway that is involved in cancer or through genetically engineered embryos that have a cancer-related phenotype. As an example of the former, overactivity of the hedgehog pathway is associated with basal-cell carcinomas and several other important types of cancers. A steroidal plant alkaloid cyclopamine that blocks the hedgehog pathway has been shown to kill tumor cells *in vitro* and to reduce cancer growth and metastases in mouse studies. In a recent pilot study, we have screened just over one-hundred randomly selected plant extracts and have identified one that causes cyclopia in the developing embryo (EE Patton and AL Harvey unpublished). This extract has been subfractionated and a group of four molecules still holds the potent activity. The identification of the active substance is underway, but it is distinct from cyclopamine. In the other approach, transplantable cancer cell lines have been generated in zebrafish [30] allowing the development of potential screening models for leukemias and hepatic cancer [31,32]. Antiangiogenic effects can also be detected using assays with zebrafish embryos in which the endothelial cells express green fluorescent protein (GFP) to make visualization of effects easier [33,34].

Currently, many screening campaigns are directed against particular protein kinases that are believed to play roles in various cancers, as well as other molecular targets, such as topoisomerases and histone deacetylase [26,35]. Plant-derived compounds have also been found to block activation of the NFκB pathway, which is critical for development of many cancer cells (e.g., [36,37]).

14.3.2 Diabetes and Metabolic Diseases

Many traditionally used herbal remedies have been suggested as being beneficial in patients with diabetes or have been shown to lower blood glucose in animals [38–40]. Since antidiabetic drugs can work through several different mechanisms of action, different assays are required for screening purposes. These should cover: insulin release (via effects on β-cell function or on β-cell numbers), insulin action (including effects on glucose uptake and on insulin's signaling pathway), carbohydrate metabolism, and lipid metabolism. While there are many examples of particular plant extracts being studied for effects on one or a few of such mechanisms, there are few examples of a comprehensive panel of assays being used or of mass screening of a large collection of natural products [39].

14.3.3 Antimicrobial Drug Discovery

With the continued emergence of microbes with resistance to current antibiotics, there is an urgent need to find new antibacterial and antifungal agents with novel mechanisms of action [41,42]. Since growing common strains of bacteria and fungi is relatively straightforward, many laboratories throughout the world have been screening plant extracts for antimicrobial effects. These simple assays should be coupled with assays that will reveal selective toxicity for microbes: for example, active antimicrobial samples should have little or no toxicity to mammalian cells in culture or to erythrocytes. Some workers infer the absence of mammalian toxicity from negative results in brine shrimp lethality assays [43], but the predictive value of the brine shrimp is questionable.

Screening assays based on monitoring growth or viability of target organisms *in vitro* may only uncover compounds that act by a limited range of mechanisms. Additional assays that probe the infectivity or virulence of the organism would appear to be needed. Ideally such assays would be conducted *in vivo*, and for this reason, suitable model organisms are being used, including the nematode worm (*C. elegans*) and the zebrafish (*Danio rerio*) [44]. Assays can be conducted on nematodes with several human pathogens: *Enterococcus faecalis, Staphylococcus aureus, Pseudomonas aeruginosa, and Candida albicans* [45–47], but it should be remembered that nematodes have an innate immune system but lack an adaptive immune system. Published assay methods could be adapted for semiautomated large-scale screening. For example, single worms in 384-well plates can be visualized by the DNA stain Sytox Green, and the standard MTT assay is known to work with nematodes. Bacteria can be visualized by engineering them to express the bright red color DsRed (GFP gives a larger background interference) so it should be possible to monitor the extent of infection and the viability of the nematodes simultaneously. Zebrafish have both innate and adaptive immune systems so that they could be more relevant to human diseases. There are reports on use of zebrafish with *Salmonella typhimurium, Bacillus subtilis*, *Listeria* species, and *Streptococcus pyogenes* [44,48]. Zebrafish may also be an *in vivo* model for tuberculosis because they can be infected by *Mycobacterium marinum* [49]. This organism causes typical granulomas in zebrafish embryos and exists within macrophage-like cells. Functional screens have identified small molecule inhibitors of the production of virulence factors by *Vibrio cholerae* [50]. There are several reports on compounds that inhibit a key process (type-III secretions) in *Yersinia*, and *Chlamydia* species, and *P. aeruginosa*, and other publications describe small molecules that inhibit the virulence-associated organelles called pili in *E. coli* (for review, see [51]). However, there do not appear to be any broad screening of collections of natural products in such assays.

14.3.4 Anti-Inflammatory Drug Discovery

There are many studies of plant extracts and plant-derived compounds as free radical scavengers and antioxidants, mostly with the assumption that such activities may be useful as an anti-inflammatory mechanism. There are simple assay methods such as the quenching of the free radical, 2,2′-diphenyl-1-picrylhydrazyl (DPPH). However, many of the active compounds discovered in this way are poorly bioavailable, and it is not clear if such compounds would be able to affect the levels of free radicals inside cells. Intracellular effects can be monitored in cells through the use of membrane-permeable probes such as 2′,7′-dichlorofluorescein diacetate (DCF-DA) (see, for example, [52]). It should be noted that plant extracts are not necessarily always antioxidant: there is evidence that they can be pro-oxidant and that this may result in anti-inflammatory effects (e.g., [53,54]).

Other assays for potential anti-inflammatory compounds are based on detecting effects on various inflammatory pathways, such as the release and action of tumor necrosis factor (TNF) α [55] and inflammatory cytokines [56], and activation of the NF-κB pathway [57].

14.4 CONCLUSIONS

There are many opportunities for drug discovery based on compounds from plants. The vast majority of plants have not been studied for biological activity, and there are a great many assays available that are relevant to diseases for which better medicines are required. Screening assays that have been developed for other purposes can generally be adapted for use with plant-derived samples. However, care still needs to be taken to avoid distractions from false positive "hits": demonstrating a clear sigmoidal concentration–response relationship and verifying hits in a similar assay that uses a different readout (e.g., fluorescent rather than chemiluminescent) usually sorts the spurious from the real hits. It is hoped that the examples provided in this chapter will help stimulate productive drug discovery screening from plants.

REFERENCES

[1] Malone, E.H., Robichaud, R.C. (1962). A Hippocratic screen for pure or crude drug materials. *Journal of Natural Products, 25*, 320–322.

[2] Irwin, S. (1968). Comprehensive observational assessment: Ia. A systematic, quantitative procedure for assessing the behavioural and physiological state of the mouse. *Psychopharmacologia (Berl.) 13*, 222–257.

[3] Giacomotto, J., Ségalat, L. (2010). High-throughput screening and small animal models, where are we? *British Journal of Pharmacology, 160*, 204–216.

[4] Collins, J.J., Evason, K., Kornfeld, K. (2006). Pharmacology of delayed aging and extended lifespan of *Caenorhabditis elegans*. *Experimental Gerontology, 41*, 1032–1039.

[5] Pun, P.B., Gruber, J., Tang, S.Y., Schaffer, S., Ong, R.L., Fong, S., Ng, L.F., Cheah, I., Halliwell, B. (2010). Ageing in nematodes: Do antioxidants extend lifespan in *Caenorhabditis elegans*? *Biogerontology, 11*, 17–30.

[6] Tang, S.Y., Halliwell, B. (2010). Medicinal plants and antioxidants: What do we learn from cell culture and *Caenorhabditis elegans* studies? *Biochemical and Biophysical Research Communications, 394*, 1–5.

[7] Bowman, T.V., Zon, L.I. (2010). Swimming into the future of drug discovery: *In vivo* chemical screens in zebrafish. *ACS Chemical Biology, 5*, 159–161.

[8] Kokel, D., Bryan, J., Laggner, C., White, R., Cheung, C.Y., Mateus, R., Healey, D., Kim, S., Werdich, A.A., Haggarty, S.J., Macrae, C.A., Shoichet, B., Peterson, R.T. (2010). Rapid behavior-based identification of neuroactive small molecules in the zebrafish. *Nature Chemical Biology, 6*, 231–237.

[9] Peal, D.S., Peterson, R.T., Milan, D. (2010). Small molecule screening in zebrafish. *Journal of Cardiovascular Translational Research, 3*, 454–460.

[10] Crawford, A.D., Esguerra, C.V., de Witte, P. A. (2008). Fishing for drugs from nature: Zebrafish as a technology platform for natural product discovery. *Planta Medica, 74*, 624–632.

[11] Vlietinck, A.J., Screening methods for detection of biological activities of plant preparations. In: Bohlin, L., Bruhn, J.G., editors. *Bioassay Methods in Natural Product*

Research and Drug Development, Kluwer Academic Publishers, Dordrecht, The Netherlands, 1999, pp. 37–52.

[12] McGovern, S.L., Caselli, E., Grigorieff, N., Shoichet, B.K. (2002). A common mechanism underlying promiscuous inhibitors from virtual and high-throughput screening. *Journal of Medicinal Chemistry*, *45*, 1712–1722.

[13] Jones, W.P., Kinghorn, A.D., Extraction of plant secondary metabolites. In: Dey, P.M., Harborne, J.B., editors. *Natural Products Isolation*, Springer, Berlin, 2005, pp. 323–351.

[14] Zhu, M., Phillipson, J.D., Greengrass, P.M., Bowery, N.E., Cai, Y. (1997). Plant polyphenols: Biologically active compounds or non-selective binders to protein? *Phytochemistry*, *44*, 441–447.

[15] Kool, J., Lingeman, H., Niessen, W., Irth, H. (2010). High throughput screening methodologies classified for major drug target classes according to target signaling pathways. *Combinatorial Chemistry High Throughput Screening*, *13*, 548–561.

[16] Rishton, G.M. (2008). Molecular diversity in the context of leadlikeness: Compound properties that enable effective biochemical screening. *Current Opinion in Chemical Biology*, *12*, 340–351.

[17] Zhu, Y., Zhang, Z., Zhang, M., Mais, D.E., Wang, M.W. (2010). High-throughput screening for bioactive components from traditional Chinese medicine. *Combinatorial Chemistry High Throughput Screening*, *13*, 837–848.

[18] Lin, L.G., Xie, H., Li, H.L., Tong, L.J., Tang, C.P., Ke, C.Q., Liu, Q.F., Lin, L.P., Geng, M.Y., Jiang, H., Zhao, W.M., Ding, J. (2008). Naturally occurring homoisoflavonoids function as potent protein tyrosine kinase inhibitors by c-Src-based high-throughput screening. *Journal of Medicinal Chemistry*, *51*, 4419–4429.

[19] An, W.F., Tolliday, N. (2010). Cell-based assays for high-throughput screening. *Molecular Biotechnology*, *45*, 180–186.

[20] Bickle, M. (2010). The beautiful cell: High-content screening in drug discovery. *Analytical and Bioanalytical Chemistry*, *398*, 219–226.

[21] Michelini, E., Cevenini, L., Mezzanotte, L., Coppa, A., Roda, A. (2010). Cell-based assays: Fuelling drug discovery. *Analytical and Bioanalytical Chemistry*, *398*, 227–238.

[22] Landry, Y., Gies, J.P. (2008). Drugs and their molecular targets: An updated overview. *Fundamental and Clinical Pharmacology*, *22*, 1–18.

[23] Chico, T.J., Ingham, P.W., Crossman, D.C. (2009). Modeling cardiovascular disease in the zebrafish. *Trends in Cardiovascular Medicine*, *18*, 150–155.

[24] Macrae, C.A. (2010). Cardiac arrhythmia: *In vivo* screening in the zebrafish to overcome complexity in drug discovery. *Expert Opinion in Drug Discovery*, *5*, 619–632.

[25] Etzion, Y., Muslin, A.J. (2010). The application of phenotypic high-throughput screening techniques to cardiovascular research. *Trends in Cardiovascular Medicine*, *19*, 207–212.

[26] Harvey, A.L., Cree, I.A. (2010). High-throughput screening of natural products for cancer therapy. *Planta Medica*, *76*, 1080–1086.

[27] Cree, I.A., Glaysher, S., Harvey, A.L. (2010). Efficacy of anti-cancer agents in cell lines versus human primary tumour tissue. *Current Opinion in Pharmacology*, *10*, 375–379.

[28] Lucas, D.M., Still, P.C., Pérez, L.B., Grever, M.R., Kinghorn, A.D. (2010). Potential of plant-derived natural products in the treatment of leukemia and lymphoma. *Drug Targets*, *11*, 812–822.

[29] Zon, L.I., Le, X. (2008). Potential of zebrafish for anticancer screening. *Expert Opinion in Drug Discovery*, *3*, 1451–1460.

[30] Mizgireuv, I.V., Revskoy, S.Y. (2006). Transplantable tumor lines generated in clonal zebrafish. *Cancer Research*, *66*, 3120–3125.

[31] Mizgirev, I.V., Revskoy, S. (2010). A new zebrafish model for experimental leukemia therapy. *Cancer Biology and Therapy*, *9*, 895–902.

[32] Mizgirev, I.V., Revskoy, S. (2010). Generation of clonal zebrafish lines and transplantable hepatic tumors. *Nature Protocols*, *5*, 383–394.

[33] Nicoli, S., Presta, M. (2007). The zebrafish/tumor xenograft angiogenesis assay. *Nature Protocols*, *2*, 2918–2923.

[34] Tran, T.C., Sneed, B., Haider, J., Blavo, D., White, A., Aiyejorun, T., Baranowski, T.C., Rubinstein, A.L., Doan, T.N., Dingledine, R., Sandberg, E.M. (2007). Automated, quantitative screening assay for antiangiogenic compounds using transgenic zebrafish. *Cancer Research*, *67*, 11386–11392.

[35] Kingston, D.G., Newman, D.J. (2005). The search for novel drug leads for predominately antitumor therapies by utilizing mother nature's pharmacophoric libraries. *Current Opinion in Drug Discovery and Development*, *8*, 207–227.

[36] Chaturvedi, M.M., Sung, B., Yadav, V.R., Kannappan, R., Aggarwal, B.B. (2010). NF-κB addiction and its role in cancer: 'One size does not fit all'. *Oncogene*, *30*, 1615–1630.

[37] Prasad, S., Phromnoi, K., Yadav, V.R., Chaturvedi, M.M., Aggarwal, B.B. (2010). Targeting inflammatory pathways by flavonoids for prevention and treatment of cancer. *Planta Medica*, *76*, 1044–1063.

[38] Prabhakar, P.K., Doble, M. (2008). A target based therapeutic approach towards diabetes mellitus using medicinal plants. *Current Diabetes Reviews*, *4*, 291–308.

[39] Harvey, A.L. (2010). Plant natural products in anti-diabetic drug discovery. *Current Organic Chemistry*, *14*, 1670–1677.

[40] Qi, L.W., Liu, E.H., Chu, C., Peng, Y.B., Cai, H.X., Li, P. (2010). Anti-diabetic agents from natural products: An update from 2004 to 2009. *Current Topics in Medicinal Chemistry*, *10*, 434–457.

[41] Di Santo, R. (2010). Natural products as antifungal agents against clinically relevant pathogens. *Natural Products Reports*, *27*, 1084–1098.

[42] Donadio, S., Maffioli, S., Monciardini, P., Sosio, M., Jabes, D. (2010). Sources of novel antibiotics: Aside the common roads. *Applied Microbiology and Biotechnology*, *88*, 1261–1267.

[43] Nisar, M., Tariq, S.A., Marwat, I.K., Shah, M.R., Khan, I.A. (2009). Antibacterial, antifungal, insecticidal, cytotoxicity and phytotoxicity studies on *Indigofera gerardiana*. *Journal of Enzyme Inhibition and Medicinal Chemistry*, *24*, 224–229.

[44] Kurz, C.L., Ewbank, J.J. (2007). Infection in a dish: High-throughput analyses of bacterial pathogenesis. *Current Opinion in Microbiology*, *10*, 10–16.

[45] Moy, T.I., Conery, A.L., Larkins-Ford, J., Wu, G., Mazitschek, R., Casadei, G., Lewis, K., Carpenter, A.E., Ausubel, F.M. (2009). High-throughput screen for novel antimicrobials using a whole animal infection model. *ACS Chemical Biology*, *4*, 527–533.

[46] Uccelletti, D., Zanni, E., Marcellini, L., Palleschi, C., Barra, D., Mangoni, M.L. (2010). Anti-*Pseudomonas* activity of frog skin antimicrobial peptides in a *Caenorhabditis elegans* infection model: A plausible mode of action *in vitro* and *in vivo*. *Antimicrobial Agents and Chemotherapy*, *54*, 3853–3860.

[47] Okoli, I., Coleman, J.J., Tampakakis, E., An, W.F., Holson, E., Wagner, F., Conery, A.L., Larkins-Ford, J., Wu, G., Stern, A., Ausubel, F.M., Mylonakis, E. (2009). Identification of

antifungal compounds active against *Candida albicans* using an improved high-throughput *Caenorhabditis elegans* assay. *PLoS One*, *4*, e7025.

[48] van der Sar, A.M., Musters, R.J., van Eeden, F.J., Appelmelk, B.J., Vandenbroucke-Grauls, C.M., Bitter, W. (2003). Zebrafish embryos as a model host for the real time analysis of *Salmonella typhimurium* infections. *Cell Microbiology*, *5*, 601–611.

[49] Cosma, C.L., Swaim, L.E., Volkman, H., Ramakrishnan, L., Davis, J.M. (2006). Zebrafish and frog models of *Mycobacterium marinum* infection. *Current Protocols in Microbiology*, Chapter 10, Unit 10B. 2.

[50] Hung, D.T., Shakhnovich, E.A., Pierson, E., Mekalanos, J.J. (2005). Small-molecule inhibitor of *Vibrio cholerae* virulence and intestinal colonization. *Science*, *310*, 670–674.

[51] Baron, C. (2010). Antivirulence drugs to target bacterial secretion systems. *Current Opinion in Microbiology*, *13*, 100–105.

[52] Carter, W.O., Narayanan, P.K., Robinson, J.P. (1994). Intracellular hydrogen peroxide and superoxide anion detection in endothelial cells. *Journal of Leukocyte Biology*, *55*, 253–258.

[53] Tang, S.Y., Cheah, I.K., Wang, H., Halliwell, B. (2009). *Notopterygium forbesii* Boiss extract and its active constituent phenethyl ferulate attenuate pro-inflammatory responses to lipopolysaccharide in RAW 264.7 macrophages. A "protective" role for oxidative stress? *Chemical Research in Toxicology*, *22*, 1473–1482.

[54] Gutteridge, J.M.C., Halliwell, B. (2010). Antioxidants: Molecules, medicines, and myths. *Biochemical and Biophysical Research Communications*, *393*, 561–564.

[55] Sethi, G., Sung, B., Kunnumakkara, A.B., Aggarwal, B.B. (2009). Targeting TNF for treatment of cancer and autoimmunity. *Advances in Experimental Medicine and Biology*, *647*, 37–51.

[56] Chen, X., Oppenheim, J.J., Howard, O.M. (2004). Chemokines and chemokine receptors as novel therapeutic targets in rheumatoid arthritis (RA): Inhibitory effects of traditional Chinese medicinal components. *Cellular and Molecular Immunology*, *1*, 336–342.

[57] Aggarwal, B.B., Shishodia, S. (2004). Suppression of the nuclear factor-kappaB activation pathway by spice-derived phytochemicals: Reasoning for seasoning. *Annals of the New York Academy of Sciences*, *1030*, 434–441.

15

PHYTOTHERAPEUTICS – INTELLECTUAL PROPERTY RIGHTS, GLOBAL MARKET, AND GLOBAL REGULATORY GUIDELINES

James D. McChesney, Raymond Cooper, and Kip Vought

15.1 INTELLECTUAL PROPERTY RIGHTS

The underlying concept of intellectual property (IP) rights is to provide exclusivity of ownership of products such that investments in development, manufacturing, and marketing of these products can be justified so as to obtain an adequate return on those investments. An overarching consequence of the absence of IP is that little or no investment can be expected or justified and, therefore, viable and needed medicinal therapies will not be developed, manufactured, and marketed. The result of this lack of development is that phytotherapeutics may not be available for the treatment or mitigation of human disease. This inability to obtain a return on investment (in this case because the economics of the market is thought not to allow adequate returns) is one of the historic causes for the lack of new drug development for tropical parasitic diseases. Only with the entrance of significant philanthropic investment and public sector funding, has this lack begun to be addressed. The following discussion is intended to be a brief overview of strategies for intellectual property protection and is not intended as guidance or recommendation to follow. One should seek the counsel of a licensed patent attorney for specifics and guidance relative to individual cases.

Intellectual property rights may be captured via multiple mechanisms. The most widely recognized is via formal patenting. In this process intellectual property is

Plant Bioactives and Drug Discovery: Principles, Practice, and Perspectives, Fourth Edition.
Edited by Valdir Cechinel-Filho.

codified into a patent application which then is submitted to an appropriate patent agency, usually that of the home country of the patent applicant. To obtain international patent protection of the IP, the application must either subsequently or directly be filed with the World Patent Office. Much effort has been made to harmonize patent regulations between countries so that effectively all major economic markets now adhere to and honor international patents.

The patent application must describe a discovery that is new (not previously recognized or known) and novel (not to be expected as predictable by an experienced and knowledgeable researcher, familiar with the discipline). The application must disclose the intellectual property in sufficient detail that "one skilled in the art" can reproduce the discovery. Finally, the discovery must have utility, that is, be useful or solve a recognized problem; in the case of a phytotherapeutic it will be able to diagnose, treat, or mitigate a disease.

Once a patent application is filed, it undergoes an examination by a patent examiner for evaluation to meet the criteria and tests of patentability, whether it is new and novel and has utility. Further, the detailed description and examples provided in the application must support the requested claims – the specific intellectual property whose ownership is sought by the applicant. This examination may take many months to several years to complete. During this examination, the filed application will publish and become public 18 months post the filing date. This way, progress and information in the discipline is made generally available to all researchers. If the examiner agrees, these tests are met and the applicant is the first to file an application claiming the IP (that is "has priority"), then a patent may be allowed and issue. Often during the examination of the application, the claims of the application are negotiated between the applicant and the patent office such that the claims allowed are specifically supported by the examples and information disclosed in the body of the application. These negotiations are to assure the breadth and extent of the claims that does not cover more than is rightfully the property of the applicant. Under the currently harmonized international patent guidelines/convention, a newly allowed and issued patent is valid for 20 years from the date of filing of the original application. Under specific criteria, the patent office may extend the patent period ("life") beyond the initial 20 years. These extensions are especially important should a patented drug entity become a "blockbuster," since every day of patent life is equivalent to millions of dollars in revenue for the patent holder. Indeed, these extensions often become the subject of intense litigation.

Patents in the area of phytopharmaceuticals usually take one of three forms – (1) a composition of matter patent, (2) a use patent, or (3) a process patent. Composition of matter patents are generally viewed as the strongest and preferable since they capture the actual chemical entity as the patented property and all uses of that chemical entity are thereby controlled and owned by the patent holder no matter who actually discovers them subsequently. Alternatively specific uses for a nonpatentable chemical entity or composition may be discovered and patented. However, should further uses be discovered for the composition then those may in turn be separately patented and owned by alternative owners. Consequently, use patents are viewed to be much less valuable than composition patents. Finally and particularly relevant to phytopharmaceuticals is the category of process patents. In these applications, the specifics of the process for preparation of extracts, compounded formulas, and even synthesis or purification of single chemical entities are disclosed and claimed. Since there are often multiple routes to reach a similar or even identical product, process patents are viewed as the weakest and least valuable of the categories.

An alternative strategy for protecting ownership of intellectual property is to take the route of Trade Secret. In this strategy the intellectual property is protected by holding it strictly confidential and proprietary and not revealed outside of the manufacturer. This is most often utilized for manufacturing processes and similar activities where disclosure of the specifics of a process such as required in a patent application would provide competitors a road map for alternative process development approaches. This strategy is not useful for product or composition protection since "reverse engineering" efforts will likely uncover the specifics of the material, such that a competitor would likely be able to completely or almost duplicate it.

15.2 BIODIVERSITY

Discovery of natural products as new pharmaceuticals is dependent upon evaluation of natural sources – biodiversity – organisms, whether terrestrial or marine, plants, animals, or microorganisms. With the recognition in the latter part of the twentieth century that biodiversity was fast disappearing and that ownership of that biodiversity rightly belonged to the country of origin, there has now evolved a strategy to return benefits to the source of the organism from which a new natural product drug or preparation was discovered. Further, since the selection of organisms to be investigated for a new drug was often made based upon traditional knowledge, it has been determined that such traditional knowledge be recognized and benefits returned in recognition of the important contribution that traditional knowledge makes to the discovery process. The intent of this strategy – codified in the Convention on Biological Diversity – is to provide guidance and framework to enable equitable benefit sharing with all stakeholders in the natural products discovery process.

The Convention on Biological Diversity provides for conservation of biodiversity, sustainable use of *biodiversity*, and importantly, equitable sharing of benefits from exploitation of biodiversity. The complexity of the Convention precludes a detailed discussion, but an excellent treatment can be found in Reid et al. [1]. Of particular relevance to natural product/phytopharmaceutical discovery and development are the access and benefit sharing provisions of the Convention. These have had profound impact upon discovery from nature. The intent to provide a framework for source countries to define and regulate access to their biodiversity is laudable. However, only few countries have been active and successful in doing so. More commonly, the outcome has been more confusion and less clarity. Many reasons can be cited for the lack of progress. A historical pattern of exploitation by agents (corporations and institutions) of developed countries and with of developing country's natural resources causes governments to be reticent to allow access and thus be slow or not motivated to establish clear, legal guidelines to allow investigators to access their biodiversity. In fact, this paranoia has in cases been so great that even in-country researchers may not access the biodiversity of their own country. A complicating factor is often the greatly exaggerated expectation of what the benefits to be shared will constitute. The coincidence of paclitaxel marketing approval and the ratification of the Convention on Biological Diversity suggested to officials and researchers in developing countries that huge monetary returns would be quickly realized from bioprospecting. No one was thinking that only very few drugs become blockbusters and even then, long lead times and large resource investments are needed to reach the market. In the specific case of paclitaxel, it took 30 years, two near complete abandonments, and more the $100 million of public and private investment. One may hope that the laudable intents of the Convention on Biological Diversity may be implemented by both developing countries, where the major portion of biodiversity exists, and in developed countries and their corporations and institutions, where the majority of the discovery capability resides, so that the full potential of natural products to provide new pharmaceuticals to meet the challenge of the great unmet human medical needs still extant for the benefit of all humankind.

15.3 GLOBAL MARKET PERSPECTIVES

The global market for botanical and plant-derived drugs was worth $19.5 billion in 2008. This is expected to reach $32.9 billion in 2013, for a compound annual growth rate (CAGR) of 11.0% [2]. Of these numbers, it is possible to assess a breakdown of the market into two segments: botanical drugs and other plant-derived drugs. The latter clearly claims the major share of business, generating almost the entire $19.5 billion in 2008. Botanical drugs only netted $127K in 2008; however, sales are projected to rise at a compound annual growth rate of >600%, through 2013. Some new botanical products may be approved for multiple indications during this time period. The projection is for the approved indications to reach a total of 14 by 2013 with oncology being the most prominent, followed by hormone and metabolism as a secondary therapeutic category. For this growth to take place, it will be necessary that the one current botanical product will translate into 12 approved drugs by 2013, up from the current situation (see Table 15.1).

The merits of plants and their extracts have been well known since ancient times. Evidence of plant-derived compounds taking on a more significant role in the pharmaceutical market today is the US Food and Drug Administration's carving

TABLE 15.1 Potential Approval Pipeline of Botanical Drugs

Botanical NDAs by FDA	Till Date
Approved	2
Applied	19
Planned	33

out of a new drug approval pathway for botanical drugs in 2004, and approving the first botanical based on these guidelines in 2006. Thus, much more opportunity exists within the industry of plant-derived drugs, if we consider that the number of higher plant species on earth is estimated around 350,000, with less than 10% screened. The well-known plant-derived single chemical entity drugs include paclitaxel, morphine, quinine, galanthamine, camptothecin vinblastine, vinchristine, and diosgenin.

Botanical drugs are those that are FDA approved under the botanical drug pathway; plant-derived drugs are those under a new FDA classification that are botanicals and contain a mixture of plant-derived substances.

In countries where there is a strong tradition of botanical use, there may be opportunities for more growth of botanical drugs. For example, India clearly has several advantages when it comes to herbal product development. However, data protection will have to be tightened further and efficiencies improved and more development of skills and resources in New Chemical Entity (NCE) research will be needed to reach their potential. That being said, there is a strong traditional folklore and a long history of usage in the country of Ayurvedic botanicals and herbs that may enter the nutraceutical markets worldwide. Whether the local drug authorities will move to phytopharmaceuticals remains an open question.

Realizing the mounting healthcare costs in the developed markets and the potential of the natural drugs discovery platform, regulators in some major markets may allow better access to proven safe nutraceuticals or give concessions to multiactive partially characterized phytoproducts to shorten approval time and cost, much like the German market.

The Food and Drug Administration (FDA) and European Medicines Agency (EMEA) recently provided a boost to the natural products industry by lowering the entry barriers for botanicals in comparison to chemicals and biologicals in their respective regions. These new guidelines more importantly also provide for unique guarantees of market exclusivity for botanicals as well as the acceptance of synergistic combinations of bioactives. Botanical drug products (or Herbal Medicinal Products as they are known in Europe) will ultimately compete alongside conventional drugs in the $500 billion global pharma market. Local regulations amongst the EU member states may differ slightly. Botanical drug products (BDPs) or phytopharmaceuticals should not be confused with pure isolated single compounds of natural origin or semisynthetic molecules such as paclitaxel and artemisinin. The new FDA regulation, in fact, is significant in recognizing semipurified multicompound plant or animal origin extracts. The FDA Guidance has also reduced emphasis and urgency on preclinical toxicological data as well as Phase I data especially for products which have a documented long history of use in the United States or any originating country.

India and China, with their vast libraries of natural compounds – many actively used in traditional systems and many still not codified – have clearly a natural advantage over the others. Apart from the starting material, India and China have now created an infrastructure and research capabilities to develop products up to Phase II. Indian investigators and CROs and Indian Pharma also have participated in global

Phase III trials but may look for partners to manage the full clinical development and regulatory aspects of a New Drug Application (NDA).

There are always niche segments that are not attractive enough for the large pharmaceutical companies to develop and market products to serve since these companies ordinarily are not interested in products with less than a $200 million annual market value. These could be potential targets for Indian companies, now that the financial barriers to new drug development have been significantly lowered. Indian companies in the pharma, biotech, Ayurvedic, and nutraceutical spaces have an opportunity to develop and market a BDP worldwide. Their global marketing setup continues to expand as there is considerable number of acquisitions by Indian pharma overseas.

Some success stories are presented. Two commonly used nutraceuticals have been approved as Botanical Drug Products by FDA – Veregen (also known as Polyphenon® E) and Omacor (now known as Lovaza). While Veregen is a green tea-based topical product for genital and perianal warts, Omacor is an omega-3 fatty acid product from fish oil approved for hypertriglyceridemia.

15.3.1 Veregen®/Polyphenon® E Ointment

Genital warts are benign tumors caused by HPV strains that disfigure the genital region and can be painful. It is one of the fastest growing sexually transmitted diseases occurring worldwide. The market revenues of genital warts treatments are growing by 7% annually. In North America and Europe 14 million people suffer from the disease, of which only about 3 million are currently seeking treatment. Among sexually active Americans 10% to 20% are thought to be infected with HPV, resulting in 750,000 new cases of genital warts each year in that country, making it the largest market. Existing treatments include laser vaporization, freezing, or repeated applications of cytotoxic substances. The major problems are the high recurrence rate (in 20–60% of patients), the side effects (local reactions such as itching or pain) and the duration of treatment. Approximately 30 million people worldwide are infected by these viruses (HPV type 6 or 11) that cause genital warts.

Veregen® was approved for the US market by FDA in October 2006 and is patent-protected until 2017. Veregen is promoted by PharmaDerm (a division of Nycomed US Inc.) to the Dermatology and OB/GYN specialties. Veregen® has been licensed to PharmaDerm for US marketing by a German Botanical Discovery company MediGene, AG. MediGene has made projections for peak annual worldwide sale from this single product to touch $150 million. Figures by other analysts put just the US market size for this segment alone at $200 million.

The active ingredient in Veregen® is a defined mixture of catechins extracted from green tea. Veregen is the first new treatment for this condition since Aldara was introduced in 1997. Veregen® is claimed to produce significant clearance rates in males, a population known for resistant genital warts. During clinical development Veregen® showed high and sustained efficacy with very few adverse events. Results were obtained from an international Phase III trial on 1000 patients in 15 countries. Veregen® activates the body's defense (immune modulation), inhibits major functions of the HPV, and counteracts specific changes in tumor cells (growth inhibition). This presents two targets for efficient combat against the tumor, which may open up new therapeutic approaches for other skin diseases.

Veregen® (or Polyphenon® E) was developed for commercial markets by Medigene, AG (Medigene, which possesses an innovative technology platform and core competence in anti cancer therapies. With the Veregen® approval, Medigene claims to be the first German biotech company with a proprietary drug approval in the United States. Veregen is their second drug with several more currently in clinical development.

Medigene received a $14 million milestone payment from Bradley Pharmaceuticals, Inc. (now PharmaDerm) after receiving US market authorization of Veregen®. The US commercial launch commenced in February 2009 and additional market approvals have been since received within the European Union. The PharmaDerm agreement includes minimum purchase quantities not only committed by PharmaDerm, but also co-development possibilities to explore new indications for Veregen, with marketing rights outside US remaining with the innovator. PharmaDerm estimates useful economic life of the product at 15 years, which is approximately 5 years beyond its patent life in the hope that development of generics for an array of natural compounds would be difficult. Though it is too early to predict the likely FDA thinking on "botanical ANDAs," PharmaDerm's projection may be refuted as the natural products space does offer some opportunities for copycats to eat into market share if not developed specifically as generics.

The Veregen® launch may be a relative nonevent for the pharmaceutical world, but eyes in the botanical sector will be closely monitoring the outcome. If PharmaDerm is correct and generic or copycat botanicals are proven to be difficult or costprohibitive to develop thereby preserving the Veregen® market outside its current patent life, then one can imagine botanicals receiving much broader interest.

Companies with a pipeline dedicated to botanicals include: MediGene (Germany), Reliant Pharmaceuticals (the US developer of Omacor), Phytomedics (USA), SunTen Phytotech (Taiwan), Phytopharma (UK), Phyto Health Pharma (The Netherlands), Nicholas Piramal (India), and Indigene Pharmaceuticals (India/USA). In the United States, Lupin has taken a lead with their antipsoriatic Desoris project. They have received approval from the DCGI (Drugs Controller India) to conduct a combined Phase-IIb/III clinical trial for its botanical drug LLL-3348 a once-daily 16-week oral therapy. Figures available for the total number of INDs filed so far for BDPs are upward of 300 with a large number of them originating from China and significant numbers of INDs and approved products from other markets (European, etc.).

Sales of herbal dietary supplements, the common form of botanicals utilized by consumers in the United States increased by nearly 5% and generated more than $5 billion in 2009. This was true even in the severely difficult economic down market when almost all consumer goods experienced a drop in sales. With botanicals incorporated into everything from cosmetics, dietary supplements, functional foods, regular foods, pharmaceuticals, etc, one does not know where the actual global market for phytotherapeutics falls. Thus, the projected market for 2013 of >$30 billion seems quite possible (Figure 15.1).

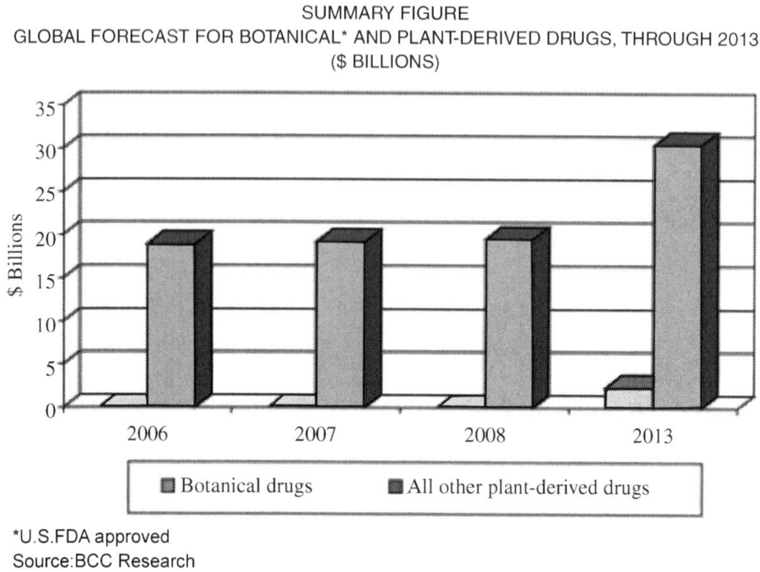

FIGURE 15.1 Global forecast for botanical* and plant-derived drugs, through 2013 ($ billions). Courtesy of BCC Research, Wellesley, MA *(See the color version of this figure in Color Plates section)*

15.3.2 Omacor

Omacor is the prescription drug derived from fish oil. The omega-3 polyunsaturated fatty acids are found in oil from certain types of fish, vegetables, and other plant sources. These fatty acids are not made by the body and must be consumed in the diet. Omacor is the ethyl ester form of these acids and works by lowering the body's production of triglycerides. High levels of triglycerides can lead to coronary artery disease, heart disease, and stroke, and is intended to be used together with diet and exercise to help lower triglyceride levels in the blood

The development of this preparation from a natural source may be model for future phytotherapeutics. Furthermore, from the marketing side, in September 2010, Pronova BioPharma ASA and Abbott Products Operations AG today announced an amendment to their existing License and Supply Agreement for Omacor. The new partnership is expected to extend sales growth and market potential of Omacor in the 69 countries covered by the agreement, including United Kingdom, Greece, Germany, Eastern Europe, the Middle East, Russia, Australia, Canada, China, and India.

This new agreement also establishes a framework for the collaborative development of new and improved omega-3 derived pharmaceutical products. Key terms of the new agreement include: the payment of an upfront incentive fee to Pronova BioPharma of EUR 19.5 million, which will be offsettable against the first EUR 19.5 million of royalties owed to Pronova BioPharma during the first 3 years of the contract. The new partnership agreement with Abbott is the culmination of this commitment to improve Omacor's market penetration and revenue growth in some of the world's fastest growing markets.

15.3.3 Generic Competition

The primary tool against generic competition will be IP as discussed previously; however, inherent complexities and variabilities in botanical composition will make it difficult to claim that a generic botanical drug is the same as the innovator product as defined in associated global regulations. Generally, generic drugs need to have the same composition of the active ingredient(s) and have the same bioavailability profile of the active ingredient(s) using the same route of administration. The specific regulations vary from region to region, but the primary consideration is the same: can the generic drug be considered a therapeutic substitution of the innovator drug product? Will the patient taking a generic drug have the same therapeutic effect and safety as the innovator product?

For complex botanical drug products that have a multicomponent active ingredient (s), the establishment of similar composition will be difficult and potentially cost-prohibitive. Likewise, this inherent complexity may make it difficult to establish bioequivalence between the generic and innovator product and present the need to perform a clinical endpoint study to confirm that the generic has the same efficacy and safety profile as the innovator drug product. Given these difficulties, the characterization of the multicomponent active ingredient may be more comprehensive than that performed for the original innovator product. The sponsor of the innovator product will likely raise these issues using formal and informal regulatory mechanisms throughout the various geographic regions in order to protect their market share.

Alternative "copycat" strategies such as the 505(b)(2) NDA mechanism in the United States and their equivalents. These strategies are not only similar to generic in that they cross-reference the market application for the innovator product for the primary establishment of the safety and efficacy, but also perform additional studies (e.g., clinical and/or animal safety studies) to further confirm the comparability of the drug to the innovator. These strategies have higher costs then generics and generally do not achieve a therapeutic equivalent code as received by generics. Therefore, the copycat product is not considered an automatic substitution for the innovator product and a sales/marketing force is needed to achieve market share, which most generic companies do not have unless there is an established branded product line within the company.

Generic competition will ultimately be dictated by the regulatory agencies and the data requirements. These issues are somewhat similar to those presented for generic biologics and it has only been recently that some regulatory agencies have begun accepting and reviewing these applications. The European Union has been on the forefront of accepting and reviewing generic biologic market application and should be expected to be more proactive for biologics. The United States is significantly lagging in outlining data requirements for generic biologics and is waiting for the congressional action, and may foreshadow what can be expected for generic botanicals.

15.4 REGULATORY PERSPECTIVES

The increased popularity and usage of herbal therapies in the larger Western markets has forced the associated regulatory agencies to address how to regulate these therapies relative to safety, efficacy, quality, and safety monitoring. The regulatory agencies of the larger Western markets such as Europe, United States, Canada, and Australia have turned their attention over the last 10 years to the incorporation and implementation of regulations specific for herbal therapies with the primary focus on traditional medicines and complementary and alternative medicines. Except for the United States, none of the regulatory agencies has attempted to address the development of either traditional or nontraditional therapies for new therapeutic indications.

Herbal medicinal products present a conundrum for regulatory agencies as they cross over between drugs, foods, and cosmetics. In most cases, it is the end use that dictates into what category an herbal product falls. Traditional herbal medicinal products are also problematic. It is presupposed that the herbal product has a common and universal origin and health benefit. Then there are issues of combining or enhancing (e.g., enriching or concentration) herbal products. The development of regulations also needs to weigh the public expectations that such products remain low cost while being considered safe and effective. Nevertheless, the four primary Western regulatory bodies have developed regulations for herbal medicinal products. The most current regulations established for herbal products are summarized in Table 15.2.

The European Union has been at the forefront of establishing regulations for herbal products with the first concerted effort made in 2001 when Directive 2001/83/EC WEU [1] was established to permit greater flexibility in the use of bibliographic data in proving the safety and efficacy of "well-established" medicines [2]. This directive maintained the requirement of a submission of market authorization applications (MAAs) for herbal medicinal products consistent with 2001/83/EC, but allowed for bibliographic data to be submitted for the establishment of safety and efficacy, either entirely or in addition to primary data. The MAA may be submitted via the centralized procedure (CP), decentralized procedure (DCP), or mutual recognition of a national market authorization (MRP). The MAAs for herbal drugs resembled generic market applications, but with comparison made to a monograph or accepted scientific data. The monographs initially originated from the World Health Organization (WHO), European Scientific Cooperative on Phytotherapy (ESCOP), and compendia. They are now increasingly coming from community herbal monographs elaborated by HMPC [3].

While this directive provided a regulatory pathway for marketing herbal medicinal products, the well-established use is not always discernable via the literature and there are products that have been well established as "traditional" therapies albeit with little data available in the literature that confirm their safety and efficacy. This issue was specifically addressed with the passing of Directive 2004/24/EC, which amends Directive 2001/83/EC and creates a simplified registration procedure for all traditional herbal medicines that do not meet the requirements for classification as well-established despite their long use [4]. To be eligible for the simplified registration, the traditional herbal medicinal product must (1) contain exclusively, one or more herbal ingredients (the addition of vitamins or minerals is allowed if their safety is established and their efficacy in ancillary to that of the herbs); (2) the indication must not require the diagnostic or therapeutic supervision of a medical practitioner; (3) the

TABLE 15.2 Current Herbal Drug Product Related Legislation

Geographic Region	Regulation	Implementation Date	Summary
Australia	Therapeutics Goods Act 1989	1991	• Establishes regulation of complementary medicines (including herbal and traditional medicines) as medicines • Agency established Australian Regulatory Guidelines for Complementary Medicines (ARGCM)
Canada	Natural Health Product Regulations	2004	• Provides regulatory pathway for market authorization of natural products (including herbal products) • Establishes requirements for the manufacture, packaging, labeling, importing and distribution of natural health products
European Union	Directive 2004/24/EC	2004 with implementation October 2005	• Amends the well established use category for traditional herbal therapies between food and full marketing authorization
			• Upgrades the EMEA herbal products working party (HMPWP) to the committee level (HMPC) • Establishes regulatory pathways for market authorizations for traditional herbal products • Establishes process for the development of industry guidelines for classification/standardization/declaration/, quality, safety and pharmacology, and efficacy of herbal products
United States	Dietary Supplement and Health Education Act (DSHEA)	1994	• Broadly defined herbs and other botanicals as dietary supplements
			• Substantially altered the definitions, standards, and mechanisms under which a claims about safety and efficacy are evaluated and enforced
	Final Rule June 2007	2007	• Establishes current good manufacturing practices in manufacturing, packing, labeling or holding operations for dietary supplements (including herbal products)

product must have a specified strength and dosage and be administered orally, externally, or by inhalation. The data supporting traditional use must be sufficient to show that they are not harmful in the specified conditions of use and the pharmacologic effects are plausible. The herbal product must have been used medicinally and safely for a minimum of 30 years and 15 years within the European Community. The directive explicitly allows nonmedicinal herbal products to be regulated under food legislation [5].

Bibliographic or expert data can be used to establish safe traditional medicinal use of the herbal medicinal product for 30 years and 15 years within the European Community. Efficacy data are replaced by the traditional use; however, a bibliographic review of safety data and an expert report are required similar to those required per Directive 2001/81/EC WEU [6]. Agencies have the ability to request additional safety data for the specific product under review [7]. MAAs are prepared in the common technical document (CTD) format and the HMPC has prepared guidelines for the specific data requirements [8].

Like the traditional herbal medicinal products, homeopathic remedies are exempted from the submission of a traditional MAA [9] and homeopathic products without therapeutic indications are afforded a simplified registration procedure [10]. Homeopathic medicinal products are defined by the Directive as any medicinal product prepared from homeopathic stocks in accordance with a homeopathic manufacturing procedure described by the European Pharmacopeia (Ph. Eur.) or a pharmacopeia of the member state [11]. A homeopathic medicinal product may contain nonherbal starting materials, which are not yet considered for the traditional use directive [3].

The homeopathic regulatory pathway to market approval is a simplified procedure where the products are assessed for their quality and safety [12]. These products are not allowed to have an indication or any therapeutic claims; therefore, they do not require the establishment of efficacy. The product may only be for oral or external use in order to qualify for the simplified procedure. In general, homeopathic medicinal products must be sufficiently dilute to guarantee their safety [13].

Per Directive 2004/24/EC, herbal medicines outside the medicinal legislation are considered as a food supplement [14]. At most, these products may require notification within the member state. These products are mainly controlled by the existing regulations within the member states, but may come under the pan-European food legislation in the future.

A summary of the regulatory pathways for different herbal products is provided in Table 15.3.

The regulations established for herbal drug products have detractors that express concern that the simplified procedure potentially allows herbal products on the market that put patients at risk. The primary concern (one that is similar across all regions) is how does one distinguish when an herbal product is an herbal drug or food supplement (or cosmetic)? This is especially the case for European countries where the majority of herbal products were classified as food supplements. This is a natural state where herbal products have dual use where nutrition/diet uses are parallel to health promotion/therapeutic uses and compounded by the fact that it is a national

TABLE 15.3 Regulatory Pathways Within the European Union for Herbal Products

Herbal product type	Regulatory Pathway	Legal Basis	Quality	Safety	Efficacy
Herbal medicinal product	Market application	Section 8(3) 2001/83/EC	Full dossier	Full dossier	Full dossier
	Market application	Section 10(1)(a) 2001/83/EC WEU	Full dossier	May be bibliographic or combination or primary data and bibliographic for confirmation of well established use	
Traditional herbal medicinal product	Registration	Section 16a 2004/24/EC	Full dossier	Bibliographic data and expert statement	Traditional use (30/15 years)
Homeopathic medicinal product	Registration	Section 14-15 2001/83/EC	Simplified	Not required	

decision on whether the product is a food or drug. This is expected to change with the implementation of the future pan-European food legislation [3]. Others note that traditional practice is not always a reliable tool for the detection of rare or late reactions, and it has limited value in predicting risks associated with nontraditional preparations or with use under nontraditional circumstances (e.g., in combination with conventional drugs). Replacing evidence of efficacy with a demonstration of 30 years of traditional use and 15 years of use within Europe is contested by those who feel that such conventional demonstration of efficacy is an equally important safeguard for patients in addition to safety assessments. Also, the relaxation of the efficacy requirements creates a major disincentive for the industry to perform clinical studies on herbal products [15,16].

The HMPC has prepared guidelines outlining the data requirements for the development and market approval of herbal medicinal products and traditional herbal medicinal products. These guidelines (summarized in Figure 15.2) address quality, safety, efficacy, pharmacovigilance, and organizational matters related to these products and reflect agency expectations and industry input. The EMEA website also includes the original concept/reflection papers for the guidelines, draft guidelines, and overview of comments received for draft guidelines. The inclusion of these additional documents gives additional perspective in the thought process that led to the final guideline.

The Ph. Eur. has incorporated a number of herbal drug related monographs and many correspond to the published HMPC guidelines. While several monographs were adopted prior to the corresponding guidelines, the HMPC took care in assuring that the two do not conflict. The Ph. Eur. listed monographs are the following [18]:

- herbal drugs;
- herbal drug preparations;
- herbal teas;
- herbal drugs: sampling and sample preparation;
- homeopathic preparations;
- determination of Aflatoxin B in herbal drugs;
- determination of tannins in herbal drugs;
- determination of essential oils in herbal drugs;
- herbal drugs for homeopathic preparations;
- extracts;
- microbiological examination of herbal medicinal products for oral use;
- heavy metals in herbal drugs and fatty oils;
- pesticide residues.

Any herbal drug, herbal tea, homeopathic preparation, or extract must comply with these monographs, as well as any applicable individual monograph.

Quality	*Safety*	*Efficacy*	*Pharmacovigilance*	*Organization*
• Declaration of herbal substances and preparations • Quality of products • Specifications • Stability testing • Purification of extracts to be considered herbal preparations • Preparation of herbal teas • Markers used for quantitative and qualitative analysis • Quality of combination products • Use of fumigants • Good agricultural and collection practices for starting materials • Quality	• Environmental risk assessment • Ethanol content • Genotoxicity testing • Risks associated with furocoumarins contained in *Angelica archangelica L.* • Nonclinical documentation for MAAs • Public statement on "CPMP List of Herbal Drugs with Serious Risks"	• Adaptogenic concept • Clinical safety and efficacy assessment • Clinical assessment of fixed combination • CTD format for MAAs	• Separate public statements on products containing: *Cimicifugae racemosae* rhizome, *Chamomilla, asarone*, Aristolochia species, pulegone and menthofuran, Capsicum/capsaicin • Public statement on the allergenic potency of products containing soya or peanut protein	• Use of CTD format in the preparation of MAAs

FIGURE 15.2 EU HMPC Guidelines Approved or in Development for Herbal Medicinal Products and Traditional Herbal Medicinal Products [17].

Another general monograph of importance is the *Substances for Pharmaceutical Use*. The monograph applies to all organic or inorganic substances that are used as active substances or excipients used in the manufacture of medicinal products for human or veterinary use. They may be obtained from natural sources or produced by extraction from raw materials, fermentation or synthesis. This general monograph does not apply to herbal related products; therefore, is a catchall for all other drug products from natural sources. This would include proprietary and generic drugs or herbal related drugs used for indications other than their traditional use [19].

It is also noted that the European regulatory agencies will review proprietary and generic drugs from natural sources against the herbal guidelines and Ph. Eur.

monographs. For example, the *Pesticide Residue* monograph will often be evoked in spite of the fact that drug substances derived from natural sources involve manufacturing processes that are often fairly involved, multistep extraction and purification procedures that are able to achieve a single chemical entity of high purity. Nevertheless, many of regulatory agencies will request evidence of purging of pesticides below certain levels either in-process or in the drug substance.

The guideline on good agriculture and collection practices was incorporated and influenced the similarly issued WHO guideline [20,21]. Up until the publication of this guideline the European regulatory agencies did not regard or enforce good regulatory or collection practices. This is in contrast to the United States where an environmental assessment is required to be included in all new and generic drug applications [22].

The Canadian Government amended their *Food and Drugs Act* in 2003 by enacting *Natural Health Products Regulations* that came into effect in 2004 [23]. The *Regulations* define Natural Health Products, product licensing, site licensing, good manufacturing practices, clinical trials, and labeling and packaging of natural health products. The *Regulations* also established a Natural Health Products Directorate (NHPD) that is charged with the ready access to natural health products (NHPs) for Canadians that are safe, effective and of high quality, while respecting freedom of choice and philosophical and cultural diversity. The NHPD reviews and approves product license applications and clinical trial applications, and prepares related guidelines. The NHPD also maintains a list of approved NHPs and approved sites for the manufacture, packaging, and labeling of NHPs. The NHPD has developed the Compendium of Monographs as a tool for the timely and efficient evaluation of the safety and efficacy of many commonly used medicinal ingredients that comprise NHPs. The NHPD has also compiled a list of acceptable nonmedicinal ingredients that may be used in finished NHPs [24].

According to the *Regulations* a natural health product is a substance listed within the *Regulations* and used in

1. diagnosis, treatment, mitigation, or prevention of a disease, disorder, or abnormal physical state or its symptoms in humans;
2. restoring or correcting organic functions in humans; or
3. modifying organic functions in humans, such as modifying those functions in a manner that maintains or promotes health.

Plant, plant material, alga, bacterium, and fungus are all listed as acceptable NHP substances. An extract or isolate from these substances is also considered an acceptable substance. NHPs may not be administered by puncturing the skin; therefore, the only acceptable routes of administration of NHPs are oral, external, or rectally [24].

The NHPD has developed a market application format specific for NHPs that are distinct from traditional drug applications and do not follow the CTD format [25]. The NHPD has also issued guidelines for the data requirements for quality, safety, and efficacy of NHPs [26,27]. Guidance is also provided that outlines the data requirements for a homeopathic medicine [28]. Once a product license is obtained the NHP is subject to manufacturing, packaging, labeling, and import for sale components of the *Regulations*.

Manufacturing is defined as a facility that fabricates or processes the NHP, but does not include growers or harvesters of the raw materials. Entities that are not considered manufacturers by the *Regulation* include health-care practitioners that compound products at the request of patients; complementary and alternative health care practitioners; and traditional aboriginal medicinal practitioners. Policies for NHP compounding are forthcoming. Site licenses must be obtained for facilities involved in the manufacture, packaging, and labeling of NHPs. Testing laboratories do not require site licenses, but the *Regulations* recommend that the testing facilities comply with good laboratory practices. Imported NHPs must also have an approved product license and be manufactured, packaged, and labeled within facilities that have a site license [24].

Manufacturing of NHPs (domestic and foreign) must comply with good manufacturing practices (GMPs) as outlined in the *Regulations*. The GMP provisions include requirements for product specifications, premises, equipment, personnel, sanitation, operations, quality assurance, product stability, record keeping, product sterility, ophthalmic use, batch sample, and recall reporting. The NHPD has issued a GMP guidelines specific for NHPs. Manufacturing sites are subject to inspection by the HFPB [29].

Clinical trials conducted within Canada that involve NHPs must be approved by the NHPD. Clinical trials are considered those involving human subjects and intended to discover or verify the NHP's clinical, pharmacological, or pharmacodynamic effects, identify any adverse events, study absorption, distribution, metabolism and excretion (ADME), and ascertain its safety and efficacy. The clinical trial sponsor is responsible for applying for clinical trial approval from NHPD. The *Regulations* outlines the obligations for the clinical trial sponsor, which includes good clinical practices (GCPs), product labeling, record keeping, information, and sample submission and adverse event reporting. Manufacturers of clinical supplies do not require a site license [24,30].

Health Canada's adverse reaction reporting system has established requirements for monitoring and reporting adverse reactions associated with the use of natural products. Product license holders are required to report serious adverse reactions and any serious and unexpected adverse reactions to their product. Likewise, clinical trial sponsors are also required to report adverse reactions and any serious unexpected adverse reactions to their investigational product [31].

In Australia, herbal medicinal products are considered "complementary medicines" along with other traditional alternative and traditional medicines as described in Figure 15.3. Complementary medicines are regulated as medicines under the *Therapeutic Goods Act 1989* [32]. The *Act* and the *Therapeutic Goods Regulations 1990* define complementary medicines as therapeutic goods consisting wholly or principally of one or more designated active ingredients each of which has a clearly established identity and a traditional use. Traditional use means use of the designated

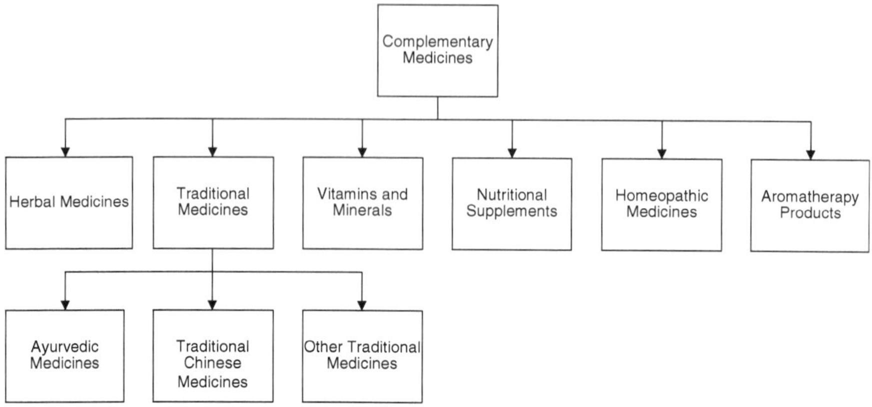

FIGURE 15.3 Australian classes of complementary medicines [34].

active ingredient that is well documented, or otherwise established, according to the accumulated experience of many traditional healthcare practitioners or an established extended period; and accords with well-established procedures of preparation, application, and dosage. Like all other medicinal products, complementary medicines must be listed in the Australian Register of Therapeutic Goods (ARTG) unless specifically exempt or excluded. A therapeutic article cannot be listed in the ARTG without the approval of an application [33].

Given the confusion and uncertainty of the regulatory process for complementary medicines, the Therapeutic Goods Administration (TGA) has developed the Australian Regulatory Guidelines for Complementary Medicines (ARGCM) to assist the industry in meeting their legislative obligations. A summary of the guidelines is provided in Table 15.4.

The premarket application is similar to the European Union requirements for the format (CTD), and type and quality of data. In fact, these guidelines reference the EU herbal medicinal product guidelines quite liberally. The guidelines also acknowledge that the bulk of the safety and efficacy data will be bibliographic. Similar to the European Union and Canada, the premarket application assessment is risk-based and assesses a number of factors including:

- the toxicity of the ingredients;
- the dosage form;
- whether the medicine is indicated for a serious form of a disease, condition or disorder, or for the treatment, cure, management or prevention of disease, condition or disorder;
- whether the use of the medicine is likely to result in significant side effects, including interactions with other medicines; and
- whether there may be adverse effects from prolonged use or inappropriate self-medication.

TABLE 15.4 Summary of Australian Regulatory Guidelines for Complementary Medicines [35]

Guideline	Description	Content
Part I	Provides guidance on the registration of complementary medicines	• Eligibility for registration • Route of evaluation • Registration process • Administrative information • Quality data requirements • Efficacy and safety data requirements • Labeling and presentation • Product information • Consumer medicine information • Changes to registered complementary medicines • Post market review • Stability of finished product
Part II	Provides guidance on listed complementary medicines	• Eligibility for listing • Listing process • Registration process • Electronic listing facility • Quality • Efficacy and safety • Labeling and presentation • Post market review • Product changes • Other requirements
Part III	Provides guidance on the evaluation of complementary substances for use in listed medicines including new substances, "switch substances" and excipients used in complementary medicines	• Eligibility • Evaluation process • Administrative information • Quality • Safety • Appendices: Example format, impurities and incidental constituents, and safety data requirements and suitability
Part IV	Provides general guidance in relation to complementary medicine modalitics such as homeopathy, traditional herbal medicine and aromatherapy. This part also provides information on exempt medicines, combination complementary/ pharmaceutical medicines and the food/medicine interface with glossary of terms	• Homeopathic preparations and mother tinctures • Traditional herbal medicines • Aromatherapy • Practitioner products • Exempt/excluded goods • Proprietary ingredients • Interface issues • Australian native and endangered species in therapeutic goods • Other legislation • Certificate of analysis – product

(continued)

TABLE 15.4 (*Continued*)

Guideline	Description	Content
		• Product specifications • Enforcement procedures • Review of decisions • Genetically modified organisms • Medicines for export • Naming new substances and terminology • Coloring permitted in medicines for oral use • Herbal ingredients – quality • Ingredients of human or animal origin • Glossary of terms, abbreviations and acronyms, and hyperlink references
Part V	Provides details of TGA policy guideline relevant to complementary medicines	• Levels and kinds of evidence to support indications and claims • Minimizing the risk of TSEs • Quantified by input • Referenced EU guidelines • Section 7 declarations • Confidentiality

Medicines that are assessed as higher risk are individually evaluated for quality, safety, and efficacy. Higher risk products approved by TGA are included on the ARTG as registered medicines. Efficacy is usually assessed by examining data from controlled clinical trials, but can be obtained also from literature reports [36,37].

Listed medicines are low risk medicines and are included on the ARTG via low-cost and streamlined electronic application and validation process. Listed medicines may only contain ingredients that have been evaluated by the TGA to be low risk, must be manufactured by licensed manufacturers in accordance with GMPs, and may be indicated for health maintenance and enhancement, or for nonserious self-limiting conditions. Listed medicines may be supplied only if they contain active ingredients permitted under Schedule 4 of the *Regulations*. Adverse drug reaction reporting follows the well-established system within Australia and sponsors of complementary medicines are obligated to report adverse reactions to the TGA [38].

Manufacturing sites of complementary medicines are required to hold a license and manufacture them in accordance with GMPs, which are consistent with those established for traditional drugs [32].

In contrast to the European Union, Canada, and Australia, the United States has very weak regulatory requirements for herbal medicinal products, which fall under the Dietary Supplement Health and Education Act (DSHEA) of 1994. A dietary supplement is defined as a product taken by mouth that contains a "dietary ingredient" intended to supplement the diet. The "dietary ingredients" in these products may include herbs or other botanicals. Dietary supplements can also be extracts or concentrates, and may be found in many forms such as tablets, capsules, soft gels, gel caps, liquids, or powders. They can also be in other forms, such as a bar, but if they are, information on their label must not represent the product as a conventional food or a sole item of a meal or diet. Whatever their form may be, DSHEA places dietary supplements in a special category under the general umbrella of "foods," not drugs, and requires that every supplement be labeled a dietary supplement [39].

Under DSHEA, a company is responsible for determining that the dietary supplements it manufactures or distributes are safe and that any representations or claims made about them are substantiated by adequate evidence to show that they are not false or misleading. Dietary supplements do not require a market application or otherwise receive a market approval. Except in the case of a new dietary ingredient, where premarket review for safety data and other information is required by law, the company does not have to provide FDA with the evidence it relies on to substantiate safety or effectiveness before or after it markets its products. FDA is not required to maintain a list of dietary acceptable dietary ingredients. Dietary supplements are significantly restricted in terms of efficacy and safety claims [39].

Because dietary supplements are under the "umbrella" of foods, FDA's Center for Food Safety, and Applied Nutrition (CFSAN) is responsible for the agency's oversight of these products. FDA's efforts to monitor the marketplace for potential illegal products (that is, products that may be unsafe or make false or misleading claims) include obtaining information from inspections of dietary supplement manufacturers and distributors, the Internet, consumer and trade complaints, occasional laboratory analyses of selected products, and adverse events associated with the use of supplements that are reported to the agency. In June 2007, FDA issued a *Federal Register Final Rule* that requires dietary supplements are manufactured in accordance with GMPs. The *Final Rule* outlines the GMP requirements specific for dietary supplements [40].

The United States Pharmacopeia (USP) has issued several additional monographs for articles of botanical origin; however, only two specifically apply to dietary supplements. The first is USP <2750>, which is the incorporation of the GMP requirements for dietary supplements consistent with the *Final Rule*. The second is USP <563> (*Identification of Articles of Botanical Origin*), which, in part, outlines the procedures for identifying source plant materials for botanical products.

The following three additional monographs apply to botanical products.

- USP <561> *Articles of Botanical Origin*
- USP <565> *Botanical Extracts*
- USP <2030> *Supplemental Information for Articles of Botanical Origin*

While these monographs do not specifically state that they apply to dietary supplements, they do apply to herbal products that are subject to USP monographs. They also apply to drug products derived from botanical sources including new and generic drugs [41].

There has been no attempt to harmonize regulatory processes for herbal drugs used for traditional indications amongst these geographic regions. A sponsor developing a traditional herbal drug intended for market in these regions would need to consult individual requirements. The regulatory processes are fairly accommodating for traditional herbal products that are well established with published monographs, but can become complex and involved for herbal products that are not traditional and have questionable safety and efficacy. A summary of the regional characteristics for herbal products is presented in Table 15.5.

While FDA has weak oversight of dietary supplements, the agency has made a concerted effort to outline regulatory pathways for botanical products intended to be marketed as a new drug or as an over-the-counter (OTC) drug. These botanical products include herbal drugs that are dietary supplements intended for new indications or extracts from herbal products. FDA has also established a review group and guidance for botanical drug products [42]. FDA has issued its first approval for a botanical product Veregen® in 2006 [43]. FDA did not develop any new regulations, but applied existing regulations for botanical drug development. Recognizing and developing a regulatory pathway for botanical drug products is unique to the United States and has not been attempted by other regulatory agencies [44].

FDA defines a botanical drug as one that is used for the diagnosis, cure, mitigates, treats, or prevents a disease; or is used to affect the structure or function of the body. The botanical drug contains ingredients that include individually or a combination of plant materials, algae, or macroscopic fungi. The product may be a solution, powder, tablet, capsule, elixir, topical, injectable, or other recognized dosage forms. Excluded from botanical products are fermentation products, highly purified, or chemically modified botanical substances, allergenic extracts, and vaccines containing botanical ingredients [44].

The most significant challenge for botanical product development in the United States is discerning whether the product is a food, cosmetic, dietary supplement, or drug. The decision tree presented in Figure 15.4 outlines the different attributes of the product in determining its category.

FDA Guidance *Botanical Drug Products* helps industry sponsors understand the data requirements for OTC products and new drug applications (NDAs). The Guidance also outlines data requirements for investigational new drug (IND) applications during different phases of development (Table 15.6). Key to the Guidance is the recognition that bibliographic data can be used for the establishment of safety during development of a botanical drug. The guidance also provides very detailed chemistry, CMC data requirements for botanical products [44].

The botanical guidance delineates IND data requirements for marketed and nonmarketed botanical drug products intended for Phase I and II clinical studies (Table). The IND data requirements for botanical drug products intended for Phase 3 clinical studies are considerably more detailed and consistent with new chemical entities. The exception is chemistry, manufacturing and controls, which is consistent with the data for Phase I and II data requirements, but with more control over the botanical raw materials. The guidance requires more details on the growing conditions, harvesting, and collection of botanical raw materials (e.g., pesticide use,

TABLE 15.5 Summary of Regulatory Requirements for Herbal Medicinal Products

	Australia	Canada	European Union	United States
Definition	Complementary Medicines	Herbal Medicinal Products	Herbal medicinal products and traditional herbal medicinal products	Dietary supplements
Market application	Pre-Market application	Product license	MAA (herbal medicinal products) or registration (traditional herbal medicinal products and homeopathic medicinal products)	Safety data required to be submitted for new dietary supplements, but no formal review procedure
Format	CTD	Form (non-CTD)	CTD (MAA) and non-CTD (registration)	Not specified
Quality data	Required	Required	Required	Not required
Safety data	Required (may be bibliographic)	Required (based on established monographs)	Required (may be bibliographic)	For new dietary supplements only
Efficacy data	Case-by-case basis	Required (based on established monographs)	Required for herbal medicinal products, but not for traditional herbal medicinal products	Not required
Pharmacovigilance	Yes	Yes	Yes	Yes
Manufacturing in accordance with GMPs	Yes	Yes (NHP-specific)	Yes	Yes (dietary supplement specific)
Labeling	Controlled for safety and efficacy claims	Controlled for safety and efficacy claims	Controlled for safety and efficacy claims	Significantly limited for safety and efficacy claims

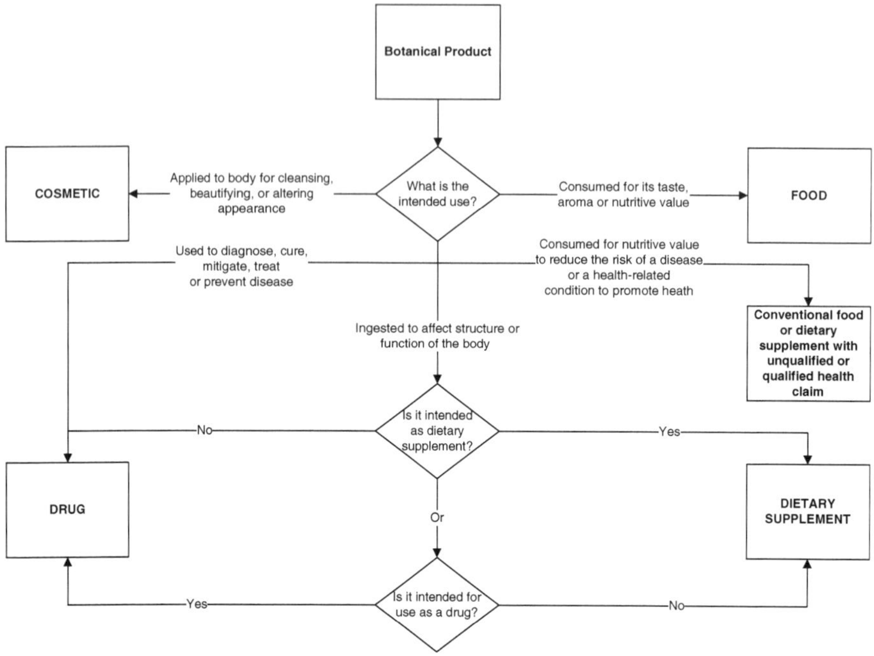

FIGURE 15.4 Development pathways for botanical products in the United States [45].

harvest locations, harvest time, plant growth at harvest, and processing of the raw materials). It is expected at Phase III that a spectroscopic and/or chromatographic fingerprint of each botanical raw material is obtained along with the chemical identity of the active constituents (or characteristic markers). Quality control tests of the raw material include chemical identification and assay of active constituents (or biological assay), heavy metals, microbial limits, residual pesticides, advantageous toxins, and foreign materials and adulterants. Additional analytical controls are also expected for drug substance and drug product.

Nonclinical safety assessments and Phase III clinical data are consistent with those required for new chemical entities. Previous human data and data from foreign clinical studies will be considered, but not expected to be sufficient by themselves to support a market application.

The environmental assessment is a key factor in determining the source and cultivation of the botanical raw materials. Unlike the majority of the synthetic compounds, herbal drugs may not qualify for a categorical exclusion from an environmental assessment. This is even the case for marketed dietary supplements that do not require environmental assessments as they are not subject to drug regulations. An environmental assessment will be required if the botanical raw materials contain a species that is listed by the Convention on International Trade in Endangered Species of Wild Fauna and Flora (CITES) or is harvested from wild specimens. FDA will review the environmental assessment to assess whether an environmental impact

TABLE 15.6 US IND Data Requirements for Marketed and Nonmarketed Botanical Products (Phase I and II Clinical Studies)

Development phase	Marketed Botanical Products	Nonmarketed Botanical Product
Chemistry, manufacturing and controls	*Drug substance*: the general method of preparation (e.g., pulverization, decoction, expression, aqueous extraction, or ethanolic extraction)	*Botanical raw material*: • identification of source material • certificate of authenticity • list of growers/suppliers and • harvesting an collection practices
	Drug product • A qualitative description of the drug product including the dosage form, route of administration, names of all ingredients, statement that the product is not adulterated with potent, toxic, or addictive botanical substances, synthetic or highly purified drugs, biotechnology-derived drugs, or other naturally derived drugs	*Drug substance* • qualitative description • quantitative description • name and address of the manufacturer • description of the manufacturing process quality • control testing container label
	• The composition or quantitative description of the finished product expressed in terms of amount per dosage unit. • Manufacturer's certificate of analysis for the clinical study product (or by cross-reference to manufacturer's application).	*Drug product:* same as for marketed product and • name and address of manufacturer • description of the manufacturing process • quality control tests description of container/closure • available stability data
Animal safety test	Animal safety test performed for crude extracts from natural sources	Not applicable
Placebo	Description of placebo components	Description of placebo components
Labeling	Copy of the container label and the immediate outer carton label of the marketed product Mock or printed representation of the proposed container label that will be provided to clinical investigators	Same as for marketed product and a quantitative description of the drug substance per dosage unit
Environmental assessment	Environmental assessment or claim of categorical exclusion	Environmental assessment or claim of categorical exclusion

(continued)

TABLE 15.6 (*Continued*)

Development phase	Marketed Botanical Products	Nonmarketed Botanical Product
Pharmacology and toxicology	Previous human experience and available animal toxicology data concerning the clinical formulation and the individual botanical ingredients within the formulation. A database search should be conducted, when feasible, to identify information relevant to the safety and effectiveness of the final formulation, individual botanical ingredients, and the known chemical constituents.	Consistent with new drugs, but reduced studies may be warranted for botanical products that are otherwise well established in other recognized markets or by compendia.
Bioavailability	Available PK/PD data	Available PK/PD data

assessment is warranted. The use of cultivated species not listed in CITES will likely qualify for a categorical exclusion from an environmental assessment [46].

FDA has reported an increase in IND applications over the last 20 years with 5–10 applications annually from 1990 to 1998; an average of 22 applications annually between 1999 and 2002; and approximately 40 applications annually from 2003 to 2007. A total of 282 applications (pre-IND and IND) has been submitted between 1999 and 2007 (see Figure 15.5). The majority of applications is submitted for cancer

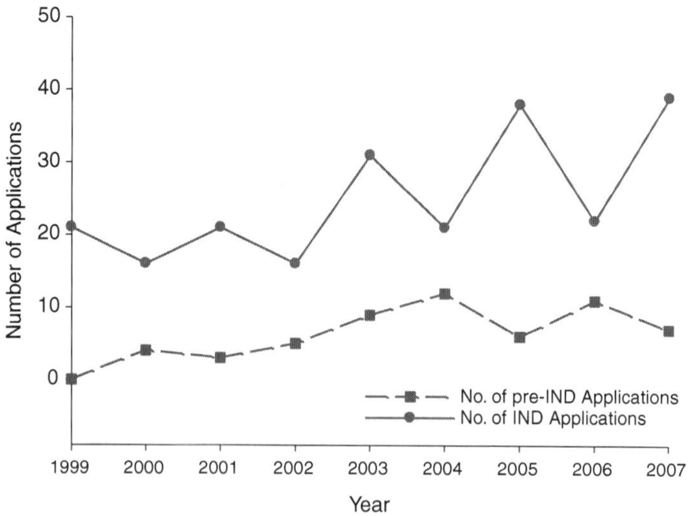

FIGURE 15.5 Number of botanical pre-INDs and IND applications submitted to FDA from 1999–2007 [43].

and related indications, followed by inflammatory and pain disorders, endocrine and metabolic diseases, viral infections, and dermatological and dental indications. Of the 282 applications, only 36% are submitted for multiple-plant combinations, which further demonstrate the difficulty in working with complex plant preparations. Forty percent of the applications have been submitted by industry sponsors while the remaining 60% were submitted by academic investigators [43].

FDA reported that the majority of the IND application relied upon previous human experience for safety and waived the need for additional safety studies. For the 225 INDs submitted between 1999 and 2007, 37 (9%) had their clinical studies placed on clinical hold for one or more serious deficiencies that could not be resolved. These applications were resolved, withdrawn by the sponsor, or remained on clinical hold. Common serious deficiencies were inadequate previous human data and/or existing animal toxicity data that were inadequate to support a clinical study; insufficient characterization of the botanical materials, drug substance or drug products; ingredients in the botanical drug product with potentially serious toxicity that had not been adequately tested in animal safety studies; quality issues such as contamination or adulteration that compromised the safety and integrity of the study; and problems with the clinical trial design, such as inadequate controls, protracted study duration not supported by prior clinical experience or animal data, and inappropriately defined trial subjects [43].

New indications for herbal drug products would be treated similarly within Australia, Canada, and the European Union. While not codified in guidance, the FDA approach to botanical drug development and utility of previous human experience are likely to be universally accepted. Therefore, one can expect a certain level of harmony in botanical drug development, but one should regard and reconcile regional differences.

15.5 CONCLUSIONS

In this chapter, we have discussed the business side of phytotherapeutics – how to secure ownership of specific products via various intellectual property strategies, current markets are and future ones and importantly, the regulatory requirements of various major market territories that lead to marketing approval with specific health benefit claims. Botanical preparations have served as the primary source of health care products for humankind since our beginnings. Even today a majority of the world's population still relies upon such materials for its primary health care. In the West, developed economies' reliance upon botanical preparations declined in importance as the pharmaceutical industry was able to bring specific single chemical entity drugs to market: these were shown to be clinically effective to treat specific disease conditions. Success in treating infectious diseases with single chemical entity antibiotics was particularly noteworthy. However, as drug resistance has become much more prevalent together with so-called life-style diseases creating a major health care issue, the value of phytotherapeutics as synergistic mixtures of active compounds is again recognized. It is only through the capture of intellectual property ownership and the diligence to gain regulatory marketing approval that phytotherapeutics useful to meet human health care needs will be available to the world's people. We hope this chapter will be of assistance to those who are working to bring this about.

REFERENCES

[1] Directive 2001/83/EC of the European Parliament and of the Council of 6 November 2001 on the Community code relating to medicinal products for human use (Official Journal of the European Communities; L311/67, 28.11.2001). Directive 2004/24/EC of the European Parliament and of the Council of 31 March 2004 amending, as regards traditional herbal medicinal products, Directive 2001/83/EC on the Community code relating to medicinal products for human use.

[2] Section 10(1)(ii) 2001/83/EC.

[3] Peschel, W. (2007)., The Traditional Herbal Medicine Directive within the European Regulatory Framework for Herbal Products. *Boletín Latinoamericano y del Caribe de Plantas Medicinales y Aromáticas*, *6*, 102–111.

[4] Directive 2004/24/EC of the European Parliament and of the Council of 31 March 2004 amending, as regards traditional herbal medicinal products, Directive 2001/83/EC on the Community code relating to medicinal products for human use.

[5] Section 16a 2004/24/EC.

[6] Section 16c(1)(c) 2004/24/EC.

[7] Section 16c(1)(d) 2004/24/EC.

[8] Guideline on the use of the CTD format in the Preparation of a Registration Application for Traditional Herbal Medicinal Products (EMEA/HMPC/71049/2007).

[9] Section 13–16 2001/83/EC.

[10] Section 14(1) 2001/83/EC.

[11] Section 1(c)(5) 2004/27/EC.

[12] Section 13(2) 2004/27/EC.

[13] Section 14(1) 2001/83/EC.

[14] 2004/24/EC.

[15] De Smet, P.A. (2005). Herbal medicines in Europe – relaxing regulatory standards. *New England Journal of Medicine*, *352*, 1176–1178.

[16] Dawson, W. (2005). Herbal medicines and the EU Directive. *Journal of the Royal College of Physicians of Edinburgh*, *35*, 25–27.

[17] The guidelines, concept papers, reflection papers and draft guidelines are provided on: http://www.ema.europa.eu/htms/human/hmpc/hmpcguide.htm.

[18] Ph. Eur. Sixth Edition, Volume 6.0.

[19] Ph. Eur. 07/2009:2034.

[20] Guideline on Good Agricultural and Collection Practice (GACP) for Starting Materials of Herbal Original (EMEA/HMPC/246816/2005).

[21] WHO Guideline on Good Agriculture and Collection Practices (GACP) for Medicinal Plants (World Health Organization, Geneva, 2003).

[22] Guidance for Industry: Environmental Assessment of Human Drugs and Biologics Applications (FDA, CDER/CBER, July 1998).

[23] Natural Health Products Regulations (P.C. 2003–847), Canada Gazette, Vol. 137, No. 13, June 2003.

[24] Overview of the Natural Health Products Regulations Guidance Document (NHPD, V 2.1, November 2003).

[25] Product License Form (V 2.0). Available at http://hc-sc.gc.ca/dhp-mps/alt_formats/hpfb-dgpsa/pdf/prodnatur/form_pl-dlmm-eng.pdf.

[26] Evidence for Quality of Finished Natural Health Products (NHPD, V 2.0, June 2007).

[27] Evidence for Safety and Efficacy of Finished Natural Health Products (NHPD, V 2.0, December 2006).

[28] Evidence for Homeopathic Medicines (NHPD, V 2.1, May 2007).

[29] Natural Health Products Compliance Guide (NHPD, V 2.1, January 2007).

[30] Clinical Trials for Natural Health Products (NHPD, October 2005).

[31] Adverse Reaction Reporting for Health Care Providers and Consumers (NHPD, January 2004).

[32] Therapeutics Goods Administration 1989 (SLI 2007 No. 161).

[33] Therapeutics Goods Regulations 1990 (July 2007).

[34] The regulation of complementary medicines in Australia – an overview. Available at http://www.tga.gov.au/cm/cmreg-aust.htm.

[35] Australian regulatory guidelines for complementary medicines (ARGCM). Available at http://www.tga.gov.au/docs/html/argcm.htm.

[36] Australian Regulatory Guidelines for Complementary Medicines (ARGCM), Part I: Registration of Complementary (June 2005).

[37] Australian Regulatory Guidelines for Complementary Medicines (ARGCM), Part III: Evaluation of Complementary Medicine Substances for use in Listed Medicines.

[38] Australian Regulatory Guidelines for Complementary Medicines (ARGCM), Part II: Listed Complementary Medicines.

[39] Dietary Supplement Health and Education Act of 1994.

[40] Guidance for Industry: Questions and Answers Regarding Adverse Event Reporting and Recordkeeping for Dietary Supplements as Required by the Dietary Supplement and Nonprescription Drug Consumer Protection Act (June 2009).

[41] USP 32 – NF 27.

[42] Available at http://www.fda.gov/AboutFDA/CentersOffices/CDER/ucm090946.htm

[43] Chen, S.T., Dou, J., Temple, R., Agarwal, R., Wu, K.M., Walker, S. (2008). New therapies from old medicines. *Nature Biotechnology*, *26*, 1077–1083.

[44] Guidance for Industry: Botanical Drug Products (CDER, June 2004).

[45] Vaccari, L.Current Regulation of Botanical Drugs in the United States (DIA 42nd Annual Meeting, June 2006).

[46] Guidance for Industry: Environmental Assessment of Human Drug and Biologic Applications (CDER/CBER, July 1998).

16

COOPERATION BETWEEN THE PHARMACEUTICAL INDUSTRY AND ACADEMIC INSTITUTIONS IN DRUG DISCOVERY

VALDIR CECHINEL-FILHO, RIVALDO NIERO, AND ROSENDO A. YUNES

16.1 INTRODUCTION

In general, research and development for the creation of new, effective, and safe drugs from natural or synthetic sources has to go through a long process, from the discovery of an active extract or pure substance to its becoming available in the market. An important point that should be emphasized is the need for heavy investments in human resources, equipment, and other expenses. While the biggest industries, generally located in the developed countries, have sufficient resources of their own to carry out this complex process, the same is not true of medium- or small-sized pharmaceutical companies, which are mainly located in less developed countries. As a solution to this problem, they often work in partnership and cooperation with research centers, usually belonging to universities, and both parties sometimes receive financial support from the government sector. This form of cooperation has shown innovative and promising results, some of which will be outlined in this chapter, together with the main advantages and difficulties relating to this form of interaction between universities and industry. It also describes the current situation of the global pharmaceutical market, with particular emphasis on phytotherapeutics, and presents the experience of Brazil – which has one of the richest biodiversity on the planet – in the development of its first phytotherapeutic drug, Acheflan.

Plant Bioactives and Drug Discovery: Principles, Practice, and Perspectives, Fourth Edition.
Edited by Valdir Cechinel-Filho.

16.2 INTERACTION BETWEEN ACADEMIC INSTITUTIONS AND THE PHARMACEUTICAL INDUSTRY

While the main role of universities is to produce and transmit knowledge to society, industry focuses its attention, in theory, on producing new and innovative products that will improve people's quality of life. In the pharmaceutical industry, interaction between universities and industry has increased in recent years, especially between smaller pharmaceutical companies, generally located in less developed countries. This interface seems to be an important strategy for meeting the economic needs of society and of an ever-weakening government. The challenges of promoting and strengthening these partnerships, for the purposes of bringing a new drug to market, involve various stages and activities, covering such matters as intellectual property rights, contract and/or agreement research, financial sources, etc. [1].

It is important to see this relationship from a perspective of the "Triple helix" concept, which was defined by a group of executives in Silicon Valley (USA) as: (1) a more prominent role for the university in innovation, on a par with industry and government in a knowledge-based society; (2) a movement toward collaborative relationships between the three major institutional spheres, in which innovative policy is increasingly an outcome of interaction rather than something that is prescribed by the government; (3) in addition to fulfilling their traditional functions, each institutional sphere also "takes on the role of the other" performing new roles beyond their traditional function. Institutions taking on nontraditional roles are viewed as a major potential source of innovation.

It is also important to observe that the economically underdeveloped condition of our region generates a "triple helix" that is different from that seen in developed countries.

Another important fact to consider is that despite their many failings, the public universities are the main domain of scientific and technological research in Latin America.

Due to several factors, including synergistic effects, the association of different mechanisms of action, molecular diversity, less collateral effects, the lower cost of research and development, credibility, etc., the search for phytotherapic drugs, in recent years, has grown more quickly than the search for synthetic drugs. This situation has led to a greater need for financial investments and heavier demands on time [2,3]. Brazil has a competitive advantage in this field due to its rich flora and biodiversity, giving opportunities for better and more rational use by pharmaceutical industries, for the development of plant-derived medicines [4,5]. The worldwide biodiversity, particularly the flora, has proven to be an important and strategic source of medicinal agents, and it is estimated that around 25% to 50% of currently marketed drugs owe their origins to natural products [6,7].

Singer and coworkers [8,9] explore the health innovation systems of some developing nations, giving special emphasis to Brazil, and demonstrate the considerable progress in this area in recent years. They also provide an in-depth diagnosis of the cooperation between universities and industry in the search for new drugs.

Although both industry and the academic stand to gain considerable advantages from this collaboration, which in theory should be reflected in society in general, there are many obstacles to a more fruitful interaction in the search for new, safe, effective and innovative drugs. Calixto and Siqueira Jr. [10] have pointed out the main barriers to the development of new drugs in Brazil, but these may equally apply to other countries:

- high risk and cost of developing new traditional drugs;
- high financial cost and little possibility of risk capital;
- long time to maturation of research and development for the projects;
- absence of formal research and development divisions in the industrial sector;
- incorporation of national companies by multi- or transnational ones;
- lack of experience in technological innovation;
- lack of researchers in industry.

Other related factors that hinder the relationship between universities and industry in the development of phytomedicines are presented by Rodrigues [11]:

- resistance of some researchers to developing projects with private companies;
- small number of industries investing in this area;
- barriers posed by the current regulatory process in terms of access to genetic resources and associated traditional knowledge.

In addition to these factors, several other important topics can be included: (1) we cannot ignore the fact that economic theorists and historians have claimed that inequality is the obstacle to prosperity and the cause of the lack of scientific and technological development [12,13] in the Latin American countries, especially Brazil, Bolivia, and Honduras, social and economic inequality are rife; (2) the economic and political elites that use large amounts of regulation to sustain their economic advantage over the rest of society – even at the cost of hindering economic growth and innovation [14]; (3) the centrality of the role of the State.

In the mid-1970s, the Codetec (Companhia de Desenvolvimento Tecnológico) was set up in Campinas (SP). Inspired by the model of Stanford University, it was linked to the University of Campinas, with the purpose of working with industry in the development of technologies. The Codetec enabled approximately 1100 mainly small- and medium-sized pharmaceutical companies, to gain commercial protection of the Federal Government for their products. However, in 1979 Unicamp suffered a major financial crisis that forced it to reduce the number of researchers, and in 1990, the Federal Government opened up foreign trade completely, leaving companies without any protection. Thus, within a few years, all those companies that were unable to compete in the international market were forced to close. In Korea, the role of the State was completely different; it created conglomerates of industries in important and dynamic areas, such as communications, television, information technology, the automobile industry, etc., protecting them until they were fully developed, then gradually and slowly subjecting them to international competition, and finally (4) the so-called paper culture, established in the academic environment, which has also hindered interaction between universities and industry. Teachers and researchers are more focused on producing and disseminating/publishing knowledge, and contributing to the training of human resources (graduate and postgraduate students), because these are important aspects for improving their curricula and thus guaranteeing approval for new projects and promoting their careers. However, in some countries, such as Brazil, this situation has changed in recent years, and the innovative activities carried out by researchers have been highly regarded and different mechanisms for encouraging work in this field have been created. For example, the Brazilian government has supported major mechanisms of incentive for the process of innovation and the subsequent transfer of knowledge and technologies to companies, with the aim of bringing new products to market. These mechanisms include the Law of Technological Innovation [No. 10,973 (02/12/2004)] and support for different organizations, such as CNPq, CAPES, FINEP, etc. Furthermore, every Brazilian State has a Foundation for Research and Innovation, or similar body, which despite their many failings, support and encourage research, development and innovation, with mechanisms for building partnerships between universities and industry.

All these factors have contributed to lessening the difference between scientific production (number of publications) and innovation (number of patents). Brazil occupied an eminent position in the worldwide ranking of scientific production (13th position) in 2010, according to the Information Sciences Institute (ISI), and is now ranked higher than some countries that have traditionally invested heavily in scientific research, such as the Netherlands, Sweden, and Switzerland. The recent Scimago Iberoamerican Ranking (SIR, 2011) placed Brazil in first position among the Iberoamerican countries in terms of the volume of scientific production published in the Scopus database. Meanwhile, The Global Innovation Index indicates that Brazil rose from 63rd position in the 2010 ranking, to 47th in 2011 ranking. In marked contrast, Swiss and Sweden are in first and second positions, respectively, and smaller countries, such as Chile and Costa Rica, occupy a higher position than Brazil in the innovation ranking. These data suggest that although Brazil has a well-consolidated and important position in the production of knowledge, it needs to improve its innovative processes. However, the above-mentioned mechanisms for strengthening cooperation between universities and industry have shown positive and promising results, confirmed by the 13 positions that Brazil rose in the ranking. But the challenges are still very great, especially due to the points already indicated, and another situation – the "brain drain" of people moving to developed countries. The vast majority of Brazilian scientists (about 70%) are in the universities, and an important strategy for retaining them here is to change this situation through more interaction between universities and industry in general. Unfortunately, this is a reality that affects many developing countries.

Leydesdorff and Etzkowitz [15] give the example of Angela Uller, Director of COPPETEC at the Federal University of Rio de Janeiro, who explained how the unit has become involved in community work. Residents of low-income slums are invited to the university to help organize cooperatives, playing a new role in the formation of low-tech service businesses. Using the interface skills gained by linking academic researchers to industry, the university brings these resources to bear in a creative way to address social problems. It is clear that the relations of university, industry and government will have great influence in the economic and social use of knowledge on the continent, and consequently, in the future of its inhabitants.

Partnership between universities and industry generates advantages, and tangible and intangible gains for both parties, and consequently, for society and the government, particularly in the area of pharmaceuticals. The advantages for the academic world are

- royalties paid to the university and researchers;
- improvements in infrastructure (laboratories, acquisition of modern equipment, etc.);
- increase in the number of researchers;
- facilities for certifying laboratories (good laboratory practices);
- spread and transfer of technologies;
- background in innovative processes;
- visibility and strengthening of university brand marketing;
- strengthening national and international relationships between scientists and universities;
- contribution to the socioeconomic development of the country;
- improvement in teaching quality;
- scholarships for graduate and postgraduate students.

The last point deserves to be better emphasized. The scholarships made possible by innovative projects between universities and industry have enabled many students to attend university, particularly private institutions, contributing to the real possibility of professional training in strategic areas, for the development of a nation.

Perhaps the main advantages for pharmaceutical companies forming partnerships with universities are significant decrease in the cost of drug research (infrastructure) and human resources, generating more competitive prices, and better financial conditions for focusing on marketing. Another interesting advantage is the accessibility of government funding. In her polemic book *The Truth About the Drug Companies*, Angell [16], an expert and highly regarded researcher and teacher at Harvard and former Editor-in-Chief of the *New England Journal of Medicine* warns of the reality and unethical means and strategies used by some power pharmaceutical industries. Concerning the partnership between universities and industry in the pharmaceutical area, she questions cooperation between public institutions and private companies. Knowledge is generated in universities through preliminary studies, and industry puts the ideas into practice, with more detailed and complex studies, such as clinical trials, after which it is transformed into a new and effective drug. This should be the natural process, but sometimes industry demands much more, looking for more profit, paying less taxes, etc., damaging the whole commodity chain and harming the people who use the drugs.

Since this book presents a specific topic relating to intellectual property, we recommend two papers as complementary reading: (1) a paper published by Jannuzzi et al. [17] that emphasizes the specific characteristics of patent protection in the pharmaceutical industry in Brazil, discussing patentable and nonpatentable inventions and the modalities and traits of patent protection and (2) a paper published by Nogueira et al. [18], which discusses the Brazilian experience of patenting bioactive substances from the biodiversity.

16.3 OVERVIEW OF THE GLOBAL PHARMACEUTICAL MARKET

In terms of productive chain, the pharmaceutical industry has distinguished itself as one of the most innovative industries, with multinational companies able to incorporate major advances from the biomedical, biological and chemical sciences into their products. In this regard, in economic terms, it is one of the most profitable industries worldwide. The worldwide pharmaceutical industry is composed of more than 10,000 companies based primarily in the United States, Switzerland, Germany, the United Kingdom, and Sweden. According to IMS (Intercontinental Medical Statistics), global pharmaceutical market increased from US$ 550 billion in 2004 to over US$ 800 billion in 2010 [19]. An estimated overview of the global pharmaceutical market (2008–2014E) is given in Figure 16.1.

In the last 5 years, global sales in the pharmaceutical industry has grown by around 40%, and the BRIC countries (Brazil, Russian, India, and China) are expected to show highest growth in the coming years, at a rate of 20% per year, contributing significantly to global growth by 2014. Cardiovascular medications account for about 12% of the market share, followed by central nervous system, dietary/metabolic, respiratory, anti-invective, musculoskeletal, genitourinary, dermatological, and cytostatic medications [20,21]. By the therapeutic category, the top five categories of medications, in

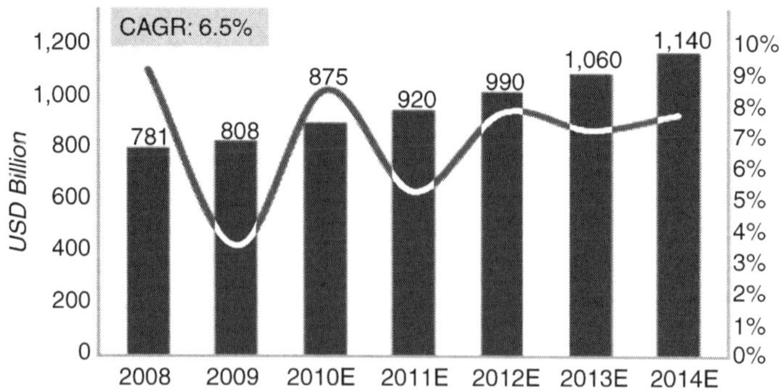

FIGURE 16.1 Global pharmaceutical market estimated. ■ Total market ■ Growth rate (*Source*: adapted from Ref. 19). (*See the color version of this figure in Color Plates section*).

terms of sales, in descending order, are: (1) cardiovascular, (2) CNS, (3) dietary metabolism, (4) respiratory, and (5) anti-invectives. These categories account for the largest volume of sales of the five largest pharmaceutical market corporations such as (1) Pfizer, (2) Merck & Co, (3) Novartis, (4) Sanofi-Aventis, and (5) Glaxo-Smith-Kline AstraZeneca [19,22]. The biggest selling drugs worldwide are Lipitor ($12 billion in sales) for lowering cholesterol, Plavix, an antiplatelet ($6 billion), Nexium, for ulcers, (US$5.7 billion); Seretide, for asthma, (US$5.6 billion), Zocor, (US$5.3 billion), Norvasc, Zyprexa, and others [23]. On the other hand, the prognosis for the future is that the only change in this scenario will be a modest increase in sales of about 5%, mainly due to the loss of ownership of patents [21].

Key factors that will influence the market, overall, are a reduction of treatment costs, of the largest markets for emerging primary treatments associated with increased uncertainty about public health, and issues related to patents. Unlike the trends of the past, the next 5 years should see a significant imbalance between the introduction of new products and loss of patents. This latter aspect will be the main factor limiting the growth of the global pharmaceutical market for mid-2013. Other key elements provided by IMS Health for the coming year are: a decrease in the contributions of the top markets (United States, Europe, and Japan), the acceleration of sales in new markets, and increased investments in generic drugs. Over the next 5 years, products that currently generate about US$137 billion in sales are expected to face generic competition. These include Lipitor®, Plavix®, Levaquin®, Xyprexa®, and Seretide®. In addition, new products that will enable innovative approaches for the treatment diseases such as osteoporosis, respiratory diseases, stroke, multiple sclerosis, and cancer are not expected to generate the same magnitude of sales [23,24].

16.4 REORGANIZATION OF THE PHARMACEUTICAL INDUSTRY

The pharmaceutical industry has evolved over time as a manufacturer of drugs, and plays an important role in the global economy. Innovations in drug discovery are the responsibility of the R&D divisions, through a process of acquiring rights to develop the drug, patent rights (for both the product and process), obtaining authority, and monopoly over the geographical extent of the market. However, with the globalization era, pharmaceutical companies worldwide are increasingly being forced to become involved in mergers and acquisitions [25,26].

No other sector has seen as many acquisitions and mergers as the pharmaceutical industry, which today faces problems of expiring patents and intense competition, with dwindling financial returns for shareholders (Table 16.1).

GlaxoSmithKline Plc (GSK) of the United Kingdom, the leading European pharmaceutical company, acquired the private biotechnology company US Reliant Pharmaceuticals for US$ 1.65 billion, one of the most important companies in the pharmaceutical industry, in 2007. This operation gave the British company access to drugs used in cardiovascular diseases, such as Lovaza (ethyl esters of omega-3). Lovaza is used in adults with very high levels of triglycerides, substances that increase the risk of coronary heart disease. Sales of this product amounted to 206 million

TABLE 16.1 Mergers and Acquisitions in the Pharmaceutical Industry Worldwide and in Brazil

Company	Headquarters	Operation Type
Abbot/EAS	USA	Abbot acquired EAS
Abbot/TheraSense	USA	Abbot acquired TheraSense
AGT Bioscience/ChemGenex Therapeutics	Australia/USA	The fusion created a new company: ChemGenex Pharmaceuticals
Amgen/Tularik	USA	Amgen acquired Tularik
Sanofi-Aventis/Medley	France/Germany	Sanofi-Aventis acquired Medley
Lilly/Applied Molecular Evolution	USA	Lilly acquired Applied Molecular Evolution
Merck & Co/Aton Pharma	USA	Merck acquired Aton Pharma
Merck & Co/Shering-Plough	USA	Merck acquired Shering-Plough
Roche/Genentech	USA	Merck acquired Genentech
Bristol Myers Squibb/Acordis	USA	Bristol Myers Squibb acquired Acordis
Mitsubishi Pharma/Green Cross Guangzhou	Japan/China	Mitsubishi Pharma acquired Green Cross Guangzhou
Fujisawa/Yamanouchi	Japan	The fusion created a new company: Astellas Pharma
Pfizer/Wyeth	USA	The fusion created a new company: Astellas Pharma
Aché/Biosintética	Brazil	Aché acquired Biosintética
Hypermarcas/Neo Química	Brazil	Hypermarcas acquired Neo Química
Biolab/Sintefina	Brazil	Biolab acquired Sintefina
Libbs/Mayne Pharma of Brazil	Brazil/Australia	Libbs acquired Mayne Pharma of Brazil

Source: Adapted from Refs. [21,27,28].

dollars in the first nine months of this same year. This process prompted the Brazilian pharmaceutical companies, which began to respond with a similar movement. In 2005, the largest national company in the industry, Aché, acquired Biosynthetic, becoming the market leader. In the same year, Biolab announced the purchase of 80% of the national laboratory, Sintefina, and Libbs acquired the operations of Australian multinational Mayne Pharma of Brazil, the leading manufacturer of oncology drugs [27,28].

16.4.1 Pharmaceutical Market in Brazil

According to forecasts for this market (IMS Health), the pharmaceutical industry will grow by 10% in 2011, and annual growth by 2015 should remain in the double digits. An overview of the pharmaceutical market in the last 9 years (2003–2011), considering only the months January to June, is shown in Figure 16.2.

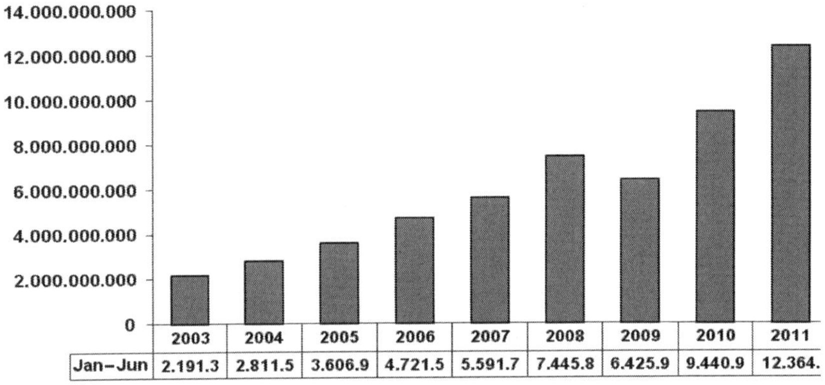

FIGURE 16.2 Pharmaceutical market in Brazil (US$) period 2003–2011 (*Source*: adapted from Ref. [32]).

The variations in sales and prices reflect a strategy adopted in Brazil since 1989. In that year, there was a series of reductions in different production phases, price controls were imposed in 1994, and there was a new law on patents in 1997. All these measures have resulted in lower prices, leading to greater competition between offers of products, and consequently, an increase in sales [25]. This growth is a clear sign of the dynamism mainly to export destinations, as demonstrated by the consistent expansion of the international pharmaceutical trade. Brazilian exports of these products have increased in recent years at an average rate of 16% per year, surpassing US$ 470 million [29]. In addition, a change in the Federal Government's industrial policy for this segment, and the creation of new products and processes, have helped boost this promising economic sector [30]. Brazil has also implemented programs to facilitate and lower the costs of customs procedures. These include the Express Customs Clearance Scheme (Blue Line) and the System of Industrial Depots under Computer Control (SIDCC). Another way to facilitate customs procedures, especially among the countries of South America, is through the adoption of common legislation on transport. Pharmaceutical industry representatives have argued that existing mechanisms are enhanced in the national territory for the MERCOSUR. One of the proposals relates to the implementation of MERCOSUR technical regulation on the practice of transport pharmaceuticals and pharmacochemicals. However discussions are ongoing in order to facilitate foreign trade operations [31].

16.4.2 The Behavior of Generic and Phytomedicines in Brazil

The sale of generic drugs is growing at a tremendous rate, and in 2007 it reached US$ 301 million, while maintaining strong growth. In 2010, the sector reached a 24.1% share in the first quarter, compared with 20.6% in the same period of 2009. The goal is to achieve a 30% market share by the end of 2012. These data are from IMS Health, an institute that audits the pharmaceutical market in Brazil and worldwide [32–34]. The class of generic drugs in Brazil was created by Law No. 9787, 1999. In 2000, the first

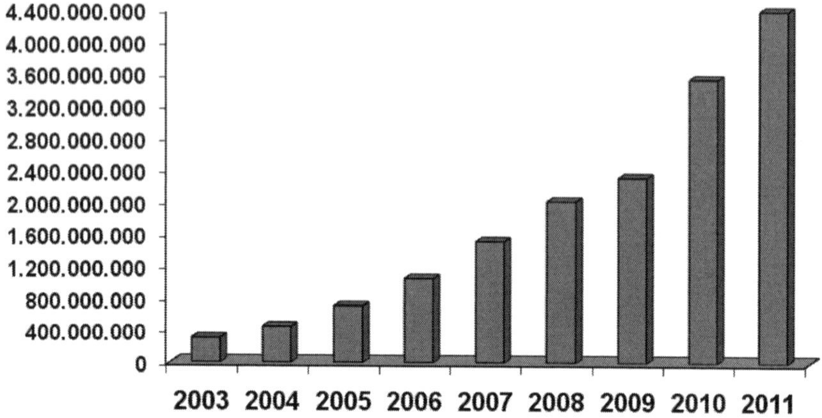

FIGURE 16.3 Generic pharmaceutical market in Brazil (US$) from 2003 to 2010 (*Source*: adapted from Ref. [32]).

products appeared on the market [35]. Figure 16.3 shows the trend in sales in Brazil for this segment of the pharmaceutical market.

A recent analysis shows that the market has been growing at a rate of 25% per year, representing 15% of sales in the pharmaceutical industry. In other words, of every 100 units of drugs sold in Brazil, 15 are generic. The industry has never experienced such a favorable scenario in Brazil. Generic drugs enjoy consumer confidence and the support of medical professionals. The introduction of generic drugs for continuous use, for the treatment of chronic diseases, has helped maintain the good performance of the sector, and these drugs now represent 48.3% of sales. While generic drugs have been growing, in general, across all therapeutic classes, the segment is expected to reach a leading position in all markets of substances intended to control chronic diseases [25,36]. Each year, over 250 million units are sold in Brazil, with revenues exceeding US$ 1.7 billion. The prediction is that Pro-Generic companies in this sector will invest US$ 350 million over the next 5 years. These funds will be invested in modernization, research and development; particularly for drugs with patents due in the coming years. However, we must emphasize that all the active principles are imported from India because the foreign trade policies meant that the incipient Brazilian industry, created between 1976 and 1990, was incapable of competing in the international market.

Another promising market in the world, which accounts for turnover of US$ 21.7 billion a year, is the herbal medicine or phytomedicine market. Today, it represents about 15% of the global pharmaceutical industry, and advances are being made toward the treatment of diseases such as vitiligo, psoriasis, and AIDS. Brazil accounts for R$550 million of this segment, and this figure is expected to double by 2012. The market is growing by about 10% each year, moving mainly toward the import industry, since much of the raw material used in our herbal medicines come from the countries of Europe and Asia [10,37].

The growth of this market is attracting investment from various industries in Brazil, competing for a share of this segment. Hence, the Aché laboratory launched Acheflan as an anti-inflammatory, investing around R$ 15 million in research and development. Other products of the Aché laboratory are Dinaton (*Ginkgo biloba*), Kamillosan (*Chamomile*), and Soyfemme (*soy isoflavones*) [38]. Meanwhile, as the demand for safety, efficacy, and quality becomes increasingly strict, more regulatory bodies and scientific studies are needed, to ensure that the population has access to quality care. In terms of expansion of the global market for products derived from plants (herbs, food supplements, cosmetics, insect repellents, dyes, etc.), Brazil, as the country with the greatest biodiversity on the planet, has an ideal opportunity to compete with much of the pharmaceutical market [39].

However, the country needs a policy with greater objectivity in the preparation of research topics of the scientific community. In this respect, a national plan – CAP-Health (Continued Acceleration Plan – Health) – was recently launched, in an attempt to boost this market. The plan is based on development and innovation in Health, which will provide an investment of R$ 88.6 billion over the next 4 years. The program features an important partnership with BNDES and FINEP that will provide an additional R$ 3 billion in the health sector to finance industry. Another point that deserves mention is the implementation of the National Policy on Integrative and Complementary (NPPIC) and the National Program of Medicinal Plants and Phytomedicine (NPMPP) linked to the Unified Health System (SHS). Both have their guidelines on the promotion and research and development of herbal medicines, emphasizing the country's biodiversity [37].

16.4.3 Acheflan: An Example of Brazilian Phytomedicine

Transforming popular knowledge of *Cordia verbenacea*, which is traditionally used to treat oedema and inflammatory processes, was the idea of Victor Siaulys, one of the founders of Aché Company. It was noted that when used as an infusion, those who drank it recovered much faster from their injuries or wounds, and this finding led the company to create, in 1989, an area of research and development of phytomedicines [38]. It was demonstrated that the active principle responsible for the anti-inflammatory action of this species was not artemetin, but α-humulene, a sesquiterpene in the essential oil. The anti-inflammatory property of α-humulene was demonstrated in preclinical and clinical experiments in different models in mice, and trials involving humans [39]. Based on the fact that the essential oils are present in the leaves, a study on the cultivation and extraction of the active principle, including its agronomic, chemical, and phytochemical development, was carried out at the Research Center for Chemical, Biological, and Agricultural (CPQBA) [40]. Parallel studies were also carried out with chemical markers, to ensure the quality control of the extract. More than 100 professionals were involved in the project between 1998 and 2004, including agronomists, biochemists, pharmaceutical, and medical professionals [40]. Preclinical studies were carried out involving pharmacological and toxicological tests in mice, followed by clinical trials involving nearly 700 patients. The product was tested in volunteers and patients with chronic tendinitis and myofascial pain, up until the final stage, with registration at Anvisa for use in the treatment of these disorders. The Aché laboratory has invested more than R$ 15 million over 7 years in research and development of Acheflan®. Right now, sales on the domestic market are a priority. However, one of the company's goals is to gain ground in the American market, which is R$ 400 million a year.

C. verbenacea DC is among the species with pharmacological applications, and has been mistakenly classified as *Cordia curassavica* (Jacq.). It is an indigenous species that grows in Brazil and is popularly known as "Erva-baleeira." In traditional folk medicine, it is used for its anti-inflammatory, antiarthritic, analgesic, and antiulcer properties [42]. Infusions of *C. verbenacea* leaves are used in the treatment of rheumatism, rheumatoid arthritis, muscle pain, prostatitis, and bruising. Pharmacological studies using extracts from the leaves have demonstrated anti-inflammatory, antimicrobial, and analgesic activities [41–43]. The principal classes of constituents have been identified as flavonoids, terpenoids, and steroids [44]. The volatile oil is rich in α-pinene (**1**), β-phellandrene (**2**), citronellol acetate (**3**), β-elemene (**4**), trans-caryophyllene (**5**), β-gurjunene (**6**), alloaromadendrene (**7**), bicyclogermacrene (**8**), δ-cadinene (**9**), spathulenol (**10**), and principally, α-humulene (**11**) (see Figure 16.4).

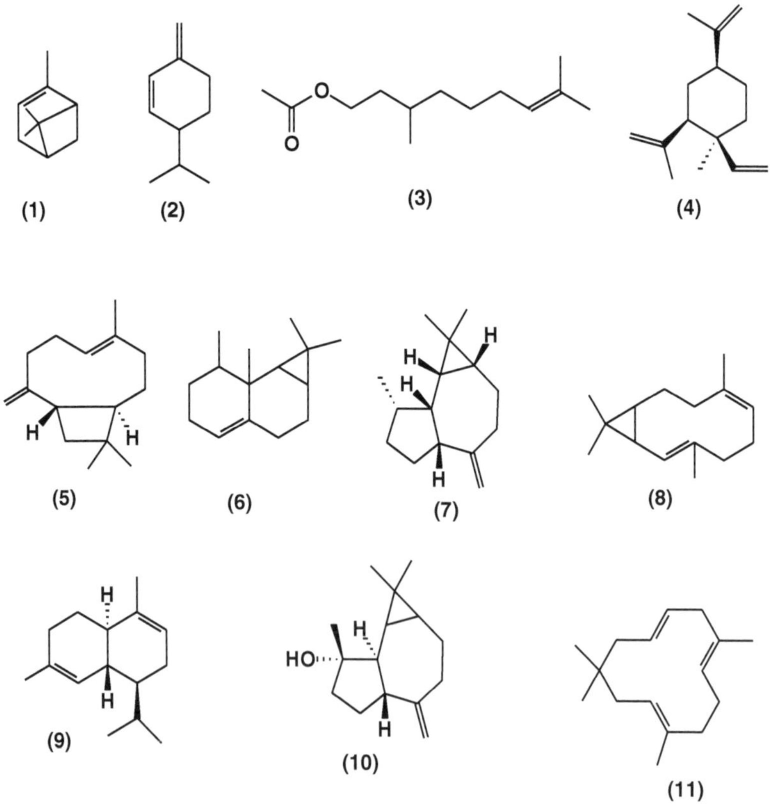

FIGURE 16.4 Molecular structures of main constituents of Acheflan (*C. verbenaceae*).

After 7 years of studies, Acheflan® was launched in June 2005. It was the first phytomedicine to be fully developed in Brazil, according to the new regulations, and was approved by the national authority ANVISA (Brazilian equivalent to the FDA). Acheflan® is formulated with *C. verbenacea* essential oil and contains 2.3–2.9% of α-humulene. It is indicated for topical use in the treatment of chronic tendonitis and myofascial pains. Other studies have been conducted in order to prove these and other activities of Acheflan® [41–44]. All these findings help explain the topical and systemic anti-inflammatory properties previously reported, and also provide support for the clinical trials conducted with the phytomedicine Acheflan®.

16.5 CONCLUSIONS

The world is going through a period of knowledge-based economic and social development that should lead to the welfare of all human beings. The role of knowledge in the economy can be analyzed in terms of a Triple helix of university-industry-government relations. Innovation, in a competitive economy, is increasingly dependent upon the Triple helix, and the pharmaceutical industries need permanent innovation because, as indicated earlier in this Chapter, there is currently an imbalance between the introduction of new products and the loss of patents. Important products will face generic competition. Companies in the pharmaceutical industry also need to produce new and better drugs in order to remain competitive. These problems have led to mergers and acquisitions between pharmaceutical companies.

In the developed countries, especially in USA, the Triple helix increasingly resembles a Development Network State in which government initiatives are critical in overcoming network failures and in providing much-needed funding for the innovation process. In this process, the roles of the government and the universities are changing; the government may participate as an entrepreneur, whether directly or indirectly, with varying degrees of involvement, while the universities have a mission of promoting economic development in partnership with industry, not only through consulting and contracting research, but through the setting up of academic research companies, the production of the knowledge, and innovative scientific development.

In Brazil and other developing countries which are the main focus of this work, the process is different. The Development Network State exists in a very incipient form; most of the industries continue producing the same drugs, without any scientific innovation; the pharmaceutical companies are mostly small, therefore they do not have their own research laboratories and cannot afford huge investments. The excessive bureaucracy, the lack of understanding on the part of the mangers, and in particular, the lack of clear and continuous political government decisions, all complicate Triple helix interactions.

However, a better understanding of the possibility of exploring natural products, and the low cost of the phytopharmaceutics, are leading to increasing collaboration between universities and industry. One notable result of such a partnership is Acheflan, a very efficient anti-inflammatory topical drug. We hope that this fact could encourage new interactions in the Triple helix of developing countries.

ACKNOWLEDGMENTS

The authors thank CNPq, FAPESC and ProPPEC/UNIVALI for their financial support.

REFERENCES

[1] Odhiambo, J. University-Industry Partnership: Meeting the Challenges of the 21st Century. Available at http://www.strathmore.edu/pdf/univ_industry.pdf, in August 15, 2011.

[2] Yunes, R.A., Pedrosa, R.C., Cechinel Filho, V. (2001). Pharmaceutics and phytotherapeutics: The need for development of the industry of phytopharmaceutics and phytotherapics in Brazil. *Química Nova*, *24*, 143–152.

[3] Yunes, R.A., Cechinel Filho, V., Novas perspectivas dos produtos naturais na química medicinal moderna. In: Yunes, R.A., Cechinel Filho, V., editors. *Química de Produtos Naturais, Novos Fármacos e a Moderna Farmacognosia*. UNIVALI, Itajaí, Brazil, 2009, pp. 9–33.

[4] Natércia, F. (2005). Parcerias e inovação impulsionam setor farmacêutico. *Inovação Uniemp*, *1*, 32–37.

[5] Villas Boas, G.K., Gadelha, C.A.G. (2007). Opportunities in the pharmaceutical industry and the local development logic based on the Brazilian biomes: The basis for a national policy discussion. *Cadernos de Saúde Pública*, *23*, 1463–1471.

[6] Cragg, G.M., Newman, D.J., Biodiversidade: um componente essencial na descoberta de fármacos. In: Yunes, R.A., Cechinel Filho, V., editors. *Química de Produtos Naturais, Novos Fármacos e a Moderna Farmacognosia*. UNIVALI, Itajaí, Brazil, 2009, pp. 51–81.

[7] Kingston, D.G.I. (2011). Modern natural products drug discovery and its relevance to biodiversity conservation. *Journal of Natural Products*, *74*, 496–511.

[8] Thorteinsdöttir, H., Quach, U., Martin, D., Daar, A.S., Singer, P.A. (2004). Health biotechnology innovation in developing countries. *Nature Biotechnology* (Suppl.), *22*, DC1–DC12.

[9] Rezaie, R., Frew, S.E., Sammut, S.M., Maliakkal, M.R., Daar, A.S., Singer, P.A. (2008). Brazilian health biotech-fostering crosstalk between public and private sectors. *Nature Biotechnology*, *26*, 627–644.

[10] Calixto, J.B., Siqueira, J.M. Jr (2008). Desenvolvimento de medicamentos no Brasil: Desafios. *Gazeta Médica da Bahia*, *78*, Suppl. 1, 98–106.

[11] Rodrigues, E. (2005). A parceria universidade-empresa privada na produção de fitoterápicos no Brasil. *Revista Fármacos e Medicamentos*, *37*, 30–39.

[12] Alesina, A., Rodrick, D. (1994). Distributive politics and economic growth. *Quarterly Journal of Economics*, *109*, 465–490.

[13] Alesina, A., Perotti, R. (2002). Income distribution, political instability and investment. *European Economic Review*, *40*, 1203–1228.

[14] Boix, C. (2009). The conditional relationship between inequality and development. *Political Science and Politics*, *42*, 645–649.

[15] Leydesdorff, L., Etzkowitz, H. (1998). The triple helix as a model for innovation studies. *Science & Public Policy*, *25*, 195–203.

[16] Angell, M.A, *Verdade Sobre os Laboratórios Farmacêuticos*. Record, Rio de Janeiro, Brazil, 2007, p. 319.

[17] Januzzi, A.H.L., Vasconcellos, A.G., de Souza, C.G. (2008). Specificities of patent protection in the pharmaceutical industry: Modalities and traits of intellectual property. *Cadernos de Saúde Pública*, *24*, 1205–1218.

[18] Nogueira, R.C., de Cerqueira, H.F., Soares, M.B.P. (2010). Patenting bioactive molecules from biodiversity: The Brazilian experience. *Expert Opinion on Therapeutic Patents*, *20*, 1–13.

[19] IMAP's (2011). *Pharma & Biotech Industry Global Report* 1–22. Available at http://www.imap.com/imap/media/resources/IMAP_PharmaReport_8_272B8752E0FB3.pdf, 4th November 2011.

[20] Hill R., Chui M. (2009). A new world of revenue growth is just breaking the horizon. Here's a first view of tomorrow's pharma. *Pharmaceutical Executive*, *29*, 1–8.

[21] IMS HEALTH. "Retail Drug Monitor," Available at www.imshealth.com, accessed on June 2011.

[22] Clinton P., Mozeson M. (2010). Available at http://amarganz.files.wordpress.com/2010/11/top-50-farmaceuticas-2009.pdf. *Pharmaceutical Executive*, 69–80.

[23] Arnold, M. (2010). Available at http://media.mmmonline.com/documents/19/outlookpt14530.pdf, accessed on June 2011.

[24] Focus (2010). *International Federation of Pharmaceutical Wholesalers*, *17*, 1–3. Available at http://www.ifpw.com/documents%5CFOCUS2010/Focus%20101007.pdf, 27th October 2011.

[25] Niero, R., Fármacos, fitofármacos e fitoterápicos: abordagem econômica e de mercado. In: Bresolin, T.M.B and Cechinel F.V, editors. *Fármacos e Medicamentos: Uma Abordagem Multidisciplinar*. Santos, São Paulo, Brazil, 2010, pp. 1–15.

[26] Anindya, D. (2011). Pharmaceutical industry and the market: The case of Prozac and other antidepressants. *Asian Journal of Psychiatry*. *4*, 14–18.

[27] Capanema, L. X. L. (2006). A indústria farmacêutica brasileira e a atuação do BNDS. *BNDES Setorial*, *23*, 193–216.

[28] Triebnigg, A. (2008). A Internacionalização da Indústria Farmacêutica Brasileira. *Interesse Nacional*, *1*, 62–67.

[29] Haag, V. E., Starosta, E. (2006). Oportunidades no mercado internacional para a indústria farmacêutica brasileira: Análise da dinâmica do setor e por produto. *Estudos Febrafarma*, *10*, 268.

[30] Galembeck, F., Santos, A. C. M., Schumacher, H. C., Rippel, M. M., Rosseto, R. (2007). Indústria química: evolução recente, problemas e oportunidades. *Química Nova*, *30*, 1413–1416.

[31] Barbosa, A.F., Mendes, R.C., Sennes R. (2007). Avaliação da política industrial, tecnológica e de comércio exterior para o setor farmacêutico. *Estudos Febrafarma*, *17*, 9–12.

[32] Sindusfarma (2011). Union of the pharmaceutical industry in the state of São Paulo. Available at http://www.sindusfarmacomunica.org.br/indicadores-economicos., 22nd October, 2011.

[33] Leite F. (2011). Available at http://www.sindusfarmacomunica.org.br/ver-destaque-da-imprensa/id/34.

[34] Scaramuzo, M. (2010). M. Available at http://www.imshealth.com/imshealth/Global/Content/Document/IMS%20in%20the%20News%20Documents/Valor%20Economico%20-%20NP.PDF.

[35] Araújo L. U., Albuquerque, K. T., Kato, K. C., Silveira, G. S., Maciel, N. R., Spósito, P. A., Barcellos, N. M. S., De Souza, J., Bueno M., Storpirtis S. (2010). Medicamentos genéricos no Brasil: Panorama histórico e legislação. *Revista Panameña de Salud Publica*, *28*, 480–492.

[36] Pereira, M. (2011). Available at http://www.istoedinheiro.com.br/artigos/61685DESTINO+PREFERIDO.

[37] Carvalho, A.B.C., Balbino, E.E., Maciel, A., Perfeito, J.P.S. (2008). Situação do registro de medicamentos fitoterápicos no Brasil. *Revista Brasileira de Farmacognosia*, *18*, 314–319.

[38] Ereno, D. (2005). Da natureza para a farmácia. *Revista Pesquisa Fapesp*, *110*, 1–3.

[39] Fernandes, E.S., Passos, G.F., Medeiros, R., Da Cunha, F.M., Ferreira, J., Campos, M. M., Pianowski, L.F., Calixto, J.B. (2007). Anti-inflammatory effects of compounds alpha-humulene and (−)-trans-caryophyllene isolated from the essential oil of *Cordia verbenacea*. *European Journal of Pharmacology*, *569*, 228–236.

[40] Calixto, J.B. (2005). Cordia verbenacea. *Arquivos Brasileiros de Fitomedicina Científica*, *2*, 5–8.

[41] Medeiros, R., Passos, G.F., Vitor, C.E., Koepp, J., Mazzuco, T.L., Pianowski, L.F., Campos, M.M., Calixto, J.B. (2007). Effect of two active compounds obtained from the essential oil of *Cordia verbenacea* on the acute inflammatory responses elicited by LPS in the rat paw. *British Journal of Pharmacology*, *151*, 618–627.

[42] Sertié, J.A.A., Woisky, R.G., Wiezel, G., Rodrigues, M., (2005). Pharmacological assay of *Cordia verbenacea*. Oral and topical anti-inflammatory activity, analgesic effect and fetus toxicity of a crude leaf extract. *Phytomedicine*, *12*, 338–344.

[43] Michielin, E.M.Z., Salvador, A.A., Riehl, C.A.S., Smânia, A. Jr, Smânia, E.F.A., Ferreira S.R.S. (2009). Chemical composition and antibacterial activity of *Cordia verbenacea* extracts obtained by different methods. *Bioresource Technology*, *100*, 6615–6623.

[44] Carvalho P.M., Rodrigues, R.F.O., Sawaya A.C.H.F., Marques, M.O.M. Shimizu, M.T. (2004). Chemical composition and antimicrobial activity of the essential oil of *Cordia verbenacea* D.C. *Journal of Ethnopharmacology*, *95*, 297–301.

INDEX

Plant Bioactives and Drug Discovery: Principles, Practice, and Perspectives, Fourth Edition.
Edited by Valdir Cechinel-Filho.

Wiley Series in Drug Discovery and Development

Binghe Wang, Series Editor

Drug Delivery: Principles and Applications
Edited by Binghe Wang, Teruna Siahaan, and Richard A. Soltero

Computer Applications in Pharmaceutical Research and Development
Edited by Sean Ekins

Glycogen Synthase Kinase-3 (GSK-3) and Its Inhibitors: Drug Discovery and Development
Edited by Ana Martinez, Ana Castro, and Miguel Medina

Drug Transporters: Molecular Characterization and Role in Drug Disposition
Edited by Guofeng You and Marilyn E. Morris

Aminoglycoside Antibiotics: From Chemical Biology to Drug Discovery
Edited by Dev P. Arya

Drug-Drug Interactions in Pharmaceutical Development
Edited by Albert P. Li

Dopamine Transporters: Chemistry, Biology, and Pharmacology
Edited by Mark L. Trudell and Sari Izenwasser

Drug Design of Zinc-Enzyme Inhibitors: Functional, Structural, and Disease Applications
Edited by Claudiu T. Supuran and Jean-Yves Winum

ABC Transporters and Multidrug Resistance
Edited by Ahcene Boumendjel, Jean Boutonnat, and Jacques Robert

Kinase Inhibitor Drugs
Edited by Rongshi Li and Jeffrey A. Stafford

Evaluation of Drug Candidates for Preclinical Development: Pharmacokinetics, Metabolism, Pharmaceutics, and Toxicology
Edited by Chao Han, Charles B. Davis, and Binghe Wang

HIV-1 Integrase: Mechanism and Inhibitor Design
Edited by Nouri Neamati

Carbohydrate Recognition: Biological Problems, Methods, and Applications
Edited by Binghe Wang and Geert-Jan Boons

Chemosensors: Principles, Strategies, and Applications
Edited by Binghe Wang and Eric V. Anslyn

Medicinal Chemistry of Nucleic Acids
Edited by Li He Zhang, Zhen Xi, and Jyoti Chattopadhyaya

Oral Bioavailability: Basic Principles, Advanced Concepts, and Applications
Edited by Ming Hu and Xiaoling Li

Dendrimer-Based Drug Delivery Systems: From Theory to Practice
Edited by Yiyun Cheng

Plant Bioactives and Drug Discovery: Principles, Practice, and Perspectives
Edited by Valdir Cechinel-Filho